AF598627

University
of Southampton

Four Centuries of Geological Travel: The Search for Knowledge on Foot, Bicycle, Sledge and Camel

IUGS/GSL publishing agreement

This volume is published under an agreement between the International Union of Geological Sciences and the Geological Society of London and arises from IUGS commission/INHIGEO.

GSL is the publisher of choice for books related to IUGS activities, and the IUGS receives a royalty for all books pulished under this agreement.

Books published under this agreement are subject to the Society's standard rigorous proposal and manuscript review procedures.

It is recommended that reference to all or part of this book should be made in one of the following ways:

Wyse Jackson, P. N. (ed.) 2007. *Four Centuries of Geological Travel: The Search for Knowledge on Foot, Bicycle, Sledge and Camel.* Geological Society, London, Special Publications, **287**.

Figueirôa, S. F. de M., Silva, C. P. da & Pataca, E. M. 2007. Investigating the colonies: native geological travellers in the Portuguese Empire in the late eighteenth and early nineteenth centuries. *In*: Wyse Jackson, P. N. (ed.) *Four Centuries of Geological Travel: The Search for Knowledge on Foot, Bicycle, Sledge and Camel.* Geological Society, London, Special Publications, **287**, 297–310.

GEOLOGICAL SOCIETY SPECIAL PUBLICATION NO. 287

Four Centuries of Geological Travel: The Search for Knowledge on Foot, Bicycle, Sledge and Camel

EDITED BY

PATRICK N. WYSE JACKSON
Trinity College, Dublin, Ireland

2007
Published by
The Geological Society
London

THE GEOLOGICAL SOCIETY

The Geological Society of London (GSL) was founded in 1807. It is the oldest national geological society in the world and the largest in Europe. It was incorporated under Royal Charter in 1825 and is Registered Charity 210161.

The Society is the UK national learned and professional society for geology with a worldwide Fellowship (FGS) of over 9000. The Society has the power to confer Chartered status on suitably qualified Fellows, and about 2000 of the Fellowship carry the title (CGeol). Chartered Geologists may also obtain the equivalent European title, European Geologist (EurGeol). One fifth of the Society's fellowship resides outside the UK. To find out more about the Society, log on to www.geolsoc.org.uk.

The Geological Society Publishing House (Bath, UK) produces the Society's international journals and books, and acts as European distributor for selected publications of the American Association of Petroleum Geologists (AAPG), the Indonesian Petroleum Association (IPA), the Geological Society of America (GSA), the Society for Sedimentary Geology (SEPM) and the Geologists' Association (GA). Joint marketing agreements ensure that GSL Fellows may purchase these societies' publications at a discount. The Society's online bookshop (accessible from www.geolsoc.org.uk) offers secure book purchasing with your credit or debit card.

To find out about joining the Society and benefiting from substantial discounts on publications of GSL and other societies worldwide, consult www.geolsoc.org.uk, or contact the Fellowship Department at: The Geological Society, Burlington House, Piccadilly, London W1J 0BG: Tel. +44 (0)20 7434 9944; Fax +44 (0)20 7439 8975; E-mail: enquiries@geolsoc.org.uk.

For information about the Society's meetings, consult *Events* on www.geolsoc.org.uk. To find out more about the Society's Corporate Affiliates Scheme, write to enquiries@geolsoc.org.uk.

Published by The Geological Society from:
The Geological Society Publishing House, Unit 7, Brassmill Enterprise Centre, Brassmill Lane, Bath BA1 3JN, UK

(*Orders*: Tel. +44 (0)1225 445046, Fax +44 (0)1225 442836)
Online bookshop: www.geolsoc.org.uk/bookshop

British Library Cataloguing in Publication Data

A catalogue record for this book is available from the British Library.

ISBN 978-1-86239-234-2

Typeset by Techset Composition Ltd, Salisbury, UK

Printed by Cromwell Press, Trowbridge, UK

Distributors

North America
For trade and institutional orders:
The Geological Society, c/o AIDC, 82 Winter Sport Lane, Williston, VT 05495, USA
Orders: Tel +1 800-972-9892
Fax +1 802-864-7626
E-mail: gsl.orders@aidcvt.com

For individual and corporate orders:
AAPG Bookstore, PO Box 979, Tulsa, OK 74101-0979, USA
Orders: Tel +1 918-584-2555
Fax +1 918-560-2652
E-mail: bookstore@aapg.org
Website http://bookstore.aapg.org

India
Affiliated East-West Press Private Ltd, Marketing Division, G-1/16 Ansari Road, Darya Ganj, New Delhi 110 002, India
Orders: Tel +91 11 2327-9113/2326-4180
Fax +91 11 2326-0538
E-mail: affiliat@vsnl.com

Contents

WYSE JACKSON, P. N. Global peregrinations: four centuries of geological travel 1

Instructions for geological travellers

VACCARI, E. The organized traveller: scientific instructions for geological travels in Italy and Europe during the eighteenth and nineteenth centuries 7

Britain

DRAKE, E. T. The geological observations of Robert Hooke (1635–1703) on the Isle of Wight 19

NICHOLAS, C. J. & PEARSON, P. N. Robert Jameson on the Isle of Arran, 1797–1799: in search of Hutton's 'Theory of the Earth' 31

Europe

KLEMUN, M. Writing, 'inscription' and fact: eighteenth century mineralogical books based on travels in the Habsburg regions, the Carpathian Mountains 49

SCHWEIZER, C. Geological travellers in view of their philosophical and economical intentions: Johann Wolfgang von Goethe (1749–1832) and Caspar Maria Count Sternberg (1761–1838) 63

TAYLOR, K. L. Geological travellers in Auvergne, 1751–1800 73

DEAN, D. R. J. D. Forbes and Naples 97

KÖLBL-EBERT, M. The geological travels of Charles Lyell, Charlotte Murchison and Roderick Impey Murchison in France and northern Italy (1828) 109

CARNEIRO, A. Sharing common ground: Nery Delgado (1835–1908) in Spain in 1878 119

WYSE JACKSON, P. N. Grenville Arthur James Cole (1859–1924): the cycling geologist 135

Greenland

WHITTAKER, A. The travels and travails of Sir Charles Lewis Giesecke 149

Russia and the Caucasus

NAUMANN, F. Alexander von Humboldt in Russia: the 1829 expedition 161

MILANOVSKY, E. E. Hermann Abich (1806–1886): 'the Father of Caucasian Geology' and his travels in the Caucasus and Armenian Highlands 177

Africa

TAQUET, P. On camelback: René Chudeau (1864–1921), Conrad Kilian (1898–1950), Albert Félix de Lapparent (1905–1975), and Théodore Monod (1902–2000), four French geological travellers cross the Sahara 183

MARVIN, U. B. Théodore Andre Monod and the lost *Fer de Dieu* meteorite of Chinguetti, Mauritania 191

Atlantic islands

WILSON, L. G. The geological travels of Sir Charles Lyell in Madeira and the Canary Islands, 1853–1854 207

PINTO, M. S. & BOUHEIRY, A. The German geologist Georg Hartung (1821–1891) and the geology of the Azores and Madeira islands 229

PEARSON, P. N. & NICHOLAS, C. J. 'Marks of extreme violence': Charles Darwin's geological observations at St Jago (São Tiago), Cape Verde islands 239

North America

SILLIMAN, R. H. Naturalists from Neuchâtel: America and the dispersal of Agassiz's scientific factory 255

ORME, A. R. Clarence Edward Dutton (1841–1912): soldier, polymath, and aesthete 271

SPALDING, D. A. E. Two Tyrrells cross the Barren Lands of Canada, 1893 287

South America

FIGUEIRÔA, S. F. DE M., DA SILVA, C. P. & PATACA, E. M. Investigating the colonies: native geological travellers in the Portuguese Empire in the late eighteenth and early nineteenth centuries 297

HERBERT, S. Doing and knowing: Charles Darwin and other travellers 311

Australasia

MAYER, W. The quest for limestone in colonial New South Wales, 1788–1825 325

OLDROYD, D. In the footsteps of Thomas Livingstone Mitchell (1792–1855): soldier, surveyor, explorer, geologist, and probably the first person to compile geological maps in Australia 343

JOHNSTON, M. R. Nineteenth-century observations of the Dun Mountain Ophiolite Belt, Nelson, New Zealand and trans-Tasman correlations 375

Japan

YAJIMA, M. Franz Hilgendorf (1839–1904): introducer of evolutionary theory to Japan around 1873 389

Geophysical travellers

GOOD, G. A. Geophysical travellers: the magneticians of the Carnegie Institution of Washington 395

Index 409

Global peregrinations: four centuries of geological travel

PATRICK N. WYSE JACKSON

Department of Geology, Trinity College, Dublin 2, Ireland (e-mail: wysjcknp@tcd.ie)

Many geologists, by the very nature of their profession tend to lead peripatetic lives: their livelihood depends on the exploitation of the products of the Earth's crust, or if they are academic geologists, teaching others about the Earth.

The origins of the profession of geologist can be traced back to central Europe in the late sixteenth century in places where mining had been a serious undertaking. By the 1800s, the science that had become known as 'geology' some 100 years earlier, began to be formulated as knowledge of the landscape and its constituents, and gradually became more widely known and understood. At this time in Europe, and later in North America, remote areas were being opened up by the cutting of canals and later by the laying of railways to facilitate economic trade. In England, William Smith (1769–1839) became familiar with a great deal of English geology as he traversed the countryside surveying canals and assessing estates for the presence of economically exploitable materials, experience and information that he encapsulated in his geological map of 1815 and in his writings on stratigraphy.

However, the potential rewards to be gained by economic exploitation of the Earth's resources are only one reason for geological travel and prospecting: other reasons for undertaking scientific travel are either unfocused curiosity or eccentricity,[1] or in contrast to this, a desire on the part of an organized individual to understand the geological framework and nature of the country to be visited. In some cases, geologists were sent abroad to study so that they might help to improve the extractive industries on their return. Similar cases of travel for education are prevalent today with many students being sent from the developing world to further their geological education.[2] In other cases, travel, that was considered by some to be a good substitute for a university education, was undertaken in order to gain experience of the natural world.[3] Sometimes scientists left home specifically to acquire good material for museum collections[4] or to obtain material for comparison with specimens known from home. In 1791, the minor cleric George Graydon (*c.* 1753–1803) visited Vesuvius and northern Italy to collect specimens so that members of the Royal Irish Academy might be able to familiarize themselves with the products of active and extinct Italian volcanoes and to compare them with products known to them from Ireland (Vaccari & Wyse Jackson 1995; Wyse Jackson & Vaccari 1997). Although this was a poor substitute for visiting the localities for themselves, the availability of collections was better than simply reading the available accounts of the volcanic regions. Geological observations were also a useful by-product of the travels undertaken by young gentlemen on the 'Grand Tour'[5] or for non-scientific reasons such as missionary outreach,[6] or even for no particular reason other than *wanderlust*. Some international expeditions were mounted with a clear scientific objective, but this might become subjugated by the nationalistic importance placed on reaching its goals. The explorer Robert Falcon Scott took three geologists (Frank Debenham (1883–1965), Raymond Edward Priestley (1886–1974) and Griffith Taylor (1880–1963)) with him on his ill-fated 1910–1913 expedition to Antarctica and a large amount of material was collected.[7] However, such scientific objectives were far from the minds of the general population and most probably the scientific community who equated success with his team reaching the South Pole first.

Whatever the reason for embarking on travels, the observer would have interpreted what he or she saw in the field in the light of their previous experience of geology and the landscape, experience that would also have been fabricated by mentors and by the availability of geological literature, either books or maps or both. It is difficult to understand fully the effect that this prior knowledge would have had in moulding the geological framework reached. It is difficult to know to what degree subliminal or conscious influences framed thinking and conceptualization. Those who lived in cities probably would have viewed the landscape differently from those who lived in rural areas. In the former, for example, the eighteenth century fabric of built-up Dublin was characterized by the use of local granite, red brick and imported limestone from the Isle of Portland, whereas the public buildings in the city of Newcastle-upon-Tyne in NE England are largely built of honey-coloured Carboniferous limestone. I am certain that few of the inhabitants of these cities recognize their characteristic geological features, but I have no doubt that it impinges on their being. Similarly,

From: WYSE JACKSON, P. N. (ed.) *Four Centuries of Geological Travel: The Search for Knowledge on Foot, Bicycle, Sledge and Camel.* Geological Society, London, Special Publications, **287**, 1–6.
DOI: 10.1144/SP287.1 0305-8719/07/$15.00

the rural region around Sydney, Australia is quite different to that around Philadelphia in the United States, which in turn is different from the Chalk Downs in SE England, and these must influence those who live there. The earliest geological travellers would have been restricted in their knowledge of foreign parts of the globe, and would not have had the benefit of the great wealth of published knowledge that is now available to the modern-day traveller. Therefore they were forced, consciously or otherwise to explain natural features seen in one country by reference to similar phenomena that they were familiar with in another. In some cases their interpretations were so conditioned by what they were familiar with at home that they misinterpreted the exotic geology. In Australia, where the distinctive *Glossopteris* fossil flora was discovered in 1804, it was invariably considered to be Carboniferous, because those palaeontologists such as Robert Etheridge Junior (1847–1920) who described it were familiar with Coal Measures flora in England and Wales. The Australian fauna is actually Permian in age (Archbold 1986). This application of English stratigraphical and palaeontological terms to non-English geology was a long-standing problem in many parts of the world and continued until the dependence on English-born geologists for survey work was broken. The imposition of Old World nomenclature was also a problem in the United States from the 1840s where some American geologists struggled to devise a local stratigraphical framework that was independent of that for Europe or that devised by the New York geological fraternity.[8]

As Martin Rudwick has written, the formulation of innovative ideas in geology is dependent on a combination of prior geological knowledge (liminal experience) coupled with new experiences gained whilst on geological travel or pilgrimage (Rudwick 1996). When the traveller returned home again he would view former field areas and geological features in a new light which transformed their geological meaning. Rudwick went on to suggest that time was necessary to allow 'the new insight to grow from its early fragile form, and this required a prolonged separation from other experts' (Rudwick 1996, p. 148). I agree with the former statement that time may be important in conceptual evolution but do not fully subscribe to his latter statement regarding separation from other geologists. I suspect that in the early days of some geological bodies such as the Geological Society of London many papers were delivered that were based on field observations made in the recent past. Obviously such papers that introduced new and or speculative ideas might have been subject to intense debate between authors and opposing protagonists,[9] who could have had major differences of opinion. Such debate could lead to either the complete suppression of the idea or the verbal reworking and subsequent publication of the innovative but initially flawed geological observations. The availability of a convivial forum for debate may in fact be as important to the propagation and acceptance of new theories as is an environment of isolation and quiet reflection and research for the development of other theories.

This volume contains twenty-eight papers that develop and expand on the many reasons for geological exploration, and which cover a broad sweep of time from the mid-1600s until the early twentieth century.

All travellers need to have some prior training in expedition techniques. An ill-prepared geologist would certainly return having failed to achieve their objective, or worse still would fail to return. Ezio **Vaccari** shows that by the 1800s a number of manuals for geologists had appeared, in which recommended equipment and techniques were illustrated and their operation discussed. The expansion of empires gave impetus for geological exploration, and the papers of Silvia **Figueirôa** and her colleagues and Wolf **Mayer** discuss scientific exploration at the opening up of the Portugese Empire in Brazil and the British Empire in Australia respectively. In the latter case the inhabitants of Sydney faced huge problems in that the location of the expanding settlement was situated in an area in which no limestone could be found. This made the building of permanent buildings difficult and it was only with the geological traversing of the Blue Mountains that a permanent source of this important commodity was located.

Some geologists do not in fact travel far from home, but still make important contributions to their science. The work of Robert Hooke (1635–1703) on the Isle of Wight and the Edinburgh-based Robert Jameson (1774–1854) on the Isle of Arran are illuminated by Ellen Tan **Drake** and by Chris **Nicholas** and Paul **Pearson**. Hooke, who has been the subject of two recent biographies (Inwood 2002; Jardine 2003), was born on the Isle of Wight and its geology must have become indelibly imprinted on to his mind. Jameson was a major advocate of the teachings of Werner. In the present study it is shown from an examination of original notebooks that Jameson was a good and competent field geologist. He made several trips to Arran on the west coast of Scotland where he described the disposition and relationships of various dykes. The nature and genesis of basalt and other igneous rocks was a topic that preoccupied geologists and men of learning for nearly two hundred years. It is well known that ideas on such rocks polarized in the 1700s with the two camps citing either water (i.e. the biblical Flood) or internal heat as being the driving mechanism for their formation. It was the work of Nicolas Desmarest (1725–1815) in central France that largely killed off the claims of the Neptunists, but as Kenneth **Taylor** discusses Desmarest's work was not in isolation and this area became a locus for geological travellers in the latter half of the eighteenth century. The papers by Marianne **Klemun** and Claudia **Schweizer** are sited in central Europe: the former describes the development of a geological framework that led to the publication of a number of important mineralogical books in the late 1700s: the latter discusses the correspondence and travels of Johann Wolfgang von Goethe (1749–1832) and Caspar Maria Count Sternberg (1761–1838) and illuminates their contrasting approach to the interpretation of similar geological features.

During the first decade of the nineteenth century, continental Europe was gripped by unrest and war. As is recalled by Alfred **Whittaker**, the former actor, musician, composer and diplomat Charles Lewis Giesecke (1761–1833) was in consequence trapped on Greenland and remained there for seven long and difficult years. During this time he assembled two collections. Much of the first was lost in 1807 during the shelling of Copenhagen and more was taken as booty and taken to Scotland. Giesecke eventually settled in Dublin and remained there for the rest of his life. Following the exile of Napoleon Bonaparte, Europe was reopened to travellers. In 1828, the threesome of Charles Lyell (1797–1875) and Roderick (1792–1871) and Charlotte Murchison (1788–1869), who unlike other wives of a similar status accompanied her husband on the undertaking, spent several months travelling through France. Martina **Kölbl-Ebert** shows how this trip was important in the geological education of both Lyell and Murchison, but also how such trips were fraught with danger and unexpected nuisances such as fleas, fever and poor weather. What is apparent from surviving letters and notebooks is that Charlotte, being fluent in French, played an important role in the logistical organization of the trip, but that she also collected fossils and sketched geological features *en route*. In 1829, Alexander von Humboldt embarked on a long and arduous journey through parts of Russia. At the time, Russia's mining industry had lagged far behind those in Western Europe in terms of technology, and von Humboldt was to examine the mining districts and see how they could be improved. This is a clear example of a geological expedition funded and carried out for purely economic reasons.

Within this volume are several papers that describe geological travels that were certainly made for academic and curiosity reasons only. Dennis **Dean** discusses the case of James David Forbes (1809–1868) who, in 1826, was drawn to Vesuvius in Italy, from where he sent home a number of letters and observations that were later published in Edinburgh. On revisiting Vesuvius some seventeen years later he found the mountain precisely as he had last seen it. Efgenji **Milanovsky** describes the remarkable career of Hermann Abich (1806–1886) who essentially gave up his academic position in order to visit the Causcasus and Armenian Highlands which had been the epicentre of a major earthquake in 1840. He became obsessed with the region and returned again and again for the next thirty years. On his death he left a large canon of geological work that included over 200 papers. Charles Lyell visited Madeira and the Canary Islands between 1853 and 1854, and was fortunate to meet up with a young German geologist Georg Hartung (1821–1891) who was staying on Madeira for the benefit of his health. The paper by Manuel **Pinto** and Annette **Bouheiry** illustrates the major contribution that Hartung made to the understanding of the geology of the island and recalls that Lyell suggested some collaborative publication; this did not happen. His trip to La Palma, the largest of the Canary Islands, allowed him, according to Leonard **Wilson**, to reject Leopold von Buch's (1774–1853) theory of crater elevation that he had proposed for its formation: Lyell recognized that the island had been built up by a series of volcanic events.

Sandra **Herbert** describes the early influences on the geological education of Charles Darwin (1809–1882) who put much of this education into practice while travelling on the HMS *Beagle*. He stopped on the Azores and 150 years later Paul **Pearson** and Chris **Nicholas** visited the island to describe and reinterpret what Darwin had seen for himself. Importantly, they conclude from their observations and comparison with written and manuscript sources that Darwin's conversion to Lyellian thinking on the voyage was not as sudden as he later suggested it had been. In following Darwin's footsteps, Pearson and Nicholas embraced the fieldwork methodology championed by David **Oldroyd** who demonstrates that, wherever possible, it is beneficial to look at the same rocks as those studied by the geologists whose history and contributions one is trying to unravel. Oldroyd himself in this volume recalls the rather turbulent, but prolific career of Thomas Livingstone Mitchell (1792–1855) who, in 1834, had the distinction of producing the first geological map of part of Australia. He also discovered fossil bones in caves in the Wellington Valley, which on their arrival in Europe caused some debate amongst the experts on their affinities. In his paper, Mike **Johnston** discusses the unravelling of another antipodean geological story: that of the Dun Mountain from where the British-born geologist Thomas Ridge Hacket (*c.* 1830–1884) described the olivine-rich rock called dunite. The discovery of chromite and copper in the region resulted in the building of the first railway in New Zealand.

Among the largest collective series of geological explorations undertaken were the various surveys in the United States. From 1836 various state surveys and from 1867 several federal geological surveys mapped and progressively worked westwards across the vast continent. Some of the most spectacular geology was encountered in Colorado where the Grand Canyon cut a swathe across the landscape, and revealed what were termed 'layercake' sequences. Clarence Edward Dutton (1841–1912) was an army man who joined the United States Geological Survey. He explored much of this area on horseback and produced a wonderfully illustrated memoir. But as Antony **Orme** discusses here, Dutton was not simply a field geologist: he investigated the Charleston earthquake of 1886 and made important and early contributions to the debate on isostasy.

Exploration can be a difficult undertaking and in the name of science men and women have travelled to some of the most inhospitable places on Earth. This volume contains accounts of expeditions to contrasting regions of intense cold and heat: David **Spalding** tells of the Tyrrell brothers, Joseph Burr (1858–1957) and James Williams (1863–1945), who found fame exploring the northern Barren Lands of the Canadian Shield, and Philippe **Taquet** documents the amazing travels by four Frenchmen in the Sahara Desert. The Canadian territories were crossed using canoe and dog sledges whereas camels or foot were the main means of transport across the sandy deserts. What is particularly remarkable about the French expeditions is that they where undertaken in areas where maps were either poor or not available: distances were computed by counting the paces of the camels. Théodore Monod (1902–2000) travelled throughout the Sahara over a period of eight decades, and as Ursula **Marvin** relates, was fascinated by the legend of a meteorite that was said to have fallen in Mauritania. Although not having to resort to camels or sledges, the British-born petrologist Grenville Cole (1859–1933) was unusual, but not unique, among geologists in preferring the use of a tricycle and later a bicycle for his travels. In 1892, he coaxed his machine along 1055 miles of bumpy rough roads in central Europe in order to investigate the volcanic districts in Hungary, and following his marriage to one of his students criss-crossed Europe on what were essentially working holidays. His appreciation of the impact of the underlying geology on the landscape is one strand of his work developed in Patrick **Wyse Jackson**'s paper.

On occasion, some geologists left their country of birth and emigrated to find opportunities for work and further research. In time, certain of them returned home, but others became fully integrated into the social and scientific fabric of their adopted countries. Louis Agassiz (1807–1873) made fundamental and pioneering contributions to palaeoichthyology and to glacial studies. Robert **Silliman** shows how, in Switzerland, he was dependent on a well-trained and skilled group of assistants, many of whom emigrated to the United States with him in or shortly after 1846. Eventually, Agassiz fell out with most of them. Some returned home to Switzerland; others remained in America where they carved out a good reputation for themselves. The westernization of Japan in the middle of the nineteenth century was encouraged by the authorities and this brought in academics from Europe and America to teach in various universities and other educational institutions. Until recently, it was thought that Edmund Morse (1838–1925) was the first to introduce evolutionary theory to Japan. However, Michiko **Yajima** refutes this and accords the honour to the German biologist Franz Hilgendorf (1839–1904) who worked in Tokyo in about 1873.

Five years later the Portuguese geologist Nery Delgado (1835–1908) travelled through parts of Spain in order to compare its geological sequences with those of his homeland. Ana **Carneiro** argues that he did this in order to decipher some of the successions in southern Portugal, but also to try to make his counterparts in the Spanish Geological Survey re-evaluate their view of adjacent geology so that it might be reconciled and consistent with that of the Portugese Geological Survey which had been published in map form in 1876.

Geophysics is a younger science than geology, but it too has its own history. Here Greg **Good** recounts the work carried out under the aegis of the Carnegie Institution of Washington between 1904 and the Second World War. Men such as Louis Agricola Bauer (1865–1932) and others crossed Africa and Asia (occasionally by being carried by porters in a sedan chair), or sailed most of the oceans of the world collecting geophysical data in an attempt to understand the Earth's magnetism.

These papers were presented at the 28th Symposium of the International Commission on the History of Geological Sciences held in Trinity College, Dublin, Ireland on 14–18 July 2003. During the course of the meeting, a field excursion visited Killiney Beach in south County Dublin where Robert Mallet (1810–1881) had carried out the first experimentally-controlled seismological explosions some 154 years earlier in October 1849. The group also visited the Victorian Cemetery at Mount Jerome where it paid homage to two of Ireland's great geologists: Sir Richard John Griffith (1784–1878) who in 1839 was the author of the first large-scale geological map of the country; and John Joly (1857–1933) who devised a scheme of colour photography, was a geochronologer, and whose book *The Surface History of the Earth* (1925) was amongst the first on tectonic geology. In the week following the symposium, a party circumnavigated Ireland and visited many sites of importance to historians of geology. These included the Giant's Causeway, a location visited by numerous visitors and men of science in the early 1800s and where the nature of basalt was disputed; Florence Court, home of the fossil-fish-collecting Earl of Enniskillen, where Agassiz, Adam Sedgwick (1785–1873) and other geologists were entertained for a week in August 1835; the Galway house of Richard Kirwan (1733–1812), the chemist and vociferous opponent of James Hutton (1726–1797); and the River Blackwater whose course through elevated anticlines was deciphered and explained by Joseph Beete Jukes (1811–1869) in 1862.

This volume draws together some remarkable stories of geological endeavour which it is hoped will be a source of inspiration to future geologists and an encouragement to historians of geology to investigate and follow in the actual footsteps or paper trail of these pioneers.

I am grateful to H. Torrens, D. Oldroyd and M. Pinto for their encouragement and help with aspects of the planning and execution of the Dublin symposium. Support was provided by Bord Fáilte, the Lord Mayor of Dublin, and by Trinity College, Dublin. I thank all the contributing authors and those who provided valuable reviews of these papers. The Geological Society of London deserves the thanks of INHIGEO for wholeheartedly embracing this project. Finally, I am indebted to my colleagues in Trinity College, Dublin, in particular G. Sevastopulo, C. Holland, G. Clayton and J. Graham for their continued support of my work. Much of this volume was edited during a period of sabbatical leave spent in late 2004 at Dickinson College, Carlisle, Pennsylvania, and I thank M. Key and his colleagues for their friendship and support during this time.

Notes

[1]Perhaps the finest case of this were the journeys in South America between 1812 and 1825 of Charles Waterton (1782–1865) who published an account of his travels in *Wanderings in South America* (1825).

[2]An early example of such an exchange took place in 1790 when two Brazilians were sent to Germany (Figueirôa 1990).

[3]This was the path followed by the English-born naturalist Robert Townson (1762–1827). See various papers in Rózsa (1999).

[4]Accounts of a number of recent expeditions undertaken for this specific purpose by staff of the Natural History Museum in London are given in Whybrow (2000).

[5]William Willoughby Cole, third Earl of Enniskillen and Philip de Malpas Grey-Egerton developed an interest in fossil fishes while on such a tour (James 1986).

[6]One such missionary was the Moravian Henry Steinhauer (1782–1818) who emigrated from England to Phildelphia in 1815; he was well versed in geology before his departure and soon after his arrival published a paper describing some Carboniferous plant fossils which were actually from England. Following his death, his valuable fossil collection eventually ended up in the Academy of Natural Sciences in Philadelphia (Torrens 1990, 2005).

[7]A total of 1919 specimens are housed in the Natural History Museum, London (Campbell Smith 1963).

[8]James Hall, with others, propagated the use of the term 'New York System' which they had established for the Lower Palaeozoic because of the difficulties of applying European nomenclature to sequences in the state. However, Hall went further and began to label the rocks of the Mid-West with the same names (Aldrich & Leviton 1987).

[9]Many such debates and episodes of infighting in the chambers of the Geological Society are recalled in Thackray (2003).

References

ALDRICH, M. L. & LEVITON, A. E. 1987. James Hall and the New York Survey. *Earth Sciences History*, **6**, 24–33.

ARCHBOLD, N. W. 1986. Nineteenth century views on the Australian marine Permian. *Earth Sciences History*, **5**, 12–23.

CAMPBELL SMITH, W. 1963. The petrography of the igneous and metamorphic rocks of south Victoria Land; based on the collections made on the British Antarctica (Terra Nova) Expedition, 1910–13. *Polar Record*, **11**, 770–771.

FIGUEIRÔA, S. 1990. German–Brazilian relations in the field of geological sciences during the 19th century. *Earth Sciences History*, **9**, 132–137.

INWOOD, S. 2002. *The Man Who Knew Too Much.* Macmillan, London.

JAMES, K. W. 1986. *'Damned Nonsense!'—The Geological Career of the Third Earl of Enniskillen.* Ulster Museum, Belfast.

JARDINE, L. 2003. *The Curious Life of Robert Hooke: The Man Who Measured London.* Harper Collins, London.

RÓZSA, P. (ed.) 1999. *Robert Townson's Travels in Hungary.* Kossuth Egyetemi Kiadó, Debrecen.

RUDWICK, M. 1996. Geological travel and theoretical innovation: the role of 'liminal' experience. *Social Studies of Science*, **26**, 143–159.

THACKRAY, J. C. 2003. *To See The Fellows Fight: Eye-Witness Accounts of Meetings of the Geological Society of London and Its Club, 1822–1868.* British Society for the History of Science Monograph **12**.

TORRENS, H. S. 1990. The transmission of ideas in the use of fossils in stratigraphic analysis from England to America 1800–1840. *Earth Sciences History*, **9**, 108–117.

TORRENS, H. S. 2005. The Moravian minister Rev. Henry Steinhauer (1782–1818); his work on fossil plants, their first 'scientific' description and the planned Mineral Botany. *In*: BOWDEN, A. J., BUREK, C. V. & WILDING, R. (eds) *History of Palaeobotany: Selected Essays*. Geological Society, London, Special Publications, **241**, 13–28.

VACCARI, E. & WYSE JACKSON, P. N. 1995. The fossil fishes of Bolca and the travels in Italy of the Irish cleric George Graydon in 1791. *Museologia Scientifica*, **4**, 57–81.

WATERTON, C. 1825. *Wanderings in South America, the North-West of the United States, and the Antilles, in 1812, 1816, 1820 and 1824*. J. Mawman, London.

WHYBROW, P. J. (ed.) 2000. *Travels with the Fossil Hunters*. The Natural History Museum, London and Cambridge University Press.

WYSE JACKSON, P. N. & VACCARI, E. 1997. *The Reverend George Graydon (c. 1753–1803): Cleric and Geological Traveller*. Royal Irish Academy, Dublin.

The organized traveller: scientific instructions for geological travels in Italy and Europe during the eighteenth and nineteenth centuries

E. VACCARI

Dipartimento di Informatica e Comunicazione, Università dell'Insubria, via Mazzini 5, 21100 Varese, Italy (e-mail: ezio.vaccari@uninsubria.it)

Abstract: The history of geological travels usually focuses on the study of individuals or groups of travellers, but it should also consider in detail some essential aspects of the meaning of geological travel, such as the beginnings of this particular kind of scientific travel, the emergence of a 'conscious' geological traveller, the concept of 'field' and fieldwork for a geologist and finally the first attempts to codify the style and the method of a geological travel. The aim of this paper is to look at possible answers to some of these questions by presenting a short outline, based on some significant examples in the historical development of a particular kind of scientific literature still little known and open to further investigation: the instructions for geological travellers, including writing by those people, generally scientists, who wanted to organize their experiences to instruct others on how to undertake geological observations methodically. From the early eighteenth century, these texts may be found within private diaries or official reports, in journals or periodicals, scholarly monographs or textbooks, and also as articles, pamphlets, booklets or even books on their own, especially in the late nineteenth century.

The period between the mid-eighteenth and mid-nineteenth century saw the development of scientific travel.[1] The appearance of the first specific instructions for geological fieldwork was clearly linked to the emergence of geology as a scientific discipline, in particular from 1760 to 1840, but also later, when this type of writing became part of the specific literature in various branches of the Earth sciences.

The process began at the end of the seventeenth century, mainly concerning naturalistic travel (the general model established among scientists) particularly in the field of botany (Sachs 1892, pp. 17–18). In 1696, John Woodward (1665–1728) published a small pamphlet anonymously, probably one of the first scientific instructions for travellers, the *Brief Instructions for Making Observations in all parts of the World and also for Collecting, Preserving and sending over Natural Things* (Fig. 1). In this work more attention was given to rocks, minerals, fossils and geomorphological features than to zoological and botanical aspects:

> Observe the several sorts of *Marls, Clays, Loams*, or other *Soils*, at the *Surface of the Earth*: And whether there be not almost every where a *Coat of one or other of these at the Surface*, whatever else lyes *underneath* [...] also take account of the several sorts of *Stone, Marble, Alabaster, Cole, Chalk, Okers, Sands, Clays* and other *Earths*: Their *Depths*; The *Thickness* of their Strata or Beds, whether level or not [...] Get an account likewise of the several *Mountains* and *Rocks*: the *Stone, Marble*, or other *matter*, of which they *consist* [...] Wherever these *Shells, Teeth, Plants Etc.*, are found the Enquirer may please to *note*, along with the *Place*, what *Sort* of Shells they are: and whether they be of the *same kinds* with *those* found upon the *Shores* of *those Parts* or not: in what *Numbers* they are found: at what *Depths*: and what Earth, Sand, or other Matter, they *contain in them* [...] Take an account of the more observable and peculiar *Diseases* of the Country [...] and of the other *Casualties*, particularly *Earthquakes* (Woodward 1696, pp. 4–8).

Woodward emphasized the importance of observing and listing all the different rock types visible on the surface and underground (with particular attention to the position and to the thickness of the strata), the various minerals and mineral veins, the 'materials' which composed the mountains, the sites where fossils may be found ('sea shells, and other Marine Bodies, at Land, in Stone [...] Trees found buried in the Earth'), as well as all the geomorphological changes due to earthquakes, volcanic eruptions and meteoric erosion. Woodward's interest for what we might now call an early 'methodology of geological observations' also emerged in a later work, the *Brief Directions for making Observations and Collections, and for composing a Travelling Register of all sorts of Fossils*.[2]

At the beginning of the eighteenth century those naturalists particularly interested in the features of the Earth's surface (not yet called geologists, but later named 'oryctologists' or 'mineralogists') were not only academics, but also ordinary 'country people', amateurs or 'curious of nature', who went into the field not only for collecting mineral or fossil specimens for their museums, but also for looking for real elements in order to understand specific geological phenomena. For example,

From: WYSE JACKSON, P. N. (ed.) *Four Centuries of Geological Travel: The Search for Knowledge on Foot, Bicycle, Sledge and Camel.* Geological Society, London, Special Publications, **287**, 7–17.
DOI: 10.1144/SP287.2 0305-8719/07/$15.00

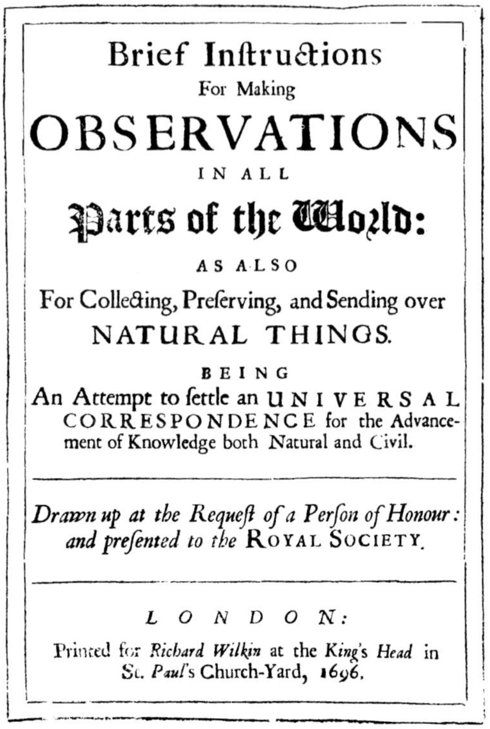

Brief Inftructions
For Making
OBSERVATIONS
IN ALL
Parts of the World:
AS ALSO
For Collecting, Preferving, and Sending over
NATURAL THINGS.
BEING
An Attempt to fettle an UNIVERSAL CORRESPONDENCE for the Advancement of Knowledge both Natural and Civil.

Drawn up at the Requeft of a Perfon of Honour: and prefented to the ROYAL SOCIETY.

LONDON:
Printed for *Richard Wilkin* at the *King's Head* in St. *Paul's* Church-Yard, 1696.

Fig. 1. Title-page of John Woodward's *Brief Instructions for Making Observations in all Parts of the World and also for Collecting, Preserving and Sending over Natural Things* (London 1696).

the different shapes and lithology of the strata connected to the origin of springs were studied in detail by Antonio Vallisneri (1661–1730), who travelled extensively through the Italian mountain chain of the northern Apennines between 1704 and 1711, mainly for geological purposes.[3]

This kind of travel, through a mountain chain and different regions, undertaken with the help of guides, was much longer and more difficult than the typical short excursion on foot that was based on the sixteenth–seventeenth century botanical model for collecting herbs for medical purposes and that had been common until the early eighteenth century. For some southern German naturalists these excursions became 'oryctological' travels in the surroundings of their native cities, or as in the case of Italian scholars from the Inquieti Academy of Bologna, who were looking for fossils in the Apennines in addition to herbs and plants (Vaccari 2000*b*). Vallisneri's travels in the Apennines might be better compared with alpine travels, such as those of Joahnn Jakob Scheuchzer (1672–1733) in the first decade of the eighteenth century (Scheuchzer 1723).

Moreover, the importance of these travels for Vallisneri prompted him to write an index of observations (*Indice di osservazioni*) in 26 points (Perrucchini 1726, pp. 404–419), where he emphasized, like Woodward, the importance of describing all the '*fossilia*', as well as the different directions of the strata,[4] their lithological content, their colours, and also the location of the mines. It was a research plan which paid attention to the structural analysis of mountains (the '*notomia de' monti*', anatomy of mountains, according to Vallisneri) based on the lithology and morphology of the strata as well as on the different kinds and locations of fossils:

> *Si noti [...] se si trovano di que' corpi marini che* diluviani, *e* antediluviani *chiamano, come giacciano, o in qual positura si trovino, se impietrati, o non impietrati, se frà sassi, o pietre rinchiusi, o fra rene, e terre, e queste di qual colore, e maniera sieno; se nel principio, nel mezzo, o nella sommità del Monte; se sopra, o molto sottoterra negli strati più fondi; se nelle miniere, o fuora; da qual parte sieno posti, se verso il più vicino mare, o verso l'interno del Monte, se sieno confuse di varie spezie, o se d'una sola, o poche; se corrispondano a quei crostacei, o pesci, o Insetti, o piante, che nel più vicino mare si trovano, o ne' lontani.*[5]

The *Indice* was originally compiled by Vallisneri as a guide for one of his pupils, Giambattista Perrucchini; however, he obtained permission to publish it because he considered it of great interest to the contemporary scientific community. It listed and explained the main elements of geological fieldwork, although, concerning the use of specific scientific instruments in the field, Vallisneri only referred to some barometers and thermometers, as well as '*altri moderni ordigni*' (other modern tools) for measuring the quality of the air.

From the middle of the eighteenth century, regional travels rapidly increased and soon became the most successful way for undertaking research in the field among the European scholars. Travels for making naturalistic observations, often mainly geological, focused on well defined geographical areas or on specific valleys, mountains or hills: in France, for example, the case of Auvergne is very significant (Taylor 2007), whereas in Italy, among the regions particularly renowned for their geological features, Tuscany was widely investigated because of its active volcanoes (Vesuvius, Mt Etna).

Within this context of regional travels, it is possible to identify the first traces of instructions for studying geological features. The florentine naturalist Giovanni Targioni Tozzetti (1712–1783) included several reports on travels throughout Tuscan in his *Relazioni d'alcuni viaggi fatti in diverse parti della Toscana*, (Targioni Tozzetti 1751–1754; on Targioni Tozzetti's travels, see Greppi 2000). He collected information on this region by employing collaborators to whom he sent a letter containing instructions on the 15 November 1751. There were 50 instructions of which 32 were of geological interest. They emphasized the need for indicating the exact position of rocks and minerals, as well as the importance of drawingsketches not only of the most spectacular geomorphological features of the Tuscan landscape, but also of the strata observed inside quarries and mines. This *Lista di Notizie d'Istoria Naturale della Toscana, che si ricercano*, a list of required pieces of information on the natural history of Tuscany, was published in 1754, together with a plan for a physical topography of Tuscany which indicated the valley as a geomorphological unit for the subdivision and the description of the territory (Targioni Tozzetti 1754, pp. 161–165, 199–210). Consequently, each Tuscan valley had to be observed and analysed according to a scheme in 24 points, which was very similar to that compiled by Vallisneri. Moreover, according to his original orographical subdivision into two units ('*monti primitivi*', primitive mountains and 'colline', hills) (Targioni Tozzetti 1754, pp. 11–29), Targioni Tozzetti required that, in each valley, 'the nature of surrounding mountains, their height, shapes, cavities, cliffs, crags, landslides, and other features', as well as 'the materials that compose the skeleton of those mountains: their amount, position, extension, subdivision, etc.' had to be observed and noted; however, at the same time, also the 'form, extension, height, and other features of the hills', together with the 'origin, nature and differences' of their 'materials' had to be studied in detail.[6] The project, outlined in the *Prodromo* but never completed, aimed to produce a complete description of Tuscany, including naturalistic, hydrographic and geomorphological features, from natural resources to human settlements and activities.

In spite of this and other examples of regional travels more 'geologically oriented', in the early years of the second half of the eighteenth century the main reference continued to be that of the naturalistic travel. Little space was given to geological matters, compared to other subjects, in the famous *Instructio peregrinatoris* by Linnaeus (Carl von Linné, 1707–1778) (Linné 1760; on Linnaeus' travels, see Sörlin 1989). In this text, in chapters 8 *Physica* and 9 *Lithologica*, Linnaeus simply exhorted travellers to climb the mountains in order to measure their height with the barometer and to compile a list of '*praecipitia*', '*fracturae*', '*species lapidum*', '*strata lapidum et terrarum*', '*terrae*', '*saxa*' and '*petrae primariae*', but also minerals and fossils '*tam rariora, quam vulgaria*', including the descriptions of the mines ('*fodinarum situs, matrix, venae profunditas pariter ac methodus fodiendi et extrahendi metalla ac acquam*') (Linné 1760, pp. 303–304, 306). Again, the content of Linnaeus' *Instructio*, which as with other contemporary instruction texts only briefly recalled the importance of collecting and carefully keeping stones, concentrated exclusively on the objects of observation rather than the practical aspects of observation and possible visual reproduction.

The mining industry also contributed to the development of the instructions for mineralogical travels, especially those that were officially organized for purposes of scientific and technical education. For example, the 1787 expedition by two Italian officers from Turin in the mining districts of Saxony, Bohemia, Hungary, Sweden, England and Scotland, a long journey with a difficult itinerary, entirely organized by the government of the Kingdom of Sardinia, was carefully prepared with very precise operative instructions for travelling and doing researches, given by their superior officer and director of the School of Mineralogy in Turin, Benedetto Spirito Di Robilant (1724–1801).[7]

These instructions were contained in an unpublished file entitled *Istruzioni per quei soggetti, a' quali la S.M. permette di viaggiare per abilitarsi nelle miniere e nella metallurgia* (Instructions for those people allowed by His Majesty to travel in order to qualify in mining and metallurgy),[8] addressed to the mineralogist Carlo Antonio Napione (1757–1827), head of the expedition and one of the best pupils of Robilant. The itinerary, the programme of studies at the mining academies, and all the visits to mines and foundries were carefully planned: the travellers were allowed to take with them only the essential scientific instruments (hygrometer, thermometer, Saxon mining compass), 'salts and fluxes' for chemical analysis in the field, notebooks and paper for drawing, weights and measures, various German volumes on mining and metallurgy, including the *Mineralogia* by the Swedish scientist Johann Gottschalk Wallerius. Their personal belongings were reduced to a minimum as their luggage would increase during the travel, with the acquisition of new mining instruments made abroad and the collection of several mineral and rock specimens. However, this travel was not intended solely as a mission for collecting pieces of useful information on the technical procedures adopted in the most renowned mining districts of central and northern Europe.

Before they left Italy, the two officers were ordered to inspect several areas in Piedmont and the Aosta valley and to prepare a detailed report (with a specific register of geo-mineralogical and mining data) that could be compared later with their travel diary. In this daily '*giornale*' the travellers were asked to note all the geological features observed during the excursions, paying particular attention to mountain structures. The content of Robilant's unpublished text of instructions is a clear example of the significant interaction between geological research and mining, as a well established relationship at the end of the eighteenth century (see also Lehmann 1759, pp. 305–313; Robilant 1790, pp. 21–22; Miché 1795).

A few years later, in 1791, the mineralogical instructions compiled by Henri Besson and Déodat de Dolomieu (1750–1801) for the expedition led by D'Entrecasteux in search of La Pérouse, emphasized the description of lithological and structural features, as well as the importance of collecting and systematically classifying rock and mineral specimens.[9] The 'mineralogist', was considered by these authors as a 'field lithologist', more than a mining expert, although he was also required to understand fully and give a complete description of the geological structures of mountains:

> Le Mineralogiste, toujours armé d'un marteau, de ciseaux ou de coins de fer qu'il porte pendus à sa ceinture, après avoir cassé quelques galets ou pierres, saura ceux qu'il doit préférer, car il faudra bien se contenter de ces fragments roulés, quand on ne peut aller chercher les roches entières dans les montagnes. [...] On observera la position des roches, les unes par rapport aux autres, si il'y a des bancs, lits ou couches bien marqués, leur différentes épaisseurs, leurs arrangements réciproques, leurs superpositions et leurs inclinaisons; bien observer vers quelle partie du ciel sont, constamment ou le plus souvent, les escarpements et les pentes les plus rapides. À la caissure et à la dégradation des grandes masses un oeil exercé peut juger, à une certaine distance, de l'espèce et de la nature des roches (Besson 1792, pp. 193–194).[10]

These texts of instructions for travellers show a good level of interest in geo-mineralogical research within those extra-European travels organized by French scientists and institutions, whereas in the great English expeditions of the same period more importance was assigned to botanical and zoological observations (Bourguet 1997, pp. 190–193).

Following the gradual establishment of geology as a scientific discipline, at the beginning of the nineteenth century the definition 'geological travel' began to appear more frequently in titles of books and articles: some significant examples of this trend are the *Viaggio geologico per diverse parti meridionali d'Italia* (*Geological travel in several parts of southern Italy*) by Ermenegildo Pini (1739–1825) or the *Geological travels* in France, Switzerland and Germany by Jean-André de Luc (1727–1817), and the *Voyage minéralogique et géologique en Hongrie* (*Mineralogical and geological travels in Hungary*), by the French mineralogist François-Sulpice Beudant (1787–1850) (Pini 1802; De Luc 1810–1811, 1813; Beudant 1822).

Together with the first definitions of geological travel, the instructions for geological travellers acquired their place within the scientific literature, although not yet in an autonomous form. Most historians have considered the *Agenda ou Tableau Général des Observations et des Recherches dont les résultats doivent servir de base à la théorie de la terre* (Fig. 2) by Horace Bénédict de Saussure (1740–1799) as the first specialized text of instructions for the travelling geologist (Saussure 1796*a*).[11] This work describes the methods and contents of a full geological investigation in twenty-three chapters, including an interesting section on instruments to be used in the field. It lists in detail the observations to be carried out along sea coasts, rivers, throughout the plains, in valleys, on erratic boulders and above all in the mountains

AGENDA,

ou TABLEAU GÉNÉRAL *des Obſervations & des Recherches dont les réſultats doivent ſervir de baſe à la théorie de la terre.*

INTRODUCTION.

§. 2304. LORSQU'ON doit contempler des objets auſſi compliqués que ceux qu'il faut étudier pour fonder ſur l'obſervation les baſes de la théorie de la terre, il eſt indiſpenſable de ſe former à l'avance un plan, de ſe preſcrire un ordre, & de minuter, pour ainſi dire, les queſtions que l'on veut faire à la nature.

" Comme le géologue obſerve & étudie
„ pour l'ordinaire en voyageant, la moin-
„ dre diſtraction lui dérobe, & peut-être
„ pour toujours, un objet intéreſſant. Même
„ ſans diſtraction, les objets de ſon étude
„ ſont ſi variés & ſi nombreux, qu'il est
„ facile d'en omettre quelques-uns : ſou-

Fig. 2. First page of H. B. de Saussure's 'Agenda ou Tableau général des Observations et des Recherches dont les résultats doivent servir da base à la Théorie de la Terre'. *Journal des Mines* (Paris 1796).

during fieldwork. The mountains were divided into 'primary', 'secondary' and 'tertiary', according to a scheme common at the end of the eighteenth century. They had to be studied with particular attention to the analysis and identification of stratified mountains and the observations on fossils had to focus particularly on the conditions of the findings (state of conservation, position, distribution, surrounding lithology). With regard to volcanoes, De Saussure recalled the differences between observations made during an eruption, on dormant and on extinct volcanoes. A long and detailed chapter was devoted to the observations inside mines: here the importance of an accurate evaluation and description of mineral veins (shape, direction, inclination, etc.) was strongly emphasized in order to identify the kind of relationship with the strata of the mountain containing the ore minerals.

It is therefore significant that the chapter on the mistakes to be avoided during geological fieldwork (*Erreurs à éviter dans les observations relatives à la Géologie*), contained a specific section on the method of observing of the strata (Saussure 1796*a*, § 2326–2327). This subject was as important for De Saussure as it had been for Vallisneri at the beginning of the century. However, the Italian naturalist had insisted only on accurate descriptions of the numerous morphologies and lithological compositions of the mountain strata, aiming to achieve a good level of systematic knowledge (Vallisneri 1715, pp. 25–29), whereas De Saussure was conscious that a wrong interpretation of the superimposition of the different kinds of strata could have determined a bad reconstruction of the history of the Earth's surface. Consequently, according to the Swiss scientist, it was no longer sufficient to observe a stratigraphical sequence only from the outside, but it was necessary to verify the continuity inside the Earth's crust (this reinforced the concept of 'geological utility' of the mines) and evaluate all the possible alterations or modifications with regard to the original formation:

Enfin, l'erreur la plus grave est celle que l'on peut commettre sur la superposition des couches. J'ai vu souvent des hommes novices dans l'étude des montagnes, croire qu'un couche reposait sur

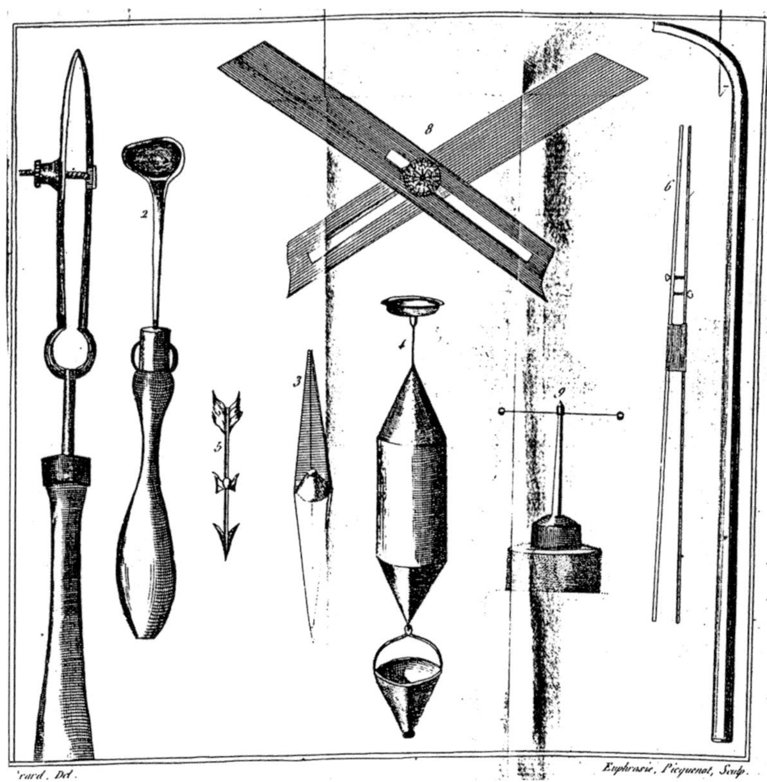

Fig. 3. Some tools and scientific instruments including a hydrostatic scale, a goniometer for measuring the angles of crystals, a small crucible for melting different substances, and some 'electrometers' for testing the electrical conductivity of minerals (Brard 1805).

une autre; un granit par exemple sur une ardoise, parce qu'ils avoient trouvé l'ardoise au bas de la montagne, et le granit dans le haut; tandis que l'ardoise n'étoit qu'appliquée contre le bas de la montagne, et que le granit au contraire s'enfonçoit dans la terre fort au-dessous de l'ardoise. Il ne faut donc prononcer qu'une couche est située sous une autre, que quand on la voit réellement s'enfoncer au-dessous d'elle. Et même lorsqu'on voit un rocher distinctement superposé à un autre, il faut examiner si celui qui est sur l'autre n'occupe point accidentellement cette situation, s'il n'a point glissé ou roulé d'une montagne plus élevée (Saussure 1796*a*, § 2326).[12]

The *Agenda* was distributed widely among European scientists because of the publication of English and German translations (Saussure 1799*a*,*b*): in particular, the instructions about the instruments were summarized by the Swiss mineralogist Heinrich Struve (1751–1826).[13]

During the first half of the nineteenth century the earliest textbooks for travelling geologists were published: however, one of these early works, the *Manuel du mineralogiste et du geologue voyageur* by Cyprien Prosper Brard (1786–1839) contained a useful synthesis of geo-mineralogical basic notions, but not yet a specific section on instructions for fieldwork (Brard 1805). Brard had included a plate with a short description on the equipment for the mineralogist on the field ('*necessaire du minéralogiste*') at the end of the volume. It included: some reagents ('*un acide habituellement nitrique, pour reconnaître les carbonates, qui s'y dissolvent; de l'ammoniaque ou alcali volatil pour reconnaître les mines de cuivre qui le colorent en bleu*'), some tools and scientific instruments such as a hammer, some pincers, a hydrostatic scale, a goniometer for measuring the angles of crystals, a small crucible for melting different substances, and some 'electrometers' for testing the electrical conductivity of minerals (Fig. 3) (Brard 1805, pp. 458–459).[14]

Although Brard's volume could not be considered a real development of De Saussure's *Agenda*, it focused on the central question of the instruments for the geologist and the mineralogists in the field. After the written rules, linked to the basic elements of knowledge of the new science of geology (having instructions that stated how to travel and what could be defined as geological travel), the concept of field and of fieldwork was

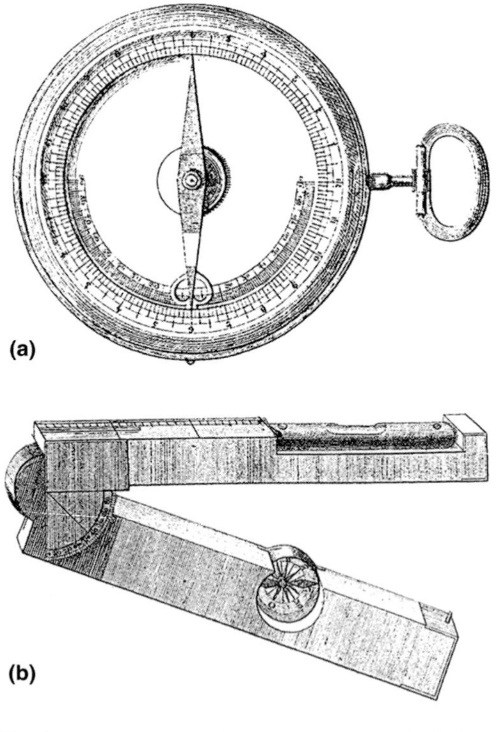

Fig. 4. Field equipment illustrated by Leonhard (1838). (**a**) A compass; (**b**) A clinometer.

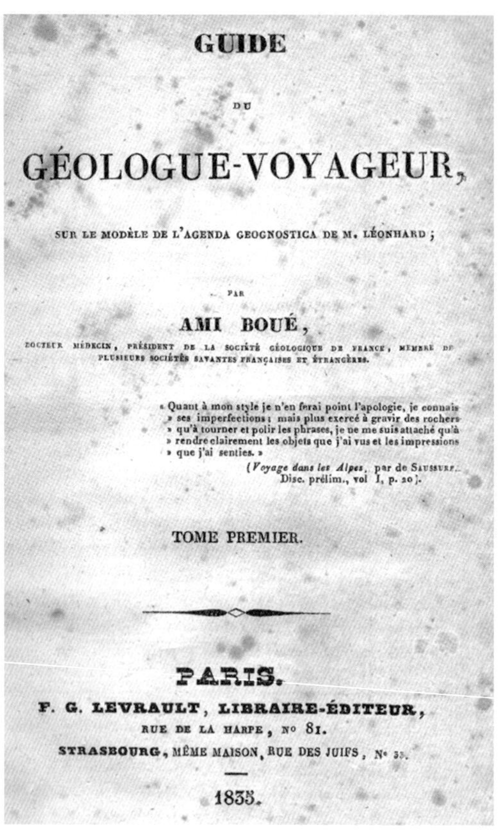

GUIDE

DU

GÉOLOGUE-VOYAGEUR,

SUR LE MODÈLE DE L'AGENDA GEOGNOSTICA DE M. LÉONHARD;

PAR

AMI BOUÉ,

DOCTEUR MÉDECIN, PRÉSIDENT DE LA SOCIÉTÉ GÉOLOGIQUE DE FRANCE, MEMBRE DE PLUSIEURS SOCIÉTÉS SAVANTES FRANÇAISES ET ÉTRANGÈRES.

« Quant à mon style je n'en ferai point l'apologie, je connais » ses imperfections; mais plus exercé à gravir des rochers » qu'à tourner et polir les phrases, je ne me suis attaché qu'à » rendre clairement les objets que j'ai vus et les impressions » que j'ai senties. »

(*Voyage dans les Alpes*, par de SAUSSURE. Disc. prélim., vol I, p. 20).

TOME PREMIER.

PARIS.

F. G. LEVRAULT, LIBRAIRE-ÉDITEUR,

RUE DE LA HARPE, No 81.

STRASBOURG, MÊME MAISON, RUE DES JUIFS, No 33.

1835.

Fig. 5. Title-page of Ami Boué's *Guide du Géologue-Voyageur* (Volume 1) (Paris 1835).

also redefined by the instruments specifically designed for travelling.

Other textbooks followed the model of De Saussure more closely such as *Agenda Geognostica* (1829) by the German geologist Karl Caesar Leohnard (1779–1862) and the *Guide du géologue voyageur* (1835–36) by Ami Boué (1794–1881) (Leonhard 1838; Boué 1835–1836). Leonhard gave a particular importance to the use of cartography and instruments in the field (Fig. 4), as well as to the practical notions for travelling ('*Art zu reisen*'), which included the knowledge of roads and transport systems, particularly in a mountain environment. Moreover, he presented a series of rules for drawing sketches and geological profiles and sections, for collecting and keeping rocks, minerals and fossils, for compiling a diary, with a complete description of the observed geomorphological

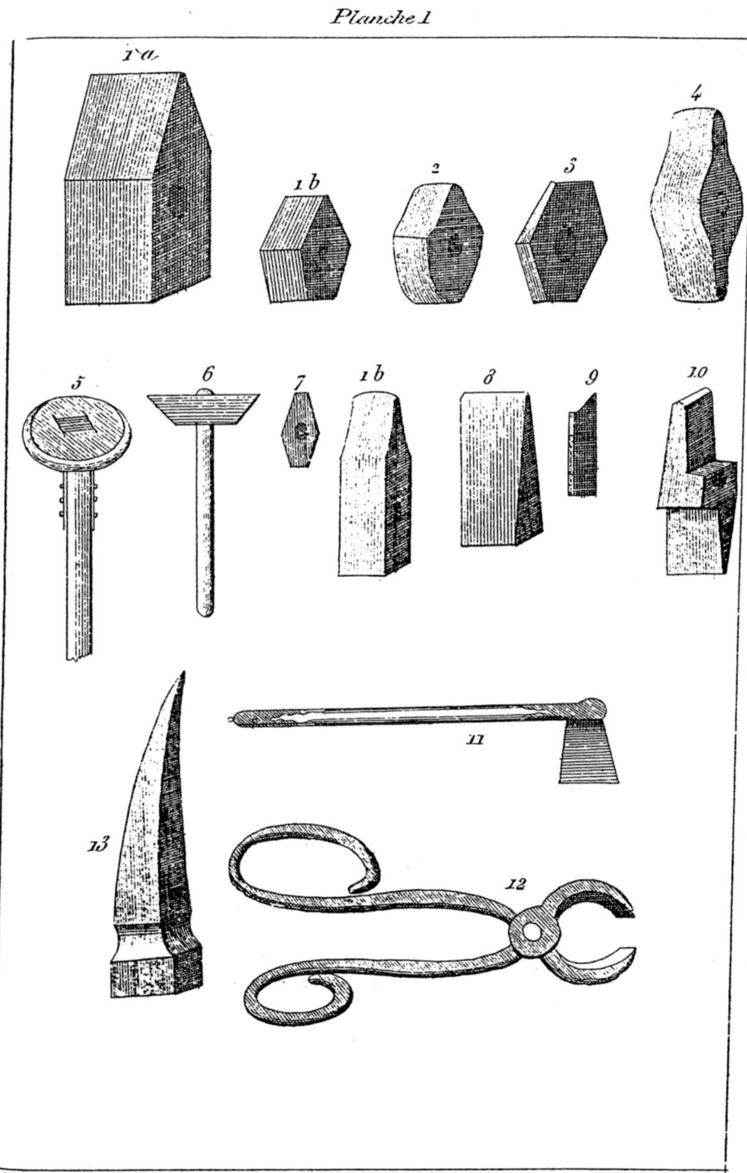

Fig. 6. Various hammers, pincers, axes and chisels. Plate 1 from Boué's *Guide du Géologue-Voyageur* (Volume 2). (Paris 1836).

features (Leonhard 1838, pp. 10–105). Ami Boué, one of the founders of the Société géologique de France in 1830 and himself a great traveller, enlarged Leonhard's work by writing a long detailed chapter to the *Préparatifs et instructions préliminaires pour les voyages geologiques* (Boué 1835–1836, **1**, pp. 9–158; on Boué, see Seidl 2002). He believed that his 'geological guide' could be read by *'amateurs de la géologie'*, although a part of the first volume and the full content of the second volume could be regarded as a complete textbook of geology. The *Guide du géologue-voyageur* (Fig. 5) followed De Saussure and Leonhard's *Agendas*, but also took into consideration other contemporary works of geological instructions, such as the *Geognostischer Katechismus* by the German mining expert Georg Gottlieb Pusch (1790–1846), which probably inspired Boué's elaboration of special geo-mineralogical signs and symbols to be inserted in topographical maps (Pusch 1819).

The *Préparatifs et instructions préliminaires* were organized into 14 chapters related to each stage of the preparation of the travel: according to Boué, the travelling geologist should first carefully study all the available maps and later the scientific works (*'ouvrages descriptifs'*), taking notes systematically. Then he gave specific instructions on the choice of tools and instruments, figured in some plates, such as several kinds of hammers, axes, pincers, a chisel for a geologist (Fig. 6), a mining compass (*'boussole du mineur'*), a clinometer, a barometer, a thermometer, and an 'aréomètre' (for measuring the specific weight of thermal or salted water) (Boué 1835–1836, **1**, pp. 28–64). The hygrometer was considered useless for a field geologist unless he wanted to make some particular meteorological observation or was interested in measuring the influence of the humidity of air on the process of decomposition of the rocks (Boué 1835–1836, **1**, p. 63). Finally, the full equipment for drawing profiles and sketches was described in great detail because Boué considered the art of drawing as *'un talent très utile au géologue-voyageur'* (Boué 1835–1836, **1**, p. 64). Among the tools listed in this chapter, a new instrument (composed of a glass slab, with a wooden frame and placed on a tripod with a direction device) was also introduced for transferring a drawing of the shape of the landscape onto a transparent sheet of paper placed above the glass.

Other chapters of the *Instructions* concern practical but no less important aspects, such as suitable clothing, special mountain-climbing equipment (such as crampons), passports and other travel documents, advice on the correct behaviour to adopt during the travel, even regarding different countries, the best periods of the year for undertaking geological excursions in different regions, instructions for collecting specimens, for keeping a journal, for surveying and for drawing a simple geological map and sections. Concerning the choice of the best places for making observations, Boué ranked the mountains highest, but also suggested that caves, quarries and mines were worthwhile visting.

The *Guide du Géologue Voyageur* can be considered a complete work for travelling geologists, with a specific long text of very elaborate and detailed instructions. This was because Boué was also trying to order and in some way propose a first codification of the activities and the correct procedures to be adopted by the geologists in the field. On the other hand, during the 1830s, the geologist was becoming a scientific, academic and professional figure. Perhaps Boué was prompted to write the *Guide* by the publication of Charles Lyell's *Principles of Geology* (Lyell 1830–1833), which were presented and reviewed at the *Société géologiques de France* by Boué himself (Vaccari 1998*a*, pp. 40–41). In effect, this essential work for the history of geology, which often referred to significant travels, also undertaken by its author, did not contain any kind of instruction for geological travellers. But from the middle of the nineteenth century many more works, published in different European languages, contributed to the filling of this gap, showing a stimulating context of interests for the meaning of geological travel and fieldwork. This great amount of scientific literature certainly deserves further attention by historians and geologists.

Notes

[1]On the extent of the scientific travel in this period, see Stafford (1984) and Kury (2001).

[2]The *Brief Directions*, written in the early years of the eighteenth century, were published posthumously in Woodward (1728, pp. 93–119). The word 'fossils' included 'every object to be found underground'. On Woodward's texts of instructions, see Torrens (1985), pp. 211–213 and Ciancio (2004), who compares them with an unpublished manuscript of geological instructions written in 1772 by the English naturalist and collector John Strange (1732–1799).

[3]On Vallisneri's geological research, see Vaccari (2000*a*, pp.166–168).

[4]*'da qual parte si pieghino, o se sieno orizzontali, o verticali, o come posti o inarcati, o colorati, ec.'* (to which side the strata bend, if they are horizontal, vertical or arched or coloured). Perrucchini 1726, p. 405.

[5]Perrucchini (1726, pp. 406–407): 'One has to take notice [...] if those marine bodies called *diluvian*, and *antediluvian* may be found, how they lie, or in which position they are found, if petrified, or not petrified, if

enclosed within pebbles, or stones, or sands, and earths, and these latter of which colour, and kind; if at the base, in the middle or at the top of the Mountain; if at the surface, or much underground in the deepest strata; if inside or outside the mines; in which side are placed, if towards the nearest sea, or towards the interior of the mountain, if they are mixed with various materials, or made of only one, or few; if they correspond to those crustaceans, or fish, or insects, or plants which may be found in the nearest or in the furthest sea'.

[6]Targioni Tozzetti (1754, p. 162): '*la natura dei monti che la circoscrivono, [le] loro altezze, curvità, caverne, scogliere, dirupi, lavine, ed altre particolarità*'.

[7]On this journey, see Vaccari 1998*b*, pp. 112–114; on Robilant, see Pipino 1999.

[8]This manuscript, kept in the State Archive of Turin, has been studied by Bulferetti (1970, pp. 12–14).

[9]See Besson (1792; the title of the manuscript of this text is *'Observations sur le choix des minéralogistes et leurs recherches pendant le voyage projeté pour la recherche de M. de la Peyrouse*': *In*: Bibliothèque centrale du Muséum d'Histoire Naturelle, Paris, Ms. 46, dossier n. 4) and Dolomieu (1791). On this expedition, see Collini & Vannoni (1996, pp. 234–238).

[10]'The Mineralogist, always equipped with hammer, chisels or iron wedges hanging from his belt, after having broken some pebbles or rocks, will know which one he must prefer, because it will be better to content himself with these rolled fragments, when one cannot go to find the whole rocks within the mountains [...]. The position of the rocks and the relationships among each other will be observed, if there are layers, beds or well marked strata, their different thicknesses, their reciprocal arrangements, their superimpositions and their inclinations; it will be as well observed toward which side of the sky may be constantly or often found the most steep slopes. Observing the fracture and the decay of big rocky masses from a certain distance an expert eye can evaluate the type and the nature of the rocks'.

[11]Also published in Saussure 1796*b* and as a booklet with the title *Agenda du voyageur géologue* (Saussure 1796*c*). Regarding Saussure and his travels, see Sigrist (2001).

[12]'Finally, the most serious error is the one regarding the superimposition of the strata. I have often seen some inexperienced men in the study of mountains who believed that a stratum was laying over another stratum; for example a granite over a slate, because thay had found the slate at the base of the mountain and granite in high places; whereas the slate was only leaning against the base of the mountain, and in contrast the granite sank into the ground well below the slate. Thus we can state that a stratum is placed below another stratum, only when we can really see the former sinking below the latter. And even when we can see a rock clearly superimposed on another, it is still necessary to evaluate if the above rock has perhaps been moved to that place accidentally, if it has slipped or rolled down from a higher mountain.'

[13]Struve (1799, pp. 33–34): 'Il semble, au premier coup-d'oeil, que plus l'on multiplie les moyens d'observation dans les voyages pédestres que l'on entreprend sur les montagnes, mieux l'on travaille au profit de la science que l'on y va à étudier' (p. 33: 'It seems, at first sight, that the more we increase the methods of observation during the mountain travels on foot the better we can work for the advantage of the science to be studied').

[14]On the instruments used by nineteenth century geologists, see Larsen (1996). On the portable equipment for chemical-mineralogical analysis, see Burchard (1993) and Gee (1989).

References

BESSON, H. 1792. Observations sur les moyens de rendre utiles les voyages des Naturalistes. Lues à la Société des Naturalistes, le 28 février 1791. *Journal d'Histoire Naturelle (Paris)*, **2**, 185–210.

BEUDANT, F. S. 1822. *Voyage Minéralogique et Géologique, en Hongrie, pendant l'Année 1818*. 4 Volumes. Verdiere, Paris.

BOUÉ, A. 1835–1836. *Guide du Géologue-Voyageur, sur le Modèle del'Agenda Geognostica de M. Léonhard*. 2 Volumes. F. G. Levrault, Paris.

BOURGUET, M.-N. 1997. Le collecte du monde: voyage et histoire naturelle (fin XVIIème siècle–début XIXème siècle). *In*: BLANCKAERT, C., COHEN, C., CORSI, P. & FISCHER, J.-L. (eds) *Le Muséum au premier siècle de son histoire*. Éditions du Muséum d'Histoire Naturelle, Paris, 163–196.

BRARD, C. P. 1805. *Manuel du Minéralogiste et du Géologue Voyageur*. Maradan, Paris.

BULFERETTI, L. 1970. I viaggi minerari di Carlo Antonio Napione 'innovatore' nel Piemonte e nel Brasile. *Rassegna Economica*, **34**, 7–31.

BURCHARD, U. 1993. Geschichte und Instrumentarium der Lötrohrkunde. *Deutsches Museum Wissenschaftliches Jahrbuch 1992/93*, 7–62.

CIANCIO, L. 2004. Raccogliere minerali e fossili nel Settecento: le 'istruzioni' di John Woodward (1696) e di John Strange (1772) a confronto. *Schede umanistiche*, **1**, 45–76.

COLLINI, S. & VANNONI, A. 1996. La Societé d'Histoire Naturelle e il viaggio di D'Entrecasteux alla ricerca di La Pérouse: le istruzioni scientifiche per i viaggiatori. *Nuncius* **11**, 227–275.

DE LUC, J. A. 1810–1811. *Geological Travels. Translated from the French manuscript*. 3 Volumes. F. C. & J. Rivington, London.

DE LUC, J. A. 1813. *Geological Travels in Some Parts of France, Switzerland and Germany*. 2 Volumes. F. C. & J. Rivington, London.

DOLOMIEU, D. DE 1791. Notes à communiquer à Messieurs les naturalistes qui font le voyage de la mer du Sud et des contrées voisines de Pole Austral. *Observations sur la Physique (Paris)*, **39**, 310–317.

GEE, B. 1989. Amusement chests and portable laboratories: practical alternatives to the regular laboratory. *In*: JAMES, F. A. J. L. (ed.) *The Development of the Laboratory. Essays on the Place of Experiment in Industrial Civilization*. Macmillan Press, London, 37–59.

GREPPI, C. 2000. Observer les montagnes d'en bas. Le rôle de Giovanni Targioni Tozzetti dans l'exploration scientifique de la Toscane au XVIIIe siècle. *In*: PONT, J.-C. & LACKI, J. *Une Cordée Originale. Histoire des Relations entre Science et Montagne*. Georg, Genève, 390–405.

KURY, L. 2001. *Histoire naturelle et voyages scientifiques (1780–1830)*. L'Harmattan, Paris.

LARSEN, A. 1996. Equipment for the field. *In*: JARDINE, N., SECORD, J. A. & SPARY, E. C. (eds) *Cultures of Natural History*. Cambridge University Press, Cambridge, 358–377.

LEHMANN, J. G. 1759. Discours sur le moyens de faire une description du monde souterrein. *In*: LEHMANN, J. G. *Traités de Physique, d'Histoire Naturelle, de Minéralogie et de Métallurgie*. Tome Premier. *L'Art des Mines ou introduction aux conoissances nécessaires pour l'exploitation des Mines Métalliques*, Ouvrage traduit de l'Allemand. Hérissant, Paris, 305–313.

LEONHARD, K. C. 1838. *Agenda Geognostica. Hülfsbuch für reisende Gebirgsforscher und Leitfaden zu Vorträgen über angewandte Geognosie. Zweite, vermehrte und verbesserte Auflage*. Mohr, Heidelberg. [1st edn Heidelberg 1829].

LINNÉ, C. 1760. Instructio Peregrinatoris quam sub praesidio D.D. Car. Linnaei proposuit Ericus And. Nordblad, Gevalia–Gestricus. Upsaliae 1759, Maji 9. *In*: LINNÉ, C. *Amoenitates Academicae*, 5 Volumes. L. Salvii, Holmiae, 298–313.

LYELL, C. 1830–1833. *Principles of Geology, being an attempt to explain the former changes of the Earth's surface, by reference to causes now in operation*. 3 Volumes. John Murray, London.

MICHÉ, A. 1795. Essai d'un Manuel du Voyageur Metallurgiste. *Journal des Mines (Paris)*, **1**, 3–25.

PERRUCCHINI, G. 1726. Continuazione dell'Estratto d'alcune Notizie intorno alla Garfagnana, cavate dal primo Viaggio Montano del Signor Antonio Vallisnieri. *Supplementi al Giornale de' Letterati d'Italia (Venezia)*, **3**, 376–428.

PINI, E. 1802. *Viaggio geologico per diverse parti meridionali dell'Italia. Esposto in lettere*. Mainardi, Milano.

PIPINO, G. 1999. Spirito Nicolis di Robilant e l'istituzione della prima accademia mineraria in Europa. *Physis*, **36**, 177–213.

PUSCH, G. G. 1819. *Geognostischer Katechismus oder Anweisung zum praktischen Geognosten für angeschende Bergleute und Geognosten*. Craz und Gerlach, Freiberg.

ROBILANT, S. B. NICOLIS DI 1790. *De l'utilité et de l'importance des Voyages et des Courses dans son Propre Pays*. Frères Reycends, Turin.

SACHS, J. VON. 1892. *Histoire de la Botanique du XVIe siècle à 1860*. Reinwald Paris. [English edition: *History of Botany* (1530–1860). Clarendon Press, Oxford (1890)].

SAUSSURE, H. B. DE 1796*a*. Agenda ou Tableau général des Observations et des Recherches dont les résultats doivent servir da base à la théorie de la Terre. *In*: SAUSSURE, H. B. DE. *Voyages dans les Alpes*, 4 Volumes. L. Fauche-Borel, Neuchâtel, § 2304–2327.

SAUSSURE, H. B. DE 1796*b*. Agenda ou Tableau général des Observations et des Recherches dont les résultats doivent servir da base à la Théorie de la Terre, par M. de Saussure, de Géneve (Envoyé par l'auteur au Conseil des Mines, pour être inséré dans son Journal). *Journal des Mines (Paris)*, **4**, 1–70.

SAUSSURE, H. B. DE 1796*c*. *Agenda du voyageur géologue, tiré du quatrième volume des Voyages dans les Alpes*. Paschoud, Genève.

SAUSSURE, H. B. DE 1799*a*. Agenda; or, A Collection of Observations and Researches the Results of Which May Serve as the Foundation for a Theory of the Earth. *Philosophical Magazine (London)*, **3**, 33–41, 147–156.

SAUSSURE, H. B. DE 1799*b*. Agenda, oder allgemeine Uebersicht der Untersuchungen, und Beobachtungen, deren Resultate zur Gründung einer Theorie der Erde nothwendig sind; aus dem Franz. (*Journal des Mines* n.XX). *Jahrbücher der Berg- und Hüttenkunde* (Salzburg), **4**, 1–70.

SCHEUCHZER, J. J. 1723. *Itinera per Helvetiae Alpinas Regiones*. 4 Volumes. P. Van der Aa, Lugduni Batavorum.

SEIDL, J. 2002, Ami Boué (1741–1881), géoscientifique du XIXe siècle, *Comptes Rendus Palevol*, **1**, 649–656.

SIGRIST, R. (ed.) 2001. *H.-B. de Saussure (1740–1799). Un regard sur la terre*. Georg, Genève.

SÖRLIN, S. 1989. Scientific travel – the Linnean tradition. *In*: FRÄNGSMYR, T. (ed.) *Science in Sweden. The Royal Swedish Academy of Sciences 1739–1989*. Science History Publications, Canton (USA), 96–123.

STAFFORD, B. M. 1984. *Voyage into Substance: Art, Science, Nature, and the illustrated Travel Account. 1760–1840*. MIT Press, Cambridge.

STRUVE, H. 1799. *Méthode analythique des fossiles, fondée sur leurs charactères extérieures*. Chez Henri Tardieu, Paris [first edition: Lausanne, 1797].

TARGIONI TOZZETTI, G. 1751–1754. *Relazioni d'alcuni viaggi fatti in diverse parti della Toscana per osservare le produzioni naturali e gli antichi monumenti di essa*. 6 Volumes. Stamperia Imperiale, Firenze.

TARGIONI TOZZETTI, G. 1754. *Prodromo della Corografia e della Topografia Fisica della Toscana*. Stamperia Imperiale, Firenze.

TAYLOR, K. 2007. Geological travellers in Auvergne, 1751–1800. *In*: WYSE JACKON, P. N. (ed.) *Four Centuries of Geological Travel: The Search for Knowledge on Foot, Bicycle, Sledge and Camel*. Geological Society, London, Special Publications, **287**, 73–95.

TORRENS, H. S. 1985. *Early collecting in the field of geology*. *In*: IMPEY, O. & MACGREGOR, A. (eds) *The Origins of Museums. The Cabinet of Curiosities in Sixteenth- and Seventeenth-Century Europe*. Oxford, Clarendon Press, 204–213.

VACCARI, E. 1998*a*. Lyell's reception on the continent of Europe: a contribution to an open historiographical problem. *In*: BLUNDELL, D. J. & SCOTT, A. C. (eds)

Lyell: the Past is the Key to the Present. Geological Society, London, Special Publications, **143**, 40–41.

VACCARI, E. 1998*b*. Mineralogy and mining in Italy between eighteenth and nineteenth centuries: the extent of Wernerian influences from Turin to Naples. *In*: FRITSCHER, B. & HENDERSON, F. (eds) *Toward an History of Mineralogy, Petrology and Geochemistry* ('Algorismus' **23**), Institut für Geschichte der Naturwissenschaften, München, 107–130.

VACCARI, E. 2000*a*. Mining and knowledge of the Earth in eighteenth-century Italy. *Annals of Science*, **57**, 163–180.

VACCARI, E. 2000*b*. Voyageurs scientifiques dans les Apennins entre le 17e et 18e siècle: perspectives géologiques. *In*: PONT, J.-C. & LACKI, J. (eds) *Une cordée originale. Histoire des relations entre science et montagne*. Georg, Genève, 160–181.

VALLISNERI, A. 1715. *Lezione Accademica intorno all'origine delle Fontane, colle Annotazioni per chiarezza maggiore della medesima*. Gio. Gabbriello Ertz, Venezia.

WOODWARD, J. 1696. *Brief Instructions for Making Observations in all Parts of the World as also for Collecting, Preserving, and Sending over Natural Things, [...] presented to the Royal Society*. Wilkin, London.

WOODWARD, J. 1728. *Fossils of all Kinds Digested into a Method, Suitable to their Mutual Relation and Affinity*. Innys, London.

The geological observations of Robert Hooke (1635–1703) on the Isle of Wight

E. T. DRAKE

College of Oceanic & Atmospheric Sciences, Oregon State University, Corvallis, Oregon 97331, USA (e-mail: drakee@onid.orst.edu)

Abstract: As Curator of Experiments at the Royal Society of London, Robert Hooke (1635–1703) was too busy to have been considered a 'geological traveller'. Yet he made fundamental geological observations whenever he did travel. He set these observations in a series of lectures he gave at the Royal Society over a period of some thirty years. These lectures were published posthumously by Richard Waller in 1705 as *Lectures and Discourses of Earthquakes and Subterraneous Eruptions*. Although his contemporary Nicolaus Stenonis, or Steno, has been recognized as the founder of geology, Hooke's more profound and compelling observations and explanations have been largely ignored by the geological community. There is also evidence that Hutton benefited considerably from Hooke's ideas.

Hooke's writings show that he derived many of his geological hypotheses from his intimate knowledge of the processes taking place on the shores of his birthplace, the Isle of Wight. This paper presents what Hooke observed and described and is illustrated with photos taken by the author on the shores of the Isle of Wight.

Few geologists are aware of the debt they owe to the seventeenth century scientist Robert Hooke. Most scientists know his name by his eponymous law but little else of his achievements. Yet his contributions to many branches of science, especially Earth science, mark him among the most original, influential, and prolific scientists in history. During Hooke's lifetime he was widely renowned for his many productive ideas and inventions. As Curator of Experiments at the Royal Society, London, he originated some of the fundamental principles governing celestial mechanics, notably the concept of centripetal force, and generously shared them with Isaac Newton, allowing the latter to perceive the gravitation problem correctly. Newton gave scant and begrudging acknowledgment of Hooke's gift to him and then only after being urged to do so by friends. Hooke proposed the oblate spheroidal shape of the Earth long before Newton claimed credit for recognizing its true shape. Hooke was the first to describe the iridescent interference colours seen when light falls on a layer of air between two thin glass plates, but ironically, these are known as 'Newton's Rings'. Hooke's other discoveries include the cellular structure of plants, micro-organisms, the giant spot in Jupiter, proof of the planet's rotation, the theory of combustion, Hooke's Law and probably also Boyle's Law. His equally numerous inventions, or improvements to instruments, include the universal joint, the air pump, many meteorological instruments (e.g. barometer, thermometer, wind gauge and hygrometer), the clock-driven telescope and telescopic sight, the sextant, the pocket watch, the iris diaphragm, the bubble level, and more. He was indeed, as his biographer Richard Waller wrote in 1705, 'an active, restless, indefatigable Genius'. In addition to all his activities as Curator of Experiments and Cutlerian Lecturer at the Royal Society in London and Professor of Geometry at Gresham College, he also designed numerous buildings and structures after the great Fire of London of 1666. Some of his structures include the famous London landmark, The Monument, commemorating the Fire, and the great dome of St Paul's Cathedral, a true engineering feat, but both are credited to Christopher Wren.[1]

Historical writings, bolstered by the power of the apple legend, have ensured that Newton's name has reigned supreme over the last three hundred years, and poets of the eighteenth- and nineteenth-century sang the praises of Newton, likening him to a god. Take, for example, the lines by Alexander Pope (1688–1744):

> Nature, and Nature's laws lay hid in night:
> God said, *Let Newton be*! and all was light.

Or these lines by William Cowper (1731–1800):

> Newton, childlike sage!
> Sagacious reader of the works of God.

Or these by William Wordsworth (1770–1850) expressing the *one* eternal mind:

> Where the statue stood
> of Newton, with his prism and silent face,
> The marble index of a mind for ever
> Voyaging through strange seas of Thought, alone.

From: WYSE JACKSON, P. N. (ed.) *Four Centuries of Geological Travel: The Search for Knowledge on Foot, Bicycle, Sledge and Camel.* Geological Society, London, Special Publications, **287**, 19–30.
DOI: 10.1144/SP287.3 0305-8719/07/$15.00

No one wrote poetry about Hooke. Such adulation of Newton was common after the seventeenth century with the consequence that the extraordinary achievements of Hooke have been overshadowed or over-looked. Although the 2003 conferences in England commemorating the tercentenary of his death in 1703 and the recent publication of several books have tried to restore Hooke to prominence, his name is still relatively obscure today.

One of the greatest pieces of geological literature, however, has been available to us since the Royal Society of London published Hooke's *Posthumous Works* in 1705. Perhaps British and continental geologists and historians of geology are familiar with this volume, but most geologists in America and possibly also in other parts of the world are not aware of its existence, far fewer have read it. Hooke gave his series of lectures to the Society, starting around 1667 and spanning a period of over thirty years. These were entitled *Lectures and Discourses of Earthquakes and Subterraneous Eruptions* and have been reprinted by the Johnson Reprint Corporation in 1969, and then reprinted again by Arno Press in 1978 under the general editorship of the late Claude C. Albritton, Jr. These lectures have now been arranged in chronological order according to Rappaport (1986), and I have transcribed them into modern type font for easier perusal and also annotated and interpreted them as well as written a biography of Hooke in my 1996 book, *Restless Genius*. His most perceptive ideas about geology were expressed in the series of lectures that ended on 15 September 1668, and in the series between 28 May 1684, and toward the end of 1687. No portrait of Hooke exists although he had sat for a portrait artist, but a representation (not a likeness) of him existed in the 'Worthies' stained glass window at St Helen's Church, Bishopsgate, London, until it was destroyed by an IRA bomb in 1992. Figure 1 shows the image of Hooke in the window before its destruction.

Fig. 1. Representation of Robert Hooke in the stained glass window named 'The Worthies Window' in St Helen's Church, Bishopsgate, London, where Hooke was buried. Unfortunately, not only was Hooke's grave lost during reconstruction of the church in the nineteenth century, but the stained glass window was destroyed from a bomb blast in 1992.

This paper shows that the shores of the Isle of Wight were the source of Robert Hooke's inspiration for his insightful ideas and conclusions about the origin and evolution of terrestrial features. Hooke's lectures laid the foundation for geology in expressing such ideas as the cyclicity of many Earth processes such as sedimentation, denudation and uplift, then erosion and again deposition, and so forth. Implicit in this process is the idea of the unconformity so important to understanding geological history from the rock strata. In spite of the general thesis that Hooke's ideas were forgotten by later geologists, I will briefly show that his ideas *were* transmitted down through the ages, especially through James Hutton. Hooke's contributions, more than those of Nicholas Steno, not only laid down the foundation of the science of geology, but many of them are still important to the current geological paradigm of plate tectonics.

Hooke's insight regarding fossils

Hooke understood the organic origin of fossils and the process of fossilization and advocated ideas about these subjects much against the prevailing thought. The latter claimed fossils to be a *lusus naturæ*, a trick of nature formed by some magical or astrological influence through a 'Plastick Vertue.' Understanding the nature of fossils led Hooke to express his ideas on biological evolution including the concept of extinction of species and generation of new ones as a result of environmental changes. He can be considered an early evolutionist, a precursor of Darwin. The subject of Hooke's remarkable ideas on evolution is treated elsewhere (Drake 1996, pp. 98–103).[2] Moreover, he immediately recognized the usefulness of fossils not only as a record of the past but he also, astoundingly, recognized the potential importance of fossils in establishing a chronology.

Hooke's concept of polar wander

Hooke also initiated the idea of polar wandering on an oblate spheroid Earth as a means for the exchange of land and sea areas. The Earth's rotation, he proposed, caused a bulge and thus greater altitude at the equator versus a flattening at the poles. He maintained that over time, a change in the positions of the poles on the Earth surface due to a change in the moment of inertia would cause different areas of bulging and flattening with the creation of new land or sea areas. The complicated subject of polar wander is treated elsewhere (Drake 1996, pp. 87–95).[3] Suffice it to mention here that polar wander was fundamental to Alfred Wegener in his hypothesis of continental drift in the understanding of the relationship among continental drift, climatic belts and biogeography. Further, the concept is an important one in the current geological paradigm of plate tectonics.

The Isle of Wight and cyclic processes

One particular line of research about Hooke was, 'What were the origins of his ideas about Earth science that seem so startling during an age when most people, even among the intelligentsia, were still steeped in medieval superstition and mysticism?' Reading his *Discourses of Earthquakes*, the answer to the source of Hooke's inspiration becomes clear. In supporting some of Pliny's writings, Hooke referred to 'an observation of my own, which I have often taken notice of, and lately examined very diligently ... which I made upon the western shore of the Isle of Wight'. Hooke's birthplace was the town of Freshwater near the western edge of the Isle of Wight where his father was the Curate of the Church of All Saints. Hooke obviously spent many happy hours as a child climbing around the western cliffs of Wight, as he had 'often taken notice' of the physical processes taking place there. Since almost his entire professional life was spent in London at the Royal Society, any other place where he could have 'often taken notice' of the changes along the shore would have been his childhood home environment. At the death of his father, because of his demonstrated artistic ability, he was sent away at the age of 13 to serve as an apprentice to the famous portrait painter, Sir Peter Lely. But because of incapacitating headaches caused by the oil paints, he was then sent to school at Westminster under the tutelage of Dr Richard Busby. During the lifetime of his mother, Hooke as an adult apparently at times had returned to his hometown because he 'lately examined very diligently' those same shores. Throughout his *Discourses* he refers to the Needles, which are some pinnacles of chalk off the tip of the western part of the island (Fig. 2). These pinnacles are the erosional remnants of the vertical arm of one of two major monoclines, the Sandown and the Brixton folds. He also repeatedly mentioned the cliffs around Freshwater Bay from which he collected fossils. In researching Hooke, therefore, I made a trip to the Isle of Wight in 1990 and again in 1996 to see for myself what could have influenced his geological thinking.

I saw firsthand what Hooke saw and marvelled at in the cliffs of his island. I could almost sense what must have impressed this highly imaginative man: I could imagine what questions must have come to his mind. How did these marine shells come to be embedded there in solid rock sixty feet above the high-water mark? What forces raised these layers of sediment up so high or tilted up almost vertical—the layers that surely were laid down horizontally by the sea?

He collected many fossils for the Royal Society which unfortunately are now lost along with his portrait as Secretary of the Society, many papers, and practically all of the hundreds if not thousands of scientific instruments and models he designed and built. It is my speculation, and I daresay that of others too, that Newton, who hated his long-time rival Hooke and who accepted the Society's presidency only after Hooke died, might very well

Fig. 2. The Needles, off the western tip of the Isle of Wight.

have had something to do with the disappearance of this massive collection. It would have been in line with Newton's personality, particularly if the missing papers had indicated the extent of Hooke's knowledge regarding universal gravitation prior to the publication of *Principia* (Patterson 1949, 1950).[4]

Hooke was an accomplished artist who illustrated and described many fossils in his collection. Luckily, some of his drawings survived, including, for example, those of ammonites published in his posthumous works; the artistry and accuracy of these drawings are to be admired even today (Fig. 3). He made what must be the first drawing of a foraminifer. Having examined it through his microscope he gave a description of this tiny shell, no bigger than the point of a pin, and noted that it resembled 'the shell of a small water-snail with a flat spiral shell: it had twelve wreathings, a, b, c, d, e, etc. [which he labelled in the drawing] all in proportion growing one less than another toward the middle or centre of the shell, where there was a very small round white spot' (Hooke 1665, Observation 11, pp. 80–81). One can almost feel the excitement of discovery Hooke felt in realizing that his microscopic shell represented such a tiny life-form. This image of Hooke looking down his microscope is not much different from a present-day micropalaeontologist bent over his or her microscope. The difference is that the scientist today usually does not have to design and build the microscope first.

Finding teeming fossil marine life embedded in sediments sixty feet above the high-water mark, Hooke concluded that these sediments deposited at the bottom of the sea had been raised to become land. He also stated that this 'raising' of the land was accomplished either 'by degrees' or more rapidly and violently by earthquakes. The word 'earthquakes' was used by him to mean any kind of earth movements, including erosion and deposition, strata pushed up by mountain-building forces or those that crack, slip, and slide due to gravity. He witnessed dramatic evidence of the dissolution and erosion of the chalky cliffs. The Cretaceous transgression of the seas deposited white chalk widely, and some 450 m of it occur on the south of Isle of Wight. The Middle Chalk reaches 80–90 m with flints appearing near the top which continue throughout the Upper Chalk. Toward the end of the Cretaceous the seas slowly retreated, leaving great areas of chalk exposed that immediately started to erode.

Huge blocks broke off the main island, some during historical and recent times. Three of these blocks were named by the local inhabitants Mermaid, Arch and Stag (Fig. 4). Two of them were known to have existed in the seventeenth century; a third, Mermaid, closest to land, toppled into the sea from the cliff sometime during the twentieth century. Arch Rock, the middle block, was still there when this picture was taken (Fig. 4) in 1990 but it was destroyed by the erosional power of the waves and disappeared in the water on the night of 24 October 1992, and is no more. Before the ravages of the sea

Fig. 3. Ammonites drawn by Robert Hooke and described in his *Discourses of Earthquakes and Subterraneous Eruptions*, published posthumously in 1705.

Fig. 4. Erosional blocks that toppled from the cliffs over the centuries, named (from left to right) Mermaid, Arch and Stag. Arch Rock disintegrated by wave action and disappeared into the sea on the night of 24 October 1992. This photo was taken in 1990 when it was still there.

had time to make Arch Rock disappear altogether, two of these features were described in an 1869 edition of *Nelson's Hand Book to the Isle of Wight*: '. . . on the eastern side of the small bay formed here by the curvature of the cliffs, [is] *The Arched Rock* [today referred to as Arch], a huge mass, originally part of the cliff, but now insulated from it at a distance of some 600 feet. The constant seas have not only effected this separation, but have beaten a way through it in the shape of a quaint Gothic-like arch. Another rude mass towers above the water close at hand.' This 'rude mass' is the block farthest from the cliff in Figure 4, called *The Deer Pond* in Nelson's handbook, now called Stag. No mention, however, is made of the huge block nearest the cliff called Mermaid, because Nelson's handbook was published in the nineteenth century when the block had not split from the cliff (Adams 1869). Mermaid broke away in the twentieth century within the memory of many residents of the Isle of Wight.

At such a rate of erosion, it is reasonable to assume that Freshwater Bay itself may have enlarged over time as a result of erosion by the sea. Some man-made features, like an old road on the east side of the bay, are now truncated by the cliff and left hanging (Fig. 5). Hooke was profoundly aware of the erosional power at play here where he had grown up and revisited to study the changes 'very diligently.' Hooke thus recognized not only the organic origin of fossils and the breaking down of the land by erosion but the re-deposition of the broken sediments, embedding of fossils, and consolidation into rock which then can be uplifted again to form land. This process described in Hooke's own words as follows demonstrates his powers of observation and shows him to be a true geologist:

I observed a Cliff of a pretty height, which, by the constant washing of the Water at the bottom of it, is continually, especially after Frosts and great Rains, foundering and tumbling down into the Sea underneath it. Along the Shore underneath this Cliff, are a great number of Rocks and large Stones confusedly placed, some covered, others quite out of the Water; all which Rocks I found to be compounded of Sand and Clay, and Shells, and such kind of Stones, as the Shore was covered with. Examining the Hardness of some that lay as far into the Water as the Low-Water-mark, I found them to be altogether as hard, if not much harder than *Portland* or *Purbeck*-stone: Others of them

Fig. 5. Road in foreground is truncated by the cliff as a result of erosion and left hanging on the east side of Freshwater Bay, Isle of Wight.

that lay not so far into the Sea, I found much softer, as having in probability not been so long exposed to the Vicissitudes of the Tides: Others of them I found so very soft, that I could easily with my Foot crush them, and make Impressions into them, and could thrust a Walking-stick I had in my Hand a great depth into them: Others that had been but newly foundered down, were yet more soft, as having been scarce wash'd by the Salt Water. All these were perfectly of the same Substance with the Cliffs, from whence they had manifestly tumbled, and consisted of Layers of Shells, Sand, Clay, Gravel, Earth, &c. and from all the circumstances I could examine, I do judge them to have been the Parts of the Neighbouring Cliff foundered down, and rowl'd and wash'd by degrees into the Sea; and, by the petrifying Power of the Salt Water, converted into perfect hard compacted Stones. I have likewise since observed the like *Phænomena* on other Shores. And I doubt not but any inquisitive Naturalist may find infinite of the like Instances all along the Coast of England, and other Countries where there are such kind of foundering Cliffs. (Hooke 1705, p. 297).

Hooke's concept of time

All the cyclic processes Hooke observed plus his other ideas involving the history of his terraqueous globe, such as polar wander and biological evolution, all taking immense amounts of time, prompted him from the start to question the validity of the narration in *Genesis* and the scriptural chronology. He asserted that Noah's Flood was nothing special, that there had been many floods in the past. Most people in the seventeenth century thought of the occurrence of fossils, and minerals in crystal form, as the result of supernatural forces. The few of his scholarly contemporaries who recognized the organic origin of fossils believed that all the superficial features of the Earth, including the occurrence of fossils left in the rocks, were the consequence of Noah's Flood alone. And furthermore, they believed that there were no seas before the Flood. John Woodward (1665–1728), whose sentiments perhaps typify those of his contemporary intelligentsia, stated that 'the terraqueous Globe is to this day nearly in the same condition that the Universal Deluge left it: being also like to continue so till the time of its final ruin and dissolution' (Woodward 1695). Hooke believed passionately that the age of the Earth must have been older than the few thousand years allowed by Archbishop Ussher whose date for the beginning of the world, 22 October 4004 BC, had been firmly established and hardly subject to debate. Hooke insisted, however, that dry land, even some very high mountains such as the Alps, because of the occurrence of marine fossils there, had at one time been under the sea. He questioned, therefore, the prevailing concept of time. He said:

I think it will be evident, that it could not be from the Flood of Noah, since the duration of that which was but about two hundred Natural Days, or half an Year could not afford time enough for the production and perfection of so many and so great and full grown Shells, as these which are so found do testify; besides the quantity and thickness of the Beds of Sand with which they are many times found mixed, do argue that there must needs be a much longer time of the Seas Residence above the same, than so short a space [of time] can afford. (Hooke 1705, p. 341).

It is inconceivable that the mathematically-minded Hooke would not have made some quick mental estimates and calculations of how long it would take for the completion of one such cycle of processes. The Isle of Wight, especially, is dramatic demonstration of the forces of nature at work. Hooke thus felt strongly that the time allowed by the Scripture simply would not accomplish all that he observed. As the religious climate of Hooke's world became more oppressive with the ascension to the throne of William of Orange and his wife Mary in 1689, Hooke realized that he needed to mollify his audience if he mentioned a time element that might offend because it exceeded the biblical chronology. For example, he said that if a duration he mentioned seemed too long for anyone to accept, one should try to think that perhaps a day at the beginning of the world might have been shorter than now (Hooke 1705, p. 322). Or, he'd refer to the writings of the Egyptians or Chinese who told of events of 'many thousand Years more than ever we in Europe heard of by our Writings.' But to placate his listeners, he'd quickly add, 'if their Chronology may be granted, which indeed there is great reason to question' (Hooke 1705, p. 327). We know, however, that he admired the Chinese greatly, for their history, their language and their inventions. He even attempted to learn to read and write Chinese. He said that the Chinese 'do make the World 88 640 000 years old.' Although Hooke might have had no difficulty thinking in terms of many millions of years for the age of the Earth, there is no evidence that he could have conceived of the 4.5 billions of years we now accept to be its age.

Hooke and Steno

Hooke's understanding of sediments deposited into 'layers' indicates that he was fully cognizant of both the concept of what became William Smith's Law of Superposition and the original horizontality of strata, both proposed by Steno (1638–1686) in his *Prodromus* (Steno 1667). I think that Hooke would have considered these simple truths as self-evident: i.e. when the bottom layer of sediments was being deposited in its watery grave, no other layer was above it, and it is a matter of common sense that anything that settles in water would be horizontal at first. Hooke's observations of the processes taking place on his shore constantly confirmed such truths. His writings show that he went beyond these simple truths to explain the cyclicity of these processes, how the loose sediments turned into rock, and why some obviously marine materials ended up high up the cliffs (Drake & Komar 1981). His ideas of rocks, minerals and fossils certainly surpassed those of Steno. Although Steno was among the few intelligentsia of that age to recognize the organic origin of fossils, he too adhered to the scriptural chronology and thought that fossils had been left by Noah's Flood. Hooke, on the other hand, discounted Noah's Flood as not very important in this respect and considered fossils as invaluable records of the past that could even perhaps give us clues toward establishing a 'chronology', one that is presumably different from the biblical version. His words were as follows:

There is no Coin can so well inform any Antiquary that there has been such or such a place subject to such a Prince, as these [i.e. the fossils] will certify a Natural Antiquary, that such and such places have been under the Water, that there have been such kind of Animals, that there have been such and such preceding Alterations and Changes of the Superficial Parts of the Earth. (Hooke 1705, p. 321)

And the following quote more than one hundred years before William Smith:

And tho' it must be granted, that it is very difficult to read them, and to raise a *Chronology* [Hooke's own italics] out of them, and to state the intervalls of the Times wherein such, or such Catastrophies and Mutations have happened; yet 'tis not impossible, but that, by the help of those joined to other means and assistances of Information, much may be done even in that part of Information also. (Hooke 1705, p. 411)

A comparison of the illustrations of crystal formation drawn by Steno and by Hooke shows that although Steno believed that crystal forms grow by surface accretion, Hooke's demonstrate that the external form is an expression of the internal arrangement of particles (Fig. 6). As has been pointed out by Cecil Schneer (1960), Hooke's diagrams express both the constancy of interfacial angles illustrated by Steno and also von Haüy's law of rational axial intercepts.

Hooke and Hutton

Hooke's ideas have been shown not only to surpass those of Steno, who is widely acclaimed as 'founder' of the science of geology, but evidence shows that they also had a profound influence on the thinking of James Hutton who is considered 'father' of modern geology (Drake 1981). Gordon Herries Davies has noticed that James Hutton's theory of the Earth, as announced to the Royal Society of Edinburgh in 1785, was almost identical to Hooke's system of the Earth (Davies 1964). This similarity in my opinion was not a coincidence. Davies believed that Hooke deserves to be

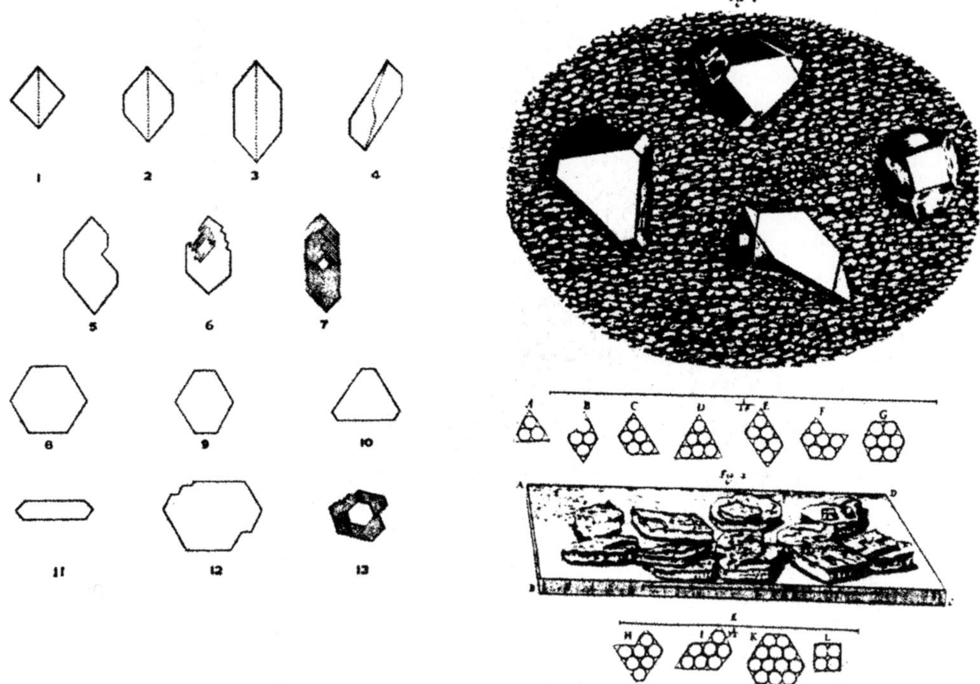

Fig. 6. A comparison of Steno's concept of the growth of crystals on the left (from *Prodromus*, 1669) and that of Hooke on the right (from *Micrographia*, 1665). Steno's diagrams show that crystal forms grow by surface accretion while Hooke's show that the external form is an expression of the internal arrangement of particles.

remembered as a precursor of Hutton. David Oldroyd (1972) has also expressed the sentiment that 'in a number of instances Hooke should receive the credit for ideas which are usually believed to have originated in the work of James Hutton'.

In spite of an interim of about fifty years after Hooke's death during which the figure of Newton dominated the intellectual scene, Hooke's ideas were transmitted by later writings demonstrating the continuity of the development of geological thought. There is convincing textual evidence that Hutton adopted many of Hooke's ideas (Drake 1981, 1996). Hutton's basic ideas are the same as those of Hooke, but he never referred to Hooke by name although in his later writings he did mention other names. In those few places where Hutton disagreed, the style of his writing is polemical. It is as if Hutton was arguing with Hooke. For example, Hooke proposed several ways in which land could be raised from its former level and conversely he assigned several ways in which land could sink. Hutton accepted the methods of raising but argued vehemently against the whole idea of land sinking. He said:

> the sinking the body of the former land into the solid globe, so as to swallow up the greater part of the ocean after it, if not a natural impossibility, would be at least a superfluous exertion of the power of nature. Such an operation as this would discover as little wisdom in the end elected, as in the means appropriated to that end; for, if the land be not wasted and worn away in the natural operations of the globe, why make such a convulsion in the world in order to renew the land? If, again, the land naturally decays, why employ so extraordinary a power, in order to hide a former continent of land, and puzzle man? (Hutton 1788)

In light of plate tectonics theory and the 'sinking' of land as a result of subduction, the lowering of the topography as a result of thermal contraction, and other sinking processes, Hooke seems indeed to have been more clairvoyant than Hutton. If physical sinking is superfluous to Hutton's theory, why did he mention it and further argue so strenuously against it? My interpretation is that Hutton studied Hooke's ideas very carefully, accepted those with which he agreed, but felt the need to strike down those with which he disagreed. Another example of Hutton's disagreement with Hooke is that of the idea of polar wander or 'axial displacement' as it was called in Hooke's time. Axial displacement or polar wander was an idea proposed by Hooke to explain the interchange of land and sea areas; the idea was ridiculed and rejected following his presentation to the Royal Society in 1687. His rationale was that if the rotation of the planet caused an equatorial bulge, then if the polar points moved to another part on the surface of the globe as a result of a change in the moment of inertia, then the equatorial bulge would eventually adjust to the new polar position as well. Such adjustment over time would cause a flattening at the new poles and a bulge at the new equator. He offered polar wander as an additional possible explanation for the redistribution of land and water, because, he claimed, there is

> a more than ordinary swelling or rising of the Sea in those Parts which are near the Æquinnoctial, and a sinking and receeding of the Sea from those which are near the Poles; so that as any Parts do increase in their Latitudes, so will the Sea grow shallower, and as their Latitudes decrease, so must the sea swell and grow high; by which means many submarine Regions must become dry Land, and many other Lands will be overflown by the Sea, and these variations being slow, and by degrees will leave very lasting Remarks of such States and Positions, in the superficial Substances of the Earth. (Hooke 1705, p. 347)

Hutton's passionate objections to this idea across a century were:

> no motion of the sea, caused by this earth revolving in the solar system, could bring about that end; for let us suppose the axis of the earth to be changed from the present poles, and placed in the equinoctial line, the consequence of this might, indeed, be the formation of a continent of land about each new pole, from whence the sea would run towards the new equator; but all the rest of the globe would remain an ocean. Some new points might be discovered, and others, which before appeared above the surface of the sea, would be sunk by the rising of the water; but on the whole, land could only be gained substantially at the poles. Such a supposition as this, if applied to the present state of things, would be destitute of every support, as being incapable of explaining what appears. (Hutton 1788, p. 222)

In spite of Hutton's protestations, this statement seems to support Hooke's idea. The idea itself, however, must have really bothered Hutton, because after his summary rejection of it, he found it necessary to continue his argument,

> But even allowing that, by the changed axis of the earth, or any other operation of the globe, as a planetary body revolving in the solar system, great continents of land could have been erected from the place of their formation, the bottom of the sea, and placed in a higher elevation, compared with the surface of that water, yet such a continent as this could not have continued stationary for many thousand year—

without, he continues, also the process of consolidation of the loose sediment—a point that is not only not denied by Hooke but originated with him. The idea of axial shift is so superfluous to Hutton's general thesis that, having brought it up, he must retreat from it by side-stepping the issue. That Hutton should refer to it at all, much less argue so strongly against it, indicates his thorough familiarity with Hooke's theory.

Hooke's importance in the history of geology

Hooke's geological contributions, like so many of his accomplishments, have suffered undeserved obscurity. Not only do his writings reveal the extent of pre-Hutton and pre-Werner geological knowledge, they contain the fundamental concepts of the pre-continental-drift/plate tectonics geological paradigm, on which many older geologists were reared.

Modern geophysicists find polar wandering, or axial displacement, a necessary part in the plate tectonics paradigm. But no one paid much attention to it in the decades after 1687 until Rudolf Erich Raspé in 1763, while reviewing and lauding Hooke's general ideas about the formation of terrestrial features, again struck down axial displacement as an idea unworthy of consideration (Raspé 1763). My research indicates that Hutton knew Raspé and in fact cited him in his writing, so that he could not have escaped knowing about Hooke's ideas in geology. Besides, Hutton's own writings testify to the fact that Hooke's 'axial displacement', an idea that Hutton need not have brought up again, deeply disturbed him and he felt the need to argue against it (Drake 1981, 1996).

Conclusion

Although Hooke's *Posthumous Works* seemed to have suffered a hiatus of about fifty years during which his ideas were not credited to him and during which the figure of Newton loomed large in the minds of men, Hooke's ideas were transmitted by later writers, demonstrating the continuity of the development of geological thought. The time element, however, was always a disturbing factor in Hooke's thinking. He could not openly proclaim a break from his religion. But all the evidence shows that he wished he could. If he had secretly thought that the Earth was unfathomably old, he was prudent enough to keep quiet about it; he had been protégé of, and championed by, wise churchmen like Bishop Wilkins and Archbishop Tillotson.

It was left to Hutton, therefore, to make that leap into immortality by stating that we find the Earth to have 'no vestige of a beginning and no prospect of an end'. Hutton's poetic dictum was wrong, of course, because our Earth had a beginning and will have an end when the Sun nears its death, but it freed geology from the biblical time constraint. Hooke's influence, however, is evident in all of Hutton's writings. It is the pages of history that have not caught up with reality after the Newton dazzle. The evidence shown by the rocks and fossils on the Isle of Wight was indisputable to the highly rational Hooke. He probably was the first true field geologist. A. P. Rossiter (1935) called him 'The First English Geologist'.

It is very encouraging that in 2003, the Tercentenary Commemoration of the death of Robert Hooke was commemorated by such prestigious organizations as the Royal Society of London, Gresham College, Oxford University (Christ Church), the Linnean Society, the London Mathematical Society, the Royal Academy of Engineering, and the Royal Microscopical Society. These groups and others have joined forces to sponsor conferences in London, Oxford and the Isle of Wight to honour the remarkable but neglected Robert Hooke. It behooves geologists and historians of geology to join these ranks, because Hooke's contributions were significant and essential to the foundation and development of the science of geology. Without a doubt, Robert Hooke was an important early thinker in the history of Earth science whose influence on later writers was far greater than hitherto appreciated. In fact, echoing the title of one of the featured sections of the journal *Geology* of the Geological Society of America, he was a super 'Rock Star'.

I thank P. Wyse Jackson not only for his fine organization of the 2003 INHIGEO meetings, but also for his helpful advice on the preparation of this paper.

Notes

[1]For details of how Hooke designed the dome for St Paul's Cathedral, see Hambly (1987). The late Dr Hambly, a distinguished London engineer, gave this 'Christmas Lecture'. A radio version of this lecture entitled 'Robert Hooke, London's Leonardo' was broadcast on BBC Radio 4 on 3 January 1987. For the story behind 'Boyle's Law' see Cohen (1964). For details of other Hooke claims, see the chapter on 'The Life of Robert Hooke' as well as other chapters related to specific subject matters in Drake (1996).

[2]The subject of Hooke's theory of evolution is discussed in Drake (1996, pp. 98–103). It is also the subject of a paper given at the Tercentenary commemoration of Hooke's death sponsored by the Royal Society and Gresham College, London, 7–9 July 2003 (Drake 2006).

[3]The subject of Hooke's concept of polar wander is discussed in Drake (1996, pp. 87–95). It is also the subject of a paper given at the Hooke commemoration at Oxford University (Christ Church), 2 October 2003 (Drake 2005).

[4]Louise Patterson argues convincingly in favour of Robert Hooke in the priority controversy of several important ideas between Hooke and Newton that resulted in the Law of Gravitation. (See Patterson 1949, 1950).

References

ADAMS, W. H. D. 1869. *Nelsons' Hand-Book of The Isle of Wight; its History, Topography, and Antiquities.* T. Nelson and Sons, London, Edinburgh and New York.

COHEN, I. B. 1964. Newton, Hooke, and Boyle's Law. *Nature*, **204**, 618–621.

DAVIES, G. L. 1964. Robert Hooke and his conception of Earth-History. *Proceedings of the Geologists' Association*, **75**, 493–498.

DRAKE, E. T. 1981. The Hooke imprint on the Huttonian theory. *American Journal of Science*, **281**, 963–973.

DRAKE, E. T. 1996. *Restless Genius: Robert Hooke and His Earthly Thoughts.* Oxford University Press, New York.

DRAKE, E. T. 2005. Hooke's concepts of the earth in space. *In*: KENT, P. & CHAPMAN, A. (eds) *Robert Hooke and the English Renaissance.* Gracewing, Leominster, 75–94.

DRAKE, E. T. 2006. Hooke's ideas of the terraqueous globe and a theory of evolution. *In*: COOPER, M. & HUNTER, M. (eds) *Robert Hooke Tercentennial Studies.* Aldershot, Ashgate, 135–149.

DRAKE, E. T. & KOMAR, P. D. 1981. A Comparison of the geological contributions of Nicolaus Steno and Robert Hooke. *Journal of Geological Education*, **29**, 127–134.

HAMBLY, E. C. 1987. Robert Hooke, The City's Leonardo. *The City University*, **2**, 5–10.

HOOKE, R. 1665. *Micrographia: Or Some Physiological Descriptions of Minute Bodies Made by Magnifying Glasses, with Observations and Inquiries Thereupon.* Jo. Martyn and Ja. Allestry, printers to the Royal Society, London. Reprinted 1966, Culture et Civilisation, Bruxelles.

HOOKE, R. 1705. (posthumously), originally presented to the Royal Society from 1668 to *c.* 1699. Lectures and Discourses of Earthquakes, and Subterraneous Eruptions, Explicating the Causes of the Rugged and Uneven Face of the Earth; and What Reasons May be Given for the Frequent Finding of Shells and Other Sea and Land Petrified Substances, Scattered Over the Whole Terrestrial Superficies. *In*: WALLER, R. (ed.) *The Posthumous Works of Robert Hooke, M.D. S.R.S., Prof. Gresh. &c. Containing his Cutlerian Lectures, and Other Discourses, Read at the Meetings of the Illustrious Royal Society*, Royal Society Secretary. Smith & Walford, London, printers to the Royal Society, 279–450. Reprinted in *The Sources of Science*, no. 73. Johnson Reprint, New York and London.

HUTTON, J. 1788. Theory of the Earth, or an Investigation of the Laws Observable in the Composition, Dissolution, and Restoration of Land Upon the Globe. *Transactions of the Royal Society of Edinburgh*, **1**, 209–304.

OLDROYD, D. R. 1972. Robert Hooke's methodology of Science as exemplified in his 'Discourse of Earthquakes'. *British Journal for the History of Science*, **6**, 109–130.

PATTERSON, L. L. 1949. Hooke's gravitation theory and its influence on Newton, I: Hooke's gravitation theory. *Isis*, **40**, 327–334.

PATTERSON, L. L. 1950. Hooke's gravitation theory and its influence on Newton, II The insufficiency of the traditional estimate. *Isis*, **41**, 32–45.

RAPPAPORT, R. 1986. Hooke on earthquakes: lectures, strategy and audience. *British Journal for the History of Science*, **19**, 129–146.

RASPÉ, R. E. 1763. *Specimen Historiæ Naturalis Globi Terraquei, Præcipue de Novis e Maris Natis Insulis, et ex his Exactius Descriptis et Observatis, Ulterius Confirmanda, Hookiana Telluris Hypothesi, de Origine Montium et Corporum Petrefactorum.* Amsterdam, Leipzig, J. Schreuder & P. Mortier. Translated and edited by IVERSEN, A. N. & CAROZZI, A. V. (1970). Hafner, New York.

ROSSITER, A. P. 1935. The first English geologist. *Durham University Journal*, **29**, 172–181.

SCHNEER, C. J. 1960. Kepler's New Year's gift of a snowflake. *Isis*, **51**, 531–545.

STENO, N. 1667. *De solido intra solidum naturaliter contento dissertationis prodromus.* Florentiæ. English version with notes by J. G. WINTER, 1916. Macmillan, New York, 169–283. Reproduced in facsimile *In*: WHITE, G. W. 1968. *Contributions to the History of Geology* 4. Hafner Publishing Company, New York.

WOODWARD, J. 1695. *An Essay Towards A Natural History of the Earth: and Terrestrial Bodies, especially Minerals: As also of the Sea, Rivers, and Springs. With an Account of the Universal Deluge: And of the Effects that it had Upon the Earth.* Printed for Ric. Wilkin at the Kings-Head in St Paul's Church Yard, London. Reprint Edition 1978 by Arno Press, Inc.

Robert Jameson on the Isle of Arran, 1797–1799: in search of Hutton's 'Theory of the Earth'

C. J. NICHOLAS[1] & P. N. PEARSON[2]

[1]*Department of Geology, Trinity College, Dublin 2, Ireland (e-mail: nicholyj@tcd.ie)*

[2]*Department of Earth Sciences, Cardiff University, Park Place, Cardiff CF10 3VE, UK*

Abstract: The Isle of Arran lies off the Ayrshire coast in the Firth of Clyde, SW Scotland. James Hutton visited Arran in August 1787 with his companion, John Clerk, and together they made the first geological investigation of the island. Hutton returned to Edinburgh satisfied that he had at last found the critical field evidence he had been searching for in support of his 'Theory of the Earth'. In June 1797, ten years after Hutton's first survey of Arran, Robert Jameson arrived on the island to investigate its geology and mineralogy. Jameson was still only 22 but had already become one of Hutton's most ardent critics. The previous year he had read two papers to the Royal Medical Society of Edinburgh refuting critical elements of Hutton's 'Theory'.

Based at Kilmichael, near Brodick, on the east coast, Jameson spent most of June, July and August systematically exploring Arran entirely on foot. On returning to Edinburgh, he combined a narrative of his observations on Arran with an earlier trip to his parents' native Shetland Islands, in what became the earliest published account of the geology of Arran; *An Outline of the Mineralogy of the Shetland Islands, and of the Island of Arran* (1798). Jameson returned to Arran for a second, shorter visit in August 1799 and subsequently published a second book, *The Mineralogy of the Scottish Isles* ... (1800), which presented a revised and updated geological description of Arran.

Jameson's handwritten journals for both his 1797 and 1799 tours of Arran survive. In this paper we draw on these first-hand accounts of Jameson's field observations and discuss them alongside his two subsequent books describing Arran's geology. The journals clearly show that Jameson was an acute observer in the field and consistently made his interpretations in accordance with Wernerian principles. Although he had read the published Volumes I and II of Hutton's '*Theory*', it appears that he had not seen the unpublished Volume III manuscript prior to either tour. Nevertheless, Jameson's exploration of the island closely follows that of Hutton's and they visited many of the same field localities. Here, we use Jameson's Arran journals to retrace his field investigations and discuss them in the light of Hutton's previous work on the island and in relation to Jameson's own Wernerian views. The difference of approach and interpretation between the two in the field on Arran mirrors the wider conflict between Plutonists and Neptunists at this time.

The Isle of Arran lies in the Firth of Clyde, just off the Ayrshire coast of SW Scotland. The island is about 32 km long and 17 km wide. Arran's topography is dominated in the north by high granite mountains, whereas to the south the scenery is more gentle and the land good for farming. Despite fleeting references to the rocks on Arran by previous visitors such as Thomas Pennant (1726–1798) in 1772,[1] no general description of its geology existed when James Hutton (1726–1797) arrived in the island in the summer of 1787. Hutton initially visited Arran for one main reason, its granite. Since the original reading of his abstract of a '*Theory of the Earth*' in 1785 (see Craig 1987), the role of granite in underpinning his '*Theory*' had become more critical. Hutton required a mechanism of uplift for strata originally formed on the seabed and large bodies of molten rock pushing up towards the surface could potentially have provided the answer. Granite had only played a minor role in Hutton's 1785 abstract as he later wrote that at that time he was:

> not perfectly decided in my opinion concerning granite; whether it was to be considered as a body which had been originally stratified by the collection of its different minerals, and afterwards consolidated by the fusion of those materials; or whether it were not rather a body transfused from the subterraneous regions, and made to break and invade the strata ... (Hutton 1794, pp. 77–78).

However, only months after the reading of the abstract Hutton embarked on a field programme in Scotland to investigate the true nature of granite. The first locality visited was the granite of Glen Tilt in September 1785 (see Craig *et al.* 1978). Following the success of this expedition he moved west the following year and toured the SW coast during September 1786, exploring the main granite plutons at the Rhinns of Galloway, Cairnsmore of Fleet and the Criffel granite at Colvend. This tour also included an aborted attempt to sail over to

From: WYSE JACKSON, P. N. (ed.) *Four Centuries of Geological Travel: The Search for Knowledge on Foot, Bicycle, Sledge and Camel.* Geological Society, London, Special Publications, **287**, 31–47.
DOI: 10.1144/SP287.4 0305-8719/07/$15.00

Arran. He returned the following year and made a successful crossing to Arran, accompanied by the young John Clerk of Eldin (1757–1832), whereupon he completed the first geological survey of the island. The main highlights of this trip have been discussed previously in Archibald Geikie's posthumous publication of Hutton's '*Theory of the Earth* ...' Volume III,[2] '*The Lost Drawings*' by Gordon Craig *et al.* (1978) and in a review of his work by G. W. Tyrrell (1950). It is beyond the aim or scope of this paper to discuss all of Hutton's fieldwork on Arran. Instead, where Jameson and Hutton visited the same locality, the significance of Hutton's earlier interpretation will be discussed here in context with Jameson's handwritten journal observations made during his geological travels of Arran in 1797 and 1799.[3]

The young Robert Jameson

Robert Jameson (1774–1854) was born in Leith on 11 July 1774. Jameson's early influences and career have been outlined and discussed at length by Jessie Sweet's work on Jameson and so need not be recounted in detail here.[4] However, in order to understand the young man who travelled to Arran in 1797 and 1799 there are a few points in his early life worth highlighting. First, it seems that from an early age Jameson was interested in natural history and also had a strong desire to travel. He read James Cook's *Voyages* as a child and collected plants and animals along the beach at Leith, no doubt watching the ships come and go in the Forth estuary. Little wonder then that he persuaded his father to let him become a mariner, only to be dissuaded by friends and to take up studies at the University of Edinburgh instead. He attended lectures there from 1787 onwards, including classics and medicine, but in 1792 and 1793 he also enrolled in the Rev. Professor John Walker's (1731–1803) class of natural history.[5] Jameson seems to have taken to this with enthusiasm and made his first sea voyage at the age of nineteen in August 1793 from Edinburgh to London, specifically to visit various zoos, institutions and museums with natural history collections (see Sweet 1963). Jameson also called on Sir Joseph Banks (1743–1820) whilst in London with letters of introduction. Banks had been the naturalist on Cook's expeditions. It is even possible that Banks was a childhood hero of Jameson's and that he epitomized Jameson's scientific career hopes and ambitions at this time.

The following year, in 1794, Jameson made his first mineralogical tour, accompanied by his younger brother Andrew (1779–1861). They travelled to the Shetland Islands, which at first glance might have seemed an odd choice. However, both Jameson's parents came from Shetland and this visit may have served the dual purpose of visiting family relatives as well as recording something of the natural history en route. Jameson had become a favourite pupil of John Walker's during this period and when Charles Hatchett (1766–1847) arrived in Edinburgh in July 1796 for a proposed tour of the Highlands, Walker offered help by sending Jameson along to accompany him over the next two weeks. The Shetland visit and the brief Highland tour appear to have been the extent of Jameson's scientific travels prior to his trip to Arran in the summer of 1797.

To be a Neptunist or a Plutonist?

Charles Darwin (1809–1882) was responsible for perhaps the most damning quote concerning Robert Jameson, which has unfortunately persisted as his main epitaph in history. In an autobiographical note in later life, Darwin wrote:

> During my second year at Edinburgh [1826–7] I attended Jameson's lectures on Geology and Zoology, but they were incredibly dull. The sole effect they produced on me was the determination never as long as I lived to read a book on Geology, or in any way study the science. Equally striking is the fact that I, though now only sixty-seven years old, heard the Professor, in a field lecture at Salisbury Crags, discoursing on a trap-dyke, with amygdaloidal margins and the strata indurated on each side, with volcanic rocks all around us, say that it was a fissure filled with sediment from above, adding with a sneer that there were men who maintained that it had been injected from beneath in a molten condition. When I think of this lecture, I do not wonder that I determined never to attend Geology (Darwin 1892, p. 15).

Darwin's description is of a man still staunchly adhering to an outdated mode of thinking. This raises important questions about Jameson's underpinning philosophy. It seems that he embraced Neptunism very early in his scientific career. In 1796 he read two papers to the Royal Medical Society of Edinburgh arguing against Plutonism, entitled '*Is the Volcanic opinion of the formation of Basaltes founded on truth?*' and '*Is the Huttonian theory of the earth consistent with fact?*'.[6] These clearly show that even whilst still a student of twenty-two, Jameson had decided which unifying theory of the Earth to reject and which to support. This begs the question as to what influences caused Jameson to make such a decision.

As not only his lecturer in natural history but also his friend and mentor, John Walker might have been expected to play a major role in determining Jameson's views on geology. However, Walker made no direct reference to either Hutton's or Werner's theories in his lectures (Scott 1966). In two previous attempts to resolve this question both Sweet and Waterston (1967) and Scott (1966) could find no unequivocal evidence that Walker was either a strict Neptunist or that he had encouraged Jameson to be one. Walker's main influence on the way Jameson thought is most likely to have been one of instilling strict scientific discipline in his pupil. For instance, in Walker's introduction to his lecture course, he wrote:

> The method of inquiry which all our ingenious Theorists of the Earth have pursued is certainly erroneous. They first form an hypothesis to solve the phenomena, but in fact the Phenomena are always used as a prop to the Hypothesis.
>
> Instead therefore of attempting to cut the gordian knot by Hypothetical analysis, we shall follow the synthetic method of inquiry and content ourselves with endeavouring to establish facts rather than attempt solutions and to try by experiments how far that method may lead us thro' the mazes of this subject (Scott 1966, p. 27).

This philosophy was clearly followed by Jameson as he performed a series of heating and cooling experiments to produce evidence as part of his investigations of basalt in 1796 (Sweet 1967).

A second major influence appears to have been the contemporary geological literature. Jameson had clearly read Abraham Gottlob Werner's (1749–1817) original journal publications as he refers to the results of Werner's travels in Scheibenberg[7] in his 1796 paper on basalts (see above). In the same paper he also quotes Richard Kirwan's (1733–1812) *Elements of Mineralogy* (Kirwan 1794) and since both of his 1796 papers attacked Hutton's ideas he had obviously also read both 1795 volumes of Hutton's *Theory of the Earth*. In the 1796 paper on basalts Jameson wrote:

> Upon such a basis is the famous Volcanic theory founded, which for many years consigned three fourths of our Globe into the hands of Pluto, until the immortal *Werner*, from a careful examination of nature, declared the absurdity of such a hypothesis (if it can be so called). He did not endeavour to confute it by subtlety of argument, but with facts, sunk it into utter oblivion. Mr Kirwan, in his admirable book of Fossils, has also given us a very able defence of the Neptunian hypothesis; which he has farther confirmed by new and important facts.

This is a telling passage from Jameson. His wording suggests that he saw in Werner's work the 'fact gathering' approach to science that had been set in him by John Walker's lectures. He uses Kirwan's more recent work as secondary supporting evidence to the Neptunian hypothesis.

One might ask why Jameson found Hutton's *Theory of the Earth* so unappealing. The answer may be simple enough. When Hutton died in 1797 only half of the originally intended '*Theory of the Earth*' had been published.[8] The first two volumes relied heavily on the published literature of other authors concerning regions that Hutton had never visited. Much of it is a literature review in search of evidence that he could use to support his '*Theory*'. Much of the *prima facie* field observations that Hutton had made since 1785 were to have formed the bulk of Volume 3, which remained in manuscript after his death. Without the mass of Hutton's own field evidence available as a published Volume 3, it must have seemed to many contempories that Hutton's *Theory of the Earth* was not supported by credible and specific detailed observation or facts. This would have been unacceptable to Jameson's strict adherence to what he had been taught by Walker.

In essence, Hutton's *Theory of the Earth* relied on the operation of several interlocking processes which still remained unobserved or unexplained in 1795, such as active uplift of mountains or intense internal heat and pressure deep within the Earth. Werner's ideas were more elegantly simplistic in that wherever strata were currently observed, they had been precipitated in that succession and orientation from a receding primeval ocean. In this respect, to a young student in 1795 it must have seemed more far-fetched to accept Hutton's theory than that of Werner.

There is perhaps one final question that should be addressed concerning the background to Jameson's tours of 1797 and 1799: why did he decide to visit the Isle of Arran? In his tour journals, Jameson does not mention his reasons for visiting Arran. It may have been at the suggestion of John Walker or it may have been on Jameson's own initiative. It is clear from Jameson's 1797 journal that he had not seen Hutton's *Mineralogical History of the Island of Arran*; which at that time lay as Chapter 9 in the unpublished Volume 3 manuscript. Hutton only makes two brief references to his work on Arran in the published Volumes of his '*Theory*...'.[9] After Hutton's death, the Volume 3 manuscript passed into the possession of John Playfair (1748–1819) who was then Professor of Mathematics at the University of Edinburgh.[10] As such, Playfair was a direct colleague of John Walker's. Not only this, but Playfair had also previously been Walker's pupil in 1782. It is possible that Playfair may have mentioned the Arran chapter or work previously done by Hutton on the island to Walker. Another possibility is that Walker or Jameson heard of it through one of Edinburgh's many scientific or social clubs.

Walker himself made many tours of the Scottish Islands between 1760 and 1785, including a formal survey of the Hebrides and Highlands in 1764 (Scott 1966). As such he must have been familiar with the potential of Arran for mineralogical investigation. Thus, there may have been an initial impetus for Jameson to visit Arran because Hutton had previously been there, or it could have been that Walker indicated it to Jameson as an island ripe for study as his next project after the Shetlands.

This question may remain unresolved. However, after returning from Dublin to Arran in early July, Jameson was joined on Arran by a 'Mr Walker', most probably his professor, and the two completed the remainder of the tour together.[11]

Jameson wrote his first geological book in 1798, combining a narrative of his 1794 Shetland tour with his 1797 trip to Arran (Jameson 1798, pp. 49–144). The text for the Arran chapters follows closely his handwritten journal, minus the more personal observations made during the tour. We are told in his obituary that the book 'soon acquired a well-merited celebrity' (Hamilton 1855) and remains the first published account of the geology of Arran. Throughout the book, Jameson used footnotes to refer his own field observations and interpretations to the Huttonian '*Theory*' and used them in evidence against it. However well this first book was received, it seems that Jameson was not altogether happy with it as it stood. He returned to Arran again the following year, 1799, for a much shorter and more focused field trip. In only ten days on the island, he targeted three main localities, all of which were crucial to the argument between Neptunism and Plutonism; the granite–schist contact in Glen Sannox, Hutton's unconformity at Lochranza and the Tormore dykes on the west coast. There can be little doubt that this second tour of Jameson's was specifically intended to visit those localities that could topple Huttonian arguments. For instance, Jameson had copied Hutton's description of the unconformity at Lochranza into the back page of his 1799 journal. Allied with field evidence from other subsequent short tours around Scotland, Jameson wrote a second account of the geology of Arran to be included in his 1800 book *Mineralogy of the Scottish Isles* (Jameson 1800). In this, many of the previous footnotes arguing against the Huttonian '*Theory*' were dropped. It seems that Jameson now felt confident enough in his observations and Neptunism that he no longer saw the need to continue arguing between the theories in print. Indeed, the book is prefaced by a short outline of Wernerian stratigraphy, setting out the philosophy by which Jameson interpreted the rocks around Scotland in the subsequent chapters. That same year, Jameson left Edinburgh to study directly under Werner in Freiburg.

Jameson's Arran journals and his approach to geological fieldwork

John Walker's lectures in Edinburgh not only covered theoretical topics but he also gave more practical help. In his introduction to the forthcoming course, he wrote that he would give:

> a general view of the distribution of the Fossils of the Globe; when I [Walker] shall suggest the form of a Mineralogical Journal to be kept by gentlemen particularly those who travel, and which will in my opinion facilitate the improvement of this science more than anything of the kind that has been attempted (Scott 1966, p. 25).

It is reasonable to assume that Jameson's journals of 1797 and 1799 closely followed the suggestions of Walker. A single handwritten journal survives of Jameson's tour through Arran in 1797. It seems that after only four days on the island, Jameson left on a boat for Ireland on 24 June. He landed back on Arran several weeks later on 18 July and made his main tour of the island (Fig. 1). The journal ends abruptly on 18 August as Jameson was gradually making his way down the west coast of the island. However, the tour continued at least for several more days as Tormore and Bennan Head are mentioned in his subsequent book (Jameson 1798, Chapter 5, p. 119). These localities would have been visited after the 18th and so it is likely that a second short journal completing the tour originally existed but has now been lost.

The 1797 journal consists of 116 pages of ink notes and occasional sketches.[12] The drawings appear to have been sketched in pencil first, as traces of the original pencil marks remain. There are also rain spots on several pages that have smudged the ink. These points, allied with references Jameson makes in his journal to notebooks, give an idea as to how Jameson approached recording his field observations.

Each day in the field Jameson generally made a looping or circular walk to cover a particular area of ground, often accompanied by a local guide. Specimens collected during the trip were stored in a chest or box that Jameson arranged to be sent around the island after him. He spent rainy days describing specimens in the box more fully. For instance, on 19 July it was 'Extremely rainy so that I could do little', so Jameson set about making a 'Description of the porphry from Cora Gills—'.[13] However, occasionally the specimen box failed to keep up with his progress around the island and his work ground to a halt: 'July 29th: Did nothing in particular—for want of my box'.[14] However, although on days of bad weather he made a formal mineralogical description of specimens in his journal, on most other days this seems to have been done in a series of notebooks that have not survived. On 27 July he was examining outcrop in Glen Rosa and wrote in the journal:

> Near the top of the hills on the North side of Glen Rosa I observed large rocks of Trap amongst the Glimmer Schiefer, bid. Note Book *W*. It appears to be entirely composed of Hornblende at least with little admixture—.[15]

This and other references to the notebooks in the journal suggest that these were used for nothing

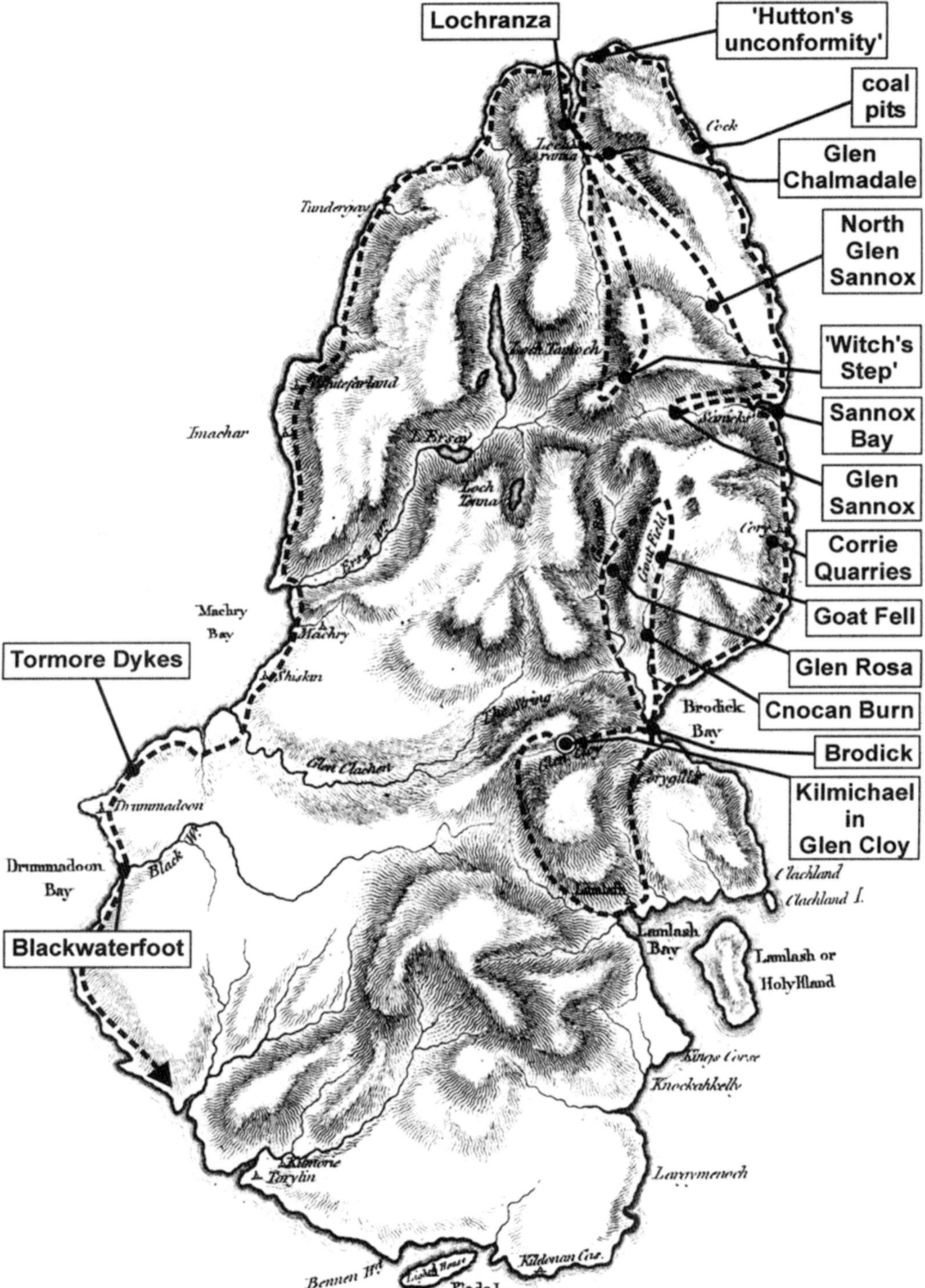

Fig. 1. The original copper-plate map of Arran produced by Jameson for his 1798 *An Outline of the Mineralogy of the Shetland Islands and of the Island of Arran*. This appears to be one of the earliest attempts to draw Arran's characteristic outline accurately. The key localities discussed here are shown in named boxes and the dotted black lines are the main routes walked by Jameson in 1797 and 1799.

more than rough jottings or hand specimen descriptions.

The main description of what Jameson saw in the field seems to have been in the journal itself which he completed religiously at the end of each day before going to bed, even when completely exhausted. On 11 August he walked and surveyed the geology from Kilmichael in Glen Cloy to Lochranza via the east coast, a distance of about 22 miles. On arriving in Lochranza at the end of the day he wrote:

... Being fatigued, wet and hungry ... made the best of our way to loch ransa. ... We now landed safe at the public house ... where we regaled ourselves with tea and, took our supper porridge. Wrote my journals, and went to the Land of *Nod*.[16]

In this respect, Jameson was employing a different approach to recording field observations from present day geologists. Whereas the usual practice today would be to record observations and interpretations in a notebook whilst still standing on the outcrop, Jameson seems to have used his journal to record the highlights of these at the end of each day. However, the rain spots on journal pages are tantalizing; on these occasions it may have been that Jameson was referring back to previously written passages in his journal whilst out in the field during a rain shower. Alternatively, he may have been actually writing the journal in ink out in the field when the weather permitted.

Jameson's second tour of Arran is contained in another single journal entitled 'Arran 1799'. As mentioned above, the tone of this second visit to the island is a little different from his first. Not only is Jameson's field work more focused, but his 1799 journal is almost completely free of the casual vignettes of Arran life that are a frequent feature of his 1797 journal.

It is beyond the scope of this paper to reproduce the Arran journals in full. Consequently, in the following sections, selected journal entries from both 1797 and 1799 will be discussed in context with localities also visited by Hutton or used by Jameson to argue against the Huttonian '*Theory*'.

Reconnaissance work: Kilmichael, 20–25 June 1797

Whatever the reasons behind Jameson's original decision to visit Arran, he set out on 17 June 1797 in a coach for Glasgow.[17] It seems likely that before the trip began it had already been arranged that, whilst on Arran, he would stay with the Fullartons of Kilmichael, in Glen Cloy near Brodick on the east coast (Fig. 1). The Fullartons were one of the oldest families on Arran having acquired titled lands on the island in the fourteenth century. With the lands went the hereditary title of Coroner of Arran. Although the head of the family in 1797 was a John Fullarton, there was also a son called John and it is not clear from Jameson's journal exactly who was his host during the visit. Jameson certainly met with a 'Miss Fullarton' in Glasgow, presumably a daughter, and they were joined there by representatives of another Arran family, the Ballantynes. After spending the 18th and 19th busy in introductions and walking out with the ladies, they all finally set sail for Arran on the morning of the 20th:

June 20th

We left Greenock this morning about 9 of clock in a boat for the Isle of Arran, where we arrived after a pretty agreeable passage about 5 in the afternoon ... The approach to Arran is extremely grand and terrific bringing to my recollection all the majestic scenery of Foula or Noss, [18] but upon a nearer approach, how diminutive do they appear, with what rugged majesty does Goatfield tower above all the other Alpine heights, like Mont Blanc amidst the sublime scenery of Switzerland.

Having landed we walked towards Kilmichael, (the seat of the ancient family of the Fullertons) where we soon arrived, found all in good health, was introduced to Mrs Fullerton and their young ladies, and soon found myself as happy as I could wish.[19]

The house at Kilmichael is reputedly one of the oldest surviving houses on Arran and is now an immaculately restored and kept Georgian Country House Hotel. The Fullarton Coat of Arms hang over the doorway and the gardens are complete with peacocks and lead to a small footbridge over the burn at or close to where an original crossing would have been when Jameson came to stay. Despite more introductions and settling in, Jameson was clearly keen to start business. Having spent his first evening on Arran making a few preliminary field observations in and around the house at Kilmichael, Jameson set out the following day to climb the highest peak on the island, Goat Fell ('Goatfield' on his 1798 map; see Fig. 1). His route from Kilmichael would have taken him into Brodick and then on to ascend Goat Fell from the south close to Cnocan Burn (Fig. 1). This route would have been the same or very close to that of Hutton's in 1787.

June 21st

Set out this morning about 11 of clock for Goatfield. In ascending the hills I found all to be sandstone for a considerable height, this was succeeded by Glimmer Schiefer, granite which last forms the summit of the hill ... Having gained the summit after considerable fatigue we had the misfortune to be surrounded with clouds ... The view however of the horrible glens underneath thro which the rain and clouds drove with fury, was inconceivably grand and stirring; what human being after witnessing such majestic works can refrain from declaring the existence of a deity, can anyone attribute such to chance; if so truth is out of the question and vanity the most deceitful of human passions leads to conclusions which have had their *own* influence on mankind. Having contemplated this scene for some time we descended which was extremely tiresome from the immense number of loose masses of granite.—We arrived at Kilmichael about 5 of clock, tired enough and very hungry.[20]

In this passage Jameson is answering a question in his journal that had obviously been occupying his thoughts. Whilst naming no names in the journal, the question would seem to have been posed in his mind by the Huttonian '*Theory*'. In considering the Hutton hypothesis that the globe had been formed by the decay of a former world, Jameson wrote in 1798 that:

This part of the Huttonian theory differs but little from that of Count Buffon: it includes, however, the question concerning the divinity of the World; a speculation fit only for fanciful metaphysicians (Jameson 1798, p. 110).

If there had been previous 'worlds' to this one, recycled by erosion, deposited at the bottom of the sea and then uplifted as mountains, was there still room for a God? Jameson's walk to the top of Goat Fell appears to have resulted in a reaffirmation of his beliefs. Having returned to Kilmichael, Jameson put the philosophical questions of the day to one side and relaxed with his new friends for the evening:

Having taken tea we walked out with the ladies across the glen to Brodic, where we had the good fortune to sup before we returned. The supper was in the true Highland style, Sowens, milk, cheese, cakes and *Whisky*. Being all seated we had an Erse grace. I ['had a great deal' *later scribbled out*] of difficulty to ['refrain from laughing' *later scribbled out*] nor was I bettered by some of my amicable friends. Having eaten most plentifully of the Sowens and such we all concluded a la Hauteur with a glass of strong whisky, and heartily pleased with a supper I will not forget for a long time.[21]

Jameson was still only twenty two and a miniature painting of him from about this period shows him to be a handsome and rather dashing young man (Sweet 1967). This anecdote and others in the 1797 journal paint a very different picture of the young Jameson from the man portrayed in later life by others such as Darwin (see above). He was clearly enjoying his company on Arran and there is a certain innocence and immaturity to his apparent excitement at drinking whisky and giggling with the others as the grace is said in Gaelic.[22] He had later gone back and scribbled out the potentially offending confession that he and the others were laughing during grace but the words are still just readable underneath the ink. This short passage also indicates another interesting side to Arran and its 'real' economy. The island, along with Ailsa Craig to the south, were havens for smugglers and illicit whisky stills. The 'Arran water' was a major supplement to crofter's income and in 1797, it is estimated that there were 50 stills in operation in the southern half of Arran alone (MacBride 1911, pp. 51–52).

On 24 June, Jameson left Arran for Ireland. He had only been on the island for three full days so it seems a little odd for him to leave so soon after arriving. It is more than likely that he had already arranged to travel to Ireland but was waiting for a suitable ship to make the passage. His journal records for 25 June:

Next morning about 12 of clock landed upon coast of Ireland upon the Isle of Magee here I saw great rocks of limestone inclined at about 12°. With immense quantities of plants, this runs in many cases amongst the Whin. We got late in the afternoon about 11 of clock to Belfast, and were well nigh being taken, having narrowly escaped the patroll.[23]

This statement suggests that Jameson's boyhood wish to become a seaman was nearly fulfilled. During 1797, the Royal Navy became engaged in various sea battles against French, Spanish and Dutch fleets and suffered two fleet mutinies at Spithead and the Nore (see Davies 2002, pp. 72–96). The Royal Navy was desperately short of able-bodied seamen during this time and used enforced enlistment or the 'press-gangs' to make up any shortfall in ship's crews. It seems that Jameson's ship landed in Belfast shortly after the 'press-gang' had passed through the docks and it is possible that Jameson would have been taken along with his vessel's merchant crew.

Glen Cloy, 20–26 July 1797

Jameson spent the next three and a half weeks in Dublin viewing the mineralogical collections of the Leskean Cabinet and meeting with Richard Kirwan and George Mitchell (1766–1803) (Sweet 1967). He sailed back to Arran and the Fullartons on 18 July via Ailsa Craig. Once safely back at Kilmichael, he set about his mineralogical survey of Arran in earnest. The most obvious place to start was the glen in which Kilmichael was situated, Glen Cloy. This valley runs more or less east–west and forks upstream to the west at the head of the valley (see Fig. 1). Jameson spent the next few days walking the hills and tracing the junction between sandstones, basalts and porphyries. On the 25 July, he wrote in his journal:

Endeavouring to discover the connection between the sandstone and Porphry at the head of Glencloy, I observed the following remarkable appearances. A large Whyn Dyke was to be seen rising through the earth from the sandstone, which penetrated the Porphry. In doing this one side of it had covered part of the sandstone along with it as both sides of the dyke were covered with sandstone breccia. The Dyke at last becomes quite narrow and drawn to a point, passing by the side of this is another Whyn Dyke which penetrated thro the Porphry passes the first one, takes a turn towards the east, where a small branch runs amongst the columnar Porphyry as may be seen in the figure (Fig. 2). The great body of the dyke continues thro the Porphyry to the top of the hill. I examined particularly the appearance of the Whyn and Porphyry at their junction, but could not perceive any difference betwixt the Porphyry here and at other parts. It is singular that the columns of Porphyry are here observed laying in very different directions and inclined at different angles of elevation . . .

N.B. The Porphyry as it approaches to the sandstone becomes more of a sandy nature.—The sandstone on one side of the dyke is very hard and compact.[24]

This is a slightly confused description of the exposure at the head of Glen Cloy although Jameson had obviously spent some time in the field trying to work out the cross-cutting relationships between the different lithologies. Jameson's rewritten version which appeared in his 1798 book is a little clearer, although he adds this disclaimer prior to providing an interpretation of his field observations:

No subject is more interesting or useful, than an examination of the relative position of strata and veins; in short, upon this is founded all our knowledge of geology. It is, however, attended with great labour and difficulty, not only on account of the many turnings and superposition of the strata, &c. but also from the frequent impossibility of tracing these strata in such a manner as to convince us of their relative position (Jameson 1798, p. 61).

The 1797 journal explanation to accompany his Glen Cloy sketch is as follows (see Fig. 2):

From all the phenomena which this interesting Glen has presented me it appears that it is composed of sandstone elevated at an angle

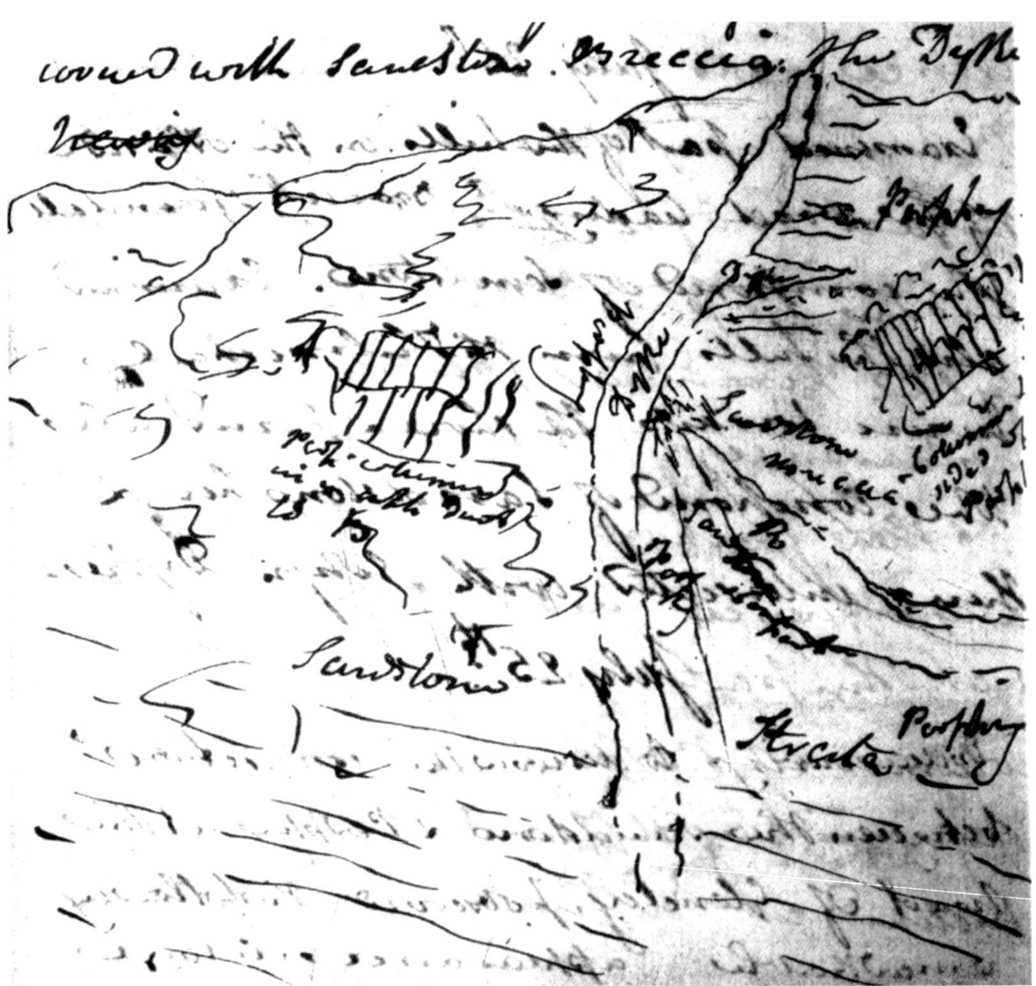

Fig. 2. Ink sketch of the locality at the head of Glen Cloy from p. 62 of Jameson's 1797 journal. Note the central basalt dyke narrowing towards the top of the hill and the patches of columnar porphyry. This sketch was simplified and produced as a copper plate for his 1798 book.

of from 12 to 30°, this sandstone forming the greatest heights and the greater solid part of the glen. The next in proportional quantity is the Hornstone Porphry, which lies upon the sandstone on the N. side; then the Sienite which we observe laid bare by the burn as it falls and forming several heights at the top of the glen and rising backwards amongst the neighbouring hills forming considerable summits.

The great flat in the amphitheatre of rocks near this top of the Glen appears as all the Glen to have been filled by sandstone which by some means or other has been carried away and has here laid bare the Sienite to which it was formerly applied as a secondary production. The next fossil is the Whyn which penetrates both Porphry, sandstone and Sienite? forming remarkable appearances connecting the sandstone and Porphry as productions of the same date and [----] doubtful that of Sienite.[25]

Jameson does not help his description by listing lithologies in order of total area of exposure to begin with, rather than in stratigraphic order. However, he seems to be suggesting that the sandstone was laid down upon the Sienite, but had been eroded away in places to form the valleys and expose the underlying Sienite. The porphyry overlies the sandstone and the two are separated by a sandstone breccia. Finally, the basalt ('whyn') veins cross-cut both the sandstone and porphyry and so post-date them. This description and interpretation of the locality in Glen Cloy is an excellent example of Jameson's keen field observations and his attempt to explain them whilst still in the field. The Glen Cloy work had taken several days and by the end of it he had produced a relatively simple stratigraphic succession that appeared to fit the available field evidence. Both Wernerians and Huttonians used cross-cutting relationships to resolve the relative age of emplacement in the same way.

There is one last point from this locality that arises from Jameson's journal entry. In the journal, he states clearly that the near vertical cross-cutting basalt, 'at last becomes quite narrow and drawn to a point' (see above). In the 1798 book this description is altered so that the dyke merely 'rises upwards getting a considerable curve; when it branches, one branch rises to the top of the hill . . .'. (Jameson 1798, p. 60). One of the characteristic features of the Wernerian 'theory of veins' was that:

All veins have been formerly open fissures . . . Thus, veins agree perfectly with fissures in their shape and position. Like them, they are seldom much contorted in their course; they wedge out at their extremities, and diminish in breadth towards their bottom; and in many cases even close completely (White 1976, p. 240).

The Glen Cloy basalt vein in Jameson's original journal description appeared to be contradicting this rule by tapering out as it progressed uphill. This would have played into the hands of Huttonians as they might have expected such a geometry if the basalt had been injected molten from below.

Goat Fell and Glen Rosa

Both Hutton and Jameson spent several days investigating the geology on and around the flanks of Goat Fell, with relatively little success. Hutton began his description of Arran by making his intentions quite clear. After taking accommodation in Brodick where the ferry from the mainland arrives, his primary objective '. . . was to examine around the mountain of Goatfield for the connection of the granite with the strata or surrounding bodies' (Hutton 1899, **3**, p. 217). Hutton's search for a 'granite–schistus' contact can be briefly summarized from Hutton (1899) as follows:

1. Hutton and Clerk climbed Goat Fell but found no contact (they did later return only to find cooling joints in the granite).
2. They then scrambled from the summit of Goat Fell down into the valley of Glen Rosa. Here they found that the sandstone was apparently conformable on the schistus, but both were inclined near vertical. There appeared to be no contact exposed in the main stream bed of Rosa Burn.
3. Abandoning the Glen Rosa and Goat Fell area, Hutton and Clerk travelled around to Glen Sannox ('South Sannox'), north of Corrie on the east coast. They proceeded up the glacial valley but again found no contact exposed in the banks or bed of Sannox Burn. Hutton states that in Glen Sannox they were '. . . equally disappointed in missing the immediate junction of the granite and schistus, the object of our pursuit, only by the space of a few yards or feet (Hutton 1899, **3**, p. 219).

Revisiting Rosa Burn and Sannox Burn today, Hutton's evident frustration is understandable. Indeed in both Glen Rosa and Glen Sannox the granite and 'schistus' can be examined *in situ* along the stream banks to within a few metres of each other, but the actual contact is not exposed.

Jameson's work in the Glen Rosa area also yielded few concrete results. Perhaps the most important entry in his journal at this point is that he was joined on Arran by a 'Mr Walker', probably Rev. John Walker, Professor of Natural History at Edinburgh. Together they climbed Goat Fell and for a second time Jameson sat at the top of the granite mountains, gazing at the scenery, and confirmed his religious faith:

July 27th

Went this morning along with Mr Walker (who had arrived in the island two days before) to Glen Rosa. Having arrived at Brodick we set forward towards the Glen and ascended the hills on the North side which were very steep and in some places rugged; having with considerable fatigue gained the summit we stood for some time admiring the grandeur of the scene. We here beheld the glen in its full extent, very deep and narrow with lofty mountains rising on both sides at an angle of 80°. Thro the middle runs a large burn Glen Rosa burn, which comes from the lofty and magnificent rocks at the top of the glen near Goatfield. The day was rainy accompanied with mist which prevented us from admiring the lofty and rugged summits of the towering hills, but the height at which we were, the great extent of the glen, its precipitous and elevated sides, the *iron like* baroness that was marked with so *strong a hand* together with the hanging clouds afforded me a scene of a most interesting nature. It is here that the philosopher in comparing the stupendous works of nature with our feeble action and cares, evidences the existence of a being, whose power we cannot estimate, whose motives we will long be ignorant of.[26]

Brodick to Lochranza via the NE shore, 11 August 1797

Jameson and Walker proceeded to walk around most of the island by way of the coast during their remaining time on Arran in 1797. They set out on 11 August for Lochranza at the northern end of the island (see Fig. 1):

Left Kilmichael this morning at 6 of clock along with Mr Walker on our walk around the island. Having passed the Castle of Brodick which is upon sandstone, we walked to the first Lime stratum before we arrived at it; observed the shore formed of sandstone which is in some places penetrated with Whyn Dykes. The sandstone is much acted upon by the sea, having a curious cavitated appearance, which is owing to the sea having washed away the softer parts of the sandstone, and left this the harder standing . . .[27]

The New Red Sandstone forms the foreshore around Brodick Bay and for much of the way up the east coast to the village of Corrie. The sandstones have developed in places a characteristic honeycomb weathering pattern that Jameson noticed and correctly attributed to differential weathering. Rather than being 'softer' parts of the sandstone though, they are patches of calcite cement more susceptible to dissolution by sea and rainwater. Jameson would have been passing down through the sedimentary succession as he walked north towards Corrie from Brodick. Once into the Carboniferous units he observed what is now termed the Corrie Limestone. This is immediately recognizable from the large productiid brachiopods, *Gigantoproductus giganteus*, that remain in life position. As part of his description, Jameson correctly spotted that the shells were all of the same type and all orientated in the same direction:

The first Quarry . . . consists of a stratum of limestone generally about 12 feet thick of a colour forming an angle of 20° and is exposed for about 50 feet. This stratum is divided into smaller ones by a [----] clay. This clay is generally 3 or 6 inches broad and contains shells dispersed in layers, (all appearing of the same species) with the convex sides downwards. The Limestone is strangely bent, in some parts assuming a waved appearance . . .

Having arrived at the Cory I observed considerable stratum of Limestone: inclined about 30 feet thick, running WNW and ESE; and is divided into stratula with clay as is the case with the other stratums . . . Here is also is a Quarry of a most beautiful white sandstone which is used principally at present for the Crinan Canal.[28]

The 'beautiful white sandstone' forms a breakwater for the small harbour in Corrie known as *Ferry Rock*. In the early 1900s the steamboats from Glasgow stood offshore at this point and lowered rowing boats to bring the passengers ashore. Jameson continued his walk with Walker by crossing Sannox Burn and continuing on to Lochranza via the raised beach around the coast. Much of the Carboniferous succession of Corrie is repeated along this stretch of shore. Nearing the most northerly point on Arran, the 'Cock of Arran', Jameson wrote:

A little to the south of the Cock is a stratum of clay slate till and is the remains of the working for coals. Here I found several pieces which have much the appearance of that of [----]. Observed two pits filled with water, which I was told were 20 fathoms deep being the length they went in search of coal; several hundred tons were thrown out and salt pans erected (of which we saw the ruins) but the seam disappeared, and so did the workers . . .[29]

Jameson reproduced this journal entry in his 1798 book, and accompanied it by a footnote explaining why the poor quality coal such as that found on the NE shore of Arran ('Blind Coal') was a critical lithology in the arguments used by both Neptunists and Plutonists alike:

Dr Hutton conceives, that this species of coal presents an irrefragable proof of the truth of his theory. Here, says he, is a coal having all the properties of that which has been submitted to the action of heat; the bitumen is separated, and charcoal remains. To the Neptunists, this affords one of the strongest facts against the theory; the separation of bituminous matter shows a want of immense compression, which is the grand fundamental basis of the hypothesis.[30]

On his tour in 1797 Jameson was unable to locate any real exposure of the coal to examine and so continued on around the coast past the Cock:

Having passed the Cock during a dreadful shower and a most miserable road, we now came to a more smooth one, when as we approached to Loch Ransa, the sandstone was to be observed joining with the schistus according to Dr Hutton, I however, did not examine this particularly at present being fatigued, wet and hungry, but made the best of our way to Loch Ransa. The evening now cleared up. The sun shone in his full splendour, presenting to us at our entrance to the loch a picture of great grandeur and sublimity.[31]

The Witch's Step, 14 August 1797

On 14 August 1797, Jameson and Walker left Lochranza and hiked to the top of 'Caime-na-Callich' ('The Witch's Step'), a peak at the heart of the granite mountains (see Fig. 1). Jameson wrote in his journal a series of key observations and conclusions from the summit:

There appears to be no regular ridge of hills in Arran, and this may be owing to the vast colossus of mountains that form its central and northern parts, taking their place from the lofty Goatfield. These mountains are all composed of Granite ... and by this height the rain has lowered them in a terrible manner, so as to give a most terrific picture of wild grandeur at the same time it shows that the rain has been the cause of the formation of vallies etc.—

N.B. The top of Caim-na-calliach has a most singular appearance. Looking like an immense colossus of stones piled above each other, by some monstrous *giant* who in former times may have waged war with the gods. These appearances show a peculiarity in the mode in which Granite decomposes, here first by exposure to the action of the air or other causes, it splits into immense irregular shaped masses, these in time fall down and then by disintegration forms Granitic sand.[32]

Jameson's conclusions here are made without our subsequent knowledge of glaciation and the effects of ice action. Explaining the extreme topography of the Arran mountains and glens presents a real problem if glaciation cannot be invoked. Jameson was forced to combine his observations into an interpretation that was something of a paradox. This was that although rain had the power to erode deep valleys, rainwater did not have the energy to transport the eroded debris very far. This passage also raises the question of geological time in a Wernerian view of the Earth; how long did it take for processes, such as erosion, to act on the rocks? In his journal entries for the Witch's Step, Jameson seems to be indicating that weathering and erosion could act relatively quickly due to the action of rain, particularly on granite.

In 1798, Jameson used his field observations and interpretation of sedimentary processes at the Witch's Step to attack the Huttonian '*Theory*':

Dr Hutton remarks, that the stony matter of this globe has been formed by the decay of a former world, whose debris has been collected by various means, at the bottom of a former ocean ....

Not a shadow of proof is brought of the debris being carried to the bottom of the ocean; on the contrary, we observe the decomposing materials of mountains filling up hollows, or forming plains. In other instances, the debris is carried to the sea shores ... where it is mixed with the waters of the ocean, and latterly thrown back upon the same or other shores, forming great tracts of land ....

Innumerable other instances might be mentioned, but these are sufficient to demonstrate the formation of land, without the supposition of the debris being carried to the fathomless depths of the ocean, and subjected to immense heat (Jameson 1798, pp. 110–111).

Whilst exploring in more detail around the summit of the Witch's Step, Jameson came across an outcrop that caused him some excitement:

Walking along the top of this awful glen, I was wonderful struck with a most astonishing appearance, a small dyke running fairly in granite .... I at first could not believe my eyes, I therefore descended I found I was right and that this was most certainly the astonishing phenomena of a Whin Dyke running fairly and decidedly in granite.[33]

If it is assumed, as we would today, that the basalt had been intruded into the granite whilst molten, then this observation by Jameson is unremarkable. However, to a Wernerian, this outcrop juxtaposed the oldest rock on Earth with one of the youngest. Using Neptunian principles, there might be two possible explanations for what Jameson saw. Firstly, that in the last stages of the 'primeval ocean's' retreat, a fissure had opened up that was so deep it had penetrated through all previous strata and into the underlying granite. Basalt had then precipitated into this fissure as a vein. The second possibility would be that in this locality none of the intervening strata had precipitated within the fissure, thus leaving the young basalt to lie directly against granite. In either case, these observations so excited Jameson that he collected specimens to show his friends. In the context of bringing youngest rocks to lie directly against oldest, he may even have seen this locality as a Wernerian equivalent of Hutton's unconformity.

The Lochranza unconformity, August 1799

On the inside back page of his 1799 journal, Jameson has jotted down a list of clothing to pack for the trip:

Trousers and draws—A waistcoat—pair shoes—2 shirts—1 pocket handkerchief—[----] [----]—razors, sulphur 4 pairs stockings—1 pair pontalons — waistcoats'

Underneath this list is written a short passage:

At Loch Ranza (see Fig. 1)—here the schistus and sandstone rise both inclined at an angle of 45°. But these primary and secondary strata were inclined in almost opposite directions, and thus met like 2 sides of a lambda being an a little disordered at the angle of their junction. From the situation of these 2 different masses of strata, it is evidently impossible that either of these could have been formed in ['their original' crossed out] that position.

These lines are copied from Hutton's *Theory of the Earth* (Hutton 1795, **1**, p. 429). It seems that at least prior to revisiting Arran in 1799, Jameson had decided to investigate this Hutton locality and written the description of it down into the back of his journal as an *aide memoire*. Jameson's own description in 1799 echoes that of Hutton:

Having taken some refreshment went to the NE point of the Glen at its entrance in order to observe the position of the strata as described by Dr Hutton in his work on the Theory of the Earth—Here observed the Micaceous schistus dipping to the SE at an angle of 45°, and these strata were in several places covered with red coloured sandstone and breccias with interposed limestone and these were at angle of 45° but dipped in a contrary direction that is to the NW—intermixed with this schistus observed a species of white foliated limestone which had a very considerable specific gravity: it had intermixed with pieces of schistus.[34]

Hutton saw the angular discordance between the two lithologies as evidence of two different cycles of deposition and uplift. However, the importance we would now place on this locality is lost in a Wernerian interpretation. The fact that the younger sandstones irregularly overlie the older 'schistus' would be expected in a Neptunist succession and consequently Jameson seems to attach little significance to this stretch of coast. More attention is paid in his subsequent book in 1798 to the cross-cutting of the schistus and sandstones by basalt 'veins' (Jameson 1798, pp. 103–104). In doing so, Jameson is stamping a Wernerian seal on what would have been seen as very much a key Huttonian locality.

Sannox and the granite–schistus contact, August 1799

On a walk from Brodick to Lochranza there is only one route which can be taken for the first part of the journey as far as Sannox Bay (see Fig. 1). However, after this point, Lochranza can be reached by either continuing to follow the raised beach around the coast via the Cock of Arran, or by cutting inland and crossing through the boggy ground of North Glen Sannox and Glen Chalmadale. Jameson chose the coastal route in 1797, possibly because he was on foot and the coastal route is less strenuous, or perhaps because he knew that there were coal workings on the coast. However, this meant that he missed the inland localities of Glen Sannox and North Glen Sannox. Ten years earlier, Hutton had also briefly visited Lochranza, but he chose to go via the glens on horseback. As he crossed the bridge in North Glen Sannox on his way north, it seems Hutton made a mental note to look further up the valley on his way back later that day:

> ... I then went forward [to Lochranza], but returning I quitted my horse, and went over the mosses and muir towards the heads of that North Sannox river which there divides into two streams. Here I had the satisfaction to find the immediate junction of the schistus with the granite, in the solid rock, exposed perfectly to view, and that in both of these rivulets, a little way above their junction. Nothing can be more evident than that here the schistus had been broken and invaded by the granite; as in this place the regular stratification of the vertical schistus is broken obliquely by the other rock ... (Hutton 1899, **3**, p. 221).

This was the critical field evidence that Hutton had been searching for. His work up to this point around Glen Rosa and Goat Fell had only been able to narrow down the contact from *in-situ* exposures to within a few yards, but had at least found loose boulders with granite veining in them up on Goat Fell.

Jameson returned to Arran in 1799 apparently determined to fill in the gaps left from his 1797 tour. This time he made sure that he took the inland path to Lochranza. Although Jameson could not have known the exact whereabouts of Hutton's granite contact localities, he visited both of them in 1799 as he walked into North Glen Sannox:

> We traversed over a hill which was mostly formed of breccia into Glen Sannicks North. In a burn at the bottom of the Glen observed the Micaceous Schistus in contact with the Granite and went down for a considerable distance and found the schistus still continues— In another Glen a little to the N of this again observed the schistus in contact with the Granite.[35]

When reproducing this paragraph of his journal for the description of North Glen Sannox in the 1800 *Mineralogy of the Scottish Isles*, Jameson makes his observations with regards the Huttonian '*Theory*' a little less equivocal:

> From this I crossed over a hill of similar rock to North Glen Sannicks. Here we observed a stream running through the glen, and in it I found the shistus in immediate contact with the granite. The shistus appeared to be a very compact micaceous rock; but the granite was not intermixed with it at the junction, nor were there any veins to be observed shooting from the granite into the micaceous rock. We now crossed over the hills into another glen, where I observed another junction of the granite and shistus, but it presented nothing remarkable (Jameson 1799 Journal, pp. 73–74).

This is a stark contradiction of Hutton's own field observations. In North Sannox Burn the thin granite veins described by Hutton can still be seen at the contact and it is remarkable that even though he had stood at the same outcrop, Jameson went out of his way to deny their existence in his 1800 *Mineralogy of the Scottish Isles*. Rather than simply write that he could not see any granite veins, he made it unambiguously clear; the granite was not mixed with the schistus at the junction and there were no veins of granite shooting into the schistus. It seems that in this particular, highly contentious, case Jameson was not prepared to entertain any suggestion of interfingering along the junction of 'primitive rock' with anything else.

The Tormore dykes, August 1799

One of the best stretches of coastline for exposing the variety of igneous intrusions to be found on Arran is along the west coast at Tormore, between Machrie Bay and Blackwaterfoot (Fig. 1). Hutton had previously seen and sketched dykes along the Ayrshire coast and Cumbrae with John Clerk, prior to his trip to Arran (see Craig *et al.* 1978). At Tormore, he only described the coastal section briefly but had clearly established in his own mind the relative ages of intrusion from cross-cutting relationships:

> ... On the west side of the island, upon the shore at Kingscove, I found it [pitchstone] traversing the strata in great dykes; sometimes it is contained in the same dyke with a whinstone, alongside

of which it runs; sometimes again the glass dyke was traversed by one of petuntze [felsite]. This last case I saw there in a glass dyke of great thickness; and it is here remarkable, that, the glass dyke being distinctly traversed by that of the porphyry, this last must appear to be of the latest date or last operation, in like manner as the whinstone dykes had appeared to be from the observations made at the south end of the island (Hutton 1899, **3**, p. 254).

For Jameson, the Tormore dykes or 'veins' provided an excellent opportunity to display the Wernerian theory in terms of the vein stratigraphy. He made a reconnaissance study of the shore in 1797, which is briefly described in his 1798 book. He returned to the area again in 1799 and this time made a more thorough survey of the Tormore shore and provided a map of it in the subsequent *Mineralogy of the Scottish Isles* in 1800. In his journal, Jameson sketched a series of detailed traverses across each vein in plan view (Fig. 3). It is now clear from comparing the journal with the 1800 book, that these sketches were later stitched together to form a composite map (Fig. 4). Jameson described the composite nature of each vein in turn from north to south along the Tormore shore in his 1799 journal:

Having taken breakfast set out for the *Tormore*. Having arrived upon the side of Machry Bay, I began an examination of this interesting section of shore.—The first object which engaged my attention was the vein of green coloured Pitchstone rising from the sea and running E & W... It is about 10 or 12 feet wide. A few yards further on there occurs another vein about 8 feet wide, it runs in common red sandstone. The sides of the veins are of white coloured sandstone next basalt, and in the middle a stone resembling Porphry (Fig. 3a).

Some yards further on another vein occurs about 7 feet wide, and is no less remarkable than the former. It runs in the common red coloured sandstone—the sides basaltes—and in the middle sandstone breccia—these matters have now the appearance of distinct strata, than the former vein, and the matters when in contact with each other are not hardened[36] (Fig. 3b).

The central zone of 'sandstone breccia' in this second vein is in fact a porphyry that has weathered to look like a sandstone and which has mixed with basalt during intrusion. The xenoliths of porphyry do indeed have the appearance of being blocks and it is entirely in keeping with Jameson's Wernerian views that he should interpret this outcrop as a fissure fill breccia. Jameson's narrative continues with a description of the next vein along the shore:

Further on another vein occurs but of far greater size than any of the former; it is about 80 feet wide and runs between NW & N and SE & S. Upon one side of the vein which runs still in red coloured argillaceous sandstone there appears a yellow coloured Porphyrous matter intermediate between Pitchstone and Hornstone—this is here about 1 foot wide—then there occurs about 2 feet of green coloured Pitchstone again about $\frac{1}{2}$ foot of Hornst—Pit matter—then a thin layer of indurated clay—then all the rest of the vein basaltes—(Fig. 3c).

As to the great vein itself which I have described in the account of Arran [1798]—I had now an opportunity of making several new observations ... As this vein of green coloured Pitchstone rises from the sea it is of considerable width ... upon the side of the vein next the sea there is a stratum that I have called sil-sandstone elevated about 60° and dipping in the same direction as the Pitchstone ... Upon the opposite side of the Pitchstone is a stratum of a substance intermediate between Pit and Hornstone ... along side this ... there is about 1 foot of basalt, which is here decomposing in balls ... Further on the vein is to be observed bounded by sandstone but this is somewhat doubtful—as the other strata matters may still be present but in such small quantity, as to be perceived—this is somewhat probable from the circumstance of the yellow horn-pit matter in vein + (Fig. 3c) being in some places only a few inches wide while in other parts is many feet—at the end of the vein nearest to Machry bay the Hornst Pit like matter appears again forming one side of the vein but the bounding substance on the other sides of the vein is not to be observed.[37]

This last passage is a crucial one, as Jameson appears to be trying to establish a common stratigraphic succession that repeats across each vein. This interpretation runs into trouble when he makes the observation that the main pitchstone vein appears in places to be bounded directly by sandstone. To explain the absence of the intervening lithologies Jameson merely points out that in a previous vein (Fig. 3c), these are seen to be variable in thickness and thus are probably present but unexposed.

Jameson's final map of the Tormore shore reproduces his sketch traverses on a much smaller scale and all together as a coherent network of veins (Fig. 4). A fascinating aspect of this map is that small piles of boulders are drawn to cover each point where the dykes should cross-cut each other. The true nature of cross-cutting relationships between dykes was recognized as providing critical field evidence by both Huttonians and Wernerians alike. Hutton wrote that he could see this relationship and establish an order of emplacement (see above). However, Jameson made it clear in his journal and on the subsequent map that exposure is lost at each dyke precisely at the point of intersection (Fig. 4). Whilst this looks a little too contrived to be true, in this particular case the field evidence seems to support Jameson rather than Hutton. There are no exposures today where one dyke can actually be seen in contact cutting across another. It can be inferred at Jameson's dyke 'P', where the baked margins of sandstone alongside the 'great vein D' are offset, suggesting that dyke 'P' has intruded at a later date and a small amount of slip has occurred during its emplacement. If Hutton really did observe an exposure showing the actual cross-cutting contact between porphyry and pitchstone, it had been covered by beach debris only a few years later and has remained so today.

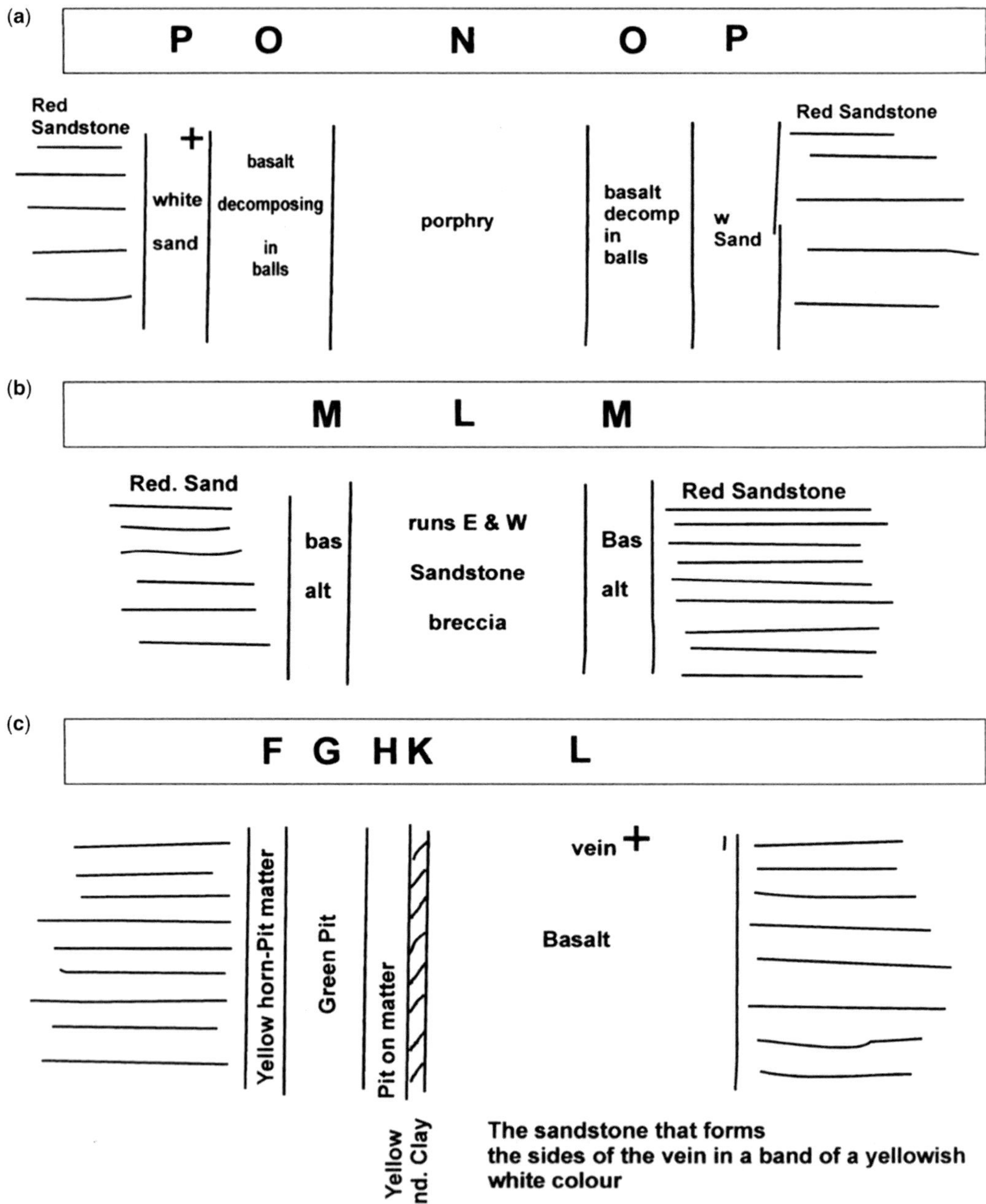

Fig. 3. Plan view sketch traverses across the main Tormore 'veins', redrawn from Jameson's 1799 journal. Note that Jameson's journal descriptions of these sketches are reproduced in the text. These sketches never appeared in print but were the vital detailed work needed for Jameson to be able to produce an overall map of the veins for publication in his *Mineralogy of the Scottish Isles* in 1800. The lettering of veins used on the 1800 map, which is reproduced here as Figure 4, has been added to the top of these sketches in rectangular boxes to allow cross-reference between the two figures.

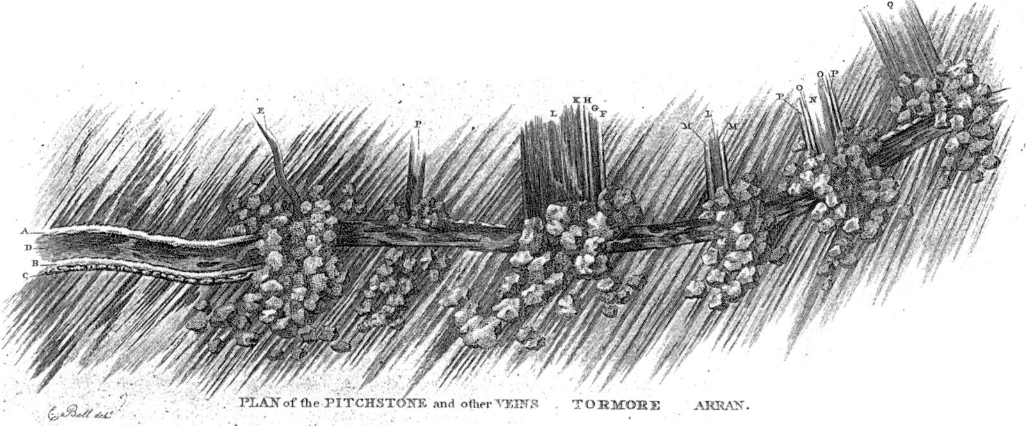

Fig. 4. 'Plan of the Pitchstone and other Veins, Tormore, Arran', from Jameson's *Mineralogy of the Scottish Isles*, 1800. Although the scale is arbitrary, the spatial relationship between dykes is accurate. Note that the low tide mark approximates to the top edge of the map and north is to the right. This remains an excellent map of the dykes along the foreshore at Tormore and can still be used successfully in the field today.

In 1893, nearly one hundred years later, John Wesley Judd (1840–1916) visited the Tormore shore (Judd 1893). The detailed description that he published of these composite dykes has led to this stretch of coast to be known colloquially as 'Judd's Dykes'. Judd wrote of Tormore:

> The first author to describe this very interesting locality was Prof. Robert Jameson ... So accurate are this plan and description that they were reproduced by Ramsay in his 'Geology of Arran', without alteration ...[38] The original plan and description by Professor Jameson still constitute an admirable guide to the position and relations of these rock masses (Judd 1893).

In fact, it seems that Jameson's original work was so good that Judd copied his methodology in describing the dykes in 1893. Each one is shown as a sketch traverse in plan view and as part of an overall map of the shore. By 1893, of course, Judd had the benefit of scale topographic maps and a modern understanding of igneous petrology at his disposal. Nevertheless, in many respects Jameson's map is superior to Judd's as it shows the composite nature of the dykes in detail.

Conclusions

For a brief period at the end of the eighteenth century, the Isle of Arran played an important role in the debate between Neptunists and Plutonists. Although Hutton produced the earliest account of Arran's geology, none of his original field data survive. Therefore, it is Jameson's Arran journals that provide the earliest surviving field description of the geology on the island. In particular, the 1797 journal is an important source of information regarding the experience of conducting fieldwork in the Scottish Islands during this period.

Jameson conducted his fieldwork with a full understanding of Wernerian and Huttonian concepts in mind. He chose to interpret his field observations using the Wernerian theory. With historical hindsight, the Neptunian theory of the Earth may seem untenable but the way Jameson interpreted his observations should not detract from the validity of the observations themselves. The Arran journals show Jameson to have been an acute and diligent observer in the field. He spent several days investigating an area until he was able to produce a simple geological history from his observations. By writing his journals each evening, Jameson was able accurately to record the work done that day and also to clarify in his mind what questions still needed to be answered during subsequent work.

It is also clear from the journals that Jameson made observations that were not always easy to explain using a strict Neptunian scheme, such as in Glen Cloy, at the Witch's Step, or Tormore. Jameson actually took a far more hard-line Wernerian view in print than he did in his journal entries. The public politics of the Neptunist–Plutonist debate appears to have overridden any ambiguity in the field evidence. Finally, the young Robert Jameson should be recognized as having been an excellent field geologist of his time and it is only because of his Neptunist views that history has not dealt with him more kindly.

The authors would like to thank the Librarians of the Special Collections at the University of Edinburgh for their help and to the Librarian for permission to publish extracts from the Jameson journals. We would also like to acknowledge all those who have helped in fieldwork on the Isle of Arran over the past thirteen years, and G. Botterill of the Kilmichael Country House Hotel, Isle of Arran, for information on the Fullarton family history and Kilmichael.

Notes

[1]Pennant (1774). The portion of text referring to Arran can be found reproduced in Whyte (1997), pp. 11–15.

[2]See the augmented reprinting of Geikie's Volume **3** as Dean (1997, pp. 191–267).

[3]Robert Jameson manuscripts Dc. 7. 126; *Journal of my Tour in 1797* and Dc. 7. 129; *Arran 1799*, held in the Special Collections of the University of Edinburgh Library. For the purposes of this paper these will be referred to simply as Jameson's '1797 Journal' and '1799 Journal'. Extracts are published here with permission of the Librarian.

[4]For a general biography of Jameson see Sweet (1976). The biographical details given in this introduction are based mainly on the obituary given by Hamilton (1855 pp. xxxviii–xli). For articles on Jameson's early career see Sweet (1963, 1967).

[5]John Walker, M.D., D.D. (1731–1803), Professor of natural history in the University of Edinburgh from 1779 until his death.

[6]See Sweet & Waterston's discussion of both papers Jameson read to the Royal Medical Society in 1796 in Sweet & Waterston (1967).

[7]Jameson quotes from Werner (1788).

[8]See Craig *et al.* 1978 and Dean 1997.

[9]Hutton (1795, **1**, pp. 429–430) briefly describes the Lochranza unconformity, and then fracturing in 'pudding-stone' conglomerates as evidence of pressure and cementation (pp. 467–468).

[10]Playfair used Hutton's unpublished mineralogy of Arran in his own research on the island during October 1797. He inadvertently refers to it in his *Illustrations of the Huttonian Theory* ... (1802). For instance, on p. 334 Playfair wrote; '...the mountain of Goatfield ... when I visited it, with a view of verifying on the spot the interesting observations which Dr Hutton had there made ...'. Hutton never referred to Goatfell in either Volumes **1** or **2** of his 'Theory ...' (1795), but he discussed it at length in his Arran chapter of Volume **3** (1899).

[11]Jameson 1797, Journal, p. 69.

[12]Jameson 1797, 1799, Journals.

[13]Jameson 1797, Journal, p. 53.

[14]Jameson 1797, Journal, p. 74.

[15]Jameson 1797, Journal, p. 73.

[16]Jameson 1797, Journal, p. 87.

[17]The first few days of the tour on Arran prior to Jameson's trip to Ireland have previously been transcribed and discussed by Sweet 1967, pp. 97–126.

[18]Foula and Noss are small islands in the Shetlands and here Jameson is clearly referring back to his previous tour of the Shetland Islands in 1794.

[19]Jameson 1797, Journal, pp. 6–7.

[20]Jameson 1797, Journal, pp. 9–11.

[21]Jameson 1797, Journal, pp. 11–12.

[22]'The Erse language' referred to by Jameson is Scots Gaelic and this was still the first language of the inhabitants of Arran at this time. Gaelic was spoken by the clergymen during services and it was taught in the schools. However, from the 1790s onwards, English gradually started to replace Gaelic across the island. James Headrick, who visited Arran in 1803 records that the inhabitants were all, 'remarkably pious and devout' (see Headrick 1807). It may be that Jameson and the others were laughing at the gravity and seriousness of the Grace's delivery or perhaps at the apparently old-fashioned way it was spoken in Gaelic.

[23]Jameson 1797, Journal, pp. 19–20.

[24]Jameson 1797, Journal, pp. 61–64.

[25]Jameson 1797, Journal, pp. 67–68.

[26]Jameson 1797, Journal, pp. 69–71.

[27]Jameson 1797, Journal, p. 80.

[28]Jameson 1797, Journal, pp. 82–83.

[29]Jameson 1797, Journal, pp. 85–86.

[30]Jameson, note 27, pp. 100–101.

[31]Jameson 1797, Journal, p. 87.

[32]Jameson 1797, Journal, pp. 104–105.

[33]Jameson 1797, Journal, p. 102.

[34]Jameson 1799, Journal, p. 25.

[35]Jameson 1799, Journal, p. 24.

[36]Jameson 1799, Journal, pp. 28–29.

[37]Jameson 1799, Journal, pp. 30–34.

[38]Ramsey's actual quote regarding the Tormore dykes and Jameson was; 'On the coast of Machrie Bay, near Tormore, there is a remarkable pitchstone vein, which has been so admirably described by Professor Jameson, that no room is now left for any additional remarks'. Ramsey then went on to reproduce Jameson's map and transcribe the text from his 1800 *Mineralogy of the Scottish Isles* (see Ramsay 1841).

References

CRAIG, G. Y. 1987. *The 1785 Abstract of James Hutton's Theory of the Earth.* Scottish Academic Press, Edinburgh.

CRAIG, G. Y., MCINTYRE, D. B. & WATERSTON, C. D. 1978. *James Hutton's Theory of the Earth: The Lost Drawings.* Scottish Academic Press, Edinburgh.

DARWIN, C. 1892. *Charles Darwin, His Life Told in an Autobiographical Chapter and in a Selected Series of his Published Letters.* D. Appleton & Co., New York.

DAVIES, D. 2002. *A Brief History of Fighting Ships.* Robinson, London.

DEAN, D. R. (ed.) 1997. *James Hutton in the Field and in the Study.* Delmar, New York.

HAMILTON, W. 1855. Anniversary address of the President of the Geological Society of London. *Quarterly Journal of the Geological Society*, **11**, xxvii–xciii.

HEADRICK, J. 1807. *View of the Mineralogy, Agriculture, Manufactures and Fisheries of the Island of Arran, with Notices of Antiquities, and Suggestions for Improving the Agriculture and Fisheries of the Highlands and Islands of Scotland.* Constable, Edinburgh.

HUTTON, J. 1794. Observations on Granite. *Transactions of the Royal Society of Edinburgh*, **3**, 77–85.

HUTTON, J. 1795. *Theory of the Earth with Proofs and Illustrations.* Volumes 1 and 2, Cadell, Junior, and Davies, London & Edinburgh.

HUTTON, J. 1899. *Theory of the Earth with Proofs and Illustrations edited by A. Geikie.* Volume 3. Geological Society, London.

JAMESON, R. 1798. *An Outline of the Mineralogy of the Shetland Islands, and of the Island of Arran.* Wm. Creech, Edinburgh.

JAMESON, R. 1800. *Mineralogy of the Scottish Isles; with Mineralogical Observations Made in a Tour Through Different Parts of the Mainland of Scotland, and Dissertations upon Peat and Kelp.* 2 Volumes. Printed by C. Stewart for B. White, Edinburgh.

JUDD, J. W. 1893. On Composite Dykes in Arran. *Quarterly Journal of the Geological Society*, **69**, 536–565.

KIRWAN, R. 1794. *Elements of Mineralogy.* 2 Volumes. 2nd edn, London.

MACBRIDE, M. 1911. *Arran of the Bens, The Glens and The Brave.* Foulis, London & Edinburgh.

PENNANT, T. 1774. *A Tour in Scotland and Voyage to the Hebrides.* B. White, Chester. [2nd volume 1774, J. Monk, London].

PLAYFAIR, J. 1802. *Illustrations of the Huttonian Theory of the Earth.* Printed for Ross and Blackwood, Edinburgh.

RAMSAY, A. C. 1841. *The Geology of the Island of Arran from Original Survey.* Richard Griffin, Glasgow.

SCOTT, H. W. (ed.) 1966. *Lectures on Geology* by John Walker. Chicago Press, Chicago.

SWEET, J. M. 1963. Robert Jameson in London. *Annals of Science*, **19**, 81–116.

SWEET, J. M. 1967. Robert Jameson's Irish Journal. *Annals of Science*, **23**, 97–126.

SWEET, J. M. 1976. Introduction. *In*: WHITE, G. W. (ed.) *The Wernerian Theory of the Neptunian Origin of Rocks.* Hafner Press, New York, pxii–xxiv.

SWEET, J. M. & WATERSTON, C. D. 1967. Robert Jameson's approach to the Wernerian theory of the Earth, 1796. *Annals of Science*, **23**, 81–95.

TYRRELL, G. W. 1950. Hutton on Arran. *Proceedings of the Royal Society of Edinburgh*, **58**, 369–376.

WERNER, A. 1788. Werners Bekanntmachung einer von ihm am Scheibenberger Hügel üder die Entstehung des Basaltes gemachten Entdeckung ... *Bergmann. Journ*, **2**, 845–855.

WHITE, G. W. (ed.) 1976. *The Wernerian Theory of the Neptunian Origin of Rocks.* Hafner Press, New York. [This is a facsimile reprint of Jameson's *Elements of Geognosy*, 1808].

WHYTE, H. 1997. *An Arran Anthology.* Mercat Press, Edinburgh.

Writing, 'inscription' and fact: eighteenth century mineralogical books based on travels in the Habsburg regions, the Carpathian Mountains

M. KLEMUN

Institut für Geschichte, Universität Wien, Dr Karl Lueger-Ring 1, 1010 Wien, Austria (e-mail: marianne.klemun@univie.ac.at)

Abstract: In the second half of the eighteenth century there was an explosion in travel literature in the German-speaking countries. Travel literature became an important medium for broadening of people's horizons. When transferring travel experiences into the written word the traveller had a wealth of literary forms at their disposal: the diary, the letter, the narrative structured according to the chronology of the journey. Moreover, topography, 'statistics' in the contemporary sense and 'apodemics' (the art of travelling) as a professional basis for travelling offered the potential for a scientific approach. The scientific interest in 'mineralogical' journeys aimed at acquiring reliable empirical data, and required formal strategies to counter the superficiality and selectivity of fieldwork. The choice of one strategy of description, and that is the thesis presented here, is based on epistemological factors. This paper looks at three works of travel literature, in which the chosen form of writing correlates with the central scientific findings of the respective authors: Belsazar de la Motte Hacquet (1739–1815), Ignaz von Born (1742–1791) and Ehrenbert von Fichtel (1732–1795) all of whom wrote 'mineralogical' books based on extensive travelling in the eastern Habsburg territories.

Travelling, writing and 'mineralogy'

In the history of science, it has generally been acknowledged that fieldwork was a predecessor to the establishment of geology as a science and it became an essential part of scientific work in the second half of the eighteenth century (Rudwick 1996, pp. 266–286). Similarly, historians of cultural–scientific travelling agree that for the aspiring bourgeoisie, travelling became a central method of acquiring world knowledge (Stagl 1995), by means of which people hoped to educate themselves. As a consequence, there was an explosion of travel literature in the second half of the eighteenth century, which occurred in a many areas of culture and knowledge (Brenner, 1989, 1990, **2**, p. 161).

Before the eighteenth century 'mineralogists' had collected rocks and minerals in the field, but the final scientific analysis had taken place indoors, in studies or collection rooms, where their finds were analysed and classified. In the 'mineralogy' of the eighteenth century, a general term used for a discipline that we now call Earth sciences, fieldwork meant that 'mineralogists' began to move about in an area and examine the rocks *in situ*, in the context of their origins. So the actual scientific work took place while travelling. The activities of the 'mineralogists' were focused on locating rocks in the field and on integrating them within the structure of a geographical distribution.

As long as scientists did not have to use a specific procedure for the recording of data, such as the structure or section of mountains that was developed only in the middle of the century, what Rudwick (1976) has called a 'visual language', the punctual form of registering data was forced into a spatial succession. Travelling became an essential part of the abstracted process of gaining knowledge, in a geographical context. The data was recorded following the route of the travels. Therefore, the system of description used by the mineralogist while travelling was important. Description as 'inscription'[1] included the articulation, rhetoric and persuasion.

My observations have shown that mineralogists used many written forms to weave their data into a story; however, is there a connection between the knowledge gained and the style of writing and selected form of 'inscription'? The reason for addressing this question is neither a literary nor a philological one, but an epistemological one. In other words: what is the connection between the packaging and the content or the form and the scientific relevance? To my mind, the first one has been neglected in favour of the second one, if not overlooked completely, mainly because too much attention has been paid to so-called 'Whiggism' (Porter 1990, p. 32–46) practised in historiography. If form takes over a function, then it is more than packaging. It is a part of the epistemological method that is analysed here. I propose that the

From: WYSE JACKSON, P. N. (ed.) *Four Centuries of Geological Travel: The Search for Knowledge on Foot, Bicycle, Sledge and Camel.* Geological Society, London, Special Publications, **287**, 49–61.
DOI: 10.1144/SP287.5 0305-8719/07/$15.00

categorization of data and its integration into a certain textual form or textual structure and into a literary defined framework does not happen arbitrarily, but that it determines scientific information. My thesis, therefore, is that appropriate, different textual forms were chosen to represent and support the scientific data collected.

Geological travellers had many text forms, those commonly used in travel literature, at their disposal. These included:

(1) a topographically-oriented overview based on '*ars apodemica*' (the art of travelling) which required a systematic procedure and observational schemas and imposed traditional patterns in order to transform the holistic encyclopaedic mode into a comprehensive synopsis;
(2) a description of a journey in the form of letters that were addressed to the reader urging him to engage in a conversation with the author; or
(3) a report or strict diary of the journey in the form of an itinerary, in which the procedure of collecting information is subordinate.

Using three books, this paper looks at the relationship between the chosen form of writing and the focus of interest of the respective authors. Ignaz von Born (1742–1791), Johann Ehrenreich von Fichtel (1732–1795) and Belsazar de la Motte Hacquet (1739–1815), each of whom wrote works on mineralogy based on their travels. All three authors were dedicated to researching the mineralogy of the Carpathian region. These three men had much more in common: they were of the same generation, they were highly productive in the Earth sciences, they played a central role in the state administration or civil service and were representatives of the group of extremely critical protagonists otherwise known as the Enlightenment movement.

Fig. 1. Ignaz von Born.

Ignaz von Born

Scientifically, Ignaz von Born,[2] (Fig. 1) was the most productive of the three. According to research on the Enlightenment, Born was considered to be amongst 'the most important personalities of the Enlightenment in the Habsburg Monarchy' (Reinalter 1991, Preface, p. 9). Born in 1742 in Kapnik, Transylvania (now Romania), Ignaz von Born studied law in Prague having left the Jesuits and subsequently attended courses for mining held by Thaddäus Peithner (1727–1792). His employment for the state began in 1769 when he became a 'Bergrat' [clerk] in the mining town of Schemnitz (Bianska Stiavnica, Selmeczbánya) in 'Oberungarn' (Upper Hungary) (now Slovakia), where the Empress, Maria Theresia (who reigned 1740–1780), had founded a mining school. By 1770, Maria Theresia appointed him to be the successor of Count Franz de Paula Colloredo (1731–1807) as 'Bergrat' in Prague, where he acted as member in the 'Oberstes Münz- und Bergamt' [Highest administrative authority on mining including the mint]. For health reasons, and in protest against the prohibition of publishing scientific or technological texts, von Born quit his post in 1772. However, in 1776 he re-entered the Habsburg service in Vienna where he classified the objects of the Imperial Mineralogical Collection and became 'Bergrat' of the 'Bergbehörde im Münz- und Bergwesen'. Born's abilities were manifested in four areas:

(1) In modern terms, he would be considered an ambitious scientific manager (Teich 1976, pp. 195–205). He worked both for the public and to professionalize the scientific fields he worked in. In Prague, he founded a private organization (1773) (von Born 1775–1784),

which called itself 'Königlich Böhmische Gesellschaft der Wissenschaften' [Royal Bohemian Society of Sciences]. Moreover, he initiated a journal of review (von Born & Trebra 1789–1790). In 1786, the 'Society for Bergbaukunde' (Fettweis & Hamann 1989, pp. 11–25) was founded, the first organization that facilitated an international exchange of technological mining knowledge. The Society had 150 members including both practical and theoretical scientists. The secrecy within mining until that time had been ordered by the state, but was undone by von Born's initiative of organizing the first international technological world congress in mining in Skleno in 1786.

(2) In connection with his post in the mining administration, von Born tried to introduce innovations that would be beneficial to the mining industry through a number of publications. He devised an original method of mining silver by means of a process of amalgamation (von Born 1786; see Teich 1975), which was used in Slovakia and Bohemia. He also developed an industrial application for bleaching organic substances using chlorine.

(3) From his interest in natural history and taxonomy, von Born developed a scheme for mineralogy, which was based on Linné's system of classification in botany. His definition of 'species' was based on chemical criteria (von Born 1772). He devised a scheme in which he categorized the groups in a pyramidal manner from 'species' into 'genera', which were grouped into 'ordines'. The latter were then divided into 'classes' (1772). In the second volume in his book (1775) he gave up this system by grouping the 'species' into strings that were then grouped side by side, thereby abolishing the original architecture of the system. In his work on the collection of the Imperial Court, von Born concentrated on the order of shells and fossils (von Born 1778). He proved to be a professional systematist when carrying out a commissioned work on the collection of Eleonore Raab, a passionate collector (von Born 1790).

(4) Von Born's social concern at a time when the conservative feudal–aristocratic conglomerate of countries was transforming into a bureaucratic, institutionally-governed homogenous state, should not be underestimated. He was prepared to air his views publicly, for example in his satirical pamphlet *Die Staatsperücke* (von Born 1773) [*The State's Whig*]. In this pamphlet, he compared British society as idealized by Alexander Pope (1688–1744) in his *Essay on Man*, with the moth-eaten aristocracy characterized by vanity in the Habsburg countries. Using Linné's system he satirically described the role of the monks. The work *Specimen Monachologiae Methodo Linnaeana* ([von Born] 1783*a*, *b*, 1784) was printed under the name Kuttenpeitscher (1783, 1784, 1786) several times featuring various titles and had already been translated into other languages by the nineteenth century.[3] From the number of editions and the audience he reached it was his most successful publication. Von Born founded five periodicals. His first was a review journal modelled on the *Göttingischen gelehrten Anzeigen*. In Vienna, he soon became the head of a Freemasons' lodge, which became an elite group due to the lack of an academy in this city. The *Journal für Freimaurer* [*Journal of Freemasons*] was complemented by a special forum for publications, the *Physikalischen Arbeiten der einträchtigen Freunde* [*The Physical Works of United Friends*] (1783–1787). The generally-accepted view that there was similarity between von Born and the figure of Sarastro (in Mozart's *Magic Flute*) is controversial; however, von Born was seen as a role-model for a wise scholar.

Travel writing about mountains

Von Born's *Travels through the Bannat of Temeswar, Transylvania and Hungary in the year 1770* was published in 1777 (Fig. 2) ([von Born] Ferber 1774). Immediately after its publication, the influential review journal *Göttingische gelehrte Anzeigen*[4] announced euphorically, that with his book von Born had laid the cornerstone for research on the composition of mountain ranges for this area. This work constituted the beginning of his geological career. In 1770, shortly before starting his job as Bergrat in Schemnitz (Slovakia), he took leave to sort out his financial affairs at home in Transylvania, which territorially belonged to the Habsburg Monarchy. His father had worked there as a mining entrepreneur and so von Born knew the area from his childhood. He travelled around Banat (Temesvar/Timisora, Oraviza/Oravica, Saska, New Moldava), which had been freed from Turkish rule and was under the rule of the Habsburg Monarchy, and reached the military border of the then Ottoman Empire. He was 'accompanied by twelve mounted mining clerks and some mountain workers on foot who had rifles,' ([von Born] Ferber 1774, p. 40) to protect him from notorious robbers ['Räuberhorden']. He reported on this situation as follows:

Des
Hrn. Ignatz, Edl. von Born,
Ritters, K. K. Berg-Raths, der Königl. Akademie der Wissenschaften zu Stockholm, der Großherzogl. zu Siena, u. der Georg. gelehrt. Gesellschaft zu Padua Mitglieds rc.

Briefe

über
Mineralogische Gegenstände,
auf seiner Reise
durch das Temeswarer Bannat, Siebenbürgen, Ober- und Nieder-Hungarn,
an
den Herausgeber derselben,
Johann Jacob Ferber,
Mitglied der Königl. Großherzogl. Akademie der Wissenschaften zu Siena, und der Ackerbau-Gesellschaft zu Vicenza und zu Florenz,
geschrieben.

Frankfurt und Leipzig, 1774.

Fig. 2. Title page of Ignaz von Born's letters.

> The unsafeness of the routes and paths made Hof Cammerrath von Hegengarthen—a man known for his affability, whose warm-heartedness I cannot praise enough—order all villages that I would pass to ensure my safety. Thus, I met in every village some forty to fifty 'Wallachs' complete with rifles and cudgels, who, led by their superior, accompanied me to the next village and who rather carried my coach on their shoulders on rocky and otherwise arduous paths than kept it in balance.[5]

Von Born was able to use his post as 'Bergrat' to his advantage and travelled part of his journey with the Imperial 'Commissarius.' ([von Born] Ferber 1774, p. 40)

Von Born's description of his travels were published in 23 letters, four years after the journey. The reason for describing the journey in letters was a political one. Since 1772, there had been an Imperial rule that prevented state-employed mining staff from publishing their work. To circumvent this rule, the Swedish scholar Johann Jacob Ferber (1743–1790), a publisher and a friend of von Born, stepped in. Von Born allegedly sent his data in letters to Ferber. In the preface of the book, the publisher stressed that the book had been printed without the consent of the author. Von Born admitted the tactics to a friend. To the Erlangen Professor Daniel Gottfried Schreber (1708–1777), he wrote: '[...] I had to pretend to be writing letters to Ferber since when I had been dismissed from my services, I had to agree that I would never publish anything on mines or mineralogy. Thus, those letters had to have another publisher.'[6]

But there were other reasons for choosing the letter-writing style for his book. A scholar, well-versed in literary genres, and part of the network of literary men who corresponded, for example, with the German poet Christoph Martin Wieland (1733–1833),[7] knew that the letter form was extremely popular in the book market. It was particularly favoured by the audience of the Enlightenment since it referred to the dominant medium of Enlightenment itself, namely communication (Stafford 1994, 1998). Moreover, it permitted the bundling of heterogeneous findings while allowing von Born's personality to show through in his writing.

Each letter was written from a different place. These were static points along his route. In his writing he looked at the past as well as into the future. He painted a picture based on the chronology of his journeys that included analysis of the region as he stayed in different places. Von Born rarely walked; generally he travelled from mining town to mining town in a horse-drawn carriage. He based himself in towns that were both administrative and mining centres, and explored the surrounding mining areas where he examined veins from which copper, iron, mercury and gold were extracted. He also wanted to learn how the areas were administered. He wrote the letters from 'Temesvar' (Timisoara), 'Oravitza' (Oravicza), 'Saska', 'Neu-Moldava', 'Dognaska' (Dognacka, Krassó-Szörenyi, Langenthal) 'Lugos', 'Nagyàg' 'Zalathna', 'Nagyàg', 'Bey Földwinz', 'Clausenburg' (today Cluj) 'Nagy-Bánya', 'Schemnitz' and the last one from 'Wien' (Vienna). Only Vienna diverged from his route. His story ended in Vienna, where he hoped it would have its most pronounced effect. He had indirectly addressed his findings about mining and the state of productivity to the Emperor who favoured the Enlightenment movement. In this final letter from Vienna, von Born expressed his regrets about the lack of possibilities for education in natural history and the low standard of mineralogical collections in this city. As for the Imperial Collection, von Born thought that it lacked any professional arrangement or taxonomic classification. He criticized the lack of local material, since 'the transition from one kind of rock to the next, in short, those mineralogical curiosities that differentiate at first sight the cabinet of a collector from that of an expert'[8] were completely missing. This comment was effective. Only two years later, von Born was appointed to work as custodian of the Imperial Collection for the Emperor.

Von Born collected many minerals in the course of his journeying. He was especially delighted with his suite of specimens, which he collected himself although he had to sell them later, for financial reasons, to London. He reminisced:

> All in all, with a view to my mineral collection, Saska may be the place where I have reaped the best harvest of my journey. Not only did I find large amounts of all kinds of copper ore—with the exception of 'Mannfeldian Kupferschiefer'[9] [which had become famous since the book of Lehmann]—but also many new hitherto unknown kinds.[10]

Von Born was not able to give a continuous overview of the mountain ranges in this large area of eastern Europe, as the relevant material turned out to be too fragmentary; the descriptions in his letters and ordered according to the places he had visited filled some of those gaps. In addition, von Born tried to address the central question that all mineralogists of the time were feverishly trying to answer: how the oldest mountain ranges had been formed. In the mines, near the ore veins, where 'hanging and lying rocks' [vertical and horizontal rocks] touched, von Born found confirmation of his original conviction that granite formed the core of mountains whereas metamorphic rocks formed the casing. His aim was to determine the superposition of fundamental rock types. He interpreted most of the sediments as 'incidental' rocks that were formed later than limestone, which he assumed to be one of the youngest rocks. Von Born concluded that there were metamorphic rocks everywhere, superimposed over the 'granitic' rocks, mostly 'Karbonatgestein' over the metamorphics. Von Born's conclusion was in accordance with that of the rest of the scientific community, who held that granite was the oldest and limestone the youngest form of rock. Younger than granite and older than limestone, von Born assumed to be the 'saxum metalliferum', by which he meant andesite and dacite, but not rhyolite (as they were later called) (see esp. Haubelt 1991; Mutschlechner 1991). His conclusion that there was a genetic continuity between andesite and ores was important for the history of petrology and was one of his most important achievements. He wanted to make his scientific findings accessible and relevant. He thought that if miners did not have any mineralogical knowledge, they would make the wrong decisions when working in the mines.

Mineralogische
Bemerkungen
von den
Karpathen.
Von
Johann Ehrenreich von Fichtel,
Kais. Königl. Gubernialrath, und Bankalgefällen-Direktor; der Gesellschaft naturforschender Freunde zu Berlin, der ökonomischen zu Leipzig, und der Sozietät der Bergbaukunde Mitglied.
Erster Theil.
Der Berg Feketehegy bey Telkebanya in Ungarn 6 Stunden von Kaschau. Zur Seite 371.
Wien, 1791
bei Joseph Edlen von Kurzbeck,
k. k. Hofbuchdrucker, Groß- und Buchhändler.

Fig. 3. Title page of Fichtel's notes about the mineralogy of the Carpathian Mountains Volume 1 (1791).

Mineralogische
Bemerkungen
von den
Karpathen.
Von
Johann Ehrenreich von Fichtel,
Kais. Königl. Gubernialrath, und Bankalgefällen-Direktor; der Gesellschaft naturforschender Freunde zu Berlin, der ökonomischen zu Leipzig, und der Sozietät der Bergbaukunde Mitglied.
Zweyter Theil.
Wien, 179
bei Joseph Edlen von Kurzbeck,

Fig. 4. Title page of Fichtel's notes about the mineralogy of the Carpathian Mountains. Volume 2 (1792), showing Hephaistos, the God of Fire.

It is not surprising that von Born's conclusions, which were in accordance with the Neptunist explanation of the formation of mountain ranges, were welcomed by Abraham Gottlob Werner in Freiberg (1745–1817), Germany, the founder of the this school of thought (see Laudan 1987, esp. p. 87ff). Werner even presented von Born's findings in his lectures (see Haubelt 1972, p. 154). Soon, von Born's publication was translated into English (von Born 1777) and a French edition followed (von Born 1780).[11] However, von Born was not a fundamental Neptunist. In his work on the mountain Kammerbühl (Komorní hurka, now in the Czech Republic), published in 1773, he defined the mountain as a burnt-out volcano for the first time and did not believe in the assumption of 'a coincidental Earth fire.' (von Born 1773).

Johann Ehrenreich von Fichtel

Eighteen years after von Born's work on the Carpathian Mountains, Johann Ehrenreich von Fichtel published his book on the Carpathian Mountains (Figs 3 & 4) (see Fichtel 1791–1792). He was interested in new explanations of the laws for formation of this mountain range. Fichtel was particularly interested in those mountain ranges that did not contain ore; he described the 'base part' as still unknown (Fichtel 1791, **1**, Vorbericht). He was looking for a 'general term, which could explain the successive change of mountain ranges on the whole', hence the volcanic interpretation of the Carpathian Mountains.

Fichtel[12] was born in Pressburg (Poszony, at that time in Hungary, today Bratislava, Slovakia) in 1732 and studied law in Hungary. After his probationary years as a solicitor, he entered the administrative services in Hermannstadt (Nagyzeben, today Sibiu, Romania), Transylvania as registrar in 1759. After the Economic Directorate of the Transylvanian Saxons had been dissolved in 1762, Fichtel moved to the Court Chamber of Accounts, the highest financial authority in Vienna, and worked there as a clerk. In 1769, he returned to Hermannstadt where, as a councillor in the Treasury of the Transylvanian Chamber, he advocated the improvement of the salt works and the promotion of the salt trade. He became a director of the Direction of the Customs Revenue between 1785 and 1787 and was finally appointed councillor of the Transylvanian Government, the highest office in Transylvania.

Although Fichtel had never had an education in mining, he taught himself and became an expert. On the urging of the *Gesellschaft Naturforschender Freunde zu Berlin* [Society for Friends of Natural History in Berlin], a society whose scope reached into eastern Europe and which was known for adopting renowned collectors as new members, Fichtel published the book *Beytrag zu einer Mineralgeschichte Siebenbürgens* [*Contribution to the Mineralogy of Transylvania*]. In this work he focused on fossil finds and local salt mines. In his *Mineralogische Aufsätze* [*Mineralogical Essays*] of 1794, in which Fichtel refined his volcanic theories on the Carpathian Mountains that had been formulated in 1791, he published a varied collection of short contributions and essays. On this topic Gábor Papp's essay on Fichtel (Papp 1998) is a valuable source of information.

In carrying out his professional duties, Fichtel travelled through Transylvania, and into Croatia and Slovenia, in order to inspect the quarantine facilities on the mountain passes along the military border of the Habsburg Empire. On such official journeys (in his travel descriptions, he quotes a period of 19 years) he randomly collected data that he eventually presented in a systematic form. Fichtel did not work and write to a planned geological itinerary. His writing was divided into a general and a specialized part. In Fichtel's own words, the first part contains the 'General description of the Carpathians, with their rocks and their remarkable curiosities occurring therein [Beschreibung der Karpathen überhaupt, mit ihren Gebirgsarten, und darinn vorkommenden Merkwürdigkeiten]'. (Fichtel 1791–1792, **1**, p. 1). He proposed studying the mountain range by approaching it from the outside and working inwards. Its structure was dictated by '*ars apodemica*'. By encircling the mountain, one eventually arrived at its centre, the volcano. According to Fichtel, many mountains were volcanic: in Lower Hungary these included the ore mountains (Erzgebirge, Kremnické vrchy, Stiavnicè vrchy), Feketehegy in the Tokaj Mountains, the Gutin (Gutai Mts), the Kele–Görgény–Hargita Mountain (Calimani–Gurghiu–Harghita) and parts of the Transylvanian ore mountains.

In this connection, it is tempting to think of Erasmus Darwin's (1731–1802) 1791 theoretical depiction of a volcano (see Guntau 1996, esp. p. 227) from a bird's eye view in order to visualize his Plutonist idea. Fichtel's writing is reminiscent of this drawing since he circles this entire area, which contains many volcanoes to arrive at his theory of the phenomenon of volcanism. The copperplate engraving on the cover of the publication illustrates his central concept. Fichtel claimed that all mountains were of volcanic origin, a theory about which he entered into heated debate.

He also argued that the Transylvanian ore veins were connected to volcanic rocks. His overview of the Carpathian Mountains, 'the surroundings of all connected Carpathian and other Mountains, which covered an area of more than 300 kilometres' (Fichtel 1791, **1**, p. 353) ended in its foothills for

Fichtel, as well as for von Born, with the 'Leopoldi Berg on the banks of the Danube River' (Fichtel 1791, **1**, p. 353) in Vienna. The view over the countries all the way to the Adriatic Sea evoked, due to the omnipresence of limestone, only 'boredom' in the passionate volcanist Fichtel. This boredom was 'interrupted shortly' (Fichtel 1791, **1**, p. 352) only three times, in Trieste, at the border between Styria and Carnia and near Graz.

In the second part of his *Mineralogical Notes*, Fichtel offered the reader a 'special treatise on volcanoes especially those of the Carpathian Mountains [*Abhandlung von den Vulkanen der Karpathen ins Besondere*]'. (Fichtel 1792, **2**, p. 415). He described his central theme as 'the simple ideas on the history of origins of the old Hungarian and Transylvanian volcanoes'. (Fichtel 1792, **2**, p. 417). This idea was the centrepiece of his description and it was no coincidence that he placed it exactly in the middle of his book. He summarized the essence of his idea:

> The old extinct volcanoes found in Hungary and Transylvania had emerged from the depth with their masses that have before been prepared in the subterranean fire chambers; they might occur alone or together with mountains of wet formation. The power of the subterranean fire caused this elevation, underneath of which the fire, in places where it was strong enough or where it met little resistance, broke through and found its way to the surface of the Earth.[13]

Fichtel distinguished two different kinds of volcanoes. He identified volcanoes that were homogenous masses and others that were made up of several layers. The first group were described as: 'those that had been elevated without any eruption, kept their entire relatively homogenous masses that had been processed in the Earth's depth, without any new addition ['Aufsätzen'] of lava.' (Fichtel 1792, **2**, p. 416f). On the other hand, he knew volcanoes that erupted periodically and thus grew through addition ['Aufsätze'] of layers of various volcanic materials. When those chasms ['Schlünde'] broke in, they were exposed to the forces of water and produced 'fake volcanic rocks'. Fichtel stressed that the elevation of the Earth's surface rather than eruptions were responsible for the formation of mountain ranges. He also offered some of his own observations as proof for this theory. Here is a sample of his line of argument:

> When nature has given away part of itself and if one follows its hints even further, one stands in the chain of natural facts. And for finding the truth, one still has to solve many a question, and often the answer to one of them confirm others. This is the situation I find myself in as I am writing these lines. I just remembered that apart from Esperies, at post station Tornye, under a huge volcano, I saw nine pyramidal beautifully overgrown pointed hills, 6 to 10 fathom [Klafter] high, just like in a garden, all placed in a straight row. The volcano had a depression, the typical sign of a former crater. The typical characteristic of all those mountains is grey, brown, light-red and light-green 'Porphyrlava' or 'Afterporphyr'. And it is there where the fire breaks through again with the eruption and its accumulation ['Anhäufung'] at the big mountain; the elevation, however, and the break of the Earth's crust without ejecting lava appears at the smaller hills.[14]

The passionate mineralogist saw himself as demiurge, drawing from a fund of arguments. In his thesis of eruption, Fichtel hoped to solve 'the knots of mineralogical disputes'. (Fichtel 1792, **2**, p. 430). He strongly opposed the contemporary, scholarly opinion that craters were necessary to prove the existence of volcanic processes. He rejected visual proof as a criterion.

The largest part of Fichtel's work was dedicated to the questions of 'what masses are elevated or ejected and what substances are volcanic products'. Fichtel defended the term 'product' against the 'mineralogical spirit of reform', a line of thought with which he addressed the opponents of his theory, the Wernerians. The cover of the book hinted at these special products of nature. It showed Hephaistos, the God of fire and master of the art of forging, in his workshop. He was the most skilful and artistic God; he had married Aphrodite and according to one myth, had fashioned the woman, Pandora. The God formed the products of nature. In two different pillars, a vertical and a horizontal pattern, there is volcanic rock. With his list of 'products', Fichtel referred to rocks that according to the Neptunists had been formed by water. He started the list with von Born's '*saxum metalliferum*', which corresponded to volcanic rocks such as andesite and dacite. Fichtel called it 'Graustein'. He also provided a detailed analysis of porphyry, basalt, and zeolith followed, all of which he described as being of genuine volcanic origin. The publication concluded with a table of products from Hungarian and Transylvanian volcanoes.

In conclusion, Fichtel's writing proceeded systematically from the general to the specific. He used a style of description that was commonly used for surveys in geography and statistics. This provided him with a traditional framework into which he tried to integrate or even harmonize his bold interpretation of the volcanic origin of the Carpathian Mountains.

Belsazar de la Motte Hacquet

Belsazar de la Motte Hacquet (1739/40–1815) (Jakob 1930; Bernleithner 1949, esp. pp. 60–63; Klemun 1988) (Fig. 5) travelled much further than the other two (Hacquet 1778–1789, 1782*a*, *b*, 1785). His journeys were very adventurous. Not much is known about his early years. Born in Leconquet (Brittany), he studied at the college of Jesuits at Pont-à-Mousson, later at the University

Fig. 5. Belsazar de la Motte Hacquet.

of Paris and then travelled to Spain and England. In the Seven-Year-War (1756–1763), he fought as a volunteer, was taken prisoner by the English and became a surgeon. He had a number of bad experiences as a prisoner in several countries until the end of the war. In 1764, he completed his education as a surgeon in Vienna, where there had been many improvements in the teaching. In 1766, Hacquet was sent to Idria (Idrija, today Slovenia) as a doctor, where as an expert in industrial medicine, he did much for the welfare of the miners and drew the public's attention to the problems of lead poisoning. From 1773 to 1787, Hacquet taught anatomy, physiology and obstetrics at the Lyceum in Laibach (Ljubljana) to students of medicine. During that period, he dedicated much time and effort to botanical and mineralogical research in the area. At last, he was able to combine his profession with his vocation and he was sent to the University of Lemberg (L'viv, L'vov, Galicia, today Ukraine) as a Professor of Natural History. After the first division of Poland (1772), this area had been put under Austrian rule. In 1805, he moved to Krakow (today Poland) as a Professor of Botany and Chemistry, where he became Dean of the Medical Faculty. When West-Galicia (that had been acquired by Austria) was returned to Poland in 1809, Hacquet had to leave Krakow and return to Vienna, where he later died.

Hacquet's passions were travelling, botany and mineralogy. He did not allow his journeys to be restricted by state borders. In the titles of his travel descriptions, he used the Latin terms for the countries, such as 'Noricum', 'Dacien' and 'Sarmatien' to define the geographical space (see Klemun 2003, pp. 25–33). His specific interest was to identify the individual character of a country or a region by describing both natural history (systematically classifying its flora, fauna, mineral and rocks) and its history (the analysis of the history of individual countries having been introduced by Göttingen Professor August Ludwig Schlözer (1735–1809) in 1769). This led to a boom in travel descriptions of particular regions. He also wondered, 'if not every region has to offer something special to the formation of mountains' (Hacquet 1785) and followed this programme as he studied the Carpathian Mountains.

Johann Gottfried Herder's (1744–1803) concept of the 'peoples close ties to the habitat they lived in' also had a lasting effect on travel literature. Hacquet 'focussed on the physical aspects of the globe, namely the groups of rocks that formed the mountain ranges,' (Hacquet 1791*a*, p. 158). He not only concentrated on rocks during his journeys, but also made several observations about the peculiarities of people and their culture, which was a form of 'statistics' (or demographics). 'Statistics' was a new science developed by Gottfried Achenwall (1719–1772). It was a collection of information on the geography, topography, political circumstances, religion, culture and the special way of life of inhabitants from a particular region. Thus, 'Hacquet's latest physical-political journeys in the years 1788 and 1789 through the Dacic and Sarmatic or Northern Carpathians,' (Hacquet 1790, **1**) contain descriptions and pictures of people in their traditional costumes as well as of typical plants.

During the eighteenth century, the postal network improved and coach transport became faster. The improvements in the road system had transformed the earlier rough winding roads into seemingly linear pathways. However, Hacquet shunned this progress. He chose to travel in the least popular way with society at that time, namely by walking. Normally only workmen, the homeless and merchants travelled in this way whereas the middle class travelled in their own coaches or with the postal system. But walking was the only feasible way of travelling when trying to explore and study inaccessible regions. This enabled Hacquet to explore the countryside independent of infrastructure and to classify the landscape according to his own ideas and perception.

Hacquet's routes followed the natural contours of an area. Thus, his written descriptions, do not appear to be systematic; instead a complex route is characterized by heights, valleys and mountain ranges. Crossing mountain ranges, walking around mountains, changing direction, moving in a zigzag manner were typical of his writing and he felt it was necessary to support the structure of the alpine area. With information on directions such as 'north north west', the reader is given hints of orientation and would thus be able to develop an awareness of spatial dimensions. Hacquet stressed that he had carried out his explorations 'along the line of the ridge of the Alpine Chain [and not only] on the slopes of the mountain ranges'. (Hacquet 1790, **1**, p. IX). Hacquet described his journeys in the only literary form possible: a descriptive narrative that followed the itinerary closely. It served as a framework for his observations, some of which were sensational, such as the one on karst landscapes. Hacquet was the first to attribute the formation of the hollow forms of karstic regions, later called doline and polje, to a collapse of washed out layers. Moreover, he based his categorization of the Noric Alps on three main zones: limestone–granite–limestone.

In his work on the Carpathians, Hacquet had to modify the principle that he had applied to the Alps of Friaul, Carnia (Slovenia), Carinthia and Salzburg, that was to 'start with the deepest layers [...] and then to climb up step by step to the peak of our Alpine Chain'. (Hacquet 1778–1789, **1**, prologue, p. XV). He started his journey in the middle of the range, on the mountain '*Pietrile rosse*', which formed a natural border to the Ottoman Empire and the Principality of Moldavia (Romania). In his second volume, published in 1791, Hacquet described areas of Moldavia in which Austria and Russia were involved in their final Turkish war (Hacquet 1791*b*). Hacquet focused his attention on the town of 'Jassi' (Iasi), in the centre of Moldavia, where a peace treaty was signed in 1792. The aim of his description was to obtain an overview of the mountain range:

This mountain range (near Jassi) is the point where the European Alpine Chain extending from west to east from Hemus [Haemus = the Balkan mountains] to the Danube River, which separates Wallachia from Banat and Transylvania, suddenly makes a sharp bend here on the Magura Mountains and changes its direction of extension from south to east, where towards the Danube River it takes the form of a regular wavelike area consisting of Moldavia, Poland or Galicia, all the way to the mountain range in the Jablunka, where Slask meets with Poland, Hungary and Moravia. Here, the mountain range ends for a short stretch transforming into a depression. Then the Bohemian Highlands [Böhmische Riesengebirge] begin which finally ends in a low mountain range extending from west to east thereby separating Saxony and Bohemia, as has already been pointed out.[15]

In his third volume, on the description of the Carpathian Mountains, Hacquet reacted to Fichtel's work, which was restricted to the southern part of the Carpathian Mountains (Hacquet 1794). He saw his description as complementary to Fichtel's pioneering work. Only in the case of Fichtel's volcanic interpretation did he disagree with the scholar:

The many important observations one has today about revolutions that the fire causes on the surface of our planet have shed much light in 'Geognosie' (although this has sometimes gone too far). Here, Mr. Fichtel has contributed his share; I and true connoisseurs of nature cannot but acknowledge his contribution in this respect; and although I do not agree with him [...] in some areas of Transylvania and the Bucovina, I hope he will not think me to be an enemy of such expansive volcanoes.[16]

Most of Hacquet's observations aimed at supporting Neptunist views, although he disliked general statements and stressed the importance of the detail. The fourth volume of his monumental work focused on the Bucovina. Hacquet concentrated entirely on the description of Galicia, the sulphurous mineral springs, the salt mine Wieliczka and the Hungarian ore mountains (Erzgebirge) (Fig. 6; Hacquet 1796).

Height played an important role in Hacquet's thinking. Triglav (Carnia, now Slovenia) and Großglockner (Carinthia, Austria) are the two highest mountains of the Noric Chain. His journey between them formed one boundary of his area of study. Hacquet used the dimension of height to create a hierarchy between different mountains. In the eighteenth century, it was the only way of differentiating mountains; after that time, individual mountains were identified (Klemun 2000, esp. p. 210f).

Differences in height were determined by various methods. For nivellement, various instruments were used as part of a geometric method of measuring height between selected points of altitude. For trigonometric height measurements a goniometer was used. Trigonometric measurements of altitude have been carried out since the eighteenth century, for example during the French expedition to Peru (1735–1743) headed by Charles-Marie La Condamine (1701–1774), during which the height of the Chimborazo was determined to be 3220 Toises.[17] In the eighteenth century this mountain was thought to be the highest in the world.

Following the invention of the mercury barometer by Galileo Galilei (1638) and Evangelista Torricelli (1643) it was realized that the instrument could also be used for calculating height. Barometric pressure increases and decreases respectively by 1 mm with a change in altitude of about 12 m. In 1676, Edme Mariotte (1620?–1684)

Hacquet's
neueste
physikalisch-politische Reisen
in
den Jahren 1794 und 95.
durch
die Dacischen und Sarmatischen
oder
Nördlichen Karpathen.

Vierter Theil.

Nürnberg,

Fig. 6. Title page of Hacquet's Journey. Volume 4 (1796).

developed a formula for calculating altitude and Johann Jakob Scheuchzer (1672–1733) applied it practically to the Swiss mountains. From then on almost all renowned astronomers and geodesists tried to improve the formula: Jacques Cassini (1733), Daniel Bernoulli (1738), Pierre Bouguer (1749) and Tobias Mayer in Göttingen (1769).

Until that time, determinations of height had been carried out only for a few peaks in the western Alps, in particular by Hacquet's idols, Johann Jakob Scheuchzer (1672–1733) and Horace Benedict de Saussure (1740–1799). The heights of the important peaks in the eastern Alps had not been determined by cartography of the Josephin land survey (see Schichl 1950, p. 178). The Genevan son of a watchmaker, Jean-André de Luc (1727–1817), who later became well known as meteorologist (De Luc 1784) geologist and constructor of scientific instruments in London, tried to carry out barometric measurements in Geneva in the 1750s. He developed a formula to correct results that offered all geodesists a relatively accurate method of determining height. Hacquet made the first measurements of the height of the Großglockner, the mountain known today to be the highest in Austria.[18] The height he calculated was close to its current measurement, although it was based merely on the height of the sun's rays. Presumably, Hacquet also carried out barometric measurements, since he recommended them to every travelling researcher of nature, and because he also owned a 'Delucian Barometer'.[19] As for the determination of the highest points of the central Carpathian Mountains, Hacquet referred to professional height measurements carried out by an army engineer. Krivàn mountain, a peak in the High Tatra, which he determined to be the highest peak, is among the highest elevations of the Tatra today.

Hacquet showed how important it was for a mineralogist to be a mountaineer. He felt that climbing was so important that he summarized the techniques he had developed during his 40 years research in an appendix to his work on the Carpathian Mountains. There, he described the 'qualities of a mountain climber', who had to have much courage and should not know any fear, and as a 'travelling researcher of nature' should 'do without those wonderful ties of love'. (Hacquet 1796, p. 345). It seemed obligatory for Haquet to produce 'sections of mountain maps'. Maps formed the basis of his work patterns and orientation and if he did not find any useful ones he devised them himself, or sharply criticized the existing ones, as in the case of Giovanni Antonio Rizzi Zannoni's (1736–1814) map of Poland, which he used for his description of Galicia.

He also criticized the fact that some nature researchers claimed to have been on a peak without ever having seen the mountain.

> Due to the description of the types of rocks at the peaks of mountains one can never be sure whether a researcher has really been there or not, since decay also brings down such rocks from the peaks. However, it is completely different with plants; there one can tell exactly how high up a botanist has gone. I have seen that many a time when someone claimed 'I have been up this or the other mountain', but the moment one has reached about one third of its height one gets another horizon, and usually also other plants.[20]

Conclusions

Fieldwork presented the early researchers with a wealth of data. These early scholars aimed to obtain reliable data in 'mineralogy'. This required formal procedures to counteract the fleeting and selective character of earlier attitudes towards fieldwork. An important criterion in the collection of data, which has been considered in this paper, is epistemology and the form of description as 'inscription'. In Hacquet's notes, the enumeration of data supported his central assumption, by describing the course of mountain chains. These descriptions followed his itinerary. The route might be interrupted so that Hacquet could climb up to a peak's summit to obtain a view of the direction of the mountain ranges, or to make important cultural observations. Von Born, on the other hand, was interested in individual mining areas and the inside of veins, and used letters to record his data. This form of description was the result of the individual places where he had written those letters. Fichtel, who formulated a new and controversial theory of volcanism in connection with the formation of the Carpathian Mountains, tried to support it by using a systematic documentation reminiscent of geography. The traditional static form lent substance to the revolutionary claim in theory.

Notes

[1]The term was coined by Kittler, F. 1990. See also the discussion in Lenoir 1998.

[2]For biographical details on von Born see: Hofer (1953) and Haubelt (1972).

[3]*La monacologia ossia descrizione metodica die frati di Giovanni Fisiolo*, Mailand 1865.

[4]Addition to the *Göttingischen gelehrten Anzeigen*, 10 September 1774, pp. 289–294

[5]Ferber 1774, p. 59 f.: 'Die Unsicherheit der Wege bewog den Hrn. Hof = Cammerrath von Hegengarthen, dessen Leutseligkeit Sie kennen, und dessen Güte gegen mich ich Ihnen nicht genug anrühmen kann, an alle Oerter, die ich vorbey reisete, den Befehl ergehen zu lassen, daß man für meine Sicherheit Sorge tragen solle. Ich traf also auf jedem Dorfe, einige vierzig bis funfzig mit Feuer = Gewehr und Prügeln versehene Wallachen an, die unter Anführung eines ihrer Vorgesetzten mich bis an das nächste Dorf begleiteten, und auf steinichten oder sonst beschwerlichen Wegen, meinen Wagen auf ihren Schultern mehr trugen, als im Gleichgewichte erhielten'.

[6]Letter from Ignaz von Born to Daniel Gottlieb Schreber (Altzedlitsch, 27 May 1773), printed in Beran 1971, p. 104: '[...] ich mußte die Einkleidung von Briefen, welche ich damals an Ferber geschrieben zu haben, vorgebe, wählen, weil man bei meiner Entlassung von meinem Dienste mir solche nur mit der Bedingnis zugestanden hat, daß ich nie etwas, so die Bergwerke oder Mineralgeschichte betrifft, drucken lassen solle. Diese Briefe mußten also einen andern Herausgeber haben.'

[7]Letters from Ignaz von Born to Christoph Martin Wieland, 2 February 1786, Autograph, Germanisches Nationalmuseum Nürnberg, ABK, Fasz. 28.

[8]Ferber 1774, p. 226: 'Übergänge einer Steinart in die andere, kurz diejenigen Mineralogischen Merkwürdigkeiten, welche bey dem ersten Anblick das Cabinet eines Sammlers von dem Cabinet eines gründlichen Kenners unterscheiden.'

[9]Mannfeldian Kupferschiefer had become famous following the publication of a book by Lehmann (1756).

[10]Ferber 1774, p. 36: 'Ueberhaupt ist, in Absicht auf meine Mineraliensammlung, Saska vielleicht der Ort, wo ich die reicheste Erndte auf meiner ganzen Reise eingebracht habe. Nicht nur alle Gattungen von Kupfererzen, die Mannfeldischen Kupferschiefer ausgenommen, sondern auch viele neue, und bisher unbekannte Arten brechen hier in Menge.'

[11]As of 1779 the Banat was placed under the administration of the then Kingdom of Hungary. Thus, the French translation did not include the Banat in its title.

[12]See *Die Siebenbürger Sachsen*. Lexikon, Innsbruck 1993, p. 124; Wurzbach 1858, **5**, Prescher & Schmidt 1993, Niedermaier 1979, pp. 1–66, esp. p. 28.

[13]Fichtel 1792, **2**, p. 416: 'Die alten erloschenen Vulkanen, welche in Ungarn und Siebenbürgen angetroffen werden, sind mit ihren, in den unterirdischen Feuerkammern vorher zubereiteten Massen aus der Tiefe hervorgestiegen, sie mögen für sich allein, oder im Zusammenhange mit bergen nasser Entstehung vorkommen. Die Macht des unterirrdischen [!] Feuers bewirkte diese Hebung, unter welcher das Feuer an solchen Stellen, wo es genugsame [genügend] Stärke hatte, oder wo es geringeren Widerstand fand, über sich durchbrach, und sich den weg auf die Erdoberfläche öffnete.'

[14]Fichtel 1792, **2**, p. 429: 'Hat sich einmal die Natur irgendwo verrathen, und verfolgt man ihren Leitfaden weiter: so stehet man in der Kette der natürlichen Begebenheiten, für eine Wahrheit noch mehrere Aufschlüsse vor sich, deren einer den andern bestätiget. So ergehet es mir, da ich dieses schreibe. Ich erinnere mich so eben dass ich ausser Esperies, bey der Poststation Tornye, 9 pyramidalische schön bewachsene, zu 6 bis 10 Klaftern hohe spitzige Hügeln, wie in einem Garten, in einer geraden Reihe, unter einem mächtigen Vulkan vor mir liegen gesehen, auf welchem letztern eine Einsattlung, das Kennzeichen eines einstmaligen Kraters, wahrzunehmen ist. Das Wesen aller dieser Berge ist graue, braune, rothlichte und grünlichte Porphyrlava, oder Afterporphyr; und hier erscheint also wieder der Durchschlag des Feuers mit der Eruption, und ihrer Anhäufung an dem grossen Berge, die Hebung aber, und der Bruch der Erdrinde ohne Fortschleuderung, an den kleinern Hügeln.'

[15]Hacquet 1791*b*, p. 93: 'Dieses Gebirg (nahe Jassi) ist der Standpunkt, wo die Europäische Alpkette, welche von dem Hemus von Westen nach Osten zur Donau hinläuft, die Walachey vom Bannat und Siebenbürgen scheidet, hier am Gebirg Magura auf einmal einen scharfen Winkel macht, und sich von Süden nach Osten wendet, wo es dann gegen letzte Gegend eine beständige wellenförmige Fläche vor sich hat, welche die Moldau, Pohlen oder Gallizien ausmacht, bis zu

dem Gebirg in der Jablunka, wo Schlesien mit Pohlen, Hungarn und Mähren zusammen komme. Hier hört das Gebirg, wie im ersten Theil erwähnt worden, in einer kurzen Strecke mit einer Einsenkung auf, dann fängt das böhmische Riesengebirg an, worauf sich solches dann endlich in ein Mittelgebirg endiget, welches vom Osten nach Westen läuft und Sachsen und Böhmen theilt, wie schon dermahlen erinnert worden.'

[16]Hacquet 1794, **3**, Vorrede, p. V and VI.: 'Die vielen wichtigen Beobachtungen, die man heut zu Tag über die Revolutionen hat, welche das Feuer auf der Oberfläche unsers Planeten verursacht, haben in der Geognosie viel Licht verbreitet (ob man zwar auch bey diesem viel zu weit gegangen ist.) In diesem hat Herr Fichtel nicht wenig beygetragen; ich und kein wahrer Naturfreund kann seine Verdienste darin verkennen, und ob gleich ich in einigen Gegenden von Siebenbürgen und der Bukowina ... nicht mit ihm übereinstimme, so hoffe ich doch, er wird mich nicht als einen Feind von so sehr ausgebreiteten Vulkanen erkennen.'

[17]1 Toise = 196.03 cm

[18]See Hacquet 1791*a*, p. 115. Here he talks about roughly 2000 Lachter.

[19]See Hacquet 1782*b*, p. 75. Here he regrets the fact that he does not have De Luc's barometer with him.

[20]Hacquet 1796, p. 345 f.: 'Mit Beschreibungen der Steinarten auf den Gipfeln der Berge ist es nicht so sicher, zu wissen, ob einer auf der Anhöhe war, oder nicht, denn die Verwitterung führt wohl auch solche von der Höhe herunter, mit den Pflanzen aber ist es ganz anders; da kann man genau wissen, wie hoch der Botaniker gekommen sey. Dies habe ich mehrmals erfahren, wo es hieß, 'ich habe diesen oder jenen Berg bereißt', aber kaum ist man auf ein drittel seiner Höhe gelangt, dann hat man einen anderen Horizont, so auch meistens andere Pflanzen.'

References

BERAN, J. 1971. Dopisy Ignàce Borna D. G. a J. Ch. D. Schreberum. *Fontes scientiarum in Bohemia florentium historiam illustrantes* **1**. Ústrêsdní archiv, Prague.

BERNLEITHNER, E. 1949. *Die Entwicklung der österreichischen Länderkunde an der Wende des 18. und 19. Jahrhunderts*. Philosophische Dissertations, Vienna.

BORN, I. VON 1772. *Index Fossilium quae collegit, et in Classes ac Ordines disposuit Ignatius S. R. I. Eques a Born. Lithophylatium Bornianum S.* Gerle, Prague. [There was also a French edition].

BORN, I. VON 1773. *Die Staatsperücke—eine Erzählung.* Vienna.

BORN, I. VON 1773. *Schreiben des Herrn Ignaz von Born an Grafen von Kinsky/Über einen ausgebrannten Vulkan bey der Stadt Eger.* Gerle, Prague.

BORN, I. VON 1775–1784 (ed.) *Abhandlungen einer Privatgesellschaft in Böhmen zur Aufnahme der Mathematik, der vaterländischen Geschichte und der Naturgeschichte.* 6 Volumes. Gerle, Prague.

BORN, I. VON 1777. *Travels through the Bannat of Temeswar, Transylvania and Hungary in the year 1770.* J. Miller, London.

BORN, I. VON 1778. *Index Rerum naturalium musei Caesarei Vindobenensis [Verzeichnis der natürlichen Seltenheiten des k.k. Naturalien Cabinets zu Vienna].* Volume 1 Testacea. Krausius, Vienna.

BORN, I. VON 1780. *Voyage minéralogique fait on Hongrie et en Transilvanie.* M. Monnet, Paris.

[BORN, I. VON] 1783*a*. *Joannis physiophili specimen monachologiae methodo Linnaena tabulis 3 aeneis illustratum, cum adnexis thesibus e pansophia*, A. Merz, Augsburg.

[BORN, I. VON] 1783*b*. *Specimen Monachologiae Methodo Linnaena.* Munich, Merz. [Later editions 1783 Vienna; 1784 London; 1802 Frankfurt].

[BORN, I. VON] 1784. *Joannis Physiophili opuscula, continent Monachologiam, accusationem Physiophili.* A. Martins, Vienna.

BORN, I. VON 1786. *Ueber das Anquicken der gold- und silberhältigen Erze, Rohsteine, Schwarzkupfer und Hüttenspeise.* Wappler, Vienna.

BORN, I. VON 1790. *Catalogue methodique et raisonné de la collection des fossiles de Mlle Eléonore de Raab.* 2 Volumes, Degen, Vienna.

BORN, I. VON & TREBRA, H. (eds) 1789–1790 (ed.). *Bergbaukunde.* 2 Volumes, Goeschen, Leipzig.

BRENNER, P. J. (ed.) 1989. *Der Reisebericht.* Suhrkamp Verlag, Frankfurt-am-Main.

BRENNER, P. J. 1990. Der Reisebericht in der deutschen Literatur. Ein Forschungsüberblick als Vorstudie zu einer Gattungsgeschichte. *Internationales Archiv für Sozialgeschichte der deutschen Literatur* **2**, 161.

DE LUC, J.-A. 1784. *Recherches sur les modifications de l'atmosphère contenant l'histoire critique du Baromètre et du Thermomètre.* Duchesne, Paris.

FERBER, J. J. (ed.) 1774. *Des Hrn. Ignatz, Edl. von Born, Briefe über Mineralogische Gegenstände, auf seiner Reise durch das Temeswarer Bannat, Siebenbürgen, Ober = und Nieder = Hungarn an den Herausgeber derselben.* Johann Jacob Ferber, Frankfurt and Leipzig.

FETTWEIS, G. B. & HAMANN, G. 1989. (eds). Über Ignaz von Born und die Societät der Bergbaukunde. *Sitzungsberichte der Österr. Akademie der Wissenschaften, philosophisch-historische Klasse*, **533**, 11–25.

FICHTEL, J. E. VON 1791–1792. *Mineralogische Bemerkungen von den Karpathen.* 2 Volumes Kurzbeck, Vienna.

GUNTAU, M. 1996. The natural history of Earth. *In*: JARDINE, N., SECORD, J. A. & SPARY, E. C. (eds) *Cultures of Natural History.* Cambridge University Press, Cambridge, 211–229.

HACQUET, B. 1778–1789. *Oryctographia Carniolica, oder Physikalische Erdbeschreibung des Herzogthums Krain, Istriens und zum Teil der benachbarten Länder.* 4 Volumes, Breitkopf, Leipzig.

HACQUET, B. 1782*a*. *Plantae alpinae Carniolicae, collegit et descripsit.* Kraus, Vienna.

HACQUET, B. 1782*b*. *Mineralogisch = botanische Lustreise, von dem Berg Terglou in Krain, zu dem Berg Glokner in Tyrol, im Jahre 1779 und 1781.* Kraus, Vienna.

HACQUET, B. 1785. *Physikalisch = Politische Reise aus den Dinarischen durch die Julischen, Carnischen, Rätischen in die Norischen Alpen im Jahre 1781 und 1783 unternommen*. Böhme, Leipzig.

HACQUET, B. 1790. *Neueste physikalisch = politische Reisen in den Jahren 1788. und 1789. durch die Dacischen und Sarmatischen oder Nördlichen Karpathen*. Volume 1. Raspe, Nuremberg.

HACQUET, B. 1791*a*. *Reise durch die Norischen Alpen/ Physikalischen und andern Inhalts unternommen in den Jahren 1784 bis 1786*. Raspe, Nuremberg.

HACQUET, B. 1791*b*. *Neueste physikalisch = politische Reisen in den Jahren 1788. 89. und 90. durch die Dacischen und Sarmatischen oder Nordlichen Karpathen*. Volume 2. Raspe, Nuremberg.

HACQUET, B. 1794. *Neueste physikalisch = politische Reisen in den Jahren 1791. 92. 93. durch die Dacischen und Sarmatischen oder Nördlichen Karpathen*. Volume 3. Raspe, Nuremberg.

HACQUET, B. 1796. *Neueste physikalisch-politische Reisen in den Jahren 1794. und 95. durch die Dacischen und Sarmatischen oder Nördlichen Karpathen*. Volume 4. Raspe, Nuremberg.

HAMANN, G. 1989. Ignaz von Born und seine Zeit, Über Ignaz von Born und die Societät der Bergbaukunde. *In*: FETTWEIS, G. B. & HAMANN, G. (eds) *Sitzungsberichte der Österr. Akademie der Wissenschaften, philosophisch-historische Klasse,* **533**, Vienna, 11–25.

HAUBELT, J. 1972. *Studie o Ignáci Bornovi*. Univ. Karlova, Prague.

HAUBELT, J. 1991. Born und Böhmen. *In*: REINALTER, H. (ed.) *Die Aufklärung in Österreich. Ignaz von Born und seine Zeit*. P. Lang, Frankfurt-am-Main, Bern, New York.

HOFER, P. 1953. *Ignaz von Born, Leben—Leistung—Wertung*. Univ. Philosophische Dissertation, Vienna.

JAKOB, G. 1930. *Belsazar Hacquet. Leben und Werke*. Große Bergsteiger V, Rother, Munich [Includes Hacquet's autobiography, pp. 223–237].

KITTLER, F. 1990. *Discourse Networks: 1800/1900*. (Trans. METTERER, M with CULLENS, C.) Stanford University Press, Stanford.

KLEMUN, M. 1988. Belsazar Hacquet—Begründer einer vielfältigen Durchforschung des Ostalpenraumes. *Carinthia* **II, 178/98**, 5–13.

KLEMUN, M. 2000. *Die Großglockner-Expeditionen 1799 und 1800. ... mit Madame Sonne konferieren*. Das Kärntner Landesarchiv 25. Klagenfurt, Verlag des Kärntner Landesarchivs.

KLEMUN, M. 2003. Raumkonzepte im Werk Belsazar Hacquets. Space concepts in the work of Belsazar Hacquet. *Hacquetia. Miscellanea Hacquet. International Conference on the Life and Work of Balthasar Hacquet*. **2/2**, 25–33. Ljubljana.

KUTTENPEITSCHER, I. L. [-BORN] 1783. *Neueste Naturgeschichte des Mönchstums*. Munich. [Later edition 1841 Bern]

KUTTENPEITSCHER, I. L. [-BORN] 1784. *Essai sur l'histoire* [...]. *de quelques especes de moines*. Monachopolis. [Later editions 1798 Paris; Johnston and Hunter, Edinburgh 1852].

KUTTENPEITSCHER, I. L. [-BORN] 1786. *Versuch einer Mönchologie nach Linnäischer Methode*. Vienna. [Later editions 1802 Frankfurt and Leipzig].

LAUDAN, R. 1987. *From mineralogy to geology. The foundation of the Science, 1650–1830*. University of Chicago Press, Chicago and London.

LEHMANN, J. G. 1756. *Versuch einer Geschichte von Flötz-Gebirgen, betreffend deren Entstehung, Lage* [...]. Klüter, Berlin.

LENOIR, T. (ed.) 1998. *Inscribing Science. Scientific Texts and the Materiality of Communication*. Stanford, Stanford University Press.

MUTSCHLECHNER, G. 1991. Ignaz von Born als Geologe, Mineraloge und Montanist. *In*: REINALTER, H. (ed.) *Die Aufklärung in Österreich. Ignaz von Born und seine Zeit*. P. Lang, Frankfurt-am-Main, Bern, New York, 117–133.

NIEDERMAIER, K. 1979. Zur Geschichte der naturwissenschaftlichen Forschungen in Siebenbürgen. *In*: WAGNER, E. & HELTMANN, H. (eds) *Siebenbürgisches Archiv* **14**, P. Lang, Cologne and Vienna, 1–66.

PAPP, G. 1998. An ardent vulcanist from Hungary. Sketches to the scientific portrait of Johann Ehrenreich von Fichtel (1732–1795). *In*: MORELLO, N. (ed.) *Volcanoes and History. Proceedings of the 20th INHIGEO Symposium*. Brigati, Genova, 505–522.

PORTER, R. 1990. The History of Science and the History of Society. *In*: OLBY, R. C. (ed.) *Companion to the History of Modern Science*. Routledge, London.

PRESCHER, H. & SCHMIDT, P. 1993. Johann Ehrenreich von Fichtel. Leben, Werk und seine Schrift Die Mineralogen gegen das Ende des achtzehnten Jahrhunderts. *Freiberger Forschungshefte D 199, Historischer Bergbau*. P. Lang, Freiberg.

REINALTER, H. 1991. Preface to: REINALTER, H. (ed.) *Die Aufklärung in Österreich. Ignaz von Born und seine Zeit*. P. Lang, Frankfurt-am-Main, Bern, New York.

RUDWICK, M. 1976. The emergence of a visual language for geological sciences. *Historical Studies in Physical Sciences* **14**, 149–195.

RUDWICK, M. 1996. Minerals, strata and fossils. In: JARDINE, N., SECORD, J. A. & SPARY, E. C. (eds) *Cultures of Natural History*. Cambridge University Press, Cambridge.

SCHICHL, E. 1950. *Die Entwicklung der physischen Geographie in Österreich bis zur Mitte des 19. Jahrhunderts*. Phil. Dissertation. Vienna.

STAFFORD, B. M. 1994. *Artful Science: Enlightenment and the Eclipse of Visual Education*. MIT Press, Cambridge, MA.

STAFFORD, B. M. 1998. *Kunstvolle Wissenschaft: Aufklärung, Unterhaltung und der Niedergang der visuellen Bildung*. Verlag der Kunst, Amsterdam.

STAGL, J. 1995. *A History of Curiosity. The Theory of Travel 1550–1800*. Studies in Anthropology & History 13. Harwood Academic Publishers, Chur.

TEICH, M. 1975. Born's amalgamation process and the international metallurgic gathering at Skleno in 1786. *Annuals of Science* **32**, 305–340.

TEICH, M. 1976. Ignaz von Born als Organisator der wissenschaftlichen Bestrebungen in der Habsburger Monarchie. *In*: AMBURGER, E., CIESLA, M. & SZIKLAY, L. (eds) *Wissenschaftspolitik in Mittel- und Osteuropa*. Camen, Berlin.

WURZBACH, C. 1858. *Biographisches Lexicon des Kaiserthums Österreich*, Volume 5. Vienna.

Geological travellers in view of their philosophical and economical intentions: Johann Wolfgang von Goethe (1749–1832) and Caspar Maria Count Sternberg (1761–1838)

C. SCHWEIZER

Institute of German Philology, University of Vienna, Dr Karl-Lueger-Ring 1, A-1010 Vienna, Austria (e-mail: c.schweizer@gmx.at)

Abstract: Two geological travellers, who clearly differ in their approach to scientific questions and their interpretation, are presented. Goethe was mainly interested in natural phenomenon and linking the interpretation of his observations in a deductive approach to a philosophically and psychologically influenced, more general view of natural mechanisms. Sternberg, on the other hand, chose an inductive approach in his conclusions from geological observations and never inclined to any philosophical explanations behind them. Goethe's natural philosophy behind his scientific concepts was strongly formed by Baruch de Spinoza (1632–1677) and led to his own subjective approach of interpreting scientific phenomena. However, Sternberg's knowledge of geognosy along the formations in the Habsburg region and in Germany is based on numerous observations at various locations and on the comparison of his findings, having a vast general knowledge on the geology of a wide area. Goethe and Sternberg's correspondence between 1820–1832 gives insight into the specific differences between these two travellers and their individual methods of geological investigation.

There was a profound change in basic geological views in the first half of the nineteenth century. During the last quarter of the eighteenth century Abraham Gottlob Werner's (1749–1817) Neptunist theory had been generally accepted, boldly claiming that all rocks were the product of crystallization and subsequent sedimentation of substances dissolved in an initial universal ocean. A primeval solid core of granite was assumed to form the ocean's ground, followed by a regular order of sedimented rock specimens (Oldroyd 1996, p. 99). However Werner had based his ideas exclusively on his geognostical observations made in Saxony. As he had not been travelling, he had never had the opportunity to compare this area with other areas in the world. In Britain, these speculative ideas were replaced by the actualistic theories of Charles Lyell (1797–1875) in his main work *Principles of Geology* (Lyell 1830–1833). Here, Lyell emphasized the necessity to find the real causes of presently occurring changes to the Earth in order to explain the changes in the past. He also approved James Hutton's (1726–1797) theory of Plutonism (Hutton 1788, 1795), after it had been re-established by John Playfair (1748–1819) in his *Illustrations of the Huttonian Theory of the Earth* (Playfair 1802). Hutton realized the importance of heat in the formation of the Earth's core. He assumed the occasional expansion of very hot material inside the Earth towards its solid core, gradually arching the Earth's crust layers. This process would sometimes even lead to bursts through the core, thus giving rise to volcanic events and pouring out the liquid magma, which would become solid by cooling (see Oldroyd 1996, p. 93). In Germany, Leopold von Buch (1774–1853) (Buch 1825) and Alexander von Humboldt (1769–1859), both defending volcanistic theories, diverged from Werner's speculative ideas, although both had been attending his lectures. Buch's theory differed from Hutton's by the assumption of strong action of fire within the Earth, leading to spontaneous eruptions of the Earth's crust by the breaking out of lava that formed the 'elevation craters' observed at the Etna, Vesuvius, Stromboli as well as the Peak of Tenerife on the Canary Islands (Oldroyd 1996, pp. 168–169. Karl Ernst Adolf von Hoff (1771–1837) supported the theory of actualism, outlined by Lyell in Britain, as stated in his main work *Geschichte der durch Ueberlieferung nachgewiesenen natürlichen Veränderungen der Erdoberfläche*(Hoff 1822–1841).

The correspondence[1] between the German poet and scientist Johann Wolfgang von Goethe and the Bohemian scientist and cofounder of the Bohemian National Museum in Prague, Caspar Maria Count Sternberg, is dominated by their common interests in science including recent botany and palaeobotany, geology, mineralogy and meteorology. It lasted from 1820 to 1832, coinciding with the first years of the modern concepts of dynamic geology. The interpretation of specifically chosen extracts[2] from these letters and from Goethe's published essays in the present

From: WYSE JACKSON, P. N. (ed.) *Four Centuries of Geological Travel: The Search for Knowledge on Foot, Bicycle, Sledge and Camel.* Geological Society, London, Special Publications, **287**, 63–72.
DOI: 10.1144/SP287.6 0305-8719/07/$15.00

article may provide an insight into the scientific approach of these two correspondents in the field of geology.

Biographical data

A few biographical details on Goethe and Sternberg should outline the cultural historical background of their enterprises as geologists.

Goethe was called to the court of Sachsen-Weimar-Eisenach by the young Duke Carl August (1757–1828) in 1775 and stayed in Weimar until his death. He was given sufficient time besides his obligations at court and his function as statesman to follow his interests in the various scientific fields mentioned above. From 1785 he frequented the Bohemian thermal spas, mostly Karlsbad, but also Franzensbad and Teplitz, where he spent much time on excursions, exploring the geology of the surroundings and collecting whole suites of rock and mineral species (Urzidil 1932). From 1821–1823 he visited Marienbad, and thereafter he did not return to Bohemia. In Marienbad he first met Caspar Sternberg in July 1822, after having corresponded with him for two years. They soon became very friendly, and their connection continued until Goethe's death in 1832.

Caspar Sternberg's life was more complex than Goethe's.[3] He was born on the 6 January 1761 in Prague, the youngest of three sons of Johann Count Sternberg (1713–1798) and Anna Maria Josepha Countess of Sternberg, born Kolowrat-Krakowská (1726–1790). His brothers, Johann (1752–1789) and Joachim (1754–1808), both followed a military career, whereas he was destined to be a clergy man. With this intention, at the age of eighteen, he was sent to the Collegium Germanicum in Rome. In 1784, he entered the clerical chapter in Ratisbon in Germany where he served under the elector-archchancellor Carl Theodor von Dalberg (1744–1817) until 1810, when the political development in Germany, strongly influenced by the Napoleonic control seizure of power, made Sternberg terminate his clerical career and return to Bohemia. There he dedicated the rest of his life to science. Self-taught, he acquired an extensive knowledge in botany, meteorology, palaeontology, geology and mineralogy. The provisional foundation of the 'Vaterländisches Museum in Böhmen' (today known as the Bohemian National Museum) in 1818, which became officially sanctioned by the Emperor Franz I of Austria (1768–1835) in 1822, could be regarded as his main achievement of public interest. Numerous publications, mostly in the fields of botany and geognosy, gave him an international reputation, and he became an honorary member of many scientific societies. Always endeavouring to promote scientific dialogue between countries, in 1832, Sternberg managed to invite the Meeting of the Association of German Scientists and Medical Doctors (Versammlungen der Gesellschaft deutscher Naturforscher und Ärzte) to Vienna and in 1837 to Prague. Both were important events, which strongly promoted the scientific reputation of the Austrian–Hungarian monarchy in Europe. Sternberg was also greatly regarded by Clemens Prince Metternich (1773–1859) and Franz I during the political era of Restoration, a time that is well known for not always promoting scientific endeavours.

Volcanism or pseudovolcanism?

When Goethe and Sternberg met in Marienbad, they undertook some trips in the company of Goethe's friend, the police inspector Josef Sebastian Grüner (1780–1864) from Eger, the Bohemian botanist Johann Emanuel Pohl (1782–1834), and the famous Swedish chemist Jacob Berzelius (1779–1848). They explored the hill Kammerbühl near Eger in the Elbogen-district in West-Bohemia, and wondered whether it was of pseudovolcanic or of volcanic origin. The geological background of the history and final solution of this question has been published in detail by Oldrych Fejfar and Fritz Steininger (1999) and by Fejfar (1999). But the fundamental problem that was addressed on this excursion was documented in Goethe's and Sternberg's letters several times in the course of the following years. These tackled the more general question of whether the Earth's crust had been created either (a) by precipitation of the basalt from the sea according to Abraham Gottlob Werner's (1749–1817) Neptunist theory; (b) by solidification of erupted lava at the surface of the molten rocks according to Leopold von Buch's (1774–1853) and Alexander von Humboldt's (1769–1859) volcanist theory; or (c) by solidification of liquid hot magma in magma chambers, as suggested by Hutton's Plutonist theory. Goethe carefully collected whole suites of rock specimens and minerals from several parts of the district around Eger and sent a documentary portion of 21 samples on 26 August 1822 to Sternberg's 'Vaterländisches Museum' in Prague (Sternberg had already left Marienbad by this date). A list was added with each sample being carefully described as: (a) natural mica schist as produced by the rock; (b) mica schist gone through fire, occurring in slags; (c) mica schist having turned redish by fire; (d) quartz externally redded' (Goethe 1809),[4] etc. He concluded from the characteristics of these suites, that Kammerbühl was pyrotypic, meaning that it had arisen under the influence of fire. Yet

Goethe did not give way to a definite approval of the volcanistic or Plutonist theory. However, in his publication of 1809 (Goethe 1820*a*), basing his views on Werner's Neptunist theory, Goethe interpreted the rocks of Kammerbühl as being volcanic products from an early phase in the Earth's history not having been erupted from deeper earth strata, but produced from molten mica schist, covered by water. He attributed the regular layers of strata at the northeastern side of the hill to a steady aqueous sedimentation process. In 1820, Goethe had reprinted this essay and added Sternberg's latest suggestion to drive a roadway under the hill in order to get some insight into the nature of its deepest strata.[5]

On 16 August 1823, with this fundamental question in mind, Goethe again sent suites of minerals and suites of rock species, this time from Wolfsberg near Czerlochin in the Pilsen district, to the Bohemian museum and mentioned them in his letter dated 10 September 1823.

It might be useful to mention two similar phenomena in Bohemia —in addition to the pyrotypic Kammerberg: first, the 'Wolfsberg' near Czerlochin; I had sent a delegate to investigate its nature. First an original rock was looked for, according to the well-thought-out demands of our old masters, and it was determined by our own persuasion, without hoping for general approval; the changes of this (original rock) were followed from the most recognizable sample to the most unrecognizable, and a thus organised collection was sent to the 'Vaterländisches Museum. [...]

The second recently discovered appearance of ancient traces of fire has been found in Altalbenreuth in the Fraisch-area, about three hours south from Eger. [...] All three of them seem to point to the fact, that their origin is *local*, as at each of the three locations different original rock species appeared altered and the products occur with different composition and appearance (Berzelius 1823. pp. 1ff).

In other words, Goethe is not claiming specificity in the rock masses involved in their changes, but is assuming the same impact on the rocks in all three locations investigated, namely fire. Heat by consequence seems to have been a secondary factor acting on already existing rocks.

Sternberg's approach to this fundamental question is not quite in conformity with Goethe's. In his letter to Goethe dated January 1824 he wrote:

In his examination of the mineral waters from Karlsbad, Teplitz and Königswart,[16] Berzelius has cut the Gordian knot and assumed all hot springs as being of volcanic origin. The correspondence of the solid components of Bohemian's springs and of the springs in the Auvergne with those from Iceland is certainly a circumstance in favour of this hypothesis, yet, their ever constant temperature, at least as far as their thermometric measurements are known, is not inclined to conform to any so far attained opinion, it might remain puzzling, as long as there is not any powerful revolution to occur and to reveal us a profound insight into the interior of the kettle of Karlsbad.[6]

Here we find the opposite to Goethe's results. Jacob Berzelius' analysis of the solid components of the springs compared at three different geographical locations seems to correspond, but their constant temperature, as Sternberg is claiming against Berzelius' interpretation, contradicts any volcanic events. Sternberg's scepticism changed in his letter dated 26 May 1824, when he wrote:

[...] also here [in Bohemia] we approached a volcano promising some information. The Calvariberg near Schlan is—as everybody knows—situated in the middle of a hard coal sandstone, it is a pillar-like basalt, the pillars are of different length and thickness, tetra-, penta-, hexagonal and are lying or standing upright, differently inclined to all directions of the wind; to the east, south and west, the sandstone is laid upon the larger basis of the basalt mountain, in the north at the stream of the suburb of Schlan, [...] the sandstone goes down with the basalt. The hard coal sandstone of this whole formation extending over several miles is white-greyish, where it meets with the basalt, it all changes colour, in immediate contact with the basalt it is brown to black, loose, easily being triturated, as it goes away from the basalt it passes all shades from brown into ochre, until it eventually regains its original colour as it leaves the basalt region. It seems rather fair to assume, that this change of colour might be caused by the higher temperature of the risen basalt, however, the isinglass appears unchanged, the subject deserves some close examination.[7]

Sternberg decided that this hill, called Calvariberg, was of volcanic origin by examining the characteristics in the arrangement of its strata and their change of colour. The aspect of strata-grouping had never been considered by Goethe, as far as the letters tell us. But the strata have been carefully examined and discussed in his essay from 1809, when he found an absolutely regular sequence of volcanic layers. However, Sternberg finally expressed his total conviction on the volcanic origin of the Bohemian hills in question in his letter dated 6 September 1824 to Goethe, which he wrote after an extended excursion in the area of the Rhine around Bonn in Germany:

The geological appearance of the quarries of Niedermennig and Mayen, the lake of Lach with its surroundings, the coherence of this area with the basalt at the left bank of the Rhine, all corresponding with the volcanic phenomena of the 'Eifel', give an illustrative example of volcanic effects in primordial periods, together with the pyrotypic, yet different, appearances in Bohemia. Also at the Rhine different modifications occur at the same location.[8]

Sternberg continued in this letter to explain the modifications of strata arrangement observed at the Eifel and compared them in detail with the ones at Kammerbühl. Goethe never went as far as this in his investigations in the early 1820s, that were confined to the hills in Bohemia mentioned in his letters. The observations, he made in 1820, on the products of Bohemian seam fires (Goethe 1820*c*) convinced him (in spite of Sternberg's investigations) of the pseudovolcanic character of the Bohemian hills in question. He found the altered and blackened rock with its highest point just below the recent surface and concluded from these findings, that the fire must have occurred, when (according to Werner) the water had already withdrawn, and was therefore to be assigned to a younger event in Earth history. To explain the blackened traces of fire, Goethe claimed that the rock must have become intermingled with combustible plant material or brown coal, which started to burn and affected the rock material all around. Thus Goethe returned to the geological views of the Bohemian mining inspector and medical doctor Franz Ambros Reuß (1761–1830), who in 1792 had attributed the pyrotypic products of Kammerbühl to pseudovolcanic events, due to burning of coal seam (Reuß 1792. pp. 1ff), and Goethe was well informed about Reuß's view. Nowadays we know that Goethe and Reuß were wrong, as Sternberg's investigations revealed in 1835, that the hill was volcanic in origin (Sternberg 1835, p. 1ff). It is Goethe's letter to Sternberg dated 14 December 1824, that reveals his deep aversion to volcanism as a theory:

> But now I have to admit, in the firm confidence in my revered friend's patience and forbearance, that I let ride my bad humour caused by von Hoff's uproarous second volume, on a whole page, which I am suppressing now, as matters of the kind might be forgiven in oral conversation, but they are not pleasant when sent into distance.[9]

In 1824, Adolf von Hoff (1771–1837) had published the second volume of his five-volume main work (Hoff 1822–1841), in which he developed his theory of actualism, saying that the forces having been active in the course of Earth history were identical to those that were active at the present time. In Hoff's second volume he claimed the Bohemian Kammerberg to be a mountain: '[...] not merely built by basalt, but also containing volcanic slags and Lava.' (see Hoff 1822–1841, p. 309). What might Goethe's 'suppressed whole page' to Sternberg be saying? It can be seen in a draft of his letter, in which he wrote:

> But now I cannot resist to talk about von Hof's [sic!] second volume, which put me into bad humour; it is just idiosyncratic with me, that I cannot bare pushing the explanation (derivation) of a phenomenon far away. God and nature have given us organs for the present, for the nearest ('das Nächste'), therefore this new plutonic kitchen will not prepare me anything savoury. As long as I am able to burn gold in my room, if constructing a powerful Volt-battery, should nature not be allowed to produce hot water up there in the rocks of Gastein?! Does the chasm need to drive a chimney right through the mountain in order to take the little amount of massive basalt up there?[10]

This undoubtedly emotional passage in Goethe's draft letter provides a hint, as to how to interpret his rejection of the idea of volcanism as such. His observations at Kammerbühl, Wolfsberg and in Altalbenreuth, mentioned in his letter of 10 September 1823 did not include his findings made on strata arrangements at the Kammerbühl in 1809, as Sternberg had carried out at the Calvariberg near Schlan. In the quoted draft Goethe was clearly pointing out, that the postulated mechanisms of volcanic effects *eo ipso* did not meet with his approach assuming the 'nearest', in other words the most obvious or the most probable natural event to explain a phenomenon. However, Goethe linked the probability of a natural event, by the actual act of interpretation to our organs, which 'God and nature have given us', thus integrating man as the *subject* into the natural context of the viewed phenomena at the *object*. We can therefore draw the conclusion that Goethe's persistent refusal of volcanism and Plutonism in his old age had its roots in his personal approach to interpretation, which exceeded purely scientific implications and strongly pointed towards a philosophically and psychologically determined intellectual and spiritual position. Being aware of this relationship between natural phenomenon and psychological constellation of its observer, Goethe had already put it into words in his essay on the Kammerbühl in 1809:

> On the occasion of such endeavours we should always consider well, that all attempts to solve nature's problems are in fact merely conflicts between intellectual power ('Denkkraft') and observation ('Anschauen'). Observation gives us an absolute notion of some performance in the past; the intellectual power, that is quite conceited, does not want to stay back, but wants to show and interpret in its own way, how the matter could and must have been performed. Yet, not feeling quite adequate, it calls the imagination ('Einbildungskraft') for help, and thus one after the other of these creatures of our mind (*entia rationis*) is produced, and it is thanks to them, that we are taken back to the observation and are approaching the matter with greater attention and for total insight (Goethe 1809).

However, Goethe underestimated in this intellectual and psychological constellation the effect of the observer's subjectivity on his intellectual power even after having returned to the phenomenon. This effect markedly applied to himself, as outlined below, when referring Goethe's natural-philosophical approach to the one of Baruch de Spinoza. Yet, in his essay *Verhältnis zur Wissenschaft, besonders zur Geologie* (1820*d*) Goethe mentioned the interference of the scientist's subjectivity with his observation of natural phenomena:

> As getting deeper into the matter, one observes, how much subjectivity really is at work in the sciences, and one does not succeed before starting to get to know oneself and one's character. [...] So I absorbed, what was appropriate to me, refused what irritated me, and as I did not need to teach in public, I taught myself in my own way without complying with anything traditional or conventional (Goethe 1820*d*, p. 215).

This essay can be regarded as a retrospective of Goethe's scientific approach that goes back to the beginning of his time in Weimar in 1775, when he started to develop deep interests in life and Earth sciences. It is remarkable that his awareness of subjectivity in scientific approaches in general and the thorough knowledge of his own character as the immediate cause of his subjectivity did not make Goethe take precautions against its overpowering effect on his scientific interpretations. On the contrary, it let him admit to accepting the involvement of his personal traits of character in the approval or disapproval respectively of particular scientific concepts. In his 1820 essay on the contemporary German geologist Karl Wilhelm Nose (1753–1835), he confirmed his opinion on the relationship between subject and object once more:

> [...] Those (other scientists) do not consider, that they are standing opposite to the object as subject, as an individual, and in spite of their presence are merely looking at the objects as well as their particular state by their own eyes and not by the general human glance (Goethe 1820*b*, p. 164).

In this context, Goethe visited and experienced Vesuvius in its active state in 1786 and was therefore familiar with the vehemence of volcanic action, as can be read in his *Italienische Reise* (1786–1787). Yet, it is symptomatic, that he was particularly inclined to accept Neptunist ideas in his old age, when Neptunism was already outdated, provided that there was no indisputable evidence for volcanic origin of the hill in question. This also partly explains Goethe's change of opinion in his judgement on the Kammerbühl. In 1809 he was still convinced of its volcanic origin, even though he attributed its eruptions to an early event in Earth's history, when the area was still covered with water, and he tried to explain the origin of the Kammerbühl, Wolfsberg and Altalbenreuth by pseudovolcanism in 1823 (Goethe 1823). It was shortly after his gathering with Sternberg, Berzelius, Pohl and Grüner at the Kammerbühl in 1822, when he became influenced by the explanations of a younger, anonymous bather, who was also interested in the geological origin of the hill near Eger. Goethe described this encounter in his article *Wunderbares Ereignis* (1823) (Goethe 1823, pp. 353–354). It happened, while Goethe had not irrevocably decided on any position in the problem. Moreover he gave a hint of his uncertainty by admitting: '[...]; but it made me think, and I of course believed to understand, that it is rather impulse then necessity determining us to stand on one side or another.' (Goethe 1823 p. 354). Here, Goethe again indicated his understanding of scientific interpretation as being the consequence of a psychologically (as outlined by the expression 'impulse') underpinned cognition process rather than a purely rational and strictly consequent derivation of an explanation from the precise observation of a given natural phenomenon. The importance of philosophically-derived, watertight evidence of the concept of nature to Goethe will be demonstrated below.

The philosophical background to Goethe's scientific approach

Hans-Ulrich Schmincke offered three options to explain Goethe's sympathy in favour of a Neptunist Earth theory: (a) his endeavour for congruence of art and science, favouring aesthetical aspects against a careful analysis of the processes of geological emergence; (b) his idea of nature presenting a whole entity; and/or (c) his lifelong aversion to chaos, tumult and disorder (Schmincke 2002, p. 119). It is obvious from the above discussion that the latter has been proven, but the other two explanations should not be regarded as alternatives, as they are causally connected to the last, following the philosophical views of Baruch de Spinoza. Also Goethe's profound dislike for tumult and violence, which characterized his whole attitude towards life in general (that he was also strongly against any kind of political revolutions, finds its explanation in his lifelong defence of Spinoza's ideas) they also markedly influenced his literary work (Jellinek 1996, pp. 19–32), which he first dealt with in his early twenties, progressively establishing his own world-view during the course of his life until old age. Spinoza should hence be regarded as having formed Goethe's philosophical and consequently rational foundations at the same time as psychologically motivated scientific modes of thought. It is therefore essential to outline the main traits of his philosophy, as far as they influenced Goethe's scientific approach.

Spinoza's philosophical achievements are all based on the religious intention to find the eternally (*sub specie aeternitatis*) good by means of mathematically deduced conclusions, founded on axioms and definitions, in order to divert the human mind from material property being regarded as merely ephemeral. Spinoza considered that the eternally good was the understanding that the world constituted a systematic entity. The insight into the actual contexts of facts should enable man to be the cause of his own behaviour and actions. Spinoza called it human liberty. These views are outlined in his *opus magnum*, the *Ethica, ordine geometrica demonstrata*, that was finished in 1675; however, its publication at the time was refused for its non-conformist religious views (Spinoza had already been excluded from the Jewish synagogue for the same reason in 1657) and it was eventually published by his friends as part of the *Opera posthuma*. It was the main source of influence on Goethe's own natural-philosophical, ethical and religious concepts. The rational, mathematical and deductive way of presenting his views allowed Spinoza to express his fundamental thought, that the organization of ratio, based on reason and conclusion, follows the organization of nature, based on cause and effect. Referring to Goethe's interpretation of natural phenomena, basically the thoughts expressed in the *Ethica*'s first two sections, *About God* and *About nature and the origin of mind*, are relevant.

In his *Ethica*, Spinoza comes from the axiom that there has to be an instance that is identical with the universe (substance, *substantia*), regarded as systematic entity as mentioned above, and at the same time causing itself (*causa sui*). He calls this instance 'God or nature' (*deus sive natura*) and defines the substance as 'what is being and is conceived by itself'. He claims that there can be only one single substance identical with God. God, who is almighty and perfect, according to Spinoza, cannot possibly be limited by borders and coincidental matters; to assume the latter would be blasphemous.

Spinoza's religious approach had a lasting effect on Goethe's own religious belief, which he dissociated entirely from Christian belief or any other faiths, that took God as an authority separated from nature for granted. By concluding that God was almighty and perfect as well as identical with nature, Goethe was convinced that nature was necessarily ideal.

Spinoza's view was deterministic, i.e. nature works out of its own necessity, and does not allow any casual events. As God is not separated from nature, he is acting out of necessity himself; consequently we live in the only possible world. Substance is determined by attributes, which according to Spinoza was 'what the intellect recognized to belong to its being' (*essentia*). Out of the substance's infinite attributes man perceives merely two: (a) the physical expansion of bodies (*corpus*), that are organized in a chain of cause and effect and (b) the action of thinking (conscience, *mens*) by imagination or ideas (*idea*), that are organized by the relation between reason and conclusion. What Goethe called 'observation' ('Anschauen'), could be considered as the act of notice associated with human sight, which is bridging the gap between Spinoza's attributes of body expansion and conscience by seeing these bodies and consequently any natural phenomena. If Goethe was pointing to the importance of the observation (Goethe 1809), he was focusing on this bridging function as the crucial act in any scientific cognition process. In this context, the two attributes of physical expansion and conscience are merely different ways of conceiving the same organization, thus forming two aspects of the same reality, which in either case is the organization of the single substance being, that is, of the universe. The attribute of the physical expansion of bodies is essentially determined by the characteristics of movement and calm, Spinoza called them infinite *modi* and defined the *modus* as 'affection of the substance or what was included in another, which it was also conceived in'. The body expansion, called finite *modi*, are composed of tiniest particles. Each body shows the tendency (*conatus*) to preserve itself, transforming various states of energy in the course of time. The attribute of conscience is determined by the intellect (*intellectus*). Spinoza explained conscience (*mens*) as imagination (*idea*), the object (*ideatum*) of which is the body (*corpus*). In order to produce imagination, bodies and their expansion need to be noticed.

In the text quoted above, Goethe emphasized (Goethe 1809) the dependence of the intellectual power (corresponding to Spinoza's *intellectus*) on the imagination (corresponding to Spinoza's *idea*), which might be regarded as the intellect's substrate and is the image of the object (corresponding to Spinoza's *ideatum*). However, philosophy is confined to theory, whereas scientific methods are based on practice. Therefore, Goethe's reflections on the function of Spinoza's items classed with the attributes of the substance could be interpreted as follows: it is the desire to prevent the intellect from drawing conclusions from faded imaginations, that makes Goethe assign to the imagination the urge to call the observer back in time to the geological phenomenon to be observed. As Goethe's geological observation was only based on the momentary perception of the Kammerbühl's strata and their qualities, he was lacking the information on strata development that represented a previous state in course of the real, therefore non-postulated and unknown chain of cause and effect leading to the strata's qualities in the present state of observation. The intellect is now apt to replace this lack of information by the imagination of facts that close the gap in the postulated chain of cause and effect or, in other words, that close the gap in the corresponding chain of reason and conclusion. The act of intellectual replacement is therefore the crucial event in the scientific interpretation of natural phenomena, and Goethe was well aware of it, when pointing to the dependence of the intellectual power on the imagination. He also bore it in mind, when claiming, that 'God and nature have given us organs for the present, for the nearest' (Sauer 1902, p. 314), a statement expressing three facts following Spinoza: (a) 'God and nature' being identities (*deus sive natura*); (b) our organs as a metaphor of our senses and intellect being a reflection of the organization in nature and, based on Spinoza's definition of the substance, irrevocably constituting part of it, and consequently; (c) nature functioning according to the intellect's most immediate, therefore most obvious and most probable way. The latter, of course, could be explained by Spinoza's view of nature's tendency to preserve itself. Considering the infinite two *modi* of the attribute of body expansion, movement and calm, it seems reasonable to assume with Goethe, that natural events, in order to preserve nature, are optimally based on the balance between the two, as nature is perfect. It is therefore not surprising, that Goethe tended to exclude excessive movement from his scientific concepts. This position was not sufficient to explain his geological position in favour of Werner's Neptunism, as Spinoza did not decide on specific maximum or minimal states of movement in natural events; he merely assigned two infinite *modi* to the finite attribute of body expansion, movement and calm. Goethe's mode of interpretation was linked to psychological influences, the existence of which he never denied, as can be proved by his statement 'As getting deeper into the matter, one observes, how much subjectivity really is at work in the sciences [...]' (Goethe 1820*d*, p. 215). Why did Goethe accept this fact and claim that 'one does not succeed before starting to get to know oneself and one's character', conclude: 'So I absorbed, what was appropriate to me, refused what irritated me, and as I did not need to teach in public, I taught myself in my own way without complying with anything traditional or conventional.' The answer returned to Spinoza and closing the circle of evidence in the light of Goethe's scientific interpretation of geological theories: object and subject reflect the two attributes of body-expansion and conscience of the substance, as thoroughly outlined above. Representing two aspects of the same organization, they must strictly correspond with each other. Goethe saw it as a *conditio sine qua non*, in order to gain insight into the attribute of body expansion being the objects, to investigate his own psychologically determined nature as the subject and consequently to find the mechanisms acting in nature. It corresponds to his own character, particularly in his old age, imagining calm, steady state modes of action and projecting this imagination into the attribute of body expansion. Goethe followed the opposite procedure from Spinoza, who claimed that the insight into the contexts of facts should enable man to be the cause of his own behaviour and actions, as stated at the beginning of this chapter. The strict correspondence of the deterministic actions of conscience to those of expanded bodies or of those in the subject to those in the object, respectively makes Goethe's reciprocal reversal of Spinoza's sentence philosophically acceptable.

Contrasting aspects: Goethe's deductive and Sternberg's inductive approach

Goethe's geological excursions in Bohemia during his last visits between 1820–1823 were only short and, with the exception of the crucial problem on the geological origin of pyrotypic hills, were limited observationally. Again we learn this from his correspondence with Sternberg. In his letter dated 20 June 1823 he casually asked: 'Where do the fossil plant printings from Falkenau belong to? I believe I recognize a great plantain and leaves similar to those from beech trees.'[11] His letter from Marienbad, 16 August 1823, sent to the 'Vaterländisches Museum' in Prague opened another topic of Bohemian geology. It focused on a fact that had been discovered by Sternberg in the previous year, namely the effect of the thermal sources of Marienbad on its basic rocks. Rock samples showing various alterations are sent together with unaltered original rock samples to the 'Vaterländisches Museum' to Prague as documentary proof of this phenomenon.[12] His letter dated 26 August 1822 reads: 'On Wednesday the 7th we drove to Schönberg, where the Kappelberg presents quite interesting things. The components of the granite in large parts side by side.'[13] Whether he was not yet sure how to interpret these granite components lying side by side, or whether he never intended to explore his findings further, we do not know. Still, his interest in geology continued until his death. His letters to Sternberg between 1824–1832 and also his diaries give sufficient evidence that he never stopped trying to keep up with the latest geological works and that he took a sincere interest in selected geognostic questions. Yet, he kept a careful position in their judgement, which he justified in his letter to Sternberg dated 8 July 1829 as follows:

> Finally I have to tell you, that I try to define my position on geology, geognosy and oryctognosy, neither polemically nor conciliatorily, but positively and individually; this is the cleverest that we can do in our old days. Sciences, we are dealing with, are disproportionately progressing, sometimes thoroughly, often hastily and fashionably, therefore we are not allowed to immediately follow on, as we have no more time to rashly stray; but in order not to stop and stay behind too far, examinations on our situations are necessary.[14]

Specifically applied to Goethe's own scientific attitude towards geology, these thoughts allowed him to retain his philosophical and psychological understanding of nature as being involved in steady, long-term but calm processes in the course of its history.

Caspar Sternberg's approach in his dealing with geology and geognosy is characterized by long-term projects. Evidence for this is given in the as yet unpublished *Ideen über naturhistorische*

Reisen in den k. k. Erbstaaten,[15] that he sent to the Bohemian magnate Franz Anton Count Kolowrat-Liebsteinsky (1778–1861) in 1817. In it he suggested a series of travels within the Habsburg region aimed at the coordination of natural historical projects on the exhaustive exploration of studies in oryctognosy (mineralogy) and geognosy along large distances of geological formations, as they connected several provinces, therefore necessitating regional geognostic investigation. At the same time he criticized the lack of attempts to relate the oryctognostic findings in the single provinces to the relevant geognostic results, claiming the correlation of mineral deposits to the relevant formations. As an example Sternberg pointed to his own findings in Bohemia, were he was able to correlate the deposition of precious metals with greywacke and of clay ironstones with the older sandstone and flinty-slate formations. Furthermore, he demonstrated that hard coal showed a regular distribution along these successions, and pointed towards a similar time of origin, whereas brown coal had to be regarded as much younger. Finally, Sternberg's suggestions to Kolowrat-Liebsteinsky encouraged the drawings of petrographic maps in order to present these oryctognostic-geognostic findings from all provinces in the Habsburg region.

Sternberg's requests to Kolowrat-Liebsteinsky should be interpreted in context with economical interests, not following purely scientific endeavours, as he realized the practical importance of finding hard coal, precious metals and gems in his own country as a means of restoring its financially poor state in the early 1800s. Sternberg invested part of his own fortune in the development of the Bohemian economy. On his Bohemian estates he built a blast furnace and employed the people of Liblín and Radnitz in his own forestry, ironworks and coal mine. In 1827, he used his influence to found a railway link between Prague and Pilsen primarily to transport the mining products. From 1826 his economic efforts were acknowledged with his presidency of the Patriotic-Economic Society of Bohemia (Patriotisch-Ökonomische Gesellschaft Böhmens), an association founded by Empress Maria Theresia (1717–1780) (Majer 1998).

Sternberg's letters to Goethe tell us, that from the scientific point of view, he endeavoured to clarify problems about the nature, geological age and mineral deposits in the widespread formations of interest. In his correspondence, he gave detailed scientific reports, almost treatises, and analyses of extended trips that led him beyond the borders of Bohemia. So in his letter of September 1822, he described the marl formation on the occasion of his excursion to the hard coal formation of Häring, Miesbach and Peißenberg in Tyrol. He found pitch coal there and analysed its fossil plants, concluding from their dicotyledonic representatives that its geological age lay between the time of origin of slate coal and brown coal, and he compared his findings with those from Switzerland, Bohemia and South Bavaria. He also analysed the seam fires of the area in detail. In his letter dated 4 August 1823 he let Goethe know:

> The main reason for my journey this year was the investigation of the salt formation in Wieliczka and of the fossil wood in the salt, another purpose was the tour through the Silesian hard coal formation.[16]

A precise report of his geognostic and palaeobotanical findings filled the two following pages. His extensive excursion in the Rhine area was mentioned above. On 1 September 1825, Sternberg informed Goethe of his tour around Istria and Illyria, again with a detailed exploration of the geognostic circumstances in mind, he outlined his suggestions to Kolowrat-Liebsteinsky in 1817. From Sternberg's geological travels, aimed at the geognostic insight into large, regional areas from various local investigations spread over the whole area in question, it can be seen that he followed an inductive scientific cognition process. His palaeobotanical findings on all his journeys were published in his main work *Versuch einer geognostisch-botanischen Darstellung der Flora der Vorwelt* (Sternberg 1820–1838), where he achieved the pioneering feat of classifying fossil plants according to Linné's system of living plants, and proved that fossil plant species lived in specific ecological conditions and in biotopically specific communities. He thus overcame the eighteenth century's idea of an antediluvian life.

Much less time has been taken up explaining Sternberg's inductive scientific approach and his way of interpretation by the conventional means of comparison than Goethe's. Unfortunately, Sternberg's scientific interests have been undermined by economical interests rather than by the defence of any philosophy, and also in the letters between Goethe and Sternberg neither Spinoza nor any other philosopher is ever mentioned. It has therefore been assumed, that Sternberg's thoughts, although a representative of enlightenment, were not influenced by philosophy. His clerical education brought him up with the conventional views of the Roman Catholic church; however, that did not determine his scientific position. However, the last sentence in his diary, dated 31 December 1837: 'Much has been given by the Lord, much has been taken, the name of the Lord be praised' (Sternberg 1909, p. 214), indicated that Sternberg still believed in God as an authority separated from nature.

Conclusions

Two geological travellers, who clearly differed in their approach to scientific questions and their interpretation, have been presented. Goethe was mainly interested in the natural phenomenon as such, and his interpretation of observations was linked to a philosophically and psychologically influenced view of natural mechanisms. In order to verify his interpretation, he went back to the phenomenon in various individual cases, following a deductive cognition process. His natural philosophy, that lay behind his scientific concepts, was strongly based on the philosophy of Baruch de Spinoza and led to Goethe's subjective approach in interpreting scientific phenomena. In contrast to Goethe, Sternberg chose an inductive approach and was never inclined to any philosophical explanations behind his observations in nature. His knowledge of geognosy along the rock successions in the Habsburg region and in Germany is based on numerous single observations at various locations and on the comparison of his findings aiming at a vast general knowledge of the geology over a wide area. These two different ways of investigating nature found their fruitful, mutual completion in Goethe's and Sternberg's correspondence from 1820–1832.

I wish to thank H. Zeman (University of Vienna, Austria) for valuable advice and O. Fejfar (Charles University in Prague, Czech Republic) for animated discussions on the geology of the Kammerbühl. This presentation has been sponsored by the Austrian Research Foundation (FWF; project 14773-G01).

Notes

[1]Sauer 1902. The letters from Goethe to Sternberg are kept in the Památník národního pisemnictví archive in Prague, those from Sternberg to Goethe are in the Goethe-Schiller-Archiv in Weimar (Germany).

[2]The extracts are quoted in the author's own English translation.

[3]The most detailed history of Sternberg's life is given in his autobiography (Sternberg 1909).

[4]When Goethe sent this article to the *Taschenbuch*'s editor, Karl Cäsar von Leonhard (1779–1862), he supported his volcanistic interpretation in an accompanying letter by Ignaz von Born's (1742–1791) article *Schreyben über einen ausgebrannten Vulcan bey der Stadt Eger*, Prag, 1773, see Engelhardt (2003).

[5]Letter Goethe to Sternberg, 10 September 1823, Eger; see Sauer (1902, p. 23); italics in the original text.

[6]Letter Sternberg to Goethe, January 1824, Prague; see Sauer (1902, pp. 84–85).

[7]Letter Sternberg to Goethe, 26 May 1824, Prague; see Sauer (1902, p. 78).

[8]Letter Sternberg to Goethe, 6 September 1824, Munich; see Sauer (1902, p. 91).

[9]Letter Goethe to Sternberg, 14 December 1824, Weimar; see Sauer (1902, p. 100).

[10]Letter (draft) Goethe to Sternberg, 14 December 1824, Weimar; see Sauer (1902, p. 314).

[11]Letter Goethe to Sternberg, 20 June 1823, Weimar; see Sauer (1902, p. 50).

[12]Letter Goethe to Sternberg, 16 August 1823, Marienbad; see Sauer (1902, p. 58).

[13]Letter Goethe to Sternberg, 26 August 1822, Marienbad; see Sauer (1902, p. 18).

[14]Letter Goethe to Sternberg, 8 July 1829, Weimar; see Sauer (1902, p. 181).

[15]Haus- Hof- und Staatsarchiv, Vienna, Sig.: Sternberg, Brasilienreise 1, Konv. 2.

[16]Letter Sternberg to Goethe, 4 August 1823, Swetla/Deutschbrod. Sauer (1902, p. 53).

References

BERZELIUS, J. 1823. Untersuchung der Mineral-Wasser von Karlsbad, von Teplitz und Königswart. *Gilberts Annalen der Physik*, **6** & **7**, 113–212.

BUCH, L. VON 1825. *Physikalische Beschreibung der Canarischen Inseln*. Königliche Akademie der Wissenschaften, Berlin.

ENGELHARDT, W. 2003. *Goethe im Gespräch mit der Erde. Landschaft, Gesteine, Mineralien und Erdgeschichte in seinem Leben und Werk*. Böhlau, Weimar.

FEJFAR, O. 1999. Brunnengast, Geolog und Spaziergänger. *In*: STEININGER, F. F. & KOSSATZ-POMPÉ, A. (eds) *"quer durch Europa". Naturwissenschaftlich Reisen mit Johann Wolfgang von Goethe*. Katalog zur gleichnamigen Ausstellung im Naturmuseum Senckenberg. Kleine Senckenbergreihe **30**, 49–75.

FEJFAR, O. & STEININGER, F. F. 1999. Johann Wolfgang von Goethe: 'Brunnengast, Geolog und Spaziergänger'—erdwissenschaftliche Beobachtungen in Böhmen. *In*: HOPPE, A. & STEININGER, F. F. (eds) *Exkursionen zu Geotopen in Hessen und Rheinland-Pfalz sowie zu Naturwissenschaftlichen Beobachtungspunkten Johann Wolfgang von Goethes in Böhmen*. Schriftenreihe der Deutschen Geologischen Gesellschaft, **8**, 7–67.

GOETHE, J. W. VON 1809. Der Kammerberg bei Eger. *Taschenbuch für die gesamte Mineralogie*, **3**, 3–24 (first print).

GOETHE, J. W. VON 1820*a*. Der Kammerberg bei Eger. *Zur Naturwissenschaft überhaupt*, **1**, 65–82 (first print).

GOETHE, J. W. VON 1820*b*. Karl Wilhelm Nose. *In*: KUHN, D. (ed.) *Goethe. Die Schriften zur Naturwissenschaft*. vollständige, mit Erläuterungen versehene Ausgabe im Auftrage der Deutschen Akademie der Naturforscher Leopoldina, **8A** (LA I 8A), Böhlau, Weimar, 157–164.

GOETHE, J. W. VON 1820*c*. Produkte böhmischer Erdbrände. *Hefte zur Naturwissenschaft* **1**, 234–238 (first print).

GOETHE, J. W. VON 1820*d*. Verhältnis zur Wissenschaft, besonders zur Geologie. *In*: KUHN, D. (ed.) *Goethe Die Schriften zur Naturwissenschaft*. vollständige, mit Erläuterungen versehene Ausgabe im Auftrage der Deutschen Akademie der Naturforscher Leopoldina, **8A** (LA I 11), Böhlau, Weimar, 215–217.

GOETHE, J. W. VON 1823. Wunderbares Ereignis. *In*: KUHN, D. (ed.) *Goethe. Die Schriften zur Naturwissenschaft*. vollständige, mit Erläuterungen versehene Ausgabe im Auftrage der Deutschen Akademie der Naturforscher Leopoldina, **8A** (LA I 11), Böhlau, Weimar, 353–354.

HOFF, K. E. A. VON 1822–1841. *Geschichte der durch Ueberlieferung nachgewiesenen natürlichen Veränderungen der Erdoberfläche*. 5 Volumes. Gotha.

HUTTON, J. 1788. On the Theory of the Earth; or, an Investigation of the Laws Observable in the Composition, Dissolution and Restoration of Land upon the Globe. *Transactions of the Royal Society of Edinburgh* **1**, 209–304.

HUTTON, J. 1795. *Theory of the Earth With Proofs and Illustrations*. 2 Volumes. Creech, Edinburgh.

JELLINEK, G. 1996. *Die Beziehungen Goethes zu Spinoza*. Wissenschaftlicher Verlag, Schutterwald/Baden.

LYELL, C. 1830–1833. *Principles of Geology, being an attempt to explain the former changes of the Earth's surface, by Reference to Causes Now in Operation*. John Murray, London.

MAJER, J. 1998. KASPAR M. GRAF VON STERNBERG (1761–1838), seine Zeit, sein Leben und sein Werk. *In*: KASPAR, M. *Graf von Sternberg, Naturwissenschaftler und Begründer des Nationalmuseums*. Böhmisches Nationalmuseum, Prague.

OLDROYD, D. 1996. *Thinking about the Earth. A History of Ideas in Geology*. Harvard University Press, Cambridge, MA.

PLAYFAIR, J. 1802. *Illustrations of the Huttonian Theory of the Earth*. Cadell and Davies, Edinburgh.

REUß, F. A. 1792. Etwas über den ausgebrannten Vulcan bei Eger in Böhmen. *Bergmännisches Journal von Köhler und Hoffmann*, **5**, 303–333.

SAUER, A. (ed.) 1902. *Ausgewählte Werke des Grafen Caspar Sternberg,* 1.Bd.:*Briefwechsel zwischen J. W. von Goethe und Caspar Graf Sternberg (1820–1832)*, Bibliothek Deutscher Schriftsteller aus Böhmen **13**. Calve, Prague.

SCHMINCKE, H.-U. 2002. Tanz auf dem Vulkan. Goethe und die Entfaltung der Vulkanologie als Wissenschaft. *In*: STEININGER, F. F. (ed.) *Senckenberg, Goethe und die Naturwissenschaften*. Kleine Senckenbergreihe **44**, 119–175.

STERNBERG, C. 1835. Rede des Präsidenten des böhm. Museums bei der 13. ordentlichen allgemeinen Versammlung, den 14. April 1835. *Verhandlungen der Gesellschaft des Vaterländischen Museums* **16**, 12–30.

STERNBERG, C. M. VON 1820–1838. *Versuch einer geognostisch-botansichen Darstellung der Flora der Vorwelt*. Fleischer, Leipzig, Sturm, Ratisbon, Calve, Prague.

STERNBERG, K. 1909. Materialien zu meiner Biographie. *In*: HELEKAL, W. (ed.) *Ausgewählte Werke des Grafen Kaspar Sternberg*; no. 2. Bibliothek Deutscher Schriftsteller aus Böhmen **27**. Calve, Prague.

URZIDIL, J. 1932. *Goethe in Böhmen*. Epstein, Vienna and Leipzig.

Geological travellers in Auvergne, 1751–1800

K. L. TAYLOR

Department of the History of Science, University of Oklahoma, Norman, OK 73019, USA (e-mail: ktaylor@ou.edu)

Abstract: Within the half-century after Guettard's epoch-making journey of 1751, geologists came to see the Auvergne region of France as a place of unusual interest for field investigation. This paper reports on an effort to catalogue instances of scientific travel in Auvergne up to the end of the eighteenth century, before observers during the first decade of the nineteenth century (such as von Buch, d'Aubuisson and Ramond) validated the establishment of Auvergne as an iconic place for geologists. In addition to those who ventured into Auvergne to investigate its geology, a significant number of the eighteenth-century observers were residents of Auvergne; these are tabulated separately from those journeying from elsewhere. Published results of Auvergne observations accomplished by 1800 suggest that the Auvergne geological phenomena were already becoming fixed as part of the geological traveller's canonical itinerary.

It also plainly appears, if I mistake not, ... that the phaenomena of recent volcanos are very little calculated to give us much instruction about the more curious igneous concretions, and the origin of vulcanic mountains in general; and that a few days tour in such countries as Auvergne, Velay, and the Venetian state are worth a seven years apprenticeship at the foot of mount Vesuvius or Ætna; where nothing but a heap of uninstructive ruins, and a sameness of phaenomena appear. And since our ideas, concerning vulcanic effects, have been almost exclusively drawn from recent volcanos, we cannot much wonder if they yet remain so imperfect. (John Strange 1775, pp. 32–33)

Vous aurés pu remarquer que l'Auvergne est infiniment plus favorable que les environs de Naples et de Rome pour étudier les opérations du feu et qu'on peut y recueillir une infinité de faits dont les caractères sont altérés. En Italie on ne se présentent [sic] pas avec cette correspondance instructive si nécessaire pour en tirer des conséquences générales. (Nicolas Desmarest to H.-B. de Saussure 1776)[1]

Que j'ai de plaisir à vous voir disposé à visiter ces régions qui certainement en Europe n'ont pas leurs pareilles. Voulez-vous voir des volcans? Choisissez Clermont de préférence au Vésuve et à l'Etna. Dans ces deux dernières montagnes, une éruption postérieure couvre les productions de celle qui l'a précédée; mais à Clermont, les laves, les courans énormes, sont à découvert depuis leur sortie du flanc de la montagne jusques dans la plaine où ils se sont arrêtées; et on découvre tous les détails, et leur nature n'est point équivoque... (Leopold von Buch to Marc-Auguste Pictet 1802)[2]

The beginnings of geological investigation in Auvergne are part of the science's historical mythology. Few places are more firmly associated in geology's historical consciousness with important facets of the science's early development than this region of south-central France. These associations are mainly with the foundations of the study of extinct volcanoes and of historical interpretations of volcanic landforms. As the epigraphs (above) illustrate, some prominent figures were prepared to proclaim Auvergne's special value for geological investigation during the last quarter of the eighteenth century. Such claims were affirmed and, if anything, strengthened during the course of the nineteenth century. Robert Bakewell, for instance, wrote in 1823:

... I felt no small degree of pleasure in finding myself in one of the most remarkable districts in Europe, placed nearly in the centre of France, and surrounded by a well-cultivated and populous country, but exhibiting incontestable proofs of a mighty conflagration, that has, at a former period, spread over many hundred square miles. The marks of the powerful agency of fire are so fresh, that the spectator might suppose in some parts it had scarcely ceased to burn; ... (Bakewell 1823, **2**, 294–295)

Somewhat later, Charles Lyell stated about Auvergne:

We are here presented with the evidence of a series of events of astonishing magnitude and grandeur, by which the original form and features of the country have been greatly changed, yet never so far obliterated but that they may still, in part at least, be restored in imagination.[3]

Near the end of the century Archibald Geikie wrote, in his appreciative chapter on Nicolas Desmarest in *The Founders of Geology*:

Among the many claims of France to the respect and gratitude of all students of geology, there is assuredly none that ought to be more frankly recognised than that, in her wide and fair domain, she possessed a region where the phenomena were displayed in unrivalled perfection ... (Geikie 1897 p. 63 [1905, p. 157])

By the close of the nineteenth century, Auvergne had become a fixed element of an international canon of places for informed geologists to see. Auvergne qualified, one might say, as an objective of geological pilgrimage.[4]

Certain aspects of Auvergne's geological story are well rehearsed in the historical literature.

From: WYSE JACKSON, P. N. (ed.) *Four Centuries of Geological Travel: The Search for Knowledge on Foot, Bicycle, Sledge and Camel.* Geological Society, London, Special Publications, **287**, 73–96.
DOI: 10.1144/SP287.7 0305-8719/07/$15.00

Leading the way is the tale of the 1751 journey during which Jean-Étienne Guettard and Chrétien Guillaume Lamoignon de Malesherbes first realized the existence of extinct volcanoes in France, in the form of the *chaîne des puys* in the vicinity of Clermont-Ferrand. This oft-told tale of discovery (over the years frequently embellished, even badly distorted) was thoroughly scrutinized 25 years ago by François Ellenberger (1978; see also Vernière 1899–1900; De Beer 1962; Michel 1971). Also conspicuous are historical accounts of the 1828 tour of Auvergne by Charles Lyell and Roderick Murchison, and the importance of that experience in particular for the development of Lyell's dynamical and historical conceptions in geology (Wilson 1972, pp. 191–200; see also Kölbl-Ebert 2007). Details of the eighteenth-century journeys in Auvergne of another distinguished geological explorer, H.-B. de Saussure, have been published by Albert Carozzi, at the same time shedding light on the work in Auvergne of other figures, notably Nicolas Desmarest (Carozzi 2000).

Notwithstanding Auvergne's familiarity as a geological setting and destination, its rise to prominence seems worthy of further study. The transformation of particular localities like Auvergne into 'iconic places' is a cultural process within the science that, to my knowledge, has not yet been subjected to close examination. This paper does not pretend to provide even the beginning of such a comprehensive examination; at most it may serve as a modest step towards a fuller understanding of how this particular district became a 'geological mecca'. This paper simply gathers the outlines of historical information about the first half-century of Auvergne's geological exploration. Brief consideration of that information may suggest useful avenues for further research regarding Auvergne's scientific enshrinement.[5]

A broad objective informing this inquiry into Auvergne's historical recognition as a special arena of geological endeavour is an effort to identify the thought-patterns of Enlightenment geologists. Records of travels to and observations in Auvergne are potentially useful as a window affording helpful perspective on how geological research worked, in this period of geology's early development. An attempt to chart the first half-century of Auvergne geological investigation is, I think, the more worthwhile if one considers that geological science during this period was in the process of developing the 'place-specificity' that was to characterize it during its nineteenth-century maturation. (A naturalist manifests the place-specific impulse in attending closely to the details that distinguish a natural feature of some particular place from another one somewhere else. In the eighteenth century, those savants who drew their ideals regarding the formulation of knowledge largely from the traditions of natural philosophy, or *physique*, tended to value place-specific information much less than generalized principles.) As will be seen in the closing remarks, most eighteenth-century naturalists were not as familiar or comfortable with notions of place-specificity taking a central role in geological science as was to become the case for many of their nineteenth-century counterparts.

Travel to (and in) Auvergne during the second half of the eighteenth century

In the decades before the Revolution, Auvergne was, according to the *Encyclopédie*, a province measuring about 40 leagues from north to south and 30 leagues from east to west, bounded on the north by Bourbonnais, on the east and SE by Forez and Velay, on the west by Limousin, Quercy, and la Marche, and on the south by Rouergue and the Cévennes or Gévaudan (Fig. 1). Auvergne was known as a region for raising livestock, for production of certain goods such as paper and fine tapestries (Aubusson), and as a net exporter of people.[6] In the 1760s, the province's population was estimated at somewhat over half a million.[7] The region was commonly divided into two distinct parts, Lower and Upper Auvergne (*Basse-Auvergne, Haute-Auvergne*). Lower Auvergne, corresponding roughly to the post-Revolutionary *département* of Puy-de-Dôme, consisted of the Limagne or the valley of the Allier, a major tributary of the Loire, and then navigable (seasonally) as far as Brioude in southern Auvergne, and also the mountainous areas to the west, including both the *chaîne des puys* and the Mont-Dore. Upper Auvergne, corresponding fairly closely to the *département* of Cantal, included the cities of St Flour and Aurillac.[8]

Auvergne generally, but particularly its substantial mountainous portion, was viewed (at least by Parisians) as rustic and off the beaten track, although Mont-Dore did boast thermal springs whose attractions for travellers dated back to Roman times. The eighteenth-century traveller faced discomforts and hazards that would be unfamiliar to a modern tourist. France at that time was far from homogeneous; the inhabitants of its different parts were accustomed to a variety of dialects, legal codes, and systems of measure.[9] A journey starting in Paris might follow a postal route traversing the provinces of Orléanais, Nivernais, and Bourbonnais, a distance of 52 *postes* or postal stations, to the northern frontier of Auvergne at Aubiat (near Aigueperse), a few leagues to the SW of Vichy.[10] Clermont was the terminus of one of the royal diligence routes from the period of Turgot's ministry (in the 1770s), although this route was one of

Fig. 1. Map of Auvergne and Lyonnais, 1748. Gilles Robert de Vaugondy (1688–1766). *Gouvernemens généraux du Lyonnois et de l'Auvergne. Par le Sieur Robert, Géographe ordinaire du Roy.* Engraved by Jean Lattré (170?–178?). Author's collection. Dimensions of original: 160 × 182 mm. The province of Auvergne is shown divided into two parts, known respectively as Upper and Lower Auvergne: *Haute Auvergne*, with its main cities of Aurillac [Orilhac] and Saint-Flour; and *Basse Auvergne*, with Clermont as the chief city. The Allier River flows north through Lower Auvergne toward its confluence with the Loire near Nevers.

relatively infrequent service. In 1795, daily delivery of the post between Paris and Clermont-Ferrand might be accomplished in four days.[11] If the roads from the north into Auvergne were not among the country's leading thoroughfares, the routes leading south, east, or west from Clermont-Ferrand were even less heavily travelled.[12] The royal government's administration was through the Intendancy in charge of the *généralité* of Riom, divided into 22 *subdélégations*. Following the example of Daniel-Charles Trudaine (le grand Trudaine, 1703–1769), who had been Intendant in Auvergne during the 1730s and as a minister in the royal government was a noteworthy advocate of improvement through cultivation of scientific knowledge, succeeding Intendants were generally open to ideas on how to apply technical and scientific talent to the cause of commercial or industrial advancement.[13]

Important resources for the serious traveller were the proper sheets of the *Carte de l'Observatoire*, known as the *Carte de Cassini*.[14] Visitors interested in the geological phenomena of Auvergne might engage local guides, and evidently often sought the company of residents with knowledge of the area and its natural history. Not least among the advantages of employing local guides was that they might help allay the fears and suspicions of a populace that could be hostile to snooping intruders. To make their way about the Upper Auvergne they might be able to use a carriage for certain routes; Saussure accomplished his journey from Geneva through Auvergne in 1776 (with his wife and two children, accompanied by a coachman and three servants) in a four-horse *berline*, an enclosed four-wheeled carriage (the maid rode inside with the family, another servant sat with the coachman, while Saussure's chief servant rode postilion) (Carozzi 2000, p. 166). But the more interesting sites for naturalists, where amenities for travellers might be scarce, generally could be accessed only on horseback, or in many circumstances, on foot (see Taylor 1994; also Vernière 1899–1900).

A half-century of observers of Auvergne geology

The inquiry involved in this paper calls for a database. A preliminary tabulation of information about Auvergne travellers is, in fact, this study's main result (see Appendix). Creating a thorough inventory of all the people who might be considered observers of Auvergne geology from 1751 to the start of the nineteenth century seemed an ambitious but not absurd ideal. However, there are several complications in compiling such a list. Not necessarily in order of importance, these difficulties are related to the problems of: (1) the definition of Auvergne itself; (2) who qualifies as a geological observer; and (3) who qualifies as a traveller. The practical resolution of some of these perplexities involves arbitrary choices.

A significant point, well understood by the French but often unclear to others, is that Auvergne encompasses only a part of the Massif Central. Within a quarter-century of Guettard's famous report on the extinct volcanoes of France, similar phenomena were being reported in neighbouring Velay, and in Vivarais, notably by Barthélemy Faujas de Saint-Fond, then by J.-L. Giraud Soulavie (Faujas de Saint-Fond 1778; Soulavie 1780–1784). Leaving aside as unproblematic the separate identification of volcanic remnants in southern parts of Languedoc (these were always treated as geographically distinct from the localities reported on by Guettard and Desmarest in the 1750s and 1760s, respectively) the greater proximity and near-contiguity of the volcanic productions in Forez, Velay and Vivarais led a number of observers to blur their geographical distinction from Auvergne.[15] As a result, in certain cases there has been a historical conflation of Auvergne geology with the entire complex of phenomena of the Massif Central. This is more apparent in some modern historical commentaries than in most contemporary scientific description, in which one generally finds comparatively fine geographical distinctions. However, the problem is compounded by historical adjustments in French political geography itself: the post-Revolutionary redefinition of political and administrative geographical units did not obliterate the old provincial identifications, but did reorganize them. Auvergne is now understood to consist of the *départements* of Allier, Cantal, Haute-Loire, and Puy-de-Dôme, which together comprise a region only generally similar to the *ancienne province* of Auvergne. Perhaps most significant is that the formerly distinct region of Velay, corresponding roughly to the modern *département* of Haute-Loire, has been folded into modern Auvergne.

A contemporary definition of the province has been used in this study, distinguishing eighteenth-century observations in Auvergne from those made in Velay or Vivarais or Forez. However, I have tried to avoid rigid exclusion of (especially) the more noted or prominently-published reports of geological features in these neighbouring areas.

As for who qualifies in this discussion as a geological observer, allowances have been made for historical changes in ideas and conventions about scientific categories, as conceived in the eighteenth century. For this paper's inventory, a broad and flexible criterion has been used in deciding whether an eighteenth-century observer or observation should be considered 'geological'. Clearly, a chemist who focused on analysis of Auvergne mineral waters belongs in the inventory.[16] Since botany was seen as the leading field science in the eighteenth century and also since botanists tended to regard examination of fossil plants as an integral part of their science (Guettard being a leading example), botanists have been included, especially when there is evidence of an additional interest in mineral phenomena. Admittedly, a few figures are included even though their geological interests were questionable (Nathaniel William Wraxall and Arthur Young, for example, the former seemingly an adventurous tourist, the latter polymathically oriented toward agriculture and industry).

A little more difficult for present purposes than determination of what is (or was) 'geological', are decisions about who was a traveller (and perhaps as a separate problem, who was a scientific traveller). As might be expected, not all the interesting or valuable geological observations were made by people who came to Auvergne from elsewhere. Indeed, one of the most notable contributions to Auvergne geology towards the end of the century was written by François-Dominique de Montlosier, an inhabitant of the area. By the 1770s, as several analysts have noted, informed travellers arriving in Clermont-Ferrand already knew of local citizens to consult and to engage as companions in exploration. Perhaps foremost among these was the Clermontois apothecary and naturalist Jean-Baptiste Mossier. Two generations of visitors relied on a Mossier, since Jean-Baptiste-Amable Mossier (the son, and a physician) carried on the tradition established by the senior Mossier; he was recommended by Vauquelin to the Swedish chemist J. J. Berzelius, during his tour of the Massif Central in 1819 (Bernhard 1989, p. 69). This inventory includes characters like the Mossiers, even though it is questionable whether their excursions from town into the neighbouring country qualify as travel. So, mindful of the difference between such local figures as Jean-François Ozy, the Mossiers or Montlosier, on one hand, and non-Auvergnats who ventured into Auvergne from outside the province, on the other, there are two distinct rosters. There may be disagreement over which roster is most appropriate for a particular individual. However, the chief objective is that the two lists should provide as comprehensive an inventory as available information permits. The risk of omission is probably greater in the case of locals, on account of the ephemerality (or non-existence) of records bearing on the activities of many provincials, especially those affected by 'papyrophobia'.[17]

This does not exhaust all the perplexities. If the Mossiers belong in a category for local naturalists of special competence, there were numerous other Auvergnats who offered hospitality to visitors, and guided or accompanied them in their excursions. Where evidence exists of their participation in observation related to geology, should they not be noted in an effort at a comprehensive tabulation? (Fouquet du Lombois and Chabrol, local field companions of Saussure in 1776, are listed here on these grounds (Carozzi 2000, pp. 189, 211)). Further, in certain cases of outside visitors who accomplished their travels with companions, some of these merit inclusion. Uninvolved family members, servants, or porters need not be placed in the inventory, but artists who made visual records of Auvergne natural features are among those who should be. Michael Shanahan, in the entourage of F. A. Hervey, Bishop of Derry, during some of his French travels, although not a trained artist (he was Hervey's architect-builder) evidently made drawings of basaltic formations, some of which were sent to John Strange. Two of the Auvergne drawings by J.-J. de Boissieu, who was commissioned to take part in the 1760s mapping project of Desmarest and François Pasumot, yielded engravings published in the collection of plates on the *règne minéral* in the *Encyclopédie*.[18] (Fig. 2 is perhaps one of the other results of Boissieu's work in conjunction with that mapping project.)

A final complication affects the chronological boundaries of the half-century chosen for this inquiry. These boundaries have been stretched slightly, in both directions. The inventory includes L.-G. Le Monnier's observations, made a dozen years before the pivotal journey by Guettard and Malesherbes in 1751, in part because of the frequency of references to them in the literature later in the century. At the other end, with several important publications (along with others not so evidently significant) appearing shortly after 1800, certain characters appear in the tabulation by waiving the closing date.[19]

Some concluding remarks

The accompanying tabulation of travellers in Auvergne represents an effort to digest information that could be used for a history of geological travel in that province. Any thorough analysis of this inventory (or of a still more comprehensive body of information for which this may possibly be used) remains to be done, and would require far more space than is available here. The following are a few impressions gained from an initial consideration of the information gathered.

Practical motivations

Among the motivations for scientific travel in Auvergne in the eighteenth century, some were practical. Although not the richest of the French provinces for mineral extraction (Alsace and Brittany come to mind as among the leading regions in

Fig. 2. View of a Rock Composed of Articulated Prisms. 1760s. Drawing by Jean-Jacques de Boissieu (1736–1810), engraving by François Godefroy (1743?–1819). Bibliothèque Nationale de France, Paris. Cabinet des Estampes. The artist Boissieu—who, along with Nicolas Desmarest, accompanied the Duc L.-A. de LaRochefoucauld (1743–1792) on his journey through Italy in 1765–1766—made illustrations of geological features in Auvergne in 1766, during the mapping project undertaken by Desmarest with the cartographer François Pasumot (see Perez & Pinault 1985–1986). Segments or 'drums' of articulated basalt prisms were sought by collectors; some brought to Paris from Auvergne in the 1760s may be seen to this day at the Muséum National d'Histoire Naturelle. Two different illustrations of prismatic basalt by Boissieu, engraved by Robert Benard, were published among the plates of the *Encyclopédie* as accompaniment to the article in which Desmarest first published his argument that these basalts are volcanic in origin (Desmarest 1768, plates VII & VIII).

eighteenth-century mining activity), Auvergne was recognized for its mineral resources. Besides the renown of some of its mineral waters, there were notable mines, for lead and antimony for instance. Contemporary digests of information on French mineral resources take note of Auvergne as a place of at least average interest.[20] In addition, at the time of his 1751 foray into Auvergne, Guettard was concerned with developing his project of a mineralogical survey of France, an enterprise manifesting contemporary belief in scientific knowledge's utility (Guettard & Monnet 1780).[21] Naturalists of a certain theoretical-mindedness did not hesitate to relate their inquiries to practical considerations; Desmarest, for example, whose experience included reporting on mining as inspector of manufactures for the neighbouring *généralité* of Limoges, took pains to report on Auvergne as a possible resource for useful materials, particularly *pozzolane* as a natural cement for construction.[22] Such practical-mindedness, consistent with the age's progressive scientific credo, was expected by the governmental patrons whose sponsorship lay behind some scientific journeys into Auvergne. As geological travellers began to visit Auvergne in greater numbers, a distinct proportion of them reported on practical knowledge of Auvergne resources. Some of these reports reflected an immediate impact in French science on the part of the École des Mines, created only a few years before the Revolution. Throughout the second half of the eighteenth century, the record of scientific investigation in Auvergne suggests that practical utility was a leading reason for seeking geological knowledge; that is, a sense of opposition or tension between theoretical knowledge and its possible uses was largely absent.

Interactions between visitors and locals

The historical fame of several of Auvergne's eighteenth-century geological visitors (Guettard, Desmarest, Saussure, Déodat de Dolomieu) has sometimes obscured an evident interaction, or interdependence, between scientific visitors and local figures such as Ozy and the Mossiers, father and son. Behind these few comparatively prominent local naturalists were a larger number of people with more modest abilities or reputations. These were individuals who in most cases had no need to be induced by visitors to cultivate knowledge of local features, and who had assembled their own collections of minerals, fossils, plants and the like.[23] Many a visitor depended on some of these men as guides and informants, also as suppliers of specimens for their collections. In return, the fact that travellers from outside, from the time of Guettard and Malesherbes onward, had initiated recognition of Auvergne's special significance as a seat of past volcanism was a source of stimulus to many Auvergnats in their pursuit of natural historical knowledge of their province. Ellenberger has shown how some of the more highly educated of Clermont's scientifically-informed intelligentsia were motivated by Guettard's 1751 discovery to develop their own sense of Auvergne's peculiar state as a museum of extinct volcanoes (Ellenberger 1978, 1994).

A cradle of geological fieldwork

Auvergne's significance as a locality during this period has much to do with field investigation, that is, with geological observations made directly at pertinent sites. Prominent issues addressed through field investigations were related to the Auvergne's rising geological profile: the discovery of extinct volcanoes in the heart of France, the recognition of columnar basalt's associations with the products of volcanic action, the mapping of the volcanic terrain, and the interpretation of these volcanic productions in geohistorical terms (a consequential step in the development of historical geology) were all based on observations made in the field. (Fig. 3 is perhaps the most conspicuous cartographic result of such systematic observation.) But did these activities exemplify fieldwork? Such a question implies a premise: that geological fieldwork (a term not then actually in use) is not an essential concomitant of all geological thinking, but has its own historical development. A swift overview of the written residue of Auvergne field observation shows that the conventions of geological fieldwork were, until the end of the eighteenth century, far from clearly defined. Relatively little survives in the way of 'field notebooks,' and some of these more nearly resemble travel itineraries than systematic records of observations. Investigators in Auvergne evidently tended to respect the idea that their general assertions should be supported (if not indeed instigated) by observations, but geological science was still in the process of evolving consensus on what features were most worth observing. A historian wishing to trace the pathways by which geological fieldwork achieved a firmer definition than simply 'observing in the field', and through which it became embedded in conventional views of how geology should be carried out, could make worse choices than to focus on how investigators in Auvergne developed their field practices.

Place-specificity, vis-à-vis *general geological problems*

A final point, where I allow my deeper motivations to surface, has to do with the degree of emphasis on place-specificity in the geological reports of Auvergne, as distinct from emphasis on organization of data in direct relation to general or theoretical issues. The observations of geological authors who visited Auvergne before 1800 appear to have been animated by an interest in one or more of a few predominant questions: among these were the origin of basalt, the causes of the prismatic forms in which basalt is found, the character of the presumed source materials of volcanic productions (which were widely thought to be altered or 'denatured' forms of such materials) and the causes of volcanic action itself. A quick and incomplete appraisal of the overall character of reporting on Auvergne geology up to about 1800, suggests that most of the observers hoped to see immediate connections between the phenomena they observed and one or more of these large, broadly causal issues. Relatively few seem to have seen their mission in terms of painstaking descriptions of geographically unique features, representative of Auvergne as a geologically individual place rather than of conveniently accessible phenomena illustrative of some problem's explanatory resolution. Perhaps it was this very sense of accessibility, in Auvergne, to phenomena that seemed to offer the promise of resolving key causal or dynamic questions that chiefly fuelled its growing status as one of geology's iconic places. If so, it suggests a paradox: the peculiar attractions of this specific locality, geologically speaking, may then have rested more on their immediate relevance to conceptual generalizations than on belief in the intrinsic scientific value of geological place-specificity.

I thank R. Rappaport for a number of valuable comments and suggestions. My thanks also to J. Ehrard for his assistance in identifying several published sources and to J. Gaudant for aid on some points of fact. I am grateful to K. Magruder for technical assistance. Research for this essay was supported by the University of Oklahoma's Research Council and Vice President for Research, and by a College of Arts and Sciences grant. Grateful acknowledgement is made as well, for generous support of special costs in publishing Figures 1 and 3, to the Vice President for Research and the College of Arts and Sciences.

Notes

[1]Desmarest, Nicolas, letter to H.-B. de Saussure, 26 December 1776. In Carozzi (2000, at pp. 198–199). [Translation: 'You will have been able to notice that Auvergne is infinitely more favourable than the environs of Naples or Rome for study of fire's operations, and that an infinitude of altered effects can be noted there. In Italy one does not encounter this correspondence which is so instructive for drawing general conclusions.']

[2]Buch, Leopold von, letter to Marc-Auguste Pictet, 2 July 1802. In von Buch (1802, p. 306). [Translation: 'What a pleasure to see you prepared to visit these regions, which in Europe certainly have no equal. Do you want to see volcanoes? Choose Clermont in preference to Vesuvius and Etna. In these latter mountains, a subsequent eruption buries the productions of the one preceding. But at Clermont the lavas, the enormous flows, are exposed from their exit from the mountain's flank to the plain where they have come to a stop; all the details are disclosed, and their nature is unambiguous . . . .']

[3]Lyell (1855, p. 127; as quoted in Scrope 1858 (reprint 1978), p. 213).

[4]For discussion of 'pilgrimage' as an aspect of geological travel, see Rudwick (1996). Rudwick (1974) discusses aspects of the growing cognizance of

Fig. 3. The Pasumot-Desmarest volcano-geomorphological map of part of Auvergne, 1774. *Map of a part of Auvergne, representing the lava flows, where basalt is found in prismatic and round forms, &c., for use in understanding the memoir by M. Desmarest on this basalt. Drawn by MM. Pasumot and Dailley, Royal Geographical Engineers.* Engraved by Guillaume De la Haye. The scale is approximately 1: 130 000. The line border measures 330 × 343 mm. (Map facsimile from an image copyright History of Science Collections, University of Oklahoma Libraries. Special thanks for assistance to Kerry V. Magruder, Collections Librarian). One of the most remarkable results of eighteenth-century scientific travel in Auvergne is this landmark effort at cartographic representation of geological phenomena. The map is the result of a collaboration between Nicolas Desmarest, whose first journey into Auvergne in 1763 showed him that the distinctively-shaped basalts must be of volcanic origin, and the cartographic engineer François Pasumot, assisted by his cartographer colleague Dailley (Taylor 1994). The map shows distribution of volcanic products (including instances of figured basalts) in the Mont-Dore district, and in other selected portions of the region near the city of Clermont (now Clermont-Ferrand). It also reflects Desmarest's recognition that those volcanic features should be understood in terms of sequential periods of volcanic eruption, and of continual alteration of the volcanic productions by erosion. Desmarest secured the support of the Royal government for this mapping project during the winter following his 1763 journey, mainly on grounds of utility in promoting knowledge about mineral resources. The bulk of the surveying was done 1764–1766. The map, reproduced here actual size, was published in 1774 accompanying a lengthy paper that Desmarest had presented in 1771 to the Royal Academy of Sciences on his investigations of basalt in Auvergne, especially basalts in the prismatic (columnar) and spherical forms that particularly caught his attention. (The publishing schedule for the Academy's annual volumes of *Histoire & Mémoires* was typically a few years in arrears.) This map surely ranks as an outstanding example of early efforts at visual representation of geological phenomena discerned through systematic fieldwork. Although the focus of Desmarest's 1771 presentation was largely descriptive and causal, on the character and volcanic origins of columnar basalt, and not on the physiography's historical features, one of the map's leading features relates directly to the historically-informed conception of the volcanic landscape about which he would soon speak. Older lava flows (*anciens courants*) are carefully distinguished from newer ones (*courants modernes*), and both of these from 'masses of old lavas melted in place', all displayed so as to show not only their distribution but also, through subtle hachuring, their topographic relations. In the closing pages of the memoir to which this 1774 map is attached (Desmarest 1774, pp. 768–775) Desmarest provided a description and explanation in which he actually distinguished four types or classes of volcanic products, instead of the three indicated in the map. Not long after this map's appearance Desmarest presented, in a showcased lecture for the Academy's annual autumn public assembly in 1775, an interpretation of the Auvergne physiography in terms of three distinct periods (*époques*) of eruption and continual erosional destruction. A summary of this talk was published (without a map) in 1779 (Desmarest 1779*a*). It was not until 1806 that Desmarest published an expanded version of this research, based on a new presentation to the Academy in 1804, with maps redrawn at a scale of approximately 1:43 000 (Desmarest 1806). Intermittent work on mapping the Auvergne volcanic terrain was a lifelong preoccupation for Desmarest from its beginnings in 1764. The final outcome of this work, still unpublished when Desmarest died in 1815, appeared in 1823 under the supervision of his son Anselme-Gaëtan Desmarest (Desmarest 1823).Historical analysts such as Martin Rudwick (1976, 2005) and François Ellenberger (1983, 1994) have noted that while the 1774 Pasumot-Desmarest map reproduced here differs in conception from what began to emerge a generation later as the classic 'geological map,' this ingeniously innovative effort was highly successful in its own way in conveying visually, with much efficiency, a great amount of information. I adopt here Ellenberger's suggestion that this be identified as a *volcano-geomorphological* map, a scientifically thematic map that is one of a kind.

Auvergne among British geologists (Daubeny, Scrope, and Lyell).

[5]A proper study of Auvergne's elevation to status as a geologically privileged locality would presumably need to draw on interpretive tools for 'place-sanctification', to be found perhaps in the literatures of cultural geography or historical sociology. I thank Bret Wallach and Robert Rundstrom, of the University of Oklahoma Department of Geography, for their suggestions on this topic; Bob Rundstrom offered the term 'iconic place', which seems to me to fit Auvergne's historical status in geology.

[6]Article 'Auvergne' (Diderot 1751). The stated dimensions of 40 by 30 leagues—or about 120 by 90 miles—seem slightly on the generous side. Toward the end of the century P.-J.-B. Legrand d'Aussy, in his book of travels in Auvergne, gave the dimensions as between 18 and 20 leagues on average east-to-west, and 40 in length north-to-south, with a total area of 651 square leagues (Legrand d'Aussy 1794/1795, **1**, 73–76).

[7]An estimate between 553 and 555 thousand is given in a 1763 almanac ([Chavagnat] 1763, p. xvi). However, according to a 1784 report for the royal administration (with the same area figure of 651 square leagues), the Généralité de Riom had a population at that time of over 681 000. See Bonin & Langlois (1989, **5**, p. 46).

[8]The political and administrative distinction between Lower and Upper Auvergne, honored over many centuries, might occasionally be challenged by an alternative distinction based on physiography. So Nicolas Desmarest, for instance, suggested in his compendium *Géographie physique* (part of the series *Encyclopédie Méthodique*), that Lower Auvergne ought naturally to be considered equivalent to the Limagne or Allier valley, thus putting all of Auvergne's mountainous regions—including the *chaîne des puys* and the Mont-Dore—in Upper Auvergne (article 'Auvergne' in Desmarest 1794–1828, II [an XII/1803], p. 878b). This suggestion does not seem to have been much followed.

[9]A useful introduction to travel in France during the last years of the ancien régime is Lough, 1987. Lough states (p. 37): 'Rare indeed were the travellers who ventured into Auvergne.' The systems of measures are treated in Alder (2002).

[10]*Calendrier d'Auvergne* [Chavagnat] (1763, p. xvii).

[11]Bonin & Langlois (1987, **1**). Diligences et Messageries Royales, pp. 46–49; postal distribution, pp. 38–41; Allier navigability, p. 25. The map on p. 48, which graphically represents the intensity of public carriage traffic throughout France in 1789, vividly conveys Auvergne's remoteness from generally frequented routes.

[12]A thorough study of eighteenth-century roadways in Auvergne is Imberdis (1967).

[13]See Gillispie (1980); also Parker (1965). An Intendant might direct subdelegates of the *Généralité* to offer assistance, for example by providing itinerant scientific observers protection from suspicious inhabitants or by furnishing lodging. Such protections were offered members of the Desmarest-Pasumot group involved in mapping part of Auvergne in the 1760s. See Taylor (1994).

[14]The Auvergne sheets can be reviewed on the internet: http://www.ifrance.com/parbelle/cartographie/auvergne.html

[15]Volcanic phenomena in southern Languedoc were reported during the eighteenth century by (among others) Montet (1766), Joubert (1779), Bertrand (1780), Joinville (1788), and Barbaroux (1788–1789). One of the observers of Forez was Passinges (1797–1798).

[16]Guettard refers for example, in his 1759 paper on Auvergne mineralogy, to analyses of the Vichy mineral waters, by Chomel [Pierre-Jean-Baptiste, 1671–1740, a physician, member of the Academy of Sciences; or perhaps his son Jean-Baptiste-Louis, 1709–1765, also a physician].

[17]The term 'papyrophobia' is borrowed from Hugh Torrens. It refers to the condition of men (and women) whose work is not easily or naturally recorded in writing. See Torrens (2002); also Torrens (1995).

[18]For Hervey see Childe-Pemberton (1925, pp. 102 ff.). (We are told, p. 103, that on other occasions Hervey employed an accomplished artist.) Boissieu (1768).

[19]An inspection of the inventory will readily show my particular reliance on the remarkable study by Vernière (1899–1900), which was based in part on unpublished sources.

[20]Observations made in 1739 by Louis-Guillaume Le Monnier (called *le Médecin* to distinguish him from his older brother Pierre-Charles Le Monnier, the astronomer), in Le Monnier 1744, republished in Gobet's 1779 compendium. Also Quériau (1748: Auvergne mineral resources discussed, pp. 24–25); the digest by Jean Hellot, in Schlüter (1750; Auvergne treated pp. 60–62); Anon. (1794); Lefebvre d'Hellancourt (1802).

[21]On Guettard, Rappaport (1972) is still the best starting-point.

[22]Desmarest's 1765 report on mining in Limousin, in Gobet (Desmarest 1779*b*). For Desmarest on *pozzolane* see Desmarest (1779*c*).

[23]Valuable information on local Auvergne collectors can be found in Paul-François Lacoste's early-nineteenth-century publications on the Auvergne volcanoes (Lacoste 1803, 1805).

Appendix

Scientific travellers in Auvergne, 1751–1800 (+): a chronological inventory, with Bibliography

Table A1. *Travellers to Auvergne from outside the province (in approximate chronological order of travel to Auvergne)*

Le Monnier, Louis-Guillaume (1717–1799) [called *Le Monnier le Médecin*]	Member of AdS 1743. His 1739 account of Auvergne natural history, including mines, in Cassini de Thury (geodesic expedition) (Le Monnier 1744*a*). Made trip up PD & to MD with Ozy. V says he made a subsequent botanical excursion in 1753. [V; DB; DSB; I; Ozy in Faujas]
Bowles, William (1705–1780) [alternate birth date given by Reynolds (1997): 1720.]	Irish-born, settled Paris 1740. Appointed in 1752 as Spanish state mines superintendent. Possibly passed through Auvergne in 1750 (as Ozy claimed). His book on the natural history of Spain refers to observation of basaltic columns in Auvergne, at Usson, near Issoire (Bowles 1775, 1776). [DB; FE; Reynolds 1997]
Guettard, Jean-Étienne (1715–1786)	Member of AdS 1743. Pioneer of mineralogical mapping. Auvergne journey with Malesherbes, 1751; 1752 report in AdS *Mémoires* [1756]. Return journey to Auvergne in 1777 (FE, 35) in company of Joubert. Travel in Vivarais as well, with Faujas, in 1770s. [V; DB; FE; DSB]
Malesherbes, Chrétien Guillaume Lamoignon de (1721–1794)	*Honoraire* in AdS 1750. Guettard's companion, 1751. Reforming minister, *directeur de la librairie*; avid naturalist. Lawyer for Louis XVI in Revolution. Letter (1769) to J.-F. Seguier on volcanoes (Bibliothèque Municipale de Nîmes, MS 417). Account of 1751 Auvergne journey published 1779. [V; Rigodon 1963; many biographical notices]
Buc'hoz, Pierre-Joseph (1731–1807)	Medically-trained, hyper-prolific author-compiler of scientific works. Published about two dozen small works on Auvergne mineralogical/geological subjects—mountains and other topographic features, mineral waters, springs, etc.; all undated. To what extent these were based on direct observation — and if so, when — is not clear. V says (17:124) he made his first botanical excursion in Auvergne in 1748, another in 1791(17:254). [V; Q]
Commerson, Philibert (1727–1773)	Physician and botanist, made botanizing tours in central France; later (1767–1769) was naturalist on Bougainville's expedition. V says he began repeated excursions in Auvergne in mid-1750s (cf. Lalande 1775, Cap 1861, Oliver 1909). [V; DSB; DBF]
Richard, Antoine (dates unknown)	Royal gardener. V states that he did the same as Commerson (i.e. herborized in Auvergne), not long after.
Montigny, Étienne-Mignot de (1714–1782)	Colleague of Guettard at AdS (member 1740). Auvergne visit prior to 1765; brought specimens to Guettard. [V; FE; Guettard 1759]
La Galissonière, Roland-Michel Barrin, Marquis de (1693–1756)	*Associé libre* of AdS 1752. Comte de La Galissonière was identified by Guettard (1759) as a supplier of information on Auvergne mineralogy (a role generically confirmed in *éloge* by Fouchy in *Histoire* of AdS, 1756, p. 151). [V; I; Groulx 1970]
Desmarest, Nicolas (1725–1815)	Inspector of manufactures from 1757. Member of AdS 1771. Travels in Auvergne from 1763. First to identify basalt as volcanic product; mapping project with Pasumot; analysis of *époques* or sequences in Auvergne volcanoes. [DSB; Taylor 1994]
Pasumot, François (1733–1804)	Geographical engineer, naturalist, antiquarian. Beginning 1764, with fellow *ingénieur-géographe* Dailley [or Dalier, forename & dates not known], prepared volcano-geomorphological map in collaboration with Desmarest (1771). [Taylor 1994, Grivaud 1810–1813]
Boissieu, Jean-Jacques (1736–1810)	Artist; sometime travel companion of Desmarest (1765–1766 in Italy, in entourage of Duc L.-A. de La Rochefoucauld); engaged 1766 to sketch Auvergne features (cf. Pasumot letters to Desmarest, 1766, Bibliothèque Municipale de Beaune, MS 304, nos. 15–16; letter from Desmarest to Duc de La Rochefoucauld, 1766, Bibliothèque Georges Duhamel, Mantes, Collection Clerc de Landresse, no. 1170). Two of his Auvergne drawings engraved & published in *Encyclopédie*. [Taylor 1994, 1995; Perez 1994; Perez & Pinault 1985–1986]

(*Continued*)

Table A1. *Continued*

Monnet, Antoine-Grimoald (1734–1817)	Originally from Champeix in Auvergne. Inspector-general of mines. Multiple Auvergne geological travels over several decades, starting as early as 1763 according to V. His *chanoine* brother, of Vic-le-Comte, was an enthusiastic mineralogical & geological observer; they sometimes went on field trips together. [DSB; much MSS material in Bibliothèque, École Nationale Supérieure des Mines]
Jars, Antoine-Gabriel (1732–1769) [also called *Jars le jeune*]	Mining engineer, *Correspondant* of AdS 1761, member 1768. Died 1769 in Clermont, on government-mandated survey of factory operations in central France. V says he got sunstroke while drawing basaltic colonnade at Saint-Arcons-d'Allier. Archives Départementales of PD. [BE; I; Garçon 2002]
Hervey, Frederick Augustus, Bishop of Derry (1730–1803)	Irish traveller in Auvergne 1770, Velay 1772. Interested in Desmarest's discoveries; he had discovered extensions of the Giant's Causeway near his Ireland home. Accompanied by son, artist(s) in his retinue, had drawings of basalt columns made, sent to his friend Strange. [Childe-Pemberton [1925]; DB]
Shanahan, Michael (dates unknown)	Hervey's architect, 'for many years the Bishop's builder and factotum' (Childe-Pemberton, **1**, p. 102), who evidently served also as tutor to his eldest son John Augustus during their travels. Some of Shanahan's 1770 drawings of basaltic formations were sent to John Strange. [Rankin 1972]
Bosc d'Antic, Paul (1726–1784)	*Correspondant* of AdS 1759. Physician, chemist, specialist on manufacturing processes (e.g. glassmaking). Lived several years in Auvergne, became member of SLC; established short-lived *école des arts et métiers* in Auvergne *c.* 1770. Some observations on Auvergne mineral resources appear in his *Oeuvres*. [V; I]
Varenne de Béost, Claude-Marc-Antoine (1722–1788) [or *Béost de Varenne*]	*Correspondant* of AdS 1752. Strange (1775, 10) refers to his observations of basalt columns in Auvergne, as reported by Sage 1772 (cf. 2nd edition, 1777, **1**, pp. 212–213). [Q; I]
Strange, John (1732–1799)	British diplomat, travelling virtuoso. In Auvergne 1773; also Velay, Lionnais. Comparisons with North Italian localities (Strange 1775). Ciancio says Strange journeyed in France at least 3 times, in 1765, 1773, and 1786–1787. [DB; FE; Ciancio 1995*a*, *b*; De Beer 1951–1952]
Wraxall, Nathaniel William (1751–1831)	Adventurous English traveller, writer. Visited Auvergne 1776. [Lough 1987, p. 317]
Saussure, Horace-Bénédict de (1740–1799)	Renowned Genevese naturalist, author of *Voyages dans les Alpes*. AdS *correspondant* (1787), *associé étranger* (1791). In Auvergne 1776, also Vivarais, with Faujas; Auvergne again 1795. [DSB; Carozzi 2000]
Aublet, Jean-Baptiste-Christian (1723–1778) [also known as *J.-B.-C. Fusée-Aublet*]	Traveller and botanist. Employed by Compagnie des Indes. V says he 'herborized, under the guidance of abbé Delarbre' in May 1776, in the vicinity of le Pont-des-Eaux and MD. [V; DBF; Q]
Mortesagne, abbé de (forename, dates unknown)	For a time resident at Pradelles, Velay. Letters in Faujas, 1778.
Faujas de Saint-Fond, Berthélemy (1741–1819)	Trained in law, with Buffon's patronage became attached to Jardin du Roi, later professor of geology at Muséum. Early supporter of study of volcanic productions of Massif Central (*Recherches*, 1778). Associated with Guettard; but Guettard broke with Faujas, enraged by his 1778 publication of Ozy's claim to priority in discovery of Auvergne volancoes. Faujas had strained relations with Desmarest. V says he did visit Auvergne (17:248, 260). [DSB]
Joubert, Philippe-Laurent de (1729–1792)	Montpellier academician, *correspondant* of AdS 1760. 1777 journey to Auvergne with Guettard. In account of Languedoc volcanic remnant (1779), said that he twice visited the volcanoes of Velay and Vivarais. [I]
Genssane, Antoine (or Antoine-François de) (died 1780) [sometimes *Gensanne*]	Mining engineer; *correspondant* of AdS 1757. Ellenberger reports his observations in Vivarais and Velay, 1777 (1994, p. 231). [BE; V; I; Q]
Adanson, Michel (1727–1806)	Noted botanist, African explorer. AdS *correspondant* 1750, member 1759. 1779 journeys included a soujourn in Auvergne, according to V in the company of Delarbre (presumably the botanist, Antoine Delarbre). [V; Lawrence 1963–1964; Chevalier 1934; Nicolas 1963; DSB; I]

(*Continued*)

Table A1. *Continued*

Lamarck, Jean-Baptiste-P.-A. de Monet, Chevalier de (1744–1829)	Member of AdS 1779. Malvezin reports he made a botanical excursion to MD and Cantal in 1779. Made return visit soon after 1800. [DSB; V; I; many biographical notices, studies]
Soulavie, Jean-Louis Giraud, abbé (1752–1813) [also called *Giraud Soulavie*]	Explored Vivarais and Velay in 1770s; ventured into Auvergne 1780 (Soulavie, 1780–1784, **3**, p. 273 ff.; also 1783). [DSB; V]
Lamanon, Robert de Paul, Chevalier de (1752–1787)	Naturalist from Salon en Provence, *correspondant* of AdS 1752. Lost on LaPérouse expedition. MS notation of his plan to make a trip to see the extinct volcanoes of Vivarais, Easter fortnight 1779 (BNF, MS f.f. 9138). Not clear whether this was accomplished. Lamanon credited by Desmarest with first recognizing freshwater formations in Auvergne (*Géographie Physique*, 'Lac'). [I; Pichard 1992; Cartwright 1997]
Simiane, Marquis de (forenames, dates unknown)	Soulavie (1783) identifies Simiane as an observer who, following long study in Auvergne, had large lava blocks transported to Paris, for examination by Cadet, Soulavie, and Marivetz.
Townson, Robert (1762–1827)	English traveller, geological author. During 1780s was in Vivarais (*Tracts & Observations* 1799, p. 149); not known if he was ever in Auvergne. [Torrens 1999]
Deluc, Jean-André (1727–1817) [sometimes *De Luc*]	Genevese naturalist-philosopher, relocated in England in 1770s. *Correspondant* AdS 1768. Brother G. A. Deluc wrote that J. A. Deluc had travelled and collected among the Auvergne volcanoes (G.-A. Deluc 1804). [DSB; V; I]
Bertrand, Philippe (also *P.-M.*) (1730–1811)	*Inspecteur général des ponts et chaussées*, author of mineralogical and geological writings, including critique of Buffon's *Époques*. Refers (Bertrand 1799) to his former repeated observations of French regions of extinct volcanoes, specifically including PD. [Q; DBF]
Besson, Alexandre-Charles (according to BE, 1725–1809; Girdlestone (1968) gives no forename, indicates born *c.* 1742)	Mining inspector, alpine travel author. Birembaut refers to him as 'nestor' (age 58) of the École des mines' class of 1783. Delarbre (1787) mentions having been at MD with him. Publications on Auvergne geology (1787, 1796). [BE; Q; Girdlestone 1968; V; Béraldi 1913, p. 151 ff.; 1917, p. 232 ff.]
Lelièvre, Claude-Hugues (1752–1835)	Mining inspector, elected to Institut 1795. G. Delarbre (1787) states Lelièvre accompanied him in mineralogical travel in Auvergne (1785) when Lelièvre was adjoint to Besson. [BE; V; I; Q]
Ramond, Louis-François (1755–1827) [also called *Ramond de Carbonnières*]	Member of Institut 1796. Claimed 1786 journey through parts of Auvergne and Velay as part of qualifications for Institut membership (Girdlestone, 312); accompanied Cardinal de Rohan as his personal secretary (V, 17:251). Named *préfet* of the *département* of PD in 1806, carried out extensive barometric altitude measurements, as well as geological observations. [DSB; V; Boyer 1930]
Legrand d'Aussy, Pierre-Jean-Baptiste (1737–1800)	Paris librarian, academician, historical writer. Books on Auvergne (1788, 1794/1795). These are popular accounts of Auvergne with strong emphasis on its attractive geographical and natural-historical features. [Q]
Baudot, Pierre-Louis (1760–1816)	Dijon antiquarian scholar, *avocat*, *substitut du procureur-général* of Bourgogne Parlement, later member of Conseil-général du Département de la Côte-d'Or, *correspondant* of Académie de Dijon. Author of 1789 MS 'Essai sur la poix minérale d'Auvergne' (24 leaves, Wellcome Institute, MS 1107). [Q]
Bournon, Jacques-Louis, comte de (1751–1825)	Mineralogist, soldier, émigré during Revolution. During prolonged sojourn in London was elected Fellow of Royal Society, was co-founder of Geological Society of London. Travelled in Forez, Auvergne, Velay, Vivarais in 1780s, notably in 1788. [DSB; V]
Young, Arthur (1741–1820)	Renowned English traveller, observer of agricultural practices. Visited Auvergne and Velay, August 1789. [V]
Duhamel, Jean-Baptiste Guillot (born 1767) [also called *Guillot-Duhamel*; *Duhamel fils*]	Inspector of mines. Son of J.-P.-F. Guillot-Duhamel (1730–1816), who was member of AdS and the Institut, and also mining inspector. Report on coal and lead extraction in Auvergne (1795). [Q; BE]

(*Continued*)

Table A1. *Continued*

Dolomieu, Déodat (1750–1801)	A leading geological traveller of late 18th century. AdS *correspondant* 1778, member of Institut 1795, author of many reports of volcanic observations. Journey in Auvergne 1797 (Dolomieu 1798). V thinks he was in Auvergne once before, in 1780 (17:256). Monnet (MS 31, École des Mines) wrote that he urged Dolomieu to see Auvergne. Dolomieu's chief focus was apparently on analysing volcanic products for purposes of fixing the nature and causes of volcanic action. [DSB; Lacroix 1921]
Fourcroy, Antoine-François de (1755–1809); **Vauquelin, Nicolas-Louis** (1763–1829); **Collet-Descotils, Hippolyte-Victor** (1773–1815)	*Fourcroy*: chemist, member of AdS 1785, Institut 1795. *Vauquelin*: chemist, student and then collaborator of Fourcroy, inspector of mines from 1794, member of Institut 1795. *Collet-Descotils:* student of Vauquelin, entered École des Mines 1794, where he eventually succeeded Vauquelin as professor, served as laboratory director. The three travelled in Auvergne in July and August 1797. (Possible motivation: to consider chemical theories of volcanic action. V says they were called to examine mineral deposits near Pontgibaud.) Letters by Fourcroy survive describing the tour (Bibliothèque Municipale de Rouen). In first decade of 19th century, Vauquelin (in collaboration with Fourcroy, also alone) published several papers on analyses of mineral productions found in Auvergne. [Smeaton 1962; Kersaint 1966; DSB]
Muthuon, Jacques-Marie (1757–1830)	Mining engineer. Said he was 'born in the middle of volcanic country' (1798), apparently meaning Velay (V, 17:256); refers (1799) to repeated collecting of volcanic products, his trajectory across the *départements* of Aveyron, Lozère, and Haute-Loire. [BE; Q]
Giroud, Alexandre (1761–1797)	Mining engineer; non-resident associate of Institut 1796. (Uncertain that he travelled to Auvergne; 1795 paper may report assay done on specimen sent to Paris.) [BE; I; Q]
Laverrière, Louis (1754–1816)	Mining engineer. Report from Auvergne 1797. [BE]
Bosc d'Antic, Louis-Augustin-Guillaume (1759–1828)	Member of Institut 1806, son of Paul Bosc d'Antic. V says he explored Auvergne, Velay, & Forez in Year IX (1800–1801). [V; I]
Breislak, Scipione (1750–1826)	Italian priest, naturalist (of German descent); specialized in volcanic study. V says he visited Auvergne summer 1801. [Cf. Breislak 1801, **1**, pp. 92, 135]
Jurine, Louis (1751–1819)	Genevese medical doctor and naturalist. Institut *associé non résidant* 1799. V says he visited Auvergne around 1801. [I]
Buch, Leopold von (1774–1853)	Freiberg-trained geologist. *Correspondant* of Institut 1815, *associé étranger* 1840. Visited Auvergne spring 1802 (Buch 1802, 1803). [DSB; I]
Cordier, Pierre-Louis-Antoine (1777–1861)	Ramond's nephew, Dolomieu's prize pupil at École des Mines, *correspondant* (1808) and member (1822) of Institut, professor at Muséum, president Conseil des Mines. Journey through Massif Central (Vivarais, Cévennes, Auvergne) in 1802 (cf. Cordier 1803). [DSB; I; Ellenberger 1979]
Lamétherie, Jean-Claude de (1743–1817) [frequently *Delamétherie*, sometimes *de la Métherie*, or *de La Métherie*]	Long-time editor of *Journal de physique*, author of works in geology, chemistry, natural history. Accompanied Cordier on journey through the Massif Central (Vivarais, Cévennes, Auvergne) in 1802 (Cordier 1803). [DSB]
Aubuisson de Voisins, Jean-François d' (1769–1841)	Freiberg-trained geologist, mining engineer. Institut *correspondant* 1821. Visited Auvergne summer 1803 (Aubuisson 1804b). (V says visited 1802.) [V; DSB; I]

[Some secondary accounts or sources are identified in square brackets. See reference list for full references.]

AdS, Académie Royale des Sciences; BE, A. Birembaut, Enseignement (1964); BMUCF, Bibliothèque Municipale et Universitaire de Clermont-Ferrand; BNF, Bibliothèque Nationale de France; DB, G. De Beer, Volcanoes (1962); DBF, *Dictionnaire de biographie française*; DSB, *Dictionary of Scientific Biography*; FE, F. Ellenberger, Précisions nouvelles (1978); I, *Institut de France. Index biographique de l'Académie des Sciences* (1979); MD, Mont-Dore; PD, Puy de Dôme; Q, J.-M. Quérard, *La France Littéraire* (1827–1839); SLC, Société Littéraire de Clermont-Ferrand; V, P. Vernière, Voyageurs et naturalistes (1899–1900).

Table A2. *Auvergnats involved in scientific observation in Auvergne (in approximate chronological order of observations)*

Ozy, Jean-François (*c.* 1711–1792) [sometimes *Osy*, *Ausi*, or *Auzi*]	Clermont chemist-apothecary, founder-member of SLC, to whom Guettard and Malesherbes carried a letter of introduction from Bernard de Jussieu in 1751. Accompanied Guettard & Malesherbes on excursions up PD & to MD. Ozy later challenged Guettard's priority. Well known to many Auvergne visitors, including Desmarest, Pasumot, Monnet, Saussure. [Multiple references to Ozy in Pasumot letters (1760s) to Desmarest: Bibliothèque Municipale de Beaune, MS 305, nos. 4, 5, 15, 20.] MSS in BMUCF on various of his *promenades*. [V; FE]
Garmage, abbé François (1718–1773) [also *Garmages*]	Chanoine & curé, St Pierre, Clermont. Founding member of SLC, before which he presented scientific and archeological discourses. President of Société d'agriculture upon 1761 founding. [DBF; V; FE, especially pp. 17–30; Mège 1884]
Grangier de Vedière, (forename & dates unknown) [V writes *Grangier de Verdières*]	Informant for Guettard, who identifies him as *Conseiller au présidial de Riom* (cf. Guettard 1755, 1759). Guettard reports his observations on tripoli quarry near Menat (West of Vichy). [V]
Guithon, (forename & dates unknown); **Alexis, R**[évérend] **P**[ère] (forename & dates unknown)	Guithon, curé of Fontanes, and Father Alexis, Capucin of Clermont, are identified by Guettard (1759) as among local observers to whom he was indebted; otherwise unidentified. [V, 17:130]
Mossier, Jean-Baptiste (1728–1809) [sometimes *Moussier*; *Mossier père*]	Chemist-apothecary at Clermont-Ferrand, from 1763. Member SLC. Often regarded as the leading local authority on Auvergne volcanoes. Testimonials by Dolomieu, Legrand d'Aussy, others. On 1776 arrival in Clermont, Saussure immediately looked up Mossier and Ozy. Published little (1804). MSS in BMUCF. Also MS 'Sur les montagnes du Montdor' 1775 (5 leaves, Wellcome Institute MS 1727/2). [Godon de St.-Memin 1804–1805; FE; Q; Carozzi 2000]
Dutour, Étienne-François (1711–1789) [also *Du Tour*, or *Dutour de Salvert*; Mège gives death date as 1784.]	Born in Riom. *Correspondant* (1746) of AdS. Provided extensive help to Guettard for his memoir on Auvergne mineralogy (1759). Acquainted with Desmarest and Pasumot (Bibliothèque Municipale de Beaune, MS 305, no. 12). Author of papers on physical questions in AdS *Savans étrangers*, also *Journal de physique*. A mentor to Romme (see below). [DBF; FE; J]
Dufraisse de Vernines, Joseph (dates unknown); **Ribauld de la Chapelle, Jacques** (dates unknown); **Pélissier de Féligonde, Michel** (1729–1767)	Members of SLC, mentioned by FE as at least occasional participants in excursions among the volcanoes. Pélissier de Féligonde was the 2nd Secretary of SLC. [DBF; FE; Mège 1884; Sauvade 1767]
Romme, Gilbert (1750–1795) [sometimes *Rome*]	Riom-born Jacobin *conventionnel*, member of the Comité d'instruction publique, advocate of the Republican calendar. [Sometimes confused with his brother Nicolas-Charles Romme (1745–1805), teacher of mathematics at a naval school in Rochefort-sur-Mer, and author of works on astronomy and navigation.] G. Romme educated in Oratorian collège of Riom. Before departing for Paris in 1774, was mentored in his scientific and mathematical studies by Dutour (above); in company of Dutour (and/or his son) and G. Delarbre (see below), made botanical and mineralogical/geological excursions. Brief return visits to Auvergne in 1786, 1788. [Galante Garrone 1971; Bouscayrol 1979; Julia 1996]
Chappel, Pierre (1738–1811)	Chemist-apothecary at Clermont-Ferrand. Member SLC, from 1777. Mège 1884 identifies him (along with Ozy & Mossier) as analyst of Auvergne mineral waters.
Mossier, Jean-Baptiste Amable (1767–1838) [*Mossier fils*]	Son of J.-B. Mossier. Physician of Clermont-Ferrand. Continued his father's practices, helping visitors. Accompanied Berzelius 1819. [Bernhard 1989; Carozzi 2000]

(*Continued*)

Table A2. *Continued*

Monnet, abbé, Chanoine at Vic-le-Comte (forename unknown; dates not known, but died between 1794 & 1800, according to Fric)	Canon at Vic-le-Comte, SE of Clermont; brother of A.-G. Monnet (with whom he sometimes did Auvergne excursions); mineralogical naturalist, member SLC. Corresponded with Guettard. Auvergne observations from at least as early as 1773. Author of many pieces (mostly unpublished MSS at BMUCF) on Auvergne volcanoes.
Fouquet du Lombois (forename, dates unknown); [or simply *M. Du Lombois*] **Chabrol** [perhaps Guillaume-Michel Chabrol, native of Riom, 1714–1792]	Two of the more prominently identified local field companions (other than Mossier and Ozy) named by Saussure in his notes of his 1776 Auvergne excursions. Carozzi (2000) identifies *Fouquet du Lombois* as 'a free-thinker nobleman of Theix' (191); V refers to him as 'un encyclopédiste convaincu' who corresponded with Saussure (17:247). Among shorter excursions, the two (Saussure & Fouquet du Lombois) were together on a trip of several days to MD. *Chabrol*, a Saussure companion for a day trip to Puy de Gravenoire, is identified as a lawyer. [Q] No indication that either was a scientifically serious observer.
Rangouse de La Bastide, Jean-Joseph-George de Leigonye, Comte de (dates unknown)	Author of 1784 work treating, in part, natural history of Auvergne. V reports that he opposed thesis of volcanization of Upper Auvergne mountains (maintained that the region is in its 'primitive state'). Argument over this with Vicomte de Sistrières-Murat; Pasumot sided with Sistrières in 1785 papers. [V]
Sistrières-Murat, François-Michel, Vicomte de (1730–1809) [also called *Desistrières*, & *Murat-Sistrières*]	According to V (17:241), in 1785 opposed anti-volcanic views of Rangouse de La Bastide (above), was supported by Pasumot. [V]
Brieude, Jean-Joseph de (1728–1812)	Aurillac physician. Published observations on thermal waters. His medical topography of Upper Auvergne (published posthumously) frequently cited. [DBF]
Montlosier, François Dominique de Reynaud, comte de (1755–1838)	Born & educated in Clermont. Deputy for nobility in Constituent Assembly; emigrated. Later retired to chateau among Auvergne puys. Prolific writer in political history (maintained that qualities of an archaeological historian are not unlike those of a geological naturalist). Long interested in geology & mineralogy (Montlosier, 1788, 1802); speaks of his youthful *courses lithologiques*. Fits Auvergne volcanoes into terrestrial natural history of vast time scale. [Many biogr. notices]
Delarbre, Antoine, abbé (1724–1807) (Q gives dates: 1724–1811) [the abbé; the botanist]	Physician & naturalist, long-time guide for Auvergne visitors (DBF); born in Clermont, returned there in 1749 after medical studies in Paris; practiced medicine at Pontgibaud; took orders, became curé at Royat 1771, then in cathedral at Clermont 1777; placed in charge of course at botanical garden in Clermont from 1781, thereafter was botanical guide for visiting scholars. V says he herborized in Auvergne from 1749. Member of SLC from 1771. A source of specimens for Guettard (1759); associated with Monnet, Saussure among others. Accompanied Fourcroy & Vauquelin on excursion to PD (according to DBF); through introduction by Broussonet, served as Arthur Young's guide in 1789 (DB). Author of natural history works on Auvergne. MSS in BMUCF. [DBF; V; Mège 1884; Q]
Delarbre, Guillaume (1754–*c.* 1789) [the mineralogist]	Evidently also a physician-naturalist. V (17:252–3) distinguishes between the *abbé [Antoine] Delarbre*, as the botanist, and *Guillaume Delarbre* the mineralogist. Legrand d'Aussy similarly indicates (1794/1795, III, 399) that the physician-naturalist was related to (thus not the same person as) the person known for his mastery of local botany. There is some difficulty as to distinguishing the work of the two Delarbres (conflated by certain sources). According to V, G. Delarbre was concerned mainly with mineralogy, author of two *mémoires* on Auvergne mineralogy. G. Delarbre (1787) refers to being at MD with Laizer and Besson, and on Auvergne journey with Lelièvre.

(*Continued*)

Table A2. *Continued*

Laizer, Louis-Gilbert, Marquis de (dates unknown) [on occasion *Lezer*, or *Lezers*]	Author of 1808 work on puy de Chopine; co-author of work on fossil pachyderm. References to his collections, and being in the field with Delarbre and Besson (G. Delarbre 1787); references by Lacoste 1805. V, 17:250 ff. [Q]
Cocq, (forename, dates unknown) [also *Lecocq*]	Cocq identified as *commissaire des poudres et salpêtres* of Clermont-Ferrand (Cocq 1805; Lacoste 1804; V, 17:258); possessor of a notable collection of volcanic productions of Auvergne. Collaborative observation with Mossier, Laizer (Cocq 1806). Regarded by Ramond (according to Boyer 1930, p. 44) as next best among local Auvergne observers after Mossier *père*. (V refers to him as 'Auvergnat d'adoption, originaire de la Lozère'.) [Q]
Ordinaire, Claude-Nicolas, abbé (1736–1809)	Chanoine, Riom; librarian, Clermont. Author of a natural history of Limagne (1787), and of *Histoire naturelle des volcans* (1802) for popular readership. (An English translation appeared in 1801, before publication of the French version!) This book's focus is on volcanoes' causes. Extent of his own observations not obvious. [Q]
Lacoste, Paul-François, abbé (*c.* 1754–1826) [also called *Lacoste de Plaisance*; his forename is mistakenly rendered as *Pierre-François* by, among other references & sources, the BNF catalog, J.-F. Michaud's *Biographie universelle*, and Q]	Professor of mineralogy at École Centrale of PD, then at Académie de Clermont. Natural history works published in Clermont from 1797; especially volcanic publications 1803, 1804; strongly drawn to Patrin's volcanological theory. An outspoken advocate for Auvergne's strong appeal for naturalists. Identifies many local individuals with whom he made excursions; names over a score of local owners of notable collections, in Clermont and other towns. Bakewell reported (1823, **2**, p. 342) that during his Auvergne travels, in the early 1820s, he had regular contact with Lacoste. [DBF; Q; Danton 1986; Gaudant 2003]
Grasset, le Chevalier [Grasset-Saint-Sauveur, Jacques (1757–1810, according to BNF catalogue)] [also *Grasset de Saint-Sauveur*]	V says (17:258) that Mossier 'parcourait les sommets et les vallées du Cantal avec un Auvergnat d'adoption, fort amateur de minéralogie, le chevalier Grasset, que les hasards de la vie militaire avaient amené à Mauriac'. [Whether this Chevalier Grasset is in fact Grasset-Saint-Sauveur is questionable; Grasset-Saint-Sauveur seems to have been a diplomat, as well as writer, but not a military man.]

[Some secondary accounts or sources are identified in square brackets. See reference list for full references.]
AdS, Académie Royale des Sciences; BE, A. Birembaut, Enseignement (1964); BMUCF, Bibliothèque Municipale et Universitaire de Clermont-Ferrand; BNF, Bibliothèque Nationale de France; DB, G. De Beer, Volcanoes (1962); DBF, *Dictionnaire de biographie française*; DSB, *Dictionary of Scientific Biography*; FE, F. Ellenberger, Précisions nouvelles (1978); I, *Institut de France. Index biographique de l'Académie des Sciences* (1979); MD, Mont-Dore; PD, Puy de Dôme; Q, J.-M. Quérard, *La France Littéraire* (1827–1839); SLC, Société Littéraire de Clermont-Ferrand; V, P. Vernière, Voyageurs et naturalistes (1899–1900).

References

ALDER, K. 2002. *The Measure of All Things*. Free Press, New York.

ANON, 1794. Apperçu de l'extraction et du commerce des substances minérales en France avant la Révolution. *Journal des mines*, **1**, (Vendémiaire An III), 55–92.

BAKEWELL, R. 1823. *Travels, Comprising Observations Made During a Residence in the Tarentaise, and Various Parts of the Grecian and Pennine Alps, and in Switzerland and Auvergne*. 2 Volumes. Longman, Hurst, Rees, Orme, and Brown, London.

BARBAROUX, C.-J.-M. 1788–1789. Description des volcans éteints d'Ollioules en Provence. *Observations sur la physique*, **33** [1788], 191–198; **35** [1789], 30–35.

BERNHARD, C. G. 1989. *Through France with Berzelius. Live Scholars and Dead Volcanoes*. Pergamon Press, New York.

BERTRAND, P. 1780. Mémoire sur les volcans de Tourves en Provence. *Observations sur la physique*, **15**, 36–38.

BOISSIEU, J.-J. DE 1768. *[Two plates in] Encyclopédie, Recueil de planches*. Volume 6. Histoire naturelle, règne minéral, 6e collection: Volcans, solfatare & Pavé des géans, plates VII (Face d'une bute toute composée de prismes articulés) and VIII (Rocher de Pereneire) Briasson, David, Le Breton, Paris.

BONIN, S. & LANGLOIS, C. (eds) 1987. *Atlas de la Révolution Française*. Volume 1, *Routes et communications*. École des Hautes Études en Sciences Sociales, Paris.

BONIN, S. & LANGLOIS, C. (eds) 1989. *Atlas de la Révolution Française*. Volume 5, part 2, *Le Territoire: Les limites administratives*. École des Hautes Études en Sciences Sociales, Paris.

BUCH, L. VON 1802. Extrait d'une Lettre de M. De Buch au Prof. M. A. Pictet, l'un des Rédacteurs de la Bibliothèque Britannique. Sur quelques observations faites aux environs de Clermont. Avec l'esquisse d'une carte. *Bibliothèque Britannique*, Sciences et Arts, **20** (An X), 306–316.

CAROZZI, A. V. 2000. *Manuscripts and Publications of Horace-Bénédict de Saussure on the Origin of Basalt (1772–1797).* Éditions Zoé, Geneva.

[CHAVAGNAT, G.] 1763. *Calendrier d'Auvergne, curieux et utile pour l'an de grace M.DCC.LXV.* M. L. P. Boutaudon, Imprimeur du Roi, [Clermont-Ferrand].

CHILDE-PEMBERTON, W. S. 1925. *The Earl Bishop: The Life of Frederick Hervey, Bishop of Derry, Earl of Bristol.* 2 Volumes. E. P. Dutton and Co., New York.

DE BEER, G. 1962 [1964]. The volcanoes of Auvergne. *Annals of Science*, **18**, 49–61.

DESMAREST, N. 1768. Basalte d'Auvergne. *In: Encyclopédie, Recueil de planches.* Volume 6, Histoire naturelle, règne minéral, sixième collection, volcans. Briasson, David, Le Breton, Paris.

DESMAREST, N. 1774. Mémoire sur l'origine & la nature du basalte à grandes colonnes polygones, déterminées par l'histoire naturelle de cette pierre, observée en Auvergne. *Mémoires de l'Académie Royale des Sciences*, année 1771, 705–775.

DESMAREST, N. 1779*a*. Extrait d'un mémoire sur la détermination de quelques époques de la nature par les produits des volcans, & sur l'usage de ces époques dans l'étude des volcans. *Observations sur la physique*, **13**, 115–126. [Fuller version presented in 1804 and published in *Mémoires de l'Institut*, 1806, **6**, 219–289.]

DESMAREST, N. 1779*b*. Des mines de la Généralité de Limoges avec les indications des carrières de pierres singulières.1765. *In*: GOBET, N. *Les anciens minéralogistes du royaume de France*, 1779, **2**, 540–553.

DESMAREST, N. 1779*c*. Lettre à M. l'abbé Bossut, de l'Académie des Sciences, sur les différentes sortes de pozzolanes, & particulièrement sur celles qu'on peut tirer de l'Auvergne. *Observations sur la physique*, **13**, 192–199.

DESMAREST, N. 1794–1828. *Encylopédie méthodique. Géographie physique.* 5 Volumes. H. Agasse, Paris.

DESMAREST, N. 1806. Mémoire sur la détermination de trois époques de la nature par les produits des volcans, et sur l'usage qu'on peut faire de ces époques dans l'étude des volcans. *Mémoires de l'Institut des Sciences, Lettres et Arts. Sciences mathématiques et physiques*, **6**, 219–289. With 4 maps.

DESMAREST, N. 1823. *Carte topographique et minéralogique d'une partie du département du Puy de Dôme dans la ci-devant province d'Auvergne où sont déterminées la marche & les limites des matières fondues & rejetées par les volcans ainsi que les courants anciens & modernes pour server aux recherches sur l'histoire naturelle des volcans. Publiée par M. Desmarest fils* [Anselme-Gaëtan Desmarest]. Ch.[les] Picquet, Paris. [About 110 × 130 cm, scale approximately 1:43 000.]

DIDEROT, D. 1751. 'Auvergne'. *In: Encyclopédie, ou dictionnaire raisonné des sciences, des arts et des métiers.* Volume 1. Briasson, David l'aîné, Le Breton, Durand, Paris, 902b.

ELLENBERGER, F. 1978. Précisions nouvelles sur la découverte des volcans de France: Guettard, ses prédécesseurs, ses émules clermontois. *Histoire et nature*, **12–13**, 3–42.

ELLENBERGER, F. 1983. Recherches et réflexions sur la naissance de la cartographie géologique, en Europe et plus particulièrement en France. *Histoire et Nature*, **22–23**, 3–54.

ELLENBERGER, F. 1994. *Histoire de la géologie.* Volume 2. *La grande éclosion et ses prémices, 1660–1810.* Technique et Documentation (Lavoisier), Paris.

FAUJAS DE SAINT-FOND, B. 1778. *Recherches sur les volcans éteints du Vivarais et du Velay.* Chez Joseph Cuchet, Grenoble; Chez Nyon aîné; Née et Masquelier, Paris.

GEIKIE, A. 1897. *The Founders of Geology.* Macmillan, London. [2nd edition, 1905.]

GILLISPIE, C. C. 1980. *Science and Polity in France at the End of the Old Regime.* Princeton University Press, Princeton.

GOBET, N. 1779. *Les Anciens Minéralogistes du Royaume de France.* 2 Volumes. Chez Ruault, Paris.

GUETTARD, J.-É. 1759 [1765]. Mémoire sur la minéralogie de l'Auvergne. *Mémoires de l'Académie des Sciences*, 1759, 538–576; pl. & map.

GUETTARD, J.-É. & MONNET, A.-G. 1780. *Atlas et Description Minéralogiques de la France.* Didot l'aîné, Paris.

IMBERDIS, F. 1967. *Le Réseau Routier de l'Auvergne au XVIIIe siècle: Ses origines et son évolution.* Publications de l'Institut d'Études du Massif Central, fascicule 2. Presses Universitaires de France, Paris.

JOINVILLE, [forename unknown]. 1788. Volcan de la Trevaresse, plus connu sous le nom de volcan de Beaulieu. *Observations sur la physique*, **33**, 24–36.

JOUBERT, P.-L. DE, 1779 [1782]. Description du petit volcan éteint, dont le sommet est couvert par le village & le château de Montferrier, à une lieue de Montpellier. *Mémoires de l'Académie des Sciences*, 1779, 575–583.

KÖLBL-EBERT, M. 2007. The geological travels of Charles Lyell, Charlotte Murchison and Roderick Impey Murchison in France and northern Italy (1828). *In*: WYSE JACKSON, P. N. (ed.) *Four Centuries of Geological Travel: The Search for Knowledge on Foot, Bicycle, Sledge and Camel.* Geological Society, London, Special Publications, **287**, 109–117.

LACOSTE, P.-F. 1803 (An XI). *Observations sur les volcans de l'Auvergne suivies de notes sur divers objets; recueillies dans une course minéralogique faite l'année dernière, an X.* Delcros, Granier et Froin, Clermont-Ferrand.

LACOSTE, P.-F. 1805 (An XIII). *Lettres minéralogiques et géologiques sur les volcans de l'Auvergne, écrites dans un voyage fait en 1804.* De l'Imprimerie de Landriot, Clermont.

LEFEBVRE D'HELLANCOURT, A.-M. 1802. Aperçu général des mines de houille exploitées en France, de leurs produits, et des moyens de circulation de ces produits. *Journal des mines*, **12**, (No. 71, Thermidor An X), 325–411. With map.

LEGRAND D'AUSSY, P.-J.-B. An III (1794/1795). *Voyage fait en 1787 et 1788, dans la ci-devant haute et basse Auvergne, aujourd'hui départemens du Puy-de-Dôme, du Cantal et partie de celui de la Haute-Loire.* 3 Volumes. Chez le Directeur de l'Imprimerie des Sciences et Arts, Paris.

LE MONNIER, L.-G. [Le Monnier le Médecin]. 1744. Observations d'histoire naturelle faites dans les provinces méridionales de la France pendant l'année 1739. *In*: CASSINI DE THURY, C.-F. (ed.) *La méridienne de l'Observatoire royal de Paris, vérifiée dans toute l'étendue du royaume par de nouvelles observations*. Chez Hippolyte-Louis Guerin, & Jacques Guerin, Paris. [Separately paginated cix–ccxxvi.]

LOUGH, J. 1987. *France on the Eve of Revolution: British Travellers' Observations, 1763–1788*. Croom Helm, London & Sydney.

LYELL, C. 1855. *Manual of Geology*. 5th edition. J. Murray, London.

MICHEL, R. 1971. Les premières recherches sur les volcans du Massif Central (18e–19e siècles) et leur influence sur l'essor de la géologie. *In: Symposium Jean Jung: Géologie, géomorphologie et structure profonde du Massif Central Français*. Plein Air Service Éditions, Clermont-Ferrand, 331–344.

MONTET, J. 1760 [1766]. Mémoire sur un grand nombre de volcans éteints, qu'on trouve dans le Bas-Languedoc. *Mémoires de l'Académie des Sciences*, 1760, 466–476.

PARKER, H. T. 1965. French administrators and French scientists during the Old Regime and the early years of the Revolution. *In*: HERR, R. & PARKER, H. T. (eds) *Ideas in History*. Duke University Press, Durham, NC, 85–109.

PASSINGES, H. 1797–1798. Mémoires pour servir à l'histoire naturelle du département de la Loire, ou du ci-devant Forez; Par le C.^{en} Passinges, professeur d'histoire naturelle à l'école centrale à Roanne. *Journal des mines*, **6**, (No. 35, Thermidor An V), 813–852; **7**, (No. 38, Brumaire An VI), 117–144; **7**, (No. 39, Frimaire An VI), 181–212.

PEREZ, M.-F. & PINAULT, M. 1985–1986. Three new drawings by Jean-Jacques de Boissieu. *Master Drawings*, **23–24**, 389–395.

QUÉRIAU, F.-G. 1748. *Discours historique, prononcé à l'ouverture de la première séance publique de la Société littéraire de Clermont-Ferrand, le jour de St. Louis, 25 août 1747*. Imprimerie de Pierre Boutaudon, Clermont-Ferrand.

RAPPAPORT, R. 1972. Guettard, Jean-Étienne. *In*: *Dictionary of Scientific Biography*. Volume 5, 577–579. Charles Scribner's Sons, New York.

RUDWICK, M. J. S. 1974. Poulett Scrope on the volcanoes of Auvergne: Lyellian Time and Political Economy. *British Journal for the History of Science*, **7**, 205–242.

RUDWICK, M. J. S. 1976. The emergence of a visual language for geological science 1760–1840. *History of Science*, **14**, 149–195.

RUDWICK, M. J. S. 1996. Geological travel and theoretical innovation: The role of 'liminal' experience. *Social Studies of Science*, **26**, 143–159.

RUDWICK, M. J. S. 2005. *Bursting the Limits of Time: The Reconstruction of Geohistory in the Age of Revolution*. University of Chicago Press, Chicago & London.

SCHLÜTER, C. A. 1750–1753. *De la fonte des mines, des fonderies, &c. Le tout augmenté de plusieurs procédés & observations; & publié par M. Hellot*. 2 Volumes. Hérissant & Pissot, Paris. [2nd edn 1764.]

SCROPE, G. P. 1858. *The Geology and Extinct Volcanos of Central France*. J. Murray, London [Arno Press facsimile reprint, 1978].

SOULAVIE, J.-L. G. 1780–1784. *Histoire naturelle de la France méridionale*. 8 Volumes. Nîmes, Belle [Volumes 1–2]. Hôtel de Venise, et chez J. F. Quillau, Marigot l'aîné, Belin, Paris.

STRANGE, J. 1775. An account of two Giants Causeways, or groups of prismatic basaltine columns, and other curious vulcanic concretions, in the Venetian State in Italy; with some remarks on the characters of these and other similar bodies, and on the physical geography of the countries in which they are found. *Philosophical Transactions of the Royal Society*, **65**, 5–47.

TAYLOR, K. L. 1994. New light on geological mapping in Auvergne during the eighteenth century: The Pasumot–Desmarest collaboration. *Revue d'histoire des sciences*, **47**, 129–136.

TORRENS, H. S. 1995. Mary Anning (1799–1847) of Lyme: 'the greatest fossilist the world ever knew'. *British Journal for the History of Science*, **28**, 257–284.

TORRENS, H. S. 2002. *The Practice of British Geology, 1750–1850*, Ashgate, Aldershot (UK) and Burlington (Vermont, USA).

VERNIÈRE, A. 1899–1900. Les voyageurs et les naturalistes dans l'Auvergne et dans le Velay. *Revue d'Auvergne*, **16** (1899), 331–348, 436–457; **17** (1900), 104–130, 239–294. [Also published separately, 1900, Typographie et Lithographie G. Mont-Louis, Clermont-Ferrand.]

WILSON, L. G. 1972. *Charles Lyell. The Years to 1841: The Revolution in Geology*. Yale University Press, New Haven.

Bibliography: Publications related to geological travel and observation in Auvergne, 1751–1800 (+)

Primary sources

AUBUISSON DE VOISINS, J.-F. d' 1804*a*. Extrait d'un mémoire sur les volcans et les basaltes de l'Auvergne. *Journal des mines*, **58**, 310–318.

AUBUISSON DE VOISINS, J.-F. d' 1804*b*. Description minéralogique du Puy-de-Dôme. *Journal de physique*, **58**, 422–427.

AUBUISSON DE VOISINS, J.-F. d' 1804*c*. Notions minéralogiques sur la phonolithe (ou Klingstein des allemands). *Journal de physique*, **59**, 367–382.

AUBUISSON DE VOISINS, J.-F. d' 1819. *Traité de géognosie, ou exposé des connaissances actuelles sur la constitution physique et minérale du globe terrestre*. 2 Volumes. F. G. Levrault, Strasbourg & Paris.

BAKEWELL, R. 1823. *Travels, Comprising Observations Made During a Residence in the Tarentaise, and Various Parts of the Grecian and Pennine Alps, and in Switzerland and Auvergne*. 2 Volumes. Longman, Hurst, Rees, Orme, and Brown, London.

BERTRAND, P. 1780. Mémoire sur les volcans de Tourves en Provence. *Observations sur la physique*, **15**, 36–38.

BERTRAND, P. 1782. *Lettre à M. le Comte de Buffon, ou critique et nouvel essai sur la théorie générale de la terre*. 2nd edition. Chez Charmet, Besançon; Chez Esprit, Paris. [1st edition 1780.]

BERTRAND, P. 1799. 'Lettre du C.^{en} Bertrand au C.^{en} Muthuon, sur ses observations volcaniques, insérées au No. XLVII de ce Journal, relativement à celles du C.^{en} Dolomieu' *Journal des mines*, **9**, (No. 53, Pluviôse An VII), 377–384.

BESSON, A.-C.? 1787. Passage de colonnes ou prismes de basalte volcanique à l'état de boules. *Observations sur la physique*, **31**, 149–153.

BESSON, A.-C.? 1789. Particularités remarquables dans quelques granits et roches primitives. *Observations sur la physique*, **35**, 121–131.

BESSON, A.-C.? 1796. Suite du tableau des mines et usines de la France: Département de l'Allier. Notice géographique. *Journal des mines*, **5**, (No. 26, Brumaire An V), 119–159. [Anonymous article; Quérard (1827–1839) identifies Besson as the author.]

BOSC D'ANTIC, P. 1780. *Oeuvres de M. Bosc d'Antic, contenant plusieurs mémoires sur l'art de la verrerie, sur la faïencerie, la poterie, l'art des forges, la minéralogie, l'électricité et sur la médecine*. 2 Volumes. Rue et hôtel serpente, Paris. [Electronic reproduction: Les archives de la Révolution Française, Bibliothèque Nationale de France.]

BOSC, L.-A.-G. 1805. Note sur un fossile remarquable de la montagne de Saint-Gérand-le-Puy, entre Moulins et Roanne, Département de l'Allier, appelé l'Indusie tubuleuse. *Journal des mines*, **17**, (No. 101, Pluviôse, An XIII), 397–400.

BOURNON, J.-L. COMTE DE 1785. *Essai sur la lithologie des environs de Saint-Etienne-en-Forez, et sur l'origine de ses charbons de pierre*. [Bibliothèque Nationale de France info.: no place of publication, no publisher].

BOURNON, J.-L. COMTE DE 1789. Mémoire sur le Pechstein et l'Hydrophane. *Observations sur la physique*, **35**, 19–29.

BOWLES, W. 1775. *Introduccion á la historia natural y á la geografia física de España*. Francisco Manuel de Mena, Madrid. [Also 2nd edition, 1782, en la Imprenta Real, Madrid; 3rd edition, 1789, Imprenta Real, Madrid.]

BOWLES, W. 1776. *Introduction à l'histoire naturelle et à la géographie physique de l'Espagne*. Transl. Vicomte de Flavigny. L. Cellot & Jombert fils jeune, Paris. [Also an Italian translation, 1783, 2 Volumes, by MILIZIA, F. Stamperia reale, Parma (not seen; in Bibliothèque Nationale de France).]

BREISLAK, S. 1801. *Voyages physiques et lythologiques dans la Campanie*. 2 Volumes. Dentu, Paris.

BRIEUDE, J.-J. DE, 1788. *Observations sur les eaux thermales de Bourbon-l'Archambault, de Vichy et du Mont-d'Or, faites dans un voyage par ordre du gouvernement, lues à la Société royale de médecine dans ses séances particulières*. Froullé, Paris. [Not seen; in Bibliothèque Nationale de France.]

BRIEUDE, J.-J. DE, 1802. *Observations économiques et politiques sur la chaîne des montagnes ci-devant appelées d'Auvergne*. Impr. de Valade, Paris, an XI. Republished from *Annales de statistique*, no. 8. [Not seen; in Bibliothèque Nationale de France.]

BRIEUDE, J.-J. DE, 1821. *Topographie médicale de la Haute-Auvergne, aujourd'hui le département du Cantal*. New edition; Imprimerie de Picut, Aurillac. [Originally from Registres de la Société royale de médecine, 1782–1783; not seen; in Bibliothèque Nationale de France.]

BRONGNIART, A. 1829. 'Volcans' *Dictionnaire des sciences naturelles*. 60 Volumes. F. G. Levrault, Strasbourg; Le Normant, Paris, 1816–1830, **58**, 334–446.

BUCH, L. VON 1802. Extrait d'une Lettre de M. De Buch au Prof. M. A. Pictet, l'un des Rédacteurs de la Bib. britann. Sur quelques Observations faites aux environs de Clermont. Avec l'esquisse d'une carte. *Bibliothèque Britannique*, Sciences et Arts, **20** (An X), 306–316.

BUCH, L. VON 1803. Observations sur les volcans d'Auvergne. Extrait d'une lettre de ce savant, à M. A. Pictet, l'un des Rédacteurs de la *Bibliothèque Britannique. Journal des mines*, **13**, (No. 76, Nivôse An XI), 249–256. [Evidently same article as Buch (1802).]

BUC'HOZ, P.-J. [Many short works on Auvergne natural history, none dated, too numerous to list here. See electronic catalog of Bibliothèque Nationale de France.]

COCQ, [forename unknown]. 1805. Notice minéralogique sur la Pinite trouvée en France, par M. Cocq, Commissaire des poudres et salpêtres à Clermont-Ferrant [sic], suivie de l'analyse de cette substance. *Journal des mines*, **17**, (No. 100, Nivôse, An XIII), 307–312.

COCQ, [forename unknown]. 1806. Mémoire renfermant des détails sur la lithologie de l'Auvergne et des environs. *Journal des mines*, **18**, (No. 114, Juin 1806), 409–434.

CORDIER, L. 1803. Lettre de Louis Cordier, Ingénieur des mines de France, à J.-C. Delamétherie. Extrait. *Journal de physique*, **56**, 220–223.

D'AUBUISSON DE VOISINS, J.-F. See Aubuisson.

DELARBRE, A. 1768. *Dissertation sur l'arcade et le mur formé par les eaux minérales de St.-Alire dans un faubourg de Clermont*. [According to Quérard (1827–1839); not seen.]

DELARBRE, A. 1785. *Essais topographiques et histoire naturelle du Mont-d'Or et de ses environs*. [According to Quérard (1827–1839); not seen.]

DELARBRE, A. 1795. *Flore d'Auvergne, ou recueil des plantes de cette ci-devant province; suivi de la description du lac de Pavin*. B. Beauvert & L. Deschamps, Clermont-Ferrand, 1796. Reissued 1797. [Not seen; 1795 original edition, according to Bibliothèque Nationale de France catalog listing for electronic reproduction. 2nd edition, 1800. Reproduced electronically 1995.]

DELARBRE, A. 1797. *Essai zoologique, ou histoire naturelle des animaux sauvages, quadrupèdes et oiseaux indigènes, d'Auvergne*. B. Beauvert & L. Deschamps, Paris. [According to Bibliothèque Nationale de France; not seen.]

DELARBRE, G. [?] 1786. Extrait d'un mémoire lu à l'Académie des sciences, sur la nature & la formation des fers spéculaires de Volvic, du Puy-de-Dôme, du Mont-d'Or, &c. *Observations sur la physique*, **29**, 119–129.

DELARBRE, G. [?] 1787. Mémoire lu à l'Académie des Sciences, sur la formation & la distinction des basaltes en boules de différens endroits de l'Auvergne. *Observations sur la physique*, **31**, 133–149.

DELARBRE, G. [?] & QUINQUET, B. [?] 1787. Mémoire sur le pechstein de Mesnil-Montant. *Observations sur la physique*, **31**, 219–223.

DELUC, G.-A. 1801. Observations sur les basaltes. *Journal de physique*, **53**, 242–249.

DELUC, G.-A. 1802. Observations générales sur les volcans. *Journal des mines*, **12**, (No. 69), 165–173.

DELUC, G.-A. 1804. Nouvelles observations sur les volcans et sur les laves. *Journal des mines*, **16**, (No. 95, Thermidor An XII), 329–354. [See also *Bibliothèque Britannique*, 1804, **26**, 336–372.]

DESMAREST, N. 1768. Basalte d'Auvergne. *In Encyclopédie, Recueil de planches*, Volume 6, Histoire naturelle, règne minéral, sixième collection, volcans. Briasson, David, Le Breton, Paris.

DESMAREST, N. 1771. Mémoire sur l'origine & la nature du basalte à grandes colonnes polygones, déterminées par l'histoire naturelle de cette pierre, observée en Auvergne. *Mémoires de l'Académie Royale des Sciences*, 1771 [1774], 705–775.

DESMAREST, N. 1773. Mémoire sur le basalte. Troisième partie, où l'on traite du basalte des anciens; & où l'on expose l'histoire naturelle des différentes espèces de pierres auxquelles on a donné, en différens tems, le nom de basalte. *Mémoires de l'Académie Royale des Sciences*, 1773 [1777], 599–670.

DESMAREST, N. 1779*a*. Extrait d'un mémoire sur la détermination de quelques époques de la nature par les produits des volcans, & sur l'usage de ces époques dans l'étude des volcans. *Observations sur la physique*, **13**, 115–126. [Fuller version presented in 1804 and published in *Mémoires de l'Institut*, 1806, **6**, 219–289.]

DESMAREST, N. 1779*b*. Des mines de la Généralité de Limoges avec les indications des carrières de pierres singulières. 1765. *In*: GOBET, N. *Les anciens minéralogistes du royaume de France*, 1779, **2**, 540–553.

DESMAREST, N. 1779*c*. Lettre à M. l'abbé Bossut, de l'Académie des Sciences, sur les différentes sortes de pozzolanes, & particulièrement sur celles qu'on peut tirer de l'Auvergne. *Observations sur la physique*, **13**, 192–199.

DESMAREST, N. 1787. Considérations générales sur le rapport des boules de lave avec les prismes de basalte articulés. *Observations sur la physique*, **31**, 65–69.

DESMAREST, N. 1794–1828. *Encylopédie méthodique. Géographie physique*. 5 Volumes. H. Agasse, Paris.

DESMAREST, N. 1811. Sur les volcans éteints de l'Auvergne. *Bulletin des sciences, par la Société Philomathique de Paris*, **3**, 213.

DESMAREST, N. 1823. *Carte topographique et minéralogique d'une partie du département du Puy de Dôme dans la ci-devant province d'Auvergne, ...* Paris.

DOLOMIEU, D. 1798. Rapport fait à l'Institut National, ... sur ses voyages de l'an V et de l'an VI. *Journal des mines*, **7**, no. 41, 385–402, no. 42, 405–421. [See also *Observations sur la physique*, 1798, **3**, 401–427.]

DUHAMEL, J.-B. G. 1795. Avantages d'une exploitation de houille et de plomb aux environs de Montaigu, département du Puy-de-Dôme. *Journal des mines*, no. 9, Prairial An III, 14–24.

DUTOUR, É.-F. 1768. Observations sur un banc de terre crétacée et de pierres branchues, qui est aux environs de Riom. Académie des Sciences. *Mémoires des savans étrangers*, **5**. [According to Quérard (1827–1839); not seen.]

FAUJAS DE SAINT-FOND, B. 1777. *Description des volcans éteints du Vivarais et du Velay, par M. Faujas de Saint-Fond: ... Ouvrage proposé par souscription.* [Advertisement for subscription for *Recherches*; a copy is bound in the copy of *Recherches* held by Bibliothèque Nationale de France, cote R.456.]

FAUJAS DE SAINT-FOND, B. 1778. *Recherches sur les volcans éteints du Vivarais et du Velay.* Chez Joseph Cuchet, Grenoble; Chez Nyon aîné; Née et Masquelier, Paris.

FAUJAS DE SAINT-FOND, B. 1784. *Minéralogie des volcans, ou description de toutes les substances produites ou rejetées par les feux souterrains.* Chez Cuchet, Paris.

FAUJAS DE SAINT-FOND, B. 1809. *Essai de géologie, ou mémoires pour servir à l'histoire naturelle du globe.* 2 Volumes. Chez Gabriel Dufour, Paris. [2nd volume in 2 parts.]

FOURCROY, A.-F. & VAUQUELIN, N.-L. 1804. Experiences comparées sur l'Arragonite d'Auvergne, et le carabonate de chaux d'Islande. *Annales du Muséum national d'histoire naturelle*, **4** (An XII), 405–411.

GIRAUD-SOULAVIE. See Soulavie.

GIROUD, A. 1795. Essai de la terre alumineuse de Royat (département du Puy-de-Dôme), envoyée à l'agence des mines par le C.[en] Châle cadet. *Journal des mines*, **2**, (No. 12, Fructidor An III), 3–4.

GOBET, N. 1779. *Les anciens minéralogistes du royaume de France.* 2 Volumes. Chez Ruault, Paris.

GODON DE SAINT-MEMIN, C. 1804–1805. Analyse du fer natif à l'état d'acier trouvé en Auvergne. *Journal de physique*, **60**, (Nivôse An XIII), 340–346.

GUETTARD, J.-É. 1751. See Vernière (1901) (secondary works).

GUETTARD, J.-É. 1752. Mémoire sur quelques montagnes de la France qui ont été des volcans. *Mémoires de l'Académie Royale des Sciences*, 1752 [1756], 27–59.

GUETTARD, J.-É. 1755. Mémoire sur le tripoli. *Mémoires de l'Académie Royale des Sciences*, 1755 [1761], 177–193; pl.

GUETTARD, J.-É. 1759. Mémoire sur la minéralogie de l'Auvergne. *Mémoires de l'Académie Royale des Sciences*, 1759 [1765], 538–576; plate & map.

GUETTARD, J.-É. 1768–1786. *Mémoires sur différentes parties des sciences et des arts.* 6 volumes. Laurent Prault/Lamy, Paris. [**1**, 1768; **2** & **3**, 1770; **4–6**: 1786. Chez Lamy, Paris.]

GUETTARD, J.-É. 1779. *Mémoires sur la minéralogie du Dauphiné.* De l'Imprimerie de Clousier, Paris.

GUETTARD, J.-É. 1782. *Minéralogie du Dauphiné. In*: LABORDE, J.-B. DE, (ed.) *Description générale et particulière de la France, t. 1* (De l'Imprimerie de Ph.-D. Pierres, Paris). [Paginated separately from the other part of the volume, which consists of a history and description of the province of Dauphiné, by Béguillet, 102 pp. Guettard's part is pp. 1–255. This includes, as the sixth mémoire, 'Sur quelques volcans éteints du Vivarais,' 33–40. This account of the province of Dauphiné reports on Vivarais journeys in 1775, in July, August, and September.]

GUETTARD, J.-É. 1958. (FRIC, R. (ed.)) Une lettre de Guettard à Monnet [the abbé] au sujet des prismes basaltiques. *Bulletin historique et scientifique de l'Auvergne*, **78**, 91–96. [The Monnet in question is the abbé and chanoine, brother of Antoine-Grimoald; The Fric article is based on an incomplete copy of the letter kept by Guettard.]

GUETTARD, J.-É. & MONNET, A.-G. 1780. *Atlas et description minéralogiques de la France*. Didot l'aîné, Paris.

HELLOT, J. 1750. État des mines du royaume, distribué par provinces. *In*: SCHLÜTER, C. A. *De la fonte des mines, des fonderies, &c. Le tout augmenté de plusieurs procédés & observations; & publié par M. Hellot*. 2 Volumes. Hérissant & Pissot, Paris. 1750–1753. **1**, 1–70. [Also in the 2nd edition, 1764.]

JARS, A.-G. 1774–1781. *Voyages métallurgiques*. 3 Volumes [**1**, 1774: Gabriel Regnault, Lyon; **2**, 1780: L. Cellot, Cl.-Ant. Jombert Aîné, L.-Alex. Jombert jeune, Paris; **3**, 1781: Didot, L. Cellot, L.-Alex. Jombert jeune, Paris.]

JOUBERT, P.-L. DE 1779. Description du petit volcan éteint, dont le sommet est couvert par le village & le château de Montferrier, à une lieue de Montpellier. *Mémoires de l'Académie Royale des Sciences*, 1779 [1782], 575–583.

LACOSTE, P.-F. 1803. *Observations sur les volcans de l'Auvergne suivies de notes sur divers objets; recueillies dans une course minéralogique faite l'année dernière, an X*. Delcros, Granier et Froin, Clermont-Ferrand, An XI.

LACOSTE, P.-F. 1805. *Lettres minéralogiques et géologiques sur les volcans de l'Auvergne, écrites dans un voyage fait en 1804*. De l'Imprimerie de Landriot, Clermont. An XIII.

LAIZER, L.-G. DE 1808*a*. Lettre sur la constitution du sol de l'Auvergne. *Journal des mines*, **23**, 407–419.

LAIZER, L.-G. DE 1808*b*. *Lettre à M. le professeur Jurine, célèbre naturaliste de Genève, membre de plusieurs sociétés savantes, etc., sur le puy Chopine, l'une des montagne volcanisées qui forment la chaîne du Puy-de-Dôme; avec la description de toutes les roches primitives ou volcaniques qu'on y rencontre*. Imprimerie de Landriot, Clermont-Ferrand.

LAIZER, L.-G. DE & PARIEU, M.-L.-P. ESQUIROU DE [No date]. *Notice sur un nouveau genre de pachyderme fossile trouvé en Auvergne, et nommé Oplotherium*. Imprimerie de Perol, Clermont-Ferrand. [8 pp. Although undated, includes textual references to other publications as late as 1833.]

LAMÉTHERIE, J.-C. DE 1802. Note sur un voyage minéralogique, fait en l'an IX, avec une classification des roches. *Journal de physique*, **55**, 129–156.

LAVERRIÈRE, L. DE 1797. Extrait d'un rapport sur les forêts et masses de houille des environs d'Issoire. Par le C.[en] Laverrière, ingénieur des mines. *Journal des mines*, **6**, (No. 36, Fructidor An V), 939–942.

LEFEBVRE D'HELLANCOURT, A.-M. 1802. Aperçu général des mines de houille exploitées en France, de leurs produits, et des moyens de circulation de ces produits. *Journal des mines*, **12**, (No. 71, Thermidor An X), 325–411. With map.

LEGRAND D'AUSSY, P.-J.-B. 1788. *Voyage d'Auvergne*. Chez Eugène Onfroy. Paris.

LEGRAND D'AUSSY, P.-J.-B. An III (1794/1795). *Voyage fait en 1787 et 1788, dans la ci-devant haute et basse Auvergne, aujourd'hui départemens du Puy-de-Dôme, du Cantal et partie de celui de la Haute-Loire*. 3 Volumes. Chez le Directeur de l'Imprimerie des Sciences et Arts, Paris. [Bibliothèque Nationale de France catalog also shows a German translation, Göttingen, 1797.]

LELIÈVRE, C.-H. 1787. Lettre de M. le Lièvre, ingénieur des mines de France, à M. de La Métherie, sur la chrysolite des volcans. *Observations sur la physique*, **30**, 397–398.

LE MONNIER, L.-G. [LE MONNIER LE MÉDECIN] 1744*a*. *Observations d'histoire naturelle faites dans les provinces méridionales de la France pendant l'année 1739. In:* CASSINI DE THURY, C.-F., *La méridienne de l'Observatoire royal de Paris, vérifiée dans toute l'étendue du royaume par de nouvelles observations*. Chez Hippolyte-Louis Guerin, & Jacques Guerin, Paris. [Separately paginated cix–ccxxvi.]

LE MONNIER, L.-G. 1744*b*. Examen des eaux minérales du Mont d'Or. *Mémoires de l'Académie Royale des Sciences*, 1744 [1748], 157–169.

LE MONNIER, L.-G. 1779. Des mines de l'Auvergne. Par M. Le Monnier, D.M.L. de l'Acad. des Sciences. 1739. *In:* GOBET, [N.] *Les anciens minéralogistes du royaume de France*, 1779, II 515–529.

MALESHERBES, C.-G. DE, LAMOIGNON DE, 1779. [His account of his journey in Auvergne with Guettard:] *In:* GUETTARD, J. É. 1779. *Mémoires sur la minéralogie du Dauphiné*. De l'Imprimerie de Clousier, Paris, cxxxix–cxliv.

MALESHERBES, C.-G. DE, LAMOIGNON DE, 1798. *Observations . . . sur l'histoire naturelle générale et particulière de Buffon et Daubenton*. 2 Volumes. Charles Pougens, Paris. An VI.

MONNET, A.-G. 1779. *Nouveau systême de minéralogie*. À la Société Typographique, Bouillon. [His 'Précis historique sur les progrès de la minéralogie en France,' pp. 1–58.]

MONNET, A.-G. 1788. Voyages minéralogiques faits en Auvergne dans les années 1772, 1784 & 1785. *Observations sur la physique*, **32**, 115–132, 179–199; **33**, 112–129, 321–339.

MONNET, A.-G. 1801–1802. Mémoire sur les petits volcans dans les anciennes montagnes volcaniques, et en particulier sur celui de la montagne de Coran, département du Puy-de-Dôme. *Journal des mines*, **11**, (No. 64, Nivôse An X), 273–278.

MONNET, A.-G. 1875. *Voyage de Monnet dans la Haute-Loire et le Puy-de-Dôme, 1793–1794*. Publié par Henry Mosnier. Imprimerie de M.-P. Marchessou, Le Puy. [Electronic document, in Bibliothèque Nationale de France.]

MONNET, A.-G. 1887. Les bains du Mont-Dore en 1786: Voyage en Auvergne de Monnet. Publié et annoté par Henry Mosnier. *Mémoires de l'Académie des Sciences, Belles-Lettres et Arts de Clermont-Ferrand*, **29**, 71–174.

MONNET (CHANOINE), 1774. Dissertation sur les débris des volcans de l'Auvergne; & sur les roches qui s'y trouvent. *Observations sur la physique*, **4**, 65–77. [Précis in *Journal encyclopédique*, 1774, **4**, 342–344.]

MONTET, [J.] 1766. Mémoire sur un grand nombre de volcans éteints, qu'on trouve dans le Bas-Languedoc.

Mémoires de l'Académie Royale des Sciences, 1760 [1766], 466–476.

MONTLOSIER, F.-D., DE REYNAUD, COMTE DE 1788. [Mus. Nat. d'Hist. Nat. copy says 1789]. *Essai sur la théorie des volcans d'Auvergne*. [No place (Paris?), no publisher]. 2nd edition. An X, 1802. Imprimerie de Landriot et Rousset, Riom, Clermont.

MONTLOSIER, F.-D. DE REYNAUD, COMTE DE, 1802. *Notice sur la pierre appelée cornéenne, ou roche de corne*. Chez Goujon, Mongie, Lenormant, Paris. An X.

MOSSIER, J.-B. 1804. Lettre de M. Mossier, contenant son opinion et celle de M. Saussure, sur l'origine de la Roche Sanadoire. *Journal des mines*, **16**, (No. 96, Fructidor An XII), 483–486. [Letter dated 1 Nivôse An 12 = 23 December 1803.]

MOSSIER, J.-B. AMABLE, 1795–1796. *Vues générales sur l'histoire naturelle des environs de Clermont-Ferrand.* Imprimerie de P. Landriot, Riom. An IV. [Not seen; listed in Carozzi (2000).]

MUTHUON, J.-M. 1798. Observations du C.[en] Muthuon, Ingénieur des mines, sur l'article du rapport fait à l'Institut national par le C.[en] Dolomieu, inséré dans le No. XLI de ce journal, qui concerne les volcans de l'Auvergne et la volcanisation en général. *Journal des mines*, **8**, (No. 47, Thermidor An VI), 869–882.

MUTHUON, J.-M. 1799. Lettre du C.[en] Muthuon, ingénieur des mines, au C.[en] Bertrand, inspecteur général des ponts et chaussées, en réponse à sa lettre insérée dans le No. LIII du Journal des mines. *Journal des mines*, **9**, (No. 54, Ventôse An VII), 439–448.

ORDINAIRE, C.-N. 1787. *Recherches sur l'ancien état de la Limagne, relativement à son histoire naturelle.* Imprimerie A. Delcros, Clermont.

ORDINAIRE, C.-N. 1801. *The Natural History of Volcanos, including Submarine Volcanos, and the Analogous Phenomena.* Translation R. C. Dallas, R. C. H. Baldwin by T. Cadell and W. Davies, London.

ORDINAIRE, C. N. 1802. *Histoire naturelle des volcans, comprenant les volcans soumarins, ceux de boue, et autres phénomènes analogues*. Levrault Frères, Paris. An X.

OZY, J.-F. 1748. *Analise des eaux minérales de Saint-Alyre.* Imprimerie de P. Boutaudon, Clermont-Ferrand.

OZY, J.-F. [Date unknown.] *Observations sur les mines de charbon & sur les anciennes saffranières de la province d'Auvergne.* [Listed as one of 16 works that could be published by Société Littéraire de Clermont-Ferrand: *Suite des mémoires de la Société littéraire de Clermont-Ferrand, qu'elle pourra donner au public lorsqu'une autorisation expresse de Sa Majesté l'aura mise en état de l'annoncer*, 1748, Imprimerie de Pierre Boutaudon, Clermont-Ferrand]

PASSINGES, H. 1797–1798. Mémoires pour servir à l'histoire naturelle du département de la Loire, ou du ci-devant Forez; Par le C.[en] Passinges, professeur d'histoire naturelle à l'école centrale à Roanne. *Journal des mines*, **6**, (No. 35, Thermidor An V), 813–852; **7**, (No. 38, Brumaire An VI), 117–144; **7**, (No. 39, Frimaire An VI), 181–212.

PASUMOT, F. 1774*a*. Observation d'un arc-en-ciel entier. *Observations sur la physique*, **3**, 416. [Observation made 1765 at summit of Mont d'Or.]

PASUMOT, F. 1774*b*. Manière dont on ramasse le grenat dans le ruisseau d'Espailly, près du Puy-en-Velay. *Observations sur la physique*, **3**, 442–443.

PASUMOT, F. 1776. Lettre à l'auteur de ce recueil. *Observations sur la physique*, **8**, 38–46. [Meteorological observations; includes elevations of Auvergne locations.]

PASUMOT, F. 1778. Mémoire sur la zéolite. *In:* Faujas de Saint-Fond, B. *Recherches sur les volcans éteints du Vivarais et du Velay*, 111–116. [A paper presented to the Academy of Sciences in 1776.]

PASUMOT, F. 1782. Mémoire sur la liaison des volcans d'Auvergne avec ceux du Gévaudan, du Velay, du Vivarais, du Forez, &c. *Observations sur la physique*, **20**, 217–223.

PASUMOT, F. 1785. Lettres sur quelques volcans de la Haute-Auvergne. *Journal général de France* [abbé de Fontenai (aka L.-A. de Bonafous) (ed.)], February, and April, 1785. [According to Grivaud; not seen; in Bibliothèque Nationale de France.]

PASUMOT, F. 1810–1813. *Dissertations et mémoires sur différens sujets d'antiquité et d'histoire.* GRIVAUD DE LA VINCELLE, C.-M. (ed.) 7 parts in 1 Volume. Paris [no publisher].

PATOUILLOT, D. 1780. Remarques sur les volcans du Vivarais & sur l'histoire de ces volcans. *Observations sur la physique*, **15**, 61–66.

QUÉRIAU, F.-G. 1748. *Discours historique, prononcé à l'ouverture de la première séance publique de la Société littéraire de Clermont-Ferrand, le jour de St. Louis, 25 août 1747.* Imprimerie de Pierre Boutaudon, Clermont-Ferrand. [31 pp. Auvergne mineral resources discussed, pp. 24–25. Quériau identified as *avocat*.]

RANGOUSE DE LA BASTIDE, J.-J.-G. DE LEIGONYE, COMTE DE, 1784. *Essai sur l'origine des fiefs, de la noblesse de la Haute-Auvergne, et sur l'histoire naturelle de cette province.* Royez, Paris. [Not seen; in Bibliothèque Nationale de France.]

ROMME, G. 1795. Vues géologiques, présentées à la Société d'histoire naturelle, dans sa séance du premier brumaire, troisième année [22 October 1794]. *Journal des mines*, **1** (No. 5, Pluviôse an III), 51–60.

SAGE, B.-G. 1777. *Élémens de minéralogie docimastique.* 2nd edition. Imprimerie Royale, Paris.

SAUSSURE, H.-B. DE. See Carozzi (secondary works).

SAUVADE, PÈRE [forename unknown]. 1767. *Éloge de M. Michel Pelissier, Ecuyer, Seigneur de Feligonde, Saulces, Le Chatelard, &c. De l'Académie de Dijon, Secrétaire perpétuel de la Société Littéraire de Clermont-Ferrand, & de l'Académie d'Agriculture de la même ville; prononcé dans la séance publique de l'Académie, tenuë le 25 août 1767, par le R. P. Sauvade, Minime, professeur de mathématique, de l'Académie de Dijon, & Secrétaire de la Société Littéraire de Clermont-Ferrand.* Chez Pierre Viallanes, Clermont-Ferrand.

SCHLÜTER, C. A. 1750–1753. *De la fonte des mines, des fonderies, &c. Le tout augmenté de plusieurs procédés & observations; & publié par M. Hellot.* 2 Volumes. Paris, Hérissant & Pissot, Paris. [Also, 2nd edition, 1764.]

SISTRIÈRES-MURAT, F.-M. VICOMTE, DE [Date unknown]. *Analyse et description topographique ... du département du Cantal.* [Publisher, place of publication unknown. Not seen; in Bibliothèque Nationale de France.]

SOULAVIE, J.-L. G. 1780–1784. *Histoire naturelle de la France méridionale.* 8 Volumes. Belle, Nîmes [volumes 1–2]; Hôtel de Venise, et chez J. F. Quillau, Marigot l'aîné, Belin, Paris.

SOULAVIE, J.-L. G. 1783. Description des couches superposées de laves du volcan de Boutaresse en Auvergne, et observations sur une planche travaillée par la main de l'homme, & trouvée sous des coulées de laves. *Observations sur la physique*, **22**, 289–294.

STRANGE, J. 1775. An Account of Two Giants Causeways, or Groups of prismatic basaltine Columns, and other curious vulcanic Concretions, in the Venetian State in Italy; with some Remarks on the Characters of these and other similar Bodies, and on the physical Geography of the Countries in which they are found. *Philosophical Transactions*, **65**, 5–47.

TOWNSON, R. 1799. *Tracts and Observations in Natural History and Physiology.* J. White, London.

VAUQUELIN, N.-L. 1805. Examen d'une pierre jaunâtre trouvée par MM. Desbassins et Godon, au sommet du Puy de Sarcouy (chaîne du Puy-de-Dôme). *Annales du Muséum d'histoire naturelle*, **6** (An XIII), 98–101.

WRAXALL, N. W. 1784. *A Tour Through the Western, Southern, and Interior Provinces of France.* Printed for Charles Dilly, London. Also a later Dublin edition, 1786. [First appeared in 1777 as part of Wraxall's *Memoirs of the Kings of France*, 2 Volumes. Printed for Edward and Charles Dilly, London. Vernière (1899–1900) says it appeared, in French translation, 1777: *Tournée dans les provinces occidentales, méridionales et intérieures de la France.*]

YOUNG, A. 1794. *Travels during the Years 1787, 1788 and 1789: undertaken more particularly with a view of ascertaining the cultivation, wealth, resources and national prosperity of the Kingdom of France.* 2nd edition, 2 Volumes. W. Richardson, London and Bury St. Edmund's.

Secondary works

BÉRALDI, H. 1913. *Le passé du Pyrénéisme. Notes d'un bibliophile. Ramond avant les Pyrénées.* Paris [no publisher].

BÉRALDI, H. 1917. *Le passé du Pyrénéisme. Notes d'un bibliophile. Le sentiment de la montagne en 1780.* Paris [no publisher].

BERNHARD, C. G. 1989. *Through France with Berzelius. Live Scholars and Dead Volcanoes.* Pergamon Press, New York.

BIREMBAUT, A. 1964. L'Enseignement de la minéralogie et des techniques minières. *In*: TATON, R. (ed.) *Enseignement et diffusion des sciences en France au XVIIIe siècle.* Hermann, Paris, 365–418.

BOUSCAYROL, R. (ed.) 1979. *Les lettres de Miette Tailhand-Romme, 1787–1797.* Aubière, Clermont-Reproduction.

BOYER, J. 1930. *Ramond et l'Auvergne.* Imprimerie Jean Vissouze, Clermont-Ferrand. [Thèse pour le doctorat (Pharmacie), Université de Toulouse.]

CAP, P.-A. 1861. *Philibert Commerson, naturaliste voyageur.* Victor Masson et fils, Paris.

CAROZZI, A. V. 2000. *Manuscripts and Publications of Horace-Bénédict de Saussure on the Origin of Basalt (1772–1797).* Éditions Zoé, Geneva.

CARTWRIGHT, D. E. 1997. Robert Paul de Lamanon: an unlucky naturalist. *Annals of Science*, **54**, 585–596.

CHEVALIER, A. 1934. *Michel Adanson: Voyageur, naturaliste et philosophe.* Larose, Paris.

CHILDE-PEMBERTON, W. S. [1925]. *The Earl Bishop: The Life of Frederick Hervey, Bishop of Derry, Earl of Bristol.* E. P. Dutton and Co., New York.

CIANCIO, L. 1995*a*. *A Calendar of the Correspondence of John Strange, F.R.S. (1732–1799).* Wellcome Institute for the History of Medicine, London.

CIANCIO, L. 1995*b*. The correspondence of a 'Virtuoso' of the late Enlightenment: John Strange and the relationship between British and Italian naturalists. *Archives of Natural History*, **22**, 119–129.

DANTON, V. 1986. Lacoste de Plaisance (abbé Lacoste). *Bulletin historique et scientifique de l'Auvergne*, **93**, 185–201.

DE BEER, G. 1951–1952. John Strange, F.R.S., 1732–1799. *Notes and Records of the Royal Society of London*, **9**, 96–108.

DE BEER, G. 1962 [1964]. The volcanoes of Auvergne. *Annals of Science*, **18**, 49–61.

EHRARD, J. 2004. Quand le Puy-de-Dôme devint un volcan. *In*: BERTRAND, D. (ed.) *Mémoire du volcan et modernité. Actes du colloque international du Programme Pluriformation 'Connaissance et représentation des volcans'.* Honoré Champion, Paris, 107–119.

ELLENBERGER, F. 1978. Précisions nouvelles sur la découverte des volcans de France: Guettard, ses prédécesseurs, ses émules clermontois. *Histoire et nature*, **12–13**, 3–42.

ELLENBERGER, F. 1979. Louis Cordier, initiateur de la pétrographie moderne. *Travaux du Comité Français d'Histoire de la Géologie*, series 1, **20** (13 pp.).

ELLENBERGER, F. 1994. *Histoire de la géologie.* Volume 2: *La grande éclosion et ses prémices, 1660–1810.* Technique et Documentation (Lavoisier), Paris.

FOUCHY, J.-P. GRANDJEAN DE, 1756. Éloge de M. le Marquis de la Galissonière. *Histoire de l'Académie Royale des Sciences*, 1756 [1762], 147–156.

FRIC, R. 1958 See Guettard (primary sources).

GALANTE GARRONE, A. 1971. *Gilbert Romme: Histoire d'un révolutionnaire, 1750–1795.* Translated from the Italian edition of 1959 by Anne and Claude Manceron. Flammarion, [Paris].

GARÇON, A.-F. 2002. Gabriel Jars, un ingénieur à l'Académie royale des sciences (1768–1769). *In*: DEMEULENAERE-DOUYÈRE, C. & BRIAN, É. (eds) *Règlement, usages et science dans la France de l'absolutisme: à l'occasion du troisième centenaire du règlement instituant l'Académie royale des sciences, 26 janvier 1699.* Tec & Doc, Paris, 237–253.

GAUDANT, J. 2003. Un observateur méconnu des volcans d'Auvergne: l'abbé Paul-François Lacoste, de

Plaisance (1755–1826). *Travaux du Comité Français d'Histoire de la Géologie,* series 3, **17**, 121–129.

GIRDLESTONE, C. 1968. *Poésie, politique, Pyrénées: Louis-François Ramond (1755–1827), sa vie, son oeuvre littéraire et politique.* Lettres Modernes–Minard, Paris.

GRIVAUD DE LA VINCELLE, C.-M. 1810–1813. Notice biographique sur M. Pasumot. *In:* PASUMOT, F., *Dissertations et mémoires sur différens sujets d'antiquité et d'histoire* [GRIVAUD, C.-M. (ed.)]. 7 parts in 1 volume. Paris, 1–14.

GROULX, L. A. 1970. *Roland-Michel Barrin de La Galissonière, 1693–1756.* Presses de l'Université de Laval, [Québec].

INSTITUT DE FRANCE, 1979. *Index biographique de l'Académie des Sciences, du 22 décembre 1666 au 1er octobre 1978.* Gauthier-Villars, Paris.

JULIA, D. 1996. Gilbert Romme, Gouverneur (1779–1790). *In*: EHRARD, J. (ed.) Volume from colloque Gilbert Romme, Riom 1995. *Annales historiques de la Révolution française*, **304**, 221–256.

KERSAINT, G. 1966. *Antoine François de Fourcroy (1755–1809): Sa vie et son oeuvre.* Éditions du Muséum national d'histoire naturelle (series D, 2), Paris.

LACROIX, A. 1921. *Déodat Dolomieu, membre de l'Institut national (1750–1801).* 2 Volumes. Perrin, Paris.

LALANDE, J.-J. L. DE 1775. Éloge de M. Commerson. *Observations sur la physique*, **5**, 89–120.

LAROUZIÈRE, F.-D. DE 2003. Le comte de Montlosier: une vision originale des volcans d'Auvergne à la fin du XVIIIe siècle. *Travaux du Comité Français de la Géologie*, series 3, **17**, 99–120.

LAWRENCE, G. H. M. (ed.) 1963–1964. *Adanson: The Bicentennial of Michel Adanson's 'Famille des plantes.'* 2 Volumes. Hunt Botanical Library, Carnegie Institute of Technology, Pittsburgh.

LOUGH, J. 1987. *France on the Eve of Revolution: British Travellers' Observations, 1763–1788.* Croom Helm, London & Sydney.

MALVEZIN, J.-E. 1879. Aperçu sur l'histoire de la botanique dans le Cantal. *Bulletin de la Société Botanique de France*, **26** [Sér. 2, t. 1], special section 'Session extraordinaire à Aurillac,' xxiii–xl.

MÈGE, F. 1884. *L'Académie des sciences, belles-lettres et arts de Clermont-Ferrand. Ses origines et ses travaux. Clermont-Ferrand, Typographie Ferdinand Thibaud.* [Facsimile edition: 1999. Imprimerie Infocampo numérique, Pau. Original 1884 edition: Typographie Ferdinand Thibaud, Clermont-Ferrand.]

MICHEL, R. 1945. A propos de la découverte des volcans éteints de l'Auvergne et du Vivarais. Notes sur deux géologues du XVIIIeme siècle: Guettard (1715–1786) et Faujas de Saint-Fond (1741–1819). *Revue des sciences naturelles d'Auvergne*, **11**, fasc 1–2, 37–53.

MICHEL, R. 1971. Les premières recherches sur les volcans du Massif Central (18e–19e siècles) et leur influence sur l'essor de la géologie. *In*: *Symposium Jean Jung: Géologie, géomorphologie et structure profonde du Massif Central Français.* Plein Air Service Éditions, Clermont-Ferrand, 331–344.

MOSNIER, H. 1895. *Le Marquis de Mirabeau au Mont-Dore (1770 et 1776).* Imprimerie G. Mont-Louis, Clermont-Ferrand. [Extrait du *Journal du Mont-Dore*, des 26, 27, 28 et 29 août 1895.]

NICOLAS, J.-P. 1963. Michel Adanson et l'Auvergne. *88e Congrès des Sociétés Savantes, Clermont-Ferrand, Actes,* part. 3, 57–67. [Not seen.]

OLIVER, S. P. 1909. *The Life of Philibert Commerson, D.M., Naturaliste du Roi: An Old-World Story of French Travel and Science in the Days of Linnaeus.* SCOTT ELLIOT, G. F. (ed.) John Murray, London.

PEREZ, M.-F. 1994. *L'oeuvre gravé de Jean-Jacques de Boissieu, 1736–1810.* Éditions du Tricorne, Geneva.

PEREZ, M.-F. & PINAULT, M. 1985–1986. Three new drawings by Jean-Jacques de Boissieu. *Master Drawings*, **23–24**, 389–395.

PICHARD, G. 1992. Robert de Paul de Lamanon (1752–1787): Entre Théorie de la Terre et Géologie. *Travaux du Comité Français d'Histoire de la Géologie,* series 3, **6**, 28–67.

QUÉRARD, J.-M. 1827–1839. *La France Littéraire, ou dictionnaire biographique des savants, ...* 10 Volumes. Firmin-Didot, Paris.

RANKIN, P. 1972. *Irish Building Ventures of the Earl Bishop of Derry, 1730–1803.* Ulster Architectural Heritage Society, Belfast.

RAPPAPORT, R. 1964. *Guettard, Lavoisier, and Monnet: Geologists in the Service of the French Monarchy.* Unpublished Ph.D. thesis, Cornell University. [UMI no. 64–13136.]

RAPPAPORT, R. 1969. The early disputes between Lavoisier and Monnet, 1777–1781. *British Journal for the History of Science*, **4**, 233–244.

REYNOLDS, G. 1997. William Bowles (1720 [sic]–1780), Eurogeologist. *European Geologist*, **5**, 67–70.

RIGODON, R. 1963. Malesherbes et l'Auvergne. *Revue d'Auvergne*, **77**, 127–132.

RUDEL, A. 1963. *Les volcans d'Auvergne.* Éditions Volcans, Clermont-Ferrand.

SMEATON, W. A. 1962. *Fourcroy: Chemist and Revolutionary, 1755–1809.* For the Author by W. Heffer, Cambridge.

TAYLOR, K. L. 1994. New light on geological mapping in Auvergne during the eighteenth century: The Pasumot-Desmarest collaboration. *Revue d'histoire des sciences*, **47**, 129–136.

TAYLOR, K. L. 1995. Nicolas Desmarest and Italian geology. *In*: GIGLIA, G., MACCAGNI, C. & MORELLO, N. (eds) *Rocks, Fossils and History.* Edizioni Festina Lente, Florence, 95–109.

TORRENS, H. S. 1999. Robert Townson (1762–1827): Thoughts on a polymathic natural historian and traveller extraordinary. *In*: RÓZSA, P. (ed.) *Robert Townson's Travels in Hungary.* Kossuth Egyetemi Kiadó, Debrecen, 19–26.

VERNIÈRE, A. 1899–1900. Les voyageurs et les naturalistes dans l'Auvergne et dans le Velay. *Revue d'Auvergne*. Volume 16 (1899): 331–348, 436–457; **17** (1900): 104–130, 239–294. [Also published separately, 1900. Typographie et Lithographie G. Mont-Louis, Clermont-Ferrand.]

VERNIÈRE, A. 1901. Notes sur les environs de Vichy et sur la découverte des volcans éteints de l'Auvergne (D'après un manuscrit autographe de Guettard, 1751). *Revue scientifique du Bourbonnais et du centre de la France*, **14**, 5–13.

J. D. Forbes and Naples

DENNIS R. DEAN

834 Washington Street, Apt. 3W, Evanston, Illinois 60202, USA

Abstract: James David Forbes (1809–1868) became a geological traveller in 1826 when, at the age of 17, he visited Naples with his family. Not content to be an ordinary, superficial tourist, he undertook sophisticated geological investigations of Mount Vesuvius and other well-known volcanic phenomena. On his return to Scotland, following a trip of some fifteen months abroad, Forbes wrote two anonymous papers based on his Italian experiences. When David Brewster, editor of the *Edinburgh Journal of Sciences*, accepted both, Forbes continued his submissions with an eight-part series entitled *Physical Notices of the Bay of Naples*, all of which appeared. These essays, summarized here, were—for such a young author—remarkably accomplished. Esteemed as genuine contributions to geological literature, they were cited approvingly over a period of twenty-five years by such prominent older geologists as Charles Lyell, K. E. A. von Hoff, Ami Boué, and Sir Henry De la Beche.

Historians of geology normally remember James David Forbes (1809–1868) (Fig. 1) for his work as a pioneering Alpine glaciologist. It is far less common to recall that his interest in glaciers was relatively late in coming. Forbes began his tragically foreshortened geological career as a teenaged volcanologist when touring Italy with his family. His only formal education to that time had been a six-month term at Edinburgh University, everything earlier having been tutorial. At the University, with which he was long associated, Forbes was influenced by the teaching of Robert Jameson. He also had a significant relationship with David Brewster, who (together with Jameson, for a time) edited the *Edinburgh Journal of Science*.[1]

The 1826–1827 itinerary

In 1826, Forbes's father Sir William (d. 4 October 1828), a wealthy Edinburgh banker trained in law, took his family on an extended European tour. His wife had died in 1810, soon after James was born. The other children were daughters Jane and Elizabeth and sons William (who would die *en route*), John, Charles, and James, the latter having postponed his second year of university to go along. Sir William, a Fellow of the Royal Society of Edinburgh, subscribed to such well known learned publications as the *Edinburgh Philosophical Journal*, edited by Jameson; and the *Edinburgh Journal of Science*, edited by Brewster. James read both journals as a child and was already a devotee of natural history before entering Edinburgh University in 1825 at the age of 16, to study law at his father's insistence. He qualified as an advocate in 1830 but never practised.

Because James kept detailed written records on a daily basis, together with dated sketches, we know the family's 1826–1827 itinerary in considerable detail (more than is reported here). They left Edinburgh by the steamship *James Watt* on 19 July 1826, arriving in London on 21 July for a stay of ten days.

On 31 July the Forbeses left London by steamship for Calais, then travelled via Montreuil, Abbeville, Amiens, Clermont, Chantilly and St Denis to Paris, where they remained from 5 to 15 August. On leaving Paris, the Forbes family continued through France and Germany to Strasbourg, Karlsruhe, Stuttgart, Ulm, Augsburg, Munich and Wasserburg. Entering Austria, they 'enjoyed the splendid scenery of the Tyrol' at Salzburg and Innsbruck. On 5 September, Forbes and his family crossed into Italy, reaching Venice on 10 September. Though they remained in Venice till 26 September, James was seriously ill for the first ten days. On 27 September they all continued south through Padua, Ferrara, and Bologna to Florence, staying from 5 to 19 October. Leaving Florence on 20 October, the Forbes family proceeded through Perugia, Spoleto and Terni to Rome, where they resided from 24 October to 9 November. Departing Rome on 10 November, they stayed overnight amidst the volcanic landscapes at Albano and Mola di Gaeta before reaching Naples on 12 November. The roads were much poorer south of Rome.[2]

The family went sightseeing in Naples from 13 November to 21 December. They were told on arrival that Mount Vesuvius, dormant since the great eruption of 1822, had begun 'smoking' again a few days before. On 14 November, Forbes wrote, 'we observed Vesuvius smoking with considerable violence' (Fig. 2). On 18 November, the

From: WYSE JACKSON, P. N. (ed.) *Four Centuries of Geological Travel: The Search for Knowledge on Foot, Bicycle, Sledge and Camel.* Geological Society, London, Special Publications, **287**, 97–107.
DOI: 10.1144/SP287.8 0305-8719/07/$15.00

Fig. 1. James David Forbes *c.* 1850, by an unknown photographer. Courtesy of the University of St Andrews Library.

Forbes family ascended Vesuvius on muleback, the young ladies riding side-saddle; they reached the edge of the crater, climbed partway down inside it, and ate a picnic lunch at the Hermitage on the way back. Three days later, James and his brothers (but no ladies) went up again to find the whole crater full of smoke and unpleasantly sulphurous fumes. Even so, the Forbes boys walked round the edge of the crater, on ground in which 'almost every little chink ... was so hot that you could not keep your hand the least time in it.' James sketched the changing cone of Vesuvius on 26, 27 and 29 November, then again on 12 December. He kept a daily journal of Vesuvian activity from 14 November to 13 December and a meteorological journal for the same period (to 15 December). On 2 December, James visited Solfatara, a still-fumarolic volcanic crater adjacent to the Phlegraean Fields. Other day trips took him to Lake Avernus, Pozzuoli, and classical sites associated with Virgil. Between 17 and 21 December he recorded visits to Pompeii, Paestum, and Herculaneum.

On 22 December the Forbeses left Naples for Rome, staying overnight at Mola de Gaeta and Albano as before. They remained in Rome from 24 December 1826 to 12 March 1827, enjoying the holiday season. But James was by now seriously interested in both geology and archaeology. On 1 March he rode out to explore the volcanic

Fig. 2. James David Forbes, sketch of Mount Vesuvius in eruption, 1827 (ink and wash, Journal I). Courtesy of the University of St Andrews Library.

nature of the Campagna. James visited the ruined imperial port city of Ostia on 6 March and on 11 March went to see 'the famous quarries of black lava with which ancient and modern Rome . . . have been paved.' On 13 March, when the family once again set out for Naples, he noted rock and soil types and landscape features along the way.

Between 16 and 22 March, the Forbeses once again went sightseeing around Naples. On the 22 James was fascinated by the ruins of Pompeii: 'the chalked advertisements, the litter of the shops, the tickets of the theatre, the gentlemen's names on their houses, the loaves, honeycombs, soap, grain, raisins, plums, nay the very fresh olives which the ancients used.' He hoped that some lost classical literature might yet be found. Further destinations included Pozzuoli and Portici. On 27 March the Forbeses sailed to the island of Ischia for an overnight stay; James stuck his thermometer into various hot springs to record their temperature. On 30 March they sailed (again overnight) to Sorrento. On 2 April, James and his brother Charles paid a third visit to Vesuvius, to see how the crater had changed since their last visit in November 1826. A perusal of the antiquities museum in Naples followed the next day; its treasures inspired James to affirm in the privacy of his diary just how meaningful the region's geology and history had become for him. On 4 April, the Forbeses left Naples for Rome, where they planned to celebrate Easter (as they had Christmas). Their homeward journey began with departure from Rome on 23 April and included Siena, Florence, Fiesole, Lucca, Pisa, Carrara, La Spezia, Genoa, Parma, Bologna, Padua, Venice, Verona and Milan. They continued through Switzerland and France to England and finally Scotland.

The series of articles and their significance

Even before leaving Edinburgh, James David Forbes had aspired to contribute his own writing to the Scottish scientific journals so highly thought of by his father. On 29 May 1826 he sent David Brewster, editor of the *Edinburgh Journal of Science*, a short item about the forthcoming conjunction of Jupiter and Venus on 31 July. Brewster accepted the note for publication without knowing who the anonymous author was. Forbes had signed himself as Delta, using a Greek letter he had learned only that year. Brewster, who knew Forbes as a student, certainly did not expect such a sophisticated contribution from anyone so young.[3]

On returning to Edinburgh University, in November 1827, Forbes took classes from Professor John Wilson (*Blackwood Magazine*'s 'Christopher North') and Robert Jameson, excelling in both. His notes from Jameson's course on natural history still survive. Forbes graduated in July 1830 and returned to Edinburgh University in 1833 as Professor of Natural Philosophy in his own right, defeating Brewster for the job. Brewster had, in the meantime, published several scientific papers by Forbes, whose identity as Delta had become known to him only in the waning days of 1828. Two of these papers related directly to Forbes's experiences in Italy. One described the family's climb up Mount Vesuvius in November 1826 ([Forbes] 1827); the other was a detailed discussion of the use of building materials in ancient Rome ([Forbes] 1828). The success of both convinced Forbes to attempt his eight-part series (Forbes 1828–1830).[4]

Physical Notices No. I

By August 1828, Brewster had received and would soon publish the first instalment of an announced series by Delta collectively entitled *Physical Notices of the Bay of Naples*. Number I, '*On Mount Vesuvius*,' dated 9 August 1828, ran in the issue for October 1828. It began with a paean to the Neapolitan region as a geological, historical, and literary wonderland still awaiting an adequate description in English. The best presently available in any language, Forbes thought, remained Scipione Breislak's *Topografia Fisica della Campania* (1797), which was available in Italian and French (1801); it promised to be a chief source and frequent irritant within Forbes's series. Devoted to topographical and scientific description, that series would combine 'a repeated survey of the locality' with 'principal works on the subject'. Other authors cited in this essay included Alexander von Humboldt, Sir William Hamilton, Charles Daubeny, Giovanni Maria della Torre, and several Italian lithologists.

Forbes began, as per his title, with Mount Vesuvius, which was in reality two separate volcanoes, Somma partially enclosing Vesuvius. The abrupt inner wall of Somma had long been a puzzle to geologists, as its lavas were traversed by numerous dykes. Forbes thought he could explain the dykes by supposing Somma to have originated as a volcanic bubble in which 'pre-existent horizontal strata of lava were upheaved by internal action' (191). When the mass cooled, it collapsed partially on itself, creating cracks and fissures to be filled later on by a different kind of lava, thereby creating the dykes.

The semicircular area between Somma and Vesuvius was called the Atrio del Cavallo because tourists making the ascent left their horses and mules at this point. Recent lavas, like those of 1822 (the last major Vesuvian eruption), had roughened this traditional resting place and left it looking unusually rugged, as Forbes had pointed out in his earlier account of the climb (Forbes 1827). Ropy lavas from the eruption of 1819 adjoined the slightly elevated Hermitage of San Salvador, where wine and other refreshments were provided by a monk in residence. Between the Hermitage and the cone lay il Cratere del Francese, so named in 1820 (when it was molten) because an unhappy Frenchman threw himself into it from despair.

Speculations on the possible origins of Somma and Vesuvius followed, together with more routine factual information. 'I have been at the summit of the volcano three times' (197), Forbes tells us (as do his diaries). He thought it obvious that 'the volcanic agency is liable to change its place and manner of acting' so as to shift the point of ejection from Monte Somma to the modern cone (197). Not all eruptions originate from the main crater, he pointed out. For example, the major eruption of 1794, which destroyed Torre del Greco, burst forth at the base of the cone. The whole mountain has been raised by subterranean heat, volcanic action having 'produced the strata on every side as far down as man has penetrated' (200). The essay included an illustrative plate showing Vesuvius in cross-section (Fig. 3). Its final topic was the various minerals produced by the mountain, which were described at length.

Physical Notices No. II

On the Buried Cities of Herculaneum, Pompeii, and Stabiae appeared in January 1829 and was again by Delta. Having noticed the topography, phenomena, and products of Vesuvius in his previous essay, Forbes now thought it appropriate to describe the most spectacular results of its volcanic agency, namely the embalming, as it were, of three whole cities and their way of life. All prior authors, whether classicists or geologists, had failed to do this topic justice. Restricting himself to physical phenomena primarily, Forbes discussed the original situation of these cities, the eruption that destroyed them, and their present condition.

Going from west to east, as in his title, Forbes began with Herculaneum, citing classical references for each. His remarks on the eruption of 79 AD derived almost wholly from the account of it by Pliny the Younger. Forbes stressed, however, that the supposed 'lava' under which the cities were buried was in reality 'tufa, volcanic dust, or decomposed trachyte' (125). By 'tufa' he meant volcanic ash now called tuff.

Sir William Hamilton (1772) had discussed the tufas of Herculaneum at length. Like Hamilton, Forbes also saw them as originating in submarine eruptions. 'All volcanoes,' he affirmed, 'hold a particular communication with the sea, which would appear to be a requisite agent in the production of their effects' (126). Alluding to his previous essay, Forbes theorized that the altered point of ejection from Somma to Vesuvius, in 79 AD, was

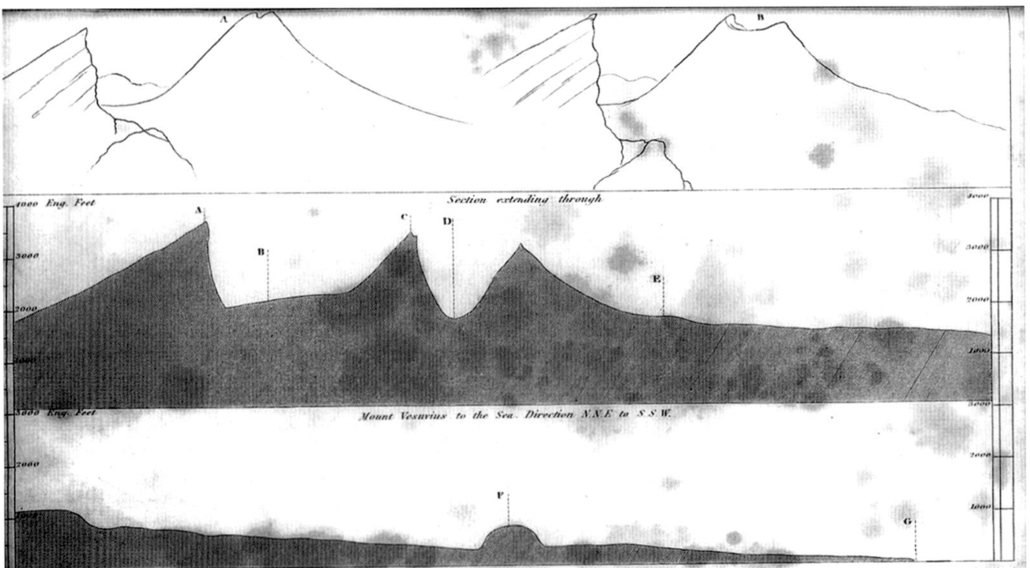

Fig. 3. James David Forbes, Plate III, Vesuvius in section, *Edinburgh Journal of Science*, **9** (1828).

Fig. 4. James David Forbes, undated sketch of Mount Vesuvius (perhaps based on a painting, Journal VII). Courtesy of the University of St Andrews Library.

due to marine influence. The crater wall of Somma facing seaward was overthrown and its debris formed the plain from which Vesuvius arose (128). 'The fall of the southern wall of the crater would bring the whole fiery deluge in the direction of the sea,' he explained, 'and, without doubt, the interment of Herculaneum was only a portion of the ravages it produced' (128).

There is also a postscript on Mount Vesuvius, in which Forbes remarked on a paper by Alexander von Humboldt concerning the structure and activity of volcanoes. Emphasizing Vesuvius, Humboldt's essay was now a chapter in his newly revised *Tableaux de la Nature* (1828), which had just appeared (Fig. 4).[5] The original paper had been given in Berlin on 24 January 1823.

Physical Notices No. III

On the District of Pausilipo and the Lago d'Agnano appeared in April 1829 and was the first 'By James D. Forbes, Esq.' Within it, he announced the topics for each of the remaining five articles (245).

Immediately west of Naples, via the delightful Strada Nuova, Pausilipo is a ridge of volcanic tufa with a flattish top and 'saddle-shaped' layering. Several fine sections of exposed strata were visible from the new road. 'The layers', wrote Forbes, 'succeed each other with great regularity and sharpness' (247). They were composed of various alternating volcanic conglomerates, pale yellow predominating. The gently undulating stratification, by no means uniform, bore 'the most irresistible marks of diluvial deposition' (247; i.e. by flood rather than fire). Just offshore, 'in the line of the ridge, and obviously separated from it either by some convulsion of nature or the slow operation of time' (248), was the small island of Nisida, which was of tufa and displayed the remains of an extinct crater. Pausilipo itself was punctured by a manmade tunnel, half a mile long, through which one could walk, though with little light. Before the Strada Nuova was constructed, this tunnel had been the only way to get from Naples to Pozzuoli—and it was still the quickest.

Enlarging on remarks in his previous paper, Forbes emphasized how the 'volcanic agency has been combined in these formations with all the peculiarity of subaqueous deposits' (251). They were, he saw, the product of submarine eruptions, even Vesuvius having 'gradually raised itself to its present character' (252) from the bottom of the sea. Though Forbes elaborated on its supposed subaqueous formation, he rejected the theory that Pausilipo had once been a volcanic crater. All such craters or crater lakes as Agnano, Avernus, Astroni, and the Solfatara, he believed, had been shaped by eruptions subsequent to their elevation out of the sea. He also believed that sea level had fallen.

Lake Agnano, Forbes noted, was never mentioned by classical authors and must have arisen during the Middle Ages. After recalling his own visit of 7 December 1826, Forbes discussed at length the infamous Grotto del Cane, that 'small emissary of carbonic acid gas' (258) in which reluctant dogs were forced to demonstrate the potency of its fumes. Though much written of since classical times, it was only a manmade hole some ten feet long and four across near Lake Agnano. Among Forbes's papers, there is a sketch of it dated 5 December 1826.

Forbes thought Astroni 'one of the best marked extinct volcanic craters in existence, and besides, one of the most agreeable spots in the whole range of the Bay of Naples' (263). With its flat bottom, precipitious walls, and several parasitic cones, it must have been actively volcanic long after the formation of the tufaceous hills and the withdrawal of the ocean. Its bedding included real lava, which led Forbes to think that Astroni might even be so old as the extinct volcanoes of the Vivarais in France (on which he would later write). At the time of Forbes's visit, it had become a royal hunting preserve, replete with evergreens, and 'a scene of the most romantic seclusion' (265).

Physical Notices No. IV

On the Solfatara of Pozzuoli appeared in July 1829. He had visited it on 2 December 1826.

Among the well-known geological features of the Bay of Naples, the Solfatara ranked second only to Vesuvius. So much had been written about it already that Forbes, as throughout his series, felt he could do little more than condense and epitomize earlier work. His modern sources now included a recently published paper on the volcanic district of Naples by G. P. Scrope (1828). Avoiding the bulky and unconnected data of other writers, he would provide only those details that an intelligent physical inquirer might reasonably wish to know. Forbes therefore described the history, structure, and phenomena of the Solfatara before concentrating on his most important topic, its products.

The Solfatara was a volcanic crater, fumarolic but not yet extinct. Classical, medieval, and later descriptions proved that its appearance had not changed in historical times. It may have erupted as recently as 1198 (rather than 1180, as Scrope had it). However, active the Solfatara may have been in past times, water within the crater was formerly more abundant, suggesting greater volcanic energy. Like others, Forbes believed that the Solfatara was connected with both Vesuvius and the sea.

By its products, Forbes meant 'such as are daily forming by the action of subterraneous volcanic agency' (134). He accepted the theories of Daubeny (1826) and Breislak (1792) that attributed volcanic action to the 'affusion of the metallic alkaline bases by sea water' (134), a chemical process he then explained. Carbonic acid gas, like that emitted in the Grotto del Cane, 'probably owes its origin to the effect of internal heat upon calcareous strata, and its prevalence in those volcanoes which are nearly extinct has been ingeniously explained by the fact that potassium, and probably the other alkaline and earthy bases when heated, decompose this gas' (135). The subsequent action of these gases accounted for all the varied products of the Solfatara.

Physical Notices No. V

On the Temple of Jupiter Serapis at Pozzuoli, and the phenomena which it exhibits appeared in October 1829. Forbes dated his visit as 24 March 1827, but there may have been earlier ones.

The so-called Temple of Jupiter Serapis stood on the shore of Pozzuoli immediately SW of the Solfatara. Though of Roman date, none of the classical writers mentioned it. The temple lay unknown until the mid-eighteenth century, when three projecting columns, hitherto concealed by shrubbery, induced an excavation and some thievery. A few inscriptions were found and various dates assigned; it seemed to be a Roman building, rebuilt atop an originally Greek foundation and completed before the close of the second century AD. Forbes accepted its traditional identification as a temple [it is actually a Roman market hall of the first century AD] and endeavoured to decide which deity it had honoured. A bearded image dug up in 1750, and now in the Naples museum, appeared to be that of Serapis, an Egyptian god of the lower world who inspired cults in ancient Greece and Rome.

Very little was known about the temple's survival and ruination during the Middle Ages. Though its submergence was sometimes attributed to the eruption of Monte Nuovo in 1538, Forbes argued for the eruption of the Solfatara in 1198 instead. Only three great marble pillars of the temple remained standing; they were those first seen in 1749. The excavation that followed their discovery was not in a quest for knowledge but rather to obtain materials for the palace of Charles III, King of Naples, at Caserta. Forbes included a detailed plan of the temple in his Plate IV and discussed its layout in his text.

The most curious feature of the three still-standing columns was that ten feet above the base of each, and corresponding exactly, there was a six-foot zone in each pillar that had been attacked and eaten into by stone-boring molluscs (*Mytilus lithophagus* Linnaeus) to a depth of as much as four inches. Even so, they avoided the hard nodules of quartz and feldspar sometimes present in the marble. The molluscs in question were no longer found in the Bay of Naples but remained abundant nearby.

The size and depth of the molluscan perforations attested to a lengthy immersion of the columns. How this came about was not easy to explain. The two most obvious theories were that the sea level had risen and fallen as indicated; or, as was more commonly maintained, that the land had been alternately lowered and elevated by earthquakes. The first explanation was unlikely because any such rise in sea level would have had to be worldwide. This was too extravagant a device by which to explain a local fact. The second explanation, though endorsed by no less an authority than John Playfair in 1802, failed to ensure that the twice-shaken columns would remain standing.[6]

There were three further theories. Spallanzani thought that the columns might have accidentally been buried in the sea, then dredged up and reused in the temple. But, as Forbes pointed out, the three zones of molluscan damage would not then have corresponded so precisely when the columns were re-erected. Such a grand temple would not have used damaged columns without at least covering them up. Some of the fragmentary columns also present had mollusc holes throughout their length and must therefore have fallen before their subsequent immersion. A second theorist, Raspe, suggested that the stone was perforated before being shaped into columns. His idea defies common sense. The latest and most farfetched theory, apparently originated by Goethe (and supported by Daubeny, among others), supposed that when the temple was covered by volcanic tufa, a small central hollow remained. Within it a salt lake formed, surrounding the columns and making the latter accessible to lithophagous molluscs, the shells of which remained in their holes after the salt lake dried up. The temple was then restored by the Romans as a ruin.[7] None of these latter theories could withstand serious scrutiny.

Forbes himself preferred the second theory, which attributed apparent changes in the waterline to alternate depression and elevation of the land. First, he reasoned, the present position of the temple, below the high-water mark, required the action of some prior cause that Goethe's lacustrine theory (the fifth above) did not include. It was well known that certain Roman ruins in the Pozzuoli area, as well as two roads, were now submerged in the ocean and could be seen through the water. Second, throughout the region there was abundant evidence of elevation. Marks of rock-eating marine molluscs abounded at Palermo, in Calabria, and on the Monte Circello, between Naples and Rome. [In No. VIII Forbes himself noted further examples in the limestone of Capri.] Third, Forbes continued, 'we have all the agents required for the accomplishment of our theory within the bounds of recorded information or the most direct analogy, without any pure assumptions whatever' (281). The *waterline* at Pozzuoli had been gradually rising since the sixteenth century—a situation precisely opposite to the rising *land* of the Baltic region. He also believed that some such changes took place suddenly. In concluding his essay, Forbes set out and replied to four objections to the theory of the elevation and depression of the land, a concept having simplicity to recommend it.

Physical Notices No. VI

On the District of the Bay of Baja [Baia] appeared in January 1830. Forbes visited this area, part of the Phlegraean Fields, on 11 December 1826, though perhaps not for the first time. With this instalment he returned to the orderly topographical survey temporarily interrupted by his essay on the Temple of Jupiter Serapis.

Two volcanic cones, Cappomazza and Monte Barbaro, adjoined the Solfatara. Though in appearance consistent with Forbes's theory of submarine eruption, they were probably created later, after Astroni and the Solfatara. Among yet more recent cones was the nearby Monte Nuovo, known to have appeared suddenly only three centuries before, on 29 September 1538. Several contemporary accounts of the event had been preserved. After two days of preceding earthquakes, a gulf opened between the village of Tripergola (situated where Monte Nuovo was) and some nearby baths; the latter made use of volcanic heat and were much frequented. Spewing forth stones, ashes, and occasionally flames, the noisy chasm approached the town and eventually destroyed much of it, throwing up a mountain in its place. The sea retreated suddenly, and a large tract of ground near the base of Monte Barbaro was elevated from the floor of the Mediterranean, much as Forbes and others had theorized in the case of the Temple of Jupiter Serapis.

The crater of Monte Nuovo, which Forbes climbed up to sketch, was as deep as the mountain was high, about 440 feet. 'The magnitude of craters,' Forbes observed, 'is almost universally in the inverse ratio of the size of the volcano' (180).

There having been no flow of lava, Monte Nuovo was constructed entirely of ejecta. Forbes supposed that its unseen base probably consisted of tufa, indicating a submarine origin. He recalled several well known examples of submarine eruptions in Greece (Santorini), Iceland, the Aleutian islands, and the Azores.

Among the immediate consequences of the eruption of Monte Nuovo in 1538 was the filling up of the Lucrene lake, originally a volcanic crater in its own right. Only a marsh now remained. Lake Avernus (Lago Averno), once artificially connected with the Lucrene, also occupied a crater. Though formerly thought to be bottomless, and fatal to birds in flight, it was at present little more than one hundred feet deep and entirely harmless. Perhaps Lake Avernus emitted mephitic fumes at one time. A most remarkable thermal spring, the Baths of Nero, was nearby. One entered, as Forbes himself did, through passages cut in the tufaceous rock. There were several similar baths nearby, together with the ruined villas of wealthy Romans who had formerly enjoyed them. The villas were built of quarried tufa mortared by pozzuolana, a local rock. The Baths of Nero, the pozzuolana quarries, the castle and village of Baja, and three temples were all situated on the western boundary of the Bay of Baja, forming a neck of land (terminated by the Capo di Miseno) that projected toward the islands of Procida and Ischia.

Physical Notices No. VII

On the Islands of Procida and Ischia appeared in April 1830. Forbes visited Ischia (only) on 27–28 March 1827, investigating its hot springs particularly.

These two islands represented the true western extremity of the Bay of Naples. Both were volcanic and had been active long before Vesuvius, a then still-dormant crater that eventually superseded them. Some think that Procida was created by an eruption of Ischia. Composed entirely of tufa interbedded with slaggy lava, Procida offered little of geological interest. Its rocks were basically the same as those of nearby Capo di Miseno. But there was a fine view of the Bay of Naples from its castle.

Mount Epomeo made Ischia less monotonous than Procida and far more interesting. Virtually the whole island consisted of rugged dells and shivered crags so untameable that travel by any means was difficult. Neither horses nor vehicles of any kind existed there, only donkeys. Much of the island remained wooded, including all but the summit of Epomeo, which was weathered and crumbling. There were also extensive vineyards, with high walls of cleared boulders and narrow lanes between. The several villages along the coast were inhabited mostly by fishermen, but the poor had only dried figs (grown on the island) to eat. Its meagre capital city was called Ischia or Celso.

In earlier times Ischia was noted for its earthquakes and eruptions. The most famous eruption took place in 1302, not from Epomeo [which is not a volcano] but from the Campo del Arso; its lavas ran into the sea near Casamicciola. The most recent earthquake took place on 2 February 1828 and devastated Casamicciola. Overall, the whole island seemed to be composed mostly of tufa. But, as Forbes admitted, 'the uncertainty of the characters of volcanic rocks is still so great that a controversy about the name or class of a rock like that of Ischia is little better than a quarrel respecting words' (335). Their modes of formation and ejection were likewise obscure and geologists remained 'wholly in the dark' about the relative ages of most volcanic rocks.

So little was known about Monte Epomeo that even its height was uncertain. Two hermits resided year-round at the top, which could be reached by a path hewn in the rock. On the morning of Forbes's ascent, poor weather obscured the otherwise magnificent view. Nonetheless, he could see several volcanic cones and the lava flows associated with them. Some of the flows contained crystals of feldspar, 'but we are obviously as yet quite ignorant of the causes of changes ... in many minerals under the action of heat' (337). Mount Taborre, between Casamicciola and Celso, was composed of a trachyte whose age had yet to be determined. Still, Forbes saw, 'it overlies a bed of clay undoubtedly not older than the tertiary series, from the shells which it contains' (368, citing Brocchi (1814), pp. 65 and 354).

Perhaps the most interesting rock on Ischia was serpentine, a beautiful shiny green substance commonly used in Naples to manufacture snuff boxes. Yet among the various Italian geologists, only Brocchi mentioned it. As pebbles of serpentine were found along the coasts of Ischia, Brocchi thought they might have been transported by the sea or used as ballast by passing ships. Forbes could not agree. He had found the same pebbles of serpentine in the bed of a small brook running down the north side of Epomeo. Though Brocchi's attempted explanation was clearly inadequate for other reasons also, the real origin of serpentine remained as yet undiscovered. Macculloch, in his *Classification of Rocks* (1821), had altered the last sheet to contradict what he had said about serpentine in the middle of the book. A dyke of serpentine in Forfarshire had convinced Lyell of its igneous origin, and Boué had similar evidence (Macculough 1821; Lyell 1825). That serpentine had been elevated from the depths of the earth by volcanic forces, Forbes concluded, was fairly certain.

Ischia, like other parts of volcanic Naples, was famous for its hot springs. Forbes studied several of them in March 1827 and reported his findings in detail. He also mentioned a lava cave on Ischia noted for its *cold* wind. Though the matter was far from settled, Forbes thought that the only possible cause had to be evaporation.

Physical Notices No. VIII

Concluding view of the volcanic formations of the district, with Notes upon the whole Series appeared in October 1830.

After an apologetic preamble to his final paper on the Bay of Naples series, Forbes lamented the inadequacy of the voluminous but unsatisfactory literature on the region that was already in existence. 'Italy,' he complained, 'has had the misfortune to be seen and described by thousands, but investigated in all her boundless store of natural science, not less captivating than her arts and antiquities—by none' (246). In the present paper he would review the geological structure of the Bay of Naples for the sake of drawing some general conclusions not possible before.

The Bay of Naples was an ideal region in which to study the various phases of volcanic activity, whether igneous, aqueous, active, quiescent, or extinct. The outline of this part of the coast of Italy was almost entirely due to the deposition of tufa on a base of older rock. 'Looking to the period when the secondary rocks were the latest in existence,' Forbes surmised, 'the primeval ocean must have washed the Apennines near Capua, fourteen miles directly landward from the present shore, and the curve of the bay extended from its present eastern point, which is composed of Apennine limestone, to a promontory to the westward .... The Bay must then have had double its present aperture' (248).

The originally contiguous limestone Isle of Capri, now insulated by an abrading current, was almost entirely composed of this earlier rock—unstratified Apennine limestone, without any admixture of volcanic rocks. The nearby Sorrento peninsula, though enclosed by limestone hills, lay on a plain consisting entirely of compact tufa. Its volcanic formations must once have been part of the Phlegraean Fields, only to be separated from them by later marine erosion. A long discussion about the origin of tufa followed. Most Italian geologists believed it to be a product of submarine volcanoes. Others advocated land eruptions or even mud volcanoes. The submarine origin of tufa seemed indisputable to Forbes because comparatively recent shells had been found in it. [As previously noted, tufa, now more commonly called tuff, is a usually stratified porous volcanic rock formed by the consolidation (fused by heat) of volcanic ash and similar debris.]

The hills of the Phlegraean Fields presented more geological variety than some authors have been willing to admit. The hill of Pausilipo, for example, with its artificial tunnel, is largely unstratified, proving that it cannot have been marine in origin. [Cf. No. III above.] For the region as a whole, Forbes supposed that one kind of pumaceous tufa had been deposited under the sea by submarine volcanoes. These, and even the upper portion of Pausilipo, were then covered by mud eruptions. Deposits from these mud eruptions, perfectly horizontal, lie in basins that are unconformable with them. In Forbes's view a second eruption of tufa from below evidently filled up fissures in the first. The craters responsible for this involved history cannot now be found.

Forbes next revealed his several differences of opinion with G. P. Scrope, who had written an important book and paper on mostly Neapolitan volcanoes (Scrope 1825, 1828). Forbes particularly chided Scrope for his rejection of the theory of elevation craters proposed by Leopold von Buch, a hypothesis that Forbes considered 'essential to the explanation of observed facts' (260). Indeed, Scrope's own theorizing seemed untenable without it. Forbes went on to explain how the dip of beds in a volcanic cone of elevation would differ from those in a cone of eruption. 'Mr Scrope's theories on this subject,' Forbes concluded, 'require reconsideration and condensation' (263).

The geological age of the tufas could be judged only by their included organic remains, which were for the most part either modern species of marine shells or others closely related to them. The oldest tufa deposits were either coeval with cenozoic rocks or younger. Even so, the trachytic rocks were younger still. Their eruption may even have caused the elevation of the submarine tufa. But different varieties probably represent very different epochs. They grade into the oldest modern lavas, of which Mount Vesuvius has left us an invaluable record. 'If, as we may hope,' wrote Forbes, 'the problem of volcanic action be one day solved, in all likelihood Mount Vesuvius will be the most fertile source of information' (265).

The final essay then concluded with a series of Notes, A–L, emending previously published instalments. Note L was just in time to notice 'Mr Lyell's excellent work on Geology' and his mention of Forbes's series within it (277–278).[8] Other geologists who cited Forbes's series on the Bay of Naples included K. E. A. von Hoff, Ami Boué, Charles Daubeny (1848), and Sir Henry De la Beche (1851).

Forbes and volcanoes after 1830

Forbes and his new wife Alicia returned to Naples in 1843–1844 as part of a year in Italy made necessary by his poor health. By that time his chief interest had become Alpine glaciers. The Forbeses left England on 14 July, reaching the continent via Dover and travelling through Ostend, Antwerp, Cologne, and Bonn. In Bonn, Forbes was confined to bed for a month with gastric fever and pulmonary inflammation. They left Bonn on 14 August for Baden-Baden, then Strasbourg, Basle, Neuchâtel, Lausanne, Bex, and Chamonix, where they stayed for ten days. Further touring in Switzerland, Germany, and then France took the Forbeses to Marsailles. They sailed to Genoa and eventually reached Naples on 24 November.[9] As Forbes remarked, 'Naples appeared to me precisely as I left it 17 years ago'. He began his geological observations a few days later. Visiting Vesuvius on 30 November, James noted the changes since he had last seen it. On 5 December he visited Astroni, Agnano and the Solfatara. Two days later he and his wife went by boat around Posilipo to Nisida; finding the volcanic formations there of interest, he sketched several sections and features. On 9 December they were at Pozzuoli, where Forbes revisited the Temple of Jupiter Serapis and again climbed Monte Nuovo to sketch the crater. He visited Mount Vesuvius once more on the 13 December, to study an exposed section, and returned to the Solfatara three days later. James showed the Naples museum to his wife on 18 December. Two days later they took a carriage to Pozzuoli and arranged for a boatman for three days to take them to Ischia. They circled both Ischia and Procida, stopping several times to go ashore. On 28 December, following the Christmas holidays, Forbes went up Vesuvius alone to investigate the older lava flows. On 30 December he and his wife visited Herculaneum and Pompeii. Forbes then revisited Vesuvius once more on 1 January 1844, when it was issuing flames but no lava. He examined other lava flows the next day. On 4 January he and his wife left Naples by steamer for an extended stay in Rome. Forbes's diary ended abruptly in mid sentence at Albano on 11 April.

Aside from his books, *Travels through the Alps* (1843) and *Norway and its Glaciers* (1853), Forbes also wrote numerous geological papers, including a series of sixteen letters on glaciers, all of which appeared in the *Edinburgh New Philosophical Journal* edited by Jameson. The sixth letter in that series compared the motion of glaciers with that of lava (Forbes 1844). Excepting a paper of his on the extinct volcanoes of the Auvergne (Forbes 1836), he had no further volcanological publications.

Conclusion

Forbes's understanding of the origin of tufa (tuff), like his advocacy of von Buch's Craters of Elevation theory, has been discredited. His use of work by other geological writers seems more like the end of an era than the beginning of one. But neither of these judgments was so obvious in his own times as each seems to us now. When Forbes wrote of the Bay of Naples, he was a very young man who had the self-confidence to cross swords with older, more established writers. Though his essays on the Bay of Naples contributed nothing of lasting significance to the emerging science of volcanology, they interested many persons in it and were thought well of by talents we now regard as superior. His contributions to glaciology, though similar in some respects, are of a higher order.

I wish particularly to thank N. H. Reid, Keeper of Manuscripts, the University of St Andrews, for his assistance throughout the latter stages of this project and my faithful typist T. Murray for her patience as new information continued to turn up. The reference staff at Northwestern University and W. Cawthorne of the Geological Society of London also made significant contributions. Versions of this essay were presented orally at the Geological Society of America meeting in Denver, October 2002, and at INHIGEO, Dublin ('Geological Travellers'), July 2003.

Notes

[1]The only modern biography is Cunningham (1990). There is also a nineteenth-century *Life and Letters* (Shairp *et al.* 1873). Robert Jameson (1774–1854) was Regius Professor of Natural History and Keeper of the museum at Edinburgh University from 1804 to 1854. He founded the Wernerian Society there and was its president for life. As a teacher of natural history, Jameson ranged widely, but in geology his thinking was generally subservient to the influence of Werner and, later, Cuvier. His open-minded journal editing provided more excitement than his lectures.

Sir David Brewster (1781–1868) was both an experimental physicist specializing in optics and a licenced preacher, though he soon abandoned the pulpit to promote the popularity of science. In addition to scientific essays and his frequent editing, Brewster wrote a biography of Newton, the first life ever to celebrate a scientific hero.

[2]J. D. Forbes, Journals, 1826–1827; Jane Forbes, Journal, 1826. Forbes was principal of St Andrews University from 1859 until his death; his papers remain there, excellently curated.

[3]In his earliest contributions to the *Edinburgh Journal of Science*, Forbes wrote on astronomical and meteorological topics.

[4]Forbes (1828–1830). His original manuscripts for parts II–VIII are preserved at the University of St Andrews. Subsequent numbers in brackets in the text refer to pages within this reference.

[5]Alexander von Humboldt, *Tableaux de la Nature* (1828), which had just appeared (Original German edition: *Ansichten der Natur*, Tübingen, 1818; second edition,

1826). English versions followed as *Aspects of Nature* (London 1849) and *Views of Nature* (London 1850).

[6]See Playfair (1802). Playfair (1748–1819) had been Professor of Mathematics (to 1805) and then of natural history at Edinburgh University. His influential reviews opposed the geological beliefs of Werner, Cuvier, and Jameson. In the passage cited, he was writing in opposition to Breislak.

[7]See Spallanzani (1798), **1**, pp. 83–89 and volumes 1–3 of Spallanzani *Opere* (1825–1826). Also Ferber (1776); in a footnote to pages 172–173, Raspe opposed Ferber's opinion that the sea level, rather than the land, had changed. Johann Wolfgang von Goethe (1749–1832) was a famous literary figure whose numerous attempts to be regarded as a serious contributor to our knowledge of the natural world have not withstood the passage of time. In his paper of 1823 on the Temple of Jupiter Serapis (opposing von Hoff (1822) *Geschichte*, **1**, pp. 455–456), Goethe tried to show that when the temple was buried, a water-filled depression probably formed in the middle, obviating the need to postulate any subsequent rise (Goethe 1823, **2**, pp. 79–88). Forbes cites Spallanzani, Raspe, and Goethe by name only (as is frequently his habit).

[8]Lyell (1830); see pp. 455–459. von Hoff, *Geschichte*, **III**, p. 329 (1834), also cites Boué (1830) **1**, p. 354, and Lyell (1830), on Forbes. De la Beche (1851) endorses all eight articles in Forbes's Neapolitan series.

[9]J. D. Forbes, 1843–1844. Journal. University of St Andrews.

References

BOUÉ, A. 1830. [Translation into French of Forbes V (1829)], *Journal de Géologie*, **1**, Paris, 354–373.

BREISLAK, S. 1792. *Essais mineralogiques sur la Solfatare de Pouzzole.* J. Giaccio, Naples.

BREISLAK, S. 1797. *Topografia Fisica della Campania.* 2 Volumes. Firenze.

BROCCHI, G. B. 1814. *Conchiologia fossile Subapennina, con osservazioin geologiche sugli Appennini e sul suolo adiacente*. Stamperia reale, Milan.

CUNNINGHAM, F. F. 1990. *James David Forbes, Pioneer Scottish Glaciologist.* Scottish Academic Press, Edinburgh.

DAUBENY, C. 1826. *A Description of Active and Extinct Volcanos.* W. Phillips, London.

DAUBENY, C. 1848. *A Description of Active and Extinct Volcanos.* 2nd edn. R. & J. E. Taylor, London.

DE LA BECHE, H. 1851. *The Geological Observer.* Longman, London.

FERBER, J. J. 1776. *Travels through Italy in the Years 1771 and 1772 ...* Translated from the German [1773] ... with Explanatory Notes ... by R. E. RASPE. L. DAVIS, London.

FORBES, J. D. 1827. Remarks on Mount Vesuvius. By a Correspondent. *Edinburgh Journal of Science*, **7**, 11–18.

FORBES, J. D. 1828. Observations on the styles of building employed in ancient Italy, and the materials used in the city of Rome. By a correspondent. *Edinburgh Journal of Science*, **9**, 30–48.

FORBES, J. D. 1828–1830. *Physical notices of the Bay of Naples* (all in the *Edinburgh Journal of Science*): No. I, **9** (1828), 189–213; II, **10** (1829), 108–136; III, **10** (1829), 245–266; IV, ns **1** (1829), 124–141; V, ns **1** (1829), 260–286; VI, ns **2** (1830), 75–102; VII, ns **2** (1830), 326–350; VIII, ns **3** (1830), 246–278.

FORBES, J. D. 1836. On the geology of Auvergne. *Edinburgh New Philosophical Journal*, **21**, 1–21.

FORBES, J. D. 1843. *Travels through the Alps of Savoy and other Parts of the Pennine Chain with Observations on the Phenomena of Glaciers.* A. & C. Black, Edinburgh.

FORBES, J. D. 1844. Sixth letter on glaciers. *Edinburgh New Philosophical Journal*, **37**, 231–244.

FORBES, J. D. 1853. *Norway and its Glaciers, visited in 1851; followed by Journals of Excursions in the High Alps of Dauphiné, Berne and Savoy.* A. & C. Black, Edinburgh.

GOETHE, J. W. 1823. Der Tempel des Jupiter Serapis zu Puzzuolo: Ein Architektonisch-Naturhistorisches Problem. *In*: GOETHE, J. W. 1823, *Zur Naturwissenschaft überhaupt, besonders zur Morphologie.* 2 volumes. Cotta, Stuttgard & Tübingen, **2**, 79–88.

GOETHE, J. W. 1824. On the geognostical phaenomena at the Temple of Serapis. *Edinburgh Philosophical Journal*, **11**, 91–99.

HAMILTON, W. 1772. *Observations on Mount Vesuvius.* T. Cadell, London.

HOFF, K. E. A. VON 1822. *Geschichte der durch Überlieferung nachgewiesenen natürlichen Veränderungen der Erdoberfläche.* Volume 1. Perthes, Gothe.

HOFF, K. E.A. VON 1834. *Geschichte der durch Überlieferung nachgewiesenen natürlichen Veränderungen der Erdoberfläche.* Volume 3. Perthes, Gothe.

HUMBOLDT, A. VON 1818. *Ansichten der Natur.* Tübingen. (2nd edn, 1826).

HUMBOLDT, A. VON 1828. *Tableaux de la Nature.* 2 Volumes. Gide fils, Paris.

HUMBOLDT, A. VON 1849. *Aspects of Nature.* Longman, Brown, Green, & Longmans, London.

HUMBOLDT, A. VON 1850. *Views of Nature.* H. G. Bohn, London.

LYELL, C. 1825. On a dyke of serpentine. *Edinburgh Journal of Science*, **3**, 112–126.

LYELL, C. 1830. *Principles of Geology.* J. Murray, London.

MACCULLOCH, J. 1821. *A Geological Classification of Rocks.* Longman, Hurst, Rees, Orme, and Brown, London.

PLAYFAIR, J. 1802. *Illustrations of the Huttonian Theory of the Earth.* Cadell and Davies, London, and Creech, Edinburgh.

SCROPE, G. P. 1825. *Considerations on Volcanos.* W. Phillips, London.

SCROPE, G. P. 1828. On the volcanic district of Naples. *Transactions of the Geological Society of London,* 2nd series, 2, 337–352.

SHAIRP, J. C., TAIT, P. G. & ADAMS-REILLY, A. 1873. *Life and Letters of James David Forbes, F.R.S., D.C.L., LL. D., Late Principal of the United College in the University of St. Andrews.* London, Macmillan.

SPALLANZANI, L. 1798. *Travels in the Two Sicilies, and Some Parts of the Appennines.* 4 Volumes. G. G. & J. Robinson, London. (Original Italian editions: 1792–1797).

SPALLANZANI, L. 1825–1826. *Opere.* 6 Volumes. Milan. (There is also a French translation).

The geological travels of Charles Lyell, Charlotte Murchison and Roderick Impey Murchison in France and northern Italy (1828)

M. KÖLBL-EBERT

Jura-Museum Eichstätt, Willibaldsburg, 85072 Eichstätt, Germany (e-mail: Koelbl-Ebert@Jura-Museum.de)

Abstract: In 1828, Charles Lyell (1797–1875), Charlotte Murchison (1788–1869) and her husband Roderick Impey Murchison (1792–1871) embarked on a long journey around Europe. The party left Paris in May 1828, travelled through the Massif Central and continued southwards. The geological programme was dedicated to stratigraphical and geomorphological observations, but sightseeing was not neglected. Published papers by Lyell and Roderick Murchison report their scientific results, but it is the unpublished journals, notebooks and letters, which illuminated their research programmes, task management and daily routine. There was an effective division of labour, which increased the scientific productivity of the trio. Lyell and Roderick Murchison decided about routes and research topics and travelled long distances on foot taking stratigraphical sections and keeping track of correlations of structures, whereas it was Charlotte Murchison's task to do much of the time-consuming fossil-hunting, sketching of landscapes and geological structures and—speaking French fluently—to visit local experts, whose expertise might add to the success of the journey. Also, the different initial expectations and working styles of Charles Lyell and Roderick Murchison become evident.

For several months in 1828, Roderick Impey Murchison (1792–1871), his wife Charlotte Murchison (1788–1869) and Charles Lyell (1797–1875) travelled together in France and northern Italy (Fig. 1). The geological programme of the journey was strenuous—dedicated to stratigraphical and geomorphological observations. However, other features such as botany, antiquities and picturesque landscapes were not neglected. Papers published subsequently by Lyell and Roderick Murchison (Lyell & Murchison 1828, 1829*a*, *b*, *c*) report on scientific results, but it is the unpublished journals, notebooks and letters that tell about the research programmes, task management and daily routine. The journey was planned expressly for geological fieldwork, the purpose of which was to educate the travellers themselves, thus being closely related to what today would be called a field trip rather than field research. There were both educational and recreational aspects to the journey; Roderick Murchison referred to this journey as their (geological) Grand Tour.[1]

Roderick Murchison had started to occupy himself with geological research about three years earlier and the journey was his first foreign geological tour. It was basically a stratigraphical interest that motivated him to undertake the journey: '[...] his leading aim had been to unravel the true order of arrangement of the rocks, and show their relation to each other and to those of other and better known regions' (Geikie 1875, p. 146). It was he who took the initiative for the tour: '[...] I induced my wife to accompany me as well as my associate[?] Charles Lyell who left his clerk Hall to take care of his chambers whilst my wife provided herself with a good stout Swiss femme de chambre & thus we started in our little green carriage'.[2]

Charlotte Murchison, being three and a half years older than her husband, was according to her friend, the celebrated scientific author Mary Somerville (1780–1872), an amiable and accomplished woman, with a solid scientific knowledge and an explicit taste for rocks and fossils. It was chiefly due to her example, that Roderick Murchison had turned his mind to geology.[3] As long as her health permitted she participated in her husband's geological work. In 1817, while travelling near Rome, she had been infected with malaria, an illness that recurred every now and then throughout her long life. As such, she was occupied especially with collecting fossils and sketching the striking geological features of the landscape (Kölbl-Ebert 1997, 2002). 'She was a good sketcher of scenery, having been taught by the famous Paul Sandby [...]'.[4] She kept her own collection of fossils, as can be seen from notes by Louis Agassiz, who clearly distinguished between the two collections of husband and wife.[5]

In 1828, like Roderick Murchison, Charles Lyell was also near the beginning of his geological career, and the journey to France and through Italy proved to be an important step in his geological education. Lyell's father expected Charles to occupy himself with his profession as a barrister

From: WYSE JACKSON, P. N. (ed.) *Four Centuries of Geological Travel: The Search for Knowledge on Foot, Bicycle, Sledge and Camel*. Geological Society, London, Special Publications, **287**, 109–117.
DOI: 10.1144/SP287.9 0305-8719/07/$15.00

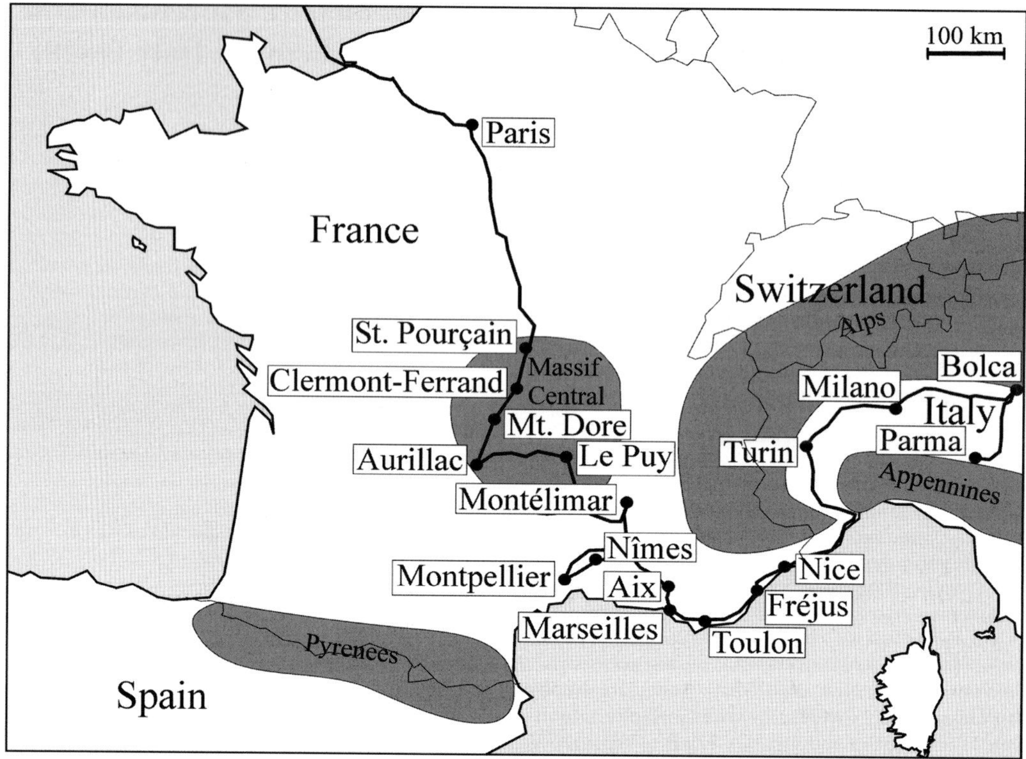

Fig. 1. Route of the French/Italian journey of Charles Lyell and the Murchisons in 1828.

(Wilson 1972, p. 187); however, in March 1828, he decided to join Murchison to collect material for a book on geology having in mind something similar to Jane Marcet's (1769–1858) *Conversations on Chemistry* ([Marcet] 1805), fancying that he could make sufficient money as an author. He came back, however, with raw-materials for a much more ambitious project. These became his *Principles of Geology* (Lyell 1830–1833) which made him one of the most prominent geologists ever (Wilson 1972, p. 186).

The fascination of the adventure of science had proven greater than that of the law: 'One must travel over Europe, to learn how completely we are in our infancy in the knowledge of the ancient history of the globe & to feel as I do now what splendid discoveries must be on the Eve of coming to light even within the time which we may hope to see'.[6]

Thus for Lyell, the journey to France marked the end of his law career and the beginning of his geological fame, as was recorded by his travel-companion Roderick Murchison, who wrote in his journal (compiled years after the journey itself):

> It was here that Lyell took the step which enabled him to rise to his present distinction.
>
> I have before said that he was a young barrister [...]; but so subsidiary did he make the study of the law to that of geology that his Clerk one Hall [...]came out to us to act as Amanuensis in Auvergne & then went home with our Collections[?].
>
> After what we had learn't of the distinction of tertiary shells, & the scheme which he had formed in Paris of systematysing their classification came vividly before him & he yearned to ransack the cabinets of Italy & visit all the principal localities for the Subapennine shells including those of Sicily. This however was not to be done without giving up a law term & then came the struggle—[...]. He had long felt very honestly[?] that he ought not to be a bother to his father who had several other children & that therefore it was his duty to try to earn a livelihood either by the law or by Science. When in this quandare [sic] & evidently[?] much perplexed he opened his mind to me & I did not hesitate a moment in counselling him earnestly to throw the law overboard & pursue his favorit [sic] career in which he was making such an excellent start & in which after an Italian & Sicilian tour he would[?] have acquired so much strength that he [illegible] publish works if [sic] importance.
>
> Thus backed up he took courage & wrote to his father informing him of the choice he had made & that he was to be thenceforward a Geologist by profession.[7]

Even if we subtract a certain amount of satisfaction on Murchison's side, who seemed to have flattered himself to have been a sort of 'midwife' to Charles Lyell's geological career, it remains as a fact that the French/Italian tour proved crucial for Lyell's firm conviction that present causes were the key to understanding the history of the earth.

Preparations

Preparations for the journey involved the collection of geological information, which formed a field guide to the localities worth seeing: 'I had made a little book full of notes from various authors before I left London. (Bakewell, Scrope, Dolomieu, Desmarest, Montlosier, Bertrand Roux (afterwards de Dome) s'Omalies [sic] d'Halley[?] Ferressac [sic] &c'.[8]

In addition, some three weeks in Paris, visiting collections and scientists before setting out further south, enabled the Murchisons to get even more information together with a practical introduction into the tertiary stratigraphy of France:

> As we had frequent access to both Cuvier & Brogniart & dined with them both and had also long colloquies with Cordier then the best mineralogical geologist in France we rapidly acquired in the various museums an amount of knowledge which qualified us to go into the volcanic region of Central France (Auvergne) & there work out in detail the real age of the deposits what had been pierced & overflowed by the volcanic rocks. A M Dusgate who had long[?] resided in Paris was of great use to us. & old Desmarest who had made maps of the country gave us good hints as well as Dufrenoy who had recently explored Auvergne & the environs. We were thus prepared to fill up the lacuna in Poulett Scrope's work which we carried with us.[9]

They also carried letters of introduction to local experts, whose expertise could be useful for the enterprise:

> Our scientific acquaintances were numerous & most civil & kind to us. The Comte de Luiger is an intelligent & gentlemanlike man—a good antiquarian, & knows the local geology of his country well. Monsr. Lecoq very agreeable in his manners & most useful as superintendent of the museum. He has begun to put it in perfect order, and his merits as a naturalist seem great & extensive—Minerals, plants, insects, all equally familiar to him. Monsr. Busset invited us to a dinner, and gave us a report only too abundant & too good for working geologists. Monr. Bouillet returned just in time to give us a sight of his museum large[?] collection of fossil bones, & the arrangement good [. . .].[10]

On the road

On 20 April 1828, the Murchisons left England for Paris. Six days later, they were engaged in geological fieldwork together with Constant Prevost in the Paris Basin to get an expert introduction into French geology.[11]

On 4 May, Lyell followed, and reached Paris on 7 May. The trio, accompanied by Charlotte Murchison's maid Annette left Paris on 11 May in a little green carriage, travelling southwards speedily until they reached St Pourçain three days later, where the geologically interesting part of the tour commenced:

> The bank of a picturesque rivulet running thro' the village, presented fine section of the Freshwater formations. *Indusia tubulata* in abundance & each tube more richly adorned with Balinia than any we have subsequently seen. The Balinia pearly white. an excelent [sic] Inn civil Landlady, & beautiful scenery.[12]

And Charles Lyell reported back home:

> We started every morning 6 o'clock—beautiful weather. I never did in my life so much real geology in as many days. . . . We have generally begun work at 6 o'clock & neither heat nor fatigue have stopped us an hour. Mrs. M. is very diligent, sketching, labelling specimens & making out shells in which last she is an invaluable assistant. She is so much interested in the affair as to be always desirous of keeping out of the way when she would interfere with the work, and as far as I yet see it would be impossible to form a better party.[13]

Admiring the 'magnificent panorama of the Puys & the valley of the Limagne',[14] they reached Clermont-Ferrand on 1 May, where Mr Hall, Lyell's London clerk joined them (Fig. 2) (see also Wilson 1972, p. 193). On 18 May, the whole party ascended the Puy de Dome. During the following week, the men went to the field regularly, often leaving Charlotte Murchison in Clermont-Ferrand, not to amuse herself but to relieve her companions from boring[15] social responsibilities and to conduct fieldwork in her own right, employing herself: 'in making panoramic sketches, receiving several of the gentry & Professors to whom we had letters in the neighbourhood & collecting plants & shells etc.'.[16] The information obtained in her fieldwork was then merged with her companion's observations, as can be deduced from notes by her husband such as this: 'On the 20th May we were making sections of the tertiary lacustrine deposits we had examined at Verduaison above Cusset & Vernet (see Blue book & Lyell's sections) & Lady M's notes'.[17]

Thus, an efficient division of labour becomes apparent, which certainly increased the scientific productivity of the trio. Whereas Lyell and Roderick Murchison decided about route and research topics and travelled long distances on foot, keeping track of correlations and bigger structures, it was Charlotte Murchison's task to do much of the time-consuming fossil-hunting, sketching landscape and geological structures and—speaking French fluently—to visit local experts, whose expertise might add to the success of the journey.

Leaving Clermont-Ferrand, the scientific party continued slowly but laboriously. On 1 June, they reached a farm in the vicinity of the Mont Dore, where the Comte de Montlosier, a local expert on natural history subjects, lived. Here they stayed for some ten days to make daily trips up to twelve

Fig. 2. The Murchisons, Charles Lyell, Lyell's clerk Hall, and an unidentified gentleman travelling in southern France in 1828 (as illustrated in Faul & Faul 1983, p. 128, fig. 8.1. Unfortunately the original source of this figure was not given in this publication).

or fourteen hours on horseback in the Mont Dore region under the guidance of their friendly host.[18]

The procedures were thus:

My wife was still in these days a good horsewoman—for that is, or was the only method by which six[?] could explore the countries well. Thus, whether we made excursions down[?] the valley of the Sioule, or got up the crater of Chaluzet or ascending to the Summit of M Dore the Pic de Sancy—descended to the basaltic filled valley of Senecterre scooped out in a great granitic plateau, or wandered over tracts of 'granite pourri' or wandered from the volcanic tract to the little cone[?] fields on the west or reached the higher & older igneous rocks of the Cantal & its 'Plomb', horses were constantly employed the carriage & Mlle. Annette going round with the baggage.[19]

The daily routine was quite strenuous: '[...] 6 hours in bed, which is all we allow, & exercise all day long for the body & geology for the mind with plenty of the vin du pays which is good here [...]'.[20] Mont Dore was still covered with snow, and the cold must have been pretty hard on Charlotte Murchison, since the fashion of those days provided only shawls and no warm coats for a lady, leading to the impression that men and women used to dress for different seasons. Although Charlotte Murchison herself did not complain in her notebook, Charles Lyell noted in a letter to his mother: 'I like it well enough, but it is certainly too early in the season to enjoy it, and Mrs. Murchison suffers from the cold and damp, though she has not often complained in this tone'.[21]

Notwithstanding the cold, Charlotte Murchison enjoyed the 'beautiful rides'[22] immensely and admired the scenery, the light and the spring flowers on the lower slopes of the mountains, a feature of the excursions, which also pleased the men:

nearly vertical strata of Trachyte, or what is more likely that this spongiferous material has in the lapse of ages been worn into these sawlike form by atmospheric causes only. The highest summits are of trachyte but the lower (as the Roc de Moumaux) are of basalt. The flowers in the greatest frequency[?] were the Narcissus boeticus, Ranunculus aconitefolium Scylla lyosotis & numerous Orchids [...] then along the rough & broken plateau de lAigle [sic] & descended down a road more fit for goats than horses in a glorious Sun set to the Baths of Mt Dòr. (about June 14. or 15^{t}. (My wife delighted in this ride & gave me the names of the wild flowers).[23]

It was these days in the lonely beauty of the Mont Dore region, where Charlotte Murchison's spirits seem to have been most exalted, and the words she wrote in her diary give a sense of her feelings towards the sublimity of nature:

to Murol to breakfast, after which we mounted to the Castle, and enjoyed a splendid view from the summit of the round Tower. To the east rich plain of the Limagne with singular tabular Puys. to the west Lac Chambon beautifully wooded and terminated by the pinnacles of the (* valley of Chaudefour) Mts. Dor. It is impossible to describe the great beauty of this view.

The oaks & Beach that form a thick forest from Murol to the Lake are worthy of Hampshire.

We returned by the valley of Chaudefour, the finest scenery we have yet beheld in this country. Enormous peaks & Dents of rock, rise thro' wood & verdure, & close in one side of the valley, whilst the snowy & rocky summits of Puy Terraud enclose the opposite one. we then reascended the Mt. Dor, and in a magnificient [sic] stream of golden sunset found the little village of the Baths at our feet. Cork screw road of descent so perpendicular that it is strange how the horses maintain their equilibrium.[24]

Around 17 June they left the Mont Dore and proceeded to Aurillac, where they stayed for a few days to take several profiles of lacustrine limestones, which led to a publication by Lyell and Roderick Murchison (Lyell & Murchison 1829*a*):

The road excellent. much the best we have travelled in France. between Mauriac & this place perpetual ascent & descent, so that there is more merit due to the engineers who have with excellent judgement continued the road. The little Inns by no means uncomfortable, [. . .] a few fleas which we probably obtained from the herds of dogs that intruded upon us at our meals at every Inn. Aurillac is a pretty chearful [sic] Town. The people well dressed, & well mannered. good horses, fine mules, & all due respect paid to the fair sex.[25]

Mrs Murchison still continued to serve as the illustrator of the party: 'When we were fairly within these freshwater deposits & their alternating & overflowing trachytic rocks my wife sketched an outline of the Cantal chain—Upper portions[?] trachyte—the lower in the vallies [sic] or sides of the mountains basaltic conglomerates, columnar porphyries'.[26]

In the Cantal, the geological trio met with women riding on horseback astride! Charlotte Murchison of course rode side-saddle as was the custom in England:

The social feature which struck us most in the Cantal was the number of well clad women on horseback astride with large[?] black petticoats falling beneath their stirrups & [illegible] with large silver buckles before & behind. The horses excited[?] & full of action.[27]

From Aurillac, the party travelled slowly, faced considerable detours, and took about two weeks to reach Le Puy en Velay, making detailed sections on the way. They left Le Puy on 11 July to start the descent through a volcanic region to the valley of the Ardèche, which led them to the river Rhône. They suffered from a sudden drop of spirits, caused by

[. . .] bad accomodation [sic] & abundance of fleas. Road barren & ugly like the worst parts of Scotland & moreover extremely bad'. But the 'descent to the valley of the Ardêche [was] very beautiful.[28]

The perils of travel

On 13 July Charlotte Murchison wrote in her notebook:

Thueyts. good village Inn. no vermin beautiful scenery but unable to sketch on visit the Gerelle[?] de l'Enfer from a lame foot. narrow escape of our lives on the descent. Postillion kicked off, just before entering Mayres, & horses ran away with us some distance.[29]

This was a gross understatement of what actually happened, and quite symptomatic for her way of keeping a diary of her journey, rarely leaving more than a few catchwords. In the words of her husband the incident sounds much more dramatic:

It was in the descent into the Ardêche amid fine groves of Chestnut trees that my wife & her maid Annette were ran away with by the wretched post-horses we had engaged. The truly awkward postilion had been thrown from his saddle by one of the horses which had been stung by a forest fly & Lyell & self were on foot examining the rocks on the side of the road when the animals freed of their manager went off at full gallop down a zig zag road cut into the side of the rocks & with a precipice on the right & scarcely any barrier. L. & self ran as hard as we could but never could reach the carriage which we saw whirling[?] from side to side & the great tall Swiss Femme a [sic] Chambre throwing about her arms & screaming whilst my wife held her down. At the village of Mayres far below the peasants seeing the Event & the [illegible] horses coming down got together & stopped them. So that when we got down we found the people all chattering away & all most anxious to get Madame into a house. after her fright. This proved to be a great Misfortune; for an awkward girl taking a great shutter from a window let it fall on Charlotte's instep & bruised all the tendons severely. This laid her up for 2 days at Thueyts and deprived her of a sight of of [sic] the beautiful scenery of the valley of the Alignon & the volcanic phenomena around Neyrac & Jaujac which L. & self explored.[30]

Already on 17 July, '[. . .] tying up Charlottes maimed foot (for she had then plenty of pluck)',[31] Mrs Murchison again joined the men on an 'expedition from Aubenas to Antraigues' to see basaltic columns.

On the next day the party crossed the Rhône by ferry boat. From the freezing cold of Mont Dore they had by now reached Mediterranean summer and during the next few days the temperature started to rise steadily, and reached 82ºF on 7 August.

The heat began to take its toll:

[. . .] I do not think the Murchisons will stand fire. Symptoms of flinching from the heat which makes scarcely any impression on me, begin to betray themselves. Thoughts of a retreat to the Alps, consultations with me whether I think it practicable to proceed farther south in the dog-days have been mooted [. . .].[32]

They now proceeded quickly via Montelimar, the Pont du Gard, Montpellier and Nîmes to reach Aix en Provence on 27 July. Their main geological concerns were now rates of sedimentation and erosion as well as signs of elevation and subsidence. They spent a whole week in Aix before continuing to Marseilles and further on to Toulon and Fréjus, in which town, on 7 August, Roderick Murchison contracted a high fever, which he himself attributed to either malaria or the effects of too much sun, and rapidly became delirious.[33] [Mrs Murchison] 'when her mind was thus recalled into exertion in attending him recovered her energy',[34] which had failed in the heat of previous days. She nursed her husband with the only thing that was at hand:

My good wife fortunately had [illegible]'s powder with her & [illegible] me with it till I fell into sleep with perspiration. When I came to myself next day I found a little French Doctor with a pigtail prescribing 'Eau de Poulet'. [...] The solemnity with which he recommended chicken broth [...] amused me. But he gave me good advice which was to move me out of Frejus [sic] as quickly as possible & in 48 hours we we [sic] ascending to purer air in the mountain of the Esterelles [sic].

I had just strength enough to gaze on the splendid prospect from the summit & to perceive dykes of porphyry whilst Lyell walking alongside of the Carriage did all the geology.[35]

The problematic health situation of both of the Murchisons required a longer halt in Nice, were they '[...] lodged in the Hotel des Etrangers and as at last'—after complaining to the landlord—'we got into rooms free from bugs we were pleased with our host M. Ferdinand'. They stayed in Nice from 9 to 27 August, although the stench of the city after a drought of five months was hard to bear.[36]

The strain begins to show

Roderick Murchison's illness had not been a sudden, simply unfortunate incident; right from the beginning of the journey it seemed to have been just a matter of time, before one of the trio would collapse under the strain, in particular the two men who egged each other on, and were unable to admit exhaustion and the need for a rest. Already in May, two weeks after the actual field-work had begun, Charles Lyell wrote to his father: 'We got under sail by six some mornings & at 5 o'c.lk some others & M. certainly keeps it up with more energy than anyone I ever travelled with'.[37] At that time, Lyell seemed to have admired his companion's energy. But soon it became apparent that Murchison could only stand the daily effort by using drugs:

Murchison must have been intended for a strong man, if the sellers of drugs had not enlisted him into their service, so that he depends on them for his existence to a frightful extent, yet withal he can get through what would knock up most men who never need the doctor. He has only given in one day and a half yet. On one occasion we were on an expedition together, and as a stronger dose was necessary than he had with him, I was not a little alarmed at finding there was no pharmacy in the place, but at last we went to a nunnery, where Mdlle. La Supérieure sold all medicines without profit [...] who hoped my friend would think better of it, as the quantity would kill six Frenchmen. M. was cured, and off the next morning, as usual.[38]

Whereas Lyell was under the impression that it was Murchison, who constantly pushed on, Murchison thought the same of Lyell:

[...] we have enjoyed uninterrupted délicieus [sic] weather & have seen every peak of every high mountain—but we now begin to be scorched, & to cob[?] out fire—when my stomach & nerves are a little disorganized I cry peccavi, & that we must not attempt the Bouche de Rhone & the Subappenines, but after all I think we shall be carried on by the Geological ardor [sic]—& Lyell is the best whipper in possible[?].[39]

Much later, while travelling to Aix en Provence, the difference in perception of the two travel-companions again became apparent, when Murchison complained that the intense heat prevented them from work:

Our journey across to Aix en Provence was most interesting, and that place offered so much that we halted a week, our work being now reduced to four or five hours in the morning, from four to nine, and a little in the evening. We hope to show you twenty or thirty species of *insects!!* from the gypsum quarries there. In this city of idleness we have been pent up during ten days, not daring to travel into Italy with these heats: it has not rained one drop here for eight months.[40]

The story reads quite differently when written by Lyell:

Mrs. M. suffers from heat & our motions have been much impeded since we reached the Rhone at Montélimart. But we have done much & never lost a whole day. We get out either on foot or caleche by 4 o'clock A.M. intending to sleep at noon, but when noon comes we are either from home or a cloud or a breeze enables us to work & so it continues till next night. If M. was not unwell about once a week I should be done up, for 5 hours sleep on hard work will not do for me, but make amends on such occasions.[41]

Obviously, in the timeless fashion of young men everywhere, Charles Lyell did not dare to admit that he needed a break too, and was secretly glad that Roderick Murchison had his little breakdowns every now and then. And even after Murchison's attack of fever at Fréjus, work continued in the field as well as at the desk, the latter pleasing Lyell especially:

Murchison has not yet regained his strength from a severe attack at Fréjus, so as to be able to take the field again. But we have been hard at work in writing, from our materials, a paper on the excavation of valleys, which is at last finished, and after two evenings' infliction, is intended to reform the Geological Society, and afterwards the world, on this hitherto-not-in-the-least-degree-understood subject. Besides this mighty operation, we have performed two jaunts with Mrs. Murchison, each at half-past four o'clock in the morning, to see certain deposits of fossil shells, and collecting these, with which she has been much pleased; and this and the cessation of eternal bustle, while the campaign was at its height, while we were as yet only crossing the Balkan, and before our descent upon these hot latitudes has restored her health and spirits, which had failed sadly. Her lord has a little too much of what Mathews used to ridicule in his slang as 'keep-moving, go-it-if-it-kills-you' system, and I had to fight sometimes for the sake of geology, as his wife had for her strength, to make him proceed with somewhat less precipitation. You may suppose it was not over prudent to attempt hard work, and only sleep, or rather to be in bed, five hours at most. I expected a break-down before.[42]

Breaking apart

Lyell and the Murchisons stayed in Nice for nearly three weeks until 27 August to restore Roderick Murchison's health, writing the paper on the excavation of valleys (Lyell and Roderick Murchison), admiring the exotic palm trees in the city (Charlotte Murchison) and '[...] making little excursions to the Environs with old Signor Risso the conchologist & examining his Quarternary, Tertiary & Numulitic deposits [...]'[43] (all three together).

The trip now progressed steadily towards northern Italy, and they admired the scenery, although troubled by a plague of mosquitoes, evoking the dangers of malaria. Turin was reached on 3 September and two days spent with detailed sections of the Cenozoic strata there. The Cenozoic kept them busy for the next two weeks. They looked for information in the libraries of Milan, saw Lake Garda, the fossil fish of Monte Bolca, the Roman amphitheatre of Verona, and a Cenozoic coral reef near Vicenza. In Padua, on 19 September, Charles Lyell took leave of Charlotte and Roderick Murchison to proceed further south to see the modern volcanoes and to study the effects of earthquakes: elevation and subsidence of land. He was encouraged to do so by Roderick Murchison, who knew Sicily and Mount Etna from his time as a soldier (Lyell 1881, **1**, pp. 200–202).

The Murchisons, meanwhile, turned north to cross the Alps, having been called back to England by an illness of Charlotte's parents. However, whilst in Tyrol, another letter reached them with good news that her parents had recovered (Wilson 1972, p. 224), and so they proceeded quite slowly again, visiting e.g. the Cenozoic fossil-lagerstaette of Öhningen [Oeningen] in southern Germany before they headed back home.

Orderly strata and a vision of a dynamic Earth

It was not only the different characters of Roderick Murchison and Charles Lyell, which added a certain thrill to the journey, they also differed in their attitude towards geology and its purposes, and thus in respect of their geological reasoning and methodology. Charles Lyell, in a letter to his father suggested a difference in expectation, at least initially, about the outcome of the tour:

The whole tour has been rich as I had anticipated (and in a manner which Murchison had not), in those analogies between existing Nature and the effects of causes in remote eras which it will be the great object of my work to point out. I scarcely despair now, so much do these evidences of modern action increase upon us as we go south (towards the more recent volcanic seat of action), of *proving* the positive identity of the causes now operating with those of former times.[44]

Whereas Lyell was obviously very much interested in processes both past and present, Murchison's interest was focused mostly on stratigraphy, initially in a rather static perception of the world, which, however, changed slowly throughout the journey:

On July the 18 [...] we stopped at the moulin de St Etienne where we met with superior strata of bluish grey limestone which by their fossils—notably Ammonites Calloviensis & Alterreyi I paralleled with the Oxford clay & Kelloway Rock of Britain. Mist & rain prevented much examination but we were struck with a modern breccia formed[?] by fragments[?] of limestone[?] & shale which tumbling down the slopes for ages had formed a mass of stratified rock 50 feet thick.[45]

While roaming the Cenozoic freshwater limestones near Aurillac together with Lyell, it seems that Murchison learned to recognize that strata were not global features of their time, but sometimes were even remnants of a small-scale, regional landscape such as a series of lakes, changing not suddenly but continuously with time:

Hence to Aurillac where we worked right hard for a week in the Fresh water & [...] we expect to be able to show you a good bottom for that lake—you will not start at this expression—you will rather rejoice when I tell you, that our numerous[?] laborious sections are bringing us to the opinion of a system of detached lakes—mind, this is between you & myself because the case will not be completely settled till we have done with this of Le Puy & summed up our comparisons [...].[46]

Realizing this was no trivial matter, but a fundamental and necessary change in understanding, the importance of which becomes apparent, to anyone dealing with visitors to a museum, who, at least in Germany, are usually quite ignorant of geology. Most of them simply find it fascinating that rocks are not global layers or shells but can be used to reconstruct changeable landscapes; landscapes, which, viewed with the time-lapsing mind of geologists, move over the Earth's surface like cumulus clouds do in the sky: moving and changing shape, sometimes swelling, sometimes dissolving and making way for other conditions.

Although Lyell was quite content with the simple idea of running water causing any amount of erosion, Roderick Murchison was uneasy with this idea—even if in the presence of Lyell he had been persuaded to agree. Years later, when he compiled his journal from his field notes, he thought it necessary to take refuge in more violent causes of change such as earthquakes:

I must if I ever have time review what Lyell (for he was the chief penman) wrote in a joint paper on the excavation of the valleys of Auvergne. For, whilst I fully admit the re-excavation of certain valleys & the clearing out of the lava & basalt which filled them, I do not believe that the original water-courses were formed by the streams, but that they originated in cracks & depressions formed by great & ancient Earthquakes, long anterior to the eruptions of the youngest volcanoes of Auvergne.[47]

Today, Charles Lyell is praised especially for his focus on geological processes. It seems, that in

Murchison's opinion, Lyell's most important achievement was his classification of the Cenozoic, although he was aware of French precursor ideas:

We were off in April & in the 26 of this month were at work in the field with Constant Prevost following his directives[?] of the Paris basin. The theoretical views of Prevost made a deep impression on Lyell who as far as I can judge imbibed[?] some of his best ideas of the operations[?] of land & freshwater attractives[?] with marine deposits from this[?] persevering & ingenious Frenchman.' [...] 'The other French authority who was of the utmost service to Lyell was the conchologist Deshayes, who who [sic] even then had described 144 species of Cerithes for the Calcaire Grossier of Paris. Deshayes had conceived the idea of distinguishing the age of Tertiary deposits by co[m]par[in]g[?] the relative number of species common to them.—I e the system afterwards arrived at by Lyell [...].[48]

Conclusion

During the early nineteenth century a large group of non-professional geologists still existed in Britain. For these people with enough money, time and leisure to study, travel and publish, geology was more or less a private interest. Therefore, privately organized field trips, like the one described, combining geological fieldwork with sightseeing, were not uncommon for the time.

For a young gentleman-geologist, hardly any formal training in geology was available, and therefore such self-organized educational field trips as the Murchison's and Lyell's journey to France and Italy, undertaken with the aid of other travellers' and the local geologists' expertise, be it written or personal, was an important factor in the training of new gentleman-geologists. This practical approach to geological education led to a smooth transition from studying geology to active research.

Because of the informal character of this gentlemanly part of early British geology, women were not excluded from participation. They were not yet opponents in the competition for jobs, but were welcomed as fellow-enthusiasts, although they rarely worked as independent scientists but rather functioned as scientific help-mates to husbands, brothers or non-related male geologists. As a result, there were many female contributors to geology in the early nineteenth century in the United Kingdom, forming a framework of assistants, secretaries, collectors, painters and field geologists to the leading figures in the geological sciences, thereby adding to and shaping their work.

Thus, Charlotte as help-mate of her husband was no singular exception. There have been other more or less productive husband–wife teams, such as William and Mary Buckland or later Charles and Mary Elizabeth Lyell. It may well be that Charles Lyell's experience with Charlotte Murchison enabled him later to expect from and to trust his own wife with being equally helpful.

I thank the staff of the library of the Geological Society of London, of the Archives Cantonales Neuchâtelois, the University Library of Neuchâtel and of Bodleian Library, Oxford for their friendly help with their treasures. I am deeply indebted to J.-P. Schaer (Neuchâtel) for saving me a lot of money and trouble, by organizing to copy local manuscript material for me. Many thanks also to P. N. Wyse Jackson (Dublin) and M.J.S. Rudwick (Cambridge) for a critical review of this paper and for improving my English. This research was financed by the Deutsche Forschungsgemeinschaft (DFG).

Notes

[1]Geological Society of London: LDGSL 789/160 (letter by R. I. Murchison to A. Geikie, 3 March 1864).

[2]Geological Society of London, Murchison Papers: M/J7, p. 21.

[3]Bodleian Library, Oxford University, Somerville Collection: Dep. c355 MSALL-2 (first draft of Mary Somerville's autobiography), p. 100.

[4]Geological Society of London: LDGSL 789/160 (Murchison to Geikie, 3 March 1864). Paul Sandby (1725–1809) was an English painter whose speciality were watercolour drawings of scenery and architecture. He worked for the Duke of Cumberland, as draughtsman for the survey of the Highlands, for Sir Joseph Banks and later as chief drawing-master for the Royal Military Academy at Woolwich (see e.g. http://91.1911encyclopedia.org/S/SA/SANDBY_PAUL.htm).

[5]Archives Cantonales Neuchâtelois, Agassiz Papers : 123/3.2.

[6]Letter Lyell to Mantell, 22 August 1928, quoted after Wilson 1972, pp. 214–215.

[7]Geological Society of London, Murchison Papers: M/J7, pp. 49a–50; the passage bears the pencil note: 'I do not think this should be published in Lyell's life-time A.G. [i.e. Archibald Geikie?]'.

[8]Geological Society of London, Murchison Papers: M/J7, p. 24.

[9]Geological Society of London, Murchison Papers: M/J7, p. 22.

[10]Geological Society of London, Murchison Papers: M/N160 (Charlotte Murchison's [C.M.] Notebook, 4 June 1828).

[11]Geological Society of London, Murchison Papers: M/J7, p. 21.

[12]Geological Society of London, Murchison Papers: M/N160 (C.M. Notebook, 14 May 1828).

[13]Letter Lyell to father, 16 May 1828, quoted after Wilson 1972, pp. 190–1.

[14]Geological Society of London, Murchison Papers: M/N160 (C.M. Notebook).

[15]See Geological Society of London: LDGSL 768/2 (letter R. I. Murchison to William Buckland, Puy en Velai, 4 July 1828).

[16]Letter Charles Lyell to his father, 26 May 1828, quoted after Wilson 1972, p. 194.
[17]Geological Society of London, Murchison Papers: M/J7, p. 23.
[18]Geological Society of London: LDGSL 768/2 (letter R. I. Murchison to Buckland, Puy en Velai, 4 July 1828).
[19]Geological Society of London, Murchison Papers: M/J7, p. 25.
[20]Letter by Charles Lyell to his mother, quoted after Wilson 1972, p. 198.
[21]11 June 1828, quoted after Lyell 1881, **1**, p. 190.
[22]Geological Society of London, Murchison Papers: M/N160 (C.M. Notebook).
[23]Geological Society of London, Murchison Papers: M/J7, p. 30.
[24]Geological Society of London, Murchison Papers: M/N160 (C.M. Notebook, 14th June 1828).
[25]Geological Society of London, Murchison Papers: M/N160 (C.M. Notebook, 18th June 1828).
[26]Geological Society of London, Murchison Papers: M/J7, pp. 30–31.
[27]Geological Society of London, Murchison Papers: M/J7, p. 32.
[28]Geological Society of London, Murchison Papers: M/N160 (C.M. Notebook, 11th July 1828).
[29]Geological Society of London, Murchison Papers: M/N160 (C.M. Notebook, 13th July 1828).
[30]Geological Society of London, Murchison Papers: M/J7, p. 35.
[31]Geological Society of London, Murchison Papers: M/J7, p. 36.
[32]Letter Lyell to his sister Caroline, 3 July 1828, quoted after Wilson 1972, p. 205.
[33]Geological Society of London, Murchison Papers: M/J7, p. 48.
[34]Letter Lyell to his father, 21 Aug. 1828, quoted after Wilson 1972, p. 211.
[35]Geological Society of London, Murchison Papers: M/J7, p. 48.
[36]Geological Society of London, Murchison Papers: M/J7, p. 49.
[37]Letter Lyell to his father, 26 May 1828, quoted after Wilson 1972, p. 197.
[38]Letter Lyell to his mother, 11 June 1828, quoted after Lyell 1881, **1**, p. 189.
[39]Geological Society of London: LDGSL 768/2 (R. I. Murchison to Buckland, Puy en Velai, 4 July 1828).
[40]Letter R. I. Murchison to Adam Sedgwick, quoted after Geikie 1875, **1**, p. 152.
[41]Letter Lyell to his sister Caroline, 3 August 1828, quoted after Wilson 1972, p. 210.
[42]Letter Lyell to his sister Eleanor, 20 August 1828, quoted after Lyell 1881, **1**, p. 197.
[43]Geological Society of London, Murchison Papers: M/J7, p. 49.
[44]Letter Charles Lyell to his father, 24 August 1828, quoted after Lyell 1881, **1**, p. 199.
[45]Geological Society of London, Murchison Papers: M/J7, pp. 37–38.
[46]Geological Society of London: LDGSL 768/2 (Geological Society of London: LDGSL 768/2 (letter R. I. Murchison to Buckland, Puy en Velai, 4 July 1828).
[47]Geological Society of London, Murchison Papers: M/J7, pp. 27.
[48]Geological Society of London, Murchison Papers: M/J7, p. 21–22.

References

FAUL, H. & FAUL, C. 1983. *It Began with a Stone. A History of Geology from the Stone Age to the Age of Plate Tectonics*. John Wiley & Sons, New York.

GEIKIE, A. 1875. *Life of Sir Roderick I. Murchison, Bart., K.C.B., F.R.S.; Sometime Director-General of the Geological Survey of the United Kingdom. Based on his Journals and Letters; with Notices of his Scientific Contemporaries and a Sketch of the Rise and Growth of Palaeozoic Geology in Britain.* 2 Volumes. John Murray, London.

KÖLBL-EBERT, M. 1997. Charlotte Murchison née Hugonin 1788–1869. *Earth Sciences History*, **16**, 39–43.

KÖLBL-EBERT, M. 2002. British geology in the early nineteenth century: a conglomerate with a female matrix. *Earth Sciences History*, **21**, 3–25.

LYELL, C. 1830–1833. *Principles of Geology, being an Attempt to Explain the Former Changes of the Earth's Surface, by Reference to Causes now in Operation.* 3 Volumes. John Murray, London.

LYELL, C. & MURCHISON, R. I. 1828. On the Excavation of Valleys, as illustrated by the Volcanic Rocks of Central France. *Geological Society Proceedings*, **1**, 89.

LYELL, C. & MURCHISON, R. I. 1829*a*. On the Tertiary Deposits of the Cantal and their relation to the Primary and Volcanic Rocks. *Geological Society Proceedings*, **1**, 140–142. [Also published in 1829 in *Annales des Sciences Naturelles*, **18**, 173–214].

LYELL, C. & MURCHISON, R. I. 1829*b*. On the Tertiary Fresh-water Formations of Aix, in Provence, including the Coal-field of Fuveau. *Geological Society Proceedings*, **1**, 150.

LYELL, C. & MURCHISON, R. I. 1829*c*. On the Excavation of Valleys, as illustrated by the Volcanic Rocks of Central France. *Edinburgh New Philosophical Journal*, **7**, 15–48.

LYELL, K. M. (ed.) 1881. *Life, Letters and Journals of Sir Charles Lyell.* 2 Volumes. John Murray, London.

MARCET, J. 1805. *Conversations on Chemistry; in which the Elements of the Science are familiarly Explained and illustrated by Experiments*. London. C. Mercer & Co.

WILSON, L. G. 1972. *Charles Lyell. The Years to 1841: The Revolution in Geology*. Yale University Press, New Haven and London.

Sharing common ground: Nery Delgado (1835–1908) in Spain in 1878

A. CARNEIRO

Centre for the History and Philosophy of Science and Technology, Faculty of Science and Technology, New University of Lisbon, 2825-114 Monte de Caparica, Portugal (e-mail: amoc@netcabo.pt)

Abstract: The Geological Survey of Portugal (Comissão dos Serviços Geológicos), was created in 1857, as a section of the Geodesic Division of the Ministry of Public Works, Trade and Industry. It benefited greatly from the workings of the ministry, which in trying to modernize the country was concerned to keep up with the latest technical and scientific developments occurring elsewhere in Europe. Since its foundation, the Geological Survey of Portugal showed a clear drive towards its participation in an international scientific dialogue and cooperation. This strategy encompassed subscription to specialized foreign books and journals; intense correspondence with foreign specialists; the regular publication of monographs and memoirs in French; occasional or permanent collaboration with foreign experts; and travelling. The main outcome of the 'travel of negotiation' undertaken in 1878 to Spain by the Portuguese geologist J. F. Nery Delgado, then adjunct to the Director of the Portuguese Geological Survey, in addition to improving relationships with the geologists of the neighbouring country, was the collection of field data that was useful for the geological characterization of the southern Portuguese regions. He was also able to negotiate and look for data which could persuade his Spanish colleagues to subscribe to interpretations consistent with the Portuguese geological map, published in 1876.

Geology is a science to which political and administrative borders are irrelevant, because the boundaries of geological units do not comply with territorial conventions or national prejudices. In addition to being intrinsically historical, geology has a spatial dimension, which makes it a transnational science (Carneiro *et al.* 2003).

The transnational character of geology led many nineteenth-century geologists to cross borders and spend considerable periods of time studying geological formations in foreign countries. At the same time, they exchanged correspondence and publications with colleagues working in distant regions, that had similar geological features. Perhaps in geology more than in other sciences, communication was vital. As early as 1878, a specialized international forum, promoting regular meetings was created—the International Geological Congress (Ellenberger 1978). This organization formally institutionalized an international dimension by standardizing geology's verbal and visual language, and promoting initiatives such as the European geological map.

The transnational and international features of geology are all the more interesting as they coexisted with the rise of nationalism in Europe, often associated with the emergence of new nations through the territorial unification of countries like Italy and Germany, or the effective occupation of overseas possessions. As Oldroyd (1996) pointed out, in the nineteenth century territory was envisaged as a resource to be conquered, dominated and exploited, since territorial expansion would bring wealth and power. Geology was thus at the heart of the governmental apparatus, as it tried to rationalize and consolidate its control over territory.

A first organization devoted to mineralogical and geological surveying was founded, in 1848, as part of the Academy of Sciences of Lisbon. However, it was officially suspended in 1855 due to structural problems, difficulties in communicating with the government, lack of staff and funds.

Meanwhile, in 1852, within the Ministry of Public Works, Trade and Industry (Ministério das Obras Públicas, Comércio e Indústria, MOPCI), an essential structure of the Liberal Regime, a Geodesic Directorate was created, with the main purpose of making geographical, chorographical, hydrological and cadastral maps (Branco 1999). The MOPCI was a large 'centre of calculation'[1] composed of a myriad of specialized 'centres of calculation' with the additional task of providing the country with an infrastructure, ranging from the construction of railways, ports, bridges and the telegraph to the establishment of the metric system, the application of statistics nationwide, and the geographical, geological and cadastral survey of the country, the latter indispensable to tax collecting.

It was in this context that, in 1857, the Geological Survey of Portugal was established as a section of the Geodesic Directorate of the MOPCI. The Survey was led by Carlos Ribeiro (1813–1882), a military engineer,[2] and by the former physician

From: WYSE JACKSON, P. N. (ed.) *Four Centuries of Geological Travel: The Search for Knowledge on Foot, Bicycle, Sledge and Camel.* Geological Society, London, Special Publications, **287**, 119–134.
DOI: 10.1144/SP287.10 0305-8719/07/$15.00

Pereira da Costa (1809–1888), then Professor of Mineralogy and Palaeontology at the Lisbon Polytechnic. Nery Delgado (1835–1908; Fig. 1), also a military engineer, was then appointed adjunct to the directors of the Survey. From 1869 onwards he became co-director with Carlos Ribeiro, and in 1882, when the latter died he succeeded him, and led the Survey until 1908.

During the directorships of Ribeiro and Nery Delgado, stratigraphy merged with palaeontology, consolidating the practice of fieldwork and of geological mapping, thereby giving rise to a geological tradition.[3]

In the context of the Portuguese Survey travelling was inscribed in a more general strategy of internationalization of the activities of the MOPCI in its effort to modernize the state. The many technical and scientific activities of the ministry were characterized by an unusual concern in keeping its enterprises in accordance with international standards. This translated into the import of technology and regular stays abroad of engineers working under its jurisdiction with the purpose of improving techniques and methods. These were regularly applied and reported in the Ministry's journal, the *Boletim de Obras Públicas e Minas* (*Bulletin of Public Works and Mining*).

Fig. 1. Joaquim Filippe Nery da Encarnação Nery Delgado (1835–1908). (Courtesy of the Historical Archive of the Institute of Geology and Mining).

Needless to say, as part of the MOPCI, the Geological Survey profited from the dynamics of this governmental institution. By 1857, there was a clear drive towards the internationalization of the geological research carried out under its aegis through various strategies such as: the publication of monographs in French, especially after the 1860s; extensive correspondence with foreign experts;[4] and travelling.[5] Undoubtedly, travel played a major role in shaping the standards of research practices (Carneiro *et al.* 2003, pp. 249–297); in negotiating interpretations of geological data with foreign experts; in consolidating abroad the reputation of the work carried out locally, notably with the participation of Survey geologists in the meetings of the International Geological Congress; and finally in attempting to systematize geological data from missions obtained in the Portuguese African colonies, either by isolated individuals or in the context of the Geographical Society of Lisbon (Carneiro *et al.* 2003, pp. 272–281).

Following Nery Delgado's appointment in 1857, he soon became one of the leading Portuguese geologists involved in a scientific international dialogue, both through his vast correspondence and his travels. His main scientific interest became the study of the Palaeozoic, and his contributions in this field earned him an international reputation. His mission in Spain in 1878 falls into the category of 'travels of negotiation', for several reasons: the Portuguese Survey was more consolidated; Nery Delgado was by then an experienced geologist thereby able to discuss as an equal his interpretations; the activities of the Spanish Survey (Comisión del Mapa Geológico) (Ayala-Carcedo 1999; Blázquez Diaz 1992), and the studies so far carried out were certainly helpful to neighbouring countries; the Portuguese geological map (in the scale 1 : 500 000) had already been published in 1876, and it was important to get the Spaniards to agree with it as much as possible, and to set the standards of future cooperation.

The context of Nery Delgado's mission in Spain

Good scientific relations between Portugal and Spain were rare in the history of science of both countries. Historically, Portugal and Spain were at odds in both wars and mutual contempt. However, by the second half of the nineteenth-century some geologists of both countries engaged voluntarily in a regular scientific exchange, the leitmotiv being the ground shared by the Iberian peoples, but on which they have indelibly marked their national and regional divisions. The relationships established between

Portuguese and Spanish geologists working in their respective national Geological Surveys call for a detailed study. They transcended strictly scientific issues to encompass a dialogue on political and economic ones, the Iberian unification being a topic of friendly discussion, though not consensual among the geologists of both countries.

Nery Delgado's visit to Spain in 1878 was preceded by an intense exchange of correspondence, which continued until his death in 1908. This correspondence shows that the relations between geologists of the Portuguese Survey and their Spanish counterparts developed in a friendly climate with views grounded in nineteenth-century Liberal ideology. They were framed in a kind of chivalric ideal of an international brotherhood of engineers, which seems to have replaced the eighteenth-century 'Republic of Letters' as an ideal realm gathering men of science and good will, regardless their nationalities or beliefs. The letter of introduction to Nery Delgado of the Spanish geologist Joaquim Gonzalo y Tarin (1838–1910), whom he was to meet during this mission, shows this spirit eloquently: 'As an engineer of the Mining Corps (...) I have the honour of addressing you with no other merit or link than that which unites the engineers of all countries'.[6]

From his mission in Spain, Nery Delgado left a written report (Nery Delgado 1879). At first, this report was meant to remain private, but Carlos Ribeiro, then the Portuguese Survey's director, decided otherwise, as Nery Delgado confided to his Spanish colleague José Macpherson (1839–1902) (Alastrué 1968; Barrera 2002*a*, *b*):

> Last year on my return from Spain, I submitted to my director an official report of what I had done, which has not really a scientific character, nor was it written to go around the world. My director thought otherwise and the report was printed. As you will probably find something that might interest you I will ask him permission to send you some copies. If you feel that I deserve that honour please send a copy to your brother Don Guillermo Macpherson and another to Don Antonio Machado [y Núñez (1812–1896)], because I do not dare to do it myself given its insignificant value.[7]

The normal procedure would have been to address a manuscript to Ribeiro, who in turn would have handed it on to the Head of the Geodesic Directorate and the MOPCI. However, he required the publication of Nery Delgado's report,[8] in this way setting a precedent, which can be seen as an important strategic move. By going public, Ribeiro was justifying the need to travel as part of the normal work of a geologist, but more importantly he was legitimizing the research carried out by him and by the geologists working under his supervision. Therefore, those in power could not ignore their role, especially as their work was being acknowledged in other countries, and similar research was being carried out and valued.

From then onwards, whenever Nery Delgado travelled abroad his reports were published (Nery Delgado 1882). Generally they are concise and accurate accounts, rich in details not only relevant to the history of Portuguese science but also to the history of European geology.

Fieldwork and theory

Nery Delgado's mission in Spain took two and a half months, that is from 28 May to 12 August 1878 and encompassed three fieldwork excursions, respectively in the provinces of Huelva (excursion 1) followed by a short stay in Madrid, then Asturias, Léon and Cantabria (excursion 2), and finally in the neighbourhood of Almadén in the province of Ciudad-Real (excursion 3).

Nery Delgado regretted the time spent travelling from one place to another and complained that storms and heavy rain prevented him from obtaining the kind of results he might otherwise have obtained. As compensation, he mentioned the cordiality which marked his personal encounters with Spanish colleagues, which he considered highly beneficial to the development of geological studies in both countries.

He then proceeded with the description of the objectives of his mission, the negotiations involved and the respective outcomes. Thus, the purpose of his visit to Huelva was to establish an agreement regarding the classification of the Palaeozoic formations located in the southern borders of the two Peninsular kingdoms; in particular, to compare the results of the research carried out in the Alentejo (southern Portugal) with those obtained previously by the Spanish mining engineer and member of the Spanish Geological Survey, Gonzalo y Tarin. The visits to Léon and Asturias were to understand the relationships between some shale formations, which were supposed to contain the 'primordial fauna' (Cambrian graptolitic shale) with contiguous Palaeozoic formations. Nery Delgado wanted to acquire practical knowledge in order to renew research on this topic carried out in Portugal, which until that point had been fruitless. At the same time, he examined sites at Colle and Almadén in the provinces of Léon and Ciudad-Real, which he considered particularly interesting for the study of both the Silurian and the Devonian. Nery Delgado wanted to collect data that would clarify the investigations on the same systems in Portugal, in particular in Buçaco (central Portugal) and Portalegre (southern Portugal). In this way, new references and means of comparison were gathered in order to describe the Palaeozoic as thoroughly as possible, which for some time—as he rightly claimed—'have deserved special attention from geologists of all countries' (Nery Delgado 1879, p. 5).

From a theoretical point of view, Nery Delgado's work at the time he visited Spain was very much based on the palaeontological interpretations of Joachim Barrande (1799–1883), a geological traveller who dedicated the last half of his life to the study of the Silurian in Bohemia. Barrande started his palaeontological career in mid-life. He had been a civil engineer and then tutor to Count Chambord, with whom he emigrated from France to Bohemia (Kriz & Pojeta 1974). According to Goulven Laurent (1987), Barrande was one of the mid-nineteenth-century palaeontologists who resisted transformism. He was a Cuvierian, because he valued precise data, based on careful observation and comparison, but he rejected a generalized catastrophism, that is the idea that there was a sudden and total extinction of a particular fauna before a younger one appeared.

Despite recognizing the existence of three faunas in Palaeozoic fossils—'primordial fauna' (Cambrian), 'second fauna' (Ordovician [which was formalized later by Charles Lapworth]) and 'third fauna' (Silurian)—Barrande formulated a theory known as the theory of colonies, which undermined the basis of Georges Cuvier's (1769–1832) and Alcide d'Orbigny's (1802–1857) catastrophism. In his studies of the Palaeozoic fossils, Barrande had observed that in some places in Bohemia assemblages of younger fossils appeared during the existence of the preceding fauna. He explained this phenomenon as being the result of migrations of organisms from one region to another, and he called such organisms 'colonies'. For him, this was a purely palaeontological phenomenon, therefore independent from stratigraphic considerations,[9] an argument which called into question the well established principle of William Smith (1769–1839) for whom the identification of strata of a particular age was associated with fossils of a specific kind (Oldroyd 1990).

Barrande's theory generated a lively controversy, that lasted for about 20 years. Among Barrande's opponents were Jan Krejci (1825–1887) of Prague, and the British geologists Charles Lapworth (1842–1920) and John E. Marr (1857–1933).[10] An alternative interpretation to Barrande's was that of Krejci who claimed that the Bohemian strata had been dislocated by earth movements, or Marr's more elaborate interpretation, arguing that these palaeontological/stratigraphic discrepancies were explained by faulting. For both, tectonic movements resulted in a confused stratigraphic sequence, whose cause was a disturbed physical structure rather than an abnormal reaction of organisms that had set out to colonize existing fauna (Oldroyd 1990, p. 224, 2000).

It is hard to ascertain Nery Delgado's position regarding the implications of Barrande's theory of colonies, particularly in relation to catastrophism, fixism, and evolution. However, when he travelled in Europe in 1881, following the meeting of the International Geological Congress in Bologna, he went to Prague and visited Barrande, just after John E. Marr had published his article opposing the 'theory of colonies' (Marr 1880). Nery Delgado's comments on these debates seem particularly interesting because he expressed the view that the theory of colonies was a strong argument in favour of Darwinian evolution, although he was well aware[11] of Barrande's rejection of evolution, which was increasingly acquiring the status of a doctrine.[12]

However, during his visit to Spain, Nery Delgado's theoretical framework was very much Barrandian, which does not necessarily mean that he endorsed creationism and fixism.

Excursion 1: Diplomacy and negotiation in Huelva

Of the various fieldtrips that Nery Delgado made while visiting Spain, the excursion to Huelva in June 1878 was the longest, lasting more than three weeks. He was accompanied by Tarin, who had a comprehensive knowledge of the province since he had produced a mining-geographic map of that region. In addition to geological fieldwork, Nery Delgado visited mines, in particular the pyrites mines of Rio Tinto and Tharsis, which were considered the most valuable of this Spanish province, and in his opinion, for their type they were probably unrivalled by any other in the world. However, mining was not a subject that interested him much personally. Although recognizing the importance of these mines, he does not provide any description or comment about their organization.[13]

The first scientific issue he discussed with Tarin was the age of the formation on which the city of Huelva was built. Huelva sits on a group of hills composed of clay and fine yellow-grey sandstone with characteristics similar to those of the shell marls of Cacela, Adiça and Mutela (Portugal). The Spanish formations were abundant in fossils, most being molluscs similar to those found in Portugal. The fossils of Huelva had been studied by Justo Egozcue y Cia (1833–1900), then Professor of Geology and Palaeontology at the Mining School of Madrid and attached to the Spanish Geological Survey, and Lucas Mallada y Pueyo (1841–1921) (Alastrué 1983), a palaeontologist serving in the same institution. Given the abundance of certain species, and the existence of others which Egozcue and Mallada thought to be characteristic of the Pliocene, they had classified the fossils of Huelva accordingly, a classification which Tarin accepted (Nery Delgado 1879, p. 7). However, Nery Delgado disagreed and contended that the fauna that inhabited Huelva during the Cenozoic was the same as had existed in Cacela and in the mouth of the Tagus River (Lisbon). The only difference, he said, was that some extra species had been found and others had vanished, which only amounted to a modification or local variation; he concludes that 'our fauna (...) corresponds perfectly to that of the Vienna basin classified as belonging to the upper part of the middle stage of the cenozoic (Upper Miocene)' (Nery Delgado 1879, p. 8).

To support his view, Nery Delgado reported that he found different types of Foraminifera in the sandstone of Huelva that were also present in the Vienna basin. This was material proof, which for the moment settled the issue and confirmed his interpretations. The outcome of this negotiation could not have pleased Nery Delgado more: the final conclusion was consistent with the classification that he and Carlos Ribeiro had made on the Portuguese geological map published in 1876 (Ribeiro & Nery Delgado 1876). Portugal, being a much smaller country, already had a geological map, whereas Spain was still struggling to produce a precise national geographic map to replace its regional maps. As in a game of chess, it was only natural that the Portuguese Survey should want to make the Spanish geological interpretations conform to its own.

Nery Delgado reported that together with Gonzalo y Tarin he had drawn an initial cross-section in Huelva, NW via Villanueva de los Castillejos and Puebla de Guzman towards Paymogo, that aimed at differentiating the relationships between the fossiliferous formations of São Domingos (Portugal), which he had classified as Silurian, the Culm[14] topping them, and with the underlying shale running northwards. However, this cross-section did not provide the expected answers (Nery Delgado 1879, p. 7).

Nery Delgado then explained that a strip running east–west in the central part of the Province of Huelva showed intense metamorphic activity, that had so deeply altered the characteristics of the original layers that it was almost impossible to differentiate them from the Azoic formation. In addition, Nery Delgado noticed that there had also been intense 'geyserian activity' (hydrothermal activity),[15] accompanied by metalliferous emissions, from which huge masses of minerals and limestone, and probably quartzite, had originated but in such an abnormal way that they did not exhibit the continuity typical of regularly formed layers (Nery Delgado 1879, p. 7). From Nery Delgado's description, he seems to have been referring to the presence of carbonatite, although he does not name it as such, as the term was only coined in 1921 by the Norwegian geologist and igneous petrologist W. C. Brögger (1851–1940).[16] Carbonatite is considered an unusual igneous rock, rich in calcite and other carbonate minerals, metals and rare earth elements. Since 1965, the magmatic origin of carbonatite has been reinforced by fieldwork and laboratory studies; however, the other hypotheses—intrusive limestone, alkaline metasomatism and recrystallization of sedimentary or metamorphic limestone, as well as hydrothermalism—remained as particular cases or complementary mechanisms (Brito de Carvalho 1978, p. 27). Nery Delgado's description seems to match the theory of a hydrothermal mechanism underlying the genesis of the then nameless carbonatite.

Nery Delgado explained that the features he had described made the study of this region so difficult that Tarin, despite his detailed investigations, was forced to paint all these layers in the geological map of this province with the colour representing the Culm. However, Nery Delgado contended, it was unlikely that this Palaeozoic formation covered such an area.

Nery Delgado was more fortunate with a second cross-section running northwards, beyond the borders of the province, and reaching the Province of Badajoz until Higuera de Fregenal. He recognized that there was a perfect replication of the strata observed in southern Alentejo (Portugal), and concluded that two seas, one Silurian and one Carboniferous had existed, either side of a barrier on both sides of which the same sediments had deposited. This explained the replication of strata, north and south of the strip of Azoic rocks, as already indicated in the Portuguese geological map, between Aldeia Nova and Corte do Pinto (Portugal), and extended into the Spanish territory.[17] He said that these ancient rocks formed a large promontory in the Silurian and the Cambrian, against which fine greywackes and greenish mudstone were deposited, covering vast areas of Portugal and Spain. He argued that later on, the formations of São Domingos (Portugal) emerged as a result of tectonic movements. They contained various specimens of Silurian fauna, which could not be found anywhere else in the Iberian Peninsula, and occasionally some graptolites, which seemed to impart a 'colonial' character to this fauna (Nery Delgado 1879, p. 8).

The same succession of fine greywackes and greenish mudstone that he had observed from São Domingos (Portugal) northwards were observed in Spain, near Molinos de San Bartolomé, five legs ESE of Barrancos (Portugal), though obscure and barely distinguishable. Nery Delgado then established some correlations between his observations carried out in Spain: the 'ampelite'[18] of the Sierra de los Cotos, which he did not examine, probably corresponded to that observed in Ensinasola; the limestone of La Serrana and Sierra de Alamo corresponded to that of Hinojales and Arroio de Molinos in the southern border of the province.

The analogies of the slate formation of the Baixo-Alentejo (Portugal) and Huelva with the Taconic of North America, that Nery Delgado had mentioned in his memoir on the Silurian of Alentejo published in 1876 (Nery Delgado 1876), were also confirmed by his Spanish friend of many years, José Macpherson, when he identified a form of *Arhaeocyathus* (*Archaeociatid*), a coral, in the limestone occurring in northern Seville. Until then, *Archaeocyathus* had only been found in the sandstone of Potsdam (Canada), but Macpherson's discovery proved the existence of Cambrian strata in the south of the Iberian Peninsula.

Macpherson's confirmation of Nery Delgado's views had a special meaning for the Portuguese geologist. Carl-Ferdinand von Roemer (1818–1891), then Professor of Geology at Breslau, who had contested Nery Delgado's classification of the strata of São Domingos (Alentejo) as being Silurian (Nery Delgado 1879, p. 9), was thus led to revise his own arguments supporting his classification of São Domingos strata as belonging to a lower division of the Culm.[19] However, Nery Delgado recognized that further studies were required in Portugal, notably in the surroundings of Barrancos, a region until that point only barely studied and difficult to explore (Nery Delgado 1879, p. 9). Nery Delgado summed up his expedition in this Spanish province:

> If my visit to Huelva did not make completely clear to me the Silurian stretch of São Domingos, and if in the research carried out by Tarin I could see no good reasons to change the results obtained in Portugal, the discoveries made near Ensinasola shed an intense light which will greatly help me in pursuing my future investigations (Nery Delgado 1879, p. 9).

He then made a short trip to Seville to meet José Macpherson, and clarify some points regarding *Archoeocyathus*. He visited the museum of the local university whose director was Antonio Machado y Núñez (1812–1896), Professor of Natural History. Nery Delgado remarked that the collections were small, but possessed invaluable samples especially molars and the maxillary bones of an elephant collected in the sandstones of the river Guadalquivir, which Machado and the brothers Macpherson (José and Guillermo) had classified as belonging to the Quaternary. Nery Delgado believed that these deposits were contemporaneous with the sandstones of the Portuguese rivers Tagus and Mondego. He also considered himself prepared to introduce corrections to the Portuguese geological map, if further studies confirmed this hypothesis:

> If that species of proboscides belongs to the Pleistocene (post-Cenozoic) and if it was effectively extracted from the sandstone deposit with quartzite pebbles there is probably a reason to correct the classification of our modern sandstone deposits, indicated in the geological map by (n^3), which will have to be divided into Pliocene and Pleistocene, when more and fruitful observations unveil the fauna which inhabited our country at the time these deposits were formed (Nery Delgado 1879, p. 10).

Nery Delgado concluded that this large sandstone deposit existing in the Iberian Peninsula should be considered a general phenomenon. He then hypothesized about its formation: it could have been formed either during the time span separating the two glaciations, or by strong currents following the thawing of the great ice cap, should the existence of these two distinct periods in the Peninsula be demonstrated.

Afterwards Nery Delgado left for Madrid, where he devoted himself to the study of the palaeontological collections of the Spanish Survey. He concentrated on those pertaining to the Silurian and the Devonian of Ciudad-Real, Asturias and Léon as a preparation for the field excursions he planned to make in the latter two provinces.

Excursion 2: Léon, Asturias and the Cantabrian mountain range

The second field excursion was to Léon, Asturias and the Cantabrian mountain range. Its purpose was to investigate the lithological characteristics and the stratigraphic relationships of the shale containing the 'primordial fauna' (Cambrian), which had not been found in Portugal at that time. However, the 'primordial fauna' had been observed in various Spanish regions, particularly in these two provinces thanks to the work of Casiano de Prado (1797–1866) (Ayala-Carcedo 1997). On this field-trip, Nery Delgado was accompanied by Lucas Mallada and Jesus Buitrago both of whom were working for the Spanish Survey and were heading to Asturias to decide on a question raised by Charles Barrois (1851–1939), who had recently contested Prado's claim of the existence and location of a Cambrian formation between Grado and Belmonte.

Prado and Edouard Verneuil (1805–1873) had discovered the 'primordial fauna' (Cambrian) in Léon and also in the centre of Asturias, between Grado and Belmonte (Prado 1850). Barrois was interested in studying this fauna, which as he argued:

> The primordial fauna is highly interesting to a great number of geologists of both worlds who attended the Exhibition in Philadelphia, and which they feel should deserve full attention in the

meeting of the International Geological Congress to be held in Paris in the next year (Barrois 1877, p. 378).[20]

Barrois added that, in France, nobody had been able to identify the 'primordial fauna' and therefore the opportunities for studying it were scarce. But, on his trip to Spain, he was also unable to find this fauna between Grado and Belmonte where Prado claimed to have found it. At first, Barrois was persuaded that the whole region was constituted by Devonian layers, but he finally located strata containing the 'primordial fauna'. However, these strata formed a regular strip in the western part of Asturias, not in central Asturias as Prado had claimed. For Barrois, the shale containing the 'primordial fauna' in western Asturias penetrated the limestone. It was overlying micaceous shale and gneiss, and covered by white sandstone and *Skolithos*, which in his view had the same stratigraphic position and the same lithological characteristics of the shale constituting the cliffs in Douarnenez Bay in Britanny. He concluded that the shale of Asturias and Britanny might be of the same age. The 'primordial fauna' of Asturias, which he hoped to be able to find in France, contained *Paradoxides*, *Conocephalites* and *Trochocystites* but differed from that of Léon because it lacked Brachiopoda (Barrois 1877, pp. 378–379).

The Spanish Survey published a translation of Barrois's paper resulting from this trip, which had been released in the *Bulletin de la Société Geólogique du Nord*, but added a note to Barrois's considerations in which it hints at a certain malaise. It reads:

The Spanish Geological Survey received the news that an engineer of the Corps of Mines, who had long lived in the Mining district of Oviedo, visited and on frequent occasions carefully observed the locations mentioned by Verneuil and Prado. In the neighbourhood of Belmonte he recognized various fossils belonging to the 'primordial fauna' and among them a *Trochocystites bohemicus barr* and the glabella of an *Elipsocephalus*. It is plausible to think that if Barrois was not able to see them, it is because geologists are not always lucky enough to find everything they seek, when they decide to undertake a quick excursion to a particular site, which is undoubtedly the case of Barrois's. The Spanish Survey therefore believes that one should suspend judgment about a particular question, until it becomes totally clear.[21]

Despite the stormy weather, Nery Delgado reported that he was able to clarify the questions that troubled him at the same time he discovered the reasons underlying what he considered Barrois's mistake, thereby providing his Spanish colleagues with arguments against the interpretation of the French geologist.

Nery Delgado pointed out that there was a quartzite mountain range in western Asturias, which the German-born mining engineer Guillermo Schulz (1800–1877) (Puche Riart & Ayala-Carcedo 2001) had classified as Devonian in his geological map. Its features corresponded to the same formation that occurred in Portugal and constituted the base of the Silurian, and contained the 'second fauna' (now known to be Ordovician). Nery Delgado described this quartzite as being either finely grained or compact, containing some mica, and being white or whitish in colour as a result of superficial alteration. Underlying this massive and thick formation, at times interrupted by thin layers of clay and clayey-siliceous, dark-greyish micaceous shale, a thick formation of clay and clay-siliceous shale more or less micaceous developed. Upwards, it effected the transition between the micaceous clay and the quartzite. At the base, there was clayey shale where various forms characteristic of the 'primordial fauna' were found, namely trilobites, belonging to the genera of *Paradoxides*, *Conocephalites* and what Nery Delgado considered a remarkable cystidium (*Trochocystites bohemicus*? *barr*), being particularly abundant in one stratum (Nery Delgado 1879, p. 16).

Nery Delgado explains that the base of this fossiliferous shale formation was composed of a thick layer of dolomitic limestone, either with a sugar-like or compact texture with spar spots with a greyish, yellowish or rosy colour, probably due to superficial alteration. In his view, this layer represented the oldest group of rocks in the region and at the same time it established the connection between the fossiliferous strip and the limestone, which in the southern part of the Cantabrian mountain range contained the same fauna. Nery Delgado argued that limestone was the key rock to understand the question because it indicated the sites of a region where the 'primordial fauna' (Cambrian) could be found:

Wherever limestone appears under the quartzite formation, there are good reasons to suppose the existence of that fauna in the intermediate shale layers (Nery Delgado 1879, p. 16).

By taking into consideration Schulz's geological sketch of the region where limestone was represented at the western end of Asturias, between Vega de Rivadeo and Santa Eulalia de Oscos, along a series of outcrops aligned with the astronomic meridian, Nery Delgado argued that one could presume the existence of the 'primordial fauna' (Cambrian) (Nery Delgado 1879, p. 16). However, that had not been Barrois's conclusion. He also said that according to Barrois's investigation, this fauna appeared instead in the borders of the province of Oviedo with Galicia.

Nery Delgado was led to conclude that similar considerations should be applied to different isolated outcrops of limestone, more or less extended but to some degree aligned and related to one another, that run parallel to the quartzite strip, crossing the province northwards in the chain separating rivers Narcea and Navia. That had been Prado's conclusion in relation to the mountain range of El Crono de Peñarubia, running northwards until reaching the sea (Nery Delgado 1879, p. 17).

Paradoxides was observed in both the western and eastern parts of the high quartzite mountain range of El Pedrorio and Peñamanteca in the neighbourhood of Belmonte. The mountains were aligned with one another, which led Nery Delgado to conclude that they were a synclinal fold, in which he had found the Devonian system, represented by alternating limestone, clay and shale, the most abundant rocks of the region (Nery Delgado 1879, p. 17).

In the valley of Santa Maria de Villandar, west of Pedrorio, the Devonian layers formed, in Nery Delgado's view, another synclinal fold, whose axis corresponded roughly to the direction of the thalweg, because the different layers were replicated in each slope, giving the impression at a first glance that they intruded under the quartzite of Pedrorio and the shale with *Paradoxides* and limestone. Nery Delgado argued that the boundary of the Devonian system lopsidedly cut the direction of the strips of 'primordial fauna' (Cambrian) and the Silurian quartzite, which was probably determined by a fault running along the foot of the mountains of Pedrorio and Siaza, the position of these layers resulting from their dislocation (Nery Delgado 1879, p. 17).

Based on these considerations, Nery Delgado claimed that he was forced to recognize that a synformal folding of the layers, running NNE along the mountain range of Pedrorio, Bejega and Peñamanteca, produced various anticlinal and synclinal folds. In his view, they explained why the Silurian quartzite of the base of the 'second fauna' (later regarded as Ordovician) came to occupy the highest points due to the greater resistance they offered to external agents and denudation. The layers of the Devonian, which by a discordant or transgressive stratification originally rested on the layers of quartzite, shale and underlying limestone, underwent the same dislocation movements as the older formations, thereby producing faults. Nery Delgado argued that these faults corresponded to the deep ravines in whose walls the lower layers were observed, showing the contact with the Devonian. He concluded that this apparent inversion had troubled Prado so much, and had mistaken the 'no less skilful geologist Barrois, 18 years later' (Nery Delgado 1879, p. 17).

From the orientation of the quartzite strips indicated on Schulz's map, Nery Delgado was in no doubt that the layers of the 'primordial fauna' (Cambrian), after inflection from NNE to SSW, inflected again southwards, southwestwards and then eastwards, penetrating the province of Léon and proceeding to Boñar and Sabero via the southern slope of the Cantabrian mountain range (Nery Delgado 1879, p. 18).

Nery Delgado and Barrois continued to exchange letters on this subject between 1879 and 1883, following the publication of Nery Delgado's report of his mission in Spain. In particular, they discussed the correlations Barrois had established between the Palaeozoic of Britanny and Alentejo (southern Portugal) by taking into account Nery Delgado's investigations in this Portuguese province.[22] Barrois, then working in the Geological Laboratory of the Science Faculty of Lille, had engaged in the study of the geology of Asturias, Galicia (Spain)[23] and Brittany published various works that he discussed with Nery Delgado.

The Cantabrian mountain range

Nery Delgado then went from Estacas to Ferredal, a place located in the parish of Quintana, where he observed the transition from the quartzite layers to the underlying fossiliferous shale, and then to the lower limestone and marbles. Ferredal is built on limestone and it was in the shale layers immediately overlying the limestone that fossils were found. Nery Delgado remarked that the transition from marble to shale was through a coarse reddish limestone similar to that of Sabero where he had found fossils. But what Nery Delgado found most striking was that the fossiliferous shale, slightly greenish and slightly chloritic, showed in some parts a similar character to the shale or slate clay of the 'colonial' horizon of Buçaco (central Portugal), especially going from Portela de Sazes to Sazes. Nery Delgado argued that this coincidence could lead one to conclude that the limestone is but a lithological accident of the shale layers, its origin being 'geyserian'. In his view, the 'geyserian activity' began to manifest before the water transported the muddy sediments, but he argued that like in Buçaco (central Portugal) it probably continued long after the sediments were deposited (Nery Delgado 1879, p. 18). Again, he seemed to be hinting at the presence of volcanic activity and the possible presence of carbonatite this time in Buçaco, which is all the more interesting as in the twentieth century its existence on mainland Portugal was absent from geological literature, mainly because calcium carbonate based rocks are invariably assumed to be sedimentary rocks.[24]

Nery Delgado contended that the study of the wide Silurian part of Asturias—a task that in his view would take a long time because the country was rough and the members of the Spanish Survey had to investigate other regions first that they barely knew—would provide invaluable data to aid the knowledge of other Palaeozoic formations. But what seemed indisputable to Nery Delgado was that when the Silurian of the Iberian

Peninsula was taken as a whole, this lower formation had undoubtedly greater thickness in the north than in the south, where it finally died out (Nery Delgado 1879, p. 18).

On the southern slope of the Cantabrian mountain range, Nery Delgado confirmed what Prado had pointed out: the 'primordial fauna' (Cambrian) appeared in narrow strips (10–40 m), essentially composed of coarse and ferruginous clayey limestone, whose red/yellowish colour could be observed especially in the upper layers. These strips were intercalated with the Devonian and the Carboniferous, and Nery Delgado remarked that their features were such that it was difficult to distinguish them from the contiguous layers containing the 'primordial fauna' (Cambrian). In his view, that was why the 'primordial fauna' of the Peninsula had not been discovered in Spain, but in Paris in the cabinets of Barrande and Verneuil, to whom Prado had sent the fossils he had collected in his field missions in the Cantabrian mountain range. Yet, Nery Delgado praised Prado's work because he had persistently followed the vestiges of these strips for more than 100 km in a very rough and so far unexplored region, thereby providing the first positive data on the existence of the 'primordial fauna' in the Iberian Peninsula. Prior to Prado's work, only the fossil fragments that he had collected in the Guadiana basin, north of Ciudad Real, had merely hinted at its existence (Nery Delgado 1879, p. 19).

Nery Delgado then observed the relationships between the 'primordial' strips of Léon and Asturias. He remarked that limestone and shale occur in both places, but in Oviedo the fossils occurred exclusively in the shale, whereas in Léon they were more abundant in the limestone. What he found most striking was that in the borders of these two Spanish provinces, or in the passage of the Cantabrian mountain range, fossils appeared both in the limestone and in the shale, which in his view proved both a petrographic and palaeontological connection between the 'primordial' strips located on both sides of that important orographic line (Nery Delgado 1879, p. 19).

However, he noted that there was a difference regarding the richness of these faunas, which resulted from more favourable conditions on the southern slope of the chain compared with the region to the north. Despite being less abundant, a greater variety of species had been discovered in the deposits of Léon than in Asturias, the absolute absence of Brachiopoda in this province being the most relevant feature. In Léon, they amounted to one third of the total of species collected. According to Nery Delgado, the composition of this fauna and the existence of certain genera linked it with the Menevian of St Davis, or associated the first stage of the 'primordial fauna of Barrande' with the fauna of northern Spain (Nery Delgado 1879, p. 19).

One of the geological cross-sections that Nery Delgado drew and found most interesting was that from Colle towards Collada de Llama, near Sabero. He drew it to observe the bedding and features of the *Posidonomya pargai* that had been discovered there. Nery Delgado had supposed that this species would be similar to the *Posidonomya becheri* of the Culm, which he had assumed to be present in Spain. He concluded instead that there was nothing common between the formation containing the first species and the particular *facies* of the Lower Carbonic, which contained the *Posidonomya becheri* in Huelva, and in the southern provinces of Portugal. He thus recognized that these species were quite different, and made sure that the conditions in which the deposits (of different age too) had been formed were also distinct. In effect, Nery Delgado did not find any Carbonic species or those of the Culm in the layers of black slate clay with many clayey-siliceous nodules, containing *Posidonomya pargai* and other Devonian species. He contended that this species was characteristic of a lower horizon than that of *Posidonomya becheri*, which until then had not been found either in the Cantabrian mountain range, or in the Peninsula (Nery Delgado 1879, pp. 19–20).

In the Spanish province of Léon, as Nery Delgado observed, the Devonian was immediately followed by the Cretaceous, with petrological features similar to those found in Portugal, namely in the Mondego valley (central Portugal), and he found this correlation in such distant places as these. In the Mondego valley, the Cretaceous was composed of a lower stage of light coloured clay, containing pebbles and small quartz rock fragments, more or less angular, of varied appearance showing either big white or yellow, purple or reddish spots of kaolin. This clay derived from the decomposition of granite rocks and led Nery Delgado to conclude that it was contemporaneous with the Cretaceous clay of Portugal. Upwards, Nery Delgado observed that there was a limestone layer, fairly clayey and coarse, with some fine and barely coherent clay layers, or marls, containing a particular fauna with species peculiar to this Spanish locality. He then correlated this limestone horizon with the limestone of Figueira da Foz (central Portugal), which, as in Léon, marked a significant change in the stratigraphic phenomena (Nery Delgado 1879, p. 20).

In Nery Delgado's view, a series of parallel faults, a common feature in Cantabria, had repeated these layers in the Middle Devonian strip where they were intercalated, placing these older rocks in a similar role to the one they played in relation to the limestone containing the 'primordial fauna' (Cambrian) (Nery Delgado 1879, p. 20).

Excursion 3: Almadén (Ciudad-Real)

Nery Delgado ended his report with his field excursion in the neighbourhood of Almadén. He emphasized that this region was famous due to its mercury mines, which supplied the world market with great quantities of this metal. However, his personal interest lay elsewhere, in the possibility they offered to geologists to study the Palaeozoic. In particular, he could observe the links of the Silurian to the system classified as Lower Devonian, which immediately followed in a seemingly concordant manner. Nery Delgado recognized that this investigation would take a long time, as the region was so difficult to explore; Casiano Prado as well as other geologists were only able to produce provisional conclusions (Nery Delgado 1879, p. 21).

One of the results that Nery Delgado obtained was that in the borders of Castella-a-Nova, as well as in Asturias and Portugal (corresponding to an area of more that 8000 square leagues) the base of the Silurian (formalized later as Ordovician) was composed of a thick quartzite formation, containing thick layers of 'bilobites' (*Cruziana*), which were strongly dislocated. They formed a mountain chain that correlated with each other in different parallel strips, thereby revealing the action of the ample movements and strong pressures they underwent (Nery Delgado 1879, p. 21).

One of these quartzite ridges ran southwards in front of Almadén and formed a slight curve opening to the village. According to Nery Delgado, a league southwards and related to the former, was a second one that represented another branch of the anticline on both sides of which were a repeat of the Ordovician and the Devonian layers. The axis of this undulation, or the valley between the two quartzite ridges, contained a thick formation of greywacke and greenish slate clay, which in the Portuguese provinces of Beira and Alentejo occurred at the same stratigraphic position, as Nery Delgado had observed (Nery Delgado 1879, p. 21).

A cross-section made in Almadén running north or northwestwards towards Chillon revealed to Nery Delgado the perfect correlation between the graptolitic shale ('colonies'), sometimes occurring in the middle stage of shale and quartzite in alternating thin layers, the trap reef contemporaneous with this shale, and the quartzite formation that constitutes the base of the Devonian (Nery Delgado 1879, p. 21). Following or contemporaneous with the genesis of quartzite, another emission of trap rocks, whose decomposition produced most of the debris found in some sedimentary layers, seemed to Nery Delgado to have coincided in time with the appearance of Brachiopoda and Bryozoa, characterizing the fauna of the Lower Devonian (Nery Delgado 1879, pp. 21–22).

In Almadén, Nery Delgado observed fossils in a quartzite layer containing many moulds of small *Nuculas*, with the same petrographical and palaeontological features of a stratum, which in Buçaco (central Portugal), were intercalated with the 'Lower Silurian' (now classified as Ordovician), corresponding to the culminating group of this division, the 'colonial' horizon. Nery Delgado contended that this observation proved that this formation and the abundance of Brachiopoda were closely associated. They had been referred to the base of the Devonian, 'to our Middle Silurian'. He argued that, perhaps, there was no reason to consider that formation to be the coeval representative of some of the upper stages of the Silurian of the Bohemia basin that are not so clearly defined outside that 'privileged region' (Nery Delgado 1879, p. 22).

Nery Delgado observed that the succession of rocks in Almadén was similar to that of Serra de Portalegre[25] (southern Portugal) with the single difference that the shale containing fossils of the 'second fauna' (now known to be Ordovician) was missing in that Portuguese region, whereas in Almadén, although poorly developed, it was clearly represented. For Nery Delgado this fact proved that the Silurian sea or gulf, where the species of the 'second fauna' would have lived, extended from Spain towards the Portuguese border, the Serra de Portalegre (southern Portugal) corresponding to the boundary of that sea. Nery Delgado argued that this reinforced his arguments in favour of the existence of an insurmountable barrier that prevented communication between the two Silurian seas of central and southern Portugal, whose deposits showed two totally distinct facies. The latter would have been closely linked to the Silurian deposits of America and northern Europe; the former showed strong analogies with those of France and central Europe. But Nery Delgado emphasized that one should not suppose that there had been a sudden transition from one of these deposits to the other. On the contrary, he believed that the Silurian strip of Serra de Portalegre (southern Portugal) somehow established the transition between the deposits of Buçaco (central Portugal) and Monfortinho (central eastern Portugal)—which have been clearly characterized—and the special facies he observed in Barrancos and São Domingos (southern Portugal) (Nery Delgado 1879, p. 22).

Nery Delgado added further hypotheses: the oscillation of the sea floor at the end of the Silurian—immediately related to the trap reef and with the appearance of hydrothermal springs—would have been responsible for the sharp discordances, and possibly the simultaneous emergence of the fauna of *Nereites* in São. Domingos (Portugal), and the graptolites in northern Huelva (Spain), Barrancos and Portalegre (Portugal). He also assumed that the quartzite with *Cruziana* was probably formed during a period of subsidence, which was preceded by slow uplift, during which deposits of the 'primordial fauna' (Cambrian) accumulated. Thus, it did not surprise him that the limestone containing *Archaeocyathus* in northern Seville, on the southern slope of Sierra Morena, was contemporaneous with the quartzite formation of Asturias, corresponding in this way to the terminus of the 'primordial fauna' (Cambrian) whose first stage of existence he believed was absent in the Iberian Peninsula, with this single exception (Nery Delgado 1879, p. 23).

Nery Delgado first classified the fauna of São Domingos as belonging to the 'Lower Silurian' (now classified as Ordovician) in 1876 (Nery Delgado 1876), but in 1899 as being Devonian (Nery Delgado & Choffat 1899). In his last memoir published in 1908, he contended that it was part of the 'Upper Silurian' (now Silurian as against Ordovician) (Nery Delgado 1908).

Impressions and souvenirs from Spain

Travelling provides references with which more authoritative comparisons can be made. In his report, Nery Delgado made evaluations of Spanish scientific practices and used them as an opportunity to criticize current Portuguese procedures.

Whilst in Madrid, he was able to appreciate the efforts of the Spanish Geological Survey, in particular the role of Manuel Fernández de Castro (1825–1895), the General Inspector of Mines presided over this institution at that time (López de Azcona 1984–1990). Nery Delgado hinted at the existence of a geological culture in Spain (Knell 2000) that joined together individuals who carried out geological research either by their personal initiative or on an official basis, a phenomenon which he could not observe in his home country:

It is really remarkable and of the highest scientific value some of the works published both by the Survey members and mining engineers working under its aegis, and by individuals devoted to geological studies, working either officially or freelance, or even by simple apostles of science (Nery Delgado 1879, p. 11).

As he well knew, in his native country, geology was only consistently practised in the context of the Geological Survey. Given the lack of 'apostles of science' (to use Nery Delgado's expression) working on a private basis, and of local scientific societies, geology emerged and was practised within a governmental framework thereby relying primarily on state economy and human resources (Carneiro 2005).

Nery Delgado recorded that the Spanish Geological Survey aimed to publish a more or less comprehensive description of a province annually, and produced two publications, the *Memorias* and the *Boletín*: the former published more detailed studies and the latter, notes, descriptions of partial and preliminary studies. At this time, the Portuguese Survey did not publish a journal, only monographs. In 1883, the Survey began to publish the first Portuguese specialized journal devoted to geology, the *Communicações da Commissão dos Trabalhos Geologicos*.[26]

The difficulties of his Spanish colleagues in carrying out their work was the object of Nery Delgado's considerations. The first difficulty was the size of the country: of 49 provinces, including the Balearics and the Canaries, eight had never been studied, for 15 the available data was very scarce, and of the remaining 26 there were some published descriptions that provided positive data to the geological description of the whole. Only upon completion of this enterprise, could a general geological map of Spain be started.[27] Nery Delgado also regretted that the petrographic study of rocks, which he considered highly useful, had not been taken up by the Spanish Survey due to lack of personnel.

As to the collections, in addition to those of the Spanish Survey, Nery Delgado examined the mineralogical collection of the Museum of Natural History of Madrid, which possessed invaluable samples, but he regretted the fact that they were displayed on aesthetic rather than on scientific criteria. He also visited the mineralogical and palaeontological collections of the Mining School of Madrid, the former classified according to Georges Dufrénoy's (1792–1857) method and the latter according to d'Orbigny's.

He was then taken to the Geographic and Statistics Institute led by General Carlos Ibañez de Ibero (1825–1891). Despite recognizing that this visit was outside the scope of his mission in Spain, Nery Delgado claimed that he could not have missed this important institution whose purposes he found similar to those of the Portuguese Geodesic Directorate. On this visit, Nery Delgado was accompanied Eduardo Benot (1822–1907), a former Minister of Development, who, in 1873, succeeded in passing the law that removed the Spanish Geographic Institute from the control of the Government, a move that pleased Nery Delgado a great deal:

Up to then the Institute had endured a difficult existence due to its dependence on the central government, which now cannot appoint or dismiss personnel arbitrarily. The hiring of staff is carried out through competition, dismissals are subject to strict rules, and the only reasons can be those directly linked to work (Nery Delgado 1879, p. 12).

Coming from a state geologist Nery Delgado's comment was all the more significant. It was a clear criticism of the imbroglio underlying the suspension of the Portuguese Geological Survey back in 1868.[28]

Nery Delgado was undoubtedly in favour of the autonomy of scientific institutions and claimed that the work of the Spanish Institute 'free from the ups and downs of politics, which in Spain like everywhere else sterilizes the most productive scientific activity, has developed rapidly and securely'. (Nery Delgado 1879, note on p. 12). On his return to Portugal, he even wrote to Eduardo Benot, asking him for details about the organization of that autonomous institution.[29]

The chromolithographic technique invented by Ibañez de Ibero to print maps at the Spanish Geographic Institute was also critically examined by the Portuguese geologist.[30] However, he did not seem enthusiastic about this method:

> I do not dare to express an opinion about the course of action taken by the wise director of the Geographic Institute. However, it seems to me that the efforts of General Ibañez—concentrating on the scientific and artistic improvement of these partial maps with the purpose of matching or even overcoming the best that has been published in more advanced countries—are not entirely profitable. It would be more advantageous if, as happens in Portugal, those efforts were directed towards a presentation of a general geographic map of Spain, precise enough to provide a basis for the work of the Spanish Geological Survey (Nery Delgado 1879, p. 15).

Nery Delgado pointed to the difficulties under which his Spanish colleagues worked, in particular the lack of an accurate general geographic map of Spain, which could have helped them to draw the boundaries of the different geological units. The fact that he sent his report to Macpherson, who most certainly publicized it, suggests that Nery Delgado seized this opportunity as an external observer to promote the needs of his Spanish colleagues, as much as he used the positive Spanish examples to express his and Ribeiro's views.

Conclusions

In the nineteenth century, travels in the context of the Geological Survey of Portugal were inscribed in the very dynamics of the Ministry of Public Works of which the Survey was part. The 'travel of negotiation' of Nery Delgado to Spain in 1878 occurred at a stage when the work and reputation of the Portuguese Geological Survey were already established. In addition to improving the relationships with the geologists of a neighbouring country, he was able to collect field data that enabled the geological characterization of the southern Portuguese regions. He was also able to negotiate and look for data with which to persuade his Spanish colleagues to subscribe to interpretations consistent with the Portuguese geological map, published in 1876.

On this account his field excursion in Huelva was significant because the fossils of Huelva had been classified by Spanish geologists as belonging to the Pliocene. Nery Delgado was able to persuade them that the fauna which had inhabited Huelva in the Cenozoic was the same found in Cacela, and in the mouth of the Tagus (Portugal). It corresponded to the fauna of the Vienna basin, and therefore belonged to the middle stage of the Cenozoic (Upper Miocene). In the sandstone of Huelva, Nery Delgado found Foraminifera, which proved his interpretations right and consistent with the Portuguese geological map. In addition, Nery Delgado contended that there had been hydrothermal activity in Huelva (central part), which he called 'geyserian activity', accompanied by metalliferous emissions, from which masses of minerals and limestone originated, thereby raising the possibility of the existence of what came to be known, in 1921 as carbonatite.

The discovery made by his Spanish friend, José Macpherson of *Archaocyathus*, a coral, in the limestone of northern Seville proved the existence of the Cambrian in the south of the Iberian Peninsula, and confirmed Nery Delgado's classification of the strata of São Domingos (Portugal), and led Roemer to revise his arguments supporting the classification of São Domingos strata as belonging to a lower division of the Culm.

In the borders of Huelva with the Province of Badajoz, Nery Delgado recognized that there was replication of the strata observed in southern Alentejo (Portugal), extending in the Spanish territory, which again was consistent with the Portuguese geological map.

Nery Delgado's field excursion in Asturias led him identify folding and faulting along the mountain range of Pedrorio, Bejega and Peñamateca, which had caused an apparent inversion of strata. Nery Delgado's conclusions about the location of the 'primordial fauna' (Cambrian) in Asturias were contradictory with those of Charles Barrois, but consistent with the observations of the Spanish geologist Prado, which certainly contributed to reinforce the position of the Iberian geologists. However, following his trip to Spain, Nery Delgado exchanged correspondence with Barrois regarding the geological correlations, which the latter established between Asturias, Brittany and Alentejo (Portugal).

In the Cantabrian mountain range, Nery Delgado compared the fossiliferous shale of that region with that of Buçaco (Portugal) and again referred to 'geyserian activity' when he claimed that the limestone in Buçaco was but a lithological accident of the shale layers (carbonatite?). In his view, there was evidence of hydrothermal phenomena before the water transported the muddy sediments, but it probably continued after the sediments were deposited.

In Almadén, Nery Delgado searched for data to reinforce his views that an insurmountable barrier had prevented communication of the two Silurian seas of central and southern Portugal. The first showed analogies with those of France and central Europe, and the latter would have been linked to the Silurian deposits in America and northern Europe.

Nery Delgado also used his travels as a subtle means of comparing and criticizing Portuguese and Spanish practices. In the national context, the publication of a report of this mission to Spain set a standard, since from that time onwards whenever he travelled his reports were published. Thus, by demonstrating that travelling abroad was part of the normal work of a geologist, he enhanced his scientific authority, and legitimized the work of the Geological Survey. At the same time, by voicing the accomplishments and difficulties of his Spanish colleagues, Nery Delgado reinforced their position and gave expression to the brotherly links uniting geologists in both countries.

The research underlying this paper was carried out in the context of the project *Laying down the Foundation Stone: 19th Century Geology in the Context of the Mining and Metals General Committee and the Geological Survey*, approved by the *Fundação para a Ciência e a Tecnologia*/Portugal, in the context of the *Programa Operacional Ciência, Tecnologia e Inovação* (POCTI) of the *Quadro Comunitário de Apoio III* (2000–2006), partly funded by FEDER. I wish to express my deepest gratitude to M. S. Pinto for his careful reading of this paper. A word of gratitude goes also to A. Dill who kindly helped me clarifying questions pertaining to carbonatite and for bibliographic guidance on this question. I am grateful to M. M. Ramalho, G. Lemeire and T. S. Mota for helping me in some questions pertaining to nomenclature. I also would like to thank P. Serrano and F. Moreira, librarians of the IGM Historical Archive, for their efficient and kind assistance.

Notes

[1]Expression coined by Bruno Latour to mean centres where specimens, maps, diagrams, logs, questionnaires and paper forms are accumulated and used by scientists and engineers (Latour 1987).

[2]Military engineers in science and technology in Portugal in the nineteenth century were trained by the Army School and/or the Lisbon Polytechnic. Given the structure of the Portuguese educational system of the time, they were the only people available in the country possessing the kind of scientific and technical training required by tasks ranging from statistical, geodesic, geographical and geological surveying to telegraphy and civil works. As far as geology was concerned, their knowledge of planning military campaigns, their acquaintance with mining and geology, their command of cartography and the practice of using explosives facilitated the organization and practice of geological fieldwork.

[3]Before the establishment of the Survey, geology was only practised in institutions for higher education—the University of Coimbra, the Lisbon Polytechnic and the Polytechnic Academy of Oporto. However, it consisted mainly of palaeontology and mineralogy still held to antiquarian practices typical of eighteenth century natural history. Students were taught geology through examples and collections of foreign countries. Despite the need for further historical research, geologists working in the context of Portuguese academia seem to have been cabinet scientists rather than field researchers, as were those working in the Geological Survey.

[4]Indeed, the Survey geologists discussed scientific questions with foreign geologists and palaeontologists rather than with their Portuguese colleagues working in higher education, whose outlook seemed to be embedded in a distinct scientific culture.

[5]This effort was complemented by the arrival in Portugal of Paul Choffat (1849–1919) who worked under contract for the Survey between 1879 and 1919 on a permanent basis and the occasional resort to foreign experts, such as the Swiss palaeontologist Oswald Heer (1809–1883) who studied the flora of the Cenozoic between 1880–1881; the Frenchman Marquis of Saporta (1823–1895) who worked on the Mesozoic flora, in 1890; C.L.P. de Loriol Le Fort (known as Perceval de Loriol) (1828–1908), a Swiss expert who analysed the Cretaceous and Jurassic faunas, in 1888, 1890 and 1896. The Survey also subscribed to various specialist foreign journals, the regular acquisition of foreign books and maps and the creation, in the 1890s, of the Survey journal, the *Comunicações dos Serviços Geológicos de Portugal.*

[6]Letter from Gonzalo y Tarin to Nery Delgado, Huelva, 26 December 1876. Historical Archive of the Institute of Geology and Mining/Portugal (IGM), Bookcase 10, Shelf 2, Box 4.

[7]Letter to José Macpherson (1839–1902), Belas, 8th March 1879, IGM Historical Archive, Bookcase 10, Shelf 1, Box 5.

[8]The official letter of Carlos Ribeiro sent on 19 October 1878 to the Head of the Geodesic Office, General Filipe Folque, reads: 'I have the honour to send to Your Excellency the report about the travel and the geological studies carried out in Spain by the Captain Engineer, adjunct to this Section, Joaquim Filippe Nery Delgado. This mission was accomplished with the intelligence and knowledge which characterize the scientific merit of this geologist. The depth and interest of this report to the geological knowledge of our Peninsula justifies, in my opinion, its publication in the format of the publications of this Section, because it

renders a good service to science and to our country'. File 'Relatórios Manuscritos', Archive of the Geographic Institute of Portugal.

[9]In Barrande's words: 'Les colonies, en général, étant des apparitions partielles et anticipées d'une faune, durant l'existence de la faune précédente, constituent un phénomène purement paléontologique, et qui, par conséquent, pourrait être complètement indépendant des phénomènes stratigraphiques, c. à d. de la nature et de la succession des roches'. (Barrande 1861). Furthermore, Barrande (1859–1860, 1881) published various memoirs defending the colonies from various opponents, notably the British.

[10]For Barrande's controversies that went on for about 20 years, see Perner 1937. A comparison of Barrande's interpretations with modern stratigraphy see Kriz & Pojeta 1974.

[11]Nery Delgado drew attention to what he perceived as a paradox in the dispute: 'It is highly remarkable that British geologists, being the most interested in supporting the theory of evolution formulated by Darwin—a theory which undoubtedly found a good basis in the theory of colonies—come now (without providing new evidence, one should say) fighting against this theory advocated by an expert who, on the contrary, does not subscribe to transformist ideas'. (Nery Delgado 1882, p. 36).

[12]Given the richness of the 'primordial fauna', and in order to be consistent with their theory, transformists suggested the existence of a series of more ancient faunas. However, as no trace of those faunas could be found in geological strata, they were led to admit that they had vanished owing to metamorphism. Barrande, in turn, responded that if geologists were unable to find fossils in the Azoic rocks, the reason was not that metamorphism had destroyed their remains but simply because they had never existed (Laurent 1987, p. 303).

[13]Also in his correspondence, his lack of enthusiasm about practical applications of geology was often expressed throughout his career. He often confided to his foreign colleagues that he could not devote the time he would like to fundamental geology owing to his official duties concerned with mining and hydrology. For example, in a letter to the Spanish geologist Lucas Mallada, Nery Delgado gave vent to his sorrows: 'As I said before, it is unfortunate that for more than one year I had to leave aside my geological work because I was assigned the mission of helping Mr Carlos Ribeiro in the exploitation of water to supply Lisbon. Now you can see my good friend that lately our geology has made little progress, not to say it has regressed. Only when our government becomes truly convinced that geological research is useful, and requires more and specialized personnel will these studies progress regularly.' After all, as a civil servant Nery Delgado was forced to comply with governmental orders, and the task of carving a place for fundamental geology in state bureaucracy proved to be a difficult one. Letter from Nery Delgado to Lucas Mallada, Lisbon, 12 June 1876. IGM Historical Archive, Bookcase 10, Shelf 2, Box 15.

[14]Continental facies of the Mississippian (Lower Carboniferous, Dinantian).

[15]Nery Delgado original expression is 'actividade geyseriana'. Nery Delgado 1879, p. 7.

[16]The reference to Waldemar Christofer Brögger's article on carbonatites usually given in scientific literature is Brögger 1921. His name is mentioned in Brito de Carvalho 1978, p. 1.

[17]Between Aldeia Nova and Corte do Pinto.

[18]Ampelite: an obsolete term for black carbonaceous or bituminous shale.

[19]Roemer wrote: 'For very good reasons you have advocated the classification of the Nereites of S. Domingos as belonging to the Silurian (...). It seems to me almost certain that the system of shale strata from which that fossil came, which is located immediately below the layers of *Posidonomya becheri*, in the province of Seville, belongs also to the Protozoic division of the Silurian'. Quoted by Nery Delgado 1879, p. 9.

[20]This is a translation in Spanish of Barrois's paper presented to the French Société Géologique du Nord.

[21]Note added by the Spanish Survey to Barrois 1877, pp. 378–379.

[22]Letter from Barrois to Nery Delgado, Lille, 19 May 1879, IGM Historical Archive, Bookcase 10, Shelf 2, Box not numbered.

[23]On the work of Barrois in Asturias see Truyols 1982.

[24]Research on carbonatites in Portugal was developed in the twentieth century, especially by geologists working in the Portuguese African colonies such as Brito de Carvalho, Matos Alves and Britaldo Rodrigues. Africa and Scandinavia are the regions where carbonatites are more abundant. Studies on the possible presence of carbonatites in mainland Portugal are to my knowledge absent (see Brito de Carvalho 1974; Matos Alves 1969, Rodrigues 1972).

[25]Serra: Portuguese word for mountain range.

[26]By then the Survey exchanged publications with 81 institutions, mainly foreign (Nery Delgado 1883–1887, Preface).

[27]The map was completed in 1889 at the scale 1:400 000.

[28]The suspension of the Survey occurred due to deep disagreements opposing Ribeiro to his co-director, Pereira da Costa (1809–1888). Costa, a former physician, Professor of Mineralogy and Palaeontology at the Lisbon Polytechnic, was in charge of the classification of fossils at the Survey. Still holding to an antiquarian technique, concentrating on cabinet tasks and resisting fieldwork due to his alleged vulnerable health, Costa was only interested in well preserved fossils, which he liked to classify and display on aesthetic rather than on scientific criteria. Costa's conceptions and susceptibility, the slow pace at carrying out his tasks, and the fact that he

appropriated Ribeiro's and Nery Delgado's work without their consent, led to deep disagreements between the two directors. However, Costa was well connected in the political sphere and friendly with Calheiros de Menezes, then Minister of Public Works. They joined forces and managed to get the Survey suspended, a situation that lasted from 1868 to 1869. As a result, the fossil and mineral collections were moved to the Lisbon Polytechnic, leaving Ribeiro and Nery Delgado with virtually nothing.

[29]Reply from Eduardo Benot, Madrid 8 August 1878, IGM Historical Archive, Bookcase 10, Shelf 1, Box 5.

[30]In Portugal, the use of lithography to print out maps was initiated in 1854 when the Portuguese Government contacted J. Lewicky, a Polish émigré living in France, who set up a lithographic workshop at the Geodesic Commission. This workshop together with that of the National Stationary Office (Imprensa Nacional), both inaugurated a tradition of lithography and chromolithography in Portuguese cartography. The first Portuguese geological map of 1876 published in the scale 1:500 000 was chromolithographed at the printing office of the Geodesic Commission (see Alegria & Garcia 1995, p. 121).

References

ALASTRUÉ, E. 1968. *La personalidad y la obra de Macpherson (1839–1902)*. Seville, Publicaciones de la Universidad de Sevilla.

ALASTRUÉ, E. 1983. *La vida fecunda de D. Lucas Mallada*, Madrid, Asociación Nacional de Ingenieros de Minas.

ALEGRIA, M. F. & GARCIA, J. 1995 (eds). *Os Mapas em Portugal, da Tradição aos Novos Rumos da Cartografia*. Lisbon, Edições Cosmos.

AYALA-CARCEDO, F. J. 1997. Casiano de Prado (1797–1864), un clásico del Naturalismo y de la Arqueología en España. *Industria y Minería*, Consejo Superior de Colegios de Ingenieros de Minas, **331**, 42–49.

AYALA-CARCEDO, F. J. 1999. De la Comisión del Mapa Geológico al Instituto Tecnológico Geominero de España: 150 Años de Geología y Minería en España. *Industria y Minería*, **338**, 29–35.

BARRANDE, J. 1859–1860. Colonies dans le bassin silurien de la Bohême. *Bulletin de la Société Geólogique de France*, **17**, 602–609.

BARRANDE, J. 1861. *Défense des Colonies*. Prague/Paris, Chez l'Auteur.

BARRANDE, J. 1881. *Défense des Colonies V. Apparition et Réapparition en Angleterre et en Ecosse des espèces coloniales siluriennes*. Prague/Paris, Chez l'Auteur.

BARRERA, J. L. 2002*a*. *Libro Homenaje al Geólogo José Macpherson*, Madrid, ILE.

BARRERA, J. L. 2002*b*. Biografía de José Macpherson y Hemas (1839–1902). *Boletín de la Institución Libre de Enseñanza (Homenaje a José Macpherson)*, **45–46**, 47–78.

BARROIS, C. 1877. Relación de un viaje geológico por España leída en la Sociedad Geológica del Norte. *Boletín de la Comisión del Mapa Geológico de España*, **4**, 373–382.

BLÁZQUEZ DIAZ, A. 1992. La Contribución Geológica del Naturalismo: Los Trabajos del Mapa Geológico Nacional. *In*: MENDOZA, J. G. & CANTERO, N. O. (eds) *Naturalismo y Geografia en España*. Madrid, 79–134.

BRANCO, R. M. C. 1999. *O Desenvolvimento da Cartografia Territorial em Portugal no século XIX. Unpublished M.A. thesis*, New University of Lisbon.

BRITO DE CARVALHO, L. H. 1978. *Carbonatitos*. Unpublished Ph.D. thesis, Lisbon, Instituto Superior Técnico.

BRITO DE CARVALHO, L. H., SAHAMA, T. G. & LOPEZ NUNES, J. E. 1974. Niobian Brookite from Mount Negoza—Carinde (Mozambique). *Revista de Ciências Geológicas da Universidade de Lourenço Marques*, **7**, 61–68.

BRÖGGER, W. C. 1921. Die Eruptivgesteire des Kristianiagebietes, IV. Das Fengebiet in Telemarken, Norwegen. *Videnskabers Selskab Skrifter Matematisk-Naturvidenskabelig Klasse*, **9**.

CARNEIRO, A. 2005. Outside Government Science, 'Not a Single Tiny Bone to Cheer Us Up!' The Geological Survey of Portugal (1857–1908), the Involvement of Common Men, and the Reaction of Civil Society to Geological Research, *Annals of Science*, **62**, 141–204.

CARNEIRO, A., AREIAS, D., LEITÃO, V. & TEIXEIRA PINTO, L. 2003. The role of travels in the internationalisation of nineteenth-century Portuguese geological science. *In*: SIMÕES, A., CARNEIRO, A. & DIOGO, M. P. (eds) *Travels of Learning. A Geography of Science in Europe*. Kluwer, Dordrecht, 249–297.

ELLENBERGER, F. 1978. *The First International Geological Congress, Paris, 1878. Episodes*, **1**, 20–24.

KNELL, S. J. 2000. *The Culture of English Geology, 1815–1851: a Science Revealed Through its Collecting*. Ashgate, Aldershot.

KRIZ, J. & POJETA, J. 1974. Barrande's Colonies Concept and a Comparison of his Stratigraphy with the Modern Stratigraphy of the Middle Bohemian Lower Paleozoic Rocks (Barrandian) of Czechoslovakia. *Journal of Palaeontology*, **48**, 489–494.

LATOUR, B. 1987. *Science in Action. How to Follow Scientist and Engineers through Society*. Open University Press, Milton Keynes.

LAURENT, G. 1987. *Paléontologie et évolution en France 1800–1860*. Paris, Editions du CTHS.

LÓPEZ DE AZCONA, J. M. 1984–1990. Mineros destacados del siglo XIX: Manuel Fernández de Castro. *Boletín Geológico y Minero*, **95–101**, 165–170.

MARR, J. E. 1880. On the Predevonian rocks of Bohemia. *Quarterly Journal of the Geological Society*, London, **36**, 591–619.

MATOS ALVES, C. A. 1969. *Estudo geológico e petrológico do maciço alcalino-carbonatítico do Quicuco (Angola)*. Lisbon, Junta de Investigação do Ultramar.

NERY DELGADO, J. F. E. 1876. *Terrenos Paleozoicos de Portugal*. Sobre a existencia do Terreno Siluriano no Baixo Alemtejo. Lisbon, Typographia da Academia Real das Sciencias, or its French version *Terrains*

Paléozoïque du Portugal. Sur l'éxistence du Terrain Silurien dans le Baixo-Alemtejo. Typographia da Academia Real das Sciencias, Lisbon.

NERY DELGADO, J. F. 1879. *Relatorio da commissão desempenhada em Hespanha no anno de 1878 por Joaquim Filipe Nery Delgado, adjunto da Secção dos Trabalho Geologicos do Reino*. Typographia da Academia Real das Sciencias, Lisbon.

NERY DELGADO, J. F. 1882. *Relatorio e outros documentos relativos à comissão cientifica desempenhada em diferentes cidades da Italia, Allemanha e França*. Imprensa Nacional, Lisbon.

NERY DELGADO, J. F. 1883–1887. Preface. *Communicações da Commissão dos Trabalhos Geologicos*, **1**, I–XII.

NERY DELGADO, J. F. 1889. *Relatorio ácerca da quarta sessão do Congresso Geologico Internacional realisada em Londres, no mez de Setembro de 1888*. Imprensa Nacional, Lisbon.

NERY DELGADO, J. F. 1908. *Système du silurique du Portugal. Etude de stratigraphie paléontologique*. Typographia da Academia Real das Sciencias, Lisbon.

NERY DELGADO, J. F. & CHOFFAT, P. 1899. *Geological Map of Portugal (1 : 500 000)*. Chez L. Whurer, Paris.

OLDROYD, D. 1990. *The Highlands Controversy. Constructing Geological Knowledge through Fieldwork in Nineteenth-Century Britain*. Chicago University Press, Chicago.

OLDROYD, D. 1996. *Thinking about the Earth: a History of Ideas in Geology*. Harvard University Press, Cambridge Massachusetts.

OLDROYD, D. 2000. (for 1998–1999) John Edward Marr (1857–1933). The foremost Lake District geologist of his era. *Proceedings of the Cumberland Geological Society*, **6**, 361–379.

PERNER, J. 1937. Les colonies de Barrande. *Bulletin de la Société Géologique de France*, **7**, 513–552.

PRADO, C. 1850. Note géologique sur les terrains de Sabero et de ses environs dans les montagnes de Léon (Espagne). *Bulletin de la Société Géologique de France*, **7**, 137–155.

PUCHE RIART, O. & AYALA-CARCEDO, F. J. 2001. Guillermo P.D. Schulz y Schweizer (1800–1877: su vida y su obra en el bicentenario de su nacimiento. *Boletín Geológico y Minero*, **112–1**, 105–122.

RIBEIRO, C. & NERY DELGADO, J. F. 1876. *Portuguese Geological Map* (1:500 000). Geodesic Office, Lisbon.

RIBEIRO, J. V. 1912. *A Imprensa Nacional de Lisbon. Subsídios para a sua história, 1768–1912*. Imprensa Nacional, Lisbon.

RODRIGUES, B. 1972. Conclusions on the distribution of alkaline and carbonatite complexes of Angola and Brasil. *Revista da Faculdade de Ciências da Universidade de Lisbon*, **17**, 89–108.

TRUYOLS, J. 1982. El Carbonífero en la obra asturiana de Barrois. *Trabajos de Geología*, Universidad de Oviedo, **12**, 7–21.

Grenville Arthur James Cole (1859–1924): the cycling geologist

P. N. WYSE JACKSON

Department of Geology, Trinity College, Dublin 2, Ireland (e-mail: wysjcknp@tcd.ie)

Abstract: Grenville Arthur James Cole (1859–1924), Professor of Geology at the Royal College of Science for Ireland, was an avid cyclist and shared this passion with his wife Blanche. Born in London, Cole studied at the Royal School of Mines and lectured for a number of years at Bedford College for Ladies. Largely concurrent with his professorship he served as Director of the Geological Survey of Ireland at a time when it was in moderate decline.

He undertook many cycling tours around continental Europe and Ireland. These trips were recounted in two early travel books. *The Gypsy Road: a Journey from Krakow to Coblentz*, published in 1894, provides a delightful account of a tour undertaken by him on a tricycle and his companion on a penny farthing, across what is now Poland, the Czech Republic, and eastern Germany. In a later slim volume entitled *As We Ride*, co-authored with his wife, a number of expeditions to France and the Balkans are eloquently described. Between 1902 and 1908 he organized a week-long geological excursion to various parts of Ireland for his students, and transport was by train and bicycle.

His cycling trips provided him with the opportunity to collect research materials, make geological observations, and to photograph features of interest. Subsequently much of this material was used in his publications and in teaching. Cole's main academic studies were in igneous and metamorphic petrology. He valued and promoted fieldwork as an essential component of geological training. He was heavily involved with professional and amateur scientific societies, and was a prolific author of both academic and popular geological papers and books.

To those who laid from west to east
The clear white road for man and beast;
To those who in our age of steel
Set the true frame and shaped the wheel;
To those whose greeting calls again
From the brown acres of the plain,
Or through the pathways of the firs,
Our equal friends and wayfarers;
To him who at each sunset post
Held wide the door, our welcome host;
To these, on many a country-side,
Our thanks be given, as we ride.[1]

Since the seventeenth century, men and women have been travelling parts of the globe in search of rocks and geological structures that could elucidate something of the Earth's history. They have travelled by many different means: James Hutton, William Buckland and early Geological Survey mappers preferred the use of horses, Roderick Impey Murchison and his wife travelled by horse-drawn coach (Kölbl-Ebert 2007); whereas the fossil fish-collectors William Willoughby Cole, the third Earl of Enniskillen (1807–1886) and Sir Philip de Malpas Grey-Egerton (1806–1881) were among those gentlemen 'geologists' who made Grand Tours of Europe during the eighteenth century. Rail travel prevailed in the early twentieth century but has now been overtaken by the motorcar. In remote areas such as Africa, geologists have resorted to canoes, or their feet. At one point towards the end of the Victorian era, cycling was very popular.

The earliest wheeled vehicles were in use in Mesopotamia around 7000 years ago, but it was not until late in the eighteenth century that self-propelled two-wheeled machines first appeared. These were the so-called 'hobby-horses', which were awkward and heavy, consisting of two large wooden wheels and with a saddle fitted to a rigid crossbeam. They could not be steered easily and were propelled by the rider running his feet along the ground in a 'Fred Flintstone' manner. These machines were particularly popular with the aristocracy of the late 1700s and early 1800s. Not content with this self-powered propulsion, early nineteenth century inventors toyed with pedals, levers and chains in an attempt to perfect efficient mechanical movement. The first recognized bicycle was built by a Scot, Kirkpatrick Macmillan (1810–1878), in 1839. A replica of his machine may be seen in the Science Museum, London. From this start evolved the hard riding 'boneshakers', which were large and heavy, like the earlier 'hobby-horses', but which were easier to ride. They came in many shapes and sizes: although two-wheeled versions were the most prevalent, three or even four-wheeled machines were not unheard of. The introduction of the boneshaker had important social implications. They were relatively cheap at £10 (in the 1860s) and so bicycling became available to the general public. Two further advances were yet to come before modern machines took shape: the first was the advent of the light triangular-framed machines

From: WYSE JACKSON, P. N. (ed.) *Four Centuries of Geological Travel: The Search for Knowledge on Foot, Bicycle, Sledge and Camel.* Geological Society, London, Special Publications, **287**, 135–147.
DOI: 10.1144/SP287.11 0305-8719/07/$15.00

typified by the 'ordinary' (or 'penny-farthing'), which was followed, in the 1880s by the chain-driven 'safety'. The second advance was the invention of the pneumatic tyre by John Dunlop (1840–1921) in 1888. This tyre is now on display in the Royal Scottish Museum. Dunlop was a Scottish veterinary surgeon working in Belfast. Soon after 1888 he moved to Dublin where his pneumatic tyres were manufactured (Cooke, undated). At last bicycling had become a comfortable means of transport and available to all.

Where do geologists fit into all this? Bicycling fever swept through Victorian Britain and many touring clubs, of which the Cyclists' Touring Club was the premier, were established. Such enthusiasm slowly spread to the United States so that by the early years of the twentieth century the pastime was as popular as in Britain (Smith 1972). Any volume of material was provided for cyclists: publications included periodicals such as *The Cycling Mercury*, *The Cyclist*, and *The Wheel World* (edited in part by Henry Sturmey noted for his gear set-up); maps on which the most suitable routes were marked;[2] guide books complete with profiles of routes showing gradients (Inglis 1896, 1897, 1908); manuals on how to repair tricycles and bicycles; and listings of the cycling clubs and hotels that welcomed touring parties (Spencer 1884). There was considerable debate about the merits or otherwise of cycling for women. This activity for females was promoted by a Miss Erskine and others (Erskine 1884, 1896, 1897; Graves *et al.* 1898) but its dangers were pointed out by others, including a Mr Ryley (1899). Catalogues advertised clothing that was both suitable for lady cyclists and which retained their modesty.

It is hardly surprising that some geologists should become committed cyclists. The bicycle widened the field area for research and allowed the rider to become personally entwined with the landscape, a feeling that could never be adequately attained by rail travel. In the United States, Helen Margaret Duncan (1910–1971) a specialist on Bryozoa and Geological Survey palaeontologist from 1942 until her death, was apparently a keen cyclist (Flower & Berdan 1977). In the British Isles, the Geologists' Association, a group of professional and amateur members, undertook a number of excursions by bicycle in the 1890s and early 1900s (Green 1989; Robinson 1990). This mode of transport was championed by J. F. Blake formerly Professor of Geology at Nottingham and by H. W. Monckton. After a short time their enthusiasm for the wheel fizzled out. However, the individuals Sydney Savory Buckman (1860–1929) who was an ammonitologist, and Grenville Arthur James Cole (1859–1924) (Figs 1 & 2) an igneous and metamorphic petrologist, were both committed cyclists. In eighteen months from the date of his purchase of his first pneumatic-tyred bicycle, Buckman covered 3000 miles, principally engaged in examining various Mesozoic successions in Normandy and around his home in the Cotswolds (Buckman 1898). Buckman, together with his wife, established the Western Rational Dress Club that advocated that women wore a knickerbocker costume. In 1900 he seriously damaged his health through over-exertion in the saddle (Torrens 2004).

Fig. 1. Grenville Arthur James Cole (1859–1924) (by permission of the Geological Society of London).

Of these cyclists it is the little-known Cole who has left most written and other material relating to his cycling activities and from this it may be seen how cycling influenced his geological and geographical work.

Biographical notes

Grenville Arthur James Cole was born in London on the 21 October 1859, the second son of John Jenkins Cole, architect to the London Stock Exchange (Wyse Jackson 1989). He was educated at the City of London School, and in his final year began to attend the lectures of Professor John Wesley Judd (1840–1916) at the Royal School of Mines. These captivated Cole, and although he was not destined to enter university, he managed to embark on an academic career in geology. Judd recognized Cole's potential and in 1878 appointed him as demonstrator to the school, a position he held for 12 years. During this period from 1886

Fig. 2. Grenville Cole and Blanche Cole (1862–1927) with a friend (standing at left) pausing during a rain shower while on a cycling trip somewhere in Ireland, *c*. 1900. Photograph by John Joly, Professor of Geology and Mineralogy, Trinity College, Dublin (1897–1933) (Joly Collection, Geological Museum, Trinity College, Dublin).

until 1890, Cole was also Head of the Geology Department at Bedford College for Ladies. Soon after he left London, a student geological society was established and named in his honour. Bedford College was later incorporated into the Royal Holloway, part of the University of London, which appropriately still awards an annual Grenville Cole Travel Bursary which funds a student to engage in research.

In 1890, Cole was appointed Professor of Geology and Mineralogy at the Royal College of Science for Ireland in Dublin (Kelham 1967) (only a 5 minute cycle from Dunlop's tyre factory), in succession to Edward Hull, and he retained this position until his death thirty-four years later. This institution had been established to provide education in science as applied to agriculture and industry and also training for teachers destined for the technical and secondary schools. Cole was later appointed as Director of the Geological Survey of Ireland in 1905 and was active in most of the societies that promoted science and natural history in Dublin. These were the Royal Dublin Society, the Royal Irish Academy, the Dublin Naturalists' Field Club, the Dublin Microscopical Club and the Irish Geographical Association. He was a member of a close-knit group of scientists who used to meet for buttered toast and tea in a refreshment room adjacent to Trinity College, Dublin.[3]

In 1896, Cole married one of his first students in Ireland, Blanche Vernon (Fig. 2), the daughter of Colonel J. E. Vernon of Clontarf Castle, a wealthy Dublin landowner, and the couple had one son, Vernon, who later joined the East African Civil Service, but subsequently practised as a medical doctor. Blanche, like her husband, was an enthusiastic cyclist, and together they travelled extensively in Europe, as well as in Algeria and Tunisia. According to the botanist Robert Lloyd Praeger, Cole's best man, after their wedding ceremony the happy couple mounted their bicycles, cycled down the aisle, and embarked on their honeymoon! (Praeger 1941, p. 272).

Towards the end of his life, rheumatoid arthritis progressively crippled Cole, but he remained characteristically cheerful and active. Grenville

Cole died on 20 April 1924 at the age of sixty-four, and Blanche only survived him by three years, dying at the same age on 17 November 1927. They are buried together in Dean's Grange Cemetery in south County Dublin.

During his career Cole received numerous awards. He was awarded the Murchison Medal of the Geological Society of London in 1909; was President of Section C (Geology) at the 1915 British Association for the Advancement of Science; was elected a Fellow of the Royal Society in 1917; and received the degree of Doctor of Science (his only degree) from Queen's University, Belfast.

Grenville Cole was a small man who stood little over five feet tall, but who had endless energy. He would frequently walk 14 miles to and from work each day, and he expected his students to work hard. He was a superb lecturer, field geologist and teacher, who had a captivating wit and a sparkle in his eyes.

Cole's geological work

At the Royal School of Mines, Judd advocated the use of practical methods (field studies, excursions and practical classes) to complement lectures. There was a shift in emphasis from the lecture theatre to the laboratory, so that by the turn of the century, practical classes, demonstrations, and field classes were soon practised in many university departments in the British Isles and on the Continent. Cole was responsible for much of the planning and teaching of the practical element of the geological courses. The job of demonstrator also allowed him time to pursue extensive research interests.

After 1890, his duties at the Royal College of Science in Dublin were to organize and deliver lectures to various groups: agricultural students, science students and student-teachers. In an average year he or his assistant gave approximately 80 lectures. His assistants at one time or another included Louis Bouvier Smyth (1883–1952), later Professor of Geology and Mineralogy at Trinity College and Isaac Swain (1874–1963), later Professor of Geology and Geography at University College Cork.

Cole's career may be divided into two periods where the thrust of his research was quite divergent. This first period was up until 1895 during which time he worked largely on petrology and the illustration of textures and relationships within igneous and metamorphic rocks; in 1889 the Geological Society of London awarded him the Murchison Fund for this work. At this time he travelled extensively in Britain, collecting Scottish basalts and Welsh volcanic rocks.

Later he became more involved in the promotion of fieldwork and the study of physical geography. He advocated the teaching of geography in universities and helped raise the profile of the subject (Herries Davies 1977). He was a prolific author of both academic and popular geological papers and books and has over 500 papers to his name (Wyse Jackson 1989).

The cycling geologist

Inspired by Judd's teaching Cole began to comprehend the importance of outdoor field studies in geology. Perhaps in response to this, he began cycling. In the preface of *Open-Air-Studies: an Introduction to Geology Out-of-Doors* which resulted from first-hand field experience he wrote 'Now-a-days, when cycling is so frequent and so free a means of travel, the geography and geology of nearer Europe have become keen realities to many of us . . . ' (Cole 1895*a*). Field studies were an aspect of the science that preoccupied his thoughts and many of his publications for the rest of his life. Cole was an avid cyclist (Wyse Jackson 1991)—this cannot be emphasized enough.

Apart from his cycling tours Cole travelled widely. He visited Canada in 1897 and South Africa in 1905 to attend the meetings of the British Association for the Advancement of Science and regularly attended when it met in Britain.

Reconstructing Cole's travels

Using various lines of evidence, I have established the routes of many of his cycling tours. Of most value in this exercise were his two travel books *The Gypsy Road* (Fig. 3) (Cole 1894*c*) and *As We Ride* co-authored with his wife Blanche (Cole & Cole 1902), as well as his geological textbooks illustrated with his own photographs and any scientific papers similarly illustrated. These include papers on Algeria (Cole 1918*a*, *b*) and Spitsbergen (Cole 1911*b*) which he visited in 1911 during an International Geological Congress meeting. An unexpected bonus, and source of valuable information about some of his European and British travels, was his collection of rocks found in the basement of the Museum Building in Trinity College, Dublin. These were still wrapped in scraps of annotated newspaper, which gave some indication of when and from where he had collected the specimens. Fortunately Cole had very distinctive, clear handwriting, and he habitually wrote in black ink, and it is not possible to mix this up with that of other hands. His student field guides produced for the student field trips, together with a recently discovered manuscript map of Ireland provided valuable information. On this dissected map he had marked many routes travelled by him

THE GYPSY ROAD
A journey from
Krakow to Coblentz
by
Grenville A·J·Cole
M·R·I·A·, F·G·S·
with
illustrations by
Edmund H·New

London & New York
Macmillan & Co
1894

Fig. 3. Title page of *The Gypsy Road* (1894).

in red ink and added some information about hotels, hostels and local geological stops.[4] Grenville and Blanche also kept a large map of Europe in their home and this was marked with the routes of their cycle tours. Unfortunately this was probably discarded after their deaths.[5]

Travels around Continental Europe

The Gypsy Road

By 1894, Cole had completed journeys on a Humber-Beeston tricycle through Italy and Bohemia in 'tandem' with a companion Gerard W. Butler who rode a penny-farthing (Fig. 4). Cole had owned his machine since 1884 and had used it on several trips. He was very proud of this tricycle and said that ownership of it opened doors throughout Europe. He had dined with the famous Hunnia Cycling Club of Pest and frequently was joined by local cycling groups. Arising from these trips came papers on volcanic rocks of Lipari (Cole & Butler 1892), Italy, the Vosges, Germany (Cole 1887) and parts of France (Cole & Gregory 1890).

In the summer of 1892, he travelled to Hungary to compare the volcanic sequences found there with those Irish examples already familiar to him (Cole 1892*b*, 1894*b*). Following this journey he wrote *The Gypsy Road*, a highly entertaining, perceptive and lucid travelogue and social commentary (Cole 1894*c*). This was described in *The Times* as being a 'breezy narrative of a cycling tour in out-of-the-way regions'.[6] It recounts a journey of 1055 miles westwards from Krakow to Coblentz, through what is now Poland, the Czech Republic, Slovakia and Germany (Fig. 5), which Cole and Butler completed in 38 days in 1892![7] In fact, not content with travelling in one direction, Cole actually cycled from the town of Zolyom over the Szturecz Mountains with another friend before rendezvous-ing with Butler in Krakow. Butler is named throughout the narrative the 'Intellectual Observer' and comes across as being a man of quiet wit and perception. Cole, on the other hand, is the extrovert of the pairing. Little is known of Butler and his background; he enjoyed painting, held a B.A. degree, and may have been a Cambridge graduate in mathematics. In response to a question from a native as if whether there were any mathematicians in England, he replied 'A few perhaps at Cambridge'. (Cole 1894*c*, p. 127). They did not find everything plain cycling. The local police in places had some reservations with their visas, they were taken for travelling salesmen in other towns, horses were scared of their machines and they had to contend with bed-bugs or being chased by geese or by fifty children—one wonders which is worse. They coped with high temperatures in Bohemia where by mid-day it was 90°F. On 19 August between noon and 7.30 pm they consumed in total 1 large flask of sparkling water, two cups of tea, 3 glasses of water, one hot coffee, two iced coffees, and ate only three small rolls and one half slice of Hausbrod. This Cole stated 'express[es] the situation better than any thermometric record' (Cole 1894*c*, pp. 135–136). In one Czech village they arrived at the inn and asked for a room for the night. Unfortunately they enquired in German and were refused. Somewhat alarmed, as it was approaching dusk and the next village was several miles away, they went to the police station where they received the same cold-shoulder treatment. When the villagers realized that they were English and not German they were immediately received with open arms. The best room in the hotel was made available to them and their conveyances were parked in another guest room. On asking why this transformation in attitude to them had taken place, they were told that although the locals could speak German, they certainly did not speak it to Germans! Elsewhere they encountered this antipathy to Germans; they were asked

Fig. 4. 'Homeward Bound': Grenville Cole (on the Beeston-Humber tricycle) and Gerald W. Butler (on the penny-farthing) cycling through Europe in 1893 (tailpiece from *The Gypsy Road*, facing p. 166. Drawing by Edmund H. New).

'Are you Christians? Or are you Germans, perhaps?' (Cole 1894*c*, p. 80).

Geologically this trip was interesting and diverse. From Poland they travelled south, passing through Myslenice, which had been visited by the mineralogist Françoise Sulpice Beudant (1787–1850) in 1818,[8] into the Hungarian alluvial plains, before crossing the limestones of the Tatra Mountains. They then entered the region of the northeastern fore-Alps and the Hungarian Alps where they cycled over Cenozoic sediments, Mesozoic limestones and occasionally encountered crystalline basement rocks. At Garam they swam in the meandering river before cycling through the rolling volcanic countryside in the moonlight. Near Hajnik they visited the ancient gold mining region of Schemnitz, which had been renowned as being the most important centre for the metal in the Austro-Hungarian Empire. Nearby the travellers collected obsidian and Cole was able to examine the local rhyolites in some detail. Thanks to an invitation by Dr Cseh, a government geologist, they were able to venture underground into the huge silver mine at Schöpferstollen where they walked through the workings for over two hours, accompanied by a mine manager and two small Slovakian boys who carried the lanterns.

They found the expansive Hungarian Plain, underlain with unconsolidated sediments, rather tedious, and Cole compared it with Salisbury Plain in England. In Hungary they followed in the tracks of the earlier traveller Richard Bright (1789–1858) who has been described as being the 'Father of Nephrology' who had been there in 1815 (Bright 1818). Once in Moravia they passed over Miocene conglomerates and noted the breathtaking local geomorphology: 'It is wonderful what amount of rock has to be cut away before you can make a decently artistic mountain'. In Prague they met with Antonín Fric (1832–1913), who worked in the Museum of the Czech Kingdom in Prague and was Director of Palaeontology there from 1880. He arranged for the travellers to visit the volcanic regions to the north of the city, and they spent a number of days examining the rocks around the vale of Algersdorf and climbed various volcanic cones such as the Geltschberg where rare exposures of white phonolite could be seen. In Britain, Cole had seen the only exposure of this rock type at the Wolf Rock in Cornwall, and rejoiced in 'the joy of riding in Bohemia from one crag to another,

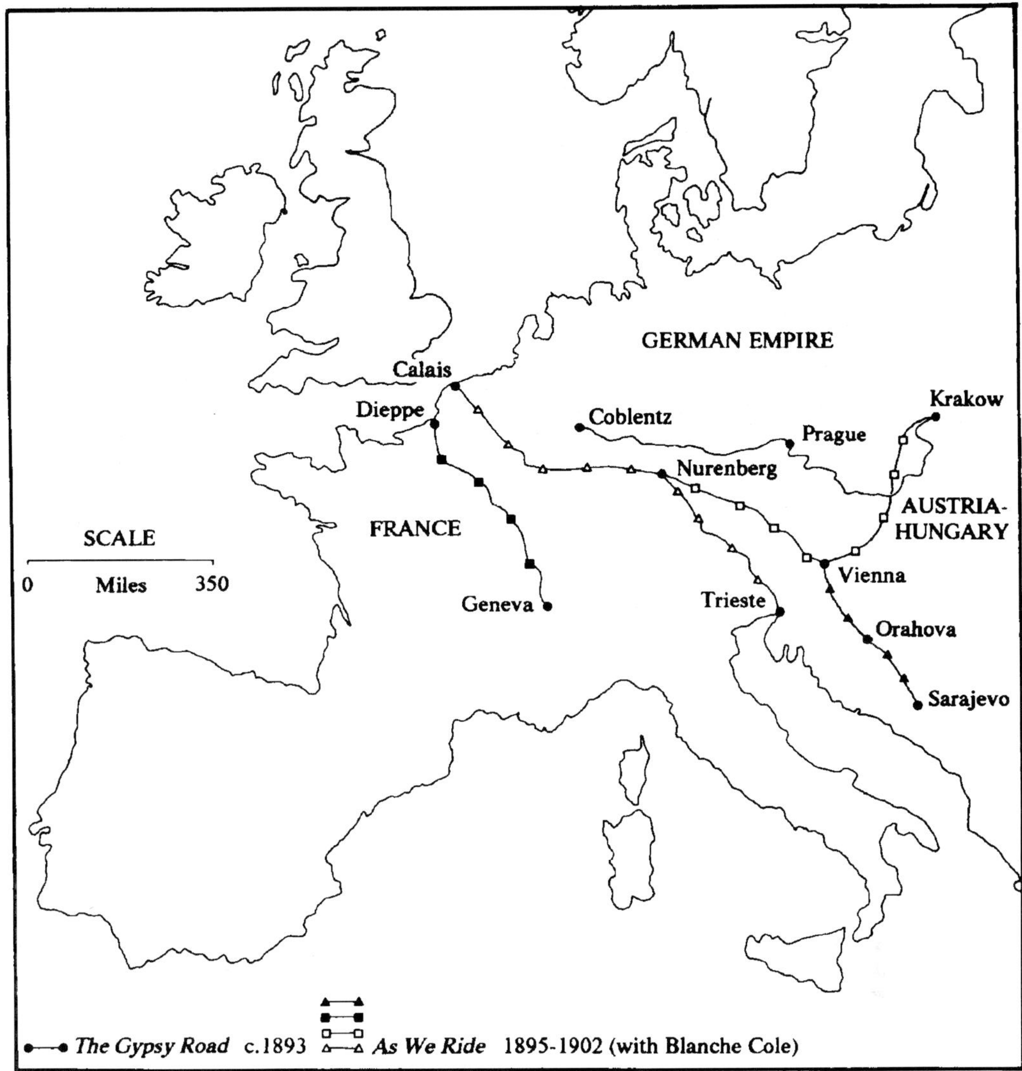

Fig. 5. Some of the routes cycled by Grenville Cole in Europe.

smiting the solid masses with the hammer, and seeing the lustrous flinty chips go skimming across the sunburnt grass' (Cole 1894*c*, p. 131). In Bohemia, Cole was in his element—wonderful rocks and wonderful cycling. Near Karlsbad they visited the volcanic cone of the Schladnigberg (Fig. 6) and traversed the deposits of Oligocene and Miocene brown coal near Dux, noting that some seams had been burning for at least 5 years and could not be extinguished, before passing out of Bohemia and into Bavaria.

By now their journey was nearing its end, and Cole having seen the volcanic deposits of Hungary could relax somewhat. He drank schnapps in a tavern in Horgenau, having been recommended to do so on medical advice by some of the elderly men of the village. He declared that his 'union with this simple country life seemed complete' (Cole 1894*c*, p. 159). However, not much further down the road in the village of Werdorf the journey nearly came to an abrupt end. Having parked their machines, Cole and Butler were relaxing when the tricycle started to freewheel downhill. It collided with a low wall, somersaulted twice and landed in a cabbage patch. The axle was bent and it seemed that the tricycle would have to be brought

Fig. 6. Cole tricycling past the volcanic cone of the Schladnigberg, near Schladnig, Germany (from *The Gypsy Road*, p. 137. Drawing by Edmund H. New from a photograph by G. W. Butler).

to Wetzler[9] to be repaired. Fortunately the local blacksmith, aided by Cole and Butler managed to take the wheel bearings apart, straighten the central axle and reassemble the mechanism. During this operation the forge was visited by a man who wished to have his horse reshod. The blacksmith dismissed him, telling him that he now ran a *Velocipedenfabrik*! The wheels ran smoothly afterwards 'and were good for another thousand miles' (Cole 1894*c*, p. 161).

Finally the pair of cyclists reached Coblentz and Cole likened them to Knights-Errant (Cole 1894*c*, p. 166); this comparison is a theme that crops up every so often in his writings. His book *The Gypsy Road* is important as it is one of the earliest travelogues of a bicycling trip and it gives an insight into eastern European lifestyles of the period.

As We Ride

In 1902, Blanche and Grenville Cole published a slim book of essays entitled *As We Ride* (Cole & Cole 1902), which gives accounts of a number of their trips through France, Germany, Poland and the Balkans. This volume, now very rare, was published for the benefit of the Royal City of Dublin hospital and it provides a glimpse of pre-First World War Europe.

In addition to containing geological insights, the essays are generally observations of the social interactions of local inhabitants and the towns through which the couple passed. One senses that Grenville had moved in his mind from hard rock geology to focus more on geomorphology. Indeed many of his publications from this period relate to the subject and its expression in Europe and to the interactions of man and his landscape (Cole 1912*a*, *b*).

Of the eight essays, half were written by Grenville and half by Blanche. She certainly did not mince her words. Writing about a visit to the Russian border town of Mlawa in the aptly-titled piece '*Defeat*' she says of the landlady of the only hotel in the town:

> She was villainously unclean, and so vacant besides that I wondered whether she were deaf, or idiotic, or both. But the length of her bill next morning in proportion to the nature of our entertainment settled this point; . . . she was by no means idiotic—far from it. (B. Cole in Cole & Cole 1902, pp. 100–101).

Rather than continue into Russia the couple retrieved their baggage and returned to Warsaw on the first available train. In *Son of Mars* she pins her political leanings to the mast: 'War is an undesirable occupation and seldom profitable; but if one must fight, please give me the French to fight with, and let the Germans fight someone else' (B. Cole in Cole & Cole 1902, p. 77). Twelve years later her sentiments counted for little.

Some of these journeys inspired Grenville to write poetry: *The Lost City* published in 1911 is his only piece of published work of this genre, and it recalls the history and evolution of a Templar hospice in the Causse de Larzac in Aveyron in SE France (Cole 1911*a*). The Coles were clearly taken with the Balkans which they visited on perhaps two occasions. Orahova, a small village now in Bosnia-Herzegovina, provided them with the title of a chapter in their book, but also and more significantly the name of their house in the fashionable district of Carrickmines in south County Dublin which they had purchased in 1902. The house is still standing and was renamed *Glenheather* by a subsequent owner.

A number of these excursions were recalled later by Cole in public lectures, including that on '*The Fringe of the Balkans*' delivered at the Royal Institution in London on Monday 8 February 1904.[10] Like *The Gypsy Road*, *As We Ride* is a fascinating commentary on a period long gone—and a reminder of the remarkable energy and close bond of Cole and his wife.

Travels in Ireland

General excursions

In 1884 Cole paid a visit to Ireland, but other than knowing it was to examine the geology of the island,[11] and that it was one of first excursions on his Humber-Beeston tricycle, nothing is known of the localities that he visited. Soon after his arrival in Dublin in 1890 on his appointment to the Royal College of Science, Cole rapidly set out to reacquaint himself with the geology of Ireland and did so largely by bicycle. In 1892, just two years after his arrival in Ireland, he penned a series of illustrated articles on the geology and scenery of Dublin that appeared in *The Irish Naturalist* (Cole 1892*a*). He was then commissioned by the owners of the North Eastern Railway in 1895, to write a similar booklet outlining the geology and scenery as seen from their railway carriages. This booklet, to accompany the photographs of the Belfast photographer Robert John Welch (1859–1936),[12] was thus the first of its kind in Ireland (Cole 1895*b*). To satisfy his fascination with volcanic rocks, he sought out intrusions of rhyolite and headed north to NE Antrim (Cole 1896). He then examined the volcanic mass of Slieve Gallion (Cole 1897). Later he carried out research on the Dalradian metamorphic belt in Donegal, produced accounts of the Donegal granite and its orbicular structures (Cole 1916) [now sadly damaged by coring] and reports on the intrusions of Tyrone and Londonderry. Later work embraced geomorphology, although he continued to write on igneous and petrological topics. One of the most important papers of his later years was that which described the course of the River Liffey, the river that bisects the centre of Dublin (Cole 1912*c*). He also put his bicycle to patriotic use: it has been reported that throughout the 1914–1918 war he rode the streets of Dublin as an army dispatch rider (Herries Davies 1995).

Leading student field excursions

Every year between 1903 and 1908 Cole took a group of his second-year agricultural students (or 'scholars' as he called them) on a week-long trip around Ireland to study geology and agricultural botany. This excursion was often held in the month of May and much of the travel was undertaken by bicycle or by train. These excursions first started in 1903 following the sanction of the scheme and agreement to fund it was given on 16 April by J. D. Daly of the Department of Agriculture and Technical Instruction, a Government office responsible for the College.[13] The cost of the first trip, which included hotels and train travel, was £5/9/11 each (approximately £340 today). It did not include the price of lunch, which was taken in the field, and which each person in the party was expected to provide for himself.

Cole expected his students to cope with the rigours of cycling field trips. As can be seen from the attractive excursion booklets that were provided for the trips, they were quite strenuous. Booklets were produced for the 1903, 1904, 1905, 1906 and 1908 trips and are similar in style, being approximately 12 pages long and A5 in size.[14] They were mass-produced on a copying machine in black, blue or purple ink and were illustrated with geological cross-sections drawn by Cole and with pasted-in portions of maps cut from the recently reissued *Murray's Handbook for Ireland* (Cooke 1902).[15] These booklets documented the routes to be covered and the features to be examined on the trips (Fig. 7).

Taking the 1903 trip as a typical example, we find that on 18 May, the group met at Broadstone Railway Station in Dublin at 6.45 am. Following departure of the train at 7.00 am they took breakfast in the 2nd class car. Five and a half hours and approximately 140 miles later, the party disembarked at the small town of Recess west of Galway having crossed over the Central Plain of

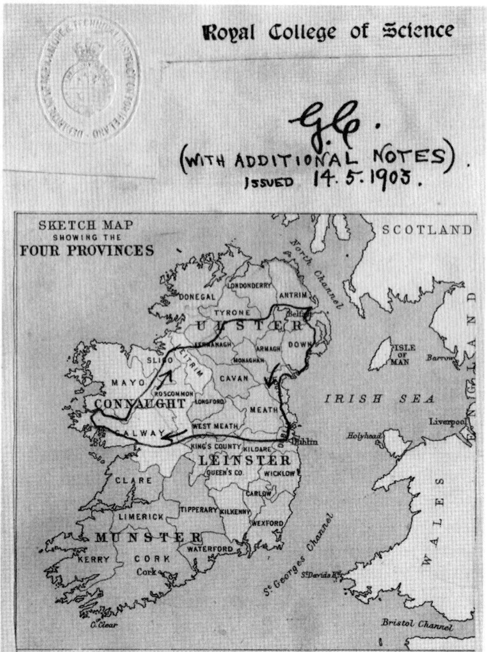

Fig. 7. The route of Cole's student geological tour around Ireland (1903) (from original booklet in the Department of Geology, University College, Dublin).

Ireland. Rather than gaze idly out of the window the students were expected to take notes on the features that they saw. From Recess, they bicycled 17 miles to Leenane where they examined Dalradian rocks and the vegetation. The next morning, they cycled 30 miles over the Silurian conglomerates and silts and Carboniferous rocks of the Mayo district. On 20 May, an early morning train carried the group northeastwards and, following a break from rail travel during which they cycled 21 miles and examined geological and botanical features of interest, they ended up in the town of Omagh. The next day they first visited Pomeroy which had yielded important Silurian fossils first described by Joseph Ellison Portlock (1794–1864) in his monumental volume on the geology of Londonderry (Portlock 1843; Tunnicliff 1980). After a 26-mile cycle over old gneissic terrain and undulating eskers, they passed Slieve Gallion and by evening had almost reached the edge of the Antrim Plateau, which geologically comprises a series of Tertiary basaltic flows. The next day, following their usual early morning rail departure, they reached Kilroot and visited an underground mine from which Triassic salt was being exploited.[16] This was followed by another 21-mile cycle into Belfast via the overlying basalt sequences exposed at Cave Hill having also examined *en route* the Triassic and Cretaceous sediments protected beneath the younger basalt. On the final day of this first excursion, 23 May, the students and Cole left Newcastle, County Down at 8.00 am and by 9.20 am had already climbed Slieve Donard, a nearby mountain. They then cycled to the small fishing port of Greencastle, loaded their bicycles on to a ferry and sailed across Carlingford Lough to Greenore. From there they took a train to Dublin and arrived in the capital at 9.10 pm.

In 1904, Cole's party of 2 staff members and twelve scholars visited Counties Tipperary and Limerick in Munster before travelling through western and northern Ireland. The following year the group had to cycle to Louisburg and climb Croagh Patrick in County Mayo (Fig. 8), before heading to the SE and the Knockmealdown Mountains in County Waterford. The cost per student for the 1905 trip was £6/7/5 excluding lunches and miscellaneous fees. The following year, in 1906, the agricultural students were joined by a number of third year natural science teachers in training and they travelled through Ireland in a double loop centred on Dublin. Following his return Cole received a letter from T. S. Gill of the sanctioning body in which he was praised for the success of the tour:

> I am glad to think that the tour was worked so satisfactorily and economically. The little hand-book is very attractive, even to a scientific ignoramus. I congratulate you and your fellow-teachers on the success of this piece of educational work, and I hope the students appreciate the great advantages which they are receiving at your hands.[17]

Fig. 8. Cole's 'Roadster' bicycle leaning against a dry stone wall next to a blackthorn tree, with the holy mountain of Croagh Patrick, County Mayo, western Ireland in the background. Note the height of the saddle: it is unusually low to accommodate Cole who was approximately 5 feet tall (from Cole 1922).

In 1907 he took his students to Tipperary again and then to see the sections in the basalt and older Mesozoic rocks in NE Ireland. He was unable to get maps for his booklet and so produced a series of handouts copied onto foolscap paper.

On 16 January 1908 he received a letter from Professor Campbell, Dean of the College, stating that the Department of Agriculture and Technical Instruction had informed him that no money would be made available to fund the field trips and that 'unless the Treasury pay ... these excursions are at an end'.[18] Cole pleaded his case to continue the excursions and even offered to use £10 from his laboratory budget to subsidize the excursions. He received permission to spend no more that £3/10/0 per head,[19] and he managed to run a 6-day excursion in March of that year. However, funding was not forthcoming the following year and the 'long excursion' disappeared from the curriculum, much to his disappointment. This must have been a serious blow to Cole, particularly as he placed huge emphasis on the educational merits of fieldwork.

Cycling and its influence on Cole's research

Cole's early petrological work often depended on materials collected during his cycling trips. By 1897 the thrust of his work altered as he began to document the importance of geological structure on the geographical landscape and he contributed short articles on the geological/landscape theme to *Nature* and *Knowledge* (Cole 1898*a–f*). He accompanied William Morris Davis (1850–1934) the American geomorphologist on a European tour in 1911 (Davis 1912). Davis, it must be said, preferred train travel to pedal power, this is hardly surprising given the expanse of the North American continent. Cole summed up his thoughts in several later books: *The Growth of Europe* (1914) and *Ireland the Outpost* (1919): 'The study of its [Europe] fundamental structure in relation to the wayward actions of its overlords is for most of us the very keystone of geography' (Cole 1919). He continued this theme in his addresses to the Geographical Societies of Britain and Ireland (he was President of both in 1919) in which he stressed the importance that geography should be introduced as a university subject reiterating again the importance of geology on human distribution (Cole 1921*a*, *b*; Herries Davies 1977).

Conclusion

Cole's scientific legacy lies in the pages of his published research output on many topics, but perhaps, of most significance on the volcanic rocks of Europe and on the foundations of its landscape. However, he also provided important social documentaries that are perhaps now of greater interest given the political changes in, and increasing awareness of Eastern Europe at the present time. His two travelogues give a valuable insight into a bygone period when the area was less turbulent.

Although Cole finally had to give up cycling late in life, due to increasing disability caused by arthritis, he never lost his love for it. In *Who's Who* for 1924 he added one poignant word to his usual entry to this annual listing of persons of influence: 'Recreation: **formerly** cycling especially as a means of travel' (Cole 1924, p. 581). Sitting in Orahova in Carrickmines in Dublin, gazing at the large wall map, he would have retraced his peregrinations across Europe and North Africa. In the Epilogue to *As We Ride* he wrote: 'Up and down, through our own Atlantic isles, or across the untracked steppes of newer lands, we ride from dark to dawn, part in memory, part perhaps in dream ...' (G. A. J. Cole in Cole & Cole 1902, p. 107).

I am most grateful to all my colleagues and friends who have aided this study: H. Torrens provided information about S. S. Buckman's cycling obsession while E. Robinson documented the role of cycling in some early excursions of the Geologists' Association. P. Kennan gave me access to the Cole-related documents held in University College, Dublin without which this study would not have been possible. I thank my wife Vanessa, who has had to 'live' with 'the cycling geologist' for nearly two decades, for her understanding and patience. This paper was completed in the leafy surroundings of Dickinson College, Carlisle, Pennsylvania during Autumn/Fall 2004. I thank M. Key and his family for their kind hospitality extended to my family during this period.

Notes

[1]Dedication in Cole & Cole (1902).

[2]*Mason & Payne's Cycling Map of the British Isles with Parts of France, Belgium and Holland.* Scale *c.* 1 : 1 584 000. London (1888), 686 × 573 mm. A series of such maps was produced by George Washington Bacon and Co., London between 1885 and 1923 and included *Bacon's Cycling Road Map: Ireland.* Scale 1 : 253 440. London (1900). It is probable that Cole owned such maps.

[3]In 1894 he published a short paper entitled '*Geologists at the Luncheon Table*' (Cole 1894*a*), about a series of marble topped tables that he and six other scientists had presented to two ladies, the Misses Gardiner. These ladies ran the *Farm Produce Depot and Refreshment Rooms* at 1–2 South Leinster Street, which was the location for their important scientific debate. The present location of the tables is sadly unknown.

[4]This quarter-inch-to-the-mile map measures approximately 2 m by 1.5 m. It is a linen-backed map and has been

divided into small sections that could be folded which allowed it to be carried easily while in transit. It was recovered in 1999 from a black plastic bag found in an attic in the Museum Building, Trinity College, Dublin. The section containing Belfast and NE Ireland is missing. The map is now in the collections of the Geological Museum, Trinity College, Dublin.

[5]I was told of this by a member of the Bewley family who lived in *Orahova* when a child. After Vernon Cole sold the house, the new owners kept the map in an outside shed before it was discarded.

[6]*The Times*, Saturday, 12 May 1894, page 8, column C.

[7]During this journey Cole used *Bradshaw's Map of Europe*; Bradshaw was the publisher of a number of highly popular maps designed with cyclists in mind.

[8]Beudant undertook his visit to Hungary at the behest of the French Government and published an account of the trip and his findings in four volumes as *Voyage minéralogique et géologique, en Hongrie, pendant l'annee 1818. Relation historique 1822*. Verdière, Paris (1822). He is noteworthy for having described the mineral species including azurite, erinite, klaprothite and smithsonite.

[9]Wetzler was a centre for high-grade precision engineering. Leitz petrological microscopes and other optical instruments used by Cole and many other geologists of the time were and continue to be manufactured in the town.

[10]*The Times*, Saturday, 30 January 1904, page 14, column A.

[11]Anon., 'Obituary. Professor Cole. Geologist and Writer', *The Times*, Tuesday, 22 April 1924, page 18, column E.

[12]Robert Welch was the foremost natural history photographer of his time in Ireland and his images illustrated many scientific papers including those authored by himself. He was an excellent conchologist. He was commissioned to photograph the Titanic and its sister ships while they were being built in Belfast, where he had his studio (Evans & Turner 1977).

[13]This letter is attached to Cole's copy of the 1903 excursion booklet (Department of Geology, University College, Dublin).

[14]Cole's copies and associated correspondence are in the Department of Geology, University College, Dublin.

[15]Cole clipped the maps out of Cooke's (1902) *Murray's Handbook for Ireland.* Originally published by John Murray, London, this was the first edition to be published by Edward Stanford. When planning the first excursion, Cole wrote to Stanford of 12–14 Long Acre, London, the well-known publisher of geological and other maps, asking him to furnish unbound sheets of these maps. Stanford's response dated 21 April 1903 indicated that he had only printed enough sheets for the binding of the volume and that no additional copies were available at that time. A subsequent postcard from Stanford to Cole, dated 23 June 1903 said that he would bear this request in mind should a further printing of the Atlas be needed. (This correspondence is attached to Cole's copy of the 1903 excursion booklet in Department of Geology, University College, Dublin).

[16]Triassic salt is still mined from underground at Kilroot and is mainly used in industry and by local authorities to grit the roads after a snowfall or heavy frost.

[17]Letter to Cole, dated 14 June 1906 (attached to Cole's copy of the 1906 excursion booklet in Department of Geology, University College, Dublin).

[18]Letter to Cole, dated 16 January 1908 (attached to Cole's copy of the 1908 excursion booklet in Department of Geology, University College, Dublin).

[19]A copy of Cole's reply to the letter of 16 January 1908 is given on the reverse side of the sheet of paper. Sanction that he may proceed with the trip, but under financial constraint, came on 24 February 1908 in an almost unreadable letter penned by a civil servant or College administrator (this attached to Cole's copy of the 1908 excursion booklet in Department of Geology, University College, Dublin).

References

BEUDANT, F. S. 1822. *Voyage minéralogique et géologique, en Hongrie, pendant l'année 1818. Relation historique 1822*. **4**. Verdière, Paris.

BRIGHT, R. 1818. *Travels from Vienna through Lower Hungary; with some Remarks on the State of Vienna during the Congress in the Year 1814*. Archibald Constable, Edinburgh.

BUCKMAN, S. S. 1898. *Proceedings of the Cotteswold Naturalists' Field Club*, **12**, 217–238.

COLE, G. A. J. 1887. The rhyolites of Wuenheim, Vosges', *Geological Magazine*, **24**, 181–182.

COLE, G. A. J. 1892*a*. County Dublin, past and present. *Irish Naturalist*, **1**, 9–14, 31–36, 53–57, 73–76, 90–95. [Reprinted 1892 with different pagination as *County Dublin, Past and Present.* Webb & Sons, Dublin].

COLE, G. A. J. 1892*b*. The variolite of Annalong, Co. Down. *Scientific Proceedings of the Royal Dublin Society*, **7**, 112–116.

COLE, G. A. J. 1894*a*. Geologists at the Luncheon Table. *Irish Naturalist*, **4**, 41–44.

COLE, G. A. J. 1894*b*. On variolite and other tachylytes at Dunmore Head, Co. Down. *Geological Magazine*, **31**, 220–222.

COLE, G. A. J. 1894*c*. *The Gypsy Road: a Journey from Krakow to Coblentz.* Macmillan & Co., London and New York.

COLE, G. A. J. 1895*a*. *Open-Air-Studies: an Introduction to Geology Out-of-Doors.* Griffin and Co., London.

COLE, G. A. J. 1895*b*. *Scenery and Geology in County Antrim.* R. Carswell and Son, Belfast.

COLE, G. A. J. 1896. The rhyolites of the County of Antrim. *Scientific Transactions of the Royal Dublin Society*, **6** (Series 2), 7–114.

COLE, G. A. J. 1897. On the geology of Slieve Gallion, in the County of Londonderry. *Scientific Transactions of the Royal Dublin Society*, **6** (Series 2), 213–248.

COLE, G. A. J. 1898*a*. The Floor of a Continent. *Knowledge*, **21**, 25–28.

COLE, G. A. J. 1898*b*. Structure of Ireland. *Knowledge*, **21**, 74–78.

COLE, G. A. J. 1898*c*. The Mourne Mountains. *Knowledge*, **21**, 121–124.

COLE, G. A. J. 1898*d*. An Old-World Highland. *Knowledge*, **21**, 170–173.

COLE, G. A. J. 1898*e*. An Esker in the Plain. *Knowledge*, **21**, 217–220.

COLE, G. A. J. 1898*f*. Volcanoes in the North. *Knowledge*, **21**, 266–268.

COLE, G. A. J. 1911*a*. The Lost City [The King (1158); The City (1450); The Watchman (1500); The City (1911); The Causse du Larzac (1911)]. *Irish Review*, **1**, 322–324.

COLE, G. A. J. 1911*b*. Glacial Features in Spitsbergen in relation to Irish Geology. *Proceedings of the Royal Irish Academy*, **29B**, 191–208.

COLE, G. A. J. 1912*a*. *The Growth of Europe*. Williams and Norgate, London.

COLE, G. A. J. 1912*b*. *Ireland: the Land and the Landscape*. Educational Company of Ireland, Dublin.

COLE, G. A. J. 1912*c*. The Problem of the Liffey Valley. *Proceedings of the Royal Irish Academy*, **30B**, 8–19.

COLE, G. A. J. 1916. On the mode of occurrence and origin of the orbicular granite of Mullaghderg, Co. Donegal. *Scientific Proceedings of the Royal Dublin Society*, **15**, 141–158.

COLE, G. A. J. 1918*a*. On Lecturing with the Lantern. *Department of Agriculture and Technical Instruction for Ireland Journal*, **19**, 14–24.

COLE, G. A. J. 1918*b*. *On Lecturing with the Lantern*. T.H. Mason, Dublin.

COLE, G. A. J. 1919. *Ireland the Outpost*. Oxford University Press.

COLE, G. A. J. 1921*a*. Why we teach geography. *Journal of Education and the School World*, **53**, 25–27.

COLE, G. A. J. 1921*b*. Why do we not teach geography? *The Geographical Teacher*, **10**, 276–279.

COLE, G. A. J. 1922. Geology. *In:* FLETCHER, G. (ed.) *Connaught*. Cambridge, Cambridge University Press, 41–60.

COLE, G. A. J. 1924. Entry in *Who's Who*. A. & C. Black, London, 581.

COLE, G. A. J. & BUTLER, G. W. 1892. On the lithophyses in the obsidian of the Rocce Rosse, Lipari. *Quarterly Journal of the Geological Society of London*, **48**, 438–446.

COLE, G. A. J. & COLE, B. 1902. *As We Ride*. Royal City of Dublin Hospital, Dublin.

COLE, G. A. J. & GREGORY, J. W. 1890. The variolitic rocks of Mt. Genevre. *Geological Magazine*, **24**, 140–141.

COOKE, J. (undated). *John Boyd Dunlop*. Dreoilín.

COOKE, J. 1902. *Murray's Handbook for Ireland*. 6th edition. Edward Stanford, London.

DAVIS, W. M. 1912. A geographical pilgrimage from Ireland to Italy. *Annals of the Association of American Geographers*, **2**, 73–100.

ERSKINE, F. J. 1884. *Tricycling for Ladies: or, Hints on the Choice & Management of Tricycles: with Suggestions on Dress, Riding & Touring*. Iliffe & Son, London.

ERSKINE, F. J. 1896. *Bicycling for Ladies*. Iliffe & Son, London.

ERSKINE, F. J. 1897. *Lady Cycling*. Walter Scott, London.

EVANS, E. E. & TURNER, B. S. 1977. *Ireland's Eye: The Photographs of Robert John Welch*. Blackstaff Press, Belfast.

FLOWER, R. H. & BERDAN, J. M. 1977. Memorial to Helen M. Duncan 1910–1971. *Geological Society of America Memorials*, **5**, 3.

GRAVES, H., HILLIER, G. L. & SUSAN (HARRIS), COUNTESS OF MALMESBURY. 1898. *Cycling, Part 1—General and Mechanical; Part 2—Cycle Racing; Part 3—Cycling for Women*. Lawrence & Bullen, London.

GREEN, C. P. 1989. Excursions in the past: a review of the Field Meeting Reports in the first one hundred volumes of the *Proceedings*. *Proceedings of the Geologists' Association*, **100**, 17–29.

HERRIES DAVIES, G. L. 1977. The making of Irish Geography II: Grenville Arthur James Cole (1859–1924). *Irish Geography*, **10**, 90–94.

HERRIES DAVIES, G. L. 1995. *North from the Hook: 150 Years of the Geological Survey of Ireland*. Geological Survey of Ireland, Dublin.

INGLIS, H. R. G. 1896. *The 'Contour' Road Book of Scotland*. Gall & Inglis, Edinburgh.

INGLIS, H. R. G. 1897. *The 'Contour' Road Book of England*. Gall & Inglis, Edinburgh.

INGLIS, H. R. G. 1908. *The 'Contour' Road Book of Ireland*. Gall & Inglis, Edinburgh.

KELHAM, B. B. 1967. The Royal College of Science (1867–1926). *Studies*, **56**, 297–309.

KÖLBL-EBERT, M. 2007. The geological travels of Charles Lyell, Charlotte Murchison and Roderick Impey Murchison in France and northern Italy (1828). *In*: WYSE JACKSON, P. N. (ed.) *Four Centuries of Geological Travel: The Search for Knowledge on Foot, Bicycle, Sledge and Camel*. Geological Society, London, Special Publications, **287**, 109–117.

PORTLOCK, J. E. 1843. *Report on the Geology of the County of Londonderry, and of Parts of Tyrone and Fermanag*. Andrew Milliken, Dublin and London.

PRAEGER, R. LL. 1941. *A Populous Solitude*. Hodges Figgis, Dublin.

ROBINSON, E. 1990. 'Clarion o'er the dreaming earth': a personal review of the G.A. *Circular* since 1858. *Proceedings of the Geologists' Association*, **101**, 101–117.

RYLEY, J. B. 1899. *The Dangers of Cycling for Women and Children*. Henry Renshaw, London.

SMITH, R. A. 1972. *A Social History of the Bicycle, its Early Life and Times in America*. American Heritage Press.

SPENCER, C. 1884. *The Cycle Directory: Containing an Alphabetical List of all Cycling Clubs & Unions, and the Touring Club*. Cassell & Co, London.

TORRENS, H. S. 2004. The life and work of the geologist S.S. Buckman. *Proceedings of the Cotteswold Naturalists' Field Club*, **43**, 25–47.

TUNNICLIFF, S. 1980. *A Catalogue of the Lower Palaeozoic Fossils in the Collection of Major-General J.E. Portlock, R.E., LL.D., F.R.S. etc*. Ulster Museum, Belfast.

WYSE JACKSON, P. N. 1989. On Rocks and Bicycles: a biobibliography of Grenville Arthur James Cole (1859–1924) fifth director of the Geological Survey of Ireland. *Bulletin of the Geological Survey of Ireland*, **4**, 151–163.

WYSE JACKSON, P. N. 1991. The cycling geologist. *Cycle Touring and Campaigning*, **4**, 151–163.

The travels and travails of Sir Charles Lewis Giesecke

A. WHITTAKER

British Geological Survey, Kingsley Dunham Centre, Keyworth, Nottingham NG12 5GG, UK (e-mail: awhi@bgs.ac.uk)

Abstract: Sir Charles Lewis Giesecke (1761–1833) was the Professor of Mineralogy at the Royal Dublin Society between the years 1814 and 1833. He was born in Augsburg, Germany, and over his lifetime pursued two careers. The first career was in the German theatre of the late eighteenth century, where, as well as acting and singing, his stage activities entailed travelling through Germany, Austria, and other parts of central Europe, together with the writing of original plays and operas for the theatre. Additionally, during his theatrical career he was engaged not only in stage management but also in the adaptation and translation of many varied works for the theatre especially in the genres of comedy, parody and travesty. At intervals he also worked as a journalist or critic, covering theatrical performances in various cities. Most importantly, for twelve years or so he was a senior member of, and official playwright to, the famous Freihaus Theatre company in Vienna, a company run by the impresario Emanuel Schikaneder, who together with the composer Mozart, conceived and staged the celebrated and seminal opera '*The Magic Flute*' in 1791. Giesecke played the role of the First Slave in this opera and was stage manager for many of its performances. Later in his life, while visiting Vienna in 1818, and during his professorial appointment to the Dublin Society, Giesecke was reported to have claimed authorship of important parts of the opera's famous libretto.

In 1794, Giesecke began to revive an earlier interest in mineralogy that he had first developed while at university in Göttingen. In 1800, he obtained a mineral dealer's licence in Vienna and undertook a remarkable series of journeys which covered much of German-speaking Europe and Scandinavia. After visiting Berlin in 1801 and acquiring the title of Bergrat or Mining Advisor to the Prussian Court, he travelled to Freiberg to meet the famous German mineralogist Abraham Gottlob Werner. Immediately after this visit Giesecke travelled to the Erzgebirge region probably specifically to examine the well-known columnar basalt locality of Scheibenberg which Werner had recently reclassified in his '*Kurze Klassification*' and which became increasingly important in the emerging discipline of stratigraphical geology. After his northern European travels, Giesecke then went to the Faeroe Islands and eventually to Greenland where he spent seven years, marooned, mainly as a result of the Napoleonic wars. Despite great hardship, Giesecke traversed and studied much of the west coast of Greenland, observing and recording the mineralogy and geology. Unluckily, his specimens together with the rest of his belongings were destroyed in the British bombardment of Copenhagen in 1807. However, some of his mineral collections were taken as booty and eventually offered for sale in Edinburgh, where they came to the notice of Thomas Allan who bought the material. On his return to Europe in 1813, Giesecke called in on Allan and their meeting soon resulted in Giesecke applying for, and being appointed to, the recently-advertised professorship in Dublin. Giesecke's geological work in Greenland launched an extremely successful scientific career and provided European geologists with essential information on the geology and mineral resources of this formerly barely-known territory. In addition to contributing significantly to scientific research, some of it with considerable economic potential, his first hand knowledge of Greenland proved to be of great practical use in the search for the elusive North West Passage, which European naval authorities were keen to explore.

Karl Ludwig Giesecke (1761–1833) was born as Johann Georg Metzler in Augsburg, Bavaria, Germany on the 6 April 1761. He was the second son of an Augsburg protestant master tailor named Johann Georg Metzler, and his wife, Sibylla Magdalena Götz. The precise identification of Giesecke's family origins only became known with certainty in 1910 with the publication of biographical details by the Danish geologist Steenstrup (1910). It was through the chance discovery of a letter from Giesecke's sister in Germany to the Danish authorities, written in 1810, enquiring about the welfare of her brother whom she knew to be in Greenland, that Giesecke's relationship with the Augsburg Metzlers was firmly established. Before the publication of the 1910 account there had been much confusion about his precise origins and identification (Scouler 1834; Wyse Jackson 1996; Whittaker 2001) for several reasons: 1) his early adoption of a pseudonym; 2) his travels and

From: WYSE JACKSON, P. N. (ed.) *Four Centuries of Geological Travel: The Search for Knowledge on Foot, Bicycle, Sledge and Camel.* Geological Society, London, Special Publications, **287**, 149–160.
DOI: 10.1144/SP287.12 0305-8719/07/$15.00

frequent change of location; and 3) because of his assumption of two totally unrelated careers. The confusion was compounded by Giesecke's apparent reluctance to discuss his early career on the stage after he became established as the Professor of Mineralogy to the Royal Dublin Society. The attitude of contemporary academic scientists to the stage is encapsulated by the remarks of Ignaz von Born on hearing that his scientific protégé Friedrich Keppner (1745–1820) had done the reverse of Giesecke by leaving a very promising scientific career for the theatre. Born's comments are paraphrased as follows:

> Mr Keppner has sacrificed all for the theatre. A theatrical poet will surely not provide a permanent living for an honest man (Whittaker 2001).

With regard to his origins, it has recently been suggested that Giesecke may even have been an illegitimate member of the Habsburg royal family (Kirchmayer 2001), which could perhaps explain the source of his university funding.

Although little is known of his early life, the young Metzler was a bright and intelligent schoolboy. This is apparent from the comments of his schoolmaster at the St Anna Gymnasium, Augsburg, Hieronymus Andreas Mertens, who gave him an excellent commendation, before he went to Göttingen University in 1781. This laudation was copied by the young Giesecke into his autograph album (Stammbuch), which is one of two such books now preserved in the National Library of Ireland, Dublin (Waterhouse 1936). At Göttingen University, Giesecke studied law from 1781 to 1783. It was while he was a student there that Giesecke took an early interest in mineralogy by attending some of the lectures of the famous German naturalist Blumenbach; this we learn from the application for the Dublin professorship in 1813. With regard to the two albums, we also know that the young Metzler had adopted his pseudonym (both first names as well as surname) as early as September 1781, because the introductory page to the album gives his name in Latin as 'Carolus Ludovicus Metzler cognomine Giesecke. After his appointment to the Dublin professorship and his becoming a member of the Danish chivalric order of the Dannebrog in 1814 (later raised to Commander in 1817); he occasionally called himself 'Carl Ludwig Metzler von Giesecke'.

It is not known why he chose these particular pseudonyms. Before the Steenstrup account it was commonly believed that he had adopted his mother's maiden name. Despite some later views that he may have had a patron named Giesecke who may have provided some funding for his university education, it seems more likely that, given his subsequent first career, and as an artistically inclined young man, he may well have been an admirer of the Klopstockian poet and writer Nicolaus Dietrich Gieseke. This is partly supported by some of the early entries in his album from his stays in Bremen that definitely have a Klopstockian, sentimental flavour to them (Waterhouse 1936).

Giesecke's autograph albums provide the main source of information about his subsequent movements (Whittaker 2001). From September 1800 the autograph entries are numerous and it is possible to follow Giesecke's travels virtually from day to day. Because of the numbering of the two preserved albums it seems possible that there were originally five albums. Various authors have also surmised that albums 1 and 2, if they ever existed, were entirely confined to Giesecke's schooldays, but that album 4 may have covered two large gaps which would coincide with his career as an actor, librettist and writer, together perhaps with the signatures and comments of Mozart, Schikaneder, Schack, Gerl, and other well known personalities from the time he spent on the stage. It is known that Giesecke left Schikaneder's company (Freihaus Theatre) in August 1800, but it now seems likely that he left clandestinely. Apparently the Viennese City records show that in 1801 distraint was levied on such effects as Giesecke had left behind for non-payment of his rent. Amongst the catalogue of the books and papers sold was 'An album with some pictures ...' It seems possible therefore that missing album 4 may still exist, forgotten in some attic, or in some private or public library in Austria.

Giesecke and the theatre

The earliest entries in album 5 date from March, May and July 1781 in Bremen, and it is clear that he left the St Anna Gymnasium at Michaelmas 1781. Album 3 was started in September 1781 when Giesecke gained numerous entries from schoolteachers and other Augsburg dignitaries. After the date of his first Christmas vacation from Göttingen spent in Augsburg, there are ten entries, in Album 5, made in Bremen, dating from January and February 1782, and prompting the question as to why Giesecke was in Bremen and not at his university in Göttingen. This may have been because he had relatives in Bremen, although one writer, named Castle (1946), stated that Giesecke was at this time fascinated by the beautiful actress Felizitas Abt, wife of the actor-manager Carl Friedrich Abt, and followed her to Bremen at Michaelmas 1783.

Giesecke was very much a polymath; he became a talented playwright, linguist, actor, singer, poet, translator and freemason. Eventually, he became fluent in German, Italian, French, Danish, English, and Greenlandic (Inuit). By 1789 Giesecke had joined the Schikaneder company at the Freihaus Theatre, Vienna, with which he remained for many years. He mostly wrote comedic works for the stage with this company, as well as playing mainly small character roles. He wrote his first opera libretto for the Freihaus company, which was an adaptation of Wieland's fairytale story 'Oberon', with music by the well known composer Paul Wranitzky (1756–1808), and in which Giesecke also played the minor role of Osmin. The opera, first performed on 7 November 1789, was a great success. The events surrounding Giesecke's contribution to this opera are somewhat typical of the travails surrounding Giesecke's activities, in that apparently the work is similar to a slightly earlier existing text of 'Oberon' by another writer. Although this has given rise to accusations of plagiarism by some commentators evaluating his literary work at a much later date, it is well to remember that plagiarism was not uncommon in those days, and that there were doubtless good reasons for taking advantage of existing material. In this particular case, with a premiere for 12 July 1789, there was little enough time to prepare and rehearse a programme for the coming autumn season. Like all such enterprises the main concern of the Freihaus Theatre was to attract large enough audiences to earn income; thus it was imperative for the individual performers to be acutely aware of the urgency and discipline of meeting production and performance deadlines. Schikaneder, as director and impresario, had full confidence in Giesecke's ability to produce a good, usable script efficiently, so as to provide the theatre company with fresh material. This would be absolutely essential, especially for the launch of Schikaneder's first season at the Freihaus.

In addition, Giesecke also edited two or more theatrical journals concerned with critical review of theatrical events at various theatres.

Giesecke in Vienna: theatre and science

While in Vienna, the composer Wranitzky and Giesecke joined the freemasonic lodge Crowned Hope, whose leader was Ignaz von Born (1742–1791), the doyen of central European science in the fields of mineralogy, chemistry and metallurgy. This masonic lodge served the function of a scientific academy in Vienna at that time and produced scientific publications amongst other works. The lodge published work by von Born, and other geoscientists of the day (Weisberger 1993).[1]

Given Giesecke's earlier interest in mineralogy from his days in Göttingen University, it is easy to see how the prevailing scientific atmosphere in Vienna in the late 1780s may well have reawakened his latent interest in the natural sciences, which came into full flower in 1794. Numerous leading mineralogists were associated with the Crowned Hope Masonic Lodge. Apart from von Born, one of the most distinguished of these was the Swedish mineralogist and protogeologist Torbern Bergman (1735–1784) who was extremely influential in European scientific circles, and greatly influenced not only von Born, but also Abraham Gottlob Werner of the Freiberg Mining Academy. According to Deutsch (1952), from 1790, Giesecke was associated with Anton David Steiger Edlen vom Amstein, a mineralogist and a favourite of Ignaz von Born.

Meanwhile Giesecke continued his work with the Schikaneder Company and in 1791, he contributed to Mozart's opera *The Magic Flute* (Whittaker 1998) as stage manager and prompt and played the non-singing part of the First Slave. He probably contributed to the libretto. Various parts of alchemical lore and hermetic philosophy were passed on, historically, in the guise of mythology and folklore, such as the myth of 'Jason and the Argonauts' and Virgil's '*Aeneid*'. In addition to providing some of the philosophical infrastructure for freemasonry, mythology and folklore have always been a rich source of material for stories, and fairytales. The collection of German fairytales by the German contemporary writer Wieland provided material for many stage performances that were extremely popular in Vienna at the time. Indeed, Giesecke as writer, stage performer, apprentice scientist and freemason was almost uniquely placed to exploit this fusion of esoteric ingredients, given his interests and activities in Vienna during the last decade or so of the 1700s (Fig. 1). The story of The Magic Flute describes the alchemical process (Whittaker 1998). The original libretto shows throughout, the precise choice of words which correspond exactly with the historical vocabulary of alchemy, and the way these have provided the basic storyline; thus, the 'serpent', which represents the alchemical *prima materia*, the silver javelins of the Three Ladies representing stibnite, and the 'Doves of Diana representing silver provide examples of this. The mineral stibnite (antimony trisulphide) represented by the silver javelins of the Three Ladies, is used to provide the 'star regulus of antimony' which reduces the iron and prepares the alchemical vessel for a process which according to the alchemists was accompanied by descriptions and symbols of trees and flowers (commonly roses). What the old alchemists were probably observing was fractal growth of materials deriving from the chemicals and metals being used in the alchemical vessel.

Fig. 1. Image of Giesecke in his acting career (courtesy of Bavarian State Library).

Acting roles

After playing the parts of young men in the early part of his acting career, subsequent to 1798 he played mature or scheming old men. This information comes from theatre tickets and playbills in the Vienna City Library and from Franz Trau in Vienna. Also there is much useful information from the handwritten records of Leopold Edlen von Sonnleithner, in the Archiv der Gesellschaft der Musikfreunde in Vienna.[2]

2 January 1790 he played Obristen Graf von Bembrok in the Comedy '*Graf von Waltron*' by H. F. Möller.

6 May 1790 he played Osmin in '*Oberon*', music by Wranitzky and libretto by Giesecke.

4 June 1790 he played Broemser in '*Gisella Broemserin*' by Simmler.

25 June 1790 he played the Burgermeister in the historical play '*King Attila*' by E. Schikaneder.

30 September 1791 he played the non-singing role of the First Slave in '*Die Zauberflöte*' by Mozart.

11 February 1792 he played Hauptmann von Allsing in '*Der Fähndrich*' by F. L. Schroeder.

28 June 1799 he played Faust's father in '*Fausts Leben, Thaten und Hoellenfahrt*' a romantic play by M. Voll with music by Johann Georg Lickel.

28 July 1799 he played Oberaufseher Mustapha in the comedy '*Der Russe in der Turkei*' that Professor Cowmeadow in Berlin wrote from the English.

As a stage writer Giesecke specialized in comedy, travesty and parody using material from, for example, '*The travestied Aeneas*' a comedy in rhyming verse given first at a benefit performance for Giesecke himself on 13 August 1799. Giesecke's play '*There are still faithful wives after all!*', described as a 'play in three acts, adapted from a true story', was produced in Vienna in 1790. It is of special interest because of its possible autobiographical input by Giesecke. In the play is a character, Law Licentiate Metzler (Giesecke's birth name), who was the University friend of the leading role Freyberg (name of both the character and place name of the mining academy). In this play Metzler is a quiet person, upright and caring, who was involved in an unhappy love affair, and hence slightly misogynist. The play provides several reverberations with Giesecke's surmised relationship with the beautiful Felizitas, as well as providing several possible links (via misogyny) with *The Magic Flute*. Part of the play's text, according to the musicologist Irmen (1991, 1996) are further resonant of famous quotes from '*The Magic Flute*'. For example, Freyberg's cry 'for me there is no help but death and despair', and references to, 'roses flower among the thorns'.

Further work as a stage writer

After the 5 January 1789 performance of '*Die Erbschaft*' by Grafen Bruhl at the Freihaus, the net profit was donated to the purchase of wood for the poor of the city of Vienna with Giesecke's appearance dressed as a beggar reciting a self-composed thanksgiving poem. Giesecke also wrote Gelegenheitsgedichte (occasional poems), as noted by Blümml (1923).

1. '*Lied eines alten Meistersangers an seine lieben Landsleute*' written by K. L. Gieseke, set to music by Joseph Swanenberg in 1797.
2. '*Volkslied bey der Wiederkehr der deutschen Streiter gesungen*' written by Karl Ludwig Gieseke and set to music by Joseph Swanenberg in 1797.
3. '*Der Asche des Herrn Feldmarschalls Grafen von Clerfayt geweiht von der Wiedner Burgergemeinde*' written by Karl Ludwig Gieseke.
4. '*Flehen um Hilfe*' a cantata written by Karl Ludwig Gieseke, Mitglied des K. K. privil. Schikanederischen Theaters. In Musik gesetzt von Herrn Ignaz Ritter von Seyfried, Compositeur des K. K. privil. Wiednertheaters, 17 März 1799.

His stage works had at their core many comedic elements that were characteristic of the contemporary Viennese stage. The following whimsical short epigrams from 1789 illustrate this aspect of his creative work in verse (Blümml 1923).

Die Mode

Was vorher dem Straüssen den Hintern bedeckt,
Wird dann auf den Kopfputz der Damen gesteckt.

Fashion

What formerly ostriches' backsides would decorate,
Will next make a headdress for ladies most delicate.

Auf einen sterbenden Saeufer

Er hat beym Wein gegeizt staets nach den ersten Zuegen,
Dafür muss er so frueh in letzten Zuegen liegen.

To a dead Boozer

With wine he was well placed to the first draught,
But through that he'll soon be breathing his last.

Nicknames

Apparently, already by 1800, Giesecke was known as 'Dietrich der Schwarzwälder' (Dietrich the Black Forester) among a merry, cheery, comedic bunch of individuals. These names continued during his second visit to Vienna to present his Greenland specimens after his Dublin appointment; with membership in 1818–1819 of the Wiener Kunstler-Club (Viennese Artists' Club) 'Die Ludlamshöhle' (Ludlam's Cavern), he went under the name of 'Harpun, der Robbe Gieseke' (Harpoon the seal), he also had another nickname 'Oberschoppe' ('Chief Sheep') of the 'Wildensteiner Chivalric Order of the Blue Earth'. This was in Burg Seebenstein in Lower Austria that belonged to the popular Archduke Johann, the founder of the famous Joanneum Museum in Graz.

Yet another work of Giesecke's is known, a comic romance entitled '*Der Wildensteiner*' given at the Castle Seebenstein and pledging the loyalty of the Seebensteiners to the Austrian Kaiser and his daughter Rosenmund. This comic romance was given at the Schloss Seebenstein in February 1802 but is thought to have been written earlier (at least by 1800) because it recalls the style of Giesecke's chivalric plays, singspiels (opera with spoken dialogue) and parodies at the Freihaus Theatre.

In a more serious vein, on 18 November 1791, only three weeks before Mozart's death, the new temple of the lodge 'New Crowned Hope' was inaugurated; Mozart conducted his *Kleine Freimaurer-Kantate* (K 623), written specially for the occasion. For many years the words were thought to have been written by Schikaneder, but Giesecke is now considered by most scholars to be the author of the text (Robbins Landon 1988).

By the 13 August 1799 Giesecke had left Vienna behind, and his professional life was subsequently dominated by scientific activities, although he remained in close contact with many of his artistic friends and colleagues.

At the end of January 1801 in Leipzig, and en route to Berlin, Giesecke met Franz Anton Hoffmeister (Austrian music publisher, composer and kapellmeister) the first publisher of a considerable amount of Mozart's music. Hoffmeister wrote several singspiels, the most successful of which was one entitled '*Der koenigssohn aus Ithaka*' The Prince from Ithaka) to a libretto by Schikaneder, first produced at the Freihaus Theatre on 27 June 1795, but later performed in Budapest, Prague and Warsaw. Giesecke knew Hoffmeister as a result of their collaboration over a heroic-comic opera '*Die Belagerung von Cythere oder: Die macht der liebe*' (*The Siege of Cythere*, or *The Power of Love*) with libretto adapted from the French by Giesecke to the original score of Gluck, with new musical sections by Hoffmeister, first performed at the Freihaus on 19 January 1796. Mozart also wrote variations on a theme by Hoffmeister, and was a subscriber to a monthly publication of piano works launched by the publisher. In addition, Hoffmeister also published musical works by Beethoven, Haydn and Pleyel.

Giesecke in Berlin: science

With his arrival in Berlin on 1 February 1801, it is clear that Giesecke's journey after he left Vienna, had not been haphazard wandering, but a carefully planned route to meet the relevant people for furthering his education, and for collecting and negotiating mineral sales. In Berlin, Giesecke met 15 people, many of them potentially influential to his interests. Between his early arrival in February and his departure in June 1801, he met with: Professor Paul Louis Simon (1767–1815) (Mining Inspector, Professor at the Berlin Mining Academy); Professor Martin Heinrich Klaproth FRS (1743–1817) (Berlin Academy, mineral analyst, mineral collector and famous Professor of Chemistry), Friedrich Wilhelm Siegfried (1734–1809) (wealthy mineral collector), Dietrich Ludwig Gustav Karsten (1768–1810) (senior mining advisor, mineral collector) and Professor Johann Georg Ludwig Manthey (1769–1842) (Professor of Chemistry in Copenhagen). While in Berlin it is likely that he attended some of Karsten's lectures on mineralogy and helped him by organizing and cataloguing his mineral collections (Hoppe 1991). Giesecke also sought, unsuccessfully, to become a member of the Berlin Natural History Society, despite presenting it with gifts of minerals and a book. Giesecke clearly got on well with Karsten, who probably advised Giesecke to visit Werner in Freiberg, before extending his activities. In Berlin he collected minerals for the Royal Prussian collection and began to use the title 'Royal Prussian Mining Commissary and Journeying Mineral Dealer'. It was a title that Giesecke used a great deal from then onwards and one that greatly facilitated his new career. The matter is discussed by Hoppe (1991) and by Whittaker (2001).

Giesecke and Werner

In mid 1801 Giesecke was in Freiberg where he met the famous mineralogist and geognosist Abraham Gottlob Werner (1749–1817) who wrote in Giesecke's album:

'The study of Nature is Instructive for Heart and Spirit and inexhaustible'.

Presumably Giesecke held long discussions with Werner and doubtless attended many of Werner's celebrated lectures at the Freiberg Mining Academy. Giesecke's sojourn in Freiberg and the Erzgebirge region was particularly important in demonstrating how his Greenland scientific work and results managed to be fully up to date within the prevailing geological paradigm, not only in terms of Werner's mineral system, but also within the developing Wernerian ideas on geognosy and geological sequence.

Werner, in addition to introducing his successful 'Mineral System' or taxonomic classification of minerals, was also strongly influenced in his wider geological work by Torbern Bergman, the famous Swedish mineralogist. Critical of his latest ideas, Werner had relocated some basalt occurrences in the construction of his later (second) system of geognostics which were known as the Flötztrappgebirge ('formation' number 11 of the Flötzgebirge). They had been examined by Werner at the Scheibenberg locality and assigned a more realistic position in Werner's sequence of rock formations, but still believed by Werner to be deposits of aqueous origin. While in the Freiberg region in 1801 Giesecke probably went to visit the Scheibenberg locality to familiarize himself with the place where Werner first recognized that the columnar basalts were not of similar geognosy to his primitive rocks, but part of his Flötz division.

Giesecke's further travels

Later that year Giesecke made his way southwestwards via Bayreuth, Bamberg and Würzburg until he eventually arrived in Frankfurt in 1802. In this region he met many professors of mineralogy or chemistry, or mining advisors and engineers, but also with singers and actors. In Bremen he met individuals from the Olbers family, an intellectual family actively promoting the theatre in Bremen and presumably the members of the Abt Company. Both of the Olbers had been students at Göttingen University and one was a member of the masonic lodge 'Golden Hart'.

From Bremen Giesecke travelled on to Hamburg and Lübeck where there was mention of a 'long geological voyage' yet to be undertaken by Giesecke. References in the albums to a 'long geological voyage' increase as Giesecke travelled along the north German coast from Rostock, Stralsund, Greifswald, and back to Hamburg in 1803 with mention of an 'Appalachian visit' and an 'American journey'. These references to America rather than to Greenland simply reflected the poor state of knowledge of the geography of these Arctic regions in the early nineteenth century, when people were uncertain as to whether it was part of the American continent or separate from it.

Giesecke then travelled in Denmark, Sweden, the Faeroe Islands (Jørgensen 1996, probable routes around Faeroe Islands p. 158) and Norway collecting and selling minerals, but also applying his scientific expertise to the curation and description of other people's mineral and rock collections as well as his own. In 1805, while Giesecke was waiting for permission from the Danish court to travel to Greenland, he made a journey to the Faeroe Islands. In the Danish Record Office (Rigsarkivet) are various documents and letters pertaining to his journey to the Faeroe Islands. Some of these papers, written in German hand, are kept at the Geological Museum in Copenhagen. In August 1805, Giesecke arrived in the Faeroe Islands where he was a guest of the Reverend Joachim Begtrup. Giesecke stayed there until 14 September 1805, which enabled him to make five journeys and visit most of the islands. The bills from the expenses for these journeys suggest that he visited at least fourteen of these islands including the most remote and inaccessible. He missed visiting only four of the islands. To show how fastidious his work was, he spent six days visiting eight localities. Apparently the minerals he collected were mostly zeolites. Jørgensen's impression of Giesecke's geological work was that he was a conscientious, modest and patient person (Jørgensen 1996). These and similar personal qualities are repeated by several of Giesecke's contacts and coworkers during his lifetime.

His first stay in Copenhagen (October 1803) had echoes of his Berlin visit in that he deliberately met some very influential people including the Keeper of the Danish Royal Privy Archive, the Royal Prussian Mining Adviser, D. Friedrich Münter and

Gregorius Wad (Professor of Zoology and Mineralogy). Friederich Münter (1761–1830) who was a prominent German freemason, historian, and antiquary was based in St Petri's Church in Copenhagen, which served as a German cultural enclave and meeting place for visiting German and Austrian intellectuals. Münter eventually became Professor of Theology in Copenhagen and Bishop of Seeland, and was not only the same age as Giesecke, but also shared an interest in the romantic poet Klopstock. In 1781, Münter travelled extensively in Europe as part of his University studies, became a member of the Illuminati (in 1783), and resided in Göttingen (another strong Illuminati centre) while at the university there in October 1781. Perhaps this is where Münter first met Giesecke, because the latter left all his books and mineral collections in St Petri's Church Copenhagen while carrying out his Greenland research and exploration. Münter also had interests in the Rosicrucian aspects of freemasonry, which Giesecke may also have shared. However, it was through the Illuminati connection that Münter became very friendly with Ignaz von Born and his daughter Maria ('Mimi') during his visit to Vienna in 1784.

Between 1803 and mid-1804 Giesecke, travelled to Gothenburg, Stockholm, Uppsala and Falun. In Uppsala he curated the royal mineral collection and was described as 'one of Germany's mineralogical collectors and a particularly knowledgable man' resulting in Giesecke (in April 1804) being made a member of the Royal Academy of Sciences in Uppsala, together with the legendary French anatomist, Georges Cuvier.

In late 1805 Giesecke collected minerals from the Faeroes and visited the Norwegian mining districts of Christiansand, Arendal and Kongsberg.

Giesecke in Greenland

On 31 May 1806 Giesecke arrived in Greenland after a difficult six-week sea journey. Conditions in Greenland were extremely harsh; long journeys had to be made by Inuit boat (umiak) usually rowed by six women, but most commonly he had to travel by sledge or on foot. According to Scouler (1834), Giesecke used to show, with pride, in his later lectures, a splendid prism of rock crystal, of nearly half a hundred weight, that he had carried home on his shoulder from a fissure in which he found it, nine miles away in the mountains.

As early as 25 May 1807 he wrote to Münter:

> I have already lived through one winter in this great stony, frozen theatre and play comedy in the morning with gusto, with which tomorrow morning I go to sea ... I am now busy doing field work where I hew and slog from morning to night. Stone collecting and distributing has its moment says wise Soloman our protective patron ... I come with luck to Copenhagen so as to disperse over the whole world (Whittaker 2001).

During his seven-year stay in Greenland he visited most parts of both east (Giesecke 1824) and west Greenland coasts. He visited all parts of the west Greenland coast from Cape Farewell in the south at latitude 69° north, as far as 76° north[20] Giesecke's study of the geology of Greenland is described in his travel diary (Giesecke 1910; Whittaker 2001).

The mineral cryolite was very rare, although it was already being traded as early as the beginning of the eighteenth century amongst the native Inuit people of the west coast of Greenland. They used the cryolite as sinkers or weights for their fishing lines and nets, because the mineral is relatively heavy, very soft, easy to shape and drill into. The first samples were brought to Europe via Copenhagen as curiosities by Danish missionaries in the middle part of the eighteenth century. It was analysed chemically and established as a mineral species by Klaproth of Berlin in 1800.

Giesecke located the major deposit at Ivigtut in 1809. However, although the mineral was known to contain aluminium and fluorine no quantitative chemical analyses were carried out until several years later. The Ivigtut location is still very attractive to professional and amateur collectors alike, because the fluoride mineral group is represented there by about 16 different mineral species. Given the contemporary knowledge of geoscience, Giesecke provided first class descriptions demonstrating that he had an excellent understanding of Greenland's landscape and geology (Figs 2 & 3). A good example is provided by the Disko Island area (Giesecke 1823), where modern maps allow clear and easy recognition of Giesecke's locations as described in his travel diaries (Giesecke 1910; Whittaker 2001). The accuracy of his observations and understanding are easily confirmed by much later work, including the most recent.

The worst event of his stay in Greenland was the arrival of the news that his existing collections, books, records, and belongings at the St Petri Church vicarage in Copenhagen had been destroyed during the British bombardment of the Danish capital city (2–5 September 1807).

This may have been the first serious setback for Giesecke, although as time went on, there were other 'travails': Giesecke suffered several accidents which in the extreme climatic and weather conditions of the Arctic region led to lameness and pulmonary problems in later life (Scouler 1834); and the Napoleonic wars, which marooned him in Greenland for seven years. In addition to the loss of his main scientific collection in Copenhagen he also lost a set of specimens on a ship sailing from

Fig. 2. Annotated drawing by Giesecke of the Cenozoic basalts overlying the Precambrian basement gneiss near Godhavn (Giesecke 1823; Whittaker 2001).

Greenland to Denmark, which was commandeered by a British vessel and diverted to Leith, near Edinburgh. The wars also interrupted the shipping traffic and regular supplies from Denmark to Greenland.

However, Giesecke had the ability to survive, to overcome such difficulties, and to continue with his work, such as his long-range correlation of the tertiary basalts, from central Europe to Greenland. It is possible that his travel diary

Fig. 3. Holstenborg District, showing an approaching sailing ship and some of the snow and ice covered structural features (joints and faults) of this part of Greenland (Steenstrup & Kornerup 1881).

contains the first ever recognition of pillow lavas (Giesecke 1910).

Despite this successful long-range correlation, Giesecke was not always accurate. On his return journey to Europe in 1813 (Thursday 9–10 September) in the North Sea region, he recorded the presence of the Flötztrappformation in the north Shetland Islands of Foula and Fair Isle. This wrong attribution is somewhat surprising, given that he was familiar with the geology of the Faeroe Islands and the Scheibenberg locality in central Europe which both contain the columnar Cenozoic basalts. However, he had just experienced a very long, arduous, stormy sea voyage, with no opportunity to land and examine the Shetland rocks, which are now known to be of Dalradian or Devonian age.

Giesecke in Dublin

Giesecke's luck was about to change dramatically. His set of specimens, which had been captured and diverted to Leith (where they were at first considered to be 'rubbish' as they had no labels) had been bought by Thomas Allan (Farrer & Farrer 1968), Scottish banker and mineralogist, who had managed to find their source via the Danish botanist Morten Wormskiold (1783–1845) (Sweet 1972). Thus on arrival in Leith in October 1813, Giesecke, fresh off the ship from Greenland and without funds, was invited to stay with Allan as an honoured guest in the fashionable quarter of Edinburgh, where his scientific standing was recognized and his portrait was painted by the famous Scottish portraitist Sir Henry Raeburn. After a few weeks Giesecke (Fig. 4), at the suggestion of Allan, applied for the recently established Professorship of Mineralogy at the Royal Dublin Society that had been advertised in Edinburgh (as well as Vienna and Stockholm). Giesecke was appointed to this professorship in 1813 having beaten off competition from such well-known geologists as Thomas Weaver, Robert Bakewell and James Millar (Herries Davies 1983).

Shortly after his Dublin appointment he travelled to Denmark to wind up his affairs there and was made a member of the Danish Order of the Dannebrog (Steenstrup 1910; Whittaker 2001). After this, he used the courtesy title of 'Sir Charles Lewis Giesecke'. On his return to Dublin he began work on the preparation and delivery of his lectures, together with the curation of the museum's mineral specimens.

In 1817, he left Dublin again to begin a long journey to the continent of Europe to purchase specimens for the Royal Dublin Society and made further useful contacts. He visited the Royal Society of London, the Geological Society of London, then Copenhagen, whence he travelled south to Vienna via Hamburg and Göttingen. During this visit he presented 832 specimens to the Austrian Emperor, valued at 1500 ducats. The Emperor gave Giesecke 1000 ducats to cover expenses and a splendid golden snuffbox set with diamonds. His return journey to Dublin included visits to Munich, Augsburg, Stuttgart, Strasbourg, Cologne and London. While in Strasbourg he renewed his contact with the Director of the Austrian Imperial Natural History Museum, to whom he had earlier sent a portion of the Tipperary meteorite (which had fallen in 1810). This opened the way for Giesecke to send a gift of mineral specimens to the poet and mineralogist Goethe, who was also a freemason, with a detailed knowledge of alchemy and a close interest in *The Magic Flute* (Whittaker 2001). This later resulted in correspondence with Goethe which covered observations and discussions on meteorology rather than mineralogy. The contact provided the opportunity for many Irish travellers to visit Goethe in subsequent years, as well as providing Giesecke with a membership diploma of the Jena Mineralogical Society. In reciprocal fashion Goethe was made an honorary member of the Royal Irish Academy in 1825 (Whittaker 2001).

Fig. 4. Portrait of Giesecke by Sir Henry Raeburn (courtesy of the Royal Dublin Society).

Giesecke was responsible for the Royal Dublin Society's museum which held large mineralogical collections (Wyse Jackson 1996), to which he donated 415 of his Greenland specimens, on his appointment as Professor of Mineralogy. He was responsible for the famous Leskean Mineral Cabinet; this cabinet contained over *c.* 7500 species of minerals, catalogued by Karsten, according to Werner's System. He added to the Irish collections, visiting and collecting minerals from various parts of Ireland, leaving accounts of his mineralogical excursions to Donegal in 1826 (Giesecke 1826), Donegal, Mayo and Galway in 1828 (Giesecke 1828) and Antrim in 1829 (Giesecke 1829). Giesecke was responsible for many new publications, including '*A descriptive catalogue of minerals in the Museum of the Royal Dublin Society to which is added an Irish Mineralogy. Dublin 1832*' (Giesecke 1832).

Giesecke died suddenly on the 5 March 1833. Unfortunately, his will and list of belongings are no longer available, so we have no idea of what books he might have owned, or what extra interests he may have had. Everyone who knew him spoke highly of his character and personality. In the *Proceedings of the Royal Dublin Society* the Assistant Secretary announced to the Vice-President the lamented death of their highly talented and esteemed Professor of Mineralogy and Keeper of the Museum, Sir Charles Giesecke:

> The high sense they entertained of the long tried talents as a Scientific Professor, and the amiable manners and character as a gentleman of the late Sir Charles Giesecke. Resolved that as a mark of respect to the memory of Sir Charles Giesecke the Museum be closed for one fortnight.

At this time, there was a lady living in London, Miss Hutton, who remembered Giesecke very well, because he had often visited her father's house. Miss Hutton spoke of Sir Charles in the following way:

> He was a singular man, very shy, I fancy (Steenstrup 1910).

He was also mentioned in a contemporary magazine. *The Dublin University Magazine* published in serial form a story (fiction) entitled '*The Confessions of Harry Lorrequer*' written by Charles Lever (1806–1872), which was later published as a novel in 1839 (Lever 1839).[3] In the chapter entitled 'A day in Dublin', two real-life people are mentioned. Sir Charles Giesecke was there described as 'one of the most modest and retiring men in existence'. When Lady Morgan (Sydney Owenson, author of the '*Wild Irish girl*') comments at a 'waltzing party' that he isn't dancing because he doesn't like women, he replies (in this book), that he kept four women in Greenland, but then he has to dispel her shock by adding that they were for rowing his umiak (or Inuit boat), much to the amusement of Harry and others within earshot. Does this suggest perhaps that Giesecke was already known in Dublin society as a misogynist, as well as a shy man and a ready wit?

Giesecke's travails re-emerged some twenty years after his death in 1833 with the publication of Cornet's book on '*The Opera in Germany* 'in 1849, in which it was reported that during his visit to Vienna in 1818, at a social gathering of old theatrical friends, Giesecke had claimed to have been responsible for important parts of the libretto of the *Magic Flute*. It is impossible to verify what was actually said at this meeting, although proof exists that Giesecke was stage manager and prompt at the Freihaus Theatre, with his own handwriting on the official prompt book used for many of the performances. Already by the mid-nineteenth century many of the leading Mozart authorities had accepted Giesecke's reported claim, for example Jahn, who in 1856 published the main authoritative Mozart biography. However, the following century brought about a further claim for the 100% Schikaneder authorship. Particularly in the early part of the twentieth century there was a long campaign against Giesecke's involvement in *The Magic Flute*, and to discredit him personally; criticisms by Schikaneder's biographer, Egon Komorzynski (1948), ranged from branding Giesecke a liar, plagiarist, swindler and cheat, to strong criticism of the poetic quality of the opera. Unfortunately, the documentary evidence that would refute accusations of unauthorized use of official titles has been lost, though Giesecke was accepted by his professional contemporaries. The pro-Schikaneder (and thus the anti-Giesecke party) comprised mainly the arts community, whereas the pro-Giesecke party comprised mainly scientists.

Conclusions

Karl Ludwig Giesecke or Sir Charles Lewis Giesecke, as he was known in the British Isles, was very unusual for his time in that he followed two totally different careers. With regard to his theatrical activities pursued in the early part of his life, he was employed as a playwright, stage poet, actor, singer, stage manager, musician and composer. His libretti were set to music by some of the best musical composers[2] of his day. In addition, he was a linguist, in German, Danish, Italian, Inuit, English and Hungarian, who translated many stage texts, including the first renditions into German of Mozart's operas, '*The Marriage of Figaro*' (1792) and '*Cosi fan Tutti*' (1794) from the original Italian, and is judged to have had considerable involvement in the preparation of the libretto of *The Magic Flute*.

However, it was in his second chosen career that Giesecke achieved a well-deserved reputation as a geological practitioner and secured his place in the history of geoscience, especially in the fields of mineralogy and geology. With regard to mineralogy specifically, he discovered many new species including sapphirine, sodalite, arfedsonite, allanite, eudialyte and gieseckite. Cryolite (an important source of aluminium) was discovered in September 1806 at the famous locality of Ivigtut. He was also first to visit the well known tourmaline occurrence at Ameralik Fjord in 1808. He also clearly understood the overall general geological structure of Greenland well, including its complex Precambrian geology (using present day nomenclature). Already by the years 1807, 1812 and 1823 he had recognized the Wernerian geognostical principles and 'formations' such as the Urgebirge, Ubergangsgebirge, Flötzgebirge, Flötztrappgebirge, and the Gebirgsformation. He also identified granites, gneisses and syenite. He made substantial contributions to the geography of Greenland including place names and cartography and to the understanding of the ethnography and linguistics of the Greenlandic Inuit people (Giesecke 1824; Jørgensen 1996). As a scientist he had great patience, tenacity and courage.

One final, important, and significant aspect of Giesecke's work was the contribution he made to the eventual discovery of the elusive North West Passage (Sweet 1974). This hinged mainly upon Giesecke's detailed knowledge of the west coast of Greenland (Giesecke 1861, 1910). In fact Giesecke was consulted many times by senior mariner explorers such as Franklin and the Scoresbys, in particular, William Scoresby junior (1789–1857) (Sweet 1974), a famous sailor with much knowledge of the Polar Seas and whaling. This Scoresby is particularly interesting in a geological context because of his close friendship and association with Professor Robert Jameson (1774–1854) Professor of Geology at Edinburgh University and a former student of Werner at the Freiberg Mining Academy. Jameson was also the founder of the Wernerian Society in Edinburgh, and also a distinguished and very influential geologist in his own right. According to Sweet (1967), Jameson was apparently the first choice for the geology chair at the Royal Dublin Society, but for some unknown reason he declined it. Giesecke's appointment to the Dublin post in 1813 predated the appointment of Sedgwick at Cambridge in 1818 as Woodwardian Professor of Geology, and Buckland at Oxford in 1814.

Thus, in essence, Giesecke is best described as a polymath, professional in each of his chosen careers. He left a considerable heritage in both fields, although more to science than to the theatre. Unlike his work for the stage, his scientific work has stood the test of time extremely well. Yet it might be said that if he were responsible for large parts of the libretto of *The Magic Flute*, then his words are likely to have been heard by more people than any other scientist!

I am grateful to E. Tillmanns, J. Meyer, E. Lebensaft and G. Kurat, all of Vienna for much help in the preparation of this account and to W. Postl and B. Moser of the Joanneum Museum, Graz, Austria. Thanks are also due to K. Secher of Copenhagen, M. Kirchmayer of Heidelberg and J. Reimer of Denmark. I am grateful to colleagues in the British Geological Survey for facilitating the work. The account is published with the approval of the Director of the British Geological Survey (NERC). Thanks are also due to my wife Jane and my daughter Clare for much help with the preparation of this paper and to P. Wyse Jackson of Trinity College, Dublin.

Notes

[1]These included Peter Pallas (Geology of Russia and Siberia 1783), Joseph Raab (Geological Survey of Galicia [1782] and its minerals, published 1784), Karl Haidinger (accompanied Raab to Galicia in 1782 and published an account in 1785), Andreas Stütz (carried out mineral surveys of Austria in 1783), Johann B. Ruprecht (metals and minerals in Hungary, and well known chemist of Schemnitz Academy, 1784) and Johann Müller (mineral surveys in Transylvania, plus experiments on antimony and bismuth, 1785).

[2]Materialen zur Geschichte der Oper und des Balletts in Wien. II. Theater auf der Wieden und an der Wien, im Archiv der Gesellschaft der Musikfreunde in Wien.

[3]Lever 1839, *The Confessions of Harry Lorrequer*, p. 158. Chapter XX. Giesecke appears in chapter XX entitled 'A day in Dublin'. Harry Lorrequer is a Subaltern in a marching regiment (infantry). He is aged 20 something and the book starts in the year 181-. The story is appropriately a comedy, riotous in places, verging on slapstick. Reminiscent of Giesecke who played Hamlet on the stage, Lorrequer played the role of Shakespeare's Othello in private performances in Cork, and a small troupe from the regiment played at garrison balls, where the players were known as the 'corps dramatique'. Amongst these extremely amusing adventures figure the Mozart operas '*Don Giovanni*' and '*Figaro*'. Interestingly, Lorrequer is also the leader of the orchestra. Lady Morgan (Sydney Owenson) also featured in the novel and is mentioned in a poem about Dublin on page 87. Lady Morgan was a novelist from a theatrical family background and well-known member of Dublin society.

References

BLÜMML, E. K. 1923. *Aus Mozarts Freundes—und Familien Kreis.* Wien, 70–89.

CASTLE, E. 1946. Aus Goethes mineralogischer Korrespondenz. Karl Ludwig Metzler von Giesecke. Der angebliche Dichter der 'Zauberflöte'. Chronik Wiener Goethe-Verein, 48–50, 84–90.

DEUTSCH, O. E. 1952. Der rätselhafte Giesecke. *Die Musikforschung*, **5**, 152–160.

FARRER, W. J & FARRER, K. A. 1968. Thomas Allan, mineralogist: an autobiographical fragment. *Annals of Science*, **24**, 115–120.

GIESECKE, C. L. 1823. On the mineralogy of Disko Island. *Transactions of the Royal Society of Edinburgh*, **9**, 263–272, pls. 15–17.

GIESECKE, C. L. 1824. On the Norwegian settlements on the Eastern coast of Greenland, or Osterbygd, and their situation. *Transactions of the Royal Irish Academy*, **14**, 47–56, pl. 13.

GIESECKE, C. L. 1826. *Account of a Mineralogical Excursion to the County of Donegal*. Royal Dublin Society, Dublin.

GIESECKE, C. L. 1828. *Second Account of a Mineralogical Excursion to the Counties of Donegal, Mayo and Galway*. Royal Dublin Society, Dublin.

GIESECKE, C. L. 1829. *Account of a Mineralogical Excursion to the County of Antrim*. Royal Dublin Society, Dublin.

GIESECKE, C. L. 1832. *A Descriptive Catalogue of a New Collection of Minerals in the Museum of the Royal Dublin Society, to which is Added an Irish Mineralogy*. Royal Dublin Society, Dublin.

GIESECKE, C. L. 1861. Catalogue of a geological and geographical collection of minerals from the Arctic regions, from Cape Farewell to Baffin Bay. Lat 59°14′N to 76°32′N. *Journal of the Royal Dublin Society*, **3**, 198–215 (edited posthumously by Samuel Haughton).

GIESECKE, C. L. 1910. Bericht einer mineralogischen Reise in Grønland. *Meddelelser Om Grønland*, **35**, 1–478.

HERRIES DAVIES, G. L. 1983. *Sheets of Many Colours. The Mapping of Ireland's Rocks*. Royal Dublin Society, Dublin.

HOPPE, G. 1991. Karl Ludwig Giesecke (1761–1833) und Berlin. *Aufschluss*, **42**, 53–63.

IRMEN, H.-J. 1991. *Mozart: Mitglied geheimer Gesellschaften*. Prisca, Essen.

IRMEN, H.-J. 1996. *Mozart's Masonry and the Magic Flute*. Prisca, Essen.

JØRGENSEN, G. 1996. Charles Lewis Giesecke, Professor of Mineralogy in Dublin: a fascinating character in the geological history of the Faeroe Islands and Greenland. *Irish Journal of Earth Sciences*, **15**, 155–160.

KIRCHMAYER, M. 2001. Ist Karl Ludwig Giesecke, geburtig aus Augsburg, Mitautor von Schikaneder's Libretto zu Mozarts 'Zauberflöte'? Auslese: zum Jahreswechsel 2000/01. Frieling & Partner, Berlin, 197–204.

KOMORZYNSKI, E. 1948. *Der Vater der Zauberflöte Emanuel Schikaneder*. Paul Neff Verlag, Wien.

LEVER, C. 1839. *The Confessions of Harry Lorrequer*. Ward, Lock & Co. Ltd, Dublin.

ROBBINS LANDON, H. C. 1988. *1791 Mozart's Last Year*. Thames and Hudson, London.

[SCOULER, J.] 1834. Biographical sketch of Sir Charles Lewis Metzler von Giesecke, late Professor of Mineralogy to the Royal Dublin Society. *Dublin University Magazine*, **3**, 161–175, 296–306 (Published anonymously).

STEENSTRUP, K. J. V. 1910. Einleitung und biographische Mitteilungen über K.L. Giesecke. *Meddelelser om Grønland*, **35**, i–ixxxvii.

STEENSTRUP, K. J. V. & KORNERUP, A. 1881. Beretning om Expeditionen til Julianahaabs Distrikt i 1876. *Meddelelser om Grønland*, **2**, 1–26.

SWEET, J. M. 1967. Robert Jameson's Irish Journal, 1797. *Annals of Science*, **23**, 97–126.

SWEET, J. M. 1972. Morten Wormskiold: Botanist (1783–1845). *Annals of Science*, **28**, 293–305.

SWEET, J. M. 1974. Robert Jameson and the explorers: the search for the north-west passage. Part 1. W Scoresby (junior), C. L. Giesecke, M. Wormskiold & J. Ross. *Annals of Science*, **31**, 21–47.

WATERHOUSE, G. 1936. Sir Charles Giesecke's Autograph Albums. *Proceedings of the Royal Irish Academy*, **43**, 291–306.

WEISBERGER, R. W. 1993. *Speculative Freemasonry and the Enlightenment: a Study of the Craft in London, Paris, Prague, and Vienna*. East European Monographs, Boulder. Columbia University Press, New York.

WHITTAKER, A. 1998. Mineralogy and Magic Flute. *Mitteilungen der Österreichischen Mineralogischen Gesellschaft*, **143**, 107–134.

WHITTAKER, A. 2001. Karl Ludwig Giesecke: his life, performance and achievements. *Mitteilungen der Österreichischen Mineralogischen Gesellschaft*, **146**, 451–479.

WYSE JACKSON, P. N. 1996. Sir Charles Lewis Giesecke (1761–1833) and Greenland: a recently discovered mineral collection in Trinity College, Dublin. *Irish Journal of Earth Sciences*, **15**, 161–168.

Alexander von Humboldt in Russia: the 1829 expedition

F. NAUMANN

Chemnitz University of Technology, Department of History, D-09107 Chemnitz, Saxony, Germany (e-mail: friedrich.naumann@phil.tu-chemnitz.de)

Abstract: Alexander von Humboldt was one of the most important figures in history in general and particularly in the history of natural sciences. He was one of the last universal scholars and his contributions have been appreciated ever since. Even today people are fascinated by his achievements as a natural scientist, explorer, researcher, adventurer, geologist and humanist. Von Humboldt is widely recognized as the founder of modern regional studies. He explored the then unknown Central American countries, studied the geographical distribution of plants for an entire continent and started the popularization of science. In doing so he anticipated an ecological view of nature and an understanding of the impact of technology on society.

From the beginning, von Humboldt attempted to understand, reflect and explain nature as a whole. Therefore, his famous volume '*Kosmos*' bears the subtitle '*Outline of a physical survey of the World*'. If nothing else, this work was the result of his almost unbelievable enthusiasm to devote his life to the study and interpretation of nature regardless of the necessary effort and exertion required.

From his childhood days, von Humboldt felt the urge to travel to faraway regions and countries seldom visited by other Europeans. He wanted to visit and explore Asia, particularly India and then Siberia, and the New World so that he could compare the Andes with the Himalayas, the Venezuelan Llanos with the Siberian steppes. Unfortunately, he was prevented from going to India. But a quarter of a century after finishing his journey to America (1799–1804) he received an offer from the Russian minister of finance, Count Georg of Cancrin, to lead a nine-month expedition through the Russian empire. The expedition started out in April 1829 accompanied by the mineralogist Gustav Rose and the botanist Christian Gottfried Ehrenberg. In late December von Humboldt returned to Germany.

Not only had the team covered a distance of 19 000 kilometres but they had also made a wealth of invaluable scientific discoveries. At the same time, relationships were established with Russia which proved very favourable for subsequent explorers.

Since 1993, the German Association of Graduates and Friends of Moscow Lomonosow University (DAMU) has been active in reviving traditional Russian–German relationships by recalling and acknowledging German explorers and researchers in Russia. To date five excursions following von Humboldt's footsteps into Russia have taken place.

Establishment of scientific education and research in Russia

The current international political situation renders it more important than ever to revive scientific and cultural traditions, as doing so helps to re-establish ties between countries and brings people together. In that sense German–Russian relations are based on a long tradition connected with the Russian recruitment of German experts such as engineers, artillery designers, craftsmen and even politicians. Under the reign of the great reformer, Peter I (1682–1725) as well as under Katharina II (1729–1796), a vast number of foreigners travelled to Russia. This was an essential step in order to achieve the gigantic task of transforming the Tsarist empire into an influential European power. Two important aims in this process were the development of the Russian mining sector and the sciences. Up to this point the predominating Byzantine influence in combination with the isolation of the country had, over centuries, resulted in little advance in technology and enlightenment. As a result, the centres of science in western and central Europe became focal points for the development of the science and technology. Peter I established an Academy of Sciences, thus following the established examples of Naples (1433), Florence (1474–1521), Florence (1582), Rome (1603), Schweinfurt 'Leopoldina' (1652), London (1662), Paris (1666) and Berlin (1700). The move was based on the ideas Gottfried Wilhelm Leibniz' (1646–1716) had laid down in a pamphlet, wherein he demanded that education and science must serve the national interest. Consequently, the systematic development of natural resources and the geological exploration of the Russian Empire became a primary goal for the sciences. An Academy of Science was set up in the young city of St Petersburg in 1725 that was closely modelled

From: WYSE JACKSON, P. N. (ed.) *Four Centuries of Geological Travel: The Search for Knowledge on Foot, Bicycle, Sledge and Camel.* Geological Society, London, Special Publications, **287**, 161–175.
DOI: 10.1144/SP287.13 0305-8719/07/$15.00

on the Paris Académie des Sciences. It included a university (for public lectures), a secondary school, a library, a museum (Kunstkammer) and a printing workshop. In its early days, the Academy was dominated by German scholars but soon attracted the attention of French, English, German, Danish and Swedish scientists who, in close co-operation with Russians and Ukrainians, forged a national academic network.

The thorough exploration of the country, especially of the distant and unknown regions of Siberia, which served as a link between Russia and China, had already started in the middle of the seventeenth century. But towards the end of that century and the beginning of the eighteenth century the rate of exploration increased. Step by step, the extensive perennial expeditions documented the vast country. The wealth of gathered material in the fields of natural science, archaeology and linguistics revealed a rich history. Two explorers of the earlier period made major contributions: Daniel Gottlieb Messerschmidt (1685–1735) travelled through Siberia between 1720 and 1727 and Vitus Bering (1680–1741) led two Kamchatka expeditions from 1724 to 1729 and 1733 to 1743.

These great expeditions gave science in Russia a huge impetus and, most notably, fostered the emergence of new disciplines. At the same time numerous artefacts were deposited in the Kunstkammer which formed the basis of a scientific collection. Michail Vasil'evic Lomonosov (1711–1765), in particular, campaigned for more expeditions so that scientific information might continue to be available. This great Russian scholar had enjoyed a modern education in humanities and natural sciences in the cities of Marburg and Freiberg and had achieved a high level of scientific expertise. His time in Freiberg was mainly associated with the well-known chemist Johann Friedrich Henckel (1679–1744) who had opened up a chemical–metallurgical laboratory for teaching purposes in 1733 on behalf of the Freiberg mining authority (Oberbergamt). His laboratory soon became a high profile educational establishment where both Henckel and his students gave presentations, classified minerals, and rocks and taught applied chemistry and metallurgy. The reputation of this 'domicile for educated miners' spread not only to America, China, Hungary, Sweden, Norway and Switzerland but also to St Petersburg. Lomonosov was actually accompanied to Freiberg by Dimitrij Ivanovic Vinogradov who, after his return to Russia, went on to solve the problems of producing porcelain at the Russian Imperial Porcelain Manufactory in 1744.

When Lomonosov returned home, he became a junior member of the Academy of Sciences and eventually a full member in 1745. He paid special attention to the Kunstkammer, the much praised 'temple of sciences' where he organized and catalogued the collections. Under his guidance a catalogue in Latin was compiled that listed all the mineral specimens (including those from Saxony) and this was published in 1745. A decade later Lomonosov participated in the foundation and development of Moscow University which still proudly bears his name. His commitment to science and in particular his publications that included *First Basics of Metallurgy and Metal Processing* published in five volumes, and *Russian Mineralogy* elevated him to a prominent place among the best known European scientists. Alexander S. Pushkin (1799–1837) wrote with high regard of Lomonosov that 'He founded the first university. To be precise, he himself was the university'.

After Lomonosov's death academic expeditions to continue the exploration and economic development of the country became more numerous, particularly amongst academics. The most important of these expeditions was led by Peter Simon Pallas (1697–1770) between 1768 and 1774 who travelled through the Urals, the Caspian Sea, the Altai before eventually reaching the distant Lake Baikal. The results and findings of this long expedition were published in a two-thousand page travelogue and in *Flora Rossica* which elevated Pallas to a central position in Russian science (Pallas 1771–1776, 1784–1788). Pallas also left extremely valuable plant collections which today form part of the botanical collections of the Natural History Museum in London. Science and technology was boosted significantly by the expeditions as the geographical, mineralogical, botanical, zoological and ethnological material collected by scholars of the St Petersburg Academy (for some a 'scholars paradise'), became an important resource for future research. As well as this, the country was given a stimulus to prosper economically, socially and culturally.

At the turn of the nineteenth century, a fundamental restructuring of the Russian system of education and research took place with the foundation in 1802 of the Ministry of National Education in St Petersburg. This ministry took over all responsibility for the co-ordination of organized education and science in the country. Furthermore, a number of new universities were founded in order to spread the educational responsibilities that had, until that time, been resting solely on the Academy. Political changes were also of huge influence as the country's territory and significance had increased considerably after the victory over Napoleon. As a result, economic reconstruction was helped by the technological advances of industrialization that had spread from England. During the reign of Nikolaus I (1825–1855) his capable

Minister of Finance, Hesse-born Russian Georg Graf Cancrin (1774–1845) was able to regain the confidence of international business in the Russian market by taking austere saving measures. To foster connections with international trade, he recommended that agriculture and the building trade should be encouraged, the industrial sector was protected by duties, and new educational institutions (i.e. a technical school and a mining institute) were built. Unfortunately, the iron fist of the 'Autocrat' also reinforced 'autocracy, orthodoxy and nationalism' which left severe problems, such as serfdom, unsolved for decades.

Following the order of the Tsar, Count Cancrin invited Alexander von Humboldt (1769–1859) to go on an expedition through Russia. This move had long been overdue as von Humboldt at that time was already regarded as an outstanding scholar, especially after his spectacular journey to the New World undertaken between 1799 and 1804. With his nomination as an honorary member of the St Petersburg Academy in 1818 he became linked to Russian science circles which helped guarantee the success of the expedition. There was a high expectation that the expedition would be successful; and it was well supported. At the same time, von Humboldt was able to fulfil a childhood dream of being able to travel through Russia.

Alexander von Humboldt: a biographical sketch

Born on 14 September 1769, von Humboldt (Fig. 1) spent his childhood and youth at Tegel Manor, then far outside the Berlin city walls. He was educated by private tutors who taught him mainly the humanities and philosophy. He was given a sense of practicality and organization without which his cosmopolitan career would probably never have materialized. Von Humboldt regarded higher education as a free formation of one's intellect and character. He started his studies at the Viadrina in Frankfurt/Oder before moving to the Georgia Augusta in Göttingen and the Commercial Academy in Hamburg. He took courses for nine months at the Bergakademie Freiberg, a period which could be regarded as the climax of his studies, since it was there that he acquired the fundamentals of mining and metallurgy as well as an excellent knowledge of the basic principles of natural sciences.

His decision to study mining presented von Humboldt with enriching experiences as he mentions in his autobiography:

> In order to learn the practical part and to perfect my skills under the great Professor Werner, I went to Freiberg in 1791 for a year. The work in the mines there strengthened my body considerably. Knowing just how much I will have to rely on by physical strength one day, I tried by all means to toughen myself up and to get used to deprivations ... Every single day I get up at 5 a. m. and, since all the mines are from a half to three quarters of an hour away from Freiberg, walk directly to the mines to proceed underground. For about 5 hours I am busy below the surface ... At about 11 or 12 o'clock I climb out of the mine and then the whole afternoon is occupied with taking courses.
>
> There has never again been a time in my life, when I have been this busy. My health has suffered a lot although I did not actually fall ill. But on the whole I am very glad, that I am engaged in a profession that, in order to love it, one has to engage in it passionately. And I have never before worked with such ease (von Humboldt 1987, p. 140).

Fig. 1. Alexander von Humboldt in his youth.

Abraham Gottlob Werner (1749–1817) and his teaching of mineralogy, geology and mining techniques had an enormous influence on von Humboldt's professional life. Werner's 'indescribably vitalising mental abilities'—as expressed by his contemporaries—drew students from all over the world to the old mining town. Charles Lyell (1797–1875) justly called him 'the great oracle of geology'.

Further parts of the educational programme included the teaching of mathematics, physics, technical mechanics, theoretical and applied mine surveying, drawing, metallurgy, metallurgical chemistry and assaying. They were taught by prominent scholars such as Johann Friedrich Lempe (1757–1801), Alexander Wilhelm Köhler (1756–1832), Johann Friedrich Freiesleben (1747–1807),

Johann Friedrich Wilhelm Charpentier (1738–1805) and Christlieb Ehregott Gellert (1713–1795).

Von Humboldt developed a small botanical garden down a Freiberg mine on which he conducted selected studies. This study, the first on a subterranean flora, was published in 1791 in Latin as *Flora subterranea Fribergensis*. On 18 February 1792, von Humboldt left Freiberg, ending, arguably the most important chapter of his early life.

Before setting off into the 'whole wide world' he decided to work for the Prussian Mining and Metallurgy Administration where he was given various responsibilities. For example he administered Upper Franconian mining in Ansbach near Bayreuth. Being appointed assessor and then rapidly promoted to various ranks within the mining authority such as Oberbergmeister, Bergrat and Oberbergrat, in recognition of his industrious and conscientious work, did not however, entirely satisfy von Humboldt. He wanted to 'live for the study of nature' and so his employment within the Prussian administration can only be regarded as a temporary step towards achieving this goal.

Before von Humboldt started out on his Russian journey, forty years passed that were filled with curiosity, the urge to explore, a passion for science, daring adventures, and unequalled scientific insights. His longing for remote and unexplored countries had been with him since his early youth. At first, the possibility for travel had only been a vision and the list of places that he wished to visit included active volcanoes such as Vesuvius, Stromboli and Etna as well as Upper Egypt, Syria or Palestine and, he thought about making a circumnavigation of the world. A programme of this size needed proper preparation, and this prompted von Humboldt to become familiar with various techniques and disciplines. He learnt how to fix his position astronomically, how to measure magnetic intensity and inclination, and how best to undertake geological, zoological and botanical studies including collecting plants. This approach corresponded with the classical trinity of researching, experiencing, and enjoying nature. However, by doing so he did not feel absolved from a duty to speak out for humane working and living conditions: 'Although it is a pleasure to enlarge the body of our knowledge by discovery, it is an even greater and much more human joy to invent something to preserve an industrious mankind, to improve and perfect important exchanges' (Beck 1961, p. 57).

Humanity as a guiding idea and its harmony with nature had always been von Humboldt's maxim that he aimed to follow at large and in detail. After all other earlier plans had failed, the opportunity for a journey through the Americas offered a totally different challenge and a chance to achieve his noble goals free of any restrictions. This journey had been preceded by a decision of the Spanish court to grant von Humboldt access to all Spanish possessions in America and the Indian Ocean (Mariana and Philippine Islands) following the mediation of the Saxon envoy, Phillipp Baron of Forell (1758–1808). The necessary official orders by the authorities followed soon, and these permitted the 'free use of all kinds of equipment for astronomic and geodetic purposes, the measurement of mountain elevations, the collection of natural specimens and investigations of any kind leading to an enhancement of science'. Thus, the first goal von Humboldt had been longing for so desperately was reached. And so in May 1799, together with the French botanist Alexandre Aimé Goujaud Bonpland (1773–1858), he embarked on a great adventure, his journey to the New World.

The programme for this expedition, known also as 'the second, the scientific, discovery of America' was enormous, almost unmanageable. The explorers spent five years and two months travelling through fascinating regions that today form the countries of Venezuela, Cuba, Ecuador, Peru and Mexico. Even today, 200 years later, von Humboldt still ranks as the most widely known and respected German throughout the whole of the Americas on account of this journey.

In 1805 he returned to Bordeaux, and later to Paris 'richer arguably than any other traveller before in collections, but most of all in observations in the vast field of natural sciences, geography and statistics'. Full of pride, he reported to his brother Wilhelm von Humboldt (1767–1835):

> My fame is more than ever. It is a kind of enthusiasm, and people are racking their brains as I often must explain astronomical, chemical, botanical and astrological matters to the tiniest detail during the meetings. All the members have been looking through my manuscript drawings and collections. They are convinced that every single aspect has been so thoroughly dealt with, as if I had dealt with one alone. Above all Berthollet and Laplace, so far my opponents, have become the most enthusiastic. The National Institute is packed every time I read (von Humboldt *et al.* 1993, p. 212).

Von Humboldt stayed on in Paris for 20 years to process and publish the scientific results of the journey. He was supported by numerous scholars to sort and organize the results in the fields of astronomy, chemistry, meteorology, zoology, botany and mineralogy; this task would have been impossible for von Humboldt alone. The outcome was the compilation *Reise in die Äquinoctial-Gegenden des Neuen Kontinents* (*Voyage to the equinoctial regions of the New Continent*), which was published in French in 1807. It consisted of 20 folios and 10 quartos and included 1425 engraved and in-part coloured charts, maps and pictures. It also contained the botanical yield of the journey, the 6000 new plant species. The actual travelogue, titled *Rélation historique*, consisted of four thick quarto volumes. A number of books followed on the subjects of ethnography, zoology, comparative anatomy, botany, geography and physiology of plants, physics, geology, mineralogy, astronomy and climatology. And eventually, the researched material became incorporated into his later works such as *Kosmos*.

Before he considered travelling through Russia and Siberia, von Humboldt had planned an expedition to upper India, the Himalayas and Tibet. The Central Asian mountain range, the 'roof of the world' had been an area that he had long wished to visit. He kept in contact with orientalists and began to learn the Persian language. It was at this time, in 1812, that Tsar Alexander I offered him the opportunity to participate in an expedition from Siberia via Kashgar and Jarkand to the Tibetan plateau. This was a proposal that von Humboldt at once embraced enthusiastically. However, the project was cancelled due to the outbreak of war between France and Russia and this deprived von Humboldt of the 'fine chance to compare the geognosy of the Himalayas and Kuen-lun with the Andes' (von Humboldt 1859). He was bitterly disappointed and began to consider going to Mexico instead.

The Russian-Siberian expedition

A new period in von Humboldt's eventful life seemed to materialize in 1827 when Tsar Nikolaus I requested a survey on the advantages and disadvantages of introducing platinum rubles. Since von Humboldt knew little about the Russian deposits of precious metals, it was suggested that he undertake a special expedition to learn more about them. Useful hints, suggestions and support for the development of Russia were expected of von Humboldt. After all, he was recognized as a famous scholar. Although the century before had seen great strides in its economy, in many parts of the country outdated and inefficient technology still dominated. That was particluarly true for the raw material sector that was dependent on mining and metallurgy. It was paralysed in a state of semi-feudal production methods and therefore in dire need of modernization.

The aim and terms of the expedition were established by the minister of finance, Cancrin, in a *Promemoria* which stated that:

> It is totally up to you in which direction you are going to travel. It is the sole desire of the government to foster the sciences. You shall be as helpful as possible to Russian mining and industry.

Von Humboldt wrote back to Cancrin:

> I do have the duty and the wish to serve your state as far as my limited knowledge on mining and technology allows by producing verbal and written reports—provided that it concerns objects rather than people. Your every single wish I will gladly fulfil since it is going to serve my own intellectual interests in nature as well as favour Russian companions most as I would like to learn the language of the country, without which one always stays a stranger to the lives of the common people (Schmidt 1924, p. 189).

As humble as von Humboldt's demands may have been, the Russian government financial and material aid helped to guarantee the success of the journey. For travel expenses from Berlin to St Petersburg 1200 ducats were paid, for the voyage itself 20 000 paper rubles. Furthermore, carriages, postal horses and military police were provided. Military protection and accommodation was provided to an extent that almost embarrassed von Humboldt: 'All the time there were greetings and provisions by policemen, administrators, Cossacks and guards of honour. Unfortunately, there was almost no moment of my own, no move possible without being guided like an ailing man' (Klenke 1876, p. 284).

On 12 April 1827, von Humboldt left Berlin and travelled towards St Petersburg. Only a few weeks before, he had been promoted by the Prussian Government to the rank of 'Wirklicher Geheimer Rath mit dem Prädicate Excellenz', a position equal in status to a Government minister. Von Humboldt was accompanied by Christian Gottfried Ehrenberg (1795–1876) and Gustav Rose (1798–1873). Ehrenberg was a doctor, zoologist and botanist who was an experienced explorer and who had already undertaken a six-year journey in Egypt. He was given responsibility for zoological and botanical studies and also acted as the doctor to the expedition. Rose was very experienced in mining activities. He had studied mineralogy, geology and chemistry and had been working as curator of the mineral collection at the Berlin University. Rose was in charge of evaluating the geological and mineral conditions and carrying out the chemical analysis of minerals. Von Humboldt reserved for himself the overall description of physical geography of the country, which included geomagnetical and astronomical observations.

This was not an expedition aimed at discovering new territories, as they repeatedly followed and crossed the tracks of earlier explorers who had already covered many parts of this vast continent and had returned with chests filled to the brim with 'treasures'. Georg Wilhelm Steller (1709–1746) and Georg Adolph Erman (1806–1877): Siberia; Vassilij Nikitic Tatiscev (1686–1750), Benedikt Franz Johann Hermann (1755–1815) and Gregor von Helmersen (1803–1885): Urals; Pjotr Ryckov (1712–1777) and Peter Simon Pallas: South Russia; as well as Johann Georg Gmelin (1709–1755) and Friedrich August Gebler (1782–1850):

Altai; all undertook travels in these regions. Instead, von Humboldt focused on the Urals and Altai regions, and later on the Caspian depression, with special emphasis on the rich ore deposits and centres of iron, silver and copper mining and processing. Peter I's decision to invite hundreds of Saxon miners and metallurgists into the country to establish a strong mining industry had already paid off. Also the entrepreneurial Demidov family was very successful, having founded the first factory for production of scythes in the Urals in 1735. They had modelled their newly founded mining and metallurgical companies on the most recent methods and technologies. Regarding the production of iron, the previously leading producers England and Sweden, had been overtaken during the second half of the eighteenth century. For silver, the situation was similar. Within a couple of years, the Altai mountain range had been turned into the most important supplier of silver and even competed with the 'classical' deposits situated in the Saxon Erzgebirge, the famous 'ore mountains'. However, by the early nineteenth century the region had started to lag behind the highly developed central and western European countries. The industry was badly affected by having old, worn out machinery, exhausted deposits and an outdated organization of production methods. Understandably, it was hoped that von Humboldt would find a remedy for these predicaments.

A further problem was the discovery of new gold deposits in the Urals and Siberia, even more so since the Siberian gold trade had just been established in 1829. As early as 1840, Russia ranked top among the world's gold producers, with 90 placer deposits being exploited in the Urals alone. In addition, there was an almost legendary richness in precious and gem stones in the Ural and Altai regions, and this formed the basis of a highly developed, eventually world famous, arts and crafts industry. The renowned Beresovsk gold mine was already known too. New platinum deposits of similar scale were discovered, that is at Nizni-Tagilsk, and immediately exploited. Von Humboldt had already seen similar, mainly sedimentary, deposits in America and was now given the chance to compare the conditions of their genesis. Further discussions 'on site' focused on geological questions and the technologies of extraction and drainage.

Preparations

Von Humboldt's work had begun in St Petersburg although he had been delayed by social obligations and attentions by the Imperial family, among them Charlotte of Prussia (1798–1860), Alexandra Fjodorovna, daughter of Friedrich Wilhelm III (1770–1840) and wife of Nikolaus I. He eventually began the research programme with astronomical, barometric and magnetic observations and used this as an opportunity to test the scientific equipment he had been provided with. He studied the magnificent collection of minerals in St Petersburg, and scrutinized specimens of those species that he felt would later become important during the expedition. Among them were gold and platinum nuggets, rare Siberian minerals and cut precious stones. Models of machines, mines, gold washings and the like drew von Humboldt's attention and these enabled him to study the specific conditions that he would encounter on his journey in advance.

Moscow University impressed him with its mineral collection of 8000 specimens that had been bought from von Humboldt's school friend, Johann Carl Freiesleben (1774–1846), a Freiberg mining official or 'Berghauptmann'. Unfortunately, the number of minerals had been reduced drastically as a consequence of the Napoleonic war thus reducing the opportunity of studying local conditions more closely.

The expedition begins

Passing Nizni Novgorod, Kazan and Perm, the expedition reached the Urals, the first low mountain range of the long journey. The stop in Jekaterinburg (Sverdlovsk) proved particularly productive since these was a mining office there. The guests were given the honour of visiting the mint, which was the central collecting point for all the gold from the Urals, cutting works for precious stones, various private collections, placer gold deposits and a number of local mines and pits. Yet, they also witnessed several mines that 'bore the marks of maladministration' as a contemporary report states.

Considering the mineral wealth in the Urals, each of the following locations provided new challenges for the professionals: the Gumezevsk pit near Gornozit, well known for its rich deposits of malachite-, corundum- and emery-containing marble and serpentine; the copper mines of Nizni-Tagilsk and Tura; the inexhaustable iron supplies from Mount Blagodat near Kuzva;[1] the iron works of Alapajevsk. They also noted various placer deposits of zirconium, sapphires, rubies, pleonaste, topaz, beryllium, octahedrite, amethyst and garnet. Outstanding emeralds were discovered in the Tokovaja mine with the most splendid example weighing in at 2.25 kg; this was transported to Moscow where it remains today in the Kremlin (gems collection). It was fortunate that the first Russian diamond was found and classified during their time in the Urals as von Humboldt could meet the high hopes of the Tsar. When setting out for the expedition he had jokingly assured the empress that he would not 'return to the monarch without any Russian diamonds'.[2]

The following journey eastward was no less exciting. The first stop was Tobolsk, the West Siberian capital at the banks of the River Irtyz. It was planned to be the 'turning point' of the journey, but von Humboldt decided to deviate from the original plan and extend the expedition in the direction of the Altai. This 'small expansion' by 3168 km (2970 werst in units used at the time in Russia)[3] promised considerable scientific gains and above all, raised the possibility of finding some rare flora and fauna. They were further tempted by the opportunity to compare the Altai to the Andes as the rather gentle and monotonous Urals did not hold very much in this respect. Thus they reached the Siberian city of Barnaul, in spite of clouds of insects and many complaints about Siberian plague and anthrax in the Barabinsk steppe. Barnaul was the central trading point for Altai silver. Twice a year, caravans transported 100 pud of silver (16 metric tons) over 4800 km to St Petersburg. Von Humboldt wrote to his brother Wilhelm that they were 'as far east of Berlin as Caracas lies to the west!' (von Humboldt 1880, p. 197). The total output of all mines in the region, however, was ten times higher. Reports, when production was at its highest, show that a miner could extract as much pure silver in one day as fits into a gloved hand. In the Altai itself the group was impressed by the area around Kolyvan with 'granite rocks of most peculiar shape, which suddenly and directly rose from the steppe ... as small, individual altars, others like far away walls and ruins of ancient castles' (Rose 1837–1842, **1**, p. 524).

This description referred to the strange woolsack-style weathering of granites. The description probably followed Hans Michael Renovantz (1744–1798) from Dresden who had been appointed Senior Russian Imperial mining officer of the Kolyvan State (Russisch Kayserlicher Oberbergmeister vom kolywanischen Staat). In 1778, Renovantz was sent off by the Imperial Cabinet to compile a comprehensive report on the ore supplies of the Altai and the Smeinogorsk region. His reserve calculations for the Kolyvan-Voskresensk mining district proved standard-setting and made him famous among scientists. His fundamental publication *Mineralogisch–geographische und andere vermischte Nachrichten von den Altaischen Gebürgen Russisch Kayserlichen Antheils* (*Mineralogical–geographical and further news from the Altai mountains' Russian-imperial share*) (Fig. 2) published in 1788 contained valuable information on the early period of Altai mining. Kolyvan became famous for its outstanding stone cutting trade. Its artistically high quality products, such as splendid jadeite vases and monumental dishes, found their way into St Petersburg's Hermitage, the palaces of the king of Prussia and into the majestic halls of the Tuileries in Paris.

Not far from Kolyvan (Fig. 3) was the silver and gold pit of Smeinogorsk, from which more than 11 tons of silver had already been extracted since 1736. Von Humboldt's expedition continued past Riddersk and Syrjanovsk near the impressive Cholsun mountains, where more mines were visited and geological studies conducted. The impressive Altai mountains, exceeding a height of 4000 m, left a big impression on the travellers and prompted von Humboldt to write to Cancrin: 'The Urals may be of high importance in respect to mining but the real pleasure of an Asian journey was only brought about by the Altai'.[4]

Another important event occurred where the River Irtysh crossed the Russian–Chinese border. Here they met a Chinese border patrol. First they exchanged information, and drank tea in a yurt. Later on Tshing-fu, the highest ranking attending Noion (officer), presented von Humboldt with a hand-signed volume of *The History of the Three Realms*, one of the best known classical Chinese novels. Of the presents given in exchange, the Chinese were surprised by von Humboldt's pencil, a writing implement they had never seen before.

On their westward return, the travellers were stirred by the sight of Kirghisian hordes, a nomadic ethnic minority people on the southern Russian border. The southern Urals formed the last station of the mining-related programme. The mining centres of Miass, Zlatoust and Kyztym were inspected as well as the Ilm mountains to the north that were rich in precious stones. Von Humboldt celebrated his 60th birthday in Miass with a party arranged by the local mining office. Still under the spell of his trip, von Humboldt evaluated the Russian expedition in a letter to Cancrin which stressed its importance. This letter was significant in the way he wanted this journey to be perceived—being of scientific and economic importance in contrast with his American expedition which appealed more to popular culture and science.

The expedition proceeded via Orenburg and Samara; it followed the Volga and passed the German township in Sarepta near Volgograd, and continued on to Astrakhan. This region situated on the Caspian Sea was populated by a number of different ethnic groups, who engaged in viticulture and fishery, oriental crafts and trades. The meetings with Armenian, Persian, Bucharian, and Indian merchants and with Kalmyk, Turkmenian and Kirgisian leaders became special moments for the travellers. Part of the schedule of the research programme was to determine the altitude (which had been disputed) and puzzling variations of the water level of the Caspian Sea, as well as examining the altitude of the whole Caspian Lowlands. It was

Mineralogisch = geographische
und andere vermischte
Nachrichten
von den
Altaischen Gebürgen
Russisch Kayserlichen Antheils
von
H. M. Renovantz,
Russisch Kayserl. Oberbergmeister vom kolywanischen Staat, Inspector und Lehrer der Bergwerkswissenschaften bey der Kayserl. Bergschule zu St. Petersburg, ordentlichen Mitgliede der Societät der Bergkunde, ingleichen der Russisch Kayserl. freyen oeconomischen Gesellschaft, der Kayserl. Academie der Wissenschaften zu St. Petersburg Correspondenten.

Mit Kupfern.

Reval, 1788.
gedruckt mit Lindforsischen Schriften auf Kosten des Verfassers.

Fig. 2. Title page of Renovantz's 1788 book about the Altai.

Fig. 3. Wool sack weathering at Kolyvan Sea (recent photograph).

considered that the results might yield geotectonic (as was understood at that time) and geographical information. Von Humboldt reported to his brother: 'It is a highlight of my life to have seen this inland sea with my own eyes and to have collected its artefacts'.[5] Finally, he travelled back to St Petersburg via Caricyn, Voronez, Tula and Moscow.

The intermediate stop in Moscow turned out to be yet another high point, as the Society of Natural Scientists, which was part of the University, felt obliged to organize an especially grandiose reception for the *Prométhée de nos jours*. The participants were suitably chosen: high-ranking and highly decorated politicians as well as the honourable group of Professors. Not lacking a certain touch of irony, the well-known Russian publicist Alexander I. Herzen (1812–1870) described the scene:

This reception of von Humboldt in Moscow and at the University was truly no small change. The General Governor, several military and civil authorities, the Senate—they all turned up: ribbons over their shoulders, full uniform, the professors belligerently marching in step with a tricorne under the arm. Von Humboldt had not anticipated all this. He had turned up in a blue tailcoat with golden buttons and he was quite stunned. Traps had been set out for him right from the entrance corridor towards the Society of Natural Scientists hall: a principal here, a dean there, a newly appointed professor here, a veteran there, who had already finished his work and was therefore speaking rather slowly. Each one greeted von Humboldt in Latin, German, French—and all this within these terrible stony tubes they call corridors, in which you can't stop for a minute without catching the flu for a month. Von Humboldt listened to all of them attentively and answered every question. I am convinced that all the savages he had encountered, no matter if red- or copper-coloured, had not caused him as much distress as this Moscow reception did. Von Humboldt wanted to talk about his observations with the magnetic needle, compare meteorological notes he had made in the Urals to Moscow data. Instead, the Principal was about to show him something plaited out of Peter I's hair. Ehrenberg and Rose finally managed to find an opportunity to talk about their discoveries.[6]

Von Humboldt had hardly any opportunity to make contact with his Russian colleagues, as he was constantly being bothered by state authorities or had to dedicate his attention to German scientists working in Moscow. The event was also taken as an opportunity to grant Rose and Ehrenberg high honours: they were awarded the St Anne Medal, 2nd class, for their contributions to research in the Russian Empire.

Leaving Moscow, they travelled on to St Petersburg which they reached on 13 November 1829; this date marks the end of the expedition. They remained for four weeks before they returned to Berlin, and the time was filled with various social obligations which turned out to be the climax of the social scene in the Tsar's capital city. Von Humboldt, who had been awarded the St Anne Medal, 1st class, took the opportunity to expound the results of his expedition at the Petersburg Academy in front of an invited audience. He explained the motivation of an explorer and developed theories on how science and technology in Russia should be continued. Political problems were not supposed to be a topic of these lectures, and prior to embarking on the expedition, von Humboldt had promised to refrain from discussing such matters. In spite of this, he used his audience with the Tsar to support the causes of people who had either been exiled for political reasons or who were victims of persecution. The group of explorers finally left St Petersburg on 15 December and arrived in Berlin on 28 December 1829.

Scientific results of the 1829 expedition

The epic journey was impressive, first of all, by its sheer dimensions. In a period of only twenty-three weeks they covered 15 468 km (14 500 werst) in Russia, on foot, on horseback, by coach and by ship. Including the Berlin to St Petersburg leg, the total distance travelled amounted to approximately 19 000 km. On average, they travelled approximately 300 km per day, and their horses were changed every 20 to 30 km. In total 568 stations were passed at which horses were switched 12 244 times. There were 53 river crossings, among them the Volga ten times, the Kama two times, the Irtyz eight times and the River Ob twice.

Originally, the Russian government had planned to provide 10 000 rubles, but had doubled the amount when they realized von Humboldt's poor financial situation. At the end of the journey, 7050 rubles were left which von Humboldt generously donated for an expedition to the Urals which was at the planning stage. Within one year such a journey was started by Gregor von Helmersen and Ernst Hofmann (1801–1871). As a symbolic gesture the Tsar presented von Humboldt with a sable fur and a vase made of Aventurine some 2 m high, that was later displayed in the Royal castle in Berlin.

With this journey over, von Humboldt's ambitions had been fulfilled and he was at the height of his fame. However, the processing of the findings and scientific results of the journey had to be finished. The material was intended to form the basis of his planned 'outline of a physical survey of the world' which would be published as *Kosmos*. Therefore, the earlier explorations in Asia and America demanded a balanced and detailed comparison. The expectations of the scientific world provided a further impetus, as more than any other previous expedition to the eastern hemisphere before, von Humboldt's journey was regarded as exemplary. The acclaim of the international community was high, and a reviewer in the Göttingen periodical *Gelehrten Anzeigen* (*Scholars Review*) wrote in 1833:

> The journey undertaken by A. von Humboldt through Siberia and to the Caspian sea, accompanied by Mr. Ehrenberg and Mr. Rose, belongs undoubtedly to the most noticeably scientific enterprises of recent times. Never before has a voyage of such dimensions, exclusively committed to scientific purposes, been completed in such a short time. And never before has such a fast expedition produced results of such high value. This could only be accomplished by a person able to combine an excellent sense of observation and deduction with the vast pool of his own knowledge and expertise. But even being bestowed with such a high degree of dedication would not have been sufficient to achieve those major goals of von Humboldt's undertaking. It took Imperial support and two talented and wise companions to participate in the observations and examinations, it also took the eagerness of authorities and privateers to assist the celebrated explorer in every possible way (Schmidt 1924, pp. 195–196).

Von Humboldt first attended to Russian mining. He criticized the lack of a workforce and hence low productivity, reprimanded the system of serfdom and recommended that they use free labour and to raise the 'human state'. Many proposals proved to be sensitive since they touched the social order in Russia, and his suggestions fell on deaf ears or were buried beneath bureaucratic obstacles as the feudal upper class was naturally not supportive of substantial reforms. He also had to take into consideration the family bonds between the Russian and Prussian monarchs that did not allow for any harsh criticism. Von Humboldt had to harbour his criticism of the Russian social order in silence. During the expedition he had corresponded with Cancrin about this sensitive matter, and his writings published after the journey also showed how he considered the plight of mine workers.

Fortunately, there were no such obstacles in the field of science, and Rose and Ehrenberg were actively encouraged to analyse and evaluate the expedition for its promised exceptional scientific results. Von Humboldt started work immediately and in 1831 published *Fragments de géologie et de climatologie asiatiques* in two octavos, which was also published in a German edition later that year. The much more comprehensive travelogue *Asie centrale ... Recherches sur les chaines des montagnes et la climatologie comparée; par A. von Humboldt* came out in three volumes in

1843 in Paris. It consisted of 1800 pages, various charts and one map. The German edition appeared in Berlin in 1844 as *Central-Asien. Untersuchungen über die Gebirgsketten und die vergleichende Klimatologie von A. Humboldt.* The translator and editor Wilhelm Mahlmann (1812–1848) in his preface praised von Humboldt's 'comprehensive abilities in all areas of human knowledge' and, above all, his talent to 'penetrate the intertwined branches of natural science from different angles, to understand the influence of nature on the life and fate of peoples and to set it all out in an easily understandable way not thought possible'. Furthermore, he noted that von Humboldt had the rare gift to 'collect, look through and systemize a chaos of facts and then to raise them to universal ideas and notions wherein all details met as rays do in a focal point' (von Humboldt 1844, pp. xi–xii).

This work, too, became a literary success in which von Humboldt gave a description of the physiography of the Russian continent, and kept his promise to 'subserve science, especially geology and the current fruitful discipline of geomagnetism'.[7] With this publication, he provided an overview of all the regions that the expedition had visited. He interpreted different natural conditions, and addressed particular problems of orography, morphogenesis, altimetry,[8] hydrology, meteorology, climatology, and geomagnetism. His examination of the physical topography and the subsequent description of it provided important data for the development of a countrywide network of meteorological and geomagnetical stations. The data collected from these stations since their establishment to the present day have proved extremely valuable in understanding recent climatic changes. Considering and acknowledging all his achievements, it seems legitimate to label von Humboldt as a pioneer of geomagnetism.

The actual summary of the expedition was written by Rose and published between 1837 and 1842 in two volumes entitled *Reise nach dem Ural, dem Altai und dem Kaspischen Meere auf Befehl Sr. Majestät des Kaisers von Russland im Jahre 1829 ausgeführt von A. Humboldt, G. Ehrenberg und G. Rose. Mineralogisch-geognostischer Theil und historischer Bericht der Reise* (Fig. 4).[9]

Shortly after the journey, Rose began to summarize a number of mineralogical details for the series *Annalen der Physik und Chemie* edited by Johann Christian Poggendorff (1796–1877). This series was published in ten parts from 1824 onwards and allowed Rose to begin compiling the results of the expedition. The diaries of all the people who undertook the expedition, which contained the geognostic, astronomical, magnetic and meteorological observations, formed the basis for these publications, and were augmented with information gleaned from the maps, books and manuscripts that had been given to them by the Russian authorities.

Rare mineral specimens and rock samples had comprised a valuable part of their baggage or were sent on later to Berlin. They were packed in crates and cases and were ultimately destined for the Royal Collection in Berlin. Today they are on display in the mineral collection of the Berlin Museum of Natural History. Since there had been no access to any laboratory equipment during the expedition, chemical and crystallographic analysis had to be carried out in Berlin so that precise identifications of the specimens could be made. Rose involved not only his students but also his brother Heinrich Rose (1798–1873) in this extensive work. They compared their material with that described in older travelogues such as those

Reise

nach dem

URAL, DEM ALTAI

und dem

KASPISCHEN MEERE

auf Befehl Sr. Majestät des Kaisers von Russland im Jahre 1829 ausgeführt

von

A. von Humboldt,
G. Ehrenberg und G. Rose.

Mineralogisch - geognostischer Theil und historischer Bericht der Reise

von

G. Rose.

Berlin, 1837.

Verlag der Sanderschen Buchhandlung.
(C. W. Eichhoff.)

Fig. 4. Title page of Rose's 1834 book about Urals and Altai.

written by Pallas or Gmelin (Pallas 1771–1776, 1784–1831, 1811–1831; Gmelin 1770), as well as in the latest issues of the Russian mining journal *Russisches Bergwerks-Journal*. Rose and his colleagues compiled a travelogue that contained so much detailed and comprehensive mineralogical data of the travelled regions, that some scientists regarded it as an official mineralogy of Russia.

Rose described 93 mineral families and 110 species in his *Systematic survey of the Urals minerals and rock formations* (*Systematische Übersicht der Mineralien und Gebirgsarten des Urals*) that was published in 1842. Among them were eleven previously unknown minerals; four of which he gave names to honour people he had met in Russia who had been most supportive of the expedition; Perovskite (after Vice President Lev Aleksejevic von Perovski (1793–1866) of St Petersburg), Tscheffkinite (Konstantin Vladimorovic Cevkin was commander of the Imperial Mountain corps in St Petersburg), Mengite (discovered by J. Menge of Miass, southern Urals) and finally Cancrinite (after Cancrin about whom Rose wrote in 1839: 'I am pleased to have the honour to enter the name of such an outstanding statesman into mineralogy, a man who has brilliantly rendered Russian mining and mineralogy a great service') (Rose 1837–1842, **1**, p. 524).

The majority of the mineral and rock samples collected by Rose during the expedition are today in the Berlin Museum of Natural Sciences and form a valuable part of the holdings of Russian material in the museum. Those collections acquired earlier include the so-called 'Russian Collection' of 3081 specimens that had been presented to the Royal mineralogical cabinet on the occasion of a state visit by Tsar Nikolaus I to Berlin in 1803; the collection of Oberbergrat August Friedrich Alexander von Eversmann (1759–1837) who had mainly collected in the Zlatoust area (southern Urals) (Hermann 1790); approximately 100 rock samples from the Jekaterinburg region which were presented by Berghauptmann B. F. J. von Hermann who had also written an early description of the Urals; and various minerals and stones from the Urals that Menge had collected during his Siberian journey in 1825 and 1826.

Rose had studied this material intensively prior to embarking on the expedition and had acquired a good knowledge of the geology and mineralogy of the areas he was due to visit. This special collection in Berlin was later to be enriched by the samples of our expedition and additional specimens. The 1829 collection demonstrates the early scientific co-operation between Russia and Germany and it documents the great value of cross-border research (see also Rose 1837–1842).

Along with mining and metallurgical information von Humboldt and Rose gave advice on how to improve the technological processes and conditions of the Russian mining and extractive industries. The importance of this part of the expedition to the Russian Crown can be seen from the fact that it had sent out the mining official Nicolai Stepanovic Mensenin to accompany von Humboldt and Rose. Mensenin had acquired a reputation of being a mining expert and had published a number of scientific works on mining technology and history, both theoretical and applied. A man of some ability he was of great use to the expedition as he spoke German and French. A short anonymous contemporary report, although certainly authored by him, and entitled *On the Journey of Mr. von Humboldt in Russia* appeared in *Gorny Zurnal* in 1830 (Mensenin, N. S. 1830). This was adapted from his official report submitted to Cancrin, his superior.

Whereas the scientific results in the fields of geology and mining and in geography and cartography had met von Humboldt's expectations, the botanical and zoological results were less satisfying. One reason may have been the 'tiring monotony' (von Humboldt) of the Urals. It was only later on that the 'beautiful and copious flora of the Altai' (von Humboldt) offered some compensation for the paucity of the former. Ehrenberg's main task lay in the identification of plants and animals, and in palaeontological studies. He brought a substantial herbarium and a zoological collection back from Russia. Parts of this collection ended up in various institutions including the Museum für Naturkunde of the Berlin von Humboldt-Universität, the Botanical Museum of the Freie Universität Berlin, the Musée National d'histoire Naturelle in Paris and the Berlin-Brandenburgischen Akademie der Wissenschaften.

For von Humboldt himself, all the collections were of value in so far as they gave him an abundance of information that he could use in supra-regional as well as global and continental descriptions and comparisons, in other words it provided the means for him to see the breath of nature in the whole. Ehrenberg concentrated on his Siberian studies by producing an account of the distribution of different species of infusoriae (microscopic animals). Elaborating on this he later made significant advances in the study of micro-organisms and is often regarded today as the founder of microbiology and micropalaeontology. His publications relating to the expedition included a special study of the Siberian tiger and his compendium of the *Mikrogeologie* (Ehrenberg 1873–1875).

Von Humboldt was regarded very highly for his 1829 Russian–Siberian expedition; even higher than for his earlier expedition to the Americas which was longer. In a letter to Cancrin dated 15 September 1829, he wrote: 'This year has become the most important of my restless life' (von Humboldt 1869, p. 92). The scientific results of this journey may also be viewed as providing a substantial body of data that was later used by him in his great work, *Kosmos* (Fig. 5), in particular in the portion entitled *Outline of a physical survey of the World*.

The 1829 journey also paved the way for expeditions to come, for instance those undertaken by the brothers Adolf (1829–1857), Hermann (1826–1882) and Robert Schlagintweit (1833–1885) in the years 1854 to 1857 through India and the Himalayas. Sadly, von Humboldt was never able to travel through Central Asia, but he nevertheless took the opportunity to compare the results of his studies and information known from the great Asiatic mountain ranges in his *Asie centrale* (von Humboldt 1844, pp. xi–xii). Many of his conclusions in this volume were hypothetical but he

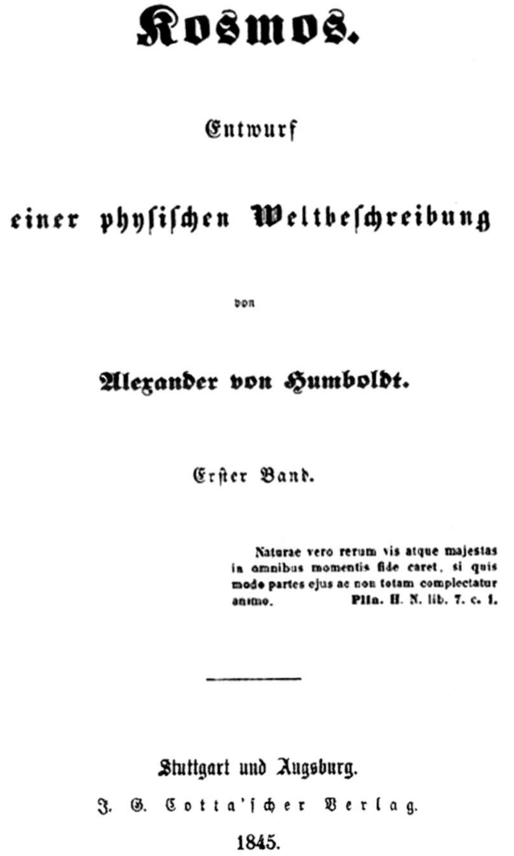

Kosmos.

Entwurf

einer physischen Weltbeschreibung

von

Alexander von Humboldt.

Erster Band.

Naturae vero rerum vis atque majestas in omnibus momentis fide caret, si quis modo partes ejus ac non totam complectatur animo. Plin. H. N. lib. 7. c. 1.

Stuttgart und Augsburg.

J. G. Cotta'scher Verlag.

1845.

Fig. 5. Title page of von Humboldt's great work *Kosmos* (1845).

undisputedly contributed significantly to the exploration of central Asia relief. Until the end of the nineteenth century, his work and the *Geology of Asia*, written by the Berlin geographer Carl Ritter (1779–1859) and published between 1832 and 1849, remained the only comprehensive accounts of physiography of the Asian continent.

Finally, the scientific contacts with Russia made during the 1829 expedition formed the basis of a prolific bilateral relationship that continued until von Humboldt's death. Since Russian science was still in need of foreign assistance, von Humboldt felt a responsibility to help. He continued to give support to prospective young scientists, retained his membership of related societies, wrote various editions of scientific books and assisted in establishing a countrywide system of meteorological and geomagnetical stations. He was also instrumental in the foundation of the central observatory for physics, the Physikalisches Zentralobservatorium, in St Petersburg in 1849. He remained in contact with Cancrin, who continued to apply von Humboldt's proposals to the benefit of his country (Suckow 1999).

Ongoing assessment of von Humboldt's work in Russia

Russian–German scientific relations date back to the eighteenth century. More than almost anything else, the contributions of German researchers in Russia are worth recalling. Von Humboldt stands out because he became an intermediary between different cultures, and promoted international cooperation at an early stage. Furthermore, humanism, active peace policies, ecological thinking and responsible use of precious natural resources were high on the agenda of his life and work. Von Humboldt's work in these areas requires reassessment in the light of a radically altered world from that in which he travelled, and this has not yet been done sufficiently. Numerous archives, as well as the letters exchanged between von Humboldt and correspondents from all over the world, require evaluation and interpretation. Some important contributions to this work have been made by both the Alexander-von-Humboldt-Forschungsstelle der Belin-Brandenburgischen Akademie der Wissen schaften (Alexander-von-Humboldt Scientific Research Centre, part of the Berlin-Brandenburg Academy of Sciences) and some historians and scientists.

A recent initiative in this reassessment was undertaken by the German Association of Graduates and Friends of Moscow Lomonossow-University (DAMU) which is based in Berlin. Since 1993, this association has been working hard to evaluate and document the work of German scientists in Russia in order to revive the traditions of German–Russian relationships. A retracing of the 1829 expedition is an important element in this research, and underlies a research programme that has spanned several years. Named *On Alexander-von-Humboldt's trail in Russia* the research project aims to collect authentic material at the places visited and described by von Humboldt, on the people he met, and the scientific and social environments and their effects. So far researchers engaged in this work have visited places, localities and buildings described by von Humboldt (if they still exist), and have gathered photographic and other documentation about them. They have located and recorded relevant original documents in museums and archives, have visited libraries such as that in Novosibirsk which holds large stocks of German specialized literature, and have carried out landscape, vegetation, botanical, ethnographic and economic studies of relevant regions.

So far, five expeditions have been undertaken during the course of this work to the following areas: Altai (1994), Urals (1995), south of Russia (1997), western Siberia (1999) and the Urals (2002).

Scientific colloquia held in Berlin after the first three expeditions have provided a forum for the presentation of the scientific results. During the fourth expedition, scientific presentations were given in Novosibirsk, Omsk and Barnaul in order to exchange experiences and results with local scientists on von Humboldt and to stimulate regional research programmes. In the course of the presentations, not only was von Humboldt's journey discussed, but also contemporary national, political and ecological problems. The idea of connecting people has been fruitful, and all participants of the expeditions have had the opportunity to gain insights into most recent ethnic, cultural, scientific and political problems of the regions that von Humboldt travelled through.

The last expedition in 2002 undertaken during UNESCO'S designated 'Year of the Mountains', took place under its protection. It started with a scientific conference in Jekaterinburg and later moved to regions in the northern and southern

Fig. 6. DAMU Expedition group on the Mount von Humboldt in the Northern Urals 2002.

Urals. A very special occasion was the first ascent of Mount von Humboldt (Fig. 6). On a joint initiative of DAMU and the Sverdlovsk branch of the Russian Geographic Society, and the subsequent decision of the Russian Government in May 2000, this 1410 m high peak in the northern Urals was named after Alexander von Humboldt. It is now a permanent reminder of an outstanding German scholar and his important role in the exploration of Russia.

Notes

[1]A memorial used to exist on the top of that mountain, commemorating the Vogul Cumpin, who had pointed out this mountain to the Russians. But instead of being rewarded as had been promised, he was burned alive as a traitor by his compatriots.

[2]The diamond was incorporated in the Royal Mineralogical Collection in Berlin. Currently it is displayed in the cabinet of the Berlin Museum of Natural Sciences (Naturkundemusem) that is dedicated to Gustav Rose and the Siberian journey.

[3]1 pud = 16.58 kg; 1 werst = 1.06678 km.

[4]Von Humboldt to Cancrin, Omsk 27 August 1829 (von Humboldt 1869). Berlin 10 January 1829, p. 88.

[5]Letters Alexander's von Humboldt, a. a. O., p. 202.

[6]From a description by the well-known Russian publicist Alexander Herzen. Quoted in Anon. [Mensenin, N. S.] (1830).

[7]From the homage to His Highness, the Emperor of Russia (see von Humboldt 1844, Preface).

[8]Measure of height (lat. *altum*-height, greek *metrein*—measure).

[9]Translation: *Journey to the Urals, the Altai and the Caspian Sea on orders of His Highness the Emperor of Russia from the year 1829 executed by A. von Humboldt, G. Ehrenberg and G. Rose. Mineralogical-geognostical part and historical account of the Journey.*

References

ANON. [MENSENIN, N. S.] 1830. On the Journey of Mr. von Humboldt in Russia. *Gorny Zurnal* **2**, 229–263.

BECK, H. 1961. *Alexander von Humboldt*. Franz Steiner Verlag, Wiesbaden.

EHRENBERG, C. G. 1873–1875. *Über die seit 27 Jahren noch wohl erhaltenen Organisationspräparate des mikroskopischen Lebens. Berlin 1862. And: Mikrogeologische Studien, über das kleinste Leben der Meeres-Tiefgründe aller Zonen und dessen geologischen Einfluß und Fortsetzung der mikrogeologischen Studien als Gesamt-Uebersicht der mikroskopischen Palaeontologie gleichartig analysirter Gebirgsarten der Erde und specieller Rücksicht auf den Poycystinen-Mergel von Barbados*. 2 Volumes. Voss Gmelin, Berlin.

GMELIN, S. G. 1770. *Reise durch Rußland zur Untersuchung der drey Natur-Reiche*. Kayserlichen Academie der Wissenschaften, St. Petersburg.

HERMANN, B. F. J. 1790. *Statistische Schilderung von Rußland, in Rücksicht auf Bevölkerung, Landesbeschaffenheit, Naturprodukte, Landwirthschaft, Bergbau, Manufakturen und Handel*. Kayserlichen Academie der Wissenschaften, St. Petersburg. St. Petersburg.

HUMBOLDT, A. VON. 1844. *Central-Asien. Untersuchungen über die Gebirgsketten und die vergleichende Klimatologie*. Carl J. Klemann, Berlin.

HUMBOLDT, A. VON. 1859. *Reise in die Aequinoctial-Gegenden des neuen Continents*. Stuttgart, 1859.

HUMBOLDT, A. VON. 1869. *Im Ural und Altai. Briefwechsel zwischen Alexander von Humboldt und Graf Georg von Cancrin aus den Jahren 1827–1832*. F. A. Brockhaus, Leipzig.

HUMBOLDT, A. VON. 1880. *Briefe Alexander's von Humboldt an seinen Bruder Wilhelm. Hrsg. v. d. Familie von Humboldt in Ottmachau*. J.G. Cotta, Stuttgart.

HUMBOLDT, A. VON. 1987. *Aus meinem Leben. Autobiographische Bekenntnisse*. Urania-Verlag, Leipzig.

HUMBOLDT, A. VON., HAMEL, J. VON., TIEMANN, K.-H. & PAPE, M. (eds) 1993. *Über das Universum. Die Kosmosvorträge 1827/28 in der Berliner Singakademie*. Frankfurt-am-Main and Leipzig.

KLETKE, H. 1876. *Alexander von Humboldt's Leben und Wirken, Reisen und Wissen. Ein biographisches Denkmal*. Lipsia, Leipzig.

PALLAS, P. S. 1771–1776. *Reise durch verschiedene Provinzen des russischen Reiches*. 3 volumes. Kayserlichen Academie der Wissenschaften, St Petersburg.

PALLAS, P. S. 1784–1788. *Flora Rossica*. Kayserlichen Academie der Wissenschaften, St Petersburg [later editions to 1831].

PALLAS, P. S. 1811–1831. *Zoographia Rosso-Asiatica*. Kayserlichen Academie der Wissenschaften, St Petersburg.

ROSE, G. 1837–1842. *Reise nach dem Ural, dem Altai und dem Kaspischen Meere auf Befehl Sr. Majestät des Kaisers von Russland im Jahre 1829, ausgeführt von A. Humboldt, G. Ehrenberg und G. Rose. Mineralogisch-geognostischer Theil und historischer Bericht der Reise*. 2 Volumes. Verlag der Sanderschen Buchhandlung, Berlin.

SCHMIDT, C. W. 1924. *Alexander von Humboldt. Sein Leben. Seine Werke*. Die Buchgemeinde Verlags GmbH, Berlin.

SUCKOW, C. 1999. Dieses Jahr ist mir das wichtigste meines unruhigen Lebens geworden. Alexander von Humboldt's russisch-sibirische Reise im Jahre 1829. *In*: (ed.) *Alexander von Humboldt—Netzwerke des Wissens*. Kunst- und Ausstellungshalle der Bundesrepublik Deutschland, Bonn, 161–177.

Hermann Abich (1806–1886): 'the Father of Caucasian Geology' and his travels in the Caucasus and Armenian Highlands

E. E. MILANOVSKY

Faculty of Geology, Moscow State University, 119899 Moscow, Russia (e-mail: milan@geol.msu.ru)

Abstract: Hermann Willhelm Abich (1806–1886) spent more than thirty years travelling and investigating the geology of the Caucasus, Transcaucasian and Armenian highlands. He was appointed Professor of Mineralogy at Dorpat University but soon afterwards began his travels. He investigated the volcanic processes of this region to determine the reasons for its tectonic and seismic instability. He later focused on Georgia with its coal deposits, and then studied the oil and gas fields and mud volcanoes on the Apsheron peninsula near Baku in Azerbaijan. He also studied the effects of the last glaciation on the topography of the region.

The famous nineteenth century European geologist Hermann Willhelm Abich (1806–1886)[1] was born and educated in Germany. Almost half of his life was spent in Russia and during more than 30 years he travelled tirelessly through different regions of the Caucasus, Transcaucasian and Armenian highlands, where he gathered data on their geological structure and development. He made a considerable contribution to our knowledge of the geology of the Caucasus, and his Russian contemporaries named him 'the father of Caucasian Geology'.

Hermann Abich: early life and geological education

Hermann Abich (Fig. 1) was born on 11 December 1806 in Berlin, the capital of Prussia, to Willhelm Abich, a mining engineer, and his wife—the daughter of the outstanding German chemist Martin Klaproth (1743–1817). Her brother, Hermann's uncle, Heinrich Julius Klaproth (1783–1835) was a well-known traveller, orientalist, ethnographer and expert in the knowledge of Caucasian peoples. He was responsible for stimulating his nephew's interest in the history and the natural history of Caucasus, while Albich's father fostered his son's interest in geology and the mining industry.

At first, Hermann entered the law faculty of Heidelberg University, but after two years he realized his mistake and transferred into the philosophical faculty of Berlin University in order to study geology, mineralogy, geography and chemistry. There he came under the guidance of some great naturalists, particularly the geologist Leopold von Buch (1774–1853), the geologist and geographer Alexander von Humboldt (1769–1859) and the geographer Karl Ritter (1779–1859). He also studied philosophy and history, attending the lectures of Georg Hegel (1770–1831) and Leopold Ranke (1795–1886) respectively. The geological ideas of von Buch and von Humboldt impressed him, and he was especially interested in their theory on the craters of uplift. In order to graduate from Berlin University in 1831 Abich had to defend his doctoral thesis in Latin: '*De spinello*', which had become the basis for four monographs and papers (Abich 1831) published in German, Latin, French and Russian between 1831 and 1834.

Following the recommendation of von Buch, in 1831 Abich left for Italy and during the following three years studied active and extinct volcanoes, including Vesuvius, Stromboli on the Aeolian Islands in the Tyrrhenian Sea, and Etna on Sicily. He later described their structure, volcanic activity and history in a special monograph issued in 1841 in Braunschweig (Abich 1841) and in a series of ten papers, published in German, French and Swedish between 1835 and 1843.

Travels in the Caucasus

At the end of the 1830s, Abich, who was already well known as an excellent geologist and mineralogist in western Europe, was invited by the Russian government to become a Professor of Mineralogy at Dorpat University (= Tartu) in what is now Estonia. This was the only university in Russia where the training was carried out in the German language. But he soon moved when in 1840 when a large earthquake struck the region around the Ararat volcano in the Armenian volcanic highland. This seismically-active region is centred on the borders between Russia, Persia (Iran) and Turkey, and the staff of the Russian Corps of Mining

From: WYSE JACKSON, P. N. (ed.) *Four Centuries of Geological Travel: The Search for Knowledge on Foot, Bicycle, Sledge and Camel.* Geological Society, London, Special Publications, **287**, 177–181.
DOI: 10.1144/SP287.14 0305-8719/07/$15.00

Fig. 1. Hermann Willhelm Abich (1806–1886).

Engineers invited Abich to determine the nature and cause of this tectonic instability which was thought to be either seismic or volcanic. In 1843, after a period of serious scientific preparation Abich compiled a special report that contained some new geological data on the region, as well as recommendations regarding the formation of a scientific expedition to study the Transcaucasian volcanic formations, an expedition in which he was prepared to participate. At the same time, he delivered a lecture at Dorpat University entitled 'On the geological structure of Armenian Highland' where he summarized the new facts and ideas of his colleagues on the geology of Caucasus, and it was published soon afterwards in the form of small book (Abich 1843). In 1844, the Russian government sent Abich on an official trip to study the geological event of 1840 that was centred on Ararat.

During 1844 and 1845 he ascended Ararat volcano several times (the last visit was particularly successful), and he was so strongly carried away by a desire to continue his field studies of the geology of the Transcaucasian and Caucasian area, that, instead of returning to Dorpat University, he decided to devote his remaining life to travelling in the Caucasus. At that time, this extensive and beautiful mountainous region was still unexplored geologically.

During the next 30 years, until his resignation in 1876, he travelled in many different parts of the region from the coasts of the Black Sea and Sea of Azov in the west to the coast of the Caspian Sea in the east.

The results of his field observations and his scientific ideas concerning different aspects of Caucasian geology were described in more than 200 publications, including some monographs that appeared in Russian, German and many other languages.

At the beginning of the 1840s, when Abich commenced his research in the Caucasus, from the point of view of geology, its vast territory represented a '*tabula rasa*'. Many places, particularly the western and eastern parts of the Great Caucasus, which were populated with freedom-loving Muslim peoples and tribes, were the locations of longstanding military conflicts between them and troops of the Russian Government, and so were inaccessible to scientists. However, it was possible to enter from the North Caucasian plain into the Transcaucasian regions of Georgia, Azerbaijan and Armenian highlands by using the military-Georgian road that crossed the Great Caucasian mountains between Vladikavkaz in the north and Tiflis in the south. Another route took travellers round the western and eastern ends of Great Caucasian ridge along the coasts of Azov, Black and Caspian seas.

After more than 30 years of travelling and geological research, much of which had been undertaken by Abich himself, almost all parts of the Caucasus had been described geologically.

The first period of Abich's geological journeys through the Caucasus began in 1844, when he travelled from St Petersburg to the North Caucasus. There he was astonished by the beauty and grandeur of the main Caucasian ridge, which he crossed between Vladikavkaz and Tiflis, and went through Georgia and Armenia to the Great Ararat volcano. This long expedition was completed in 1850 when he returned to St Petersburg. After research on the volcano of Mount Ararat, Abich studied other extinct volcanoes including Alagez (Aragats) situated in the Armenian highlands (Fig. 2). He then focused on the central part of Georgia with its coal deposits, and on Black Sea coast of Georgia. Finally, he studied the eastern part of Azerbaijan, including the oil and gas fields and mud volcanoes on the Apsheron peninsula near Baku.

Because of his very long absence from Estonia, Abich was removed from the staff of Dorpat University; but instead he was employed by the Corps of Mining Engineers, which was under the control of the Caucasian Deputy Count M. S. Vorontsov, who held Abich's scientific and practical capabilities in high regard. He had written to von Humboldt about Abich: 'He was very useful for us not only as the author of fundamental scientific works, but also in practical relation

Fig. 2. Canyon-like valley of Aparan river on the south-eastern slope of Alagez (Aragats) volcano near Echmiadzin in the Armenian highlands. Drawing by H. Abich (from Abich 1882).

for the satisfaction of needs of our country, particularly for successful search of coal, peat, salt, etc.' In 1849, Abich in the company of Russian troops commanded by General Eristavi, crossed the Great Caucasian ridge from the south to the NW to the highest mountain of Caucasus, Elbrus volcano (5.6 km high) and there he began his geological research on the northern slopes of the Great Caucasus.

The first region which he had studied was dotted with young laccoliths (Beshtau and many others) that were intruded into Tertiary sedimentary rocks. Numerous springs of curative mineral waters, on which were centred several famous spas, were known as the 'Caucasian Mineral waters'. During his first geological expedition Abich together with his companion, the mining engineer V. V. Sokolov, completed a major study of the stratigraphy, tectonics, magmatism, mineral deposits, as well as the geography of many earlier unknown Caucasian regions. He described their results in about 30 papers published in German and French journals that were systematically translated into Russian by the geologist A. D. Ozersky. Ozersky was responsible for translating into Russian Roderick Impey Murchison's (1792–1871) classic volume on the geology of Russia, which was published in *Gorny Journal* and in other Russian editions. Abich also regularly informed the Corps of Mining Engineers on the progress of his Caucasian work and to their headquarters in St Petersburg and in Breslau sent many dozens of boxes containing rocks, minerals and fossils for chemical, mineralogical and palaeontological investigation.

In August 1850, after his return to St Petersburg, Abich temporarily left Russia and spent 10 months travelling in Germany, Belgium, France and England. He gave lectures and engaged in interesting and successful discussions in scientific societies, and met with outstanding European geologists including his beloved teacher and friend von Humboldt. In order to help his talented pupil and disciple, von Humboldt persistently applied to the Head of Staff of the Corps of Mining Engineers of Russia asking that Abich be given a long period of leave and a grant to enable the publication of his principal scientific work: 'Extensive H. Abich's works in the Caucasus and Armenia deserved acknowledgement everywhere, and Mr. Abich had occupied one of the primary places amongst European geologists'. After waiting two years von Humboldt's request was granted.

On Abich's return to Russia in May 1851 he visited St Petersburg and shortly after that returned to the Caucasus where he remained until 1853. In this short period he studied the Tertiary deposits in the basin of Sea of Azov (from the city of Taganrog in the north to Taman peninsula in the south). Then he went to Tiflis where he was obliged to spend several months because of a long illness. In the spring of 1852 he resumed his geological study of Kertch peninsula in the eastern part of Crimea and then went again to Tiflis and further through Armenia up to Persian boundary. During this journey he established and mapped most of the Tertiary deposits in the Transcaucasus.

In the spring of 1853 he left Tiflis for a long trip to western Europe, in order to continue his medical treatment and to make contact with his foreign scientific colleagues. Soon afterwards, he learned that the St Petersburg Academy of Sciences had elected him a full ordinary member in recognition of his valuable geographical and geological investigations in the Caucasus.

After his return to Russia in 1854 he was sent to the Caucasus for two more years. However, because he had to prepare his great work *Comparative Geological Analysis of Main Features of Caucasian, Armenian and North-Persian Mountains as a Basis of Geology of Caucasian Countries* for publication (the so called '*Prodromus*') (Abich 1859) he was forced to spend a great deal of time away from Tiflis in St Petersburg. Finally in 1857, he was obliged to move from Tiflis to the capital of Russia. '*Prodromus*' was published in 1859 and represented the first fundamental general and systematic description of the relief and geology of the Caucasus region. This work was critically acclaimed by many foreign and Russian geologists, in particular, by Professor G. E. Tshurovsky, the head of geological school of Moscow University, who discussed it in his book *Geological Essay of Caucasus* in 1872. Soon after his move to St Petersburg, Abich married Adelaide Hess, the daughter of the famous Russian chemist and member of St Petersburg Academy of Science, H.I. Hess.

After completion of his '*Prodromus*', which had taken him four years to prepare (1854–1858), he decided in spite of his age to embark on yet another expedition to the Caucasus. This can be considered his third expedition to the area. He explored the Northern slope of the Great Caucasus (the area between Elbrus and Cazbek volcanoes with Mesozoic and Cenozoic rocks, and springs in the Caucasian Mineral Waters region); examined the southern slope of the Great Caucasus (Ratcha, Lechkhum, Svanetia, Shemacha) and its western and eastern ends (Kertch, Taman and Apsheron peninsulas); and spent time in the Minor Caucasus and on the Armenian volcanic highland, where he examined the volcanoes of Great and Small Ararat. This period of travel was extensive and prolonged, with the exception of some short interludes, and lasted eighteen years until 1875. In 1866 at the age of sixty he was elected an honorary member of St Petersburg Academy of Sciences. He was also a member of many foreign and Russian scientific societies, including the Moscow Naturalist Society, which was the oldest such body in Russia. Abich's creative scientific activity had not diminished with age. During these eighteen years he had published about 120 original new papers and books, all of which demonstrated continued and sustained scholarship.

The last years in Vienna

In 1876, Abich retired on a full pension from his geological job after thirty-five years. He left St Petersburg and spent the last ten years of his long and fruitful life in Vienna, the capital of Austro-Hungarian Empire. This city was one of main centres of European geology, and he intended to devote his retirement to the preparation and publication of a final synthesis of his work in the Caucasus.

Unfortunately, this programme was only partly realized. The first two volumes of this series entitled *Geological Explorations in Caucasian Countries* were published (Abich 1878, 1882) during his lifetime. For these he was presented with the Konstantin gold medal of Russian Geographic Society. The third volume was published posthumously in 1887 and contained a short biography of Abich written by his wife Adelaide and his long-standing friend the great Austrian geologist Eduard Suess (1831–1914) who was also responsible for editing the volume (Abich 1887). The second and third volumes were subsequently translated into Russian by B. Z. Kolenko and published in Tiflis in 1899 and 1902 in an abridged form under the title *Geology of Armenian Highland. Western and Eastern parts. Orographic and Geological Description.* The fourth and last volume remained unfinished by Albich before he died; Suess in fact used a great deal of this material in his famous work *Die Antlitz der Erde* (1883–1885).

Hermann Abich's scientific abilities were highly regarded in Russia and in western Europe. He died on 1 July 1886 as a result of appendicitis. According to Abich's will his body was cremated, and the urn containing his ashes was buried in the grave of his mother in Koblenz in Germany.

Conclusions

Hermann Abich spent over forty years studying the geology of the vast and extremely complicated territory of the Caucasus. He had a natural gift for geology, persistence, good health and endurance, an excellent education and professional training at Berlin University. This, together with an independent study of Italian volcanoes, prepared him for his diverse geological investigation of the Caucasian region. He was ideally prepared for a full and detailed study. In addition, for many years he devoted himself only to his beloved science and long journeys, and avoided marriage until he was 50 years old. All these natural talents, together with his excellent gift as an artist of geology and landscape and his acquired knowledge and principles of life, led to him becoming an important figure in the Earth sciences, and his research brought great benefits to Russia and its multinational society. Abich, however, was not a pioneer in new scientific methodologies but tended to follow the older more orthodox teachings of his mentors. For all of his life he adhered to the theories of von Humboldt and von Buch on the craters of uplift as the main cause of mountain building. Because of this he was most interested in the role volcanic processes played in the development of structures seen in the Armenian highlands (Milanovsky 1998). He was also a pioneer in the scientific study of the present and Quaternary glaciation of the Caucasus (Milanovsky 2000).

Note

[1]A relatively complete biography of Hermann Abich was published by E. Suess and A. Abich-Hess as the preface to the third volume of H. Abich's *Geologische Forschungen in Kaukasischen Ländern* (1887). See also Volkova & Tikhomirov (1959) for a comprehensive assessment of his scientific activity and achievements.

References

ABICH, H. 1831. Chemische Untersuchung des Spinells und der Minerale von analoger Zusammensetzung. *Annalen der Physik*, **23**, 11, 305–359.

ABICH, H. 1841. *Geologische Beobachtungen über die vulkanischen Erscheinungen und Bildungen in Unter- und Mittel-Italien.* **1**, Lief. 1–2. Braunschweig.

ABICH, H. 1843. *Über die geologische Natur des Armenischen Hochländes.* Fest Rede. Dorpat, 68 S. 1 Karte.

ABICH, H. 1859. Vergleichende geologische Grundzüge der Kaukasischen, Armenischen und Nordpersischen Gebirgen als Prodromus einer Geologie der Kaukasischen Ländern. *Memoirs de l'Academie des sciences de St Petersburg, Science, mathemathique et physique*, Part 1, series 6., t. **7**, 359–534.

ABICH, H. 1878. *Geologische Forschungen in den Kaukasischen Ländern. T. 1. Eine Bergfauna aus der Araxesenge bei Djoulfa in Armenien.* Commission bei A. Hölder, Wien, VII, 126 S. 11 Tafeln.

ABICH, H. 1882. *Geologische Forschungen in den Kaukasischen Ländern. T 2. Geologie des Armenischen Hochlandes, 1. Westhälfte.* Commission bei A. Hölder, Wien, 472 S. 19 Tafeln.

ABICH, H. 1887. *Geologische Forschungen in Kaukasischen Ländern. T 3. Geologie des Armenischen Hochlandes, 2. Osthälfte.* Wien, S. 20 Tafeln mit Forwozt bei E. Zuess and A. Abich.

MILANOVSKY, E. E. 1998. Main development stages of volcanological researches in Russia. *In:* MORELLO, N. (ed.) *Volcanoes and History.* Proceedings of the 20th INHIGEO Symposium, Napoli-Eolie-Catani. Genova, 351–370.

MILANOVSKY, E. E. 2000. The Plio-Pleistocene glaciation in eastern Europe, Siberia and Caucasus: evolution of thoughts. *Eclogae Geologicae Helvetia*, **93**, 379–394.

SUESS, E. & ABICH, A. 1887. Preface, S. I–XII. *In:* ABICH, H. (ed.) *Geologische Forschungen in Kaukasischen Ländern.* Volume 3. Osthälfte, Wien, (S. I–XII).

VOLKOVA, S. P. & TIKHOMIROV, V. V. 1959. [Life and works of H. W. Abich. *Essays on the History of Geological Knowledge.*] Volume 8, Moscow, 177–238 [in Russian].

On camelback: René Chudeau (1864–1921), Conrad Kilian (1898–1950), Albert Félix de Lapparent (1905–1975) and Théodore Monod (1902–2000), four French geological travellers cross the Sahara

PHILIPPE TAQUET

Muséum National d'Histoire Naturelle, 8 rue Buffon, Paris 75005, France (e-mail: taquet@mnhn.fr)

Abstract: From 1920 to 1990, these brave geologists travelled through the Sahara from Mauritania to Libya and from Algeria to Niger. During these hikes across thousands of kilometres, often in very difficult conditions, they were able to trace the main features of the geology of these desert regions, they established stratigraphical sections of the main sedimentary provinces, discovered volcanic and eruptive complexes and drew geological maps of large areas.

Today, helicopters, four-wheel-drive vehicles, satellite observations and global positioning systems allow people to visit the most remote regions of the Sahara safely; however, geologists, naturalists and explorers like Chudeau, Kilian, De Lapparent, Monod made the most of their observations and discoveries thanks two essential auxiliaries: the camel and the goatskin bottle.

The portraits and the principal contributions to the geology of the Sahara of these four pioneers are presented here with maps of their itineraries.

The immense desert countries of the Sahara were first crossed by nomads travelling for trade. The most famous Islamic traveller was Abou Abd Allah Mohammed Ibn Battuta (1304–1377) who, at 22 years old, left Tangier to explore the World, the Middle-East, Asia and finally Saharan Africa. Leaving Morocco on 18 February 1352, he went to Niger and returned two years later to Fez. His was the only description of a crossing of the huge desert produced during the Middle Ages (Rouch 1958). It was not until the 1550s that Leon the African, another Muslim traveller (who converted to Christianity) produced his descriptions of Africa.

The first great crossing of the Sahara made by a European traveller took place in 1825 and 1826 by the Englishman Major Alexander Gordon Laing. Leaving Tripoli, he arrived in Timbuktu on 18 August 1826, but he was murdered on the way back (Maugham 1961; Monod 1977). The following year, the Frenchman René Caillé (1799–1838) crossed the Sahara from south to north via Timbuktu. On the east side of the Sahara, in Tripolitania, Denham, Oudney and Clapperton left Tripoli and journeyed to Chad lake from 1822–1825); then the German Heinrich Barth, accompanied by the geologist Oterweg, joined the James Richardson expedition in 1850 and travelled to Tripoli, Agades and Kano; of the three companions, only Barth returned from this tour in autumn 1855. Barth was the first to make geological observations and to note the presence of granites and sandstones in the central Sahara. Other such intrepid explorers included the French Henri Duveyrier (1840–1892), who from 1859 to 1862, visited the borders of the Southern Tunisia, of Tripolitania and Algeria, and in 1864 published a memorable work *Les Touaregs du Nord*, the first part of which was devoted to physical geography and to geology. In 1864, the German explorer, Gerhard Rohlfs (1831–1896), crossed the Sahara from west to east from Morocco to Tripolitania.

The French conquest of the city of Algiers in 1830 marked the beginning of the scientific explorations of the Sahara in the south of Algeria. Geological studies started in 1844 and were led according to the pacification of the Southern territories and were focused on two large projects: the creation of a large inland Saharan sea and the building of trans-Saharan railway.

The inland sea of the Sahara

Due to Captain Roudaire's instigation, military men, business men and adventurers rushed to undertake a mammoth project: to open a passage between the Mediterranean sea and the region of the 'Chotts' of South Algeria and Tunisia. This region constitutes a large depressed area south of the Atlas mountains that is below the sea level; the promoters hoped that digging a canal from the Gabes gulf would create a large inland sea, that would be fertile and prosperous area, in addition to quietening the nomadic and quarrelsome populations (Letolle & Bendjoudi 1997). This project, supported by Roudaire and Ferdinand

From: WYSE JACKSON, P. N. (ed.) *Four Centuries of Geological Travel: The Search for Knowledge on Foot, Bicycle, Sledge and Camel*. Geological Society, London, Special Publications, **287**, 183–190.
DOI: 10.1144/SP287.15 0305-8719/07/$15.00

de Lesseps, the famous promoter of the Suez Canal, was particularly useful to the interests of speculators but it was completely unrealistic; this arid region was sparsely populated and the supporters of the project, against the advice of the geologists, such as Pomel (1877), had not considered the existence of a barrier not far from the coast, the Gabes threshold, that had to be crossed. They were also ignorant of the role of evaporation and of the precipitation of the mineral salts which undoubtedly would have transformed the Inland Sea of the Sahara into a giant salt marsh. Nevertheless from 1850 to 1875, the idea of an Inland Sea nourished passionate controversies and discussions. In 1905, the famous novelist Jules Verne published a story inspired by this project (Verne 1905).

In contrast to this French project, a British one was proposed, to create an inland sea near the West African coast, not far from Tarfaya. In 1877, the Englishman D. Mackenzie settled in a fortress called Victoria; however, the Reguibat nomads murdered him in 1880 and the project never succeeded.

The trans-Saharan railway

The idea of linking Algiers to Niger by a railway crossing the Sahara from north to south generated a project as utopian as that of an inland sea. Colonel Flatters was in charge of the study of the layout of the future line. He set out with a party of soldiers and engineers to study its feasibility but during its second expedition, the party was killed by the Touareg on 16 February 1881 after reaching the southern edge of the Hoggar mountains. The nomads, having fed the explorers poisonous dates, cut the throats of nearly all the members of the expedition (Decraene & Zuccarelli 1994). This tragedy had enormous repercussions in France; there is a monument inside the Montsouris Park in Paris.

This drama, and the rebellion of the members of the Ouled Sidi Cheikh tribe in the west of Algeria (from 1864 to 1883) brought a halt to exploration and research. These started again in 1898 with the Foureau-Lamy mission who succeeded in reaching Chad after crossing the Sahara and making several geological observations and collecting some fossils (Foureau 1904–1905).

Flamand made the first great synthesis of the geology of the south of Algeria (Flamand 1911). Flamand explored the Algerian Sahara up to In Salah as a geologist, under the protection of a military column of 140 men of the camels' corps, but with the desire to lead a peaceful and scientific expedition. However, after an ambush, the column, under the orders of the captain Theodore Pein captured the city of In Salah by force on 28 December 1899, breaking the promise made to Flamand by the military commander. This event caused a serious dispute. Following the return to Algiers, there was a duel between Flamand and Pein, fortunately without dramatic consequences for the two antagonists (Decraene & Zuccarelli 1994). The law of 24 December 1902 established the autonomy of the South Territories and a geological survey was founded with Flamand as deputy director. In 1911, the latter published an comprehensive volume entitled: *Recherches Géologiques sur le Haut Pays de l'Oranais et le Sahara (Algérie et Territoires du Sud).*

René Chudeau (1864–1921)

In 1904, the geographer E. F. Gautier asked a young 40-year-old associate to join him, a geologist, who had graduated in natural sciences. At the age of 26, René Chudeau had been in charge of lectures at the University of Besançon before being removed from his position abruptly by the University council who disapproved of the life he shared with a suspected prostitute. Chudeau tried to join Gautier at Timimoun in May 1905. The project to establish a telegraphic line between In Salah and Timbuktu led Chudeau to go to the south and to cross the Sahara from Zinder and the Chad Lake. He returned to France by Bamako and Dakar in Senegal 18 months after his departure having covered distance of 7500 km which he had walked without an escort. He mapped the districts he crossed, made a geological survey and took numerous observations (Bourcart 1925). Chudeau discovered the first dinosaur bones in Niger at the bottom of the Tiguidi cliff, south of Agades. He published the results of his work in two volumes—a synthesis full of new geological observations, the first of which was written in collaboration with Gautier: *Le Sahara Algérien* and the second written alone: *Le Sahara Soudanais.*

Fascinated by Africa, Chudeau started again in 1908 for Mauritania and in 1909 travelled in the Timbuktu region. At the end of the same year, he crossed the Sahara again from north to south with the commission of the trans-Saharan railway; he returned in 1913 and 1914 to Timbuktu and visited the Taoudeni basin, discovering unknown ancient volcanoes. In all his trips Chudeau walked a total of 18 000 km through unexplored countries of the Sahara (Fig. 1) (Chudeau 1907*a*, *b*, *c*, *d* 1909). Chudeau demonstrated the presence of Cretaceous rocks in the south of the Sahara which are widely represented in the north, and that the former, lower Cretaceous rocks are overlain in the Zinder region by Upper Cretaceous successions. He was interested in the Quaternary de-
actual geological phenomena such

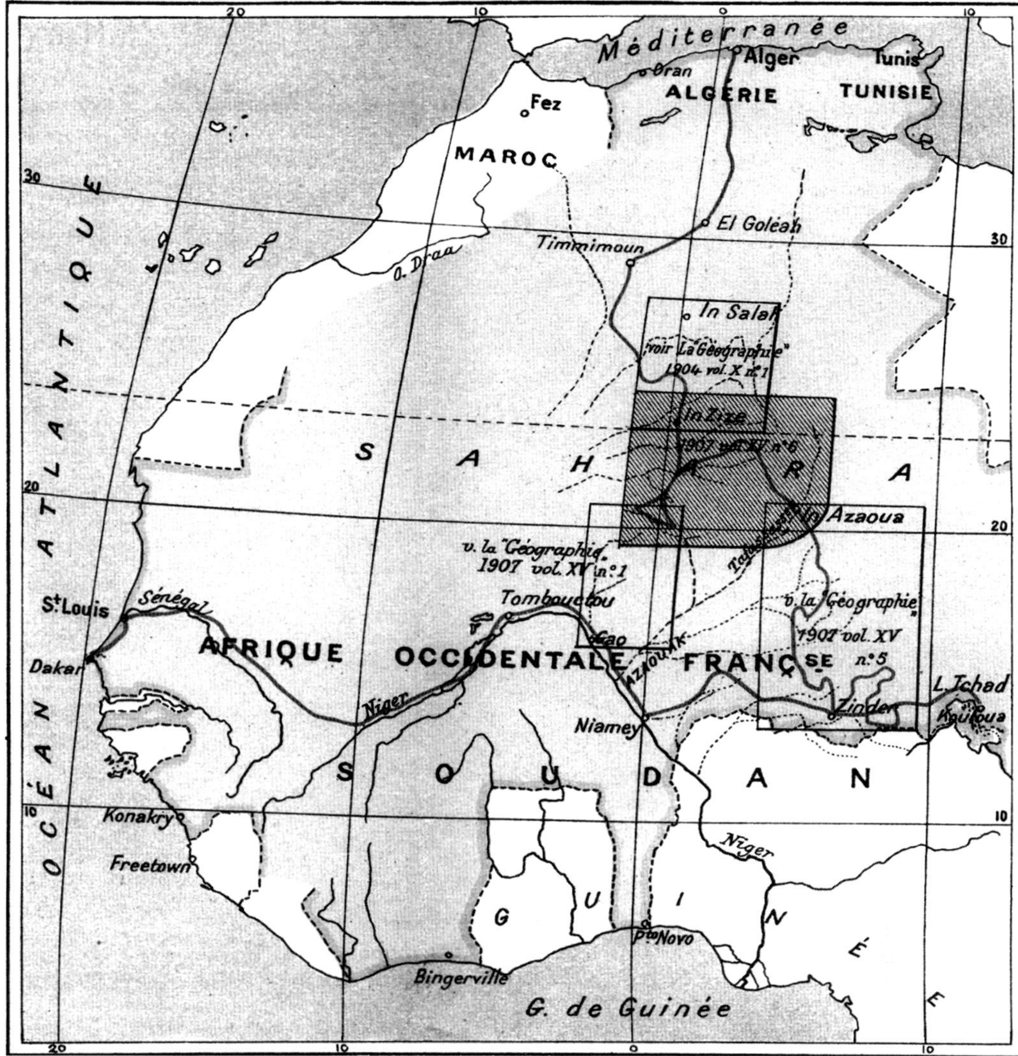

Fig. 1. The itineraries of René Chudeau through the Sahara (after Chudeau 1907*a–c*).

of dunes, the patina of the rocks, and in processes of aeolian erosion.

Back in Paris, Chudeau was obliged to accept a very modest job as a consulting engineer at the industrial bank of China. The bank subsequently went bankrupt and he was dismissed. Chudeau died miserably and alone. Ironically the day after his death, he was elected to a position at the Muséum National d'Histoire Naturelle, the institution where his collections remain to this day.

Conrad Kilian (1898–1950)

François Theodore Conrad Kilian was born on 25 August 1898 at the Château des Sauvages, near Lamastre in the Ardeche, France. His father, who lived in Strasbourg, was of Huguenot extraction and had an older Irish ancestory and links with Cuvier's family. His father was Wilfrid Kilian a great specialist of alpine geology and a member of the Academy of Sciences. Conrad was educated late under the tutelage of Charles Deperet in Lyon. In October 1921 he was engaged by a businessman living in Constantine (Algeria) who wanted to find the legendary locality of the emerald layer of the Garamantes, the first inhabitants of the Sahara. Kilian was employed as a prospector and embarked on his survey in the company of an older warrant officer of the camel corps. The emeralds were in fact green feldspars, amazonite, and after many incredible mishaps, Kilian, who was abandoned by his military companion, completed his exploration of the Tassilis des Ajers and of the central Hoggar. He developed a clear and excellent understanding of the main features of the geology of the central Saharan massif and of its aureoles. He discovered graptolitic shales *in situ* and realized that the Tassilis could be divided into two distinct units. The internal Tassilis he regarded as being 'pre-Gothlandian' [Silurian] comprising Lower Ordovician sandstones, whereas the external Tassilis consisted of upper Devonian sandstones. They were separated from each other by Silurian shales containing graptolites. Looking for a clear unconformity between the sandstones at the bottom and the crystalline basement rocks, Kilian also demonstrated that the basement was Precambrian and not simply pre-Silurian. The presence of a conglomerate allowed Kilian to separate this Precambrian sequence into two series (which he named Suggarian for the base and Pharusian for the top). These significant discoveries made by a young geologist who was only 22 years old at the time were submitted in 1922 to the Academy of Sciences in the form of a short note entitled *Aperçu général de la structure des Tassilis des Ajers* and then again to the International Geological Congress of Geology in Brussels in 1925 in a paper entitled *Essai de synthèse de la géologie du Sahara sud-constantinois et du Sahara central.* The information that these contained was so new that Wilfrid Kilian, Conrad's father asked a practised geologist, Jacques Bourcart, to verify his son's logs! (Bourcart 1924).

Fascinated by the Sahara, Kilian returned again in 1926 (Fig. 2), when he crossed the desert on a white camel dressed like a Touareg, he was preceded by a squire carrying a banner, and proclaimed himself as sovereign explorer (Boissonnade 1971). During this trip he made numerous discoveries; he established the existence on the Saharan platform of a '*Continental intercalaire*' containing fossil reptiles and fishes, a sequence situated between the marine upper Carboniferous (Namurian to Gzelian depending on the region) and the marine Middle Cretaceous (Upper Cenomanian), a concept still used today (Lefranc & Guiraud 1990). It was also Kilian who defined a '*Continental terminal*' of Upper Cretaceous and Cenozoic age.

Kilian explored the confines of the Italian Fezzan and Algeria, in a no-man's land where he discovered unknown mountains that he named Monts Doumergue; he offered these to France and caused a diplomatic crisis. Back in France he published details of his fieldwork and prepared the two first sheets of the International geological map of Africa, the second being entirely his own work (Bourcart 1951; Furon 1955).

Kilian was rewarded with the gold medal of the French Geographical Society and he was invited to London in May 1932 by Sir Francis Rodd to deliver a lecture on his Saharan explorations to the

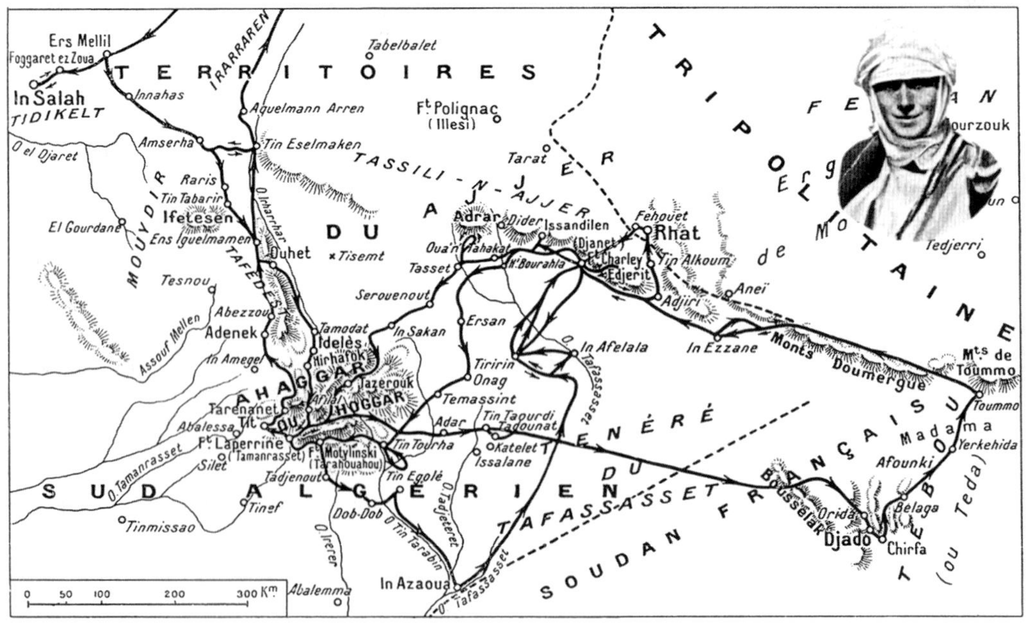

Fig. 2. Portrait and itineraries of Conrad Kilian through the Sahara (after *L'Illustration*, 1929).

members of the Royal Geographical Society. At the end of an intense and brief life, he was worn out by emotional disappointments, and by political and administrative wranglings in France over his opinions for the delimitation of the borders between Chad, Libya and Algeria (these took place against the background of intense prospecting for the first time for oil in the region). Kilian died in a mysterious and strange way in a hotel in Grenoble on 30 April 1950. He was found covered with blood and hanged at the espagnolette of his room window. There is uncertainty surrounding the circumstances of his death; some people feel that it was suicide, but others think he was murdered.

Albert Félix de Lapparent (1905–1975)

Albert Felix de Lapparent was born at Mont-Dieu in the Ardennes on 9 September 1905. His grandfather, Albert Auguste de Lapparent was a geologist who had founded the laboratory of the Catholic Institute in Paris in 1876 and became perpetual secretary of the Academy of Sciences in 1907. His uncle, Jacques de Lapparent was a well-known petrographer and mineralogist who taught in Strasbourg and Paris;

Albert Felix de Lapparent was ordained a priest in 1929. Following family tradition, he then engaged in geological studies and prepared a thesis on the sedimentary geology of Provence between Var and Durance. For this thesis he was awarded the Cuvier Prize of the Academy of Sciences in 1939. While continuing his geological work in France, De Lapparent, began, following the suggestion of the great Saharan geologist Nicolas Menchikoff, to study the stratigraphy of the Mesozoic Saharan basins. In 1946, he began a long period of exploration of huge monotonous regions travelling by camel. He undertook these trips with one or two guides, three or four camels and little food and endured conditions that were described as torrid during the day and icy during the night (Bordet 1977).

De Lapparent gave a detailed account of the stratigraphy of the Gourara, of Touat and Tidikelt. He began to collect vertebrate bones discovered during his Saharan trips, and he immersed himself in the study of the dinosaurs of the Sahara so he could develop a better understanding of the stratigraphy of the Continental Intercalaire. He travelled through the Sahara eight times (Fig. 3): southern Tunisia in 1951 and 1952; Tamesna (Niger) in 1953; Niger (Agades and Zinder) in 1954; Chad (Tibesti, Ennedi and Borkou) in 1955; Algeria in 1959; Hoggar, Edjelé and Fort Flatters in 1958 where he was the first to interpret the geology of the Edjele region where later the first oil field inside the Algerian Sahara was discovered. I was lucky to be able to invite him to Niger in 1966 and to go with him to the Tademaït plateau in Algeria in 1970.

In 1960, de Lapparent published a memoir devoted to the dinosaurs of the Continental Intercalaire of the Sahara (Lapparent 1960), a memoir which marks the starting point of the studies on the terrestrial vertebrate faunas of the Mesozoic of the Sahara. He was the first to discover the bones of a giant crocodile in the cuttings of an underground irrigation canal (called a foggara) in Aoulef, South of the Tademaït. This crocodile was named in 1966 *Sarcosuchus imperator* by de Lapparent's niece, France de Broin and by the present author. *Sarcosuchus imperator* the skull of which is 170 cm long, measured 11 metres in length and is the biggest crocodile to have been found so far.

On Christmas Eve 1948 de Lapparent almost lost his life in the Sahara desert when he was travelling in the Azaoua region on the borders between Tunisia and Libya. He fell from his camel and dislocated his shoulder. His guide then roped him onto the back of his camel and hurried to Fort Polignac, a trip that still took several days! de Lapparent was then taken by car to Rhat, but he still had to wait three days before being taken by aeroplane to Tunis where he was finally operated on. He never fully recovered from this mishap and remained partially paralysed for the rest of his life.

All the archives and field notes of A. F. de Lapparent are today preserved in the Geological Institute Albert de Lapparent in Cergy Pontoise near Paris in France.

Théodore Monod (1902–2000)

Théodore Monod was born in Rouen on 9 April 1902. He was descended, on his father's side, from five ministers. He began to visit the Jardin des Plantes and the Museum National d'Histoire Naturelle in Paris when he was five years old. He wrote his first paper *Une relation zoologique et botanique d'un voyage dans le Midi de la France* in 1914 when he was only 14 years old.

Initially, Théodore Monod was tempted to study theology, but finally he chose natural sciences, obtaining his Master's degree from the Sorbonne in 1921. He was awarded a scholarship in the Museum National d'Histoire Naturelle and was elected as assistant at the laboratory of fish and colonial animals. After his first trip to Mauritania, Théodore Monod didn't embark on the boat back to France but preferred to travel to Senegal overland on camel back. From that time onwards, his research centred on two areas: both

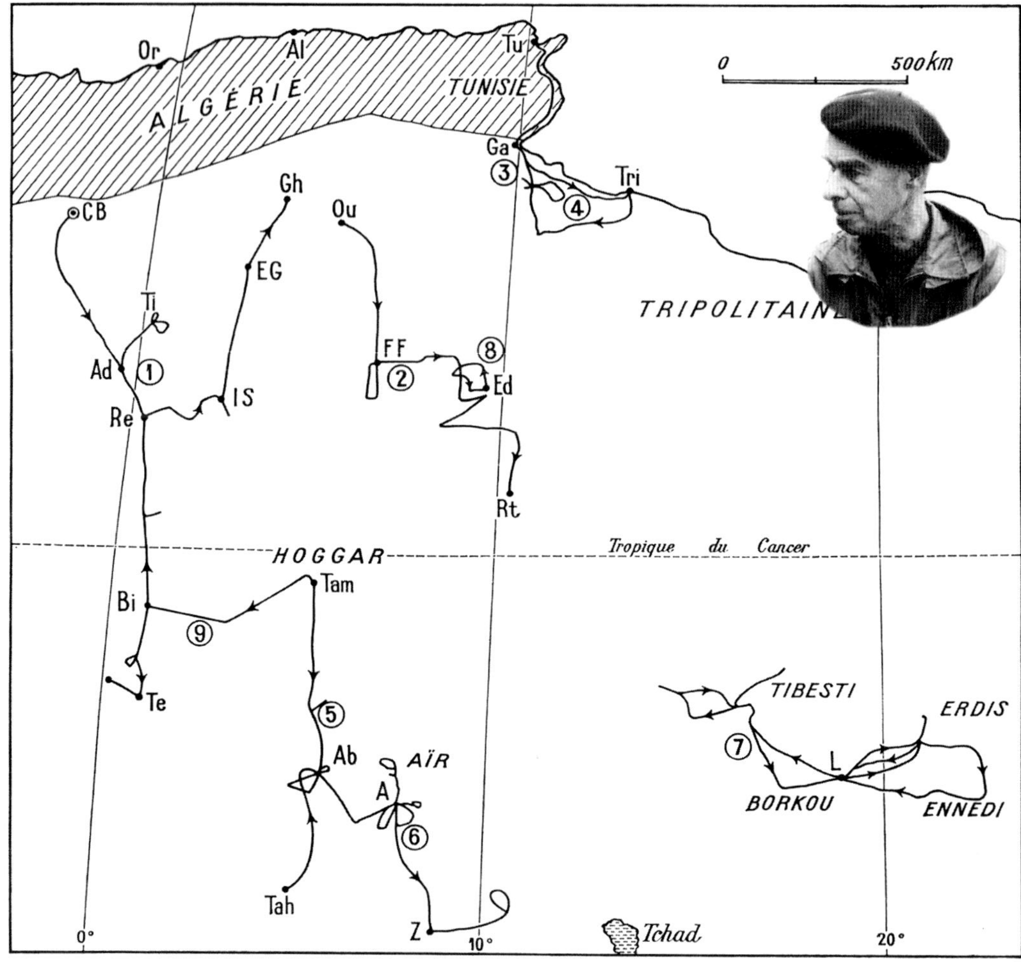

Fig. 3. Portrait and itineraries of A. F. de Lapparent through the Sahara (after Lapparent 1960).

'oceanic' (the vertical marine that yielded fish and crustaceans) and the horizontal sandy deserts in which he found rare treasures in the arid environments. Switching research between these two different environments, from the sea to the desert, from the desert to the sea, from one ocean to the other, was the idea of Théodore Monod, and so the pattern of work continued until his death at the age of 95 years. Théodore Monod crossed the Sahara in all directions, mainly on foot, in severe conditions, with some companions and few camels (Fig. 4).

In 1927, Théodore Monod participated as a naturalist on the Augieras–Draper expedition, organized by the French Geographical Society, and crossed Africa from Algiers to Dakar via Hoggar. In 1938, he was asked to manage the Institut Français d'Afrique noire which had just been established in Dakar, and he made this institution (which still exists today as the Institut fondamental d'Afrique noire) a remarkable centre for scientific research in Africa. He was Director of this Institute for 26 years until 1964. During the Second World War he patrolled along the borders of the desert between Chad and Libya, during which time he was elected as Professor of the Museum National d'Histoire Naturelle and Director of the laboratory of fisheries. After the war he divided his time between the exploration of the Sahara, undertaken during long and intrepid camel trips, mainly in the Majâbat al-Koubrâ on the borders of Mauritania and Mali, and the exploration of the Atlantic Ocean. For the latter he participated in the first dives of the bathyscaph of Professor Piccard. This was whilst working in the laboratories in Dakar and Paris.

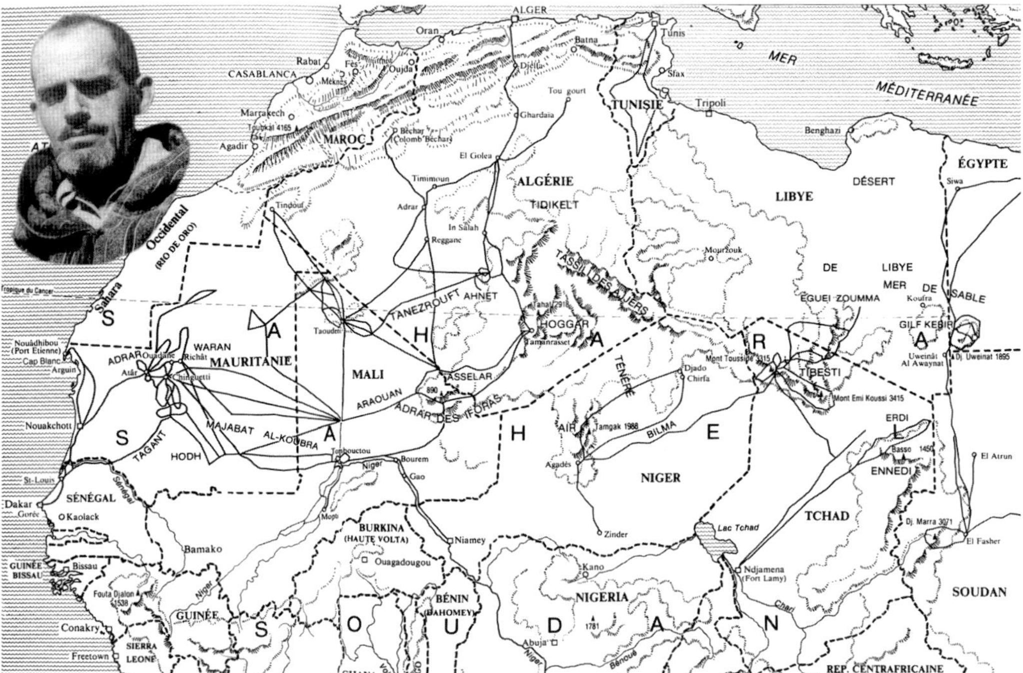

Fig. 4. Portrait and itineraries of Théodore Monod through the Sahara (after Monod 1937).

Théodore Monod was also an excellent Saharan geologist and palaeontologist: as early as 1930, when he was doing his military service as a second class cameleer in the camel corps, he discovered an intermediate series between the Precambrian and the Palaeozoic of the NW Hoggar (Monod 1931–1932), which he named purple series and which became famous, because it illustrated the first known example of Panafrican molasse. These were an extraordinary accumulation of sediments that were derived from an enormous mountain chain, as high as the Himalaya, but now eroded. He then studied the cliffs of the Mauritanian Adrar, describing its stratigraphical succession, and reported on Precambrian stromatolites, (calcareous masses with laminar structures that result of the consolidation of fine mud by blue algal filaments associated with bacteria). He had a passionate interest in the concentric complex of the Richat in Mauritania which he had discovered in 1934. This was a strange ring-shaped geological structure, composed of Precambrian sediments and containing siliceous rocks made of primitive unicellular micro-organisms. In 1934, he also discovered Silurian graptolitic rocks in the Tassilian aureoles of Ahnet, and soon afterwards in 1937 wrote an account giving a synthesis of the structure of the West Sahara. His findings are valid today and are still used on all the recent geological maps. Following these studies he discovered Carboniferous plants in the Westphalian of Taoudeni, and Devonian faunas with goniatites and fusulinids in the erg Chech.

Although a man of science, Théodore Monod was also a humanist (Taquet 2002). Throughout his life, he fought for a more fraternal world and for a humanity that was finally humanized. He was also a talented writer and knew how to make his geological knowledge and scientific questions more accessible, explaining for example the geological formation of the Sahara by way of a cookery recipe (Monod 1937) and publishing books such as *Méharées* or *L'Hippopotame et le Philosophe* with which he became very popular. He received the gold medal of the French Geographical Society, the gold medal of the Royal Geographical Society and the gold medal of the American Geographical Society. He was commander of the Legion d'honneur, commander of the National Order of Senegal and commander of the Merite Saharien.

Conclusion

Today, helicopters, four-wheel drive vehicles, satellite observations, global positioning systems allow anybody to walk in the most isolated parts of the Sahara. However, there are still some dangers inherent in these immense desert spaces and travellers still die of thirst every year in the Sahara.

But as the computer replaces the geological hammer, the motor vehicle replaces the camel and the aluminium can replaces the goat skin, it is right to pay tribute to some of the pioneers, to those geologist travellers, who at enormous cost, covered thousands of kilometres, hiking or on camelback for weeks or months, in order to collect samples, to survey outcrops, to draw maps and to trace the main lines of the geology of the Sahara.

References

BOISSONNADE, E. 1971. *Conrad Kilian. Explorateur souverain.* France Empire Ed., Paris.

BORDET, P. 1977. Albert F. de Lapparent (1905–1975) Notice biographique. *Mémoire Hors-Série de la Société Géologique de France*, **8**, 7–18.

BOURCART, J. 1924. *Un voyage au Sahara. Note préliminaire sur les résultats géologiques de la mission Olufsen au Sahara.* Publication du Comité de l'Afrique française.

BOURCART, J. 1925. René Chudeau (1864–1921). Notice nécrologique. *Bulletin de la Société Géologique de France 4*, **25**, 449–467.

BOURCART, J. 1951 Conrad Kilian (1898–1921). Notice nécrologique. *Bulletin de la Société Géologique de France*, **6**, 303–312.

CHUDEAU, R. 1907*a*. D'Alger à Tombouctou par l'Ahaggar, l'Aïr et le Tchad. *La Géographie,* T.XV, **6**, 261–336.

CHUDEAU, R. 1907*b*. L'Aïr et la région de Zinder. *La Géographie,* T.XV, **6**, 321–336.

CHUDEAU, R. 1907*c*. D'In Zize à In Azaoua. *La Géographie*, T.XV, **6**, 401–420.

CHUDEAU, R. 1907*d*. Excursion géologique au Sahara et au Soudan (Mars 1905–Décembre 1906). *Bulletin de la Société Géologique de France*, **IV-7**, 319–346.

CHUDEAU, R. 1909. *Le Sahara soudanais.* Colin, Paris.

DECRAENE, P. & ZUCCARELLI, F. 1994. *Grands Sahariens à la découverte du 'désert des déserts'.* Denoël, Paris.

FLAMAND, G. B. M. 1911. *Recherches géologiques et géographiques sur le haut pays de l'Oranais et sur le Sahara (Algérie et Territoires du sud).* Rey ed., Lyon.

FOUREAU, F. 1904–1905. *Documents scientifiques de la missions saharienne (Mission Foureau-Lamy) d'Alger au Congo par le Tchad.* Masson, Paris.

FURON, R. 1955. Histoire de la Géologie de la France d'Outre-Mer. *Mémoires du Muséum national d'histoire naturelle.* Nlle série, série C., Sciences de la Terre, T.V.

LAPPARENT, A. F. DE, 1960. Les Dinosauriens du ⟪Continental Intercalaire du Sahara Central⟫. *Mémoires de la Société Géologique de France.* (Nouvelle série). Mémoire **88A**.

LEFRANC, J. P. & GUIRAUD, R. 1990. The Continental Intercalaire of northwestern Sahara and its equivalents in the neighbouring regions. *Journal of African Earth Sciences*, **10**, 27–77.

LETOLLE, R. & BENDJOUDI, H. 1997. *Histoires d'une mer au Sahara. Utopies et Politiques.* L'Harmattan. Paris.

MAUGHAM, L. R. 1961. *The Slaves of Timbuktoo.* Longman, London.

MONOD, T. 1931–1932. L'Adrar Ahnet. Contribution à l'étude physique d'un district saharien. *Revue de Géographie Physique et de Géologie Dynamique*. T. IV. fasc. **2**, 107–150; fasc. **3**, 223–262; T.V, fasc. **3**, 245–297.

MONOD, T. 1937. *Méharées.* Je Sers ed. Paris

MONOD, T. 1977. Le dernier voyage de Laing. 1825–1826. *Bibliothèque d'Histoire d'Outre-Mer.* Nouvelle série, Travaux, **2**, 1–126.

POMEL, A. 1877. Le projet de mer intérieure et le seuil de Gabès. *Revue Scientifique*. 2nd series, **19**, 433–440.

ROUCH, J. 1958. L'Afrique. *In*: ROUCH, J. (ed.) *Histoire universelle des Explorations.* Nouvelle Librairie de France, F. Sant'Andrea Paris, **4**, 155–187.

TAQUET, P. 1997. Au temps des crocodiles mésozoïques sahariens. *In*: BILLARD, R. & JARRY, I. (ed.) *Hommage à Théodore Monod naturaliste d'exception.* Muséum national d'Histoire naturelle. Archives, 57–72.

TAQUET, P. 2002. La vie et l'œuvre scientifique de Théodore Monod. *In*: *Académie des Sciences et Institut de France*. Discours et Notices biographiques. T.V, 181–186.

VERNE, J. 1905. *L'Invasion de la Mer.* Hetzel. Paris. [New edition Cérès, 2003].

Théodore Andre Monod and the lost *Fer de Dieu* meteorite of Chinguetti, Mauritania

U. B. MARVIN

Harvard-Smithsonian Center for Astrophysics, Cambridge, Massachusetts, 02138, USA (e-mail: umarvin@cfa.harvard.edu)

Abstract: Théodore Monod (1902–2000), of the Muséum National d'Histoire Naturelle in Paris, was a natural scientist with an extraordinarily wide range of interests and expertise. His early researches were chiefly in marine zoology, which he continued to pursue throughout his career. However, in 1923 he began to work in the Sahara Desert and travelled through it for thousands of kilometres on foot and camel-back collecting samples for the museum and keeping detailed journals of the geology, palaeontology, flora, fauna, prehistoric artefacts, and the customs and cultures of the peoples he met. Monod published nearly 700 technical articles in scientific journals and numerous books for general readers. He won innumerable honours in France and internationally, but only one of his books, few of his articles, and no biographical accounts of Monod's activities have been published in English. This paper focuses on his search for the *Fer de Dieu*, an iron meteorite said to be 40 metres high and 100 metres long, lying in the desert of Mauritania. He found no trace of it, but the meteorite remains legendary in the history of meteoritics.

Monod's early years and some highlights of his career

Born in Rouen on 9 April 1902, Théodore Monod was five years old when his family moved to Paris. Shortly thereafter, his mother began taking him to the Jardin des Plantes and the adjacent Muséum National d'Histoire Naturelle on visits that inspired him to become a naturalist. Monod attended the Ecole Alsacienne in Paris and received his Baccalauriat at the age of 16. Thereupon, he faced a difficult choice between pursuing studies in the natural sciences or in theology. Being descended from a succession of five Protestant pastors, and having a highly distinguished father then officiating at the Oratoire du Louvre, Théodore was sorely tempted to take up the family calling. Nevertheless, in 1918 he entered the Sorbonne to pursue studies in natural sciences. By 1921 he had completed his courses in zoology, botany, and geology and entered the Muséum National d'Histoire Naturelle to work toward his doctorate. In January 1922 he was named as assistant in a laboratory devoted to studies of fish and fisheries in France's overseas possessions.[1]

After making three short cruises, Monod's first major research assignment took him to Port-Étienne (today's Nouâdhibou) at the northwestern corner of Mauritania where he remained from December 1922 to October 1923. He then seized an opportunity to return to France via Dakar by joining a caravan of 19 men, one woman, and ten camels travelling southward as far as Saint-Louis d'Sénégal. The trip of some 800 km lasted from 14 October until 12 December 1923, during which Monod walked or rode a camel for up to ten hours a day, making diversions to examine the geology or the plant and animal life and to meet with local people.[2] Captivated by this experience, he developed a passionate attachment to deserts, which he saw as another kind of ocean with oases as ports and caravans as navies. In his lifetime, Monod trekked for many thousands of kilometres through north and central Africa and to places as distant as Ethiopia, Yemen, and Madagascar, but his deepest loyalty was to '*le vrai Sahara*'.[3] He became particularly attached to the Adrar highlands of central Mauritania which he referred to as his 'diocese'.[4]

Monod returned to France early in 1924 and resumed research on his thesis. Once again, however, his work was interrupted by a year-long assignment that took him back to Africa to study the aquatic fauna in the principal rivers of Cameroon. From September 1925 to September 1926 he travelled through Cameroon on foot and by pirogue and tipoye (litter) across the steppes and through the jungles from the coast of Equatorial Guinea to Lake Chad. Later that year, back in Paris, he earned his Ph.D. with his principal thesis, of nearly 700 pages in two volumes, on the *Gnathiidae*, a family of miniscule crustacean parasites on fish. His new data and complete reclassification of this group of isopods brought him recognition as the world authority on them. He also presented a secondary thesis, as was required at that time, on estuarine life in the River Seulle, which enters the sea in lower Normandy.

From: WYSE JACKSON, P. N. (ed.) *Four Centuries of Geological Travel: The Search for Knowledge on Foot, Bicycle, Sledge and Camel.* Geological Society, London, Special Publications, **287**, 191–205.
DOI: 10.1144/SP287.16 0305-8719/07/$15.00

In October 1927 Monod returned to Africa on a five-month expedition financed by a wealthy American, W. P. Draper, and led by M. Augiéras. The Société de géographie recommended Théodore Monod and his colleague, Wladimir Bresnard, as naturalists on the trek. Starting from Algiers, they travelled southward through the Hoggar to remote Tamanrasset and thence southwestward through the French Soudan (Mali) to Timbuktu and on to Dakar. At Asselar, near the southwestern border of Algeria with present day Mali, Monod and Bresnard excavated *l'homme d'Asselar*, a fossilized skeleton of a Neolithic human who had drowned in a lake and been preserved in soft mud which had filled the cranium. Today the skeleton is exhibited in Paris in the Palaeontology section of the Muséum National d'Histoire Naturelle (Vray 1994, p. 155).

In July 1928, four months after his return to Paris, Monod met Olga Pickova (1900–1980), a young Jewish woman of Czech origin, with whom he soon fell in love. In his journal he wrote that it was unbelievable that he had fallen in love but even more unbelievable that she loved him too (Vray 1994, p. 183). They were formally betrothed on 30 September 1928, but had to delay their marriage while he endured his military service, which was especially onerous because Monod, at heart, was a pacifist and antimilitarist. However, his experience of the desert won him an assignment as a saharien 2nd class with the camel corps in the far reaches of the Tidikelt-Hoggar of Algeria. There, he speculated that he probably was the first European, and certainly the first geologist, to cross the Ahnet mountain chain in which he collected samples and kept detailed notes on the geology. He also sketched some fine rock engravings, of elephants, hippos and other creatures, of a type he found in several localities in the Sahara.

On 22 March 1930, Théodore and Olga were married in a civil ceremony that was consecrated two days later in a service at the Oratoire du Louvre. His family welcomed Olga among them. Théodore was particularly pleased when his father read the consecration in Hebrew. Théodore and Olga had three children, a daughter, Béatrice, born 13 June 1931, and two sons, Cyrille, born 8 February 1933, and Ambroise, born 8 March 1938. In later years, Ambroise accompanied his father on one of his long treks, and Cyrille edited, posthumously, a collection of his father's letters.[5]

At Dakar, on 14 July (Bastille Day) 1938, Monod founded *l'Institut français d'Afrique Noire* (*l'IFAN*). He served as its director until 1965, initiating a wide diversity of research projects on the continent and offshore. He moved his family to Dakar where Olga had an office at the Institute. During World War II, Monod denounced the German occupation of France on Radio-Dakar and aided the efforts of the resistance in which Olga was particularly active. 1943 was a dark time for them: Théodore's father died in Paris and Olga lost her entire family in Auschwitz and Terezín—except for one sister who had gone to Palestine in 1928 (Vray 1994, p. 287). In that same year, however, Théodore was made a full professor at the Muséum in Paris and he held that position until he became an honorary professor on his retirement in 1974.

Meanwhile, on 26 October 1948 this indefatigable explorer accompanied professor Auguste Piccard (1884–1962) in the first dive of the Bathyscaphe FNRS-2 that Piccard had designed. This attempt carried them to a depth of what Monod described as '25 000 millimetres'. Indeed, that dive had to be aborted at 25 m, but after numerous later ones had been successful, Monod accompanied an experienced diver to 1400 m on 22 April 1954. Monod was fascinated by the opportunities presented by this means of exploring the ocean depths, while maintaining his own allegiance to the deserts (Vray 1994, pp. 310–312).

The passages Cyrille published of his father's letters show that the frugal life he led in the desert suited Monod perfectly. A natural philosopher of deep spirituality with a profound reverence for life, Monod became a vegetarian and an ardent pacifist. He once wrote that the Christian era ended on 6 August 1945, with the dropping of the atomic bomb on Hiroshima. During the final 17 years of his busy life, when he had become internationally famous and showered with honours, Monod wielded his influence by holding an annual three-day, 'publicity-seeking' fast on the anniversary of that date, to call attention to the sheer inhumanity of the event and to urge France and the other powers to abandon their nuclear programmes.

The *Fer de Dieu* of Chinguetti: 1924–1934

The '*Fer de Deu*' was an immense iron meteorite alleged to lie about 45 km SW of the village of Chinguetti in the Adrar desert of Mauritania (see map, Fig. 1). The world at large first heard about this giant iron meteorite on 28 July 1924, when Alfred Lacroix (1863–1948), the distinguished mineralogist at the Muséum d'Histoire Naturelle, who also was the permanent secretary of the Académie des Sciences, read a paper to the Académie about a small 4.5 kg meteorite that he had received from a geologist, Henry Hubert (1879–1941), the colonial administrator at Dakar.[6]

Lacroix identified the specimen as an iron meteorite unlike all others then known. It was about 5 cm thick and nearly rectangular, with its

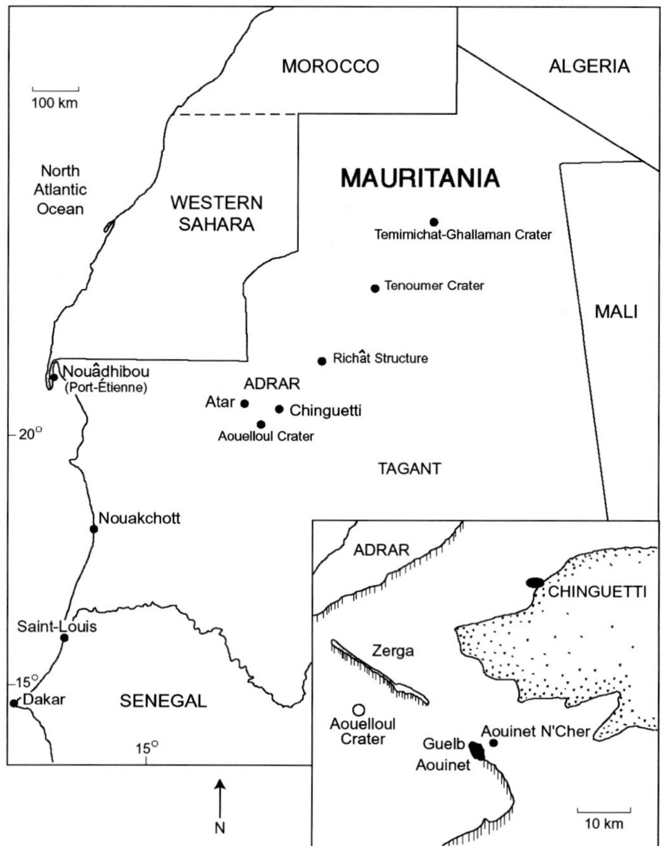

Fig. 1. Locality map of Mauritania showing the principal cities and topographic features discussed in the text. The inset shows the great sand dunes encroaching on Chinguetti, a part of the Adrar highlands, the Aouelloul crater and Zerga, the nearby ridge for which a small stony meteorite, found on the crater rim, is named. It also shows Aouinet N'Cher, a temporary waterhole, and guelb Aouinet, the promontory on a ridge that Monod concluded was mistaken for a giant iron meteorite. (Based partly on the sketch map of Fig. 13 in Monod & Zanda 1992).

largest face measuring 16 × 9.5 cm (Fig. 2). Lacroix described the meteorite as a breccia consisting of grains of metallic nickel–iron, making up 80% of the volume, admixed with silicate minerals, primarily hypersthene with lesser amounts of olivine and feldspar, making up 20%. At that time, only five so-called stony-iron meteorites were known worldwide, and their mineralogy and metal-silicate proportions were very different from those of the meteorite from the Adrar. Today, stony-iron meteorites are still rare, constituting fewer than 1% of the meteorites that have been collected having been seen to fall.

Henry Hubert had not collected the Adrar meteorite himself. He said it had been given to him in the summer of 1917 by Army Captain Gaston Ripert (1881–1957), who was passing through Dakar *en route* to a post in Cameroon. Ripert told Hubert he had found the meteorite in

Fig. 2. Chinguetti, the mesosiderite which gave rise to the legend of the giant iron. The whitish constituent is metallic nickel–iron, which occurs in rounded aggregates and as small grains scattered throughout the specimen. The dark constituents of all sizes are mainly pyroxene. (The specimen measures approximately 16 × 9 cm. Photo from the Muséum National d'Histoire Naturelle; used with permission.)

1916 while he was serving at Chinguetti as the resident administrator and commander of the camel corps. He described the circumstances under which he had collected it and asked Hubert to send the meteorite to Dr Lacroix in Paris.

Hubert, a former student of Lacroix, fulfilled his obligation tardily. The specimen arrived at the Museum and was accessioned into the collection in March 1921. This delay of more than four years remains puzzling, although Ripert, himself, suggested in 1932 that it may have been due to all the dramas that were being played out in France between 1917 and 1921. Lacroix initiated the long process of having the meteorite sliced into three parallel slabs, one of which was used, in part, for mineralogical and chemical analyses. Another portion presumably was sent back to Dakar at Hubert's request, although neither the specimen nor any records pertaining to it have survived. Three more years passed until in July 1924 Lacroix presented his results to the Académie des Sciences. He opened his report with Hubert's written account of Ripert's story as told to him in 1917 (Lacroix 1924):[7]

On a New Type of Meteoritic Iron Found in the Desert of the Adrar in Mauritania

Monsieur, the administrator Ripert, formerly commandant of a platoon of the camel corps of Mauritania, has made in the desert of the Adrar an important observation which has been transmitted to me by M. Henry Hubert, along with the specimen described in this note. It is an iron of extraterrestrial origin.

This specimen was recovered approximately 45 km southwest of Chinguetti, and west of the guelb Aouinet. It lay isolated on an enormous metallic mass measuring about one-hundred metres on one side and forty metres high; it jutted up in the midst of dunes covered with desert plants, the sba [sbot]. It has the form of a parallelepiped, compact, without fissures. The visible surface is dominantly vertical like a cliff, but the wind-blown sand has hollowed it out at the base such that the summit overhangs it; the portion open to aeolian erosion is polished like a mirror. The sand has accumulated against the opposite face and covered it entirely, which prevents an evaluation of the third dimension. One corner of the summit of the mass is bristling with small needles that the Maures have tried to carry away, but the malleability of the metal only permits them to bend them. Some smaller blocks are scattered in the vicinity.

After describing the mineralogy and chemistry of the small specimen, Lacroix closed his report with the following observation:

The questions of composition and of structure are not the only particularities that call attention to the iron of Chinguetti; if, in effect, the dimensions given by M. Ripert are exact, and there is no reason to doubt them, the metallic block constitutes by far the most enormous of known meteorites; if we can admit that the third dimension, which could not be measured, is equal to the smallest of the two others, it would measure no less than 160 000 cubic metres.

Such a declaration, by such an authority, quickly generated international excitement. Scientists and laymen alike imagined a fabulous iron meteorite lying in the remote desert of Mauritania. Expectations of finding it were high. A meteorite 40 m high and 100 m long should be hard to miss in any terrain.

The search began in September 1924 in response to an official request from Lacroix to the Governor of Mauritania, who passed it on to the chief of the Chinguetti district, who, in turn, assigned one Lieutenant Bonnin to locate the meteorite. Lt Bonnin explored the area of great dunes to the south of Chinguetti, and the plateaus to the SW, and returned on 28 October 1924, with no sighting to report. He pointed out that the area to the SW consists of the so-called 'reg', a wide, rocky expanse of almost unbroken relief, which allows a very extensive panorama in all directions. Any giant meteorite should be readily visible on the reg, but he saw none there. To the south and SE of Chinguetti lies the 'erg', a landscape of dunes, many, he said, with heights similar in scale to the sought-after meteorite. Nevertheless, Bonnin was convinced that even if it lay among the dunes, as Ripert had said it did, the huge metallic mass could not have escaped his detection.

Four years later two more searches took place. From late December 1928 to mid-January 1929 Emile Bruneau de Laborie, a seasoned explorer of the Sahara, searched diligently for the mass of iron and found no sign of it. That same year Lt Maurel mounted one more fruitless search. Later in 1929, Auguste Chevalier, a professor who was travelling from Paris to St Louis, carried a letter from Lacroix to M. Choteau, the Lieutenant Governor of Mauritania. Lacroix asked Choteau to arrange for an investigation that would yield the following items:

1. A sketch giving the exact position and orientation of the meteorite;
2. Photographs taken from all sides;
3. Several kilos of samples to be sent to the Museum; and
4. Designation of the meteorite as a natural monument to remove it from all depredations.

Choteau replied that Dr Malavoy, director of the Department of Mines of French West Africa, was already in the process of studying the geology and mineralogy of the country and would be interested in searching for the meteorite.

In February 1930 Jean Malavoy (1903–1945), looked for the iron in the areas to the SW, south, and SE of Chinguetti but failed to find it. On 24 February 1930, Malavoy wrote to his predecessor, Henry Hubert, saying that the affair constituted '... *a gigantic humbug, voluntary or not*' (Monod

& Zanda 1992, p. 30). Malavoy then wrote to Lacroix assuring him that there was no giant meteorite; the only known meteorite from the Adrar was the small specimen he already had examined.

Also in 1930, Malavoy sent three official letters, signed by the Governor General of French West Africa, to Gaston Ripert, asking for more details. The following year, on 8 February 1931, Ripert replied to the Governor General from Cameroon where he was managing a coffee plantation. He said his memory had become so vague that he could furnish no more precise details than he had given to M. Hubert in 1917. Meanwhile, he had lost track of the notes he had written when he returned to Chinguetti after collecting the small meteorite.

The following year, however, Ripert, then residing in Nice, added some new details to the story in a reply he wrote to a letter from Professor Jean Bosler (1878–1973), the director of the Observatory at Marseille. Ripert's letter, dated 6 October 1932, included the following remarks:

> . . . In order to satisfy the urgent demands of my guide, I had taken with me neither compass nor any material that would permit me to make notes of any measurements whatsoever; also, I could remain only a very short time, because of the guide's haste to leave the spot; so my observations were extremely cursory. I was able to state the meteorite formed a sort of cliff . . . facing approximately southwest, while the northeastern side was completely buried in sand . . . . The upper surface was strongly polished by windblown sand. I found the small block up there, rounded on all its edges, that I delivered to M. Hubert. On one corner of the summit, facing west, I believe, erosion had exposed metallic needles sufficiently thick so that I could not break them or remove them. The metallic nature of the rock cannot be in doubt; as shown by examination of the block I gave to M. Hubert. I tried to detach one of the needles by striking it very hard with the small block. The block shows traces of the shock at each point where it struck . . . . The surface appearance of the mass was in no way comparable to that of the blackish, polished surface of the rocks that are found on the 'reg' and on the sandstone plateau of the Adrar. On my return, I made some notes of my observations, while my recollections still were fresh; but I have since moved about so much that I do not know where any of my notes are . . . (Monod & Zanda 1992, p. 50).

Ripert added that the natives had seemed anything but anxious for the existence of the iron to become known to Europeans. He realized this when he overheard some of the camel drivers at Chinguetti talking among themselves about the *Fer de Dieu* ('the iron fallen from the sky'). He asked about this from the head man of Chinguetti, Idi Ahmed Ould Seïn, who at first vigorously denied its existence. Only after long conversations did Sidi Ahmed consent to lead him, secretly, to the spot on condition that he would take nothing with him that might enable him to describe the exact position of the meteorite or to make notes. In January 1917, some months after their trip, Sidi Ahmed died suddenly, possibly through poisoning. Ripert said he did not wish to establish any connection between these facts, but he felt absolutely certain that the position of the meteorite was well known to all the village natives and, more particularly, to the nomads. He suggested that the giant meteorite, which was half-covered with sand in 1916, might be wholly covered by then (1932), but, even so, he was astonished it had not been found by the civilian and military expeditions. Perhaps the searches had been blocked by a conspiracy amongst the natives.

On 19 November 1932, Ripert paid a visit to Professor Bosler at Marseilles. On this occasion, Ripert told Bosler that he had made a slip when he said the iron lay to the SW of Chinguetti when, in fact, it lay among the dunes to the SE. He also remarked that the journey to the iron took about ten hours by camel, including what seemed to be several detours; he and his guide had left in the evening and arrived early the next morning. Bosler was most impressed with Ripert whom he judged to be a man of unquestionable honesty and good faith.

Gaston Ripert, the original source and a principal player in this story, was born in Oran, Algeria, in 1881, and remained in North Africa until 1902 when he entered the military school at Saint-Cyr in France. He obtained degrees in mathematics and natural science and he eventually learned to speak English, German, Italian, Spanish and Arabic plus several African dialects. In 1904, he graduated from Saint-Cyr as a second lieutenant of the colonial infantry and spent the next nine years at posts in Senegal and the Ivory Coast. In 1912 he was named a Chevalier of the Légion d'honneur.

Ripert left the army in 1913 and entered the colonial administration. However, with the onset of World War I on 2 August 1914, Ripert was mobilized and sent to the front. He was wounded twice, first in France and later, more seriously, in the Dardanelles campaign. There, he was cited for the excellence of his leadership as commander of his troops and decorated with the *Croix de guerre with palms*. In 1915, Ripert was discharged from the military hospital at Toulon as a 40% invalid and assigned to service in North Africa, which included two short tours of duty at Chinguetti in 1916 and 1917. Ripert's integrity was never questioned by anyone, not even by Malavoy, who doubted the existence of the giant iron meteorite. This faith in Ripert weighed heavily in prolonging the search and making the problem of the Chinguetti iron so difficult to resolve.

In September 1932, the Meteor Commission of the International Astronomical Union, meeting at

Cambridge, Massachusetts, passed a resolution urging the government of France to find the giant meteorite whatever the cost. The next day, an article by the astronomer, James Stokley (1900–1989), appeared in *The New York Times* of 6 September 1932, under the following headlines:[8]

ASK FRANCE TO HUNT AFRICAN METEORITE

Astronomers at Cambridge Meeting stress Importance of a Sahara Expedition

MILLION-TON MASS FELL

Magnetic Survey Suggested for Locating Phenomenon Described Some Years Ago.

TO DISPROVE VAPOR THEORY

Polar Year Stations Are Asked to Attempt to Determine Amount of Meteorite Dust Present.

Professor Bosler, present at the meeting, recommended that a magnetic survey should be made of the type used to locate ore deposits. He said the site could easily be reached from Port-Étienne with no danger to a scientific party except from brigands, and that the French Army, that was already in the region, would provide protection. Bosler pointed out that exploration of the Chinguetti area would be much easier than that of the great meteor explosion of 30 June 1908, in Siberia, which could be reached only through miles of forest where mosquitoes were especially bad.

Meteorite craters in the 1930s

Professor Bosler's remarks scarcely began to describe the difficulties of entering and working in the remote Tunguska region of Siberia, where Leonid A. Kulik (1883–1942) had led three expeditions in 1927, 1928 and 1929–1930 to the site of the explosion of a brilliant fireball that had been witnessed streaking northwestward for some 1500 km over central Siberia. Kulik discovered there a forest of huge trees that had been felled with their trunks fanning radially outward from a boggy area dotted with numerous depressions. He believed the depressions were meteorite craters and he had one of them drained without finding any trace of meteoritic materials. Nevertheless, Kulik believed it was imperative to continue the search and he called for an aerial survey of the Tunguska area—which finally took place in 1938. Meanwhile, many astronomers, including Bosler, agreed that the search should go on but were somewhat intimidated by Kulik's reports of the arduous efforts required to penetrate the devastated area and perform fieldwork there. A search of the region near Chinguetti would, indeed, be easier than that. (To this day, no meteorite fragments have been found at Tunguska, but calculations indicate that an immense explosion took place above ground demolishing whatever was left of the incoming body, whether it was a fragment of a comet or of a stony meteorite.)

Perhaps the most urgent voice raised at the IAU meeting in 1932 was that of Dr Charles P. Olivier (1884–1975), Professor of Astronomy and Director of the University of Pennsylvania's Flower Observatory. Olivier stressed the importance of finding the African iron, particularly because it would emphatically disprove theoretical calculations that a mass of this size hitting the earth would instantly be dissipated into vapour. The spreading of such views on high scientific authority, he said, has caused a complete cessation of work to find the large mass that was believed to have caused the meteor crater in northern Arizona.

Olivier was referring to calculations made in 1929 by Dr Forrest Ray Moulton (1872–1952), a mathematician at the University of Chicago, at the request of Daniel Moreau Barringer (1860–1929), the president of the Barringer Crater Company.[9] As early as 1906, Barringer had staked a mining claim on the Arizona crater, and after extensive drilling with no showings of ore, he still expected to locate a massive iron meteorite buried beneath the crater floor. Moulton calculated that the explosive impact of a large iron meteorite, striking the Earth at cosmic velocity, had excavated the crater and vaporized much of the metal. Barringer died shortly after receiving Moulton's assessment. Whether or not the report was a causal factor, Barringer's death did, indeed, lead to a cessation of work on the crater.

Until 1928, Meteor Crater was the only natural depression in the world for which an impact origin had been proposed. Although some scientists favoured such an origin for it, most geologists, especially in America, opposed it. In 1896, Grove Karl Gilbert (1843–1918), the chief geologist of the US Geological Survey, had reported his investigations of the crater and pronounced it to be volcanic (Gilbert 1896). Few geologists were prepared to challenge his authority. However, between 1928 and 1933, three more suspect impact craters, associated with iron meteorites, were discovered on three continents: in 1928, Daniel M. Barringer, Jr (1900–1962) identified the first one at Odessa, Texas; in 1932, A. R. Alderman found a group of 13 craters associated with iron meteorites, and with fragments of silica glass in the largest one, at Henbury, Australia; and in 1933, Harry St John Bridger Philby (1885–1960) reported finding two craters, rimmed and floored by black glass with a small iron meteorite nearby at Wabar in the Empty Quarter of Arabia. The presence of iron meteorites associated with these craters, provided strong circumstantial evidence of their impact origin.[10] Not until the 1960s would shock effects in the target rocks be developed as a reliable criterion for an impact origin of craters and structures that lacked meteorites.

Meanwhile, scientists faced the conundrum posed by large iron meteorites that lay on the surface of the Earth with no associated craters. Typically, meteorites orbiting at cosmic velocities begin to decelerate due to friction with air molecules as soon as they enter Earth's upper atmosphere. Most of them slow down to the velocity of free-fall and then plummet to Earth making nothing more than small pits in the soil. Large bodies, weighing ten tons or more, are more likely to retain some of their cosmic velocity and to collide with the Earth in an explosive, crater-forming impact. How, then, was it possible to account for certain huge irons that showed no signs of having disrupted the landscape where they fell. There was the 31 ton Ahnighito iron that lay on the rocky west coast of Greenland with no sign of a crater when it was first seen in 1897 by Lt Robert H. Peary; and the 70 ton Hoba iron, first recognized as a meteorite in 1920, that lay on undisturbed ground in the desert of Namibia (SW Africa). In addition, there was the alleged million-ton iron in the Adrar desert near Chinguetti in Mauritania.

In 1951, two astronomers (Richard N. Thomas, of the University of Utah and Fred L. Whipple, of Harvard University) published calculations showing that a huge iron could make a soft landing if it were to approach the Earth tangentially, at an angle of 1° or less, and follow a long trajectory until it decelerated to a terminal velocity of less than 1 km/sec. Without mentioning Chinguetti by name, they predicted that meteorites *very much larger* than Hoba could land without causing explosions (italics in original) (Thomas & Whipple 1951).

More than two decades later, in 1975, Robert F. Fudali, of the Smithsonian Institution, and Dean R. Chapman, of the NASA Ames Research Center in California, published calculations showing that a huge iron, plunging through the atmosphere at an angle of 3° or less, could strike Earth's surface, ricochet, land again, and slide to its resting place. For an iron the size of Chinguetti, they estimated the probabilities of such an event as being between 0.1–1%, but not zero (Fudali & Chapman 1975). (They said nothing about deep grooves it might make as it scoured its way across the ground.)

Sixteen days after Stokley's article appeared in *The New York Times*, a copy of the paper reached Jean Malavoy in Dakar. He responded with the following letter, which appeared in *The New York Times* of 10 October 1932:[11]

To the editor of The New York Times,

I have just learned of the article published in the Times of Sept. 6 entitled: 'Ask France to Hunt African Meteorite,' by James Stokley.

This meteorite is unfortunately a huge hoax. It was mentioned in 1920 to my predecessor by a soldier on duty in these desert regions. This soldier told him where this meteorite was, its dimensions (100 metres in length, 40 metres in width (sic), and probably 40 metres in height), and gave him a 4-kilogram sample of the meteorite. The latter, examined by M. Lacroix, a French expert, was found to be very interesting and of a species entirely new. Moreover, M. Lacroix asked the government that searches be made in order to find again this gigantic meteorite. Several officers, aided by numerous natives, to whom big rewards were promised, searched in vain for several years. I went through this region in 1930 without success.

Finally, I am convinced that the only extant meteorite is the specimen studied by M. Lacroix. What the soldier had taken for the principal mass of the meteorite was in reality a sandstone plateau, of which there are many in this region. These pieces of sandstone are very hard and are covered with a black film of iron oxide (desert varnish); we also find a little laterite. Any one who knows nothing of geology can, in utterly good faith, take it for meteoritic iron.

This is what happened: the soldier picked up the real meteorite (4 kilos, about 8.80 pounds) at the top of a plateau and believed that the entire plateau was of the same sort. I believe that it is useless to search further; it would be to lose much time and money.

J. Malavoy, Chief of the Geological Services of French Western Africa. Dakar, Senegal, Sept. 22, 1932.

Monod's investigations of the *Fer de Dieu*: 1934–1939

By 1934, the time had come for an investigation of the giant meteorite of Chinguetti by Théodore Monod, who went to Mauritania on behalf of the Muséum d'Histoire Naturelle in Paris. Monod was happy to go. He had spent four years leading a sedentary life in the laboratory, and welcomed the chance to return to Africa.

In preparing for his mission, Monod asked Professor Bosler for Gaston Ripert's addresses in Cameroon and in Nice. He then wrote to Ripert, asking for more details of his story, and he also wrote to Malavoy. Malavoy was the first to reply. In a letter dated 26 January 1934, he repeated his charge that the iron was a giant humbug and said it was useless for Monod to search for it. Malavoy regarded the malleable metallic needles Ripert had described on the summit of the iron mass as the result of an imagination overheated by the Saharan sun. He implied no bad faith on Ripert's part, but left him looking foolish.[12] Perhaps, we should note that 'malleable' was Lacroix's term, not Ripert's. But the description of the needles as 'malleable' lent some credence to the story because meteoritic Ni–iron is malleable and needles of it most likely would bend without breaking as Ripert said they did.

Monod left Paris on the evening of 13 March 1934, with two objectives: first, to go to Chinguetti to look for the 'giant, colossal, fabulous' iron meteorite said to be 100 m long by 40 m high; second, to go on to Asselar to study the rock formations where he and Bresnard had collected the fossilized human skeleton in 1927. When he arrived at Dakar, he was welcomed by the elder of his two brothers, Sylvain Monod (1896–1987), who was serving there as administrator of the colonies. At Saint-Louis, Monod encountered Captain G. de Linares, who had written to him on 4 May, too recently for him to have received the letter, about a blacksmith (unnamed) who had told him that at Chinguetti the blacksmiths used metal from the *Fer de Dieu*. The man went on to describe a block of iron as large as his house, around which lay small blocks. He added that the natives always avoided the site when accompanied by Europeans because they did not want infidels to know about it. Captain de Linares could add no details about its location, but his report did show that the story of the *Fer de Dieu* was a familiar one in the region.

On 5 May 1934, Monod arrived at Atar, the largest city of the Adrar, about 50 km from Chinguetti. There, he met with the commander of the military post and two of his lieutenants. One of them, Lt Larroque, told him of an old holy man who claimed to have seen the iron in 1913. Nothing came of this lead. Monod talked with one Baseïd, who had been in Chinguetti in 1916 and 1917 and had served as the principal interpreter to Ripert. Baseïd claimed to have accompanied Ripert on all of his local trips, but he had never heard him speak of the meteorite. Ripert, he said, had first arrived at Chinguetti on 26 March 1916, and remained there until 1 September. He then travelled southward to Tagant with a nomad group and stayed there for five months. On 23 January 1917, Ripert returned to Chinguetti and remained there until 28 May 1917. Ripert then left, by way of Dakar, for a post in Cameroon. Sidi Ahmed Ould Zeïn died in January 1917. Any night-time trip from Chinguetti to the iron meteorite would have taken place in the late winter or early spring of 1916 during Ripert's first stay there. Bazeïd was astonished to hear that Sidi Ahmed Ould Zeïn had guided Ripert to the meteorite because Sidi Ahmed was a villager, ignorant of the by-ways of the countryside. Clearly, Bazeïd doubted the existence of the meteorite, and he also doubted that the local people were keeping it a secret. He thought they simply knew of no such an object in their vicinity.[13]

After examining the local geology, Monod went on to Chinguetti where he arrived very early in the morning of 5 June, a day of overcast sky and blowing sand. Monod found Chinguetti to be the most 'Saharan' village of the Adrar. He described it in his journal as a classical oasis of the type popularized on coloured postcards—with a wadi and palm trees and an immense dune in the background. Chinguetti had an illustrious history that began toward the end of the thirteenth century when it served as the capital of the Maurs (Moors). It lay on a principal route of the salt caravans that travelled between Timbuktu and Saint-Louis, and the cities of Morocco. It also served as a gathering place for the annual pilgrimages to Mecca. At its height in the eighteenth century, Chinguetti had 12 mosques, 25 madrassas, five great libraries full of manuscripts, and a distinguished population of scholars and poets. Decline set in during the nineteenth century and by the time of Monod's visit in 1934 the population had dwindled to only a few thousand. Most of the historic buildings lay in ruins, and the old quarter was beginning to fill with sand. Today, one of the libraries survives and Chinguetti still ranks as the seventh most holy city of Islam. In 1996 it was listed as a World Heritage site in the hope of restoring some of its historic buildings and finding ways to halt the encroachment of the sands.

On his arrival, Monod posted copies of the following notice, in French and Arabic, at various locations in and around Chinguetti:[14]

To the residents of the region of Chinguetti.

The resident, chief of the subdivision of Chinguetti, makes known to all that the government wishes to acquire information on a very large rock fallen from the sky in the region of Chinguetti a very long time ago. This rock consists of iron and is as large as a house; at the summit, it bristles with many points that cannot be broken off but only bent over. The rock lies to the east or southeast of Chinguetti, among dunes covered with sbot; it is called 'the rock that fell from the sky', or the '*fer de Dieu*'. The government has sent to the region Monsieur Monod to study the matter, and it announces that the person who will guide Monsieur Monod to the rock, and he finds it, will receive from the Bureau of the Residence, the sum of one thousand francs.

This was the first time that a large sum of money had been offered publicly through official channels for information leading to the location of the meteorite. On 5 June 1934, Monod wrote in his Journal:

That will make some noise around the village: who knows? A happy chance may, one day or another, reveal whether the people are pretending or are simply ignorant of the location of that pebble (Monod 1997, p. 160).

He added that although Ripert's story was quite romantic and filled with inconsistencies (which he felt certainly were involuntary on Ripert's part), people in France believed in that 'accursed' rock, as did people in the entire world, except in Chinguetti where nobody said a word about it.

In Chinguetti, Monod interviewed Abd er Raman Ould Seïn, a brother of Sidi Ahmed, who had no memory of Sidi Ahmed's making such an overnight trip. Monod questioned a blacksmith, Abdallah Ould Beida, who had lived in Chinguetti for 53 years. The blacksmith said: '*My eyes have not seen it; my ears have not heard of it*'. The chief of the Laghlal tribe, Mohammed el Béchir, told him he knew absolutely nothing about the rock and added:[15] '*If you will search until the end of the world you never will find anyone who knows of this rock*'. Given such encouragement, Monod set out on 9 June to begin his own search.

With temperatures soaring, and beset by a series of violent sandstorms, Monod travelled through the countryside mostly to the SE of Chinguetti collecting samples and keeping notes of the stratigraphy, topography, prehistoric relics, and rock paintings and inscriptions. He noted that the prevailing winds of the Adrar blew from the north and NE, so he surmised that Ripert, without a compass, had become confused when he reported that the wind-polished cliff faced SW.

On 18 June, Monod returned to Chinguetti to ship boxes of specimens to Paris and to interview more people. Everyone with whom he spoke claimed total ignorance of the giant iron, and all struck Monod as being perfectly sincere. He learned that Mraitéré, an aging concubine of Ripert's, still lived in the city, and interviewed her in mid-July, when he returned from another foray into the desert. She said that aside from his trip with the nomad group, Ripert never passed a night outdoors. She had no knowledge of the story of the meteorite (Monod & Zanda 1992, p. 78).

Months later, Ripert responded to Monod from Cameroon in a letter dated 24 October 1934. He said, in part:

> . . . I know that the general opinion is that the stone does not exist: that, to some, I am purely and simply an imposter who picked up a metallic specimen on the 'reg' of Mauritania . . . that, to others, I am a simpleton who mistook an exposure of sandstone, blackened and polished as is so often the case in the country, for an enormous meteorite. I shall do nothing to undeceive them. I care very little for what any of them may think. I know only what I saw, because I saw it, and I could probably have found the mass again if I have been directed to do so when my visual memory and my age still permitted. And now my only chagrin is that I can no longer help you in your search (Monod & Zanda 1992, p. 59).

In reviewing the situation toward the end of 1934, Monod wrote that for the fifth time since 1924 careful searches had proved to be unproductive. He concluded, tactfully, that the dossier on the meteorite of Chinguetti, without being definitely closed, had largely evolved toward a negative solution (Monod & Zanda 1992, p. 79). But to Monod, himself, the case was closed, and he said as much in 1937 in the first edition of his popular book, *Méharées* (camel treks). There, he referred to it as the famous—but hypothetical—giant meteorite, the existence of which might be theoretically possible but, in his view, was most improbable. However, he admitted the fascination it held for those who hoped to discover such a colossal mass—by far the largest meteorite of the entire world.

The case was not closed for everyone, however, especially not for Professor Olivier and the Meteor Commission of the International Astronomical Union, which met in Stockholm in 1938 and, once again, passed a resolution urging the French government to find the meteorite. In later years, Monod remarked that these resolutions caused bad feeling among the French toward Americans, who seemed to believe they could easily find the iron if they were to take up the search themselves.

On 5 May 1938, the geologist, Andre Pourquié, flew over the Chinguetti region to search for the giant iron at the behest of the Société Française d'Exploration Minière. He noticed the feature known locally as 'the hole of Aouelloul', which lies about 40 km SW of Chinguetti. Thinking it might be a meteorite crater he visited it on the ground later that year and described it to Théodore Monod. World War II began in the autumn of 1939 and scientific explorations ceased in the Adrar, as they did in much of the world.

Post-war investigations of the *Fer de Dieu*

Monod's first post-war stirrings of interest in the Adrar region arose in 1950 when he arranged to make a flight over the Aouelloul crater and to investigate it in the field. By that time, he had the advantage of carrying a regional map[16] and some photographs made from the air in 1949 and 1950 by Lt Jacques Gallouédec, an officer of the French Air Force. At the crater, Monod found numerous fragments of black silica glass lying on the outer slope of the southeastern portion of the rim. By analogy with the black glasses reported at the Henbury and Wabar craters, he presumed (correctly) that these were impact glasses and sent them to the British Museum of Natural History [now the Natural History Museum, London] for comparative analyses. The results showed small quantities of nickel in the glass whereas the Ordovician sandstones in which the crater was excavated contained no nickel at all. Monod revisited Aouelloul in March 1951, and found additional glass fragments within the crater as well as outside it. This, he wrote, considerably strengthened the case for an impact origin. Later in 1951, Monod and Pourquié published a detailed description of the Aouelloul crater, with a plane-table map and photographs taken from every angle, including overlapping ones that created panoramic views (Monod & Pourquié 1951).

Monod wrote that the discovery of the Aouelloul crater, probably of impact origin, lying in the vicinity of the legendary iron of Chinguetti prompted him to reopen this subject, which he had previously regarded as closed. Once again, he wrote to Jean Bosler and to Gaston Ripert in hopes of learning something new from them. Bosler replied on 4 April 1951, expressing full confidence in Ripert and his story. He said that Ripert's long letter and his visit to him did not seem to offer the least possibility of trickery. Bosler reminded Monod that if M. Ripert had not been perfectly sincere, and sure of his facts, he would have taken good care not to visit with him and expose himself to the 'indiscreet fire of my questions'. He added that it was enough to talk with Ripert for a few moments to feel certain of the kind of man with whom one was dealing.[17]

On 6 June 1951, Ripert replied to Monod that if he indicated earlier that the huge mass lay to the SW of Chinguetti it certainly was a slip on his part (one which he already had corrected during his visit to Bosler); that zone, he said, consists of gravelly surfaces on bedrock, whereas the area in which he found the meteorite was one of high dunes lying SE of the city. To Monod's query about the whereabouts of the notes Ripert made on his return from the expedition, Ripert replied that he had no notes. He had made the traverse blindly, under the leadership of Sidi Ahmed, to whom he had promised not to take notes, and he regarded a promise to an indigenous person as being as valid as one to a European. As we observed earlier, Ripert had previously told Bosler (and he had also written to the Governor General of French West Africa) that he had made notes on his return while his memory was still fresh, but had lost track of them. We only can guess whether it was his original explanation or his new one that was accurate. Ripert added with some bitterness:

> I have been cruelly punished for my indiscretion in talking to my friend Hubert confidentially in a friendly style ... I strongly desire that in your geological exploration of the Adrar you will find the meteorite, and show that I am neither a 'farceur' nor an imbecile.[18]

Monod found Ripert's 'slip' of saying 'southwest' instead of 'southeast' to be totally incomprehensible. But, being much involved in other lines of research, he did not personally mount a new search for the iron at that time.

In 1952 Monod reviewed the history of the search in an article titled: *Le probléme de la météorite de Chinguetti*. In it he listed all of the puzzling inconsistencies in the story, but he pointed out that if, perchance, the huge mass lay among the dunes of Ourane, ten hours by camel to the SE of Chinguetti, it would be in territory that was very difficult to explore. A random search there would be useless. Therefore, before giving up all hope of finding the iron, he recommended that magnetic surveys should be made on the ground and in the air, if airborne magnetometers should prove to be feasible (Monod 1952).

In 1954, the American natural historian and meteoriticist, Lionel F. Brady (1880–1963), translated Monod's 1952 article into English and published it in Volume 1 of *Meteoritics* under the title: 'The Problem of the Chinguetti [French West Africa] Meteorite' (Brady 1954). In a companion paper, Lincoln Lapaz (1897–1985), the editor of *Meteoritics*, and his daughter, Jean, reviewed the history of the Chinguetti iron and printed their own translation of Lacroix' paper of 1924 (Lapaz & Lapaz 1954). They argued that the metallic character and dimensions of the mass as described by Lacroix' should be accepted as valid. They sharply criticized the searches that had been made and Monod's doubts about the iron that he had expressed in *Meharées* in 1937. Lincoln and Jean Lapaz expressed full confidence that the giant iron meteorite soon would be found. Seven years later, however, in their book *Space Nomads* (Lapaz & Lapaz 1961), they made no mention at all of the giant iron of Chinguetti. Clearly, they must have lost their faith in its existence during those intervening years.

In 1989, Monod remarked that between 1934 and 1987 all was quiet with respect to the giant meteorite. The Americans were calm, the French had abandoned the search, and the Bedouins never spoke of it (Jarry 1990, p. 171). Perhaps he was right with respect to the giant iron, but in the 1950s Monod, himself, made at least one little-known visit to the region to look for meteorites.

In 1954, William A. Cassidy, then a research assistant at the Institute of Meteoritics in Albuquerque, wrote to Monod asking for his advice in arranging for a visit to the Aouelloul crater.[19] Cassidy, who held the first Fulbright Fellowship to be awarded in meteoritics, had just completed a year-long investigation of tektites and impact craters in the outback of Australia and wished to visit Aouelloul on his way home. Monod graciously proffered his aid, and when Cassidy reached Dakar, Monod gave him an introductory letter to the Commander of the Camel Corps at Nouakchott, the capitol of Mauritania.

Members of the Corps drove Cassidy and a Swiss companion, Ernesto Släpfer, into the desert and left them at the crater with a tent, food, water, a stove, and all the necessities for a week or more of camping. One of the cameliers told Cassidy that Monod previously had commandeered the entire camel corps to search that area for meteorites.

He had lined up the men, side-by-side, a few metres apart, and had them walk slowly across the desert leading their camels. Each time one of them saw a strange rock, he signalled to Monod who rode over to examine it. Unfortunately, this exhaustive approach yielded no meteorites.

At their camp, Cassidy and Släpfer rented camels from the local Arabs and explored the area. From the crater rim they could see the black profile of guelb Aouinet, 20 km or so to the SE. It looked rather like a 100 m meteorite partially buried in sand so they rode over and climbed it. It lies about 35 km SW of Chinguetti (as the crow flies), where Ripert originally said he saw the huge iron meteorite and collected the small one.

Mauritanean craters

Having collected the glass fragments that eventually would guarantee an impact origin for the Aouelloul crater, Monod took an interest in other craters. In 1954, he presented a paper entitled '*Accidents circulaire où cratériformes*' at the 19th International Geological Congress in Algiers (Monod 1954). ('Accidents' are any features that interrupt regional structures without obvious cause). Monod described five Mauritanian craters: Aouelloul, Tenoumer, Temimichat Ghallaman, Richât, and Semisayat. Although one investigator, A. Allix (1951) had suggested that the large Tenoumer crater, 1.8 km in diameter (about 50% wider than Meteor Crater in Arizona) might be of impact origin, Monod concurred with the general consensus that, with the exception of Aouelloul, all of these craters were caused by terrestrial explosions. For Aouelloul, he advanced the cautious statement that a meteoritic origin could not be excluded. In 1966 such an origin would be fully confirmed when Edward Chao and his colleagues, of the US Geological Survey, found minute spherules of metallic nickel–iron in the Aouelloul glass (Chao *et al.* 1966).

On 4 October 1957, the Space Age dawned with the orbiting of Sputnik I, and suddenly, meteorites ceased to be natural curiosities of limited interest and became high priority objects of research among scientists interested in them as pieces of other planets and as missiles orbiting through space. By then, large areas of the world had been photographed from the air and scientists pored over the maps searching for circular features to be investigated on the ground. Organizations in several countries compiled lists of circular craters or scars designating them as being of proved, probable, or possible impact origin.

In 1966, Monod and C. Pomeral published a field and laboratory study of the Tenoumer crater (Monod & Pomerol 1966, p. 165). They identified siliceous rhyodacitic lavas in dykes outside the rim and concluded that the crater was a small volcanic caldera in granitic country rock. By that time, studies of special features created in quartz and other minerals by instantaneous impact shocks were well advanced. In 1971, such features were reported in a suite of samples from the Tenoumer crater that Monod had supplied to a team of investigators led by Bevan French of the NASA Johnson Space Center. French and his colleagues showed that Monod's 'rhyodacitic lavas' were, in fact, impact melt rocks (French *et al.* 1970).

Meanwhile, in 1969, Robert S. Dietz, of the National Oceanographic and Atmospheric Administration, and two colleagues found evidence of an igneous origin for both the Richât, a strikingly ringed crater frequently used as a landmark by pilots during World War II, and the Semisayat dome (Dietz *et al.* 1969). Thus, of Monod's five Mauritanian 'cratériform accidents', two were of impact origin and two were volcanic. That left the crater of Temimichat Ghallaman, with its dramatically hexagonal rim, *c.* 670 m across, the origins of which remain uncertain. However, three of the craters, Aouelloul, Tenoumer, and Temimichat Ghallaman, lie on a straight line that trends northeastward across the desert for 600 km. In 1971, W. A. Cassidy, then of the University of Pittsburgh, and R. Fudali conducted gravity surveys of these three Mauritanian craters and suggested, without conclusive proof, that all them may be impact craters excavated by three fragments of a single body, which broke up during a long, low-angle flight through the atmosphere (Fudali & Cassidy 1972).

In 1973, Raleigh Drake, a member of a Smithsonian team searching for impactite fragments found a small, 76 g stony meteorite on the rim of the Aouelloul crater. His discovery was particularly surprising because the stone was mostly covered by a fusion crust that clearly distinguished it from all the country rocks. Furthermore, the meteorite lay on the southeastern portion of the rim which already had been picked-over by searchers for glass fragments on their hands and knees. From the first, it seemed doubtful that the small stone could have been a fragment of the body that excavated the crater. These doubts were confirmed when radiometric age-dating, reported in 1976 by Robert Fudali and Philip Cressy, showed that the stone fell to the crater rim less than 350 000 years ago but the impact event that formed the crater occurred about 3.1 million years ago (Fudali & Cressy 1976). Inasmuch as its presence at the crater was strictly coincidental, the small stone was named Zerga for a prominent ridge nearby.[20]

Final searches for the *Fer de Dieu*

In 1987, Monod was 85 years old and had failing eyesight. However, he possessed a curiosity which he said left him defenceless, and he travelled to Mauritania to examine the dunes SE of Chinguetti where a semi-circular feature had been reported in 1980 by Jacques Gallouédec, an aviator who was familiar with the topography of Mauritania. Monod realized that this might be a meteorite crater located approximately where Ripert had, on second thought, placed his giant iron. So he travelled there with Bruno Lamarche, an ornithologist from Nouakchott. They found nothing meteoritic.

Monod went back to the Adrar in January 1988, with a film director, Karel Prokop, who was equipped with a global positioning system. To Monod the GPS was a fabulous device, well known to mariners but ignored by camel drivers of the boundless, sandy deserts. Prokop recorded Monod's activities in the field on that excursion and on a later one they made in December of the same year. From his films, Prokop produced two television specials titled: *The Old Man and the Desert* and *The Old Man, the Desert, and the Meteorite*, to much acclaim in France.

On their second trip, Monod and Prokop travelled southwestward, from Chinguetti in the direction originally indicated by Ripert. Monod searched for more samples of impact glass around the Aouelloul crater, and, after 10 days in the area, *VOILA*: there it was! A large black butte on the horizon about 40 m high and 100 m long—the proportions originally reported by Captain Ripert. Monod had seen it many times before, but from this particular viewpoint he suddenly concluded that the feature Ripert had visited in 1916 was guelb Aouinet, itself. Ripert, he decided, had mistaken this sandstone and quartzite butte, covered with a black patina, for a huge mass of metallic iron (Monod & Zanda 1992, p. 93).

Monod pondered one final question: if this was their destination, why did Sidi Ahmed Ould Zeïn lead Ripert to guelb Aouinet? Perhaps it was to show him some iron ore. Malavoy had mentioned occurrences of a little laterite in the sandstone plateau in his letter of 22 September 1932 to the *New York Times*. And on this trip, in December 1988, Monod found lateritic iron oxide at the foot of guelb Aouinet. From this material it would be possible to extract metal. Such practices were common in other parts of the Sudan and blacksmiths may have pursued them here in ancient times, or even in the recent past. Monod had stated in his article of 1952 that he felt certain that iron smelting had not occurred in this area, but now he suggested that searches be made for evidence of such activities. If this were the answer, the trip by night and the guide's urgent desire to leave at dawn may simply have been because the sun would become too hot later in the day. Interestingly, Ripert never mentioned it, and it seems that no one ever asked him about the trip back to Chinguetti, which would have extended their absence up to twenty hours and made their outing doubly hard to conceal. And most of the return trip would have been made under the full glare of the Saharan sun regardless of how early they had started. Ripert's silence on this part of the story led Baseïd to speculate that the visit to the meteorite had been made as a side trip on the long trek with the nomad group to Tagant. But there were no reports to confirm Ripert's leaving the main party for any significant amount of time.

By the late 1980s, a new generation of scientists wanted to see the area for themselves, and Monod was glad to lead them back to his '*diocèse*'. Early in 1989, Monod returned to Chinguetti accompanied by four leading meteoriticists: Paul Pellas, the curator of meteorites at the Muséum d'Histoire Naturelle, Claude Perron, also of the Museum, P. Pontes of the University of Paris, and Christian Koerberl, of the University of Vienna. No one seriously expected to see the large iron, but they searched, without success, for additional meteorites like the small mesosiderite Ripert had collected.

Once again, Monod decided that the case was closed, but, since the iron lived on in the minds of many astronomers and geochemists, he felt it was necessary to set the record straight. In June 1989, Monod declared to the Académie des Sciences in Paris that the mystery of the giant meteorite of Chinguetti had, without doubt, been solved. Ripert had mistaken guelb Aouinet for a giant iron meteorite. In August, he repeated his declaration to the Meteoritical Society at its annual meeting in Vienna. Monod (1989) published his final word on this subject in an article titled, 'New data on the problem of the Chinguetti giant meteorite of the Adrar, Mauritania' in *Comptes Rendus* of the Académie de Sciences:

> The existence of a giant meteorite in the Adrar of Mauritania, largely accepted since 1924, must now be abandoned. There was a mistake on the nature of the rock of a butte that is entirely sedimentary with no trace of metal.

Monod's conclusion found support from the work of two meteorite specialists, Sara Russell, of the Natural History Museum, London, and Phil Bland, of the University of California, who visited the Adrar carrying a magnetometer sensitive to changes in the Earth's magnetic field. It registered no significant variations. While there, Bland searched in vain for objects fashioned from meteoritic nickel–iron being used by the local peoples.

Subsequently, a team of investigators led by K. C. Welton at the University of California, and including Bland and Russell, reported that its content of cosmogenic isotopes showed that the small stony-iron meteorite collected by Ripert had orbited through space as part of a body with a radius no larger than about 80 cm, and that it landed on the Earth about 18 000 years ago (Welton *et al.* 2001). These findings revealed the small mesosiderite to be an individual meteorite and ruled out any possibility that it could have been split off from an immense mass of the same composition, as had been maintained by Ripert.

According to Claude Perron, the small meteorite in Paris shows none of the pits or scratches that Ripert had said it did when he used it to pound on the metallic needles on the large mass. This brings up the question of what those metallic needles were? Monod wrote in 1952 that it would be psychologically improbable for Ripert to have invented his stories, and he specifically referred to the needles as details he felt would be would be inexplicable if Ripert had not actually observed them. To Malavoy, however, the needles lent an air of fantasy to the story. He did not hesitate to call them products of Ripert's overheated imagination. No one published a natural explanation of the needles, such as one suggested to the author by William Cassidy in 2003. Cassidy said he had seen eroded irons in which projections of the Ni-rich metal, taenite, were left standing after the wearing down of the adjacent Ni-poor metal, kamacite. The taenite 'spikes' bend rather than break. Cassidy's explanation would be the most satisfactory one, by far, if the giant iron actually existed.

In December 1991, Monod returned to guelb Aouinet (Fig. 3) with Brigitte Zanda, a meteorite specialist at the Museum. After their trip, Monod and Zanda (of the Muséum national d'Histoire naturelle, Paris) co-authored their book, *Le Fer de Dieu: Histoire de la Météorite de Chinguetti*, that appeared in 1992. It is the definitive history of this quest based on Monod's field observations and their extensive archival research. Towards the end of the book they pose a series of questions and fashion two possible scenarios from the answers. Their basic question is: Do we accept Ripert's story, or do we not? Alfred Lacroix accepted it uncritically, as did Charles Olivier and other members of the International Astronomical Union. Jean Bosler accepted it after conducting a two-hour interview with Ripert. Monod willingly took up the search, knowing that others had failed, in the expectation that, if the giant iron lay anywhere within a 45 km radius of Chinguetti, his offer of a rich reward would tempt someone from the local population to lead him to it. When all searches failed, on the ground, from the air, and with sensitive magnetometers, Monod concluded that there was no giant iron in the Adrar and that Ripert, instead of wholly fabricating his story, had mistaken guelb Aouinet for a mass of metal.

Fig. 3. Théodore Monod near the summit of guelb Aouinet in January 1992 (Courtesy of the photographer, Brigitte Zanda).

However, after seeing guelb Aouinet for themselves both Brigitte Zanda and Claude Carron found it almost impossible to accept it as the landmark described by Ripert, unless, as Zanda put it, he saw it at night, at a distance, in a fog (Padirac 2002, p. 12). Guelb Aouinet is not a parallelepiped but a peak at one end of a long ridge, and, according to both Zanda and Carron, it is fully recognizable at any distance as consisting of the local sandstone and quartzite. (Perhaps we may speculate that it was only after the dimming of his own eyesight that Monod identified guelb Aouinet with the giant iron.) Zanda became convinced of Ripert's competence in geology when she found his notes, stored in the archives in Paris, accompanying a rock collection he had submitted to the Museum. Without fully believing that the giant iron exists, she would have liked to find a logical explanation for Ripert's story.

Leonard J. Spencer (1870–1959), a former curator of meteorites at the British Museum (Natural History) once suggested there may have been a misprint of m for cm in Lacroix' account: instead of measuring 40 × 100 metres, the iron actually measured 40 × 100 centimetres (Hey 1966, p. 105). That idea never gained acceptance although it offers solutions to several problems. An iron that size easily could be an unfissured parallelepiped displaying a vertical face with a scooped-out base, a high polish on surfaces exposed to wind erosion, and protrubing spikes of taenite on an upper corner. Ripert could have picked up the small meteorite which, by coincidence, lay on top of the larger one without having to make an arduous climb to the summit of a 40 m cliff with his guide continually badgering him to leave the area. After they left, the iron could have been covered completely by great thicknesses of sand that shielded it from magnetometers, particularly if it lay in the dunes to the SE of Chinguetti. If there were a misprint, when did it take place: in the original letter that Hubert sent to Lacroix? Or in Lacroix' report? Either way, Ripert had several opportunities to correct an error of such magnitude in his interview with Bosler and in his correspondence with Monod, but he did not do so.

Perhaps another reason why the idea of a misprint has not been welcomed is that it demolishes the fable of the gigantic iron, which most of us would like to be true. Many retellings of the story invoke the spirit of the Arabian Nights: the secretive trip without a compass or measuring devices, the massive cliff of polished iron, and the subsequent poisoning of the guide. One recent version strikes the right chord by describing how, in the first light of dawn, the enormous metallic mass appeared on the horizon like a huge ship with polished metal sides, sinking in a sea of sand (Padirac 2002, p. 9). Compared with this, Ripert's descriptions are quite matter-of-fact. But our imaginations still nourish the image of that enormous iron meteorite lost in the desert near Chinguetti.

If Monod's conclusion was correct, and there is no giant iron, numerous questions remain unanswered. Where did Ripert collect the small meteorite? Did he pick it up on top of guelb Aouinet? Did he find it on another trip and mix up his samples? Did he, in fact, give that specimen to Hubert? Or did Hubert mix up samples and send Lacroix one of his own? We never shall know. Today, that small meteorite is listed under the name 'Chinguetti' in the authoritative *Catalogue of Meteorites* published by the Natural History Museum, London (Grady 2000, p. 144). The *Catalogue* describes the small mesosiderite and then perpetuates the story of the large iron: 'An enormous mass, said to be 100 m long and 45 m high, with other smaller masses, was found in the desert of Adrar, about 45 km SW of Chinguetti ... there are doubts about the existence of the huge mass ...'

In 1999, Monod was preparing for one more trip to Chinguetti when he suffered a crippling stroke at the age of 97. Théodore Monod died at Versailles on 22 November 2000. Over the years, his attitude toward the giant meteorite of Chinguetti had softened somewhat. On his trip with Brigitte Zanda, Monod stood near the summit of guelb Aouinet (Fig. 3) and asked:

> Do we ever have to abandon all hope? Is it not perhaps a good thing that by refusing to give in to the evidence, the dreams that lie half awake in us all may persist? (Monod & Zanda 1992, p. 116).

I wish to thank B. Zanda for an electronic copy of her photograph of Théodore Monod on guelb Aouinet and for her permission to use it. I also thank W. Cassidy for information he provided to me, and my reviewers for very helpful suggestions after their critical readings of my manuscript.

Notes

[1]Vray (1994) has been used a principle source of information on Monod's life and career.

[2]Monod described his first caravan trek in *Maxence au Désert* (Monod 1995). In the volume, he called himself 'Maxence' and told the story in the third person.

[3]For a lyrical expression of his responses to deserts as types of seas, see Chapter 1 in Monod's book *Méharées* (1937).

[4]A photograph of Monod riding a camel in his '*diocèse*' may be found in Hureau 2002, p. 32. This book was written to accompany an exhibit of the same name held at the Muséum from May to October of 2002, the 100th anniversary year of Monod's birth.

[5]Cyrille Monod 1997, *Les Carnets de Théodore Monod.*

[6]Monod & Zanda (1992) is the most complete and authoritative source of information on the search for the giant meteorite.

[7]Unless otherwise noted, English translations from the French were made by the author.

[8]The headlines and the text of J. Stokley's article are reproduced from *the New York Times* of 6 September 1932 as Figure 10 in Monod & Zanda 1992, p. 34.

[9]Moulton's report is included as Appendix B in Abrahams (1983), pp. 296–309.

[10]For a review of early crater discoveries, see Marvin (1986, 1999).

[11]Malavoy's letter of 5 September 1932 is reproduced as it appeared in *The New York Times* of 10 October 1932 by Monod & Zanda (1992, Fig. 11, p. 35).

[12]See Monod & Zanda 1992, pages 34–35 for passages from Malavoy's letter of 26 January 1934 to Monod; and Figure 12 for an extract in Malavoy's handwriting.

[13]For Bazeïd's testimony about his association with Ripert, see Monod & Zanda 1992, p. 68–69.

[14]The text of Monod's offer of 1000 francs for information leading to his discovery of the giant meteorite is given by Monod and Zanda (1992, p. 11), and also on the back cover of their book.

[15]Monod & Zanda 1992, p. 70, quote this statement from Monod's Journal. Elsewhere, Monod, himself, and other authors have substituted '*the last judgment*' for '*the end of the world*': (e.g. Vray 1994, p. 213; Monod 1997, p. 160; Monod 1989, p. 126). Since both Christians and Moslems anticipate the last judgment at the end of the world, we cannot be certain which expression Mohammed el Béchir actually used.

[16]American Flight Chart FC-62 (1947).

[17]This passage from Bosler's letter of 4 April 1951 to Monod is quoted in L. F. Brady's English translation of Monod (see Brady 1954, p. 314, note 9).

[18]Passage from Ripert's letter to Monod (in Monod & Zanda 1992, p. 125, note 23).

[19]Cassidy, William. 2003. Personal communication.

[20]The Zerga meteorite is an ordinary chondrite (LL6) described by Grady (2000, p. 543).

References

ABRAHAMS, H. J. (ed.) 1983. *Heroic Efforts at Meteor Crater, Arizona*. Farleigh Dickinson University Press, Madison and Teaneck, New Jersey.

ALLIX, A. 1951. Note et correspondance à propos des cratères météoritiques. *Revue de Géographie de Lyon*, **26**, 357.

BRADY, L. F. 1954. The problem of the Chinguetti [French West Africa] meteorite. *Meteoritics*, **1**, 308–314.

CHAO, E. T. C., DWORNIK, E. J. & MERRILL, C. W. 1966. Nickel–iron spherules from Aouelloul glass. *Science*, **154**, 759–765.

DIETZ, R. S., FUDALI, R. F. & CASSIDY, W. A. 1969. Richât and Semsiyat domes (Mauritania): not astroblemes. *Bulletin of the Geological Society of America*, **80**, 1367–1372.

FRENCH, B. M., HARTUNG, J. B., SHORT, N. M. & DIETZ, R. S. 1970. Tenoumer Crater, Mauritania: Age and petrologic evidence for origin by meteorite impact. *Journal of Geophysical Research*, **75**, 4396–4406.

FUDALI, R. F. & CASSIDY, W. A. 1972. Gravity reconnaissance at three Mauritanian craters of explosive origin. *Meteoritics*, **7**, 51–70.

FUDALI, R. F. & CHAPMAN, D. R. 1975. Impact survival conditions for very large meteorites, with special reference to the legendary Chinguetti meteorite. *Smithsonian Contributions to the Earth Sciences*, **14**, 55–62.

FUDALI, R. F. & CRESSY, P. J. 1976, Investigation of a new stony meteorite from Mauritania with some additional data on its find site: Aouelloul Crater. *Earth and Planetary Science Letters*, **30**, 262–268.

GILBERT, G. K. 1896. The origin of hypotheses, illustrated by the discussion of a topographic problem, *Science*, new series, **3**, 1–24.

GRADY, M. M. 2000. *Catalogue of Meteorites*. 5th edn. The Natural History Museum, London. Cambridge University Press, Cambridge.

HEY, M. H. 1966. *Catalogue of Meteorites*. 3rd edn. The British Museum (Natural History), London.

HUREAU, J.-C. 2002. *Le Siècle de Théodore Monod*. Muséum National d'Historie Naturelle, Actes Sud.

JARRY, I. 1990. *Théodore Monod*. Plon, Paris.

LACROIX, A. 1924. Sur un nouveau type de fer météoritique trouvé dans le desert de l'Adrar en Mauritanie. *Comptes Rendus de l'Academie des Sciences*, Paris, **179**, 303–313.

LAPAZ, L. & LAPAZ, J. 1954. The Adrar (=Chinguetti), Mauritania, French West Africa, Meteorite. *Meteoritics*, **1**, 187–196.

LAPAZ, L. & LAPAZ, J. 1961. *Space Nomads: Meteorites in Sky and Laboratory*. Holiday House, New York.

MARVIN, U. B. 1986. Meteorites, the Moon, and the history of geology. *Journal of Geological Education*, **34**, 140–165.

MARVIN, U. B. 1999. Impacts from space: the implications for uniformitarian geology. *In*: CRAIG, G. Y. & HULL, J. H. (eds) *James Hutton—Present and Future*. Geological Society, London, Special Publications, **150**, 89–117.

MONOD, C. (ed.) 1997. *Les Carnets de Théodore Monod*. Le Pré aux Clercs, Paris.

MONOD, T. 1937. *Méharées: Exploration au Vrai Sahara*. Je Sers, Paris [and all later editions through 1989].

MONOD, T. 1952. Le problème de la météorite de Chinguetti. *Bulletin de la Direction des Mines du Gouvernement Général de l'Afrique Occidentale Française*, No. 15, **II**, 407–410.

MONOD, T. 1954. Sur quelques accidents circulaire où cratériformes du Sahara occidental. *Comptes rendus International Geological Congress*, 19th, Algiers, 1952, Part 2, 85–93.

MONOD, T. 1989. Donnes nouvelles sur le problème de la météorite geante de Chinguetti, Adrar (Mauritainie). *Comptes Rendus Académie des Sciences*, Paris, **309**, series 2, 547–552.

MONOD, T. 1995. *Maxence au Désert: Journal de Route d'un voyage en Mauritanie de Port-Étienne a Saint-Louis (14 octobre–12 novembre 1923)*. Terres d'Aventure, Actes Sud, Arles.

MONOD, T. & POMEROL, C. 1966. Le Cratére de Tenoumer (Mauritanie) et ses laves. *Bulletin de la Société Géologique de France*, series 7, **8**, 165.

MONOD, T. & POURQUIÉ, A. 1951. Le cratère d'Aouelloul (Adrar, Sahara occidental). *Bulletin de l'Institute Français d'Afrique noire*, **13**, 293–304.

MONOD, T. & ZANDA, B. 1992. *Le Fer de Dieu: Histoire de la Meteorite de Chinguetti*, Terres D'Aventure, Actes Sud, Arles.

PADIRAC, D. 2002. Chinguetti: The Quest for 'The Other Meteorite'. English translation by Sally Sutton. *Meteorite*, **8**, 8–12.

THOMAS, R. N. & WHIPPLE, F. L. 1951. The physical theory of meteors. II. Astroballistic heat transfer. *Astrophysical Journal*, **114**, 448–465.

VRAY, N. 1994. *Monsieur Monod: Scientifique, Voyageur et Protestant*. Actes Sud, Paris.

WELTON, K. C., BLAND, P. A. *ET AL*. 2001. Exposure age, terrestrial age and pre-atmospheric radius of the Chinguetti mesosiderite: Not part of a much larger mass. *Meteoritics & Planetary Science*, **36**, 939–946.

The geological travels of Sir Charles Lyell in Madeira and the Canary Islands, 1853–1854

L. G. WILSON

797 Goodrich Avenue, Saint Paul, Minnesota 55105, USA (e-mail: wilso004@umn.edu)

Abstract: Throughout his life, Lyell travelled extensively, always as a keen observer. He viewed the Earth's geological history as continuous with and subject to the same processes of change as at present. Leopold von Buch's theory of craters of elevation contradicted Lyell's view of Earth history. Thus Lyell travelled to Madeira and the Canary Islands in 1853 to see von Buch's evidence. Lyell found the islands formed by a long series of volcanic eruptions, not by the single explosive upheaval that von Buch had described. Nevertheless, Lyell still accepted Léonce Élie de Beaumont's claim that lava flows could not form compact rock on steep slopes. In 1855, Lyell learned from Eilhard Mitscherlich that on Stromboli contemporary steeply inclined lava flows were forming solid rock. In 1857, Lyell went to Sicily where unmistakable evidence contradicted Élie de Beaumont. In the walls of the Valle del Bove, steeply inclined layers of lava were intersected by dykes that pointed towards a former centre of eruption at Trifoglietto, later buried by volcanic rocks emitted from the present centre of eruption at the summit of Etna, proving that the Valle del Bove could not have originated as a crater of elevation.

Lyell the traveller

For Charles Lyell (1797–1875) travel was part of his life from an early age. Although born in Scotland, he grew up in Hampshire on the edge of New Forest. His father made annual trips to Scotland to supervise affairs on the family estate of Kinnordy in Forfarshire. When Lyell was seventeen and had just completed school, he accompanied his parents to Scotland to spend almost three months at Kinnordy. During that time, he joined his father and William Jackson Hooker (1785–1865) on plant-collecting expeditions, Mr Lyell being an amateur botanist. No doubt the interaction with men in the scientific community piqued his interest, for in 1817, Lyell spent three weeks as a guest of Dawson Turner (1775–1858) at Yarmouth where he examined the geology of the Yare delta. Later that year, Lyell and two college friends visited the west coast of Scotland to see the basalt pillars on the island of Staffa and Fingal's Cave, rendered famous by Sir Joseph Banks's (1743–1820) description of them (Banks 1772, **1**, 299–309).

Lyell with his parents and sisters made an extended carriage tour through France and Switzerland and into Italy as far as Florence during the summer of 1818. In the Swiss Alps, Lyell observed the effects of advancing glaciers and of an extraordinarily violent flood at Martigny caused by the sudden release of water from a lake dammed up by a glacier on the Dranse River (Lyell 1830–1833, **1**, pp. 194–196). While a law student at Lincoln's Inn in 1822, Lyell began to study the geology of the SE of England, an interest he shared with Gideon Mantell (1790–1852), a surgeon at Lewes in Sussex. Lyell's geological pursuits took him on various journeys into Sussex and Kent and to the Isle of Wight. In 1823, he spent more than two months in Paris, where he became acquainted with Georges Cuvier (1769–1832), Alexander von Humboldt (1769–1859) and Constant Prevost (1787–1856). Von Humboldt was famous as a world traveller and Prevost had travelled widely throughout Europe. On his return home, Lyell began to compare the ancient freshwater formations of the Paris and Hampshire basins with the modern sediments forming in Scottish lakes. The types of rocks and the fossils they contained were similar. Lyell demonstrated that no fundamental difference existed between ancient and modern lake sediments (Lyell 1826).

When George Poulett Scrope's (1797–1876) *Geology of Central France* was published in 1827, Lyell wrote an extended review of it. The book so aroused Lyell's interest in the extinct volcanoes of France that in the summer of 1828 he travelled with Roderick Murchison (1792–1871) and Mrs Murchison to see these ancient volcanoes (Lyell 1827; Kölbl-Ebert 2007). During two months of intense geological work in Auvergne, Velay, and Vivarrais, Lyell and Murchison found decisive evidence for the repeated excavation of river valleys by running water. The frequent similarity of the Cenozoic freshwater formations to much older Mesozoic formations, such as the New Red Sandstone (Triassic) or the Chalk (Cretaceous) in England, also struck them forcibly. Lyell was

From: WYSE JACKSON, P. N. (ed.) *Four Centuries of Geological Travel: The Search for Knowledge on Foot, Bicycle, Sledge and Camel.* Geological Society, London, Special Publications, **287**, 207–228.
DOI: 10.1144/SP287.17 0305-8719/07/$15.00

impressed by the evidence for the slow accumulation of laminated marls, hundreds of feet thick, their paper-thin foliations each representing a single year's accumulation of the cast off valves of the crustacean *Cypris*. The *Cypris* valves, together with caddis fly larva cases, or 'indusiae', found in strata of 'indusial' limestone, showed that the strata had been deposited in clear, calm lake water. At Aix-en-Provence, Lyell and Murchison examined another Cenozoic freshwater formation containing an abundance of fossil insects. A few miles away at Faveau, they found a much older Secondary (Mesozoic) freshwater formation, containing shells and seams of coal, lying beneath Secondary marine strata.[1] The coal deposits, accumulated on land, provoked Lyell, in writing to Gideon Mantell, to exclaim 'so much for the aboriginal universal ocean!' He concluded by adding that 'one must travel over Europe to learn how completely we are in our infancy in the knowledge of the ancient history of the globe'.[2]

Excited by the geology that he and Murchison had seen in France, Lyell decided to continue south through Italy to see active volcanoes: Vesuvius at Naples and Etna in Sicily. Murchison urged him to go, even though Murchison himself could not make the journey. They travelled together as far as Padua in northern Italy. There on 26 September they parted, the Murchisons to travel north through the Tyrol, Lyell to go south.

At Nice, Lyell had learned that the strata along the French coast contained many shells belonging to living Mediterranean species. In Italy, the Subapennine formation, a Cenozoic formation, contained a similar proportion of living species among its fossil shells. At Florence, Lyell learned that the travertine strata of the Arno Valley constituted a still younger freshwater formation overlying the Subapennine strata. The fossil shells and Chara species of the travertine all belonged to living species. Since in France the freshwater strata had undergone uplift and been inclined by volcanic activity, Lyell anticipated that in the active volcanic region around Naples he would find more recently elevated strata. That is exactly what he did find. On the island of Ischia, just NW of Naples, high on the sides of the extinct volcano Mount Epomeo, Lyell collected thirty species of fossil sea shells from clay strata that all belonged to living Mediterranean species. The shells showed that volcanic eruptions on Ischia had been accompanied by large-scale elevation. Lyell thought of the Mediterranean marine fauna as relatively recent. His impression of the recentness of the elevation of Ischia was strengthened by the evidence of the Temple of Serapis on the shore of the Bay of Baiae at Pozzuoli, near Naples. During the historical period, the Temple of Serapis had undergone both subsidence and re-elevation. The striking phenomena of the temple, together with the burial of Herculaneum and Pompeii by volcanic eruptions, convinced Lyell of the power of contemporary volcanoes and earthquakes to produce geological changes. As he continued his journey through Sicily, Lyell came to recognize that contemporary causes of change had been active for an extremely long period of time.

After sailing from Naples to Messina, Lyell rode on muleback south along the east coast of Sicily. From Zafferana he rode into the Valle del Bove, the great depression on the eastern side of Mount Etna. Its surrounding cliffs provided a section into the heart of the great volcano, showing that it had been built up by a long series of volcanic eruptions. In clay strata at the base of Etna, forming the foundation upon which the great volcano rested, Lyell obtained fossil shells of modern Mediterranean species, many retaining their original colours. From the immense number of volcanic eruptions required to produce the massive bulk of Etna, which had grown relatively little during historical times, Lyell realized that Etna must have erupted over hundreds of thousands of years. Yet, the modern shell species of the Mediterranean were still older. Etna gave Lyell an intimation of how much greater the vista of past geological time might be than he, or any of his contemporaries, had yet imagined.

After studying Etna, Lyell rode southward over the limestone plateau of the Val di Noto. Its hard white rock, containing only casts of shells, reminded him of the English Oolite, a Jurassic formation. At Syracuse, Lyell found, in soft blue clay strata, older and lower than the Val di Noto limestone, beautifully preserved shells of living Mediterranean species. The Syracuse fossils so startled him that he thought he must have mistaken the order of the formations. After criss-crossing southern Sicily, he finally found the whole succession of Sicilian strata revealed in a great escarpment at Enna in the interior of the island. Although many of his observations during his tour through France and Italy caused Lyell to question the geology he had been taught, the discovery that a hard limestone containing casts of shells was geologically recent truly astonished him. 'All idea of attaching a high antiquity to a regularly stratified limestone, in which the casts and impressions of shells alone were discernible vanished at once from my mind'.[3] The fossil shells from Sicily were almost entirely living species. By contrast, the fossils of the Subapennine beds of Italy included about 40 per cent of extinct species. Deeply impressed by all that he had learned during his eight-month journey through France and Italy, and especially by the revelations of Sicilian geology, Lyell wrote from Naples to Roderick Murchison describing the geological principles that he intended to incorporate in the book he was planning. These principles were 'that *no causes whatever* have from the earliest time to which we can look back, to the present, ever acted, but those *now acting*; and that they never acted with different degrees of energy from that which they now exert'.[4]

Lyell had begun his tour confident that the catastrophist assumptions of Georges Cuvier and of English geologists like William Buckland and Adam Sedgwick were mistaken. They drew a sharp contrast between a former world, characterized by the tumultuous upheaval of mountains, tremendous volcanic eruptions, and great floods, and the tranquil order of the present world characterized by very slow and moderate geological change. Lyell had found that Cuvier's and Alexandre Brongniart's (1770–1847) distinction between ancient and modern lake deposits was illusory. In Auvergne, he saw that Cenozoic freshwater formations had accumulated gradually over a very long period of time. In southern Italy and Sicily, volcanic activity and earthquakes had produced significant geological changes even within the historical period. Nevertheless, the volcanic activity that had built up Mount Etna and elevated southern Sicily had occurred gradually over many hundreds of thousands of years. From his travels through France and Italy and especially in Sicily, Lyell's initial belief hardened into an unshakeable conviction that through an indefinitely long geological past the Earth had been essentially as it was at present—subject to constant, imperceptibly slow change, the effects of a stable system within the Earth and on its surface. He realized that for a geologist, travel was indispensable. From Naples on his return from Sicily, he wrote to Roderick Murchison: 'We must preach up travelling ... as the first, second, and third requisites for a modern geologist, in the present adolescent state of the science'.[5] At Rome, ten days later, he was even more emphatic, writing to Murchison: 'We want nothing short of a radical reform in geology & we shall have one soon if honest men will travel & write & travel again!'[6]

Lyell did travel and write and travel again. In July 1830, after seeing the first volume of the *Principles of Geology* through the press, he set out for Spain to study the extinct volcanoes of Catalonia, similar to those of central France, and the geology of the Pyrenees.

Lyell and Mary Horner (1808–1873) were married at Bonn in 1832. Their wedding trip through central Europe included the Rhine and Rhone river valleys, Switzerland, and northern Italy. As they travelled, the Lyells called upon geologists whom Lyell knew personally or by reputation. At Gurnigel, near Bern, they met Bernhard Studer (1794–1887). The two men discussed Studer's 1827 paper showing that the gneiss and schist strata of the Stockhorn in the northern Alps were Mesozoic strata, altered by heat and pressure, and not Transitional (Studer 1827, p. 9). In the Rhone Valley, Lyell met Jean de Charpentier (1786–1855), superintendent of the salt mines at Bex. Charpentier had written a geological description of the Pyrenees and had made a study of the erratic boulders found in the Alps and the Jura, concluding that the boulders had been transported by glaciers. At Geneva, the Lyells called on the Swiss geologist Louis Albert Necker (1786–1861). With the aid of Necker's directions, Lyell found the granite veins at Valorsine extending into fossil-bearing sedimentary strata that Necker had described in 1826. The granite had transformed the sandstone and conglomerate strata next to them into gneiss. Above the sandstone and conglomerate was slate and above the slate an unaltered limestone containing ammonites (Necker [de Saussure] 1826). Lyell's encounters with Swiss geologists and Alpine geology in the summer of 1832 profoundly influenced his understanding of the effects of heat and pressure on sedimentary strata. In 1833, he introduced the term metamorphic to describe stratified crystalline rocks, reflecting their origin as sedimentary strata, transformed by heat and pressure (Lyell 1830–1833, **3**, pp. 374–375).

In 1833, Lyell again travelled on the Continent. At Heidelberg he met the mineralogist Carl Cäsar von Leonhard (1779–1862). He also met geologists or studied collections of fossils at Stuttgart, Göppingen, and Bayreuth. On his return through Belgium, Lyell stopped at Liège to see Philippe Charles Schmerling's (1791–1836) collection of animal and human bones obtained from caves along the Meuse River.

Lyell made his first trip to Scandinavian countries in 1834. In Sweden and Denmark he examined evidence that the land along the shore of the Baltic was rising. Lyell found that the reported rise of the land was real; a mark in a rock cut at sea level in 1820 was, by 1834, more than five inches above the water. Lyell also found fossil shells of living Baltic species in gravel beds thirty feet above the sea, additional proof that the land had risen. At Copenhagen he met Johann Georg Forchhammer (1794–1865) with whom he examined the Danish chalk formation.

The following year Lyell returned to the Continent, spending most of his time in Switzerland. At Porrentruy, he called on Jules Thurmann (1804–1855), an authority on the Jura Mountains, who explained their structure to Lyell (Thurmann 1833). Lyell spent several weeks studying the Bernese Oberland around the Lake of Thun and from there proceeded to the Jungfrau. At an elevation of nine thousand feet, he found limestone strata containing ammonites alternating with strata of crystalline gneiss, confirming, he thought, that the gneiss had undergone metamorphosis.

In 1836 Lyell, visited the Isle of Arran in the Firth of Clyde to study the relationship of granite to stratified rocks. The next year he went to Norway to expand on that study at Christiania (Oslo). Leopold von Buch (1774–1853) had described the granite at Christiania as overlying 'Transition' strata (von Buch 1811), and Lyell wanted to see this for himself. The Norwegian mineralogist Baltasar Keilhau (1797–1858) took Lyell to see the junctions of the granite with the stratified rocks. Keilhau told Lyell that although the granite did extend slightly over the edges of the strata at some points, it had clearly intruded from below as molten rock.[7]

The common theme of Lyell's travels to Switzerland, Norway, and Arran was to study the metamorphosis of sedimentary strata by granite intrusions. He discussed such intrusions in the *Principles* and developed the subject further in 1838 in the *Elements of Geology* (Lyell 1838, pp. 487–528). In the second edition of the *Elements* in 1841, he enlarged his treatment of metamorphosis, adding a north–south geological section of the Isle of Arran (Lyell 1841, **1**, pp. 324–437).

In 1838 and 1839 Lyell confined his travels to England to study the Crag formation, of Miocene age, in Suffolk and Norfolk. In 1840, he went to France to examine the Faluns of Touraine, also Miocene, seeking to correlate the Faluns with the Crag. In 1843, he again spent more than a month in Touraine collecting fossils from the Faluns.

During the 1840s, Lyell's travels were dominated by his visits to America in 1841–1842 and 1845–1846. These extended visits were made possible by invitations to give Lowell Institute lectures at Boston, each series with a generous stipend. Upon his arrival in America in August 1841, Lyell travelled across New York State in the company of James Hall (1811–1898) to see formations described by the New York State Survey and to visit Niagara Falls. Henry Darwin Rogers (1808–1866) guided him over Appalachian ridges to see the Pennsylvania Coal formations. After delivering the Lowell lectures in Boston during the autumn, Lyell went south to study the Cenozoic formations of the Atlantic coastal plain in Virginia, the Carolinas, and Georgia. In the spring he returned north, crossed the Appalachians into the Ohio Valley, and from Ohio travelled through the Great Lakes and down the St Lawrence River to Quebec. During the summer of 1842 he visited Nova Scotia to study the Coal formation at Pictou and the great fossil cliff at Joggins.

After giving the second series of Lowell Institute lectures in the autumn of 1845, Lyell travelled once more to the south. From Savannah, he crossed Georgia and Alabama by railway and stagecoach to Montgomery, descended the Alabama River to Mobile by steamboat, stopping at Claiborne to study the Eocene fossil cliff. After visiting the Mississippi Delta, he travelled up the Mississippi and Ohio Rivers by steamboat, stopping at various places to collect fossils and at New Madrid, Missouri, to see the effects of the 1811–1812 earthquake. Lyell and Mary returned to England in June 1846, some nine months after their departure.

In 1850, Lyell travelled through Germany and in 1851 he spent six weeks intensively collecting fossils among the Cenozoic formations of Belgium, seeking to relate them to the Faluns of Touraine and the English Crag. In 1852, during Lyell's third visit to America to give another series of Lowell lectures, he and John W. Dawson (1820–1899) discovered a fossil reptile at Joggins, Nova Scotia, the first from a Carboniferous formation in North America, and a fossil *Pupa*, the first fossil land shell from a Carboniferous formation anywhere. In 1853, Lyell paid a fourth brief visit to America as a member of the Royal Commission representing Great Britain at the opening of the New York Industrial Exhibition.[8] On his return home, Lyell began to plan a trip to Madeira and the Canary Islands.

Craters of elevation

Lyell's chief aim was to see the Canary Islands, particularly Tenerife and La Palma, which had provided Leopold von Buch with illustrative examples of 'craters of elevation'. In Tenerife, von Buch had identified the circle of escarpments enclosing the area called Las Cañadas at the foot of the great Pico de Teide as the periphery of a former crater of elevation. von Buch thought that the layers of lava and other volcanic rocks in the precipices surrounding Las Cañadas had been poured out originally in horizontal layers on the sea floor. He proposed that the accumulated layers of volcanic rock had subsequently undergone uplift caused by a sudden explosion that had left an immense crater. Within that crater of elevation, Pico de Teide had been thrust upward as a solid mass of trachyte. von Buch distinguished the crater of elevation enclosing Pico de Teide from the small volcanoes in the lower parts of Tenerife. These he considered to be 'cones of eruption', that is, ordinary volcanoes. On La Palma, the Caldera de Taburiente, six miles in diameter and five thousand feet deep, provided von Buch with an even more vivid example of a crater of elevation.

Lyell had opposed the theory of craters of elevation in all seven editions of the *Principles of Geology*. Among other difficulties, he noted that von Buch described the precipices enclosing the Caldera in La Palma as composed entirely of volcanic rocks. Yet if the lava layers had been poured out beneath the sea, they should be interspersed with sedimentary beds containing seashells. Furthermore, according to von Buch, the layers of basalt in the walls of the caldera were not shattered and broken as they ought to have been if they had been thrown up in a tremendous explosion.

Despite criticisms by Lyell and in France by Constant Prevost (1787–1856), von Buch had continued to insist on the validity of craters of elevation. In Paris, Léonce Élie de Beaumont (1798–1874) attached the craters of elevation theory to his theory of the sudden elevation of mountain chains. In 1834, Élie de Beaumont and his colleague Armand Dufrénoy (1792–1857) accompanied Leopold von Buch to southern Italy to examine the active volcanoes Vesuvius and Etna. von Buch decided that both had originated as craters of elevation (Buch 1836*b*). Impressed by the discontinuity between the ancient structure of Etna revealed in the walls of the Valle del Bove and the modern summit cone, Élie de Beaumont agreed with von Buch that Etna had originated as a crater of elevation, arising suddenly at the same time as the equally sudden rise of the Alps and the Pyrenees. Such events ushered in 'the present state of things' on Earth (Élie de Beaumont 1836).

In support of the theory of craters of elevation, Élie de Beaumont argued that a liquid lava flow could not solidify into a continuous layer of compact rock on a slope inclined at more than three degrees. In the walls of the Valle de Bove, lava layers dipped at angles of twenty-seven to thirty degrees. Élie de Beaumont's explanation was that the layers had been tilted from their original position. In support of his argument, he measured the inclinations of some thirty lava flows on and around Mount Etna and gave a detailed description of Etna, accompanied by drawings.

Although Lyell disbelieved entirely in the theory of craters of elevation, he was not prepared to dispute Élie de Beaumont's opinion that lava flows could form continuous sheets of compact rock only on horizontal or very gently inclined surfaces. It seemed intuitively probable. Liquid lava would flow down steep slopes too rapidly to retain its continuity. Nor would Lyell deny that elevation and disturbance might accompany volcanic activity. In 1828 on the island of Ischia, he had found beds of clay containing Mediterranean shells and, in Sicily, similar shells several hundred feet above the sea on the eastern slope of Etna. Furthermore, Lyell had shown that the whole of southern Sicily had been elevated thousands of feet within the period of living Mediterranean species, the elevation being accompanied by intermittent volcanic activity. Where Lyell differed from Élie de Beaumont was that he thought such elevation had occurred in a slow, gradual, manner as a result of ordinary earthquakes and ordinary volcanic eruptions.

Despite Lyell's criticism, the theory of craters of elevation became widely accepted among geologists, especially after 1836 when von Buch published a French translation of his descriptive account of the Canary Islands (von Buch 1836*a*). Lyell could not combat Buch's theory successfully until he had examined for himself the Canary Islands from which von Buch had drawn his examples. En route to the Canaries, he would stop at Madeira, another volcanic island. von Buch had visited Madeira briefly in 1815 but did not use it as an example (von Buch 1825, 1826). James Smith (1782–1867) of Jordan Hill, near Glasgow, had published a brief, but accurate, account of the geology of Madeira. Smith described Madeira as composed of several thousand feet of volcanic rocks, poured out on land during the Tertiary period. Among the lavas, basalt prevailed, but tuffs and conglomerates also formed a large proportion of the volcanic rocks. Since all had been poured out on land, Madeira could not be a crater of elevation. The central mountains of Madeira were the ruins of volcanoes once much higher. The Curral de Baixo (Grand Curral), Smith thought, was the ruin of a great crater, but one formed on land. So closely did the Curral resemble the older parts of Tenerife and other Canary Islands that Smith rejected the idea that they too could be craters of elevation. Smith also mentioned the few instances of sedimentary rocks on Madeira: at São Vicente a limestone containing corals and fossil shells, at São Jorge a coal bed, and at Caniçal sand dunes containing land shells (Smith 1839–1842 [1841]).

Madeira

Lyell arrived at Funchal (Fig. 1) on 18 December 1853, along with Mary Lyell, her maid-companion Antonia Schmidt, her sister and brother-in-law, Frances and Charles Bunbury (1809–1885), and Frances Bunbury's maid, Mrs Rennie. He intended to spend a relatively short time there, perhaps a month, before going on to the Canary Islands. In fact, he spent two months in Madeira. Repeatedly the island surprised and perplexed him. The day after Lyell and his party were landed from a small boat on the beach at Funchal and taken in a brightly-painted ox-drawn sledge to a Portugese pension, Lyell went to call on Major Antonio Azevedo, a

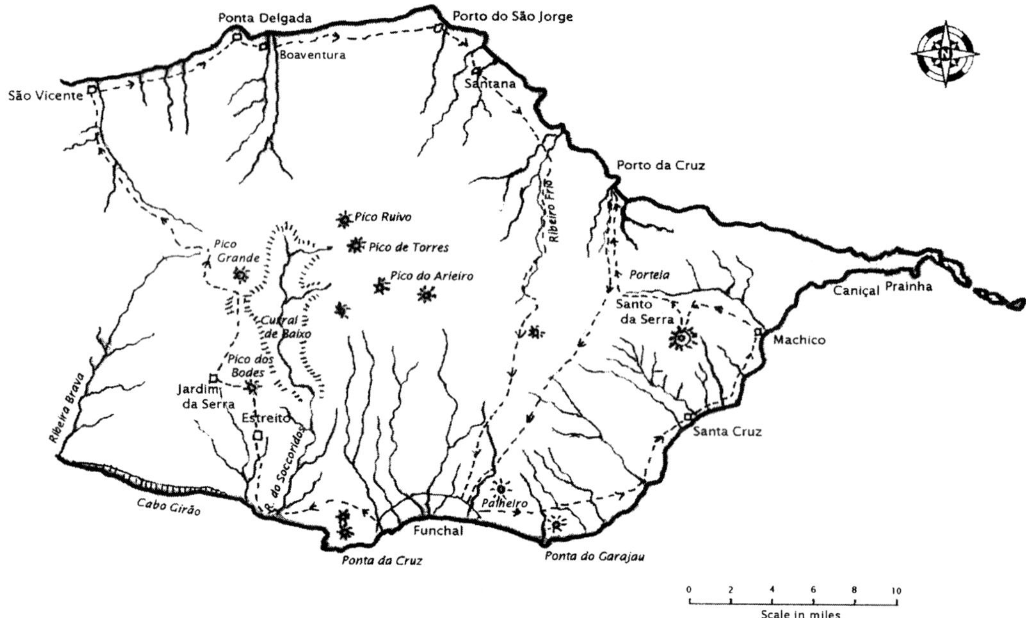

Fig. 1. Lyell and Hartung's travels in Madeira, 1854.

military engineer in charge of public works in Madeira. Azevedo was thoroughly familiar with the geology of the island and willing to help Lyell. He also told Lyell about Georg Hartung (1821–1891), a German living at Funchal, who was likewise interested in geology. Lyell went immediately to call upon Hartung whom he found to be a man in his thirties, the son of a printer at Königsberg. A consumptive, Hartung had spent three winters in Madeira for his health and had filled his time by studying the natural history and geology of the island (Pinto & Bouheiry 2007). He jumped at the chance to work with Lyell, and since he had become fluent in Portugese, he proved a valuable companion and guide. He was even willing to accompany Lyell to the Canary Islands.

In central France, George Poulett Scrope had described lava flows that had descended river valleys. The rivers had then cut new channels through the lava to reveal in the banks, beneath a layer of basaltic lava, the former gravel bed of the river. In 1828, Lyell and Murchison confirmed and extended Scrope's observations. Hartung told Lyell that in Madeira he knew of no such water-worn gravels beneath layers of lava. In 1850 Hartung's friend Oswald Heer (1809–1883), who had come from Zurich to spend the winter in Madeira for his health, had discovered a volcanic tuff containing charred tree branches in a ravine near Funchal. The charred wood indicated that the tuff had formed on land.[9]

Hartung took Lyell first to Cabo Garajau, a promontory some four hundred feet high at the eastern end of the bay of Funchal. The steeply sloping lava layers of Cabo Garajau puzzled Lyell. He thought they must be part of an anticline. An alternative possibility was that Cabo Garajau was part of a volcanic cone. In that case the inclination of the layers might be due partly to upheaval and partly to their deposition on the sides of the cone.[10] Clearly, Lyell was constrained by Élie de Beaumont's theory that layers of compact lava could form only on horizontal or gently sloping surfaces. If lava layers were inclined at twenty to thirty degrees, they must have been tilted. It was the same assumption as geologists applied to sedimentary strata; namely, that they were laid down originally in horizontal layers and, when found inclined, had been disturbed from their original position. Nevertheless, as Lyell explored Madeira over the next two months, he found that some layers of volcanic rock were inclined simply because they had flowed down an inclined surface. Even as he and Hartung returned from Cabo Garajau to Funchal, Lyell saw among lava layers that they passed examples of quaquaversal dips that to him were very perplexing.[11] Another puzzling feature was that, despite the frequent occurrence in Madeira of lava layers inclined at more than twenty degrees, the shoreline of the island showed no signs of upheaval: no coastal plain, no raised beaches, no lines of inland cliffs.

On the western outskirts of Funchal in the Pico da Punta da Cruz, Lyell and Hartung found beds of cinders inclined as steeply as twenty-six degrees.[12] The volcanic rocks in Pico da Punta da Cruz were exactly like those in the extinct volcanoes of Auvergne in central France.[13] 'When the slope exceeds 8° here', Lyell noted, 'there is much scoriaceous & largely vesicular lava & mixed slag—'.[14] He had with him a copy of Élie de Beaumont's 1835 monograph on Etna that he consulted frequently, striving to reconcile the structure of Madeira to Élie de Beaumont's opinion.[15]

Accompanied by Azevedo and Hartung, Lyell sailed in a small boat along the coast west of Funchal to study Cabo Girão, a sea cliff some nineteen hundred feet high, providing a section through a former volcano. It showed two clusters of dykes, some of them extending vertically more than a thousand feet. From each cluster of dykes, layers of lava and tuff sloped away on either side. The generally vertical position of the dykes indicated 'that the scoriaceous lavas & the lapilli or tuffs must have been originally inclined, for if they had owed their present tilt to mechanical derangement, the dykes w^{d}. have been inclined'.[16] Hartung suggested that the great cape was a model in miniature of the whole of Madeira, its clusters of dykes marking centres of eruption, like those of old volcanoes in the centre of the island. That day the sea was rough. All three men became seasick. Lyell made some crude sketches, but with the tossing of the boat and his seasickness, he could not take in all that the cliff showed, nor make good drawings. He needed to return on a calmer day, which he did (Fig. 2).[17]

In the steep sides of the lower valley of the Ribeira dos Soccoridos, which drained the Curral de Baixo in the centre of Madeira, Lyell observed what appeared to be continuous layers of basaltic lava descending with a gentle slope from the interior of the island towards the sea.[18] Hartung told him that in Madeira layers of lava tended to slope downward from the centre of the island. If the lava layers coming from the centre of Madeira were continuous sheets of basalt, Lyell thought they might have flowed from volcanic eruptions in the centre before the smaller volcanoes along the coast became active. But, the wedge-shaped lava layers visible in the Cabo Girão and the steeply inclined layers of cinders and tuff, suggested that lava flows coming from the interior must have occurred more or less when the smaller volcanoes near the coast were active.[19] East of Funchal in a ravine at Santa Cruz, Lyell found a section through a basaltic lava flow that had thickened or thinned as it filled the ravine along which it had flowed, indicating that the lava flow had poured over a landscape already deeply eroded.

At the end of December, two days of torrential rain transformed the streams descending from the mountains to the sea through the city of Funchal into raging torrents of muddy water. The rushing

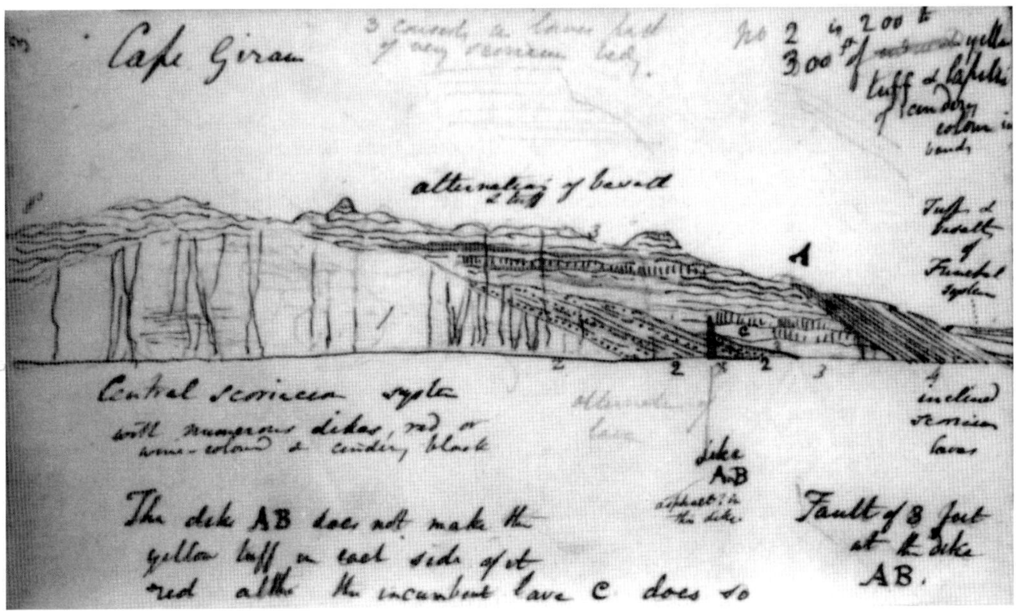

Fig. 2. Lyell's sketch of Cabo Girão, Madeira, from the sea (Notebook 191:3, Kinnordy MSS).

water rolled large boulders noisily along stream beds, demonstrating vividly to Lyell how the streams had carved out the ravines in which they flowed.

At the beginning of January 1854, Lyell and Hartung spent three days on the island of Porto Santo, thirty miles NE of Madeira, to which they sailed in the British warship HMS *St Jean d'Acre* through the kindness of its captain. Porto Santo was a volcanic island, formed much like Madeira by volcanic eruptions on land. On the small island of Baixo off the west end of Porto Santo, they found a fossil coral reef, resting on volcanic tuff. Baixo provided clear evidence of the elevation of formerly submarine volcanic rocks.

After their return from Porto Santo, Lyell and Hartung set out with Major Azevedo on an extensive exploration of Madeira. Riding westward from Funchal, they crossed the Ribeira dos Soccoridos and then turned inland into the high country to the summit of Pico dos Bodes, one of a series of volcanic cones extending from the southern coast towards the centre of the island. After three nights as guests of Henry Veitch, a former British Consul at Funchal, in his country house at Jardim da Serra, Lyell and his party, which now included Mary Lyell and the Bunburys, rode through the mountains along an old packhorse trail to the north side of the island. At one point the path led for more than a mile, across a knife-edge ridge. On either side the ground dropped away some two thousand feet. The little cavalcade paused to admire the view. They were almost in the centre of Madeira. Below them to the east lay the vast cavity of the Curral de Baixo. On its farther side, rose the highest mountains in Madeira, Pico Ruivo, Pico dos Torres, and Pico do Arieiro, their tops castellated with multitudes of protruding dykes. To the west lay the great valley of Serra de Água. In front of them rose the mass of Pico Grande with its thick, almost horizontal layers of lava. At the far end of the ridge a large dyke rose vertically in the mountainside. Excited by what he saw, Lyell astonished his companions and guides by standing on the edge of the precipice, Charles Bunbury recalled, 'with his glass to his eye, expounding the structure of the rocks in a sort of extempore lecture, seemingly perfectly forgetful that a slight movement would send him down several hundred feet' (Bunbury 1894, **3**, p. 66).

The great valleys on either side had been excavated out of the soft volcanic tuff by running water. Dykes of harder rock had preserved the narrow ridge on which they stood. In the mountains some of the layers of basalt were nearly level; others were steeply inclined. Layers of lapilli tended to be inclined more steeply than those of basalt. The central and highest portions of Madeira, with many dykes, had been the site of as many points of eruption (Fig. 3). In Pico Grande, an upper mass of basalt several hundred feet thick overlay soft tuffs and lapilli. The removal of the tuff by running water would undermine the overlying layers of basalt except, as in the ridge on which they stood, where the hard dykes resisted erosion.[20]

At São Vicente on the north coast of Madeira, Lyell and his colleagues examined a bed of limestone, about eleven hundred feet above the sea, containing Cenozoic fossil shells and corals. The elevated limestone strata proved that the north coast had undergone considerable elevation. The fossils also offered a means of dating that phase of volcanic activity.[21]

In a cliff at Boaventura on the north coast, Lyell found a section through a volcanic cone, buried under a thousand feet of lava layers, evidently the result of lava flows from the interior of the island. Lava flows from the central volcanoes had overwhelmed and buried smaller cones near the sea. By contrast, the volcanic cones around Funchal must have arisen after the activity of the central volcanoes had subsided. Denuding forces had subsequently dissected the Madeiran complex of volcanic rocks, streams of running water had excavated deep, narrow valleys, and along the coast waves had cut back the rocks to create great precipices.

Although Lyell continued to assume that steeply inclined lava layers had been disturbed, he was finding more and more evidence in conflict with that assumption. In the bank of the Ribeira dos Marinhos at Boaventura, he found a compact lava inclined at an angle of twenty-seven degrees. 'If we attribute the inclinat.[n] to upheaval or dislocation', he noted, 'we are met with the difficulty that other beds above and below are gently inclined or horizontal'.[22] At São Jorge he found a similar contradiction in a layer of lava dipping at twenty-five degrees, whereas other layers above and below did not.[23]

At São Jorge, Lyell was interested mainly in a well-known bed of lignite that cropped out on a hillside about nine hundred feet above the sea. Above the lignite was a layer of tuff about four feet thick and above it again Lyell found a bed of clay containing an abundance of fossil leaves. Most of the leaves were those of ferns, but some were of myrtles and laurels. All belonged to species still living in Madeira. They showed that the ancient Madeiran volcanoes had poured out their lavas over a forested land surface much like that of modern Madeira. The ravine at São Jorge proved incontestably that Madeira did not consist of layers of volcanic rock poured out beneath the sea and later uplifted.[24]

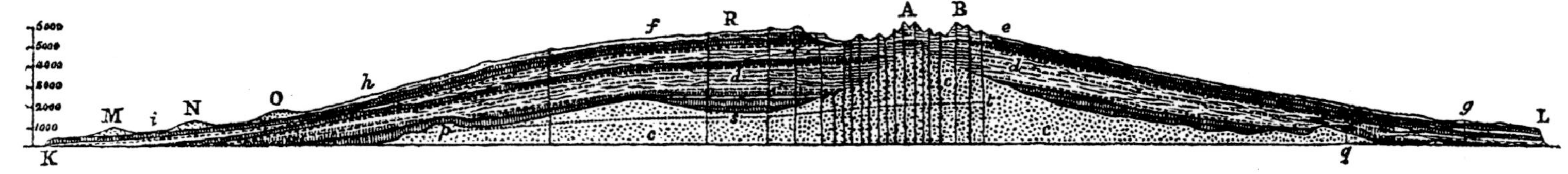

Length of section twelve miles. Drawn on a true scale of heights and horizontal distances from the observations of C. Lyell and G. Hartung, 1853–4.

A. Pico Torres (or Pico do Gatto), about 6050 feet high.
B. Pico Ruivo; the highest mountain in Madeira; about 6060 feet high.
c. Scoriæ, agglomerate, lapilli, tuff, and ejectamenta, with some highly scoriaceous lava.
d. Alternations of lava with tuff and lapilli, or with parting layers of red clay (laterite). Under this same head of "alternations" must be included all the beds between R and *s*.
e. Commencement of more highly inclined lavas on north side of Madeira; slope usually 10 degrees.
f. Commencement of more highly inclined lavas on southern slope, usually at angle at 15 degrees.
g. Dike of Jogo da Bola, in Ribeiro S. Jorge.
h. Slope of beds 15 degrees, occasionally but rarely 20 degrees.
i. Slope or dip of lavas 5 degrees.
K. Point da Cruz, near Funchal.
L. Point S. Jorge, on north coast.
M. Pico da Cruz, 843 feet high; modern cone.
N. Pico S. Martinho, 1100 feet high.
O. Pico S. Antonio, 1440 feet.
p. Buried cone in Ribeiro do Torreão.
q. Lignite and leaf-bed.
R. Pico S. Antonio, 5706 feet high.
p, *s*, *t*. Line below which the rocks are not exposed to view. All below this line is given conjecturally.

1 — 2 — The beds indicated by the sign No. 1. consist of lavas more or less stony, under which occur red clays or laterites, probably ancient soils (see p. 475.), represented by the interrupted lines, No. 2. These red bands, as well as the lavas, No. 1., are very numerous in nature, and for want of space a few only are introduced into the diagram.

Fig. 3. Lyell and Hartung's section of Madeira from north to south (Lyell 1855, p. 517).

After they returned from the north coast, Lyell and Hartung spent several days in the Curral de Baixo. Lyell decided that the Curral was not a crater, as James Smith had thought, but simply a valley created by running water. Here, deep within the volcanic heart of Madeira, there was no sign of marine sediment. The great mass of the island must have been formed by volcanic eruptions above the sea. On either side the mountains bristled with protruding dykes, showing the intense eruptive activity that had once occurred there.

From his two months in Madeira, Lyell concluded that when the island was growing as a result of frequent eruptions, most of the rainfall would be absorbed by the porous volcanic rocks, as it was today on Mount Etna. After eruptive activity subsided, streams would begin to flow from the water-soaked mountains and would rapidly erode the soft tuff and lapilli.

Lyell remained uncertain about the role of uplift in causing the inclination of lava beds and the elevation of Madeira, but even if no elevation had occurred, by volcanic eruptions alone the island would have achieved an estimated height of forty-five hundred feet above the sea. Successive periods of volcanic activity had been separated by long intervals of rest. Some volcanoes remained extinct for ages before others erupted elsewhere.

Fig. 4. Lyell and Hartung's travels in Gran Canaria, March 1854.

Yet throughout the whole long period of volcanic eruptions, the types of lava, scoriae, and tuff remained the same.[25]

Canary Islands

At Funchal on 17 February, the Lyells and Bunburys together with their maids and Georg Hartung boarded the British steamer *Severn* bound for Tenerife. On landing at Santa Cruz in Tenerife, they were quarantined for cholera for three days in a house outside the town, obtained for them by the British consul. The day after their release, the Lyells and Georg Hartung seized the opportunity to take the Cadiz steamer to Gran Canaria. The Bunburys remained in Tenerife, but moved from Santa Cruz to Puerto de la Orotava (now Puerto de la Cruz) on the northern coast of the island.

On landing at Las Palmas in Gran Canaria (Figs 4 and 5), Lyell and his party resorted to the English Inn. The British consul, Mr Houghton, offered Lyell every assistance, putting him in touch with a Spanish engineer, Pedro Maffiotte, who was interested in geology. Gran Canaria was very different from Madeira. The flat-roofed houses, interspersed with palm trees, against a background of bare, dry hills, gave it a decidedly African appearance. Geologically, Gran Canaria was also unlike Madeira. It showed clear signs of uplift. A narrow coastal plain extended around much of the island. Near Las Palmas, former sea cliffs nearly three hundred feet high bounded the coastal plain on the landward side. The cliffs showed horizontal layers of stratified tuff, lava, and a conglomerate containing fossil sea shells. From the conglomerate, Lyell and Hartung collected more than sixty species of shells that proved the formation to be upper Miocene, like the elevated limestone at São Vicente in Madeira. Just north of Las Palmas, between the cliff and the sea-shore, Lyell also found an elevated beach about twenty-five feet above the sea. With the help of Pedro Maffiotte, he collected more than fifty living species of sea shells; not all of them still lived in the waters immediately surrounding Gran Canaria, some occurred on the coast of Africa, others in the Mediterranean. From the fossils, Lyell concluded that volcanic activity in Madeira and Gran Canaria had begun in the upper Miocene and continued beyond the Pliocene (i.e. into the Pleistocene) period (Lyell 1865, pp. 668–669). Above Las Palmas, among hills covered with yellow-flowered euphorbias, Lyell and Hartung visited the Caldera de Bandama, an extinct volcanic cone with a large and well-preserved crater.

Maffiotte accompanied Lyell and Hartung on an exploration of the island that lasted several days. From Las Palmas they rode south along a dry coastal plain. Near Jinámar they crossed a lava flow, as black and sterile as when it first descended from an inland volcano. As they approached Telde,

Fig. 5. View of Las Palmas, Gran Canaria, from La Isleta (Notebook 194:84, Kinnordy MSS).

they rode up onto a plateau formed of layers of yellow tuff and dark basalt. Further south they descended again onto the coastal plain, now quite desert-like, dotted with large euphorbias. To Lyell the vegetation seemed distinctive, unlike anything in Europe.[26] As they continued south across the plain, mountains rose on their right. From Maspalomas at the southern tip of Gran Canaria, they rode up the desolate and barren Barranco de Fataga towards the central mountains. In the cliffs on either side, columnar basalt rose layer upon layer to heights of several hundred feet. At the head of the barranco, they rode through pine trees into the Barranco de Tirajana and continued upward into a high irregular valley, Las Calderas Altas. On the pine-covered ridges, massive volcanic dykes protruded like pillars. From the town of Tejeda, set among such dykes, they descended the northern slope of Gran Canaria through San Mateo and back to Las Palmas. Lyell's initial impression of the central mountains was that Gran Canaria was an upheaved mass. He was uncertain where the principal centre of eruption had been.[27] Gran Canaria possessed an abundance of volcanic cones, many with their craters intact.

Lyell and Hartung made another expedition, this time through Teror on the north slope and across a high ridge, or cumbre, to Artenara on the western slope. Thence they returned through Tejeda to Las Palmas. In Gran Canaria, as in Madeira, the lava layers in the high central mountains were almost horizontal.

On 9 March, after two weeks in Gran Canaria, the Lyells and Georg Hartung sailed to Tenerife to rejoin the Bunburys at Puerto de la Orotava. Two days later Lyell and Hartung sailed from Puerto de la Orotava in a two-masted barkentine, the *Zurbano*, bound for La Palma, the westernmost of the Canary Islands. Mary Lyell and her maid, Antonia Schmidt, remained with the Bunburys.

As the *Zurbano* sailed in toward Santa Cruz on the east coast of La Palma, the forested mountains of the island reminded Lyell of Madeira. Lyell and Hartung spent their first three days on La Palma exploring the ravines in the mountainside behind Santa Cruz (Fig. 6). Leopold von Buch considered these deep barrancos on the outer slope of the Caldera to be fissures formed in the sides of the mountain at the time when La Palma was raised from the sea bed (von Buch 1836*a*, p. 284). Nevertheless, the ravines became larger as they descended, showing that they had been excavated by the streams running in them. The layers of compact basalt interspersed with layers of tuff in their sides reminded Lyell of similar formations in the ravines of Madeira. Near the sea in the Barranco de la Madera, the lava layers were inclined gently at about five degrees. Three miles up the ravine, the layers were more steeply inclined, at ten to fifteen degrees, and a few dykes intersected them. Near the crest of the ridge that formed the rim of the Caldera, the layers became even more steeply inclined and there were numerous dykes. The increased frequency of dykes towards the centre of the island was similar to their distribution in Madeira.

Having acquainted themselves with the Santa Cruz area, Lyell and Hartung set out to explore the great Caldera de Taburiente (Figs 7 and 8). In a heavy drizzle, they rode through thick laurel woods across the central ridge of the island. On its western side, they caught their first glimpse of the caldera through a gap in the hills. Near the town of Los Llanos, they rode down a steep trail into the valley of the Barranco de las Angustias, which contained the stream draining the caldera. They forded the stream and followed a winding road up the farther side to a terrace, three thousand feet above the barranco, at the base of the wall of precipices enclosing the Caldera. They rode along the terrace through pine trees to the small farm of Tenerra. Behind Tenerra rose the NW wall of the caldera, more than two thousand feet high, displaying in its face horizontal layers of basaltic lava and tuff, intersected by many vertical dykes. To the south and east of Tenerra lay the enormous cavity of the caldera.

The Caldera de Taburiente astonished Lyell and Hartung. Leopold von Buch and others had described it as a simple basin. Instead, it contained a rugged terrain of pine-covered ridges and ravines. In the central area several pyramidal shaped rocks, covered with pine trees, rose more than a thousand feet. To the SW rose Pico de Bejenedo, separated from the rest of the encircling precipices by a gap, called La Cumbrecita. Water from various barrancos united at Dos Aguas to form the stream that flowed down the Barranco de las Angustias to the sea, descending about fifteen hundred feet in four and a half miles.

From Tenerra, the steep ridges and ravines appeared to have been shaped largely by running water. They were carved out of lower volcanic rocks, quite different from those of the precipices of the upper encircling wall. The precipices showed layers of volcanic agglomerate, scoriae, lapilli, and basaltic lava, dipping outward at angles of 10–28°. From the base of the precipices to the bottom of the caldera, the rocks consisted of coarse breccias and agglomerates, tea-green and sometimes light yellow in colour in contrast to the brown, lead-coloured or reddish tints of the basaltic precipices above. The lower rocks were so interlaced by dykes, extending in every direction and frequently winding sinuously, that Lyell and Hartung could not determine their general dip.

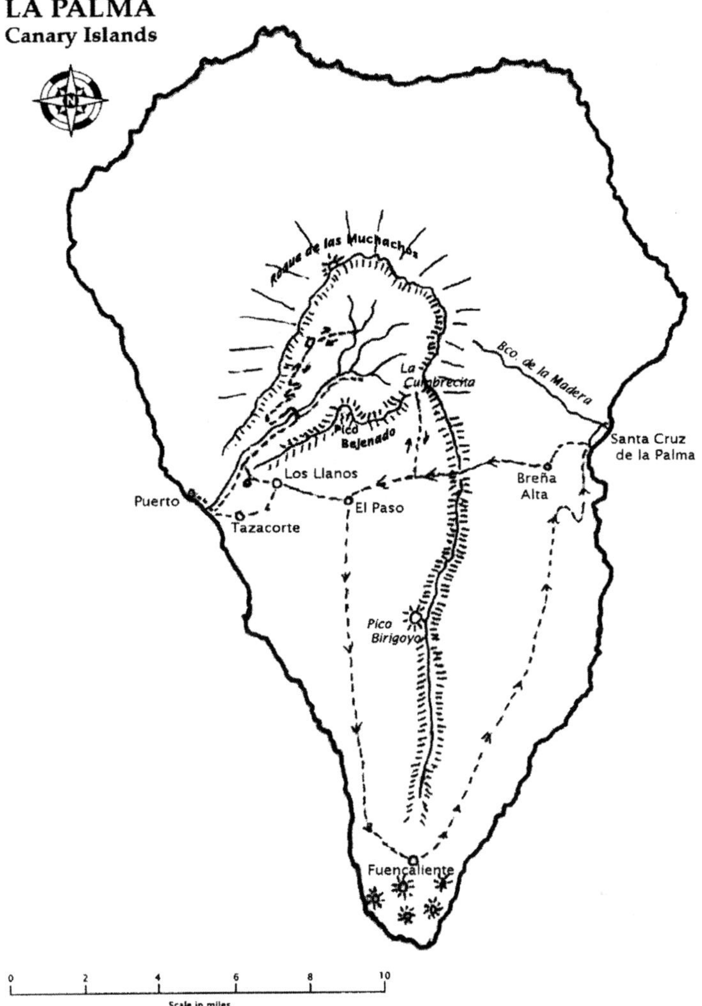

Fig. 6. Lyell and Hartung's travels in La Palma, March 1854.

Only in the Barranco de las Angustias did they detect a general southerly dip. The lower rocks seemed to have been produced by very gaseous volcanic eruptions, but they were so much altered by the multitude of dykes that they were frequently crystalline (Fig. 9; Lyell 1855, pp. 502–503).

Next morning, Lyell and Hartung followed a path leading from Tenerra down into the Barranco de Taburiente in the middle of the caldera. As they descended, the Roque de los Muchachos, the highest point on the north wall of the caldera came into view. Looking up and down the caldera, Lyell saw that its shape was not circular, but elongated from east to west like a normal mountain valley.[28] Leopold von Buch had not mentioned the great rock pyramids in the bottom of the caldera. They consisted of the same layers of volcanic rock as those in the precipices several thousand feet above them. Lyell speculated that the rock pyramids were detached portions of the upper precipices that had broken off and sunk down to their present position.[29] On their return to Tenerra that afternoon, Lyell, ever more aware of the difference between the lower volcanic rocks of the caldera and those of the upper precipices, drew a sketch of the precipices behind the hacienda, with vertical dykes, but fewer than those among the lower volcanic rocks of the caldera.[30]

On 21 March, Lyell and Hartung returned on horseback through the pine trees and down the winding road to the stream at the bottom of the Barranco de las Angustias, observing with new

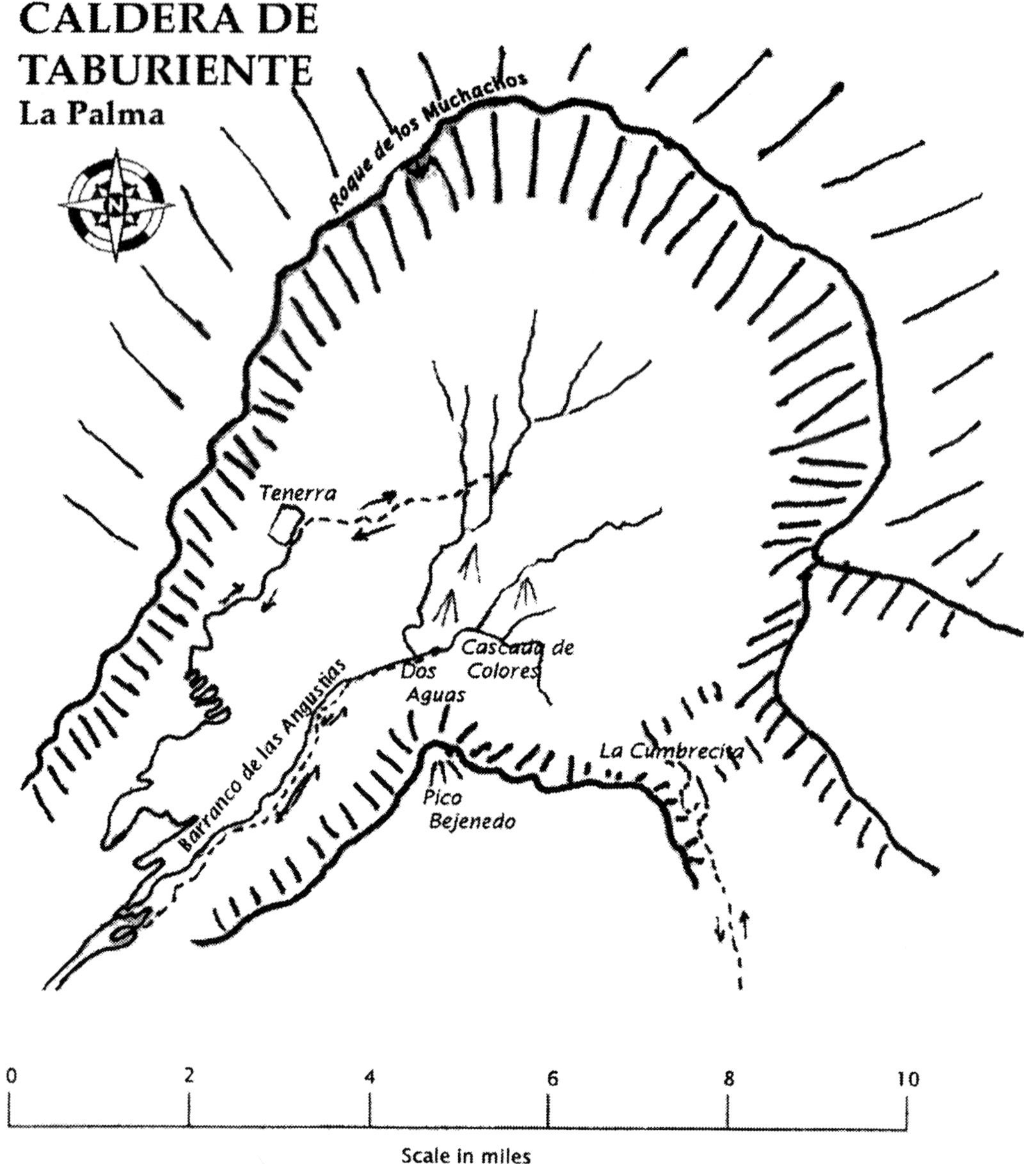

Fig. 7. Lyell and Hartung's travels in Caldera de Taburiente, March 1854.

interest the lower rocks they passed, so different from those in the precipices behind Tenerra.[31] At the bottom of the barranco they followed a trail alongside the stream up to Dos Aguas, the meeting place of two streams coming from the various barrancos. From Dos Aguas they followed up the stream to the right into the Barranco de Limonero. In the banks of the stream, the layers of rock dipped at various angles to the SW. At a small waterfall,[32] Cascada de Colores, named for the bright orange-coloured dykes in its face, the dykes appeared to form most of the mass. Above the waterfall, the dip of the rock layers changed to the east and slightly to the north, opposite to the dips below the cascade and the dips of the layers in the face of Pico Bejenedo directly above them. The dykes of the cascade with rock layers dipping away in all directions thus represented a former centre of eruption, possibly the centre that had given rise to Pico Bejenedo.

Fig. 8. Lyell's first view into the Caldera de Taburiente, La Palma (Notebook 195, p. 182, Kinnordy MSS).

After spending the night at La Vina, a farm in the caldera near the upper end of the Barranco de las Angustias, Lyell and Hartung rode down the barranco to the sea. On either side they observed thick beds of gravel interspersed with layers of basaltic lava. The sides of the barranco revealed a history of successive lava flows, separated by long intervals during which the stream deposited gravel. The gravel was entirely river gravel, its depth of several hundred feet showing the enormous quantity of material that had been carried down from the caldera. The total would be far greater than the gravel itself, because the stream would have carried fine mud and sand, derived from soft volcanic rocks, directly into the sea. The gravel beds also showed that formerly the stream had flowed at a higher level. They persuaded Lyell that the caldera had been hollowed out chiefly by erosion from running water. Only the precipices of its upper wall, Lyell thought, might have been formed by subsidence of volcanic magma within the heart of the volcano.

Lyell and Hartung made their last observation of the Caldera de Taburiente from La Cumbrecita, the gap between the high east rim and Pico Bejendo. It coincided with the boundary between the upper and lower groups of volcanic rocks. To the right extended the precipices of the east wall of the caldera, whereas on the left rose Pico Bejenado, its layers of lava dipping to the SW.

During ten days in La Palma in 1815, Leopold von Buch appears to have spent two days exploring the caldera. From Argual he descended into the Barranco de las Angustias and followed the stream as far as Dos Aguas. He mentioned alternating beds of basalt and brown tuff in the sides of the barranco but said nothing about river gravel. He described numerous dykes, that he called veins (filons) that formed a network among the rocks of the caldera floor. He noted the absence of basalt in the bottom of the caldera and its presence in the upper precipices, but he did not otherwise distinguish the lower volcanic rocks from the upper ones of the precipices (von Buch 1836*a*, pp. 277–278). He also viewed the caldera from the eastern rim where it appeared to him to be simply a vast abyss. He failed to note the rock pyramids rising from the bottom of the caldera. From Dos Aguas, where the pine-covered pyramids themselves blocked his view, he could not easily see them. From the rim of the caldera, they would appear to blend into the pine-covered ridges.

On 25 March, Lyell and Hartung began their exploration of the southern part of La Palma. From El Paso, they rode south along the western slope of the island to Fuencaliente through a landscape of black, sterile lava flows and numerous volcanic cones. The volcanic peaks on their left often reached heights of five to six thousand feet. South of Las Manchas they rode over lava flows inclined at angles on average of seven degrees.[33] One lava flow had descended a slope of thirteen degrees.[34]

At Fuencaliente, near the southern tip of La Palma, Lyell and Hartung visited a then quiescent

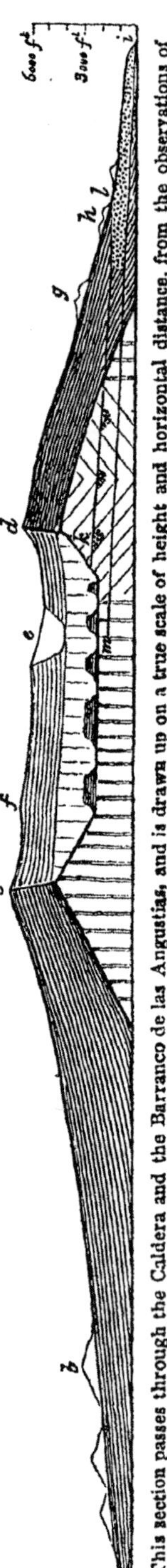

This section passes through the Caldera and the Barranco de las Angustias, and is drawn up on a true scale of height and horizontal distance, from the observations of C. Lyell and G. Hartung. 1854. (See Map, p. 498.)

a. Barlovento Point, see map, fig. 642., p. 498.
b. One of several cones, S.S.E. from Barlovento Point.
c. Pico de la Cruz, 7730 feet high, forming part of the northern boundary of the Caldera.
c, d. The Caldera.
d. The summit of the mountain called Alejanado, 6210 feet high, forming the southern wall of the Caldera.
e. The Cumbrecito, or higher opening into the Caldera.
f. Pico de Cedro, 7470 feet high; the highest point on the eastern margin of the Caldera.
g. Lateral cone on the flanks of Alejanado.
h. Cone of Argual.
i. Cliff of Tazacorte.
k, l. Old inclined water-line, marked by upper limit of gravel or conglomerate.
m, i. Level of the river or torrent of the Barranco de las Angustias.

The stronger lines in this diagram express that part which alone falls into the line of section; the fainter lines that portion of the Caldera which is in perspective, and could be seen by a spectator standing on the west side.

Fig. 9. Lyell and Hartung's Section through the Caldera de Taburiente, La Palma (Lyell 1855, p. 501).

volcano, about 350 feet high, with a circular crater at its top. On the sides of the cone were lava channels and tunnels. They returned along the east side of La Palma to Santa Cruz.

During his fortnight's intensive study of La Palma, Lyell had discerned successive episodes of volcanic activity in the island. The lower volcanic rocks of the caldera represented the earliest phase of volcanism, the precipices of the upper wall, a second phase. The volcanoes of the central ridge of La Palma south of the caldera were a third phase, and the historically active volcanoes around Fuencaliente constituted a fourth, contemporary, stage of activity. In 1855 Lyell described the structure and volcanic history of La Palma in his *Manual of Elementary Geology* (Lyell 1855, pp. 498–512). He did more than disprove Leopold von Buch's interpretation of the caldera as a crater of elevation. Lyell described accurately the broad outlines of the volcanic history of La Palma. In 1993, in their geological map of La Palma, the Spanish geologists Lavarro Latorre and Coello Bravo described the structure and successive episodes of volcanic activity of the caldera in terms similar to those of Lyell. Between episodes of volcanic activity were intervals of calm during which erosion by water removed large quantities of material. Radioactive dating showed the age of the basal complex in the floor of the caldera to be three million years, and the wall of precipices above to be the remnants of a great volcano that was active two million years ago. Lavarro Latorre and Coello Bravo had an advantage over Lyell and Hartung in that they could study the inner structure of La Palma from water galleries dug through the mountains for irrigation purposes (Lavarro Latorre & Coello Bravo 1993). Nevertheless, Lyell and Hartung showed that in La Palma volcanic activity had occurred in successive episodes separated by long periods of rest. La Palma showed, Lyell wrote, that 'the activity of subterranean heat may often be persistent for more than one geological period in the same place, relaxing perhaps its energies for a while but then breaking out afresh with an intensity as great as ever' (Lyell 1855, p. 504).

Tenerife

On 27 March, Lyell and Hartung sailed from Santa Cruz de la Palma, landing at Puerto de la Orotava in Tenerife that evening. Pressed for time before their departure on the next monthly steamboat bound for England, Lyell pushed himself to see as much of Tenerife's geology as possible. After a bare two-days rest, at three o'clock in the morning on 30 March, he and Hartung set out to study the high country around Pico de Teide (Fig. 10).

Charles Smith (1804?–1885), an Englishman resident for many years in Tenerife, accompanied them. Smith was a skilled botanist, thoroughly familiar with the island.

As they ascended the mountainside, Lyell noted the successive plant zones through which they climbed. Tenerife was then largely stripped of its pine forests, but the yellow-flowered Teide sticky broom (*Adenocarpus viscosus*), which had grown beneath the pines, formed a distinct zone. Still higher, the slopes were covered with white broom, or retama (*Spartocytisus supranubius*).

In late morning on the Alto de Guamasa, above Las Cañadas, Lyell sketched Pico de Teide and Las Cañadas (Fig. 11). From the Alto they rode among low hills and across the dry bed of a lake to the foot of La Fortaleza, a northern outlier of the ring of precipices surrounding the base of Pico de Teide. An inclined path along the face of La Fortaleza took them to Degollada del Cedro, a saddle in the rim of the precipice with a grove of pine trees. There at an altitude of eight thousand feet they stopped at half past one to rest. From Degollada del Cedro they descended the mountainside to Icod de los Vinos and thence back to Puerto de la Orotava.

Two days later, Lyell and Hartung rode from Puerto de la Orotava across the central ridge of Tenerife to Güimar on the eastern side of the island. The next day they rode to Santa Cruz to rejoin Mary Lyell and the Bunburys who had travelled there directly from Puerto de la Orotava. On 8 April, the whole party boarded the steamer *Severn*. Georg Hartung disembarked at Madeira, and the others continued on to England.

In 1851, in the third edition of the *Manual*, Lyell discussed La Palma briefly on the basis of Leopold von Buch's account in his description of the Canary Islands. In the fourth edition published in 1852, he left the discussion unchanged (Lyell 1851, pp. 390–394). On the steamboat returning to England, Lyell began to draft his discussion of volcanic mountains anew, beginning with La Palma.

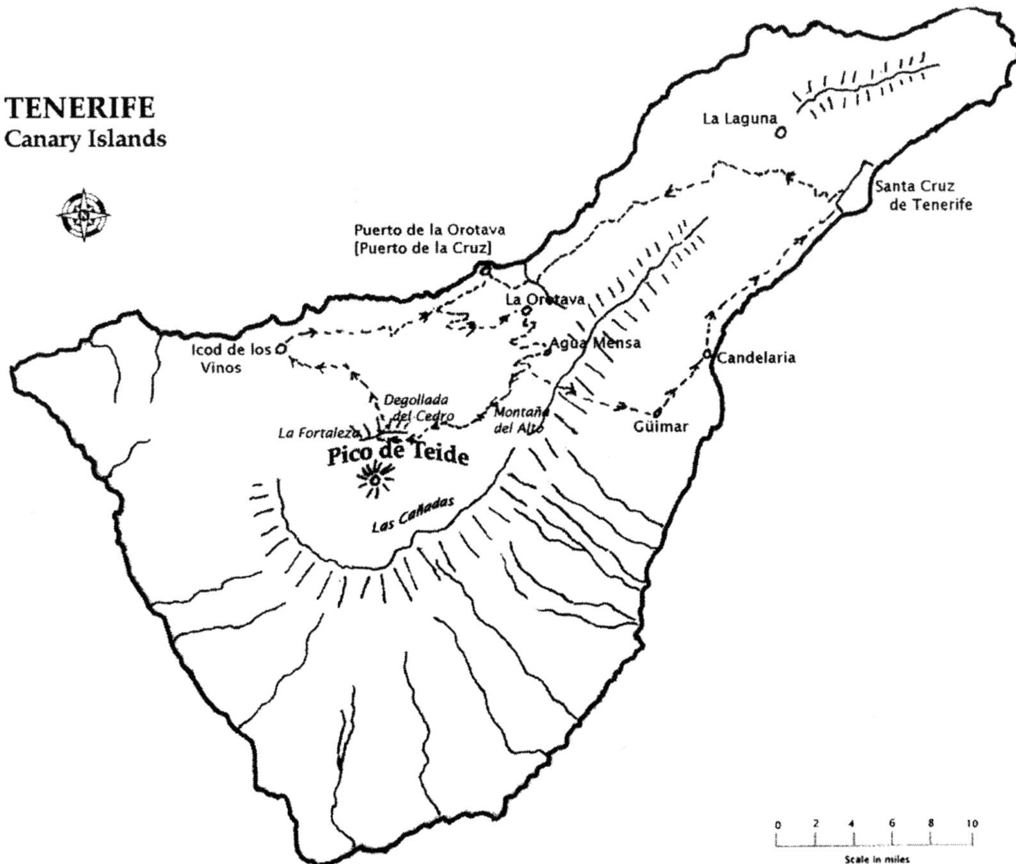

Fig. 10. Lyell and Hartung's travels in Tenerife, March 1854.

Fig. 11. Lyell's sketch of Las Cañadas, Tenerife (Notebook 196, p. 20, Kinnordy MSS).

He repudiated the theory of craters of elevation, as he had done since 1830 (Lyell 1830–33, **1**, pp. 386–395). He questioned to what extent the position of the layers of lava and scoriae composing a volcano was simply the result of successive eruptions and what contribution 'subterranean movements' had made to the same result. 'Both agencies have doubtless been at work but in what degree has each exerted its power'.[35]

In his notes, Lyell began with a discussion of Madeira before moving on to the Canary Islands, an order that he reversed in the fifth edition of the *Manual*. In Madeira, he described the chain of volcanoes that had erupted along the central axis of the island, their centres marked by a larger proportion of scoriae, tuffs, breccias, and agglomerates to layers of solid lava. At a greater distance from the eruptive centres, the proportion of solid lava to loose volcanic rocks was greater. All of the volcanic eruptions in Madeira had occurred on land. All of the stone fragments in the tuffs, breccias, and agglomerates deep in the heart of the island retained their angularity. They were not water-worn and contained no sea shells. At São Jorge, the leaf bed showed that the thousand feet of lavas and tuffs overlying it had been poured out on land. Madeira had been produced by a long succession of volcanic eruptions interrupted by intervals of rest, often prolonged. Later eruptions might have resulted in the disturbance and upheaval of layers of lava produced in earlier eruptions. Such upheaval might incline the lava layers at higher angles than those at which they had flowed initially.[36]

In suggesting the possible upheaval of lava layers after they were formed, Lyell was not in any sense accepting the theory of craters of elevation. That theory required the accumulation of layers of lava in a horizontal position on the sea bottom before they were raised up in a sudden paroxysmal explosion. The theory of craters of elevation had no place for the gradual accumulation of volcanic rocks on land as a result of successive volcanic eruptions.

Lyell speculated that when the central volcanoes of Madeira were in eruption, outflows of lava may have occurred at the same time on their sides and near the sea just as, when Mount Etna was in eruption, outflows of lava sometimes occurred from its satellite volcanoes on lower slopes. 'Hence', observed Lyell, 'when the mountain is rent & distended during lateral eruptions, the fissures may be filled with melted matter injected from below & changes of form & position may be superinduced in the older parts during the growth of the newer . . .: '[37] Such movements might accompany each volcanic eruption, but individual movements would be small.

After summarizing in his notes what he had learned of the structure of Madeira, Lyell decided that in the *Manual* he would begin with an account of La Palma, then describe Leopold von

Buch's theory of craters of elevation, and end with a description of Madeira.[38] He noted that the volcanic phenomena of La Palma were much more complicated than von Buch had assumed. Its broad outline was that of a great volcanic cone formed by many centres of eruption. At various times, volcanic eruptions may have generated movements in the accumulated layers of lava, scoriae, and tuff. 'von Buch was probably correct', noted Lyell, 'in assuming that the whole—at least the greater part of the force which tilted the beds was posterior in date to the original flowing of the lavas, but he was in error in imagining that it was posterior to the injection of the dykes'.[39] The injection of dykes was the cause of whatever uplift and tilting of the layers of lava that had occurred. Lyell still assumed that most inclined layers of lava had been uplifted and tilted after they had solidified, because of Élie de Beaumont's insistence that lava flows could not solidify into compact rock on slopes of more than three degrees.

Lyell was also influenced by James Dwight Dana's (1813–1895) account of the great active volcanoes of Mauna Loa and Mauna Kea in the island of Hawaii. When lava rose in the crater of Mauna Loa, fourteen thousand feet above the sea, the immense hydrostatic pressure of the liquid lava produced eruptions at lower levels on the sides of the mountain, indicating that the lava had forced its way through cracks in the accumulated solid layers of volcanic rock to create dykes (Dana 1849).

As he reflected on his observations in Madeira and the Canary Islands, Lyell was clearly concerned to explain the inclined position of layers of lava and scoriae. In numerous instances, layers of lava simply possessed the inclination of the slopes down which they had flowed. In other instances, where the geological evidence was not compelling, Lyell followed Élie de Beaumont's dictum.

In revising the *Manual of Geology* for a fifth edition through the autumn and winter of 1854–1855, Lyell's ideas changed. At the end of his account of the theory of craters of elevation, Lyell wrote:

> I have alluded to an opinion entertained by some able geologists, that no lava can acquire any degree of solidity if it flows down a declivity of more than three degrees. This doctrine I believe to be erroneous. The lava which has flowed from the cone of Llarena near Port Orotava in Teneriffe, is very columnar in parts, and yet has descended a slope of six degrees. Another stream of recent aspect near the town of El Passo, in Palma, has a general inclination of ten degrees ... (Lyell 1855, pp. 507).

In the *Manual*, Lyell included two sections through the caldera of La Palma drawn to scale by Georg Hartung, portraying the structure and volcanic history of the island (Fig. 2). Lyell also showed that the precipices encircling Las Cañadas at the base of Pico de Teide were the remnants of a larger volcano that had once existed about where Pico de Teide now stands. He described Madeira in a manner similar to La Palma with a north–south section across the island, drawn to scale, showing the concentration of dykes in the central region and some of the younger volcanic cones near the south coast (Lyell 1855, pp. 515–522).

The day after Lyell completed the preface to the fifth edition of the *Manual*, he and Mary departed for Berlin to visit Mary's sister Leonora, who, the year before, had married Georg Pertz, the Prussian royal librarian. At Berlin, Lyell talked with the crystallographer Eilhard Mitscherlich (1794–1863), Professor of Chemistry in the University of Berlin. Mitscherlich told Lyell that during the previous summer he and two colleagues had visited southern Italy and the Aeolian Islands. On the island of Stromboli they had seen modern lava flows forming thick layers of compact rock, a hundred feet in width, on slopes of fifteen, twenty, and even twenty-nine degrees. Mitscherlich's account of lava flows on Stromboli was a revelation to Lyell. His efforts to explain steeply inclined layers of lava in Madeira and the Canary Islands had been unnecessary.[40] Élie de Beaumont's claim that lava flows could not form thick sheets of compact rock on steep slopes was simply not true. Lyell decided that before he and Hartung revisited the Canaries, they should study the active volcanoes of southern Italy and Sicily.

At the Geological Society of London in April 1856, George Poulett Scrope showed that on five occasions over the preceding century an explosive eruption had produced a large crater at the summit of Vesuvius. In the intervals between explosions, smaller eruptions had gradually filled the crater until lava flows spilled over its lip (Scrope 1856). Scrope's demonstration that Vesuvius had not formed as a crater of elevation, but was the product of a long succession of volcanic eruptions of varying intensity, strengthened Lyell's desire to revisit Vesuvius and Etna.

In October 1857, Sir Charles and Lady Lyell arrived at Naples. With the aid of Italian geologists, Lyell studied Vesuvius, which was then in eruption, half of the mountain covered with a mantle of fresh lava. From Naples they went to Catania in Sicily, where Lyell spent two weeks examining lava flows from Etna. In various places, a lava flow had passed over the rim of a cliff and down steep slopes of 26° to 38°, yet had solidified as a thick layer of compact rock. In the Valle del Bove, Lyell found lava flows from the 1852 eruption of Etna that had solidified into compact rock on slopes of 44° to 50°. From the floor of Valle del Bove, Lyell ascended a two thousand foot precipice to the upper slopes of Etna, climbing over the surface of a lava flow fifteen feet thick inclined at an angle of about 35°. Such lavas solidified into thick layers of compact rock on very steep slopes.

In September 1858, Lyell returned to Sicily to spend five weeks studying the structure of Etna. Twice he descended from the upper part of Etna into the Valle del Bove to study the layers of lava and scoriae in its walls. The sections of lava flows in the walls of the Valle del Bove showed that ancient lava flows were identical in all respects to modern ones. Such facts contradicted Élie de Beaumont's assertions about the lavas of Etna that had remained unchallenged for almost a quarter century (Wilson 1998*b*, pp. 21–37). Most devastating to the theory of craters of elevation was Lyell's argument for the former existence of a principal centre of eruption at Trifoglietto in the Valle del Bove that had subsequently been covered by lava flows from the modern centre of eruption on Etna. 'If such be the structure of Etna ...', wrote Lyell, 'we must abandon the elevation-crater hypothesis; for although one cone of eruption may envelope and bury another cone of eruption, it is impossible for a cone of upheaval to mantle around and overwhelm another cone of upheaval ...' (Lyell 1858, p. 761).

Conclusion

Charles Lyell's four months in Madeira and the Canary Islands exerted a profound influence upon him. In 1828, he had studied the extinct volcanoes of central France and in Italy the active volcanoes of Vesuvius and Etna. He had examined the extinct volcanoes at Olot in Catalonia in 1830 and those of the Eifel district in the lower Rhine Valley in 1831. Thus when he sailed from Southampton in December 1853, Lyell was familiar with a broad range of volcanic phenomena. Nevertheless, Madeira astonished and puzzled him. The sheer scale and complexity of the island's volcanic activity—the enormous production of volcanic rocks, the long succession of eruptions at many centres throughout the island, the immense denudation of volcanic rocks revealed in the deep river valleys, the dykes protruding from the eroded mountain tops, and the stupendous coastal cliffs—all revealed an indefinitely long series of past volcanic activity stretching back to the Miocene, but now entirely quiescent.

Geologically, Gran Canaria was a repetition of Madeira, although superficially the two islands appeared quite different. Gran Canaria was bare and arid; Madeira was green and lush. Gran Canaria showed uplift around the coast; for the most part Madeira did not. The barrancos of Gran Canaria were narrower and shallower than the deep river valleys of Madeira, but the high mountains with protruding basaltic dykes were similar in the two islands, as were the smaller volcanic cones near the coast.

La Palma, although green and forested like Madeira, was rendered unique by the Caldera de Taburiente. The caldera revealed a history of at least two great episodes of volcanic activity, separated by an interval of rest. Along the central ridge of La Palma extending south of the caldera were volcanoes that had been active within historical times, indicating a third period of volcanic activity extending to the present.

Finally, in Tenerife Lyell saw that the great arc of precipices, enclosing Las Cañadas around the base of Pico de Teide, were remnants of a once larger volcano that had arisen roughly on the same site as Pico de Teide and that had later disappeared. Lyell did not speculate on the cause of its disappearance. He may have thought it had been eroded away during a long period of rest between episodes of volcanic eruptions or that it had been partly blown away, like Vesuvius, in an explosive eruption. Lyell did speculate that if Pico de Teide continued to erupt, its lava flows and other emissions would eventually fill Las Cañadas. Lava flows from Pico de Teide could then descend to the sea on its south and east sides as they did to the west (Lyell 1855, pp. 513–555).

The difference between Charles Lyell's observations in 1854 in Madeira and the Canary Islands and those of Leopold von Buch in 1815, thirty-eight years before, is striking. Lyell was interested in the geological history of the islands; von Buch was not. For von Buch, the islands had no history; they had sprung from the sea as craters of elevation. He mentioned the presence of augite, olivene, peridote, and feldspar in various lavas, but speculated little on the origin of the rocks containing them. He did not explain or justify the theory of craters of elevation, so much as take it for granted. He assumed that the difference between the circle of cliffs around Las Cañadas on Tenerife, or the caldera on La Palma, and a volcanic cone of eruption was obvious. It needed no explanation; at any rate he provided none. Lyell, by contrast, sought to explain the striking features of Madeira and the Canary Islands in terms of their historical origin and development. He saw that history with remarkable clarity and completeness. More recent geologists have filled out details but have not fundamentally altered Lyell's interpretation of the geology of Madeira and the Canary Islands.

Although Lyell had always rejected the theory of craters of elevation, he had not disputed Élie de Beaumont's argument that lavas could not form continuous layers of solid rock on slopes greater than three degrees. Lyell believed Élie de Beaumont to be an accurate observer, and Élie de Beaumont claimed that his opinion was based on the examination of many lava flows on and around Mount Etna. Therefore, even when Lyell observed inclined lavas that could not have been uplifted, he viewed them as special cases. Where circumstances did not rule out uplift, Lyell was prepared to believe that inclined lava might have been uplifted. Always in his mind was the possibility that earthquakes might accompany volcanic activity and that the infusion of dykes into volcanic mountains would disturb existing layers of lava to some degree. When at Berlin in 1855, Eilhard Mitscherlich described to Lyell modern lava flows on Stromboli solidified into sheets of solid rock on slopes as steep as 29°, Lyell was shocked. The Stromboli lavas meant that Élie de Beaumont was completely wrong. When Lyell returned to Sicily in 1857 and 1858, he found that Élie de Beaumont had misstated facts on the structure of Mount Etna and its lavas. Not only was the theory of craters of elevation a delusion, but the geological evidence that Élie de Beaumont had presented to sustain it was completely false.

Notes

[1]Lyell's and Murchison's observations on their tour through France are described in detail in Wilson (1972, pp. 190–266) and discussed by Kölbl-Ebert 2007.

[2]Lyell to Mantell. 22 August 1828, Mantell MSS, Alexander Turnbull Library, Wellington, New Zealand.

[3]Lyell 1830–33, **3**, x–xi. For a detailed account of Lyell's observations during his 1828 tour see Wilson (1972, pp. 218–261).

[4]Lyell to Murchison, 15 January 1829, Murchison MSS, Geological Society of London. In Lyell 1881, **1**, 234–235.

[5]Lyell to Murchison, 12 January 1829, Murchison MSS. In Lyell 1881, **1**, 232–234.

[6]Lyell to Murchison, 22 January 1829, Murchison MSS. The sentence was not included in the letter as printed in Lyell 1881, **1**, 239–242.

[7]Professor Geir Hestmark of the University of Oslo is preparing a paper on Lyell's 1837 visit to Christiania. He has identified the sites to which Keilhau guided Lyell. He finds that they remain unchanged today.

[8]Lyell's American travels are described in detail in Wilson (1998*a*).

[9]Lyell, Notebook 185, pp. 88–89. Sir Charles Lyell's scientific notebooks number 266; he kept them from the beginning of his scientific career in 1824 until his death. They are the property of Lord Lyell of Kinnordy.

[10]Lyell, Notebook 185, p. 101.

[11]Lyell, Notebook 185, p. 102.

[12]Lyell, Notebook 185, p. 106.

[13]Lyell, Notebook 185, pp. 110–111.

[14]Lyell, Notebook 186, p. 6.

[15]Lyell, Notebook 185, p. 69; 186, p. 75.

[16]Lyell, Notebook 186, p. 45.

[17]Lyell, Notebook 186, p. 45.

[18]Lyell, Notebook 186, p. 3.

[19]Lyell, Notebook 186, pp. 53–54.

[20]Lyell, Notebook 188, pp. 71–73.

[21]Lyell, Notebook 188, p. 78.

[22]Lyell, Notebook 189, p. 19.

[23]Lyell, Notebook 189, p. 35.

[24]Lyell, Notebook 189, p. 43.

[25]Lyell to Horner, 21 February 1854. Kinnordy MSS.

[26]Lyell, Notebook 193, p. 131.

[27]Lyell, Notebook 193, p. 194.

[28]Lyell, Notebook 194, p. 214.

[29]Lyell, Notebook 194, pp. 222–225. Cf. Lyell 1855, p. 502.

[30]Lyell, Notebook 195, p. 8.

[31]Lyell, Notebook 195, p. 12.

[32]Lyell, Notebook 195, p. 18.

[33]Lyell, Notebook 195, p. 70.

[34]Lyell, Notebook 195, p. 78.

[35]Lyell, Notebook 197, pp. 19–20.

[36]Lyell, Notebook 197, p. 28.

[37]Lyell, Notebook 197, p. 30.

[38]Lyell, Notebook 197, p. 36.

[39]Lyell, Notebook 197, pp. 37–38.

[40]Lyell to Horner, 24 March 1855. Kinnordy MSS.

References

BANKS, J. 1772. Account of Staffa. *In*: PENNANT, T. (ed.) *A Tour in Scotland and Voyage to the Hebrides*. 2nd edn. 2 Volumes. B. White, London.

BUCH, L. VON. 1811. Voyage dans les environs de Christiania en Norwège. *Journal de Physique*, **73**, 289–291.

BUCH, L. von. 1825. Note sur l'île de Madère. *Annales des Sciences Naturelles*, **4**, 14–21.

BUCH, L. VON. 1826. Observations made during a visit to Madeira, and a residence in the Canary Islands. *Edinburgh New Philosophical Journal*, **1**, 380–385; **2**, 73–86.

BUCH, L. VON. 1836*a*. Description physique des Îles Canaries, suivie d'un indication des principaux volcans du globe. (Trans. C. Boulanger) F. G. Levrault, Paris and Strasbourg.

BUCH, L. VON. 1836*b*. On volcanoes and craters of elevation. *Edinburgh New Philosophical Journal*, **21**, 189–206.

BUNBURY, C. J. F. 1894. Journal, February 27th 1875. *In*: BUNBURY, F. J. (ed.) *Life, Letters and Journals of Sir Charles J. F. Bunbury, Bart.* 3 Volumes. Privately printed, London.

DANA, J. D. 1849. *Geology*. Volume 10 in United States Exploring Expedition (1838–1842). *United States Exploring Expedition*. G. P. Putnam, New York.

ÉLIE DE BEAUMONT, L. 1836. Sur la structure et sur l'origine du Mont Etna. *Annales des Mines*, 3rd series, **9**, 175–216, 575–630; **10**, 351–370, 507–576.

KÖLBL-EBERT, M. 2007. The geological travels of Charles Lyell, Charlotte Murchison and Roderick Impey Murchison in France and northern Italy (1828). *In*: WYSE JACKSON, P. N. (ed.) *Four Centuries of Geological Travel: The Search for Knowledge on Foot, Bicycle, Sledge and Camel.* Geological Society, London, Special Publications, **287**, 109–117.

LAVARRO LATORRE, J. M. & COELLO BRAVO, J. J. 1993. *Caldera Taburiente, Geological Map.* Ministerio de Medio Ambiente, Madrid.

LYELL, C. 1826. On a recent formation of freshwater limestone in Forfarshire and on some recent deposits of freshwater marl. *Transactions of the Geological Society of London*, 2nd series, **2**, 72–96.

LYELL, C. 1827. Art. IV. *Memoir on the geology of central France; including the volcanic formations of Auvergne, the Velay and the Vivarrais, with a volume of maps and plates.* By G. P. SCROPE, F.R.S., F.G.S., London, 1827. *Quarterly Review*, **36**, 437–483.

LYELL, C. 1830–1833. *Principles of Geology.* 3 Volumes. John Murray, London.

LYELL, C. 1838. *Elements of Geology.* John Murray, London.

LYELL, C. 1841. *Elements of Geology.* 2nd edn. 2 Volumes. John Murray, London.

LYELL, C. 1851. *Manual of Elementary Geology.* 3rd edn. John Murray, London. [The pagination is identical in the fourth edition].

LYELL, C. 1855. *Manual of Elementary Geology.* 5th edn. John Murray, London.

LYELL, C. 1858. On the structure of lavas which have consolidated on steep slopes; with remarks on the mode of origin of Mount Etna, and on the theory of 'Craters of Elevation'. *Philosophical Transactions of the Royal Society of London*, **148**, 703–786.

LYELL, C. 1865. *Elements of Geology.* 6th edn. John Murray, London.

LYELL, K. M. 1881. *Life, Letters, and Journals of Sir Charles Lyell, Bt.* 2 Volumes. John Murray, London.

NECKER [DE SAUSSSURE], L. A. 1826. Sur les filons granitiques et porphyriques de Valorsine et sur le gisement des couches coquillières des montagnes de Sales, des Fizs, et de Platet. *Bibliothèque Universelle des Sciences, Belles Lettres et Arts*, 33–92.

PINTO, M. D. & BOUHEIRY, A. 2007. The German geologist Georg Hartung (1821–1892) and the geology of the Azores and Madeira islands. *In*: WYSE JACKSON, P. N. (ed.) *Four Centuries of Geological Travel: The Search for Knowledge on Foot, Bicycle, Sledge and Camel.* Geological Society, London, Special Publications, **287**, 229–238.

SCROPE, G. P. 1856. On the formation of craters and the nature of the liquidity of lavas. *Quarterly Journal of the Geological Society of London*, **12**, 326–350.

SMITH, J. 1839–1842 [1841]. On the geology of the island of Madeira. *Proceedings of the Geological Society of London*, **3**, 351–355.

STUDER, B. 1827. Remarques géognostiques sur quelques parties de la chaine septentrionelle des Alpes. *Annales des Sciences Naturelles*, **11**, 5–47.

THURMANN, J. 1833. *Essai sur les Soulèvemens Jurassiques du Porrentruy. Nancy* [*Mémoires de la Société des Sciences de Nancy*, Book 2] **1**.

WILSON, L. G. 1972. *Charles Lyell: the Years to 1841: the Revolution in Geology.* Yale University Press, New Haven and London.

WILSON, L. G. 1998*a*. *Lyell in America: Transatlantic Geology, 1851–1853.* Johns Hopkins University Press, Baltimore.

WILSON, L. G. 1998*b*. Lyell the man and his times. *In*: BLUNDELL, D. J. & SCOTT, A. C. (eds) *Lyell: the Past is the Key to the Present.* Geological Society, London, Special Publications, **143**, 21–37.

The German geologist Georg Hartung (1821–1891) and the geology of the Azores and Madeira islands

M. S. PINTO[1] & A. BOUHEIRY[2]

[1]*Centro de Estudos de História e Filosofia da Ciência e da Técnica, Universidade de Aveiro, 3810 Aveiro, Portugal (e-mail: mspinto@csjp.ua.pt)*

[2]*Hauptstrasse 96a, CH-8246 Langwiesen, Switzerland*

Abstract: In 1853 and 1854, the German geologist G. Hartung (1821–1891) was living in Funchal, on the Portuguese island of Madeira. During this time Charles Lyell visited the archipelago to carry out the fieldwork that led the famous British geologist to write *On the Geology of Some Parts of Madeira*, published in the *Quarterly Journal* in 1854, and to make abundant references to the geology, morphology, palaeontology and living flora and fauna of the islands in the sixth edition of the *Elements of Geology* (1865) and in the 1868 edition of the *Principles of Geology*. Lyell was accompanied by Hartung during the visit to Madeira and to the Canary Islands. Hartung left on him a deep and lasting influence, so that Lyell wrote that his German colleague had 'proved a most active fellow-labourer'. Lyell liked to talk to young geologists from whom he felt 'old stagers' had much to learn, and that was probably the case with Hartung.

Hartung visited the Azores archipelago in 1857 and in 1860 produced a book (*Die Azoren in ihrer äusseren Erscheinung und nach ihrer geognostischen Natur*) and an atlas with very fine plates of views of the volcanic landscape of S. Miguel Island. In 1864, he published *Geologische Beschreibung der Inseln Madeira und Porto Santo* on the geology of Madeira and Porto Santo that was the result of his ideas and field observations made from 1850 to 1854. Both books have palaeontological sections written by German-speaking authors. *Betrachtungen über Erhebungskrater, ältere und neuere Eruptivmassen nebst einer Schilderung der geologischen Verhältnisse der Insel Gran Canaria*, published in 1862, also refers to Madeira and the Azores.

Hartung's work on these archipelagos was important from both a geological and a historical point of view, and he became involved in the discussion of Leopold von Buch's 'upheaval' and Charles Lyell's 'upbuilding' theories.

This is an account of the geological work carried out in the mid-1800s by the German naturalist Georg Hartung (1821–1891) in the Portuguese archipelagos of Madeira and Azores in the Atlantic Ocean. It focuses on the importance of Hartung's work in advancing the scientific understanding of such volcanic islands, and also on his contribution to the debate on the so-called Erhebungskrater ('craters of elevation') or upheaval theory of Leopold von Buch (1774–1853), and the opposite theory of upbuilding, promoted by Charles Lyell (1797–1875).

Hartung also visited the Spanish Canary Islands in order to discover more about their geology and the volcanic processes that were responsible for their origin and physical development, and to compare them with Madeira and the Azores.

The Madeira archipelago of which Madeira and Porto Santo are the main islands with Funchal being the capital city, and the Canary archipelago which comprises the main islands of La Palma, Gomera, Hierro, Tenerife, Gran Canaria, Fuerteventura and Lanzarote, both lie off the northwestern part of the African coast. They are relatively close to each other (some 650 km apart) but in the mid-1800s there were poor links between them, such that Hartung sometimes had to travel from one island group to the other via mainland Europe. The Azores archipelago which comprises the islands of S. Maria, S. Miguel, Terceira, Pico, Faial, S. Jorge, Graciosa, Flores and Corvo is located in the middle of the North Atlantic, some 1300 km west of Lisbon, and again Hartung had to travel from Europe in order to visit the islands. This created severe problems in carrying out geological work, and certainly was expensive. He paid for the travelling and living expenses himself and received no funds for his work from governmental or institutional sources. A case *d'amour à l'art*, one might say.

This bio-bibliographic sketch is based mainly on a reading of the prefaces of his books and on letters that he exchanged with Charles Lyell. Unfortunately, we did not have access to letters from and to other scientists (such as the Swiss palaeontologist Oswald Heer (1809–1883) from Zurich), with whom Hartung worked and corresponded.

From: WYSE JACKSON, P. N. (ed.) *Four Centuries of Geological Travel: The Search for Knowledge on Foot, Bicycle, Sledge and Camel*. Geological Society, London, Special Publications, **287**, 229–238.
DOI: 10.1144/SP287.18 0305-8719/07/$15.00

Bio-bibliographic sketch

Georg Friedrich Karl Hartung was born on 13 July 1821 in Königsberg, (now Kaliningrad, then in Prussia) and died, unmarried, in Heidelberg on 28 March 1891. He was the first child of Georg Friedrich Hartung (1782–1849) and Anna Maria Sophie Greis (1797–1870). His father, a freeman and a municipal councillor of Königsberg owned a prosperous printing and publishing house in the city, and also edited the traditional liberal Königsberger Journal *Hartung'sche Zeitung*. A year before his death he left his older son his house, and the firm was later passed on to Georg's younger brother Hermann (1823–1901) who had a degree in mathematics from the University of Königsberg. These family assets were almost certainly the source of funds for Hartung's travels and it is possible that he received some practical preparation in the printing and publishing business that allowed him to make a living. In 1871 the family sold the printing house.

It is not clear from his writings whether or not he had an interest in science before going to Madeira in 1850, but he never received any preparation in order to become a naturalist. He attended the Insterburg Classical Grammar School for seven years, and in 1838 when he was 16 years old had qualified for entry to university. Rather than follow a university career immediately, he probably joined his father's firm, where his ability to execute fine drawings was most useful (Wilhelm 1997).[1]

Hartung's artistic capabilities were amply demonstrated in his books, and his progress in adapting to be a geological artist was praised by Charles Lyell in 1854 and again in a letter dated January 1855: 'He draws tolerably, and improves in this daily'. What is surprising is that no representation of Hartung is known to exist except for a small watercolour that depicts him and Charles Lyell during one of their geological excursions in Tenerife (Lyell 1881; Wilson 1998).

Lyell described Hartung as being 'very zealous', 'helpful', 'a most active fellow-labourer' and an 'apt scholar'. It is not surprising that Lyell got on well with Hartung; it was well known that Lyell was sympathetic to and liked to talk to younger geologists from whom he felt 'old stagers' like himself had much to learn; Hartung was no exception (Lyell 1881; Macomber 1993; Wilson 1998).

It was most probably after his arrival in Madeira, where from 1850 he had begun to spend the winter to escape the harsh climate of his native land, that he became attracted to natural history at the relatively advanced age of 29 years. Hartung lived in Funchal, where he shared a house with Oswald Heer, who had been advised by von Buch to stay there to recover from a severe, dangerous cough. This suggests that Hartung had also travelled to Madeira for treatment which, at the time, was common among Europeans suffering from chest diseases. This supposition is supported in letters from Hartung addressed to Lyell in which he wrote that suffered from catarrh and usually got much better after staying in Madeira. His interest in natural history was fostered largely due to the influence of Heer, and included not only geology but also botany, entomology and, according to Lyell, agriculture (Hartung 1857*a*, *b*, 1860*a*, *b*, 1862, 1864; Lyell 1881). His preparation as a field geologist resulted from his practical interest in that science and from accompanying Earth scientists, like Heer and Lyell, on field trips around the island. From his observations, conversations and the exchange of letters with these and many other scientists, and from readings books by von Buch, Lyell, Charles Darwin and others he read the conceptual models put forward on the formation and development of the volcanic islands that he visited, and more importantly began to develop his own models.

Hartung was concerned that he had not received any formal education in geology, and so in 1855 he thought of reading geology at the University of Heidelberg for at least one semester. However he missed the beginning of the academic year and so decided instead to have private lessons in geology and mineralogy, and these continued intermittently because of his travels to the islands, in that and in the following year.[2]

In October 1862, according to a very short paragraph in one of his letters addressed to Lyell, he received the degree of Doctor of Philosophy from the University of Königsberg, most probably in recognition of his work in the volcanic islands of the Atlantic.[3] This explains why on the title page of his book on Madeira and Porto Santo (1864) is printed '*Dr. G. Hartung*' and probably explains why he is referred to as 'geologist' and 'Dr. h. c.' (*honoris causa* doctor) in a recent publication for German-speaking visitors to Madeira (Wilhelm 1997).[4] It also explains why Heer referred to him in the preface of his 1877 book *Flora Fossilis Arctica* as '*my friend Dr. G. Hartung*' (Heer 1877).[5] Curiously, for no apparent reason, Charles Lyell addressed him as 'Professor George Hartung' in a letter dated October 1856 (Lyell 1881).

Hartung spent much of the early 1850s engaged in his geological travels to the Atlantic islands, but dedicated most of the later 1850s and 1860s to writing up and publishing the results of his observations and ideas (Hartung (1857*a*, *b*, 1860*a*, *b*, 1862, 1864; Hartung & Arlett 1858; Fritsch *et al.* 1867).[6] He published articles first on the Canary Islands in 1857 and 1858, then on the Azores: *Die Azoren in ihrer äusseren Erscheinung und nach* *ihrer geognostischen Natur* (1860); then on the various archipelagos (*Betrachtungen über Erhebungskrater, ältere und neuere Eruptivmassen nebst einer Schilderung der geologischen Verhältnisse der Insel Gran Canaria* (1862); then about Madeira and Porto Santo: *Geologische Beschreibung der Inseln Madeira und Porto Santo* (1864), where he had started his geological activities some thirteen years before and where he had worked accompanying Lyell some eleven years before; and finally in 1867 about Tenerife. Only one scientific manuscript by Hartung is known to exist, on the geology of the Terceira island (Azores) written in French in 1857; it is in the library of the Instituto Geológico e Mineiro (formerly the Geological Survey of Portugal) in Lisbon.

In later years Hartung also published works on his travels in Scandinavia, which he undertook on account of his interest in glacial geology which he had first developed in the Azores. There, according to letters dated 1858 addressed to him from Lyell and from Darwin to Lyell and to J. D. Hooker, he had found erratic boulders and plants showing that icebergs had been stranded there (Hartung 1877*a*; Hartung & Dulk 1877; Lyell 1881).[7]

Hartung's travels around various Atlantic islands

That Hartung should have travelled to the Portuguese island of Madeira in 1850, is not surprising on two accounts: first it was a well-known haven for those suffering ill-health, and secondly the German-speaking community on the island was a relatively large one, in the second half of the nineteenth century, because many were associated with the export of embroidery and Madeira wine to Europe through Hamburg (Wilhelm 1997).[8]

Hartung accompanied Heer for some five to six months on his excursions in the island. Heer, who had travelled from Switzerland to Madeira in 1850, specialized in palaeobotany and entomology and was also interested in geology. He became a full professor at the prestigious University of Zürich and also at the Eidgenössische Technische Hochschule (ETH) in the same city (Wilhelm 1990, 1997).

In April 1851, Hartung and Heer departed the island for Cadiz, Spain, and from there they travelled to the Canary Islands where they visited Lanzarote, Fuerteventura and Gran Canaria. It is believed that Hartung spent the winter of 1851 to 1852 there (Wilhelm 1990).

In 1852 Hartung was again in Madeira where he spent that and the following winter, according to Charles Lyell. Both Hartung and Lyell worked together for two months during the winter of 1853–1854, and Lyell benefited greatly from his company since the German naturalist was able to speak Portuguese and had a broad knowledge of the topography and the geology of the island (see maps that accompany Wilson's paper in this volume for routes and places visited). These observations made in Madeira were to be included in a book by Lyell that would contain a report of the result of their joint observations; a plan was outlined by Lyell in a letter dated January 1855 in which he also mentioned the German's good work on the island (Lyell 1881). No report, as a separate piece within a book, was ever published, although Lyell did write a paper on the geology of Madeira that was presented in 1854 to the Geological Society of London (Lyell 1854)[9] and in 1855 prepared a manuscript with notes on the visit to the island with Hartung (Fig. 1).[10]

In 1854, Hartung spent a month with Lyell visiting the Canaries, where they visited Tenerife and Palma and he again returned to Gran Canaria, where he had the opportunity to hear Lyell's thoughts and ideas on the local geology of the islands and on volcanic mountains in general. That helped Hartung understand what he had observed in all the islands, and clarified what had not been clear from reading von Buch's concepts. He spent the summer of 1854 in Europe, where he prepared geological sketches for Lyell, and travelled around in Germany and Switzerland where he met with Heer in Zürich; in November he returned to Madeira (Lyell 1881; Wilhelm 1997). He worked there and in Lanzarote and Fuerventura until April 1855 and used an interesting method of transport which he described thus: 'I may mention the way in which I travelled in these islands. The mode of conveyance is on camels, which are the most ungeological animals you can imagine'.[11]

In May 1855, Hartung travelled from Madeira to London where he spent several weeks with Lyell. The visit had both social and scientific purposes, but not much is known about it. He then went to Germany where he stayed for the rest of the year, travelling to various parts including Darmstadt, Königsberg and Heidelberg. During this time he saw to some family business, worked on his geological samples, prepared geological sketches and took private lessons. In a letter dated December 1855 he described what he did in terms of advancing his geological education during this tour of England and Germany:

> As for my studies, I arrived somewhat late to take full advantage of the 'Semester' as they call the lectures of one half year. However I was not sorry for that, because I had had such a rare occasion to learn much more in a short space of time during those weeks which I spent in London. But at the same time I was anxious to get if possible as much as could be given in one semester, notwithstanding all that, which you teached (*sic*) me. I contrived therefore

Fig. 1. Hartung and Lyell with their guides at Degolliado di Cedro, Tenerife, about 2.00 pm on 30 March 1854; looking SE over las Canadas. Watercolour by Georg Hartung. (Courtesy of Malcolm Lyell; from Wilson 1998, p. 32, Fig. 7.)

to arrange a 'Privatissimum' with Dr. Leonhardt ... who is lecturing to me alone. He ... has a very fine collection of specimens at his disposition. Every time he places before me a great number of rocks, which I may observe at my leisure, asking particular questions.[12]

In December 1855, Hartung rented a house in Heidelberg which he retained until his death, and spent most of 1856 working on his samples from Madeira and the Canary Islands. He again attended private lessons, this time in mineralogy with Dr Leonhardt. The second half of the year was dedicated to the preparation of the planned trip to the Azores. He kept in touch with Heer and also with Robert Bunsen (1811–1899) who was going to analyse the samples from S. Miguel.[13] He also met Lyell, according to a letter sent from Munich and dated October 1856 in which the British geologist addressed him as 'Professor George Hartung' (Lyell 1881).

In December 1856, Hartung travelled to Madeira, where he stayed until April 1857 and from Madeira he went to the Azores where he remained from the middle of April until the end of August. There Hartung experienced serious difficulties in travelling around, not only between islands but also in each of the nine islands that compose the archipelago.[14] He exchanged letters with Lyell in that year and again in 1858 about Lyell's 'upbuilding' theory and von Buch's 'upheaval' theory, and also noted in some letters his having found fragments of certain rocks that Lyell believed to have been transported by ice (Lyell 1881).

In the first half of 1858 Hartung commuted frequently between Darmstadt and Heidelberg to study his rock and fossil samples, and to discuss them with Professor Heirich-Georg Bronn (1800–1862), a professor at Heidelberg University. In September he started visiting districts in Germany in which volcanic rocks cropped out, and he made contact with several German scientists interested in volcanism. In the following year and the first half of 1860 the preparation of his book on the Azores took a great deal of his time, and it was published in June. During the remainder of 1860 he travelled through several volcanic areas of Germany; he also continued to study his samples from Madeira and Porto Santo. In 1861 he visited central Europe and later that year and in 1862 he visited the Auvergne. His letters show that he could not help but compare these very old volcanoes with the recent ones that he had seen in the islands.[15]

After 1857, Hartung did not return to Madeira, the Canary Islands or to the Azores. Even while he prepared his papers and books for publication, and the geological and topographic maps of Tenerife, which were assembled in conjunction with his colleagues Karl Fritsch (1838–1906) and Wilhem Reiss (1838–1908), he saw no reason to return to make any additional local observations.

From the middle 1870s he began to travel to Scandinavia and continued to visit this region until 1880. In Volume 4 of Heer's immense *Flora Fossilis Arctica*, a work of 6 volumes plus a supplement that integrated contributions of a number of co-authors, Hartung contributed a short nine page report on the coal seams at Andö, but it is not obvious that he had participated in the fieldwork that was carried out between 1869 and 1875 under the direction of a mining expert (Hartung 1877*b*).

Hartung's publications on Madeira and the Azores

A main feature of Hartung's books related to the geology of Madeira and the Azores (and his book on the island Gran Canaria has to be included in this category) is that each one of them, although focusing on one of the island groups, refers to the others, and points out their similarities more than their differences.

Another feature is that the books deal with physiographical aspects of the islands and the field relationships of the rocks; these were normally depicted in very fine drawings and in paintings of landscapes and geological sections. Petrographical and fossil studies, the latter by respected palaeontologists, like Heirich-Georg Bronn and Karl Meyer-Eymar (1826–1907), also make up a significant portion of the books.

In all of the volumes, Hartung follows Lyell's concepts on volcanic processes relating to the origin and evolution of the volcanic islands, that is he adheres to the 'upbuilding' theory. The following is a quotation of a precise and concise description by Leonard Wilson of aspects of that theory and also of the work carried out in Madeira by Lyell and Hartung:

Craters of elevation

In 1835 ... von Buch, ... Élie de Beaumont and ... Dufrénoy launched a concerted attack on Lyell's theory centred upon the question of the age of the Mount Etna. In opposition to Lyell [they] insisted that Etna was quite young. It had arisen suddenly, they said, as a crater of elevation ...

During an extended visit to the Canary Islands in 1815 Buch decided that the great caldera on the island of Palma had been formed by a sudden upheaval of accumulated layers of lava, poured out originally in a horizontal position on the sea floor. In 1825 Buch published the theory at Berlin in his book on the Canary Islands and in 1836, to make the theory of craters of elevation better known, he published in Paris a revised and expanded French translation of his Canary Islands book ... The same year Élie de Beaumont published ... a long paper ... on the structure and origin of Mount Etna ... [where] ... he insisted that the lavas must have been poured out on a nearly horizontal surface and upheaved later ...

In 1848 ... Lyell was knighted, but it was not until 1854 that. . .he was able to visit Madeira and the Canary Islands. . . In Madeira he met a young German, Georg Hartung, who joined him in studying the geology of the island. Lyell and Hartung found that Madeira consisted entirely of lavas and basalts produced by a long series of volcanic eruptions of centres on the island. Major cones of eruptions in the centre of the island had buried smaller cones of eruption around its periphery. The many streams of water flowing down from the central mountains had cut deep narrow valleys, providing sections through the accumulated lavas. Frequently they included sections through old volcanic cones buried under later flows of lava from higher centres of eruptions. Such sections revealed multitudes of small vertical dykes, marking channels where molten rock had risen through layers of lava and scoriae ... They showed also that lavas could form sheets of compact rock on steep slopes, and those had not been disturbed since they solidified. The valleys showed Lyell how much more powerful rivers were in eroding the land than he had thought previously ... From fossil plants in a bed of clay among volcanic rocks at Sao Jorge, hundreds of feet above the sea, Lyell discovered that when Madeira was only a half its present size, it was already covered with plants. The additional lavas that had built up Madeira had therefore flowed from the centres of eruption on land. They had not been poured out over the sea bottom. Some volcanoes had become extinct before others began to erupt, but the whole island was a product of a long succession of volcanic eruptions (Wilson 1998).

Hartung's book (1864) on Madeira and Porto Santo relied essentially on the work carried out in 1853 and 1854 by himself and Lyell, but, as the German naturalist worked there on his own after that field season, the book has a personal touch seen, for instance, in his fine drawings of landscapes and geological sections.

An initial section on submarine morphology of the surrounding area of Madeira, based on soundings made available to Hartung, was followed by a description of the subaerial landscape. A long discussion on the effects of sea erosion on the creation of the abrupt sea cliffs followed, and was linked to a discussion of the relative ages of the volcanic masses exposed on the cliffs. He also addressed the importance of the process of deposition of volcanic products relative to the processes of formation and accumulation of organic deposits seen in the island. From this Hartung deduced that fluctuations of the ground either downwards or upwards may have occurred. This section was followed by another lengthy discussion on the effects of running water, and of running water combined with natural trough formation on the cutting of the steep valleys seen in Madeira. This led him to propose a classification of the valleys into two types. Directions, sizes, shapes and slopes observed in ridges, plateaux, valleys, elevations and depressions in general were discussed and tentatively interpreted in terms of old and recent external geodynamic processes, which led to the conclusion that valley formation and evolution, whatever the valley type, had been due mostly, if not completely, to the action of running water. Petrographical and field descriptions of the main rock types were given, as well as their distribution and modes of occurrence and jointing. The study of the internal structure of the mountains preceded a study of their formation and shaping. For the former purpose, he subdivided volcanic products into two main groups: (a) recent, external layers comprising volcanic cones (classified according to their shapes and relative ages) and lava flows (phonolites, dolerites basalts, trachytes, trachydolerites) with which some landforms (channels, arches, grottos) were considered; and (b) old, deep layers composed of hyperstenites, diabases, melaphyres, syenites and porphyres, that, being older than the features described under the grouping (a), could not have their geological age determined. The study of the formation and shaping of mountains was accomplished by considering first the main mass of the mountains and then the peripheral mountain ridges of Madeira. The most important conclusion was that the Madeira and Porto Santo islands had grown gradually upwards from a wide base of sea floor as a consequence of repeated eruptions with deposition, accumulation and superposition of volcanic products, the young eruptions forming the main mass of the mountains most probably having taken place more or less in the middle of the island. These eruptions in the centre of the island had buried smaller peripheral cones of eruption.

The final chapters written by Hartung are about Porto Santo Island and an extended abstract with concluding remarks. The book ends with a study by Karl Mayer of the fossils not only from Madeira and Porto Santo but also from Santa Maria, one of the islands of the Azores. The bibliography was not thoroughly prepared by Hartung and did not include several papers and books published between 1818 and 1861 by Portuguese, British, German and Austrian authors about the geology and palaeontology of Madeira.

Following Lyell's visit to the Madeira archipelago made with Hartung in the winter of 1853–1854, his references to and writings on the geology changed not only in their content but also in their length. The 1846 edition of his *Principles of Geology* contained a short, one page conjectural account of its geology; his 1857 *Manuel de Géologie Élémentaire* and the sixth edition of *Elements of Geology* (1865) contained many pages on the volcanic aspects of the islands; whereas he devoted almost a complete chapter to the living and fossil flora and fauna of the archipelago in the 1868 edition of his *Principles of Geology*. According to Hartung, Lyell had not described the geology of the island in the 1855 edition of his *Manual of Elementary Geology* because he had insufficient field information about a very complex structure. Supplementary data were in fact obtained in 1854 by Hartung who sent them to Lyell.[16]

Hartung's book (1860) on the Azores had a similar structure to that of his earlier book on Madeira and Porto Santo. The first part, describing the general geography, morphology and vegetation cover of the islands, was followed by an overview of the geology, general information on historical eruptions and earthquakes, then by a study of the fossils of Santa Maria, this time by H.-G. Bronn, and finally by a geological description of each one of the islands, in which he stressed the similarities and differences among them. An appendix included chemical analyses of lavas, and comments on the lack of limestone deposits. The figures in this book and the accompanying atlas, some of which were in colour, were fine illustrations, and the geological sections were very clearly drawn. From the book, the 1857 manuscript and the letters addressed to Lyell, it seems that Hartung worked in the Azores to his full capabilities, knowing what to look for and where to be able to draw conclusions about the relationship between the structure of the islands and their formation and relief. The main conclusion, again, was that the mountain chains of the Azores could not be interpreted as the result of upheaval but as the result of a gradual accumulation of volcanic products.

Another important idea of Hartung's regarding circular valleys that he had observed in the Azores, was that the structures might in fact be craters of explosion. He documented these ideas in a letter to Lyell dated 10 October 1857:

> Since I have seen the Caldeiras (*sic*) of the Azores I feel sure that explosions on a very great scale must have constituted a power which produces immense circular valleys. I do not mean to say that I think it is impossible that a valley like the Caldera of Palma should have become excavated chiefly by aqueous erosion after one, two, or more rivulets have been produced, by which the water got access to the scor. format. [scoriae formation?]. But the outward physiognomy of the Azorean Caldeiras must induce the observer to speculate on the possibility of the pre existence of circular depressions, caused by explosions.[17]

A few pages later, in the same letter, Hartung presented a genetic classification of the valleys of the Madeira, the Canaries and the Azores: 'Inter colline valleys', 'Valleys of aqueous erosion' and 'Valleys produced by explosion'.

Hartung's book on the geology of the Gran Canaria island *Betrachtungen über Erhebungskrater, ältere und neuere Eruptivmassen nebst einer Schilderung der geologischen Verhältnisse der Insel Gran Canaria* published in 1862 contains many references to the Azores and Madeira, although from the title of the book and certain chapters (About the origin of the Palma caldera; The geology of the Gran Canaria island; About the old and the young eruptive masses; and On the origin of the local elevations) it would be impossible to know this. This is probably his most interesting book; in it Hartung compared the geological aspects of the three archipelagos and made a synthesis of his main ideas. There is a curious aspect related to it. In a letter addressed to Lyell dated November 1860, Hartung referred to a 'small pamphlet' that he had considered writing on the question of the so-called craters of elevation comparing the volcanic phenomena observed in several regions of Central Europe with those seen in the Atlantic islands that he knew so well. About two years later, in October 1862, in another letter also addressed to Lyell, Hartung announced that he had just published a pamphlet in Leipzig in which he addressed some theoretical questions on volcanism. The title of the pamphlet was *Betrachtungen über Erhebungskrater, ältere und neuere Eruptivmassen nebst einer Schilderung der geologischen Verhältnisse der Insel Gran Canaria*, and it was a book of some 115 pages!

Conclusion

Hartung was a man who first developed an interest in natural history, particularly in geology, when he was nearly thirty years old, thanks to his meeting with the Swiss naturalist Oswald Heer. Heer, like Hartung, had travelled to Madeira for health reasons, not to study science, but their meeting and friendship sparked off in Hartung an interest that was to yield some important contributions to the understanding of the geology of the island.

Hartung also had the rare chance of getting acquainted with Charles Lyell, one of greatest geologists of the time, whom he accompanied on field trips in Madeira and the Canary Islands and from whom he got not only a precious practical preparation in field geology, but also a conceptual model for the physical development of the Atlantic volcanic islands. Lyell once addressed Hartung as he being his 'first pupil in the volcanic line' (Lyell 1881) and it has to be concluded that was perfectly true by reading, in the numerous letters addressed to Lyell, the detailed descriptions of the observations made by the German naturalist and the strong argument that he used in favour of the upbuilding theory.

The question of the publication in co-authorship, by Lyell and Hartung, of a paper on Madeira is a topic that is tentatively raised in some of the Hartung's letters to Lyell including that of 14 September 1857:

If I could only show you my sketches and describe the configuration of the different islands, you would (I am sure) publish the Madeira paper without hesitation and even some doubtful passages, which are however so necessary that they could hardly be omitted, would loose at once they seeming ambiguity.

Was he still thinking of Lyell's promise, made in 1855, to send him a book containing the report of the result of their joint observations? In fact a short report of such observations, with Lyell as the sole author, had been published in 1854 in the *Quarterly Journal* of the Geological Society of London, under the title *On the geology of some parts of Madeira* (Lyell 1854). In writing that letter in 1857, Hartung was probably thinking of a different kind of account, a larger one and with his sketches, of their joint observations.

Through Lyell, the German geologist became involved in the discussion of Leopold von Buch's 'upheaval' and Lyell's 'upbuilding' theories. In the foreword of his 1862 book Hartung wrote that the evidence of occurrence of elevated craters had become so fragile that a totally different interpretation about craters like the Caldera of Palma, in the Gran Canaria island, should not be regarded as offensive to those honoured men who had defended the upheaval theory (Hartung 1862). This is the expression of a true gentleman scientist.

As the Prussian geologist and palaeontologist Curt Gagel (1865–1927) rightly pointed out, Hartung's very detailed, prolix descriptions of his observations in Madeira must be considered not only as a reaction to von Buch's theory, but also the result of the influence of Lyell's fascinating personality (Gagel 1913). In a letter addressed to Hartung, dated 16 October 1857, sent from Naples, Lyell wrote 'Lady Lyell desires her kindest regards, and begs me to say she is sorry you are not with us. So am I, but every day makes me feel that you would only have been confirmed in the true doctrines, which you have gradually and not hastily embraced ...' (Lyell 1881). He was certainly referring to Hartung conversion from von Buch's catastrophism ideas to his uniformitarianism concepts applied to volcanism.

Hartung contributed to the progress of the geological knowledge of Madeira, the Azores and the Canary Islands in collaboration with respected palaeontologists whose names he included with his on the books that he authored. These men included Heinrich-Georg Bronn, who, according to Goulvent Laurent (1997), was one of the greatest palaeontologists of the time, and Karl Mayer-Eymar, who became professor at the Eidgenössische Polytechnical School and at the University of Zürich. Among the fossils of Madeira described by others some such as *Cardium hartungi* Bronn, *Janthina hartungi* Mayer and *Ilex hartungi* Heer have names given in honour of the German naturalist (Zbyszewski *et al.* 1975).

Late in his life, when he was some 55 years old, Hartung changed direction in his research, and began to carry out studies on glacial geology in

Scandinavia, a subject that he had touched on in the Azores where he had made some discoveries that had interested Lyell and Darwin.

How well known are his works on the history of geology of Madeira and the Azores archipelagos? Apart from the fact that his books can only very rarely be found second-hand (partly explained by the small numbers of copies that were printed and sold as Hartung complained in a letter to Lyell dated October 1862)[18] two or three other facts reveal that his scientific output has been forgotten. One of his books, *Betrachtungen über Erhebungskrater, ältere und neuere Eruptivmassen nebst einer Schilderung der geologischen Verhältnisse der Insel Gran Canaria* is not listed in bibliographic references to the geology of Madeira and of Azores although it discussed many aspects of the subject (Rodrigues & Aires-Barros 1992; Pinto 1998). Even Hartung's books on the Canary Islands were not referred to in a recent, extended bibliography about the archipelago (Cruz 2001). German scientific encyclopaedias and dictionaries make no reference to Hartung.[19] The manuscript relating to the island of Terceira, found by one of the present authors (MSP) in Lisbon in the library of the Instituto Geológico e Mineiro, has never been cited in the literature.

Nevertheless his geological sections of the Azores islands are correct[20] and the interpretation of the geological evolution of Madeira as given by Gagel, a critic of Hartung's work, is, according to him, closer to the Lyell–Hartung model than to the monogenetic volcano model of the German volcanologist Alphons Stübel (1835–1904). Stübel's model was proposed after he had worked in Madeira in 1863 (Stübel 1910).

It seems that the only author who paid attention to Hartung's valuable work was Mitchell-Thomé who considered his books on the geology of Madeira and of the Azores to be classic geological works (Mitchell-Thomé 1976). The early study of these Atlantic islands was important in the formulation of ideas about the origin and evolution of volcanic islands in general. Hartung's unpublished letters to Lyell, which are detailed and well illustrated, are precious documents for the history of research on the geology of such islands. Hartung was a true geological traveller who certainly does not deserve to fall into oblivion, and whose publications should be regarded as classics.

We thank M. Kölbl-Ebert for her help in providing information and material not available in Portugal. The authors are also deeply grateful to L. Wilson for having made available copies of Hartung's letters to Lyell, the originals of which are located in the Special Collections Division of Edinburgh University Library. P. Wyse Jackson deserves special thanks for his careful 'editorial handling' of the draft.

Notes

[1]See also: www.familysearch.org

Allgemeine Deutsche Biographie 1875–1912, herausgegeben durch die Historische Commission bei der Königl. Akademie der Wissenschaften. 56 Bände. Duncker & Humblot, Leipzig (Digitales Register—www.ndb.badw.de); Krollmann 1974, p. 253.

Verein für Familienforschung in Ost- und Westpreussen e.V., Hamburg 1980. Schriftenreihe Quellen- Material-Sammlung 1. Die Kartei Quassowski, Buchstabe H. Zusammengestellt von Helmut Zipplies. Im Selbstverlag des Vereins, Hamburg, p. 118; see also Die Abiturienten des Königlichen Gymnasiums und Realgymnasiums in Intersburg, 1861–1910 nebst einem Anhange: Die Abiturienten vor 1861. Von Prof. N. Biesenthal. Interburg. Dr Albert Bittners Buchdruckerei. 1910.

Stadt Heidelberg, Germany, Archiv—Letter, 25 August, 2003; see also *Heidelberger Zeitung* 31 March 1891.

[2]Hartung, G. 6 June 1854 to 23 May 1868. Collection of letters addressed to Sir Charles Lyell, in Edinburgh University Library Special Collections Division.

[3]Hartung, G. 6 June 1854 to 23 May 1868. Collection of letters to Lyell.

[4]Wilhelm recorded that he had received directly from Königsberg University the indication that, according to the university records, Hartung had received a *h. c.* degree of Doctor of Philosophy.

[5]Hartung wrote a whole section in Heer's book: *Ueber die Pflanzen-Versteinerungen von Andö in Norwegen—I. Schilderung des Fundortes und der Lagerungsverhältnisse* (Hartung 1877*b*).

[6]Hartung (1857*a*): Hartung stated here that he was from 'Konigsberg (Prusse)'. He wrote in the preface that he started travelling to Madeira in 1850, that he was forward to live in the same house as Prof. Oswald Heer, from Zurich, whom he accompanied for 5 months in his excursions on the island and with whom he also had travelled home through the South of Spain. He also mentioned that he had not received training to become a naturalist, but for a different profession.

Hartung (1857*b*): This book contains an appendix *Verzeichnis der aufgefundenen Tiere und Pflanzen. 1. Landmollusken, mit Bemerkungen über die Mollus kenfauna der canarischen Inseln überhaupt, von Herrn Professor A. Mousson*, in which Albert Mousson (1805–1890) of Zürich, described some varieties of snails that Hartung had collected.

Hartung & Arlett (1858).

Hartung (1860*a*): Hartung wrote in the preface that he had been attracted by the wonderful natural phenomena seen in Madeira when he had the good fortune of spending some months there accompanying Prof. Heer. From visits to Madeira and to the Canary islands made in Lyell's company, Hartung got introductory views on Lyell's ideas on the formation of the islands that helped him to understand what he saw in nature, contrary to what had happened with von Buch's ideas. He mentioned that Lyell intended to publish a book about Madeira and the Canary islands, more detailed than the 1855 edition of the *Manual of Elementary Geology*. About the Azores he mentioned difficulties in travelling there and stressed the similarities and differences between these islands and Madeira. Based on Lyell's ideas he tried to explain the relationship between the internal structure of the islands and the formation of mountain ranges. He thanked the various scientists who had contributed to the book. And finally he wrote about his attempt to describe the Azorean plants comparing them with the vegetation cover of Southern Europe, and of the Madeira and the Canary islands.

Hartung (1860*b*): This volume contains a map and 19 fine drawings, 3 of them in colour.

Hartung (1862): In the foreword Hartung wrote that the evidence of occurrence of elevated craters had become so fragile that a totally different interpretation about craters like the Caldera of Palma, in the Gran Canaria island, should not be regarded as offensive to those honoured men who had defended the upheaval theory. Then he gave an explanation why Lyell had not described the geology of the island in the 1855 edition of his *Manual of Elementary Geology*: insufficient field data (that both had gathered together) about a very complex structure. Hartung then explained that he tried to combine M. Reiss' study of diabases and lavas of the Palma island with von Buch's field observation and with his own observations in order to shed more light on the geology of the Gran Canaria. The importance of comparing the volcanic rocks composing the various Atlantic islands with similar rocks occurring in Germany was then described. Finally the origin of volcanic mountains was discussed.

Hartung (1864): Hartung stated in the preface that, having been encouraged by Prof. Heer to study some aspects of the geology of Madeira, he had the chance of confirming the origin of some agglomerates supported by Heer against prevailing opinions and that Lyell, in 1853, had been able to give a full explanation of the phenomenon. He then reported that Lyell's short visit to Madeira had been followed in 1854 by new works in Madeira, Lanzarote and Fuerteventura carried out by himself, Hartung, in order to compare old and recent lavas. The results of such works had been sent to Lyell who planned to publish in co-authorship with Hartung a geological description of Madeira and Porto Santo, but lack of time had prevented the British geologist from preparing a draft and so Hartung had decided to publish his material alone, stating that under Lyell's supervision it would have received a true appreciation. He also referred to his travels in the Azores and in several German regions to observe recent and old volcanoes. Finally, he thanked several scientists who had collaborated in the preparation of this book.

Fritsch, Hartung & Reiss (1867): Hartung wrote in the foreword that he had only visited the most important places in Tenerife, whereas the other authors had visited the whole island. Hartung had planned to visit this island once more, but it is not clear whether or not he managed a second visit: one in 1854 with Lyell and another one related to the preparation of this book.

[7]http://darwin.lib.cam.ac.uk

[8]Hartung, G. 6 June 1854 to 23 May 1868. Collection of letters to Lyell; A. Vieira, personal communication, 2003.

[9]This paper was presented on 22 March 1854, and was read by Leonard Horner. In it Lyell wrote that he had been favoured in nearly his excursions by the company of Hartung, who had proved a most active fellow labourer.

[10]www.nahste.ac.uk/GB 0237 Sir Charles Lyell Gen. 118 Lyell 2/59–76.

[11]Hartung, G. 6 June 1854 to 23 May 1868. Collection of letters to Lyell.

[12]Hartung, G. 6 June 1854 to 23 May 1868. Collection of letters to Lyell.

[13]Hartung, G. 6 June 1854 to 23 May 1868. Collection of letters to Lyell.

[14]Hartung, G. 6 June 1854 to 23 May 1868. Collection of letters to Lyell.

[15]Hartung, G. 6 June 1854 to 23 May 1868. Collection of letters to Lyell.

[16]Hartung, G. 6 June 1854 to 23 May 1868. Collection of letters to Lyell.

[17]Hartung, G. 6 June 1854 to 23 May 1868. Collection of letters to Lyell.

[18]Hartung, G. 6 June 1854 to 23 May 1868. Collection of letters to Lyell.

[19]M. Kölbl-Ebert, and M. Guntau, personal communication, 2003.

[20]V. H. Forjaz, personal communication, 2003.

References

CRUZ, C. M. G. 2001. *The Origin of the Canary Islands: a Chronology of Ideas and Related Concepts from Antiquity to 2000.* International Commission on the History of Geological Sciences, Sydney.

FRITSCH, K., HARTUNG, G. & REISS, W. 1867. *Tenerife geologisch topographisch dargestellt. Ein Beitrag zur Kenntnis vulkanischer Gebirge: eine Karte und sechs Tafeln mit Durchschnitten und Skizzen nebst erläuterndem.* J. Wurster, Winterthur.

GAGEL, C. 1913. Studien über den Aufbau und die Gesteine Madeiras. *Zeitschrift Deutschen Geologischen Gesellschaft*, **64** (1912), 344–491.

HARTUNG, G. 1857*a*. Die Geologischen Verhältnisse der Inseln Lanzarote und Fuertaventura. *Neuen Denkschriften der allgemeinen Schweizerischen Gesellschaft für die gesammten Naturwissenschaften*, **15**, 1–168. Zürich.

HARTUNG, G. 1857*b*. *Die geologischen Verhältnisse der Inseln Lanzarote und Fuertaventura.* Basel.

HARTUNG, G. 1860*a*. *Die Azoren in ihrer äusseren Erscheinung und nach ihrer geognostischen Natur. Mit Beschreibung der fossilen Reste von Prof. H. G. Bronn*. Engelmann, Leipzig.

HARTUNG, G. 1860*b*. *Die Azoren in ihrer äusseren Erscheinung und nach ihrer geognostischen Natur. Atlas enthaltend neunzehn Tafeln und eine Karte der Azoren*, Engelmann, Leipzig.

HARTUNG, G. 1862. *Betrachtungen über Erhebungskrater, ältere und neuere Eruptivmassen nebst einer Schilderung der geologischen Verhältnisse der Insel Gran Canaria. Mit zwei Karten und fünf Tafeln*. Engelmann, Leipzig.

HARTUNG, G. 1864. *Geologische Beschreibung der Inseln Madeira und Porto Santo. Mit dem systematischen Verzeichnisse der fossilen Reste dieser Inseln und der Azoren von Karl Mayer*. Engelmann, Leipzig.

HARTUNG, G. 1877*a*. *Die skandinavische Halbinsel. Eine geologische Skizze*. Habel, Berlin.

HARTUNG, G. 1877*b*. *Ueber die Pflanzen-Versteinerungen von Andö in Norwegen—I. Schilderung des Fundortes und der Lagerungsverhältnisse)*. *In*: HEER, O. (ed.) *Flora Fossilis Arctica—Die Fossile Flora der Polarländer (Vierter Band)*. J. Wurster & Co., Zürich.

HARTUNG, G. & ARLETT, R. N. 1858. *Geologische Karte der Inseln Lanzarote und Fuerte Ventura* 1:250 000.

HARTUNG, G. & DULK, A. F. B. 1877. *Fahrten durch Norwegen und die Lappmark*. Kröner, Stuttgart.

HEER, O. 1877. *Flora Fossilis Arctica—Die Fossile Flora der Polarländer (Vierter Band)*. J. Wurster & Co., Zürich.

KROLLMANN, C. 1974. *Altpreussische Biographie, herausgegeben im Auftrage der Historischen Kommission für ost- und westpreussische Landesforschung*. Elwert, Marburg.

LAURENT, G. 1997. Paléontologie et évolution: état de la question en 1850 d'après l'œuvre de Heinrich-Georg Bronn. *In*: GOHAN, G. (ed.) *De la Géologie à Son Histoire: Ouvrage edite en hommage à François Ellenberger, 175–188*. Commission des Travaux d'Histoire Scientifique (Ministère de l' Éducation Nationale de France), Paris.

LYELL, C. 1854. On the geology of some parts of Madeira. *Quarterly Journal of the Geological Society of London*, **10**, 325–328.

LYELL, K. M. (ed.) 1881. *Life, Letters and Journals of Sir Charles Lyell, Bart*. Volume 2. John Murray, London.

MACOMBER, R. W. 1993. Lyell, Sir Charles. *The New Encyclopaedia Britannica* **7**, Micropaedia, 585–586.

MITCHELL-THOMÉ, R. C. 1976. Geology of the Middle Atlantic Islands. *Beiträge zur Regionalen Geologie des Erde*. Gebrüder Borntraeger. Berlin and Stuttgart.

PINTO, M. S. 1998. Effects of eruptions on society—the case of the Azores Archipelago. A brief historical account. *In*: MORELLO, N. (ed.) *Volcanoes and History*. Brigati, Genoa, 565–580.

RODRIGUES, B. & AIRES-BARROS, L. 1992. A geoquímica dos vulcanitos do continente português e das ilhas atlânticas dos Açores, Madeira, Selvagens, Cabo Verde e S. Tomé e Príncipe: síntese e perspectivas. *In: História e Desenvolvimento da Ciência em Portugal no Séc. XX*. Publicações do II Centenário da Academia das Ciências de Lisboa, Lisboa, 873–900.

STÜBEL, A. 1910. *Die Insel Madeira. Photographische Wiedergabe einer Relief Karte zur Erlanterung das Vulkanische Baues dieser Insel*. Museum für Länderkunde, Leipzig.

WILHELM, E. A. 1990. Visitantes de Língua Alemã na Madeira (1815–1915). *Revista Islenha*, **6**, 48–67.

WILHELM, E. A. 1997. *Visitantes e Escritos Germânicos da Madeira, 1815–1915*. Direcção Regional dos Assuntos Culturais, Funchal.

WILSON, L. C. 1998. Lyell: the man and his times. *In*: BLUNDELL, D. J. & SCOTT, A. C. (eds) *Lyell: the Past is the Key to the Present*. Geological Society, London, Special Publications, **143**, 21–37.

ZBYSZEWSKI, G., FERREIRA, O. V., MEDEIROS, C., AIRES-BARROS, L, SILVA, L. C., MUNHÁ, J. M. & BARRIGA, F. 1975. *Carta Geológica de Portugal na escala 1:50 000. Notícia explicativa das folhas ≪A≫ e ≪B≫ da Ilha da Madeira*. Serviços Geológicos de Portugal, Lisboa.

'Marks of extreme violence': Charles Darwin's geological observations at St Jago (São Tiago), Cape Verde islands

P. N. PEARSON[1] & C. J. NICHOLAS[2]

[1]*Department of Earth Sciences, Cardiff University, Park Place, Cardiff CF10 3YE, UK (e-mail: Paul.Pearson@earth.cf.ac.uk)*

[2]*Department of Geology, Trinity College, Dublin 2, Ireland*

Abstract: The first stop on Charles Darwin's famous voyage around the world in HMS *Beagle* was at Porto Praya (Praia), the principal town on the island of St Jago (São Tiago) in the Cape Verde archipelago. From 16 January to 8 February 1832, Darwin enjoyed his first substantive opportunity to study the natural history of an exotic place. Darwin himself regarded this occasion as a significant turning point in his life because, according to his autobiography, it was here that he decided to research and publish a book on the geology of the places visited on the voyage. He also recalled that it was here, the very first port call, that convinced him of the 'wonderful superiority' of Charles Lyell's uniformitarian geology over the doctrine of successive cataclysms that he had been taught in England. Later commentators have generally accepted this account, which is significant for understanding the intellectual background to the *Origin of Species*, at face value. In this paper we reconstruct some of Darwin's observations at St Jago based on his contemporaneous notes and diary, and in the light of our own visit made in January 2002. We find little evidence to substantiate the claim that he interpreted the geology in Lyellian terms at the time. Instead, he formulated a theory involving a great cataclysm to explain the dramatic scenery in the island's interior. He speculated that a torrent of water had carved the main valleys of the island, leaving deposits of diluvium in their beds. It is indisputable that Darwin came to embrace gradualist thinking enthusiastically during the voyage. Some of his observations made on St Jago, especially relating to uplift of the coast, were instrumental in this change of view, but the conversion was gradual, not sudden. His later published works make no mention of his original catastrophist interpretations.

Catastrophies, Lyell, and Darwin's recollections of St Jago

In a volume devoted to geological travellers, it is worth noting that Charles Darwin (1809–1882) was one of the most widely travelled of all geologists in the early nineteenth century, having circumnavigated the globe aboard HMS *Beagle* (1831–1836) (Herbert 2007). Because of his later renown as an evolutionist, a great deal of attention has been paid to the zoological observations he made on the voyage, but from the outset, geology was his main occupation. At 1383 folios, his geological manuscripts from the voyage outnumber his zoological notes by a ratio of about four to one. Unlike the zoological notes, these manuscripts have not all been published, although they contain much of interest.

Darwin was 22 years old when the *Beagle* sailed. He was by no means a finished naturalist, but by the standards of the day he was well educated in geology (Secord 1991). At Edinburgh, he had attended lectures on the subject and went on day trips in the field. At Cambridge he had received more detailed instruction from his mentor John Stevens Henslow (1796–1861), and in the summer of 1831 he had acted as a field assistant to Adam Sedgwick (1785–1873) who was at that time conducting his classic research on the strata of North Wales. He had also undertaken geological field excursions on his own (Secord 1991; Roberts 2000; Lucas 2002*a*, *b*; Herbert & Roberts 2002).

One of the great theoretical issues of the time, even if it was only beginning to be recognized as such, was whether the geological record was best interpreted as a series of catastrophic events, or whether slowly acting processes of erosion, deposition and deformation could account for geological appearances (Hallam 1989). For many years, it had been believed by naturalists that the most recent in a succession of global catastrophes was a major flood or deluge, which had left a layer of superficial deposits known as 'diluvium'. The reverend dons of Oxford and Cambridge (Henslow and Sedgwick included) were among those who had equated this deluge with the Noah's flood, taking satisfaction in the apparent convergence of geology and scripture, at least in this respect.

By the time Darwin sailed, however, research on the 'diluvial' deposits had made this theory less and

From: WYSE JACKSON, P. N. (ed.) *Four Centuries of Geological Travel: The Search for Knowledge on Foot, Bicycle, Sledge and Camel.* Geological Society, London, Special Publications, **287**, 239–253.
DOI: 10.1144/SP287.19 0305-8719/07/$15.00

less tenable (Herbert 1991). Sedgwick formally renounced it in his Presidential Address to the Geological Society in 1830 (Sedgwick 1830). One of its chief proponents, William Buckland (1784–1856) at Oxford, was to do the same in 1836, the year of Darwin's return (Buckland 1836). It is not clear what Henslow's position was in this period, but Darwin certainly knew that the diluvial theory and the geological evidence for the Flood was a contentious issue.

Closely related to this catastrophism debate was the issue of the uniformity or otherwise of different rock types in different geological periods. An important standard work that was used frequently by Darwin was *Active and Extinct Volcanoes* by Charles Daubeny (1795–1867) (Daubeny 1826). Sedgwick recommended it to him in 1831 during preparations for the voyage (Burkhardt & Smith 1985), and it was to prove useful in its descriptions of igneous rocks and minerals. Daubeny (who was a colleague of Buckland's at Oxford) took the standard view that different rocks, such as different types of lava, were characteristic of different periods in earth history, and that geological processes had been acting with diminishing intensity up to the present day. By contrast, in 1830 Charles Lyell (1797–1875) published the first volume of his famous *Principles of Geology* (Lyell 1830), which not only repudiated the diluvial theory, but went much further in arguing that processes now operating, working at the same intensity, were sufficient to account for all geological phenomena, and that the same types of rocks have been produced throughout Earth history. Lyell disparaged the tendency of geologists to invoke great catastrophes in an *ad hoc* fashion to explain appearances in the rocks and landscape. Instead he argued on methodological grounds that explanations based on known processes were always preferable. He employed the Newtonian notion of the vera causa—the natural cause—as the yardstick by which valid geological theories should be measured.

Although Lyell was widely respected among the geological elite, they regarded his theorizing with suspicion. Looking back on this controversy, Darwin related in his autobiography that 'the sagacious Henslow, who, like all other geologists believed at that time in successive cataclysms, advised me to get and study the first volume of the *Principles*, which had then just been published, but on no account to accept the views therein advocated' (Barlow 1958). During the voyage, however, Darwin became a convert to Lyellian geology—even a 'zealous disciple', as he himself declared in 1834 (Burkhardt & Smith 1985). On his return to England, he became Lyell's most valuable ally in the scientific debates of the late 1830s, and the two men became lifelong friends. Secord (1991) has remarked that it was 'astonishing that he should have decided to align his work with Charles Lyell's controversial programme of geological reform, which had almost no followers in England'. But for Darwin's later work on evolution, it was a very important step. It has been generally acknowledged that Lyell's emphasis on the long accumulated effects of gradual changes, acting over expanded time periods, was an important component of Darwin's later theorizing.

Darwin himself implied in his autobiography (Barlow 1958) that his conversion to Lyellian geology occurred at the beginning of the voyage, in St Jago. Indeed the point is made twice, in similar terms, once in the original manuscript of 1876, and again among additions that were written in 1881, the year before his death:

> The very first place which I examined, namely St. Jago, in the Cape de Verde islands, showed me clearly the wonderful superiority of Lyell's manner of treating geology.

> I am proud to remember that the first place, namely St. Jago, in the Cape de Verde archipelago, which I geologised, convinced me of the infinite superiority of Lyell's views over those advocated in any other work known to me.

In another passage from the autobiography, Darwin elaborated on the importance of this first stop as one of the defining moments of his life:

> The geology of St. Jago is very striking, yet simple: a stream of lava formerly flowed over the bed of the sea, formed of triturated recent shells and corals, which it has baked into a hard white rock. Since then the whole island has been upheaved. But the line of white rock revealed to me a new and important fact, namely that there had been afterwards subsidence around the craters, which had since been in action, and had poured forth lava. It then first dawned on me that I might perhaps write a book on the geology of the various countries visited, and this made me thrill with delight. That was a memorable hour to me, and how distinctly I can call to mind the low cliff of lava beneath which I rested, with the sun glaring hot, a few strange desert plants growing near, and with living corals in the tidal pools at my feet.

The story of Darwin's sudden conversion in the field to Lyellian geology has been frequently related by his many biographers. However, the autobiography was not intended for publication, and in view of its parable-like nature it has been regarded with suspicion as an exact source (Ospovat 1981). Analysis of Darwin's geological notes from St Jago (Secord 1991; Pearson 1996; Keynes 2002; Sloan 2003) has revealed that he did indeed confront several issues that were important in the Lyellian debate, and because of this, most commentators have largely accepted his implied sudden conversion to Lyellian gradualism in the first days of the voyage.

We have conducted a new study of Darwin's geological work on St Jago in 1832, based on all the written sources and our own observations of the important localities made in January 2002. We agree with Oldroyd (1999)[1] that visiting the field localities can be of great help in appreciating the written texts. The rocks themselves, if one can be confident they are the same as seen by a previous observer, constitute a form of historical record in their own right. Apart from the urban sprawl immediately around Praya, the landscape of St Jago is relatively unchanged from Darwin's time. We visited at the same time of year as Darwin, which was during the local dry season. Being on the spot has helped clarify many potential ambiguities in the written record and gives a sense of scale to the phenomena in question.

In contrast to earlier writers, we have concluded that Darwin's work during his first visit to St Jago was squarely in the mode of his teachers in his invocation of sudden and violent episodes to explain geological features, including the presence of 'diluvium' on the island. It is even doubtful whether he had read Lyell's *Principles* in detail before the *Beagle* departed for Brazil. He was soon to do so, however, and it appears that he reinterpreted some of his St Jago observations in this theoretical light in the following few months and years. We note that our analysis does not strictly contradict the letter of Darwin's account. Indeed, it makes the episode more interesting as a case study in what later became known as 'catastrophist' versus 'uniformitarian' thinking in the early 1830s, when this most important debate was taking shape.

We also differ from received wisdom in one other respect, which is of interest only as a point of biographical detail. Insofar as previous writers have expressed an opinion, it has generally been stated that the 'memorable hour' in which Darwin resolved to write a book on the geology of the voyage (see the quotation above) occurred on Quail Island (now Santa Maria island, a small islet in the bay of Porto Praya) on 17 January 1832. We think it more likely that it was on the east coast of the main island on 23 January. We locate this place referred to by Darwin, which is a beautiful spot that he remembered with clarity for his whole life as marking the start of his scientific career.

Darwin's itinerary

The *Beagle* arrived in Praya, the port and capital of St Jago, on the afternoon of 16 January 1832 and departed on 8 February. It is possible to reconstruct Darwin's itinerary (Fig. 1) and geological observations from this period using several sources, both published and unpublished. The main published works are the *Journal and Remarks* (Darwin 1839) and the *Geological Observations on the Volcanic Islands visited during the voyage of HMS* Beagle (Darwin 1844), in which the first chapter is devoted to St Jago. Relevant passages from Darwin's autobiography (Barlow 1958) have been quoted above. Darwin's correspondence from the period (Burkhardt & Smith 1985) includes important letters he wrote to his father and J. S. Henslow in the months after the visit. Darwin kept a general diary throughout the voyage, in which he recorded his movements and described events (Keynes 1988). His field notebook from the period is kept in Down House, Kent.[2] His geological specimen notebook is at the Department of Earth Sciences in Cambridge University.[3] The most important document, however, is a manuscript entitled 'Diary of observations on the geology of the places visited during the voyage' (hereafter referred to as the 'geology diary') which is kept at Cambridge University Library.[4] This contains two structured accounts summarizing his observations and ideas, the first written on 18 January relating to two days' observations on Quail Island (pp. 15–20) and the second written on 8 February, or soon after, giving a similar account of what he had seen on the mainland of St Jago (pp. 21–36).

The geology diary is of particular importance because it is a contemporaneous and detailed account. Some aspects of the text, particularly the Quail Island observations, have been analysed in detail by Herbert (1991). It was written on only one side of each page, with additional notes or reflections on the back of the pages, keyed by letters to the original account. Unfortunately many of these notes are undated. Most were probably written shortly after the main text, possibly even on first reading, but some were clearly added later in the voyage.

It is obvious that Darwin, who had never before travelled abroad, greatly enjoyed his experiences on St Jago. For some time he had yearned to see the tropics, having been inspired by Alexander von Humboldt's (1769–1859) *Personal Narrative* of journeys in South America. On the very first afternoon, he visited the citadel of Praya, tasted a banana for the first time and saw oranges and coconuts growing in their native soil. As he recorded in his diary, 'I returned to the shore, treading on Volcanic rocks, hearing the notes of unknown birds, & seeing new insects fluttering about still newer flowers.—It has been for me a glorious day, like giving to a blind man eyes' (Keynes 1998). Writing to his father about the visit, he remarked that 'Geologising in a Volcanic country is most delightful, besides the interest attached to itself it leads you into the most beautiful & retired spots' (Burkhardt & Smith 1985), and to Henslow 'St Jago is singularly barren & produces few plants or insects.—so that my hammer was my usual companion; in its company most delightful hours I spent' (Burkhardt & Smith 1985).

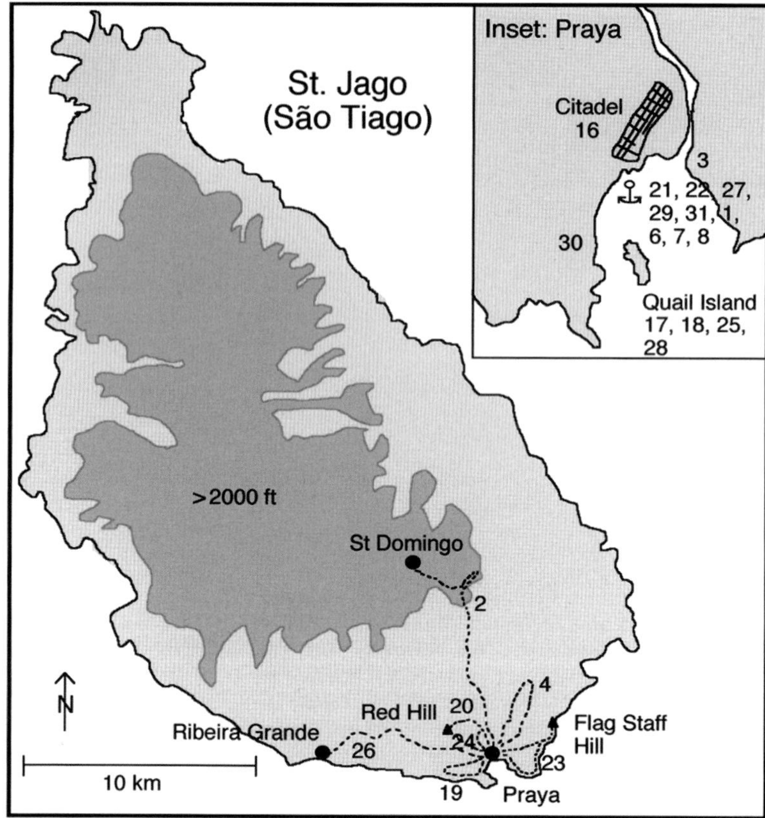

Fig. 1. The island of St Jago, drawn from a modern map, showing Darwin's itinerary during his stay. Numbers refer to days of the month, beginning with 16 January 1832 and ending with 8 February. Days spent mostly on the ship are shown in Praya harbour (see inset).

Darwin's initial strategy was to study the geology of Quail Island (now Santa Maria island) (Fig. 2). This is a rocky prominence just a few hundred metres long and less than a hundred metres wide, with a flat top and steep sides. It was selected by Captain Robert Fitzroy (1805–1865) as a secure site to construct an observatory in order to take an exact fix of the longitude using the ship's chronometers. Darwin reasoned that it would also be useful in giving him a representative section of the geology of the district, which indeed proved to be the case. This he did on his first full day in Praya, 17 January, and wrote up on the 18th after making some additional observations.

Following this, over several weeks, Darwin made a series of excursions in the surrounding area (see Fig. 1), including visits to two small volcanic cones called Red Hill (visible in Fig. 2) and Flag Staff Hill (later referred to as Signal Post Hill in his published writings). He also took two excursions by pony, the first to the old town of Ribeira Grande (Cidade Velha) on 26 January and the second to a village called St Domingo in the upland interior of the island on 2 February. As well as the geology, Darwin spent much time collecting the invertebrate fauna of the rocky shore. After a day's fieldwork he would often spend the following day aboard ship working on the geological and zoological collections before venturing out again.

We have visited the key localities and re-traced all of Darwin's excursions. Rather than attempt a long, detailed re-description of all that he saw and did, we concentrate here on issues that are of most interest regarding the interlinked theoretical issues of geological time, catastrophism, and the uplift and subsidence of the land.

Geological time

By the time Darwin received his training, there was general agreement that the previously accepted Biblical time-span for the Earth of a few thousand years was insufficient to account for the planet's history. He encountered this idea at Edinburgh, if not

Fig. 2. Quail Island (Ilheu Santa Maria) from the air, looking west Darwin worked here initially on 17 and 18 January 1832. The Red Hill volcano is in the centre distance. The urban sprawl on this side of the bay is modern.

before, and it would have been reinforced by the teachings of Henslow and Sedgwick. Nevertheless, the revived catastrophist/uniformitarian debate was related to differing concepts about how long Earth history might be. The generally accepted view was that the rock record reveals a series of more or less violent events, perhaps declining in intensity toward the present day, whereas to the uniformitarians the present was unexceptional and the Earth had always been much as it is now (Hallam 1989). The catastrophists could compress Earth history into a relatively short time-span compared to the gradualists, who required truly immense periods to accomplish multiple cycles of erosion and uplift of mountains. Lyell, and to some extent George Poullet Scrope (1797–1876), were the only serious British uniformitarians in the early 1830s.

The rocks of Quail Island, which Darwin studied on 17 January, threw up an immediate puzzle in relation to the question of geological time. The island is like a geological sandwich (Fig. 3), with volcanic rocks lying above and below a central layer of fossiliferous, calcareous sand. This Darwin interpreted as a localized beach deposit ('former beach'), although because of its broad extent, he later came to think of it as having been deposited as a layer of sediment on the shallow sea floor. It had evidently been covered by a thick sheet of basalt, which had produced some interesting heating effects at the junction. Looking over to the mainland, Darwin could see the same white band of rock occurring in the cliffs below the citadel of Praya, and all around the bay, implying that the rocks of Quail Island had at some time connected across the entire area.

The white rock is accessible in several places around Quail Island. Darwin found it to contain abundant fossil shells, and he made a collection of limpets, oysters, gastropods, and sea urchins, something which we were able to repeat with ease. He was surprised to find that an almost identical collection of shells can be made on the modern beach at his feet, implying that the limestone was not very old:

> The organic remains in their beds are certainly of a Tropical type & as far as my knowledge goes are the same as those of present day. <—I should think this old coast one of no great duration?—>[5]

Darwin noticed that some of the shells retain traces of their original pigmentation and used his

Fig. 3. Cliff exposure on the NE side of Quail Island, looking south, showing the central band of white limestone that Darwin initially interpreted as a 'former beach' resting on weathered volcanic rocks and overlain by a black basaltic sheet. Height of cliff: approximately 10 m.

blowpipe to see if the shells emitted an 'animal smell' when fused, that is, to determine if there was any organic matter present. There was not. It has been suggested (Secord 1991) that this is evidence of Charles Lyell's early influence on Darwin, but other geologists may have done the same.

The significance of this rock to Darwin was that despite having modern-looking shells in it, the band of white sandy limestone had nevertheless been covered by a thick sheet of molten basalt and after this a great deal of very hard material had been removed by erosion, separating the island from the mainland.

Another important discovery on Quail Island was an outcrop of muddy conglomerate on the west side of the island, which had been laid down after the other rocks that make up the island and contained within it reworked pebbles of them. Significantly, Darwin noted that 'It looks to me like a part of the long disputed Diluvium'.[6] Herbert (1991) has shown that Darwin's use of the term diluvium does not necessarily imply his support for the Noachian deluge. Sedgwick, for example, continued to use the term in a descriptive way for sediments of different ages, and was so doing in his trip with Darwin in 1831 (Secord 1991). Later work on the mainland of St Jago revealed to Darwin much more diluvium in the valleys, and we note that in one case he described some of it that he interpreted as having been 'deposited previous to the general upheaving of the island',[7] implying that to Darwin also, not all of it was contemporaneous. Be that as it may, such large amounts of diluvium still demanded an explanation, and as we shall see, a large flood was one possibility. At a later date, possibly on his return to Praya on the homeward bound leg of the voyage in 1836 (as discussed below), Darwin crossed out the reference to the diluvium, along with other significant passages, with the statement 'I have drawn my pen through parts which appear absurd'.[8]

On 20 January, while walking in one of the valleys behind the old town, Darwin made an unexpected discovery that prompted him to reflect on geological time. This was a mature baobab tree of the genus *Adansonia*, named by Linnaeus in honour of the explorer–naturalist Michel Adanson (1727–1806). Darwin returned with Captain Fitzroy and one of the officers to measure its height and circumference accurately. Fitzroy climbed the tree with a lead and line to confirm the measurement, a feat that was no doubt easy for a naval officer and not beneath his dignity. A sketch of the tree, probably by Fitzroy, and various trigonometric calculations can be seen in Darwin's notebook.[9]

The baobab tree was interesting, because as Darwin recorded in his diary and geological notes, according to Adanson, the largest trees reached the great age of 6000 years. The implication of this in the old Biblical chronology is that such trees might have sprouted in the first week of Creation! Darwin remarked in his diary that 'This one bears on its bark the signs of its notoriety—it is as completely covered with initials & dates as any one in Kensington gardens' (Keynes 1998). His geological notes reveal significant reflections prompted by the discovery of this tree:

> In this [one of the valleys north of Praya] grows the celebrated Baobab or Adansonia; this tree only 45 feet high, measured two feet from the ground round the solid trunk. 35.—Some of the same species in Africa were supposed by Adanson to reach the enormous age of 6000 years.—The very appearance of the tree strikes the beholder that it has lived during a large fraction of the time that this world has existed.—Of course the valley must be older & it is this one that has finally left the neighbourhood of Praya in the state we now find it.—How long a time intervened between this period and the deposition of former beach it is impossible to say.—during it three great phenomena occurred, the flowing of the lava.—the upheaving of the coast. & the great beds of diluvium collected in the older valley.—To what a remote age does this in all probability call us back & yet we find the shells [in the 'former beach'] themselves & their habits the same as exist in the present sea.[10]

This passage can be interpreted in various ways, but is probably best regarded as a rhetorical account highlighting the expanded view of geological time that was orthodox among geologists at the time. It is very unlikely that Darwin really believed that the tree had lived during a large fraction of the time that this world has existed. He was well aware that most other limestones, such as those near his home in Shropshire, contained only extinct species of fossils, and must be far older than the 'former beach' of Praya. The Praya limestone, however, must be of substantial antiquity because it predates the uplift and erosion of the island.

Despite this interpretation of Darwin's remarks, further evidence is provided in the following section that the modern concept of geological time is probably far more expanded than Darwin was contemplating. He would surely have been surprised to learn that the limestone was between two and five million years old, as modern geologists believe (see below). Hence we suggest that the fact that the Praya limestone was geologically recent encouraged him to compress the geological events of St Jago, including its uplift and the formation of the landscape, into a comparatively short time span relative to that which would be envisaged by geologists today.

Darwin's catastrophism

During the weeks of fieldwork following his initial observations at Quail Island, Darwin studied the coastal exposures in both directions and climbed most of the neighbouring hills. Even from a distance, he observed that these fall into two main classes; volcanic cones, such as Red Hill and Flag Staff Hill, and 'truncate conical hills', which he interpreted as remnants of table land left after the erosion of a network of valleys around them. The citadel rock of Praya was one of this latter class, as was Quail Island itself, although its base was below sea level.

Darwin was puzzled by the form of the valleys which he described as a remarkable feature of the island, and he expended considerable effort in studying them. The largest of those examined were behind and east of Praya, the one that runs in the direction of Trindad to the north and contained the baobab tree (although, sadly, no longer), and the valley of St Martin, which it was necessary to cross on the road to Ribeira Grande, which he did on 26 January (Fig. 4). All these valleys are similar, in that they have wide, flat bottoms with steep sides and a jumble of large and small boulders litter their floors. In places they branch and rejoin around the truncate hills. Only the valley of St Martin contained running water when Darwin visited it, and that in small amounts. So how had they been formed?

This question was a highly charged one in terms of the catastrophist/gradualist debate that was then beginning. Most geologists were inclined to think of large landscape features as having been produced suddenly, by violent events. But an alternative explanation that was gaining ground through the work of Lyell and Scrope (following James Hutton (1726–1797) and John Playfair (1748–1819) in previous generations), was the long-continued action of rivers. For example, as Lyell himself emphasized in his long 1827 review of Scrope's book on the Auvergne, Scrope had proved, contrary to other English geologists, that the 'valleys, which decidedly owe their form to the agency of water, have not been shaped out by one sudden and violent inundation, but progressively by the action

Fig. 4. The valley of St Martin on the road to Ribeira Grande, as studied by Darwin on 26 January 1832. Like others on the island, the valley has a broad flat bottom littered with boulders, and steep sides. Darwin variously speculated that the valleys had been excavated by a single torrent of water as the island was uplifted and the seawater drained away, leaving boulders of diluvium in the bed, or that torrents of freshwater from inland were responsible.

of rivers, or of such floods as may occur in the ordinary course of nature' (Lyell 1827).

Darwin, however, clearly sided with the catastrophists in his attempts to explain the St Jago valleys. As he put it, 'immense bodies of water have fairly cut through the lava & former beach. — leaving at the bottom a thick pile of large boulders'.[11] He cited three reasons for believing this. Firstly, 'I am decidedly of the opinion that these valleys were formed by great bodies of water & not by gradual effects—the mere flatness of their bottoms gives it a strong presumption'.[12] In other words, there was no incision, as would be expected from gradual cutting down by streams. Second, 'this body of water must have been rush — and not continuous streams, as the spaces between the large stones are not filled up'.[13] And third, the erosion must have happened rapidly, because it had all occurred since deposition of the Praya limestone (i.e. the aforementioned 'former beach') with its geologically modern shells—as he put it, 'continued action of water is here precluded by shells of modern date'.[14]

The issue he debated with himself was whether the erosion had been caused by the running off of seawater, as the island was bodily upheaved from below the waves (a quite amazing theory to modern eyes), or freshwater torrents from inland. He seems eventually to have inclined to the latter view because he was unable to find marine shells in the diluvium in the valleys, and such a large amount of seawater would imply that at the time of deposition, the white limestone must have originally been some way below the sea, at a depth where he thought shells might be uncommon. This he explained in an undated addition:

It certainly sounds the most probable explanation.—that when the former coast was upheaved the valleys were formed by the draining off of the sea.—but in this case we must suppose the beach abounding with shells to have been formed at a great depth.—sufficient to cover the superimposed bed of lava, & then to allow a body of water enough to make a torrent when the island was lifted up.—shells at such depths are I believe uncommon.—again we have evidence that there was one stream sufficient to cut through the lava & former beach leaving a bed for the lava posterior to the general covering.—this then is a *vera causa*.[15]

Darwin's reference to the concept of *vera causa* in this context is interesting, possibly even defiant, as his statements can hardly be interpreted as being Lyellian in spirit. He was invoking a single stream of water, either salt or fresh, of such magnitude as to sculpt almost the entire landscape that we see today, including isolating Quail Island from the mainland. The 'lava posterior to the general covering' refers to the small streams that emanate from still existing vents such as Red Hill and Flag Staff Hill. Only these, and the growth of modern vegetation (including the baobab tree), post-dated this great event. There is no apparent source for such a volume of water on such a small and dry island as St Jago, so Darwin had to 'go back to the torrents of rain that usually are said to accompany Volcanic action'.[16]

Further evidence of Darwin's 'catastrophism' comes from his observations made in the interior of the island on 2 February, when he went on horseback to St Domingo (Fig. 1). Here he described in his diary the view (Fig. 5) as seen entering the upland valley:

> On approaching St Domingo a turn in the road first showed us the background of wild peaked rocks.—their forms are most fantastic; one part looks like a castle wall, others like towers & pyramids.—Every thing betrays the marks of extreme violence: & which is better shown by the rocks being in horizontal beds (Keynes 1988).

His geological notes go further, and link the supposed cataclysm that created the interior landscape with the coastal valley network:

> St. Domingo . . . is situated in an amphitheatre of black and nearly precipitous rocks about 500 feet high.—The surrounding country bespeaks the utmost violence; pieces of rock torn apart from their beds stand like castle walls erect.—others as wild and jagged in their outlines as Mica slate give to the horizon a grand and picturesque appearance.—The effect is rendered more striking by the beautiful contrast of the valley; where oranges, banana and Cocoanuts are flourishing together in their luxuriance.. . .—I conceive it to be clear, from the pieces left standing and from the corresponding appearance on each side of the valley, that the country was originally covered with a uniform bed of this rock.—and that after being shattered by some great force: these valleys were formed by the agency of large bodies of water: To this latter force the valleys nearer the coast give abundant evidence.[17]

Fig. 5. The valley of St Domingo, which Darwin visited on 2 February 1832. Darwin interpreted the jagged scenery as having been formed in a single cataclysmic event, in which the rocks were torn apart from one another and the intervening valleys were excavated.

This explanation of the formation of the landscape is similar to that given to him the previous summer by Adam Sedgwick, who argued that the valleys of North Wales (now recognized as glacial) had been formed when the rocks on either side had been torn apart from one another, leaving diluvium in their bases (Secord 1991). Unlike many of his other observations on St Jago, it was later edited out of his published works.

Uplift and subsidence

One of the great themes of Lyellian geology was that of uplift and subsidence of the land. Lyell was keen to show that historically recorded earthquakes had resulted in measurable changes to the land's surface, and by implication that given enough time, whole continents could be raised or submerged without the agency of truly cataclysmic events, just earthquakes of the same magnitude that occur today.

The frontispiece of the *Principles* is the so-called 'Temple of Serapis' at Pozzuoli in Italy, whose appearance was something of a *cause célèbre* at the time. According to Lyell, the temple provided firm physical evidence of absolute changes of the level of the land in historical times. Originally it must have been built on dry land. The columns, however, had been drilled by marine rock-boring bivalves to about half their height, proving that they must have been under salt water for a considerable period. Since then, the temple had again been raised so that everything was now above the high water mark. As no such evidence of changes in the level of the Mediterranean was evident elsewhere, Lyell conjectured that changes in the elevation of the land must have happened, probably at the times of known historical earthquakes in medieval times. The earth movements were gentle enough, however, to leave some of the columns standing upright.

This was in stark contrast to earlier writers, such as Johann Wolfgang von Goethe (1749–1832), who had attempted to explain appearances by postulating a temporary dam of volcanic debris around the temple that had been filled by seawater in a flood which only gradually seeped away. This theory had recently been endorsed by Charles Daubeny in his book on volcanoes. Daubeny argued that changes in the height of the land were precluded by the fact that the columns had not toppled over during the violent earthquakes. Writing, it should be remembered, before Lyell, he stated that 'Had such been the case, it is probable that not a single pillar of the temple would now retain its erect posture to attest the reality of these convulsions.' (Daubeny 1826, p. 162).

Darwin considered analogous questions of uplift and subsidence in relation to the white limestone ('former beach') of Praya, which he followed in both directions along the coast. One of his most successful days of fieldwork was 23 January, when he resolved to explore the coast to the east of Praya in the direction of a hill called Flag Staff Hill (which is confusingly referred to in the published works as Signal Post Hill). Here the cliffs are particularly impressive as they are exposed to the full Atlantic swell. For several miles along the coast the limestone can be seen as a prominent band about half way up the inaccessible cliffs. Darwin found that 'the line here as observed by a Theodolite is truly horizontal'.[18]

As the limestone contains marine fossils, Darwin reasoned that either the island has been uplifted, by some great force, or the sea level had dropped. Since the latter hypothesis required a drop in the level of the entire Atlantic, Darwin (with no notion of eustasy) thought it 'clearly impossible', and in any case, the different height of the band elsewhere, where it was not horizontal, 'proves it is not subsidence of water'.[19]

> When viewing from a distance an extent of cliffs one is struck by the great force it must have required to have raised fields 2 or 3 miles broad of these rocks at least 50 feet.—(which is supposing the former coast was at the surface of the sea at high water mark!)—A considerable thickness of the lower crystalline rocks must likewise have been elevated at the same time.—Taking this into consideration it is perfectly astonishing that the force should have acted so uniformly that a spirit level with sights proved the former beach to be as truly level as the present.[20]

A note added on the back of this sheet, refers to the case of the Temple of Serapis;

> Dr. Daubeny when mentioning the present state of the temple of Serapis doubts the probability of a surface of country being raised without cracking buildings on it.—I feel sure at St Jago in some places a town might have been raised without injuring a house.[21]

It is clear from various writings that Darwin thought of the uplift as having occurred in a single event, not gradually. Although the area east of Praya had been raised flat, other parts were folded down to lower levels, and there were also fault lines or dislocations that were now marked by narrow ravines on the table-land itself. There is no evidence, as has recently been remarked (Keynes 2002), that he was suggesting 'gradualist, rather than catastrophic, subsidence and elevation of the land'. Nevertheless, in the above passage Darwin did adopt a similar position to Charles Lyell on the Temple of Serapis in that he invoked changes in the level of the land, not the sea, to account for appearances. It is therefore striking that there is no mention of Lyell, who wrote an extended account of the temple in the first volume of *Principles*. Critically, it is also unclear when the note quoted above was written. We think it plausible that Darwin had yet to read Lyell's discussion of the Temple of Serapis when he made this note, but that when he did eventually do so, he was pleased to have independently come to similar conclusions.

An analogous case can be made for another of Darwin's undated additions to his geological diary of St Jago, in which the apparently young age of the white limestone caused Darwin to confront the issue of directionality in Earth history. In Daubeny's *Active and Extinct Volcanoes*, the siliceous trachytes were said to be older than sheets of basalt, which in turn were older than basalt that had flowed in streams. Darwin, however, remarked that:

> If as I suppose the shells in the former beach are the same as now exist.—the superimposed lava comes under the class of formations of the present day.—Dr Daubeny states that those lavas, which like this, are in their composition Basaltic & form fields rather then circumscribed streams originated at the time when Tertiary formations were depositing.—does not this instance war against the rule.[22]

Again, this observation has rightly been interpreted as supporting a non-progressionist position, but there is no mention of Lyell, and the note is an undated addition. Also, at other times on St Jago, Darwin traced streams of basalt to the small craters of Red Hill, and Flag Staff Hill, which he would have seen as supporting Daubeny's classification.

The 'memorable hour'

Adjacent to the cliff line some miles north east of Praya is the aforementioned low hill known as Flag Staff Hill. Darwin climbed this on 23 January and recognized it as an eroded volcanic vent, similar to another (Red Hill, west of Praya) that he had seen a few days previously. He traced some streams of lava from it into a nearby valley, proving that the volcano was of recent origin and the lava stream was younger than the excavation of the valley.

In his diary, Darwin then tells us that he descended the cliff with some difficulty by a curious ravine at the northern end of the hill. This, we have confirmed, is still the only accessible route down to the base of the cliffs for miles in each direction. At the bottom of the cliff he traversed north and south and examined the rock exposures. He was surprised to find that the white limestone, which maintains a constant level both north and south, locally dips locally below sea level. An idea of his interpretation of appearances is given by a woodcut in his 1844 book on volcanic islands (Darwin 1844, p. 9) (Fig. 6).

It seemed beyond coincidence to Darwin that this place of deviation in the level of the limestone was exactly below the volcanic vent of Flag Staff Hill, and he thought he could use the instance to relate the phenomena of volcanism and uplift. In fact, at different times he proposed three different explanations:

1. The rocks had not been upraised beneath the hill, because the volcano had acted to relieve the pressure of uplift.[23]
2. The vent existed before the uplift and its additional weight prevented its upheaval.[24]
3. The hill had been uplifted with the rest of the landscape, but there had been subsidence around the vent after it had poured forth lava (Darwin 1844, p. 9, Barlow 1958).

In our opinion, having visited at the same localities and also viewed the area from offshore by boat, the appearances are not as simple as indicated in Darwin's diagram of 1844. But that is beside the point, as is the exact interpretation that Darwin put on his observations. What is significant from a biographical point of view is that he thought he had made a significant and original observation, however it was to be interpreted, relating uplift and subsidence to the existence of subterranean volcanic activity. Darwin had been exposed to Huttonian views of the Earth as a student at Edinburgh (Secord 1991), in which the phenomena of internal heat, volcanism and surface elevation were linked, and this example would surely have delighted Hutton. Equally, it was a discovery that could

Fig. 6. The woodcut from Darwin (1844, p. 9) in which he showed how the white limestone dips locally below sea level in an asymmetrical fashion under Flag Staff Hill. Note that the lower band (with horizontal ruling) represents the sea. A, B and C (somewhat confusingly) are the three layers of rock in the cliff section, as shown in Fig. 6. A, ancient volcanic rocks, shown with oblique stripes; B, calcareous stratum; C, upper basaltic lava, shown with dashed lines.

Fig. 7. The base of the cliff under Flag Staff hill, at the point in which the white limestone dips under sea level. This is probably the location remembered by Darwin as the place where he developed new ideas on volcanism and uplift and decided to plan a book on the geology of the *Beagle* voyage. The band of white limestone ('former beach', as in Fig. 3) can be seen half way up the cliff in the background.

easily be incorporated into a Lyellian view of the processes when Darwin came to read Lyell in detail.

At the northern end of the coast section beneath Flag Staff Hill the limestone band descends close to sea level and can be approached in some protected undercuts or caves at the bottom of the cliff (Fig. 7). We know from the notes that Darwin went to this place, and we contend that this is the most likely place that Darwin referred to in his autobiography (quoted above) as where he decided to write a book on the geology of the voyage. All previous writers of which we are aware have suggested Quail Island on the 17 January, and indeed Darwin does state in his diary for that day that it was a memorable occasion to him, that he fished in rock pools there and observed corals growing on their native rock (Keynes 1988). But at Quail Island there is no volcano, and Darwin could not have made the connection between the presence of a specific volcanic crater and the local subsidence of the land which he later remembered as being the key idea that made him think he could write a book. We note also that the Flag Staff Hill locality is particularly grand place. It has much larger, more impressive rock pools than on Quail Island, with the requisite corals, and tremendous views over the open ocean. This, combined with a sense of isolation would have appealed aesthetically to the young Darwin, who was much taken with the beauty of scenery throughout the *Beagle* years. Furthermore, when Darwin departed St Jago for the last time in 1836, homeward bound for England, he wrote in his diary, 'I confess, I feel some good will to the island; I should be ungrateful if it was otherwise; for I shall never forget the delight of first standing in a certain lava cavern & looking at the swell of the Atlantic lashing the rugged shores.' (Keynes 1988). If this certain lava cavern is the same place in which he decided to write a geology book, as has been implied (Herbert 1991), it describes well the Flag Staff Hill locality on the exposed windward coast of St Jago, but not at all Quail Island, which is in the protected harbour of Praya.

Conclusion: Darwin and Lyell

To modern geologists, the island of St Jago is a deeply eroded remnant of a large volcano, one of several that owe their origin to a mantle plume. The geological map of the island indicates that the white Praya limestone is from the Pliocene epoch (deposited between about 2 and 5 million years ago). The valleys are wadis, and owe their form to occasional flash floods from the interior. Darwin was clearly right to infer that they had been cut by large torrents of water. However, many such floods, acting over vastly extended periods of time, would be necessary to create the landscape we see today. The small cones of Red Hill and Flag Staff Hill are geologically quite recent, and indicate a late-stage reactivation of the volcanism similar to that seen on other islands in the group. No volcanic activity or large earthquakes have been recorded in historic times on St Jago, but the neighbouring volcano of Fogo (which can be seen on a clear day) is active.

With the exception of Sandra Herbert (1991), who acknowledged elements of Darwin's catastrophism, most modern commentators who have studied the geology diary have tended to accept the account of Darwin's sudden conversion to Lyellian geology at the time of his first visit to St Jago. Thus Secord (1991) wrote that Darwin had 'faced an interpretative crux at the very start of the voyage and solved it triumphantly through Lyell's style of interpretation'. For Sloan (2003), 'we find Darwin seeking naturalistic explanations for the layers of geological formations, and appealing to a gradualist, rather than catastrophic, subsidence and elevation of the land'. And according to Keynes (2002) he 'immediately fell in wholeheartedly with Lyell's gradualist and not yet generally accepted approach that geological changes resulted from slow processes operating over long periods of time'. Many more such statements could be culled from the secondary literature, and this overly simple story is in danger of becoming a historical myth.

We find no evidence of Darwin even having read Lyell in detail by the end of his visit in 1832. In describing the geology of St Jago, he referred to Charles Daubeny's *Active and Extinct Volcanoes* in four places, but not once to Lyell. But whether or not he had read the *Principles*, Darwin can hardly be described as applying gradualist thinking in his overview of the geology. We have here reproduced for the first time parts of his geology diary relating to the interior scenery and valleys that can only be interpreted as being 'catastrophic' in nature. Darwin was, however, actively questioning orthodox views, such as those of Charles Daubeny, and habitually seeking evidence to test his theories. Several of his observations from St Jago would soon make more sense to him in terms of Lyellian thinking rather than the progressionist/catastrophist school of his teachers (Secord 1991).

If Darwin cannot be considered as an instant convert to Lyellian gradualist geology at the time of his first visit to St Jago, when, where and how did the conversion happen? From the note in his autobiography regarding Henslow's suggestion to read the *Principles*, it seems probable that he had not seen the volume until at least the few months preceding the voyage. Darwin is known to have consulted with Henslow about the voyage on the 3–4 September 1831 (Keynes 2002), and it is likely that this was the time he was recommended to read the *Principles* as part of the necessary preparation.

Darwin's own copy of the first (1830) volume is inscribed as a gift from Captain Fitzroy, which implies that it was presented on or shortly before departure, and certainly not before 5 September 1831 when the two men first met (Keynes 2002). The months preceding the voyage were a hectic time for Darwin. There is no evidence that he read the volume at this time, although he could have done so while waiting for fair winds in Plymouth (Herbert 2007). For much of the journey to St Jago, Darwin was severely afflicted by sea-sickness.

The first clear evidence of Darwin having read and approved of Lyell is a letter he wrote to Henslow on 18 May 1832 from Rio de Janeiro, in which he recalls the St Jago visit:

> here we spent three most delightful weeks.— the geology was pre-eminently interesting & I believe quite new: there are some facts on large scale of upraised coast (which is an excellent epoch for all the volcanic rocks to [be] called from) that would interest Mr Lyell.' (Burkhardt & Smith 1985)

Unknown to Darwin, Henslow abstracted various passages from Darwin's letters, including this one, and read them to the Cambridge Philosophical Society, and published them as a pamphlet. (Henslow 1838). Then, more than a year later (27 November 1833), Darwin's sister Catherine wrote to him saying that:

> I hear your Theory of the earth is supposed to be the same as what is contained in Lyell's 3d Vol. (Burkhardt & Smith 1985)

Whether this refers to the reception in England of the above passage, or to some other of Darwin's letters that have been lost, is difficult to tell.

A 'conversion' to Lyellian thinking did occur during the voyage, but evidently it was more gradual than Darwin later implied. On the east coast of South America, at Bahia and then as the *Beagle* made its way further south, Darwin observed large areas of upraised granitic rocks extending over huge areas. He wrote in his geology diary:

> If this remarkable continuity was more satisfactorily proved it would be a fine instance for those who attribute the present state of the world more to great causes at distinct epochs than to a succession of smaller ones.[25]

This quotation is interesting not only in that Darwin was seemingly favouring sudden widespread uplift, but also in that it proves he was aware of the catastrophist/gradualist controversy as an active area of debate, and seeking evidence that bears on it.

By March 1834, however, after observing the evidence for uplift in Chile, he wrote to his cousin William Darwin Fox (1805–1880) saying 'I have become a zealous disciple of Mr Lyell, and I am inclined to take his views even further than the author himself' (Burkhardt & Smith 1985). This is evident in much of his subsequent work, especially his theory of coral reefs, which depended on gradual subsidence matched by the slow growth of coral on their volcanic foundations. But it is also true, as pointed out by Armstrong (1992), that even in the same month as his letter to Fox he was still contemplating great catastrophic earthquakes of a magnitude far exceeding historical ones to explain geomorphic features on the Falkland Islands.

At some point, Darwin went back to his notes and deleted passages that refer to the diluvium, and also the recent date of the white limestone of Praya. Secord (1991) has suggested that this happened quickly, but as Herbert (1991) has argued it seems more likely from the context that it was done much later, possibly in 1836 when Darwin briefly returned to the island. When he came to write up his *Journal of Researches* and his book on the *Geology of the Volcanic Islands Visited During the Voyage of HMS* Beagle, he reworked the greater parts of his notes in a Lyellian fashion, and simply ignored passages such as that suggesting the interior of the island bore the marks of extreme violence and that a single torrent of water had excavated the valleys. Instead, in 1844 he suggested that the valleys had been 'scooped out by the waves of the sea', but still recorded his puzzlement at not having found marine shells in their bottoms (Darwin 1844).

In conclusion, Darwin's autobiographical statements about his conversion to Lyellian geology at St Jago are simplistic and potentially misleading, but not literally incorrect. As the autobiography was not intended for publication, and was a brief account written long after the events described, it is perhaps not surprising to find discrepancies with his contemporary notes and papers. An analogous situation has occurred with respect to his account of the discovery of the principle of natural selection (Ospovat 1981). The boldness of Darwin's statements, however, may have led subsequent researchers to overvalue those aspects of his contemporary geology diary that might be seen as supporting an immediate Lyellian conversion, while not giving sufficient weight to the more obvious catastrophism in that document.

We are grateful to the Philip Leverhulme Prize Fund to PNP which supported our fieldwork and other researches. We thank the staff of Cambridge University Library and the Down House Museum, Downe, for assistance in locating material, and G. P. Darwin for permission to reproduce extracts from the geology diary.

Notes

[1]See also Pearson & Nicholas (1992).

[2]Darwin, C., ms. Field Notebook 1–4, Down House, Downe, Kent. Also available on microfilm. Also see notes in Keynes (1988).

[3]Darwin, C., ms. Geological specimen notebook, Department of Earth Sciences, Cambridge University.

[4]Darwin, C., ms. 'Diary of Observations on the Places Visited on the Voyage'. Cambridge University Library, DAR 32.1. Also available on microfilm.

[5]Darwin, ms. 'Diary of Observations ...', p. 16 verso. Note that angled brackets (<>) indicate a later deletion.

[6]Darwin, ms. 'Diary of Observations ...', p. 20.

[7]Darwin, ms. 'Diary of Observations ...', p. 32.

[8]Darwin, ms. 'Diary of Observations ...', p. 20

[9]Darwin, ms. Field Notebook 1–4.

[10]Darwin, ms. 'Diary of Observations ...', p. 34.

[11]Darwin, ms. 'Diary of Observations ...', p. 23.

[12]Darwin, ms. 'Diary of Observations ...', p. 33.

[13]Darwin, ms. 'Diary of Observations ...', p. 19 verso.

[14]Darwin, ms. 'Diary of Observations ...', p. 19.

[15]Darwin, ms. 'Diary of Observations ...', p. 34 verso.

[16]Darwin, ms. 'Diary of Observations ...', p. 35.

[17]Darwin, ms. 'Diary of Observations ...', pp. 31–32.

[18]Darwin, ms. 'Diary of Observations ...', p. 22.

[19]Darwin, ms. 'Diary of Observations ...', p. 24 verso.

[20]Darwin, ms. 'Diary of Observations ...', p. 24.

[21]Darwin, ms. 'Diary of Observations ...', p. 24 verso.

[22]Darwin, ms. 'Diary of Observations ...', p. 35 verso; see Pearson (1996) for a discussion of the issue of igneous rock composition and uniformitarianism in Darwin's later work.

[23]Darwin, ms. 'Diary of Observations ...', p. 30; Darwin (1844, p. 9).

[24]Darwin, ms. 'Diary of Observations ...', p. 27 verso.

[25]Darwin, ms. 'Diary of Observations ...', p. 32; quoted by Herbert (1968).

References

ARMSTRONG, P. 1992. *Darwin's Desolate Islands: a Naturalist in the Falklands, 1833 and 1834.* Picton, Chippenham.

BARLOW, N. (ed.) 1958. *The Autobiography of Charles Darwin 1809–1882, With Original Omissions Restored.* Collins, London.

BUCKLAND, W. 1836. *Geology and Mineralogy Considered with Reference to Natural Theology.* William Pickering, London.

BURKHARDT, F. & SMITH, S. (eds) 1985. *The Correspondence of Charles Darwin, Volume 1: 1821–1836.* Cambridge University Press, Cambridge.

DARWIN, C. 1839. *Journal and Remarks, 1832–1836. In*: FITZROY, R. (ed.) *Narrative of the Surveying Voyages of His Majesty's Ships* Adventure *and* Beagle;

Between the Years 1826 and 1836. Volume 3 Henry Colburn, London.

DARWIN, C. 1844. *Geological Observations on the Volcanic Islands Visited During the Voyage of HMS* Beagle, *Together with some Brief Notes on the Geology of Australia and the Cape of Good Hope*. Smith Elder, London.

DAUBNEY, C. 1826. *A Description of Active and Extinct Volcanoes; with Remarks on their Origins, their Chemical Phenomena, and the Character of their Products, as Determined by the Conditions of the Earth During the Periods of their Formation*. W. Phillips, London.

HALLAM, A. 1989. *Great Geological Controversies*. 2nd ed. Oxford University Press, Oxford.

HENSLOW, J. S. 1838. Extracts from letters addressed to Professor Henslow by C. Darwin Esq. Privately printed for Cambridge Philosophical Society. Collected Papers, **1**, 3–16.

HERBERT, S. 1968. The Logic of Darwin's Discovery. Unpublished Ph.D. dissertation, Brandeis University.

HERBERT, S. 1991. Charles Darwin as a prospective geological author. *British Journal for the History of Science*, **24**, 159–162.

HERBERT, S. 2007. Doing and knowing: Charles Darwin and other travellers. *In*: WYSE JACKSON, P. N. (ed.) *Four Centuries of Geological Travel: The Search for Knowledge on Foot, Bicycle, Sledge and Camel*. Geological Society, London, Special Publications, **287**, 311–323.

HERBERT, S. & ROBERTS, M. B. 2002. Charles Darwin's notes on his 1831 geological map of Shrewsbury. *Archives of Natural History*, **29**, 27–29.

KEYNES, R. D. (ed.) 1988. *Charles Darwin's* Beagle *Diary*. Cambridge University Press, Cambridge.

KEYNES, R. 2002. *Fossils, Finches and Fuegians: Charles Darwin's Adventures and Discoveries on the* Beagle, *1832–1836*. Harper Collins, London.

LUCAS, P. 2002*a*. 'A most glorious country': Charles Darwin and North Wales, especially his 1831 geological tour. *Archives of Natural History*, **29**, 1–26.

LUCAS, P. 2002*b*. Jigsaw with pieces missing: Charles Darwin with John Price at Bodnant: the walking tour of 1826 and the expedition of 1827. *Archives of Natural History*, **29**, 359–370.

LYELL, C. 1827. Art. IV. *Memoir on the geology of central France; including the volcanic formations of Auvergne, the Velay and the Vivarrais, with a volume of maps and plates*. By G. P. SCROPE, F.R.S., F.G.S., London, 1827. *Quarterly Review*, **36**, 437–483.

LYELL, C. 1830. *Principles of Geology, Being and Attempt to Explain the Former changes of the Earth's Surface by Causes now in Operation*. Volume 1. John Murray, London.

OLDROYD, D. 1999. Non-written sources in the study of the history of geology: pros and cons, in the light of the views of Collingwood and Foucault. *Annals of Science*, **56**, 395–415.

OSPOVAT, D. 1981. *The Development of Darwin's Theory*. Cambridge University Press, Cambridge.

PEARSON, P. N. 1996. Charles Darwin on the origin and diversity of igneous rocks. *Earth Sciences History*, **15**, 49–67.

PEARSON, P. N. & NICHOLAS, C. J. 1992. Defining the base of the Cambrian: the Hicks-Geikie confrontation of April 1883. *Earth Sciences History*, **11**, 70–80.

ROBERTS, M. B. 2000. I coloured a map: Darwin's attempts at geological mapping in 1831. *Archives of Natural History*, **27**, 69–80.

SECORD, J. A. 1991. The discovery of a vocation: Darwin's early geology. *British Journal for the History of Science*, **24**, 133–157.

SEDGWICK, A. 1830. Annual General Meeting of the Geological Society, Presidential address. *Philosophical Magazine*, new series **2**, 310.

SLOAN, P. R. 2003. The making of a philosophical naturalist. *In*: HODGE, J. & RADICK, G. (eds) *The Cambridge Companion to Darwin*. Cambridge University Press, Cambridge, 17–39.

Naturalists from Neuchâtel: America and the dispersal of Agassiz's scientific factory

R. H. SILLIMAN

Department of History, Emory University, Atlanta, GA 30322, USA (e-mail: RHSilliman@aol.com)

Abstract: When Louis Agassiz went to America in 1846, he took with him, or was soon joined by, a whole retinue of Swiss protègès and assistants who in the preceding decade had turned his scientific work into a corporate, collective enterprise. Among the new arrivals were E. Desor, A. Guyot, L. Lesquereux, C. Girard, and L. Pourtalès—a labour pool that enabled Agassiz to re-assemble the Neuchâtel 'scientific factory' on American shores. Indeed, some of the troop immediately went to work under Agassiz's supervision, plunging eagerly into the rich new fields of investigation that America afforded them. Although it was seemingly productive, by 1850 the group had dispersed.

Some of the Neuchâtel naturalists, urged on by the egalitarian politics of the day, lost patience with Agassiz's domineering ways and determined to strike out on their own, fashioning independent careers. Others, still loyal to Agassiz, left to establish themselves in regular posts. All of these young naturalists struggled to adapt to American culture and make a go of it. Although Agassiz himself was hospitably received and well provided for, his followers found that they were looked down upon because of their foreign manners, broken English, and idealized scientific pursuits that appeared lacking in usefulness to their practically-minded hosts. Nor did they have abundant prospects for earning a living, obliged as they were to compete with native-born Americans for the few full-time scientific positions then available.

Nevertheless, in the end they were remarkably successful. Lesquereux in palaeobotany and Guyot in physical geography reached the top in their fields in America. These two, along with Pourtalès, were honoured by election to the National Academy of Sciences. As for the others, Desor was launched on a productive career as a Quaternary geologist, when a family matter recalled him to Switzerland, and Girard was building an excellent reputation in ichthyology and herpetology, when he was obliged to return to Europe because of his support for the Confederacy in the Civil War. Working in geology and natural history, where broad experience and comparative observation paid dividends, the Swiss turned their European background into an advantage.

Alluding to immigrant scientists, the American geologist J. Peter Lesley (1819–1903) effused: 'Men who are forced to fly from their ancestral homes to begin a new career elsewhere, acquire rapidly by the struggle for life a noble development of all their powers; gaze upon the new world around them with new eyes; inform themselves of what would never have interested them; ally themselves with the strongest and wisest whom they find; invent enterprises, place scaling ladders against the ramparts of fame, and in the end come to be of the number of the world's rulers.' (Lesley 1883, p. 519). Surely Lesley did not think that the usual refugee arriving on American shores would respond so strenuously to exile and make of it a springboard to fame and fortune. Certainly he must have known that in 'the struggle for life', the losers are at least as numerous as the winners. But Lesley was not thinking of typical immigrants. He was thinking of men of rare character and ability like Édouard Desor (1811–1882), the subject of his biographical memoir, and perhaps others among the Neuchâtel naturalists, such as Léo Lesquereux (1806–1889), whom he knew even more intimately than Desor. These naturalists were the assistants and protègès of Louis Agassiz (1807–1873) who had followed him to America.[1] None of them enjoyed Agassiz's prominence or commanded his popular esteem, but in scientific circles, at least, they all came to be quite well known. They were a talented, energetic, enterprising band with high aspirations. Strip Lesley's statement of its romantic hyperbole and it fairly represents these young men from Switzerland. The following biographic sketches of the Neuchâtel naturalists offer a tentative description of what immigration to America meant to them and to their careers.

About a decade before he left Neuchâtel for the New World, Agassiz (Fig. 1) first surrounded himself with an entourage of scientific workers. Earlier, when his research had focused more or less exclusively on fossil and living species of fish, he had employed artists to make drawings

From: WYSE JACKSON, P. N. (ed.) *Four Centuries of Geological Travel: The Search for Knowledge on Foot, Bicycle, Sledge and Camel.* Geological Society, London, Special Publications, **287**, 255–269.
DOI: 10.1144/SP287.20 0305-8719/07/$15.00

Fig. 1. Louis Agassiz (1807–1873), as he appeared about 1844, two years before emigrating to the United States.

of his specimens but otherwise he worked alone, even if a wide correspondence and frequent travel put him in touch with Europe's leading naturalists. By 1837, however, his research projects had multiplied—in addition to fish he was studying radiates and molluscs, as well as alpine glaciers—and he was in desperate need of assistance. Accordingly, he hired a secretary, Édouard Desor, and employed as research assistants Carl Vogt (1817–1895) and Amanz Gresley (1814–1865), as well as Desor, who doubled in this capacity. A youngster, Charles Girard, was apprenticed to the assistants and served them as bootblack and errand boy. Others came into this growing organization, students like Louis François de Pourtalès and promising young naturalists like Arnold Guyot (1807–1884).

But that was not all. Agassiz acquired his own print and lithography shops to publish the books his scientific team produced. To keep the press active, he put out special works like his *Nomenclator zoologicus* (1842) and *Bibliographia zoologiae et geologiae* (1846) and produced translations of English scientific treatises such as William Buckland's *Geology and Mineralogy considered with Reference to Natural Theology* (German edition, 1838). Another facet of these operations was the production of casts of shells and other natural history specimens that could be sent to naturalists across Europe. Reflecting back on all this, the erstwhile assistant Carl Vogt wrote, 'It was, if I may so express myself, a scientific factory with community property.' (Vogt 1883, p. 16). The workers received no regular wages, contenting themselves with whatever compensation Agassiz could scrape up when there were financial needs. In this idealistic operation monetary reward was understood to take second place to the satisfaction of a disinterested pursuit of truth. Nor could the assistants deny that ultimately their work had a *quid pro quo* in the solid training in natural history they received from Agassiz. Still, the enterprise was very costly, if not in labour costs, in other ways, particularly in the publishing department. Unhappily the situation worsened over time, because Agassiz could not forego inaugurating new projects, even when old ones were not yet completed, and debt mounted. Agassiz finally had to close the print and lithography shops. Discouraged—realizing that Neuchâtel could not provide the support his numerous and far-reaching projects required—he looked elsewhere (Lurie 1960, pp. 106–121). For a long time he had dreamed of being a naturalist-explorer following the model of Alexander von Humboldt (1769–1859). Where better to realize this dream than by going to the New World, where von Humboldt's reputation had been established? As usual his aspirations were high. He would travel to America and fully master its natural history. It never occurred to him that on the other side of the ocean he could operate without assistants. The scientific factory would have to be re-assembled in America.

Before Agassiz departed there was an alarming development. From the beginning tensions existed within the 'scientific factory'. Certainly, the assistants found much that was admirable and captivating in Agassiz, and from the experiences they shared with one another, especially those they had on Alpine peaks testing Agassiz's glacier theory, they developed a kind of fraternal bond (Fig. 2). 'It was an extraordinarily stimulating life up there', Vogt wrote. 'Friends had to be acquainted with all the facts, opponents convinced, their objections refuted, their doubts set aside—how many nights did we spend there until near morning by the light of a candle in a glass, wrapped in our cloaks, with steaming grog and glowing cigars and the vale, sunk in icy silence, filled with our discussions. Swiss, Germans, French, Italians, Americans streamed by, and the closest friendships were made during a work-filled life together, which had to bring hearts closer to one another'. (Vogt 1883, p. 23). But the assistant whose effusion this was had moments of dejection, too, particularly when he fell to brooding over what he perceived to be Agassiz's theft of credit for scientific work done by himself and the other assistants. On the eve of the departure for America, Vogt decided it was

Fig. 2. 'Hotel des Neuchatelois', the first summer field station of Agassiz and his companions on the Lower Aar Glacier, 1841–1842.

time to come out from the master's shadow and stand on his own feet (Fig. 3). As he later remembered it, Agassiz in tears pleaded with him to make the trip to America (Vogt 1883, p. 24). Unmoved, Vogt quit the scientific factory and headed for Paris. Subsequently at Giessen and Geneva he distinguished himself as a zoologist and an anthropologist. Philosophically, as is described below, he put considerable distance between Agassiz and himself. What of the other assistants? In the uncertain conditions that life in America would present, could Agassiz count on their continuing loyalty and steadfastness?

Fig. 3. Carl Vogt (1817–1895).

Agassiz arrived in Boston in the autumn of 1846 (Lurie 1960, pp. 122–130). As he went about acquainting himself with the United States his experiences were all positive and gratifying. The Americans were utterly charmed by him and extended him the warmest hospitality. He took a month touring New Haven, Philadelphia, Washington, D.C. and other centres of scientific activity getting to know the country's leading naturalists and geologists. During the winter, he delivered the Lowell Institute Lectures in Boston which were a huge hit and led to numerous invitations to give paid lectures in other cities. In the autumn of 1847 he was offered a professorship in Harvard's new Lawrence Scientific School. In America, it seemed, money came easily, and he could look forward to escaping the penury that had impeded his efforts in Neuchâtel. With bright prospects for the future, he decided to become a permanent resident of the United States. It was not for himself alone that he took this step to better his circumstances. He associated a true devotion to science with monetary sacrifice. It was important to his scientific plans to have the resources to employ at least a small staff of assistants and artists.

Encouraged to cross the Atlantic by political and economic troubles in Europe, Swiss émigrés began to arrive at the three-story brick house that Agassiz rented in East Boston (Lurie 1960, pp. 130–147). Desor and Girard arrived in April of 1847, joining Pourtalès who had travelled with Agassiz. The gathering staff also included Jacques Burkhardt (1808–1867), an artist, and Auguste Sonrel, a lithographer. In the autumn of 1848 Arnold Guyot and Léo Lesquereux appeared. By then, retaining the East Boston house for another half year, Agassiz had moved his residence to Cambridge, and, as an additional place to store his mounting collections, added a bathhouse along the Charles River. Some twenty-two foreigners, virtually all with a Neuchâtel connection, now made up the Agassiz household. Their quarters had all the features of a zoological laboratory and natural history museum. The scientific factory was back in operation. Early investigations carried out by the Neuchâtel band included a twelve-hundred mile excursion through the eastern United States that Desor and Pourtalès undertook to gather evidence for the ancient ice sheet hypothesized by the glacial theory. For the first time in his career, Agassiz had ready access to fresh marine organisms, and marine zoology went to the top of his scientific agenda. For specimens, Agassiz kept an eye on the Boston fish market, and in the summer of 1847 did his own collecting at sea as a guest aboard the USS *Bibb*, a Coast Survey steamer. In 1848, obtaining help from his assistants, he co-authored with Augustus A. Gould (1805–1866) the first volume of a textbook *Principles of Zoology* and led an excursion to Lake Superior that provided the basis for another book.

Édouard Desor (1811–1882)

For a couple of years the scientific factory was thriving and productive, but by 1850 it had largely been shut down and its working corps dispersed. A crippling blow had struck the enterprise as early as April 1848, when relations between Agassiz and his secretary Desor suddenly went sour (Fig. 4). For casual acquaintances of the two men the split must have seemed astounding, since for over a decade Desor had been Agassiz's friend and collaborator and his right-hand man in supervising the scientific factory. As a Neuchâtel friend later recalled, 'Desor speaks of Agassiz as much or more than of himself. Agassiz said this; Agassiz did that. He associated with his projects; he rejoiced at his successes and recounted them, celebrated them and sang them with enthusiasm. Their life was shared; everything was half and half, ... the honeymoon, one might say, of a marriage of choice'.

Fig. 4. Édouard Desor (1811–1882).

(Favre & Berthoud 1882, p. 62). What alienated these two men from one another? Animosity over sharing credit for collaborative work was one factor in the quarrel, but it does not explain the vehemence of the charges the two hurled at one another, the blackening of one another's character that they engaged in. Three successive arbitration panels consisting of their peers in the Boston scientific community met to adjudicate their differences, and each time the panelists found in Agassiz's favour. But the fact that righteous men of the standing of the Rev. Theodore Parker (1810–1860), the noted radical Unitarian and abolitionist, took Desor's part shows that all the wrong-doing was not on one side. The upshot of the matter was that Desor was fired from his secretarial post, and the two men, nursed bad feelings toward one another for the rest of their lives and were never reconciled.

Although Agassiz's biographer Edward Lurie faults entirely what he sees as Desor's maliciousness, there are reasons to think that beyond personal animosities philosophical and ideological issues divided the two men (Lurie 1960, pp. 152–161). Desor and Carl Vogt had much in common. As college students they had both been caught up in the liberal-nationalistic politics of the Burschenschaft movement and had taken flight to avoid prosecution by the authorities. In 1837, still a fugitive and unemployed, Desor settled into a small spare room in the Vogt family home in

Berne, when Agassiz showed up wanting to hire a secretary. Desor was recommended for the post by Vogt's father and was soon at work in Neuchâtel. Carrying on Agassiz's correspondence, drafting large portions of Agassiz's books, he had a special thing about clarity, so much so, his friends joked, that he even made clear things that he himself did not understand. When in 1839 Carl Vogt completed his studies, he, too, joined the Agassiz enterprise, and for five years worked beside Desor. They were a bright, sophisticated, vocal pair with a shared sceptical, ironic, realistic outlook and a strong antipathy to authority. Daily contact between the two assistants mutually reinforced these attitudes and in the end undermined Agassiz's control over them. In the career that he fashioned after severing relations with Agassiz, Vogt became a well-known biologist and anthropologist, a champion of Darwinism, and one of the leading scientific materialists of the day bracketed with the likes of Jacob Moleschott (1822–1893) and Ludwig Büchner (1824–1899) (Gregory 1977, pp. 51–79 and *passim*). Desor remained with Agassiz somewhat longer and was never quite as radical as Vogt, but in America his friendship with Theodore Parker and Eliza Lee Cabot Follen (1787–1860), another abolitionist, who happened to be Carl Vogt's aunt, clearly placed him in the progressive, reformist camp on the side opposite Agassiz (see Desor 1861). Thus, the desertion of both Vogt and Desor split the Agassiz circle apart along ideological lines. In light of the subsequent debates between Agassiz and the Darwinians this division of the 1840s was a noteworthy harbinger of things to come.

Sacked as Agassiz's secretary in 1848, Desor became unemployed and was at first quite discouraged about his prospects. He thought of returning to Europe, but he had no money for the voyage and friends prevailed upon him to make the best of his situation in America. As one who daily took time for self examination and cultivated a Franklinian ideal of 'self-government' in his life, he clearly had the inner resources to do just that, though it was difficult to know in what direction to turn (Favre & Berthoud 1882, p. 71). In the summer of 1848, about the time Agassiz left on an excursion to Lake Superior, Desor accepted an invitation to go on board the Coast Survey steamer *Bibb* as an assistant to its commander Lt Charles H. Davis (1807–1877). In this work he had the freedom to dredge for marine life and examine sand and gravel formations configured by tides and currents. Desor made a highly favourable impression on Davis and before the sailing season was over the two had made plans to co-author an oceanographic treatise and even talked of exploring the Pacific Ocean together. The following February Desor heard from Davis that the collaborative work was off. Reflecting on the falling out between Agassiz and Desor the naval officer had decided to throw in his support with Agassiz. Believing that he had had a book contract with Davis, Desor claimed monetary damages in a legal suit he inaugurated against him (Anon. 1852). The upshot of this affair was that Desor found himself again needing to look for paid work. Since positions in natural history or geology were unlikely to open up until the summer of 1849, he passed the time classifying the molluscs and corals he had collected sailing on the *Bibb*, taking an active role in sessions of the Boston Society of Natural History, and teaching French at a girls' boarding school. He also gave a free tuition course in French to some '*grandes demoiselles*', an experience he found 'so much the easier and more agreeable as the majority of the students knew French better than I'. (Favre & Berthoud 1882, p. 64).

In the summers of 1849 and 1850 Desor secured an appointment as an assistant on the federally-sponsored geological survey of the Lake Superior Land District under the direction of John W. Foster (1815–1873) and Josiah D. Whitney 1819–1896). The conditions were primitive, and the work difficult, since 'nearly the whole of this area is an unbroken wilderness, interspersed with tangled thickets, almost impassable marshes and inland lakes, which retard the progress of the explorer.' (Foster & Whitney 1851, p. iv). To this brief description in the official report of the survey Desor added detailed testimony in a series of letters he sent back to Switzerland (Desor 1879, pp. 1–70). Here he recounts an excursion by boat (a 'mackinac') that he made in July 1850, in the company of another of the survey's assistants, Charles Whittlesey (1808–1886), into Michigan's Upper Peninsula. Desor was particularly charged with studying the drift and the alluvium, and in the survey's two-part report ('the Copper Region' and 'the Iron Region') the chapters devoted to surface deposits were written by him. That Desor had the requisite background for the work to which he was assigned was noted at the beginning of the report:

> His previous investigations of the drift in parallel latitudes in western Europe, and of glacial action as manifested in the Swiss Alps and the shoals along the coast of the Atlantic as observed by him during his connection with the Coast Survey, had qualified him to enter upon this field with every prospect of success (Foster & Whitney 1850, p. 16).

With the completion of the fieldwork for the Lake Superior survey, Desor went back to Boston for the winter and awaited the geological campaigns of 1851. This time it was the state survey in Pennsylvania under the direction of Henry Darwin Rogers

(1808–1866) that employed him. The state legislature after a delay of a decade at last authorized publication of the final report, and so Rogers re-instituted fieldwork to carry out some last-minute checks and to add a few details on the state's geology. Desor was appointed a 'geological assistant' and again, as in Michigan, given the task of investigating superficial deposits, including any traces of glacial action. Thanks to Desor's recommendation Rogers hired another Agassizite, Léo Lesquereux, for whom the Pennsylvania appointment would be the first of many he would obtain in government surveys as a palaeobotanist. Although they had seen one another at Agassiz's establishment in Neuchâtel, Desor and Lesquereux did not really become acquainted until they were brought together briefly in Boston. Initially, Desor had been quite impressed with the deaf botanist, but his admiration became even greater as he witnessed Lesquereux's unmatched expertise in the field in Pennsylvania. Taking time from his own duties, Desor generously assisted Lesquereux by making the drawings of his specimens of coal flora. Susan Lesley, describing the hotel room in Pottsville where she visited her husband J. Peter Lesley, the survey's topographer, left this sketch of the corps:

> It requires some connivance on my part to keep the room in tolerable order, which is quite necessary, as it is the rendezvous of Mr. Rogers, Désor, Lesquereux, Shaefer, and Colter, at all hours of the day when they are at work indoors, and need to consult continually with Peter, or bring him new materials for his map. . . . I see very little of the gentlemen, in any way to converse with them, they are all so busy, and so wholly absorbed in their work. But I enjoy very much seeing them come in to their evening meal, tired to death; but Desor always good-humored and full of talk, and poor deaf Lesquereux, the most interesting to me of the whole group (Ames 1909, **1**, p. 254).

By early 1852, it seemed as though Desor was beginning to carve out a career for himself in the study of American geology, but he was obliged to change direction suddenly. He received word from Switzerland that his brother Fritz, a physician, was ailing and needed him to return home. Desor promptly departed for Neuchâtel, where he remained, never to return to America. In his old surroundings he earned a living as a teacher, when in 1858, his brother died and left him a fortune. He promptly abandoned the classroom and took up an Agassiz-like role in promoting and subsidizing scientific activities in the Neuchâtel region. At Combe-Varin he maintained a rustic hunting lodge, where he hosted a steady stream of distinguished visitors including the British geologist Charles Lyell. Desor continued his own eclectic research, including, notably, pioneering work in the archaeology of the ancient Swiss lake dwellers. He also sponsored and took part in an expedition to the Sahara in an attempt to determine whether the desert had once been a sea. In his later years, with a reputation as a radical republican, he became prominent in local, cantonal, and federal politics. He died in 1882. When his friends Léo Lesquereux and Fritz Berthoud (1812–1890) reviewed his career, they felt confident that he would have succeeded very well in America, if he had remained there (Favre & Berthoud 1882, pp. 66, 74).

Charles Girard (1822–1895)

On his voyage to America aboard the *Sylvie de Grasse* from Le Havre in March 1847, Desor was accompanied by Charles Girard (Fig. 5). Girard, then in his mid-twenties, was of peasant stock from Upper Alsace. His parents had some connection with the Agassiz family, and as a teenager he had come to Neuchâtel to study at the local *collège*. When formal lessons were over, Girard became attached to the Agassiz enterprises as a kind of intern or lowly assistant, earning his keep by writing down fossil descriptions under Desor's dictation, running various errands, and shining shoes. With Agassiz's departure for America, Girard stayed behind for five months helping to pack up the natural history specimens and the books from the naturalist's personal library that were to be shipped across the Atlantic. Upon their

Fig. 5. Charles Girard (1822–1895).

arrival in the United States these items, and extensive additions to the collections Agassiz was making in America, all had to be arranged in some kind of order in the house in Cambridge or at some other location. There was plenty of work for Girard. In assisting Agassiz, as he had been doing for some eight years, he was learning a good deal of zoology, and was well on his way to becoming a naturalist in his own right. In Boston his training continued as every day he and Pourtalès took a dory out into the harbour and dredged the muddy bottom for marine animals, or at low tide scoured the beaches looking for specimens. By 1849 Girard had published papers on the fish genus *Cottus*, and was soon to release his findings on flatworms.

At this time, in 1849 or 1850, a disagreement involving Girard erupted among the Agassizites, and while Agassiz supported him in the dispute, Girard decided he had had enough of the 'scientific factory' and resigned. He remained in Cambridge long enough to bring a monograph on *Cottus* nearly to completion, and then attempted to use this work as leverage for a new job. It made a considerable impression on Spencer F. Baird (1823–1887), the newly appointed Assistant Secretary of the Smithsonian Institution, who agreed to arrange for its publication and who offered Girard a job as his principal assistant. It was a secure base in the years to come during which Girard was exceedingly productive. In Agassiz's mind, Girard's move to Washington D.C. and the alliance with his rival Baird was an act of consummate disloyalty. Writing to Baird about the *Catalogue of North American Reptiles in the Smithsonian Institution. Part I. Serpents* (1852) that Baird had co-authored with Girard, he could not resist taking a disparaging swipe at his former assistant, with whom he was still angry. 'If you had been willing to listen to my advise [*sic*] before', Agassiz told Baird, 'you should have known that Girard though capable of sustained work and endowed with considerable ability in distinguishing the peculiarities of animals, has no judgment, and is utterly unable to trace original researches without supervision. Moreover he is as obstinate as a mule, if contradicted, which makes it necessary that he should be led with a high hand and kept in an entirely subordinate position. Now this supervision of his work you have not made.... ' (Herber 1963, pp. 54–55). Agassiz never softened his judgement on Girard, and Girard saw no reason to regret his decision to leave Cambridge. The split between the two men was definitive. Vogt, Desor, and now Girard—the Agassiz circle was contracting.

Girard's work with Baird, though confined to a single decade, was extensive, adding up to nearly two hundred papers, articles and reports (Goode 1891). These were virtually all taxonomic studies in which species making up various groups of animals, primarily fish, amphibians, reptiles, and worms, were named, described and classified. The materials Girard worked with were collections of the Smithsonian Institution acquired in many federally-sponsored expeditions and surveys of the 1840s and 1850s. Often his publications were ascribed jointly to Baird, even if occasionally Girard did much or all of the work. According to G. Brown Goode, 'he held high rank among descriptive naturalists' in America. In 1854 Girard became a naturalized citizen of the United States and in the same year embarked on a course of medical studies at Georgetown College in Washington, D.C. He received his M.D. in 1856.

When the Civil War began Girard was on a visit to Paris. Sympathizing with the Confederacy, he undertook to assist the South by staying abroad and acting as a purchasing agent for the Confederate States, obtaining drugs, medical supplies and uniforms on their behalf. He also entered into a partnership with Jean Alexander François Le Mat, a New Orleans physician and inventor of the Le Mat 'revolving grapeshot pistol', with the purpose of manufacturing these guns and supplying them to the Confederate forces. In addition to providing war material, Girard championed the cause of the South. In scores of articles published in the Paris journal *Pays* he extolled the justness of the Southern cause, while heaping contempt on the North and its leaders. In 1863 he boarded a Confederate blockade-runner in England and crossed the Atlantic, before landing safely in Charleston. For some weeks he toured the Carolinas and Virginia, interviewing Jefferson Davis and other Confederate leaders and getting a first-hand view of how the South was faring under military and economic pressures. On his return to Paris, Girard published *Les États confédérés d'Amérique visités in 1863. Mémoire adressé à S.M. Napoléon III.* Here was additional propaganda, a bid by Girard to convince Frenchmen that they should tender friendship and accord diplomatic recognition to the South as an independent sovereign state. When the Civil War was over Girard paid the United States another brief visit, but finding his situation there intolerable, he returned to Paris, where he practised medicine and carried out zoological investigations until his death in 1895.

When Girard was a youth in Neuchâtel, his work in the Agassiz enterprise was supervised by Desor, who 'dealt with him firmly and reprimanded every mistake with stinging remarks' (Vogt 1883, p. 16). One can only wonder if, in America, years later Desor, the abolitionist, had an occasion to reprove Girard for his proslavery stand. In *Les États confédérés* Girard referred to slave owners as kind and patriarchal and characterized American plantation slavery as an advance for blacks over the barbaric life of their ancestors in the wilds of Africa. That slavery should be abolished by decree he dubbed abstract and unrealistic. A practical end to slavery would be by 'gradual emancipation brought on with the advances of the black race, time, the era, and civilization itself'. The belligerency of the North was an assault on the right of Southerners 'to live peaceably in the land that they owe only to their Creator and their labor' (Girard 1864, p. 141). Girard saw the planters of the South as representative of a Latin race, struggling for independence from the Anglo-Saxons of the North. These views were doubtless solidified during the summer of 1851 that Girard spent in Charleston, working with the herpetologist and ichthyologist John E. Holbrook (1794–1871). But it is likely that the core of his ideas was the doctrine of polygenesis, the separate, successive creation of human races, and belief in Negro inferiority that came straight from Agassiz. If, at least in part, different racial views encouraged the split between Desor and Agassiz, the separation of Girard and Agassiz must have had other roots. Their views on race were hardly distinguishable.

Léo Lesquereux (1806–1889)

Leo Lesquereux, the palaeobotanist and bryologist of the group, differed from Desor and Girard in not having been one of the old 'scientific factory' (Fig. 6). Nevertheless, he was born and raised in the canton of Neuchâtel, had Arnold Guyot and August Agassiz, brother of Louis Agassiz as school friends, and as a young man had been noticed and encouraged in his botanical pursuits by Agassiz. When Lesquereux was ten years old he tumbled down the face of cliff and was knocked unconscious. Miraculously no bones were broken, but he sustained permanent damage with the slight loss of hearing in one ear. After graduating from the Neuchâtel Academy he embarked on a teaching career, going first to Germany, where in Eisenach, in the Grand Duchy of Saxe-Weimar he gave tutorials in French. There, he fell in love with one of his pupils, Sophia, the daughter of General von Wolffskel von Reichenberg, commandant of Eisenach. Before they could marry he was obliged to return to Switzerland and improve his finances by further teaching. The wedding took place in July, 1829 and the couple eventually had five children. In hopes of improving his hearing, which was becoming more and more of a handicap in the classroom, Lesquereux went to Paris to get medical help. The operation was bungled, and he was left almost entirely deaf. Teaching was now out of the question. To support his family his best course of action appeared to be to return to his parents home and there, to apprentice himself to his father, learning the art of making watch springs. Lesquereux undertook to do this, while at the same time he eagerly pursued a longstanding hobby, the study of plants, particularly the mosses. His entry in an essay contest on peat bogs and methods for exploiting them won first prize and caught Agassiz's attention. Lesquereux then produced a manual on the subject for use in schools and was commissioned to travel all across Europe making a further study of peat bogs. The hopes engendered by these modest successes were suddenly dashed with the economic and political disturbances of 1847–1848. When the watch business became unprofitable and subsidies for scientific work that had come from Prussia were cut off, Lesquereux judged that he had no choice but to do what other Neuchâtel naturalists were doing and emigrate to America. Inquiring of Agassiz about the prospects of scientific work there, he was invited to come with his family and stay in Cambridge until he could find employment. With his wife and five children, confident that Divine Providence would see him through, he boarded a ship in Le Havre. Reflecting back on the voyage, he subsequently wrote, 'I came to that promised land a poor immigrant family, having an abominable voyage of sixty days in the entrepont (steerage) of a sailship, together with 300 companions of misery, the most terrible experience of my life' (Quoted in Lesley 1895, p. 209). Docking

Fig. 6. Léo Lesquereux (1806–1889).

in New York, he proceeded to Boston, where he made contact with Agassiz.

Evidently the help he got from Agassiz was not as much as he expected, but it was something. Initially, he was given work identifying plants, particularly the mosses, collected during the expedition to Lake Superior (Agassiz 1850, p. 154 n). For a time, too, he assisted Asa Gray as a curator of his herbarium at Harvard. But much brighter prospects soon emerged when he was put into contact with William Starling Sullivant of Columbus, Ohio, a wealthy businessman and the country's leading authority on cryptogams (Rodgers 1968). Lesquereux moved his family to Columbus, and for a year or two enjoyed a regular salary from Sullivant in return for his services as a collector. Subsequently Sullivant paid Lesquereux on an occasional basis, which served to supplement the income brought in by a watch and jewellery business that Lesquereux established. When he found work on the state and federal surveys as a palaeobotanist, there were welcome proceeds from that as well, even if frequent absences from home were a drawback.

Columbus was his home for his remaining forty years during which he fashioned a remarkable career. For more than two decades he worked closely with Sullivant, helping to elevate his patron's reputation as the world's leading bryologist. When Sullivant died in 1873 he left unfinished a comprehensive work on the mosses of North America. Thanks to the efforts of Lesquereux and a collaborator Thomas Potts James (1803–1882) the manual was completed and published a decade later, when it became the standard work on the subject. Meanwhile, starting in 1851, Lesquereux found frequent, if short-term, employment on the geological surveys. It was largely his extensive knowledge of coal plants that led to his appointments to the Pennsylvania, Kentucky, Illinois, and Indiana surveys, and his *Description of the Carboniferous Formation in Pennsylvania and throughout the United States* (1880–1884) was a pioneering work on the subject. In his contribution to the Hilgard survey of Mississippi he reported on Tertiary plants, and in the work he did for the US Geological and Geographical Survey of the Territories led by F. V. Hayden he studied extensively the fossil plants of the Tertiary and the Cretaceous.

It was never easy for Lesquereux. The honours he received such as election to the National Academy of Sciences and membership in a score of European scientific societies must have been gratifying. Yet work was irregular and not well paid.[2] He never received visitors in his home, which was bursting with specimens and the tools of his trade, and he never went to scientific meetings. Totally deaf, he suffered from failing eyesight as well. Still, year after year he persisted with his work.

There was never much of a chance that Lesquereux would become part of a scientific work force under Agassiz. In 1848 he was one of the team that classified specimens from the Lake Superior expedition, and over a few summers beginning in 1867 he worked at Agassiz's Museum of Comparative Zoology. However, Agassiz's projects, when they were not geological, were almost exclusively zoological, and Lesquereux did nothing but botany. Even if botanical investigations had come to occupy a larger place in Agassiz's schemes, it is not clear that a close working relationship between the two men could have evolved. Although they were the same age, they differed greatly in personality and circumstances. Featuring largely in Lesquereux's life was the fact that as a father of five he was always financially stretched and, although he understood that the rewards of science were not monetary, he could not shake off the stark realization that he had little wealth and social status. This was a preoccupation that went back many years. Lesquereux was the son of an uneducated artisan, and as a young man had gazed up the social ladder. He had made preparations to enrol at university, had tutored the well-born in French, and had taken a bride among the daughters of the aristocracy. But then the onset of total deafness dashed his high hopes. Forced out of the classroom, he took up the humble trade of watch making, which, when he contemplated his wife's origins, must have been a painful reminder of the social distance he had failed to travel. Encounters with Agassiz, who was well educated, prosperous, commanding, and lionized by all, must have been depressing. In his *Lettres écrites d'Amérique* Lesquereux charged that Americans feel incapable of making good judgements on their own and that is why they give such a warm reception to celebrities arriving from Europe. He went on, 'It's enough that a name has made a little stir on the other side of the Atlantic for him who bears it to be received as a demigod. If he is a man of science, knows English and, bending to the customs of the country, looks through the ads in the newspapers and goes from city to city giving his lectures and hawking his science as you would merchandise, in a little time he is assured of a huge collection of dollars.' (Lesquereux 1853, p. 18). No names were mentioned, but the fit with Agassiz is too close to miss, and so, too, is the resentment. *Lettres écrites d'Amèrique* was intended for the emigrating poor of Switzerland and exposed for their benefit the various forms of exploitation they could expect in their journey to 'the promised land.' Unlike Vogt and Desor, Lesqueruex was not a democrat on principle, but at a gut level he joined these other Neuchâtel naturalists in identifying with the oppressed.

Lesquereux's settlement in Ohio may have been as much flight from Agassiz's elitist Cambridge as it was the allure of paid employment.

Arnold Guyot (1807–1884)

Just as Agassiz was the foremost zoologist and Lesquereux the foremost palaeobotanist, so Arnold Guyot was the foremost geographer of nineteenth-century America (Fig. 7). In some ways the careers of Lesquereux and Guyot followed remarkably similar trajectories. They were both of Huguenot ancestry (as were Agassiz and Desor) and were born less than a year apart in small villages in the canton of Neuchâtel. When they both attended the College of Neuchâtel, they became close companions. 'Guyot and I', wrote Lesquereux, 'were for some years, brothers in study, working in common, and often spending our vacations together ... I owe much in life to the good influences of this friendship' (Lesquereux, quoted in Dana 1886, p. 312). Early in life there were prospects that they both might enter the ministry, but in each case the appeal of natural history proved stronger in their vocational choices. The scientific careers they fashioned were of comparable distinction, even though Lesquereux did not have the higher education that Guyot drew upon. They both finished their careers in America, honoured alike by election to the National Academy of Sciences.

Fig. 7. Arnold Guyot (1807–1884).

Whereas in Switzerland Lesquereux had only occasional contact with Agassiz, Guyot was an active member of the old 'scientific factory'. He had perhaps first become acquainted with Agassiz around 1825 when they were both guests of the family of Alexander Braun (1805–1877) in Karlsruhe, Germany. In the Braun home nature study was avidly pursued, and Guyot came away with positive impressions of science. His studies, however, still focused on the classics and on theology, which he pursued for two years in Neuchâtel under the supervision of the local pastors. Thereafter he continued his theological studies in Berlin for a year with Frederich Schleiermacher (1768–1834) and August Johann Neander (1789–1850). Thanks to influences stemming from Alexander von Humboldt, Karl Ritter (1779–1859) and other renowned scholars who were then in Berlin, Guyot turned from theology to the natural sciences. In 1835, having written a dissertation on the 'Natural Classification of Lakes', he was awarded a Ph.D. by the University of Berlin. He was in Paris working as a private tutor, when in 1838 he received a visit from Agassiz, who told him about his Ice Age theory and challenged him to view the evidence amidst the glaciers of the Alps. That summer Guyot spent six weeks examining glaciers in the Central Alps and made important observations on moraines, the movement of glaciers, and the banded structure ('the blue bands') of the ice. His findings were read before the Geological Society of France, but they were not published because he and Agassiz had agreed to collaborate on a book in which Agassiz would treat the glaciers, whereas his responsibility would be the erratic boulders. Although Guyot did not receive immediate credit for the discoveries he had made, he was appointed Professor of History and Physical Geography at the Neuchâtel Academy. This made him a colleague of Agassiz, with whom he took up lodgings in 1844. Between 1840 and 1847 Guyot strengthened the glacial theory through an extensive study of the distribution of erratic boulders. He was teaching at the Academy when it was shut down during the revolution of 1848. This led him to emigrate to America.

Guyot arrived in the United States in September 1848, and established immediate contact with Agassiz, who had urged him to emigrate. In his American 'debut' early in 1849 he delivered (in French) a course of lectures at the Lowell Institute published as *The Earth and Man* (1849). Here, drawing on the ideas of his Berlin mentors, von Humboldt and Ritter, Guyot presented a synthetic portrayal of the Earth as a progressive dynamic, organic entity, whose various physiographical features function as so many different shaping influences in the forward march of humankind, socially, politically, and morally. The book established Guyot's reputation as a respected geographer, and that in turn led to his arrangement with the Massachusetts Board of Education to travel across the State each summer during a six-year period speaking to teachers on the best methods of imparting geography to their pupils. Later, following up on this experience, he published a series of geography textbooks and wall maps that enjoyed a wide circulation. In 1854 Guyot moved from Cambridge to Princeton, where he accepted a chair at the College of New Jersey as Professor of Geology and Physical Geography, a position he held for the next thirty years. Here he established a natural history museum just a little before Agassiz was laying the foundations of his great museum at Harvard. Prominent among Guyot's exhibits were five thousand rock specimens that he had collected in the Alps while studying erratic phenomena. A Swiss who had never lost his love of the mountains, he spent many summers making barometric measurements to determine the heights of Appalachian peaks from Maine to Georgia. Under the direction of the Smithsonian Institution he played a leading role in efforts to establish a national system of meteorological observations. He advised on instrumentation and especially in New York and Massachusetts, helped with the locating and equipping of observation stations. In almost all of these activities there was little that Guyot had not first devised and tried out when he was a professor in Neuchâtel. In religion, too, there was continuity, as Guyot sustained the faith that had once made him a theology student. In America he often spoke out on the harmony of science and religion and when faced with the challenge of Darwinism, proposed an exegesis of the Book of Genesis that preserved Biblical truth. At his death in 1884 he was perhaps the last anti-evolutionist who remained among respected scientists in America.

Guyot's conservatism in this regard was not objectionable to Agassiz, who led the fight against evolution for over a decade. In fact after the move to America Agassiz seems to have been as close to Guyot as he was to any of his former Neuchâtel associates. To all appearances they felt toward one another only mutual admiration and respect. In age, education, and enterprise, they were near equals, and their scientific specialties (aside from their shared interest in glaciology) were divergent enough to forestall a sense of rivalry. In delivering Agassiz's eulogy before the National Academy of Sciences in 1878, Guyot had only the highest praise for his 'beloved friend and life-long associate'. Even with regard to that controversial matter of Agassiz's treatment of his subordinates in the old 'scientific factory' he had only good to report:

> In casting a glance at these varied labors, we cannot but admire the extraordinary industry displayed by Agassiz in this remarkable decade of his life, and also his marvelous power in inciting, directing, and organizing the work of the numerous helpers whom he knew how to employ. All those engaged with him caught a sparkle of his ardor and worked with a will—all drew a direct benefit from their intercourse with him. The artists felt their powers raised to a higher degree of perfection by his demands on their talent; the others learned how to see and observe, and every day he became conscious of new progress in the path of knowledge. (Guyot 1886, p. 63).

This is high praise, generous praise, when it is realized that Guyot's failure to publish his glacial discoveries in 1838 was owing to Agassiz's unfulfilled promise that the credit for these fundamental contributions would come later in a co-authored book. The fact is that forty years later Guyot had not forgotten the incident and mentioned it in his eulogy of Agassiz. He did so, however, in a thoroughly matter-of-fact way without a hint of blame. He wanted things put right, but was not interested in censuring his departed friend.

Louis François de Pourtalès (1823 or 1824–1880)

Equally loyal in his attachment to Agassiz was Louis François de Pourtalès (Fig. 8). Pourtalès

Fig. 8. Louis François de Pourtalès (1823 or 1824–1880).

was born in Neuchâtel in 1823 and attended the Neuchâtel Academy, where he was one of Agassiz's favourite pupils. During the summer of 1840, while still in his teens, he joined Agassiz and his team of alpine investigators who camped in the shelter of an overhanging boulder on the Aar glacier, the so-called 'Hotel des Neuchatelois'. Everyone had been assigned tasks, and his was to help Agassiz monitor the weather and take temperatures of the ice. After graduation from the Academy, Pourtalès undertook medical studies at the University of Bonn, but was forced by illness to withdraw before obtaining his degree. With little to do, having left school, he obtained the permission of his family to go to America with Agassiz. Settling into East Boston and then Cambridge, he was involved in the housekeeping activities of the Agassiz enterprise but had a hand in the scientific work as well, collecting marine animals from Boston harbour and assisting Desor in the search for evidence of glacial action. When Agassiz collaborated with Augustus A. Gould in writing the *Principles of Zoology*, it was Pourtalès who drew most of the illustrations.

In 1848, taking advantage of friendly relations between Agassiz and Alexander Dallas Bache (1806–1867), Pourtalès joined the United States Coast Survey, of which Bache was the Director. His tenure in this branch of government service continued for more than twenty years and provided the framework in which he built his reputation as a distinguished oceanographer. From 1854 onwards he was in charge of the Tidal Division with responsibilities that stirred his interest in the sea bottom and the life forms it harboured. Given the opportunity to study the nearly 9000 bottom samples brought up by Coast Survey vessels, he published a pioneering map charting the distribution of different sediment types along the eastern coast of the United States. He carried on deep-sea dredging with improvements of his own devising and made important discoveries. Samples he dredged from the Florida Straits between 1867 and 1869 brought up a variety of bottom dwelling invertebrates from depths of as much as 850 fathoms, which contradicted the accepted view of biologists that life does not extend below 200 fathoms. The harvest of new corals, crinoids, and holothurians brought in by the Coast Survey vessels was written up by Pourtalès in reports that were published more often than not in the *Bulletin of the Museum of Comparative Zoology*.

On the death of his father in 1870, Pourtalès inherited the title, 'Count', and his wealth. Pourtalès promptly resigned from the Coast Survey, returned to Cambridge, and attached himself to the Museum. In 1871–1872 he headed up the scientific team assembled for the cruise of the *Hassler*, a new ship that the Coast Survey had put at Agassiz's disposal. From Boston the expedition steamed south along the eastern coast of South America, through the Straits of Magellen and north to San Francisco. The dredging apparatus of the *Hassler* failed to perform as expected, but Pourtalès had the gratification of taking part with Agassiz in verifying the former existence of glaciers in the Southern Hemisphere. Back in Cambridge the plans of a new project were hatched, the creation of a summer institute for teachers of natural history. In July 1873, fifty school teachers from all over the United States gathered on Penikese Island in Buzzard's Bay, Massachusetts, to receive instruction from Agassiz and a small group of associates that included Pourtalès and Arnold Guyot, who came to give a special course of lectures. Four months after the institute adjourned for the summer Agassiz died. According to his son, Alexander, Louis had been grateful in his final years to have had Pourtalès at the Museum. 'In youth', the younger Agassiz elaborated, 'one of his favourite pupils, throughout life his friend and colleague, he now became the support of his failing strength'. (Agassiz 1905, p. 86). Nor did the devotion of Pourtalès stop there. On Agassiz's death he joined with Alexander Agassiz in picking up the reins of the Museum's administration. The latter said of Pourtalès that 'his devotion to science was boundless. A model worker, so quiet that his enthusiasm was known only to those who watched his steadfast labour, he toiled on year after year without a thought of the self, wholly engrossed in his search after truth. He never entered into a single scientific controversy, nor even asserted or defended his claims to discoveries of his own which had escaped attention.' (Agassiz 1905, p. 87). For Agassiz who had to deal with the forwardness of a Vogt or a Desor, Pourtalès might have seemed to be the ideal assistant.

Conclusion

By 1850, the Swiss émigrés had departed Agassiz's Cambridge residence, and the Neuchâtel 'scientific factory' that had so recently been resurrected in the New World ceased operations. For Agassiz this development was little more than a temporary inconvenience. He had had the help he needed during the settling-in phase of his move to America and if now he required assistance, Pourtalès was on call and American helpers could were found to augment his staff. After 1859, when the Museum was up and running, the graduate students it brought in could conveniently be tapped as assistants. But what of the Swiss who broke with Agassiz, abandoned his patronage and struck off on their own?

Having, in Lesley's imagery, 'placed scaling ladders against the ramparts of fame', the Neuchâtel naturalists were remarkably successful in their career ventures. In fact, they attained greater heights than one could reasonably have expected for a handful of immigrants coming from a small region of Europe. Agassiz became the leading zoologist of nineteenth-century America; Lesquereux, the leading palaeobotanist; and Guyot, the leading geographer. Girard ranked among the most respected herpetologists, and Pourtalès left a permanent mark on the history of deep-sea dredging and marine zoology. Desor, if he had stayed longer in America, would probably have made important contributions to Quaternary geology and the glacial theory. Agassiz, Guyot, Lesquereux, and Pourtalès were all elected to the National Academy of Sciences. It is an impressive record of achievement.

In probing how it came about, it might be appropriate to consider first some of the hurdles of life in America that immigrants had to clear. What these were in the experience of one immigrant was amply and colourfully described in Lesquereux's *Lettres écrites d' Amérique*. Contending that strife and want had rendered Europe intolerable, Lesquereux upheld the dream of America as the Promised Land, but he painted in grim colours the process of re-settlement in the New World. In the first place, he judged, Americans regarded their own God-given culture as superior to any other, and they disliked foreigners and foreign accents (Lesquereux's was that of a person who had learned English by lip reading after becoming totally deaf). For the immigrant looking to make a living as a scientist there were additional discouragements. Americans had an inordinate love of money, and they valued science only for the practical benefits it bestowed in manufacturing, commerce, and agriculture. They always wanted to know 'of what use' something was. Nor did their devotion to democracy with its levelling equality redound to the benefit of science. Though commendable for its universality, American education was uniform and geared to the least adept. Absent from the curriculum were the specialized, advanced courses requisite for the preparation of scientists. As a result, Americans, for all their practical knowledge and inventiveness, knew little of science, and had little to boast of in the way of scientific societies, museums, botanic gardens, and celebrated universities. America, or so it seemed to Lesquereux, was a scientific wasteland.

Lesquereux's charges of anti-intellectualism in American culture have been heard often enough to contain some truth, but at just the time Agassiz and his followers arrived science was enlarging its scope and reputation in American society. Scientific institutions were on the increase, and employment opportunities for scientists were growing. Teaching positions were opening up in colleges and universities, government surveys with paid positions were proliferating, and scientific consulting was getting under way. The Agassiz assistants who were desirous of fashioning independent careers generally shied away from teaching (Guyot was an exception) because of their discomfort in speaking English. But that left ample possibilities for government employment, particularly in the natural history surveys, and that was where the Neuchâtel naturalists mostly ended up. Agassiz's connection with Bache opened the doors to the Coast Survey for Desor and Pourtalès, and though Agassiz regretted it, his connection with Spencer F. Baird paved the way for Girard's employment at the Smithsonian Institution. Lesquereux found positions, first with a private patron and then with numerous government surveys, for which his uncommon qualifications in palaeobotany were his best recommendation. Desor had been looking for survey assistantships for employment when he was called back to Switzerland.

The Agassizites succeeded in joining the stream of rising professionalism in American science, but how did they overcome the hindrances described by Lesquereux and perform so well in their scientific work? Several factors of varying importance could have contributed. Perhaps Lesley was correct in suggesting that desperation builds motivation. The Neuchâtel naturalists fled terrible political and economic circumstances and believed their only hope for a decent life was to work hard and build their fortunes in America. Certainly Lesquereux found a powerful incentive to do his best and persist in his labours in his need to support five children. The same was true of Guyot, who although single until the age of fifty-nine, when he married Sarah Haines, the daughter of a former governor of New Jersey, had large household expenses because of the dependency on him of his mother, his sisters, and other relatives. By contrast, Desor, Girard, and Pourtalès were able to give their work their undivided attention because they were not married. Might religion have been a spur to their scientific work? Without pausing to critique the Merton thesis, it is worth mentioning that Agassiz, Guyot, Desor, and Lesquereux all had Huguenot ancestors, raising the possibility that some of their industriousness was rooted in the Protestant work ethic.[3] Education was doubtless important, but it may not have been a necessary ingredient in the success of the Neuchâtel naturalists.

Attending German universities, where they had considerable exposure to philosophy and science,

Agassiz and Guyot had imposing educational credentials. Vogt and Desor had university degrees as well, one in medicine and the other in law, but one cannot help wondering how much their involvement in student politics may have distracted them from their studies. Pourtalès had started on a medical curriculum, but left the university without a degree. Lesquereux and Girard had only secondary education, although belatedly, when he was in his mid-thirties, the latter obtained a medical degree. Arguably, Lesquereux, Pourtalès, and Girard did not require a higher education for the work they did in descriptive taxonomy.

But to excel in biological classification, as they did, broad experience was indispensable, and that was what the Swiss possessed in measures exceeding the experience of most of their American counterparts. When the Neuchâtel naturalists set out to describe and classify new species, they were usually in a good position to make the comparisons with related species because they knew the Old World species.[4] They also had a high standard of classification in which the taxonomist proceeds comparatively, considering the whole web of relations in which the creature under consideration participates, taxonomical, palaeontological, embryological, environmental, and so forth, precisely what Agassiz had instilled in his assistants (Winsor 1991, p. 90). The European background and training of the Neuchâtel naturalists equipped them well for descriptive biology. The familiarity with the mosses that Lesquereux acquired in studying European peat bogs, for example, hastened his becoming an expert in American bryology. Similarly, Girard blossomed in ichthyology, probably because of the extensive knowledge of fish he had absorbed as a protégé of Agassiz in Neuchâtel. Corresponding circumstances gave the Neuchâtel naturalists a leg up in other fields in which they worked. When Desor and Guyot investigated drift deposits in North America their experience in the Alps (augmented in Desor's case by explorations in Scandinavia) gave them an advantage over American scientists in picking out the traces of ancient glaciers in the landscape; nor in geography was there much of a chance that Guyot would have a native-born rival. The science hardly existed outside of Germany, and there Guyot found his teachers in Ritter and von Humboldt. To a greater or lesser extent the whole Agassiz band adopted the globe-encircling outlook of these two men. Perhaps in fostering a universal-comparative perspective that validated and juxtaposed the two worlds of their experience it was the critical factor that lifted their work beyond the commonplace.

The author is greatly indebted to P. K. Wilson of Penn State College of Medicine for prolonged, stimulating conversations on the Neuchatel naturalists. My thanks to him and to our spouses for tolerating without complaint our absorption in things Swiss.

Notes

[1]Were more space available, the list 'Neuchâtel naturalists' considered here could have been expanded to include Alexander Agassiz, the son of Louis Agassiz and a distinguished marine biologist and oceanographer, and Jules Marcou, a stratigraphical geologist and map-maker.

[2]See Lesquereux to F. V. Hayden, 12 December 1870, cited in Cassidy (2000, p. 165).

[3]c.f. Landes (1979, esp. pp. 32–39), where Protestantism and the Protestant ethic are highlighted as factors in the Swiss success in watchmaking.

[4]In Neuchâtel, of course, they did not have access to fresh specimens of living marine organisms, but in the natural history collections of Paris and other European scientific centers Agassiz had ample opportunities to study these species in the form of preserved specimens.

References

AGASSIZ, A. 1905. Biographical Memoir of Louis François de Pourtalès. *Biographical Memoirs of the National Academy of Sciences*, **5**, 79–89.

AGASSIZ, L. 1850. *Lake Superior: Its Physical Character, Vegetation, and Animals compared with those of other and similar regions. With a Narrative of the Tour by J. Elliott Cabot.* Gould, Kendall & Lincoln, Boston.

AMES, M. L. (ed.) 1909. *Life and Letters of Peter and Susan Lesley.* G.P. Putnam's Sons, New York & London.

ANON, 1852. *Trial of the Action of Edward Desor, Plff. Versus Chas. H. Davis, Deft. Before the Circuit Court of the United States for the District of Massachusetts; for Breach Of Contract to Write a Book on the Geological Effects of the Tidal Currents of the Ocean.* Stacy & Richardson, Boston.

CASSIDY, J. G. 2000. *Ferdinand V. Hayden. Entrepreneur of Science.* University of Nebraska Press, Lincoln & London.

DANA, J. D. 1886. Memoir of Arnold Guyot, 1807–1884. *Biographical Memoirs of the National Academy of Sciences*, **2**, 309–347.

DESOR, E. 1853. De l'esclavage aux Etats-Unis à l'occasion de la case de l'oncle Tom (Uncle Tom's Cabin) par mad. H. Stowe-Beecher. *Extrait de la Revue Suisse*, **16**, 23–43.

DESOR, E. 1861. Esquisse de la vie de Théodore Parker. *In*: MAYER VON ESSLINGEN, C. (ed.) *Album von Combe-Varin. Zur Erinnerung an Theodore Parker und Hanz Lorenz Küchler.* Schabelitz'sche Buchhandlung, Zurich, 309–331.

DESOR, E. 1879. *La Forêt Vierge et le Sahara.* J. Sandoz/ Neuchatel; Desrogis, Genève.

FAVRE, L. & BERTHOUD, F. 1882. Edouard Desor. Discours proncés à l'ouverture des cours de l'Académie de Neuchâtel le 12 avril 1882. *Musée neuchatelois.*

Recueil d'histoire nationale et archéologique. Supplément à livraison de janvier 1883, 28–74.

FOSTER, J. W. & WHITNEY, J. D. 1850. *Report on the Geology and Topography of a Portion of the Lake Superior Land District in the State of Michigan. Part I. Copper Lands.* House of Representatives, Washington, D.C.

FOSTER, J. W. & WHITNEY, J. D. 1851. *Report on the Geology of the Lake Superior Land District. Part II. The Iron Region.* A. Boyd Hamilton/Washington, D.C.

GIRARD, C. 1864. *Les États confédérés d'Amérique visités en 1863.* E. Dentu, Paris.

GOODE, G. B. 1891. Bibliographies of American Naturalists: V. The Published Writings of Dr. Charles Girard. *Bulletin of the United States National Museum*, **41**.

GREGORY, F. 1977. *Scientific Materialism in Nineteenth Century Germany.* D. Reidel, Dordrecht & Boston.

GUYOT, A. 1886. Memoir of Louis Agassiz, 1807–1873. *Biographical Memoirs of the National Academy of Sciences*, **2**, 39–73.

HERBER, E. C. (ed.) 1963. *Correspondence between Spencer Fullerton Baird and Louis Agassiz—Two Pioneer American Naturalists*, Smithsonian Institution, Washington, D.C.

LANDES, D. 1979. Watchmaking: A Case Study in Enterprise and Change. *Business History Review*, **53**, 1–39.

LESLEY, J. P. 1883. Obituary Notice of Edouard Desor. *Proceedings of the American Philosophical Society*, **20**, 519–528.

LESLEY, J. P. 1895. Memoir of Leo Lesquereux, 1806–1889. *Biographical Memoirs of the National Academy of Sciences*, **3**, 189–212.

LESQUEREUX, L. 1853. *Lettres écrites d'Amérique. Henri Wolfrath, Neuchâtel.*

LURIE, E. 1960. *Louis Agassiz. A Life in Science.* University of Chicago, Chicago.

RODGERS, A. D. 1968. *'Noble Fellow' William Starling Sullivant.* Hafner, New York & London.

VOGT, C. 1883. *Eduard Desor. Lebensbild eines Naturforschers.* S. Schottlaender, Breslau.

WINSOR, M. P. 1991. *Reading the Shape of Nature. Comparative Zoology at the Agassiz Museum.* University of Chicago, Chicago & London.

Clarence Edward Dutton (1841–1912): soldier, polymath and aesthete

ANTONY R. ORME

Department of Geography, University of California, Los Angeles, CA 90095-1524, USA (e-mail: orme@geog.ucla.edu)

Abstract: Clarence Edward Dutton (1841–1912) was one of several scientists who laid the foundations for modern geology from their work in North America during the late nineteenth century. Dutton was a career soldier who fought in the American Civil War and remained with the US Army Ordnance Corps to his retirement in 1901. Despite military obligations, Dutton developed a profound interest in geology and, on secondment first to the Powell Survey and later to the fledgling US Geological Survey, made important contributions to volcanic geology, seismology and physical geology. His lifelong fascination with volcanism led to improved understanding of the volcanic geology of the American West, Hawaii, and Central America. This work linked naturally with the emerging science of seismology, as reflected in his study of the 1886 earthquake in Charleston, South Carolina, and he is often credited with introducing the 'new seismology' to American scientific audiences. Awareness of volcanic and seismic hazards in turn led him to caution against a proposed sea-level canal across the Nicaraguan isthmus. His contributions to physical geology are most evident in several reports on the American West, notably the *Report on the Geology of the High Plateaus of Utah* (1880), the *Tertiary History of the Grand Cañon District* (1882), and *Mount Taylor and the Zuni Plateau* (1885). These reports, presented in colourful prose, reveal both the author's scientific acumen and his aesthetic appreciation of nature.

Whereas most of Dutton's work must now be placed in its historical context, in his recognition of isostasy, a term he coined in 1882 to reflect the debate then raging concerning Earth's crustal behaviour, his ideas were remarkably prescient. Dutton's interest in isostasy derived in part from his studies of volcanic geology, seismology, and crustal behaviour, and in part from his fieldwork in the American West. Initially, however, it was Dutton's description of the great denudation of the Colorado Plateau, rather than any isostatic implications, that influenced geomorphology during the earlier twentieth century, dominated as it was by Davis' cycle of erosion. The subsequent demise of Davisian geomorphology, and the ensuing quest for alternative models, led to a reawakening of interest in isostasy as a concept basic to the explanation of Earth's surface features. Somewhat belatedly, Dutton's concept of isostasy is once again at the centre of debate regarding denudation and crustal behaviour. Wherever these debates focus, they confront a problem fundamental to geology, geodesy and geophysics, namely the extent to which the Earth's present relief reflects a quest for balance between the subsurface forces generating uplift, subsidence and mass transfers at depth, and the climate-induced processes responsible for denudation and mass transfers of rock waste across the surface. Dutton's probing of isostasy predated modern debate in geomorphology by more than a century.

The establishment of the United States Geological Survey in 1879 brought together some remarkable scientists who had already made significant contributions to geology as members of various independent surveys in the American West. Notable among these were John Wesley Powell (1834–1902), Grove Karl Gilbert (1843–1918), and Clarence Edward Dutton (1841–1912), close friends with differing personalities and talents who combined to advance science along several fronts. The flamboyant Powell, whose reputation had been established from his first exploration of the Colorado River in 1869, occupied the uneasy interface between scientific investigation and sponsoring agencies. The more ascetic Gilbert, the consummate intellectual, grappled with relationships between form and process. The reserved but generous Dutton, the career soldier who addressed scientific questions with great literary skill, complemented the others in several areas and in one respect, his interest in isostasy, was notably farsighted.

Of this trio, Dutton is perhaps the least known, the focus of less attention during his lifetime than Powell, the subject of fewer retrospective accolades than Gilbert. And yet, for a part-time scientist whose most original work was condensed into just 15 years, Dutton travelled widely and achieved much that is worthy of analysis in the light of modern advances in the Earth sciences. This paper outlines Dutton's life and accomplishments, and then reviews his contributions to volcanic geology and seismology, before examining in greater depth

From: WYSE JACKSON, P. N. (ed.) *Four Centuries of Geological Travel: The Search for Knowledge on Foot, Bicycle, Sledge and Camel.* Geological Society, London, Special Publications, **287**, 271–286.
DOI: 10.1144/SP287.21 0305-8719/07/$15.00

his understanding of isostasy and the legacy of this work to later developments in physical geology.

Biographical sketch

Compared with many of his contemporaries, much less is known about Dutton's early and later life because most of his papers disappeared shortly after his death. In contrast, his productive middle years are well represented by his many writings and by records maintained by the various agencies for which he worked. Additionally, many salient facts and perspectives are recorded in obituaries and memoirs, in genealogies of the Dutton family, in later biographies, and in extant correspondence, most notably that with Archibald Geikie during the 1880s.

Clarence Edward Dutton was born on 15 May 1841 in Wallingford, Connecticut, the fourth of five children from the marriage in 1833 of Samuel Dutton (1806–1851), shoemaker and sometime postmaster, and Emily Curtis (1805–1875).[1] The Duttons, an extended family with roots traceable to twelfth century England, derived from immigrants to New England during the seventeenth century. Clarence Dutton graduated from Yale College in 1860 where he won a literary prize and established credentials as a mathematician and raconteur.[2] The influence of Macaulay's essays and Thackeray's novels was to resurface in his later writings. His strong physique and athletic prowess, notably in gymnastics and rowing, were to serve him well later in the field (Longwell 1958). He then spent a brief period at Yale's theology school but soon felt unsuited to the ministry.

Following the outbreak of the American Civil War in April 1861, Dutton joined the 21st Connecticut Volunteers in September 1862, fighting for the Union cause, as first lieutenant and adjutant. He was wounded at the battle of Fredericksburg in December 1862 and was promoted to captain in March 1863 (Wilson & Fiske 1887–1889). In January 1864, he transferred by examination into the regular army as a second lieutenant in the Ordnance Corps. He was subsequently engaged in conflicts at Cold Harbor, Drury's Bluff, and the Bermuda Hundred before ending the war in command of the ordnance depot of the Army of the Potomac. His elder brother, Arthur Henry Dutton (1838–1864), an engineering graduate from the military academy at West Point in 1861, rose to colonel in the 21st Connecticut Volunteers but was mortally wounded in skirmishing in the Bermuda Hundred and died on 2 July 1864. Meanwhile, Clarence Dutton married Emeline Babcock (born 1840) of New Haven, Connecticut, in April 1864, and they had two children, Marion in 1865 and Clarence Edward in 1871.

The war's end in 1865 saw Dutton posted to the Watervliet Arsenal in West Troy, New York, and he was promoted to first lieutenant in 1867. While there, he was influenced by Robert Whitfield and James Hall, who directed his attention to geology, and by Alexander Holley of the nearby Bessemer Steel Works who interested him in steel production (Diller 1911, 1913; Becker 1912).[3] His first scientific paper, *On the Chemistry of the Bessemer Process*, was read before the American Association for the Advancement of Science at their Troy meeting in 1870. He was posted to the Frankford Arsenal in Philadelphia the same year and then to the Washington Arsenal in 1871, where he remained until May 1876. Promoted to captain in June 1873, Dutton spent the remainder of his career, nominally at least, with the Ordnance Corps, from which he retired in 1901 (Fig. 1).

Dutton's transfer to Washington in 1871 was fortunate. In the aftermath of war, the capital was a hive of scientific, cultural and political activity while the distant surveys of the American West were in full flow under sometimes contentious leadership and variable funding. Dutton soon became engaged in the scientific activities of the capital, notably with the Philosophical Society of Washington. Early critiques of crustal deformation in 1871[4] and the contraction hypothesis in 1873[5] showed that he was already pondering some of the more contentious geological issues of his day. His

Fig. 1. Captain Clarence Edward Dutton, Corps of Ordnance, US Army.

enthusiasm was noted and on 15 May 1875, following recommendations from John Wesley Powell and Joseph Henry, Secretary of the Smithsonian Institution, the War Department assigned Dutton for duty with the Powell Survey. Thus began Dutton's 15-year association with the American West during perhaps the most formative period in North American geology.

Although the American West had been explored to some extent earlier, notably during the railroad surveys of the 1850s, the years immediately following the Civil War saw renewed interest in the region which focused on several major expeditions. The first of these, the *United States Geological and Geographical Survey of the 40th Parallel* (1867–73) under the direction of noted geologist and mining engineer Clarence King (1842–1901), was designed to render geological maps of the country about to be opened up by the Union Pacific and Central Pacific Railroads.[6] The other significant expeditions were the *United States Geographical Survey West of the One Hundredth Meridian* under Lt. George Montague Wheeler of the War Department (informally called the Wheeler Survey, 1867–79), the *United States Geographical and Geological Survey of the Rocky Mountain Region* under John Wesley Powell (the Powell Survey, 1867–79), and the *United States Geological and Geographical Survey of the Territories* under Ferdinand Vandiveer Hayden (the Hayden Survey, 1869–79).

These surveys were undoubtedly aided by completion of the first transcontinental railway in May 1869, but beyond the railheads in northern Utah Territory travel was challenging and accomplished mostly by saddle horse and wagon across difficult uncharted terrain. High elevations, deep canyons, and a continental climate limited most fieldwork to the summer months and even then hazards were posed by thunderstorms and flash floods. By modern standards, these were reconnaissance surveys that focused primarily on topographic mapping and resource identification. Nevertheless they generated some remarkable scientific work, including that arising from the fortunate collaboration of Powell, Gilbert and Dutton. Powell's explorations of the Colorado River in 1869 and 1871–72 had enthralled a larger American audience, whereas Gilbert, who had served with the Wheeler Survey from 1871 to 1874 before joining the Powell Survey, was acquiring a reputation for discerning science.

After helping Gilbert with the intrusive rocks in the Henry Mountains of Utah (Dutton 1877), Dutton progressed from 1875 to 1877 onto Utah's high plateaus where his survey team provided base topographic maps on which Dutton recorded the essential geology. His *Report on the Geology of the High Plateaus of Utah* appeared in 1880 (Dutton 1880*b*). He then devoted three field seasons, 1878–1880, to a geological reconnaissance of the Grand Canyon region to the south. When not in Washington, his winters were spent in the west as chief ordnance officer for the Department of the Platte which allowed him time for reflection and writing. He was also one of the co-founders of the Cosmos Club in 1878, a focus for the intellectual elite in Washington.

Meanwhile, competition for responsibilities and resources between the remaining surveys had been resolved in March 1879, after much political wrangling, by the formation of a single United States Geological Survey (USGS), with Clarence King taking office as its first Director on 24 May (King 1880, p. 3). The three surveys were formally discontinued by the US Congress on 30 June 1879, with the USGS, now charged with coordinated surveys of the west, inheriting many of their personnel and much of their scientific work. King was highly regarded but it was Powell, now briefly Director of the Bureau of Ethnography, who was instrumental in encouraging this arrangement and who succeeded King as Director on 19 March 1881 (Powell 1882, pp. xi–xvi). Dutton became head of the short-lived Division of the Colorado (1879–81), one of four regional divisions of the USGS,[7] to complete his work in the region. His remarkable study of *The Tertiary History of the Grand Cañon District* appeared in the Second Annual Report of the USGS for 1880–81 and then in the Survey's monograph series in 1882 (Dutton 1882*a*). Initially, the work of the USGS was confined to the west but, in August 1882, Congress was persuaded to extend and fund operations nationwide 'to continue the preparation of a geological map of the United States.'[8]

At Powell's request, Dutton's scientific work now refocused on volcanic geology, although he had never seen an active volcano. Powell thus made provision for him to visit the Hawaiian Islands in 1882 while Dutton's new assistant, Joseph Silas Diller (1850–1928), began reconnaissance work in the far west (Dutton 1884). Between 1883 and 1886, Dutton's team focused mainly on the volcanic terranes of the Cascade Range, on the then little known volcanoes extending from Lassen's Peak, California, northwards into Oregon and Washington Territory. In 1884, Dutton was formally given charge of the new Division of Volcanic Geology, and was elected to the National Academy of Sciences. During 1883 and 1884, he also studied extinct volcanic terranes in New Mexico Territory, access to which had been aided by completion of the second transcontinental railway in March 1881. This work appeared as *Mount Taylor and the Zuñi Plateau* in the Sixth Annual Report of the USGS in 1885 (Dutton 1885*a*).

On 31 August 1886, while Dutton was working in the Cascade Range in Oregon, a major earthquake occurred in the Charleston area of South Carolina. Dutton was invited by Powell to study the event and, reaching Charleston in October, began a thorough investigation that was to culminate in his report of 1889 (Dutton 1889*a*). The latter year also saw Dutton address the Philosophical Society of Washington *On Some of the Greater Problems of Physical Geology*, a paper in which he formally introduced the word 'isostasy' although he had in fact used it earlier.

However, Dutton's involvement with the American West was changing. In the Tenth Annual Report to the US Congress for 1888–89, Director Powell had noted that the duties of the USGS now extended to 'the investigation of the extent to which the arid regions of the United States can be redeemed by irrigation . . .' (Powell 1890, p. 3). Ever willing, on 1 July 1888 Dutton became head of the hydraulic branch of the new Irrigation Survey where he initiated studies of precipitation, evaporation, stream flow and suspended sediment discharge, and evaluated the potential for engineering works such as dams and canals (Dutton 1890).[9] However, despite Dutton's warnings, Powell continued to use some irrigation survey funds to support general topographic mapping. This brought rebuke from Congress, irrigation being a sensitive issue in Washington, and in his Twelfth Annual Report for 1890–91 Powell noted that 'a portion of the work of the Irrigation Survey was discontinued' (Powell 1891, p. 5). Dutton may have been a casualty of these events. After his report on irrigation for the 1888–89 fiscal year (Dutton 1890, p. 78–108), he made no further contributions to USGS publications.

Although Dutton may have left the USGS at his own request following the disagreement with Powell, he was evidently not happy at this turn of events. In a letter of 6 June 1891 to Archibald Geikie (1835–1924), Director General of the Geological Survey of Great Britain, Dutton said that he was again fully employed by the War Department, having been promoted to major on 1 May 1890, but expressed dismay at being transferred from Washington, D.C., to the San Antonio Arsenal, Texas, where he saw no life for himself.[10] Somewhat sadly, he told Geikie that he was considering writing a cowboy novel or possibly a historical novel about Davy Crockett.

Whether or not these novels were ever written, Dutton found new life, during furloughs from Texas, by visiting volcanoes in Mexico and Central America, and by preparing reports on the proposed inter-ocean canal through Nicaragua for the US Congress in 1892 (Dutton 1891–1892). In 1899, he was recalled to duty in the office of the Chief of Ordnance in Washington, D.C., but on 7 February 1901, at his own request, he retired from active service (Pyne 1999, pp. 172–173). His book on *Earthquakes in the Light of the New Seismology* appeared in 1904 (Dutton 1904), and in his last major paper, in 1906, he returned to explain volcanoes in relation to the recently discovered concept of radioactivity (Dutton 1906).[11] After several years of declining health, he died from arteriosclerosis at his son's home in Englewood, New Jersey, on 4 January 1912. He apparently just fell asleep. Among his last words were 'Farewell to my old friends on the Geological Survey' (Diller 1913, p. 16).[12]

Volcanic geology

Dutton's interest in volcanic geology, reflecting perhaps his early exposure to molten ores in Troy's furnaces, was stimulated by contemporary theoretical debates regarding Earth's crustal behaviour and then by familiarity with the Tertiary lavas that spewed onto the Colorado Plateau. His first field season in charge of the Division of the Colorado in 1879 had focused on the topographic survey (by John Renshawe) and geological mapping of the arid Uinkaret Plateau where perhaps 200 volcanic vents and basaltic flows, dominated by Mount Trumbull (2447 m), cover Carboniferous and Permian sedimentary rocks above the North Rim of the Grand Canyon (Dutton 1880*a*, pp. 29–31). Several of these cones are perched on the brink of the canyon and spill *coulées* of olivine basalt into the gorge and nearby valleys such as the Toroweap. To illustrate these relationships, his 1882 report on the *Tertiary History of the Grand Cañon District* included drawings of Mount Trumbull, of dykes in the canyon wall, and of recent lava flows on the Uinkaret by artist-geologist William Henry Holmes (1846–1933), and of lava falls in the inner gorge by romantic artist Thomas Moran (1837–1926) (Dutton 1882*a*, Plates XVIII–XXI). Distinguishing between the older basalt flows on the plateau and younger basalts cascading into the gorges, he inferred a relatively recent age for the incision of the Grand Canyon. In the following field season, 1880, he asked Richard Goode to complete a topographic survey of the comparable extinct volcanic field in the San Francisco Mountains just south of the Grand Canyon (Dutton 1882*b*, note 13, pp. 5–10).

Thus, when Powell asked him to direct his attentions towards volcanic geology in 1882, Dutton was no neophyte. But neither had he seen an active volcano. Indeed, his limited experience showed in a letter to Archibald Geikie in January 1881, wherein he recognized only two categories of volcanic eruption: what he termed, from the literature, the Mediterranean-type of frequent small events and cinder cones, and from his own field studies, the Rocky Mountain-type of fewer but larger eruptions over a wide area.[13] Powell therefore provided funds for Dutton to visit the Sandwich (Hawaiian) Islands from May to October 1882.

Dutton was by no means the first geologist to examine the Hawaiian volcanoes. For example, James Dwight Dana (1813–1895), later Professor of Natural History and Geology at Yale, had visited them between 1838 and 1843 (Dana 1849). Dutton spent much of his time on the big island of Hawaii, observing the active caldera at Kilauea, the legacy of fissure and crater eruptions from the frequently active pile of Mauna Loa, the olivine-rich lava flows, the dormant Hualalai, and the dissected features of extinct Mauna Kea and Kohala. He concluded that the lavas of Kilauea were generated in shallow reservoirs only 1–4 km beneath the surface. He also made briefer excursions to Maui, Kauai, Molokai and Oahu, recognizing as had others before him that volcanic activity ceased earlier as one progressed NW from Hawaii, although he erred in saying that this did not necessarily mean that the eruptions to the NW began earlier.[14] In a letter written to Dana after his return, Dutton admitted that he had made no great discoveries but that:

> I understand much better than I ever did before the action and behavior of lavas, their modes of accumulation and their methods of flowing.[15]

No doubt invigorated by his sojourn in the Hawaiian Islands, Dutton prepared to apply his new knowledge to the Cascade Range in the far west. The problem was that, although Gilbert Thompson and M.B. Kerr had begun topographic surveys there in 1882 and 1883, 'in 1884 the topographic survey and mapping of that country had not sufficiently progressed to enable systematic geological work to proceed satisfactorily.' (Dutton 1885*a*, p. 114). He also found the maps of the earlier Wheeler Survey 'wholly unsuited for geological work.' (Dutton 1888, p. 97). He therefore took the opportunity to examine part of New Mexico territory, where topographic mapping had begun under J. Howard Gore and Gilbert Thompson as early as 1881. In his subsequent report on Mount Taylor and the Zuni Plateau he focuses in part on the great volcanic pile of Mount Taylor. In a letter to Geikie in 1884, he again distinguishes between the thick older flood basalts on the plateau and the more recent lava flows flowing up to 160 km from vents on Mount Taylor.[16]

Thus, after significant delays, studies of the volcanic geology of the Cascade Range began in earnest in 1885. Dutton's new assistant, Joseph Silas Diller, had begun reconnaissance work from Mount Shasta to the Columbia River in 1883 and 1884, collecting samples for petrographic analysis and observing the relationship between the volcanic and other rocks of the region. It was not until 1885 that Dutton, now head of the Division of Volcanic Geology, began serious field studies there. His 1885 field season focused mainly on northern California, on the vast volcanic pile of Mount Shasta (4317 m), isolated from the main Cascade Range, and on Lassen's Peak (3187 m) and the Modoc volcanic plateau (Dutton 1888, pp. 99–100). He recognized the mostly andesitic and basaltic nature of the lavas, while the superposition of recent lavas on Cretaceous and early Cenozoic sedimentary rocks allowed him to establish a maximum age for later volcanism. Later in 1885 and in 1886, he followed the range northward through Oregon to the Columbia River.

During 1886, he undertook a study of Crater Lake, a mid-Holocene caldera in the Cascade Range (Dutton 1888, pp. 97–103). This involved the construction of three boats in Portland, their transport by rail to Ashland, and thence by wagon to the lip of the caldera's near-vertical walls down which they were lowered by an infantry detachment (Dutton 1888, pp. 102–103). Some 168 soundings of the lake were obtained by lowering a piece of pipe on a spool of piano wire from the stern of the *Cleetwood*, the half-ton survey boat. The maximum recorded water depth of 608 m was remarkably close to the sonar depth of 589 m obtained in 1959.

He also attempted to understand the complex geology of the Coast Ranges of northern California and Oregon which he thought were in some way related to Cascade volcanism. He stated:

> The terms dip and strike here lose all significance, for everything like system or order in the arrangement of the beds (or of what once may have been beds) has been obliterated by crushing or crumpling forces which have acted upon them. The folding and distortion have not resulted here in a systematic plication like that of the Appalachians, as if by a force acting in a single or two opposite directions, but by forces from every possible direction. (Dutton 1888, p. 100)

Today, several generations later, geologists armed with an understanding of accretionary and subduction tectonics continue to struggle with the issues posed by these ranges.

Further studies in volcanic geology were interrupted by the Charleston Earthquake of August 1886, the report on which precluded fieldwork during the 1887–88 fiscal year, and by his subsequent involvement in the fledgling Irrigation Survey from July 1888. He retained an interest after leaving the USGS, however, and was able to visit volcanoes in Mexico and Central America while on furlough from the San Antonio Arsenal.

In his oft-quoted 1889 paper, Dutton recognized three important questions in physical geology, the first of which was: 'What is the potential cause of volcanic action?' (Dutton 1889*b*). He considered the problem then unsolved and thus declined to answer. Later, in his last paper in 1906, following the discovery of radioactivity in 1896,[17] he returned to the theme. Here he states that volcanic eruptions 'are caused . . . by a development of heat, resulting from radioactivity, in limited tracts at a depth of one to three—at the very utmost not over four—miles from the surface . . .' He supported this argument from his belief in shallow reservoirs beneath Hawaiian volcanoes and from recent measurement by others of radioactivity in soils, rocks and hot springs. In seeking to lay to rest, one more time, the contraction hypothesis of Earth's history, he emphasized that the heat generated by radioactivity far exceeded that lost to the atmosphere and space through conduction and radiation. The 1906 paper was an excerpt from a book on volcanism that he was preparing but never published. In a letter shortly before his death, Dutton reaffirmed his interest in volcanicity but considered its cause still unresolved.

Seismology

Dutton's interest in seismology was probably kindled by his broader concern for questions of Earth's crustal stability and by his exposure, while in the Hawaiian Islands, to minor earthquakes associated with volcanic eruptions. Also, his report for the 1884–85 fiscal year was remarkably prescient in noting that several earthquake tremors had occurred during 1884 along the Atlantic seaboard of the United States. Dutton convened a group that included representatives of the US Naval Observatory and the Signal Service, and also William Morris Davis (1850–1934) of Harvard, to investigate the practicality of setting up an earthquake observation network (Dutton 1885*b*, p. 61). Lacking suitable instruments, they concluded that this was not feasible and that it was best to rely on a network of voluntary observers during and after specific earthquake events.

Dutton was on the Warm Spring Indian Reservation east of the Cascade Range when the Charleston Earthquake occurred on 31 August 1886 (Dutton 1889*a*, p. 209). He learned of this event, and of Powell's request that he undertake its investigation, on reaching Portland on 10 September. Dutton's understandably delayed arrival in Charleston in mid-October was hardly ideal but, fortunately, W. J. McGee (1853–1912) of the USGS and Thomas Corwin Mendenhall (1841–1924), then of the US Signal Service, had been dispatched to the scene immediately after the earthquake. With additional input from local observers Earle Sloane, G. E. Manigault, Carl McKinley and others, and with questionnaires circulated by Ensign Everett Hayden of the US Navy, Dutton was able to assemble unusually detailed facts regarding the timing, propagation and intensity of the earthquake.

One feature that makes the Charleston event particularly noteworthy is that this was the first strong earthquake to occur in the United States following the adoption of standardized clock time and time zones throughout the nation in 1883. There had of course been earlier historic earthquakes, notably in the central Mississippi Valley and California, but the Charleston event was the first substantial event whose effects could be timed over a large area. Hindcasting from available data suggests that the Charleston earthquake had a moment magnitude (M) of 7.3, the largest such event in the eastern United States in historic time (Johnston 1996).

Dutton's report on the Charleston earthquake is a remarkably fine document, considering the facilities and understanding of the time. Refining Robert Mallet's approach to the Neapolitan earthquake of 1857, Dutton developed a new isoseismal method for estimating the depth of the earthquake focus by identifying two critical points, the epicentrum and a point on a radius from the epicentrum at which shock intensities diminished most rapidly, which in turn allowed him to define an index circle of comparable points around the epicentrum. He identified two foci, one at a depth of about 20 km beneath Woodstock, the other at a depth of about 13 km beneath Rantowles (Fig. 2). Regarding the event's wider impact, he plotted questionnaire and interview data on the recently developed Rossi-Forel scale of earthquake intensity. The resulting isoseismal map showed that the event was felt as far north as New England and southern Ontario, and westward into the Mississippi valley (Fig. 3). He also improved understanding of the propagation velocity of earthquakes, which for the Charleston event he determined to be 5184 m s^{-1}. He was thus able to devote much attention to the nature and mechanisms of earthquake wave motion but, lacking information, he refrained from speculating on the cause of this event.

Although he never investigated another major earthquake, in retirement Dutton presented his knowledge in a book on *Earthquakes in the Light of the New Seismology*, published in 1904 (Dutton 1904). This book of 16 chapters defines two classes of earthquake (volcanic and tectonic), analyses seismic vibrations, propagation rates and wave forms, discusses relevant aspects of the

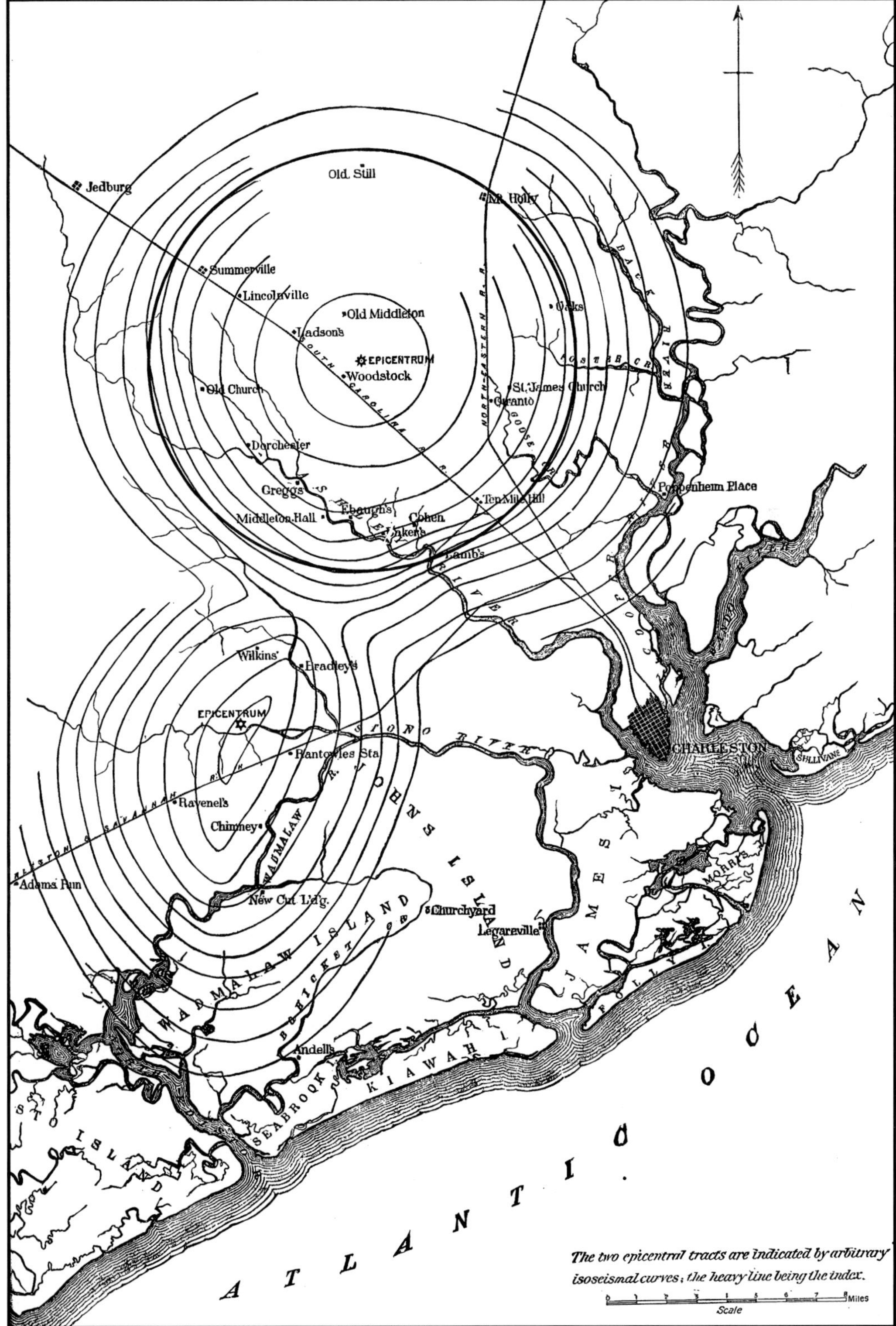

Fig. 2. The Charleston earthquake of 1886; isoseismals within the epicentral tracts (Dutton 1889*a*).

Fig. 3. The Charleston earthquake of 1886; isoseismals throughout eastern North America (Dutton 1889*a*).

Earth's crustal properties, and seeks to explain the distribution of earthquakes across the globe. He also returned, in his 1906 paper on volcanoes and radioactivity, to defend his isoseismal method for defining earthquake depth and thereby intensity, but by then the field had moved on.

Crustal behaviour, isostasy and denudation

Generations of Earth-science texts indicate that the term *isostasy* was first coined by Dutton in his 1889 address *On some of the greater problems in physical geology* but in fact he had introduced the term in 1882 in a review of *Physics of the Earth's Crust* by the Reverend Osmond Fisher (1817–1914) (Fisher 1881). Fisher, for many years rector of Harlton, Cambridgeshire, was a leading opponent of the then prevailing contractional hypothesis of mountain building, a view in which he found a ready ally in Dutton. In a footnote to the 1882 review, Dutton stated that:

> In an unpublished paper I have used the terms isostatic and isostacy to express that condition of the terrestrial surface which would follow from the flotation of the crust upon a liquid or highly plastic substratum;—different portions of the crust being of unequal density (Dutton 1882*c*).

The 'unpublished paper' to which Dutton referred was presumably to become his 1889 address to the Philosophical Society of Washington, not published until 1892, in which he formally proposed the term *isostasy* (having, on the advice of Greek scholars, changed the 'c' to an 's').

There were essentially three series of events that brought Dutton to this stage: first, the debate then raging among geologists concerning the Earth's crustal behaviour; second, Dutton's own theoretical musings derived from the literature and discussions with other scientists; and third, his field observations on the Colorado Plateau.

The debate concerning the Earth's crustal equilibrium already had a lengthy heritage in Dutton's time. Leonardo da Vinci (1452–1519) had speculated on it around 1500 (Delaney 1940).[18] Then, in a critical period from the mid-eighteenth to the mid-nineteenth centuries, a succession of European surveyors, mathematicians and astronomers became intimately involved in vexing questions regarding Earth's shape and the origins of mountains. Among these, Pierre Bouguer (1698–1758) recognized gravity anomalies in the Andes which R. G. Boscovich (1711–1787) attributed to thermal expansion of rocks at depth that generated surface uplift and subsurface voids.[19] More directly relevant to Dutton's ideas were those expressed by John Herschel (1792–1871) in the 1830s. Herschel, in a letter to Charles Lyell included in a treatise by Charles Babbage (1792–1871) (Herschel 1837), expressed his belief that sediment loading at ocean margins caused sea-floor subsidence and consequent uplift of continental margins. It is unfortunate that Herschel illustrated this hypothesis from the emergent shorelines of Scandinavia but in other respects it was an interesting notion. Shortly afterwards, the Survey of India, directed by George Everest (1790–1866), found discrepancies between the geodetic and astronomic positions of locations on the Ganges Plain, which he attributed to geodetic errors of closure. John Henry Pratt (1809–1871), then Archdeacon of Calcutta, disagreed. In a paper read to the Royal Society in 1854, Pratt suggested that the discrepancy was due to the gravitational effect of the Himalayas but he computed that such an effect was likely to be more than three times the observed value (Pratt 1855). George Biddell Airy (1801–1892), the Astronomer Royal, was not surprised at the discrepancy and reasoned from simple principles that Earth's outer layers comprised a thin crust overlying a denser fluid layer of 'lava.' (Airy 1855). He likened this relationship to timber rafts of comparable density but differing heights floating on water, the taller blocks supported by timber reaching greater depths beneath the water. In further papers, Pratt responded with his own theory by implying that elevated regions were underlain by low density rocks whereas depressed regions were underlain by high density rocks (Pratt 1864). The Pratt and Airy hypotheses were of course germane to the other contentious issue of the time, namely the origin of mountain ranges which, with some notable exceptions, most scientists of the day, including Pratt, attributed to the Earth's progressive cooling and periodic contraction through time.

All these observations and hypotheses were grist to Dutton's mill. Not yet 30 years old, he entered this debate with vigour with an address to the American Philosophical Society in Philadelphia on 7 April 1871 (Dutton 1873, pp. 70–72).[20] In this paper, published in 1873, Dutton explained regional crustal deformation in terms of changes in rock density following metamorphism. He hypothesized that uplift was caused by the expansion of rocks whose density had been lowered by water added during metamorphism. Subsidence was caused by rock contraction which reduced volume but increased density. Volcanism occurred as shallow crustal rocks became lighter when supersaturated with water and were displaced upwards through vents and fissures by subsiding heavier overlying rocks. He believed that volcanism would thus coincide with elevated regions, invoking as evidence the Rocky Mountains which had experienced scattered Miocene and Pliocene volcanism.

Dutton expanded his thesis in two subsequent papers to oppose the then prevalent contraction hypothesis of mountain building. In 1874, based on Fourier's cooling theory and Thomson's calculations, he argued that there had been little cooling of Earth's nucleus over time and therefore that contraction due to secular cooling could not explain the formation of mountain ranges, certainly not those of post-Cretaceous age (Dutton 1874).[21] In 1876, he observed that the mechanical force 'arising from contraction by cooling is many times less than the quantity required for mountain building' (Dutton 1876). He also felt that both the varying ages, dimensions and distributions of mountain ranges were inconsistent with uniform and progressive contraction of Earth's nucleus. However, he offered no theory to replace the contraction hypothesis.

By 1876, Dutton was aware of similar views held by Osmond Fisher and when Fisher's *Physics of the Earth's Crust* appeared in 1881, Dutton used his complimentary review as a forum to promote his own ideas. In his book, Fisher sought to destroy the contraction hypothesis which Dutton called 'nothing but a delusion and a snare, and that the quicker it is thrown aside and abandoned, the better it will be for geological science' (Dutton 1882*c*, p. 287). Instead, Fisher assumed a rigid crust overlying a liquid or plastic substratum on which the crust is in approximate hydrostatic equilibrium.

> The crust must be in a condition of approximate hydrostatical equilibrium, such that any considerable addition of load will cause any region to sink, or any considerable amount deduced off an area will cause it to rise. (Fisher 1881, p. 275)

Such flotation allows the crust to shift laterally. Lateral compression in turn produces narrow fold belts and crustal thickening, nine-tenths of which lie below the mean surface of the crust. Fisher was at a loss to explain lateral compression of this magnitude but inclined to invoke elastic vapours and fissures in the lower crust. This allowed Dutton in his review to explore an alternative explanation for mountain building, namely isostasy.

Neither Fisher nor Dutton were really comfortable with the reasoning behind isostasy, invoking both the Airy and Pratt hypotheses. Thus, in paraphrasing Fisher's views, Dutton states:

> The height to which the surface of the disturbed tract will rise above the mean level will be proportional to the difference of densities of crust and substratum respectively and also to the amount of local thickening of the crust by the compression. The ratio of the density of the crust to the density of substratum he [Fisher] takes at about 0.905, which is not far from that of ice to water. Hence when a disturbed tract of the crust is thickened by compression much the greater part of the thickening consists in additions to the under surface of the crust and only a small part in additions to the upper surface. The bulge downward is some nine times greater than the bulge upward. (Dutton 1882*c*, p. 286)

This invocation to the Airy hypothesis is followed by one to the Pratt hypothesis as Dutton prefaces his concept of 'isostacy' (sic) by summarizing Fisher's observation of flotation thus:

> The profiles of the earth therefore become, in their broader features, simply those which are due to flotation. The higher portions float higher because they are less dense; the depressed portions sink deeper because they are more dense. (Dutton 1882*c*, p. 289)

Meanwhile, Dutton's field studies of the American West had opened a Pandora's box of evidence to support his rejection of contraction theory and his belief in isostasy. His 1876 paper contains the first reference to the Colorado Plateau, which he had first seen during the 1875 field season, concluding from its elevated horizontal strata that the plateau was obviously not the result of contraction. But in 1885, in his report on Mount Taylor and the Zuni Plateau, he concluded that 'The mountains of the West have not been produced by horizontal compression, but by the action of some unknown forces beneath which have pushed them up' (Dutton 1885*a*, p. 198).

His field studies on the plateau, however, also exposed him to the vast amount of denudation that had been inflicted by the Colorado River and its tributaries (Fig. 4). He was particularly impressed by the evidence for massive uplift of the plateau's more-or-less horizontal rocks, and the vast thickness of Mesozoic and Cenozoic sedimentary rocks that must have been removed by erosion to expose the Palaeozoic strata in and around the canyons. Significantly, in his 1882 monograph on the *Tertiary History of the Grand Cañon District*, Dutton stated:

> If it be true that the Grand Cañon district received between the close of the Carboniferous and the close of the local Eocene 10,000 feet of deposits averaged over its entire surface, it follows that at the latter epoch the summit of the Carboniferous lay at least 10,000 feet below sea-level and was much more nearly horizontal than it is at present. And if such was its position and configuration, the great faults and displacements which traverse it must be of Tertiary age, and there must have been an enormous amount of uplifting, ranging from 12,000 to 18,000 feet, in various portions of the district. These are some of the consequences of the great denudation. (Dutton 1882*a*, p. 69)

He referred frequently to the great uplift (in the early Cenozoic) and the great denudation (in the late Cenozoic) of the Colorado Plateau (Fig. 5), mostly in the sense that uplift triggers denudation. Only in the above segment does he hint that denudation causes uplift.

Dutton's 1889 address to the Philosophical Society of Washington is a curious document. In identifying three of the greater problems of physical geology, he identifies (1) the cause of volcanic action (which he considers unsolved and therefore ignores); (2) the cause of regional elevation and subsidence (for which he finds no satisfactory explanation); and (3) 'the cause of the foldings, distortions, and fractures of the strata' (which he examines) (Dutton 1889*b*, p. 201).

Returning to earlier arguments concerning mountain building, he rejects the contraction hypothesis for want of an adequate mechanism and as unsuited to the field evidence. He then

Fig. 4. The Grand Canyon dissecting the Colorado Plateau, providing evidence of great uplift and great denudation captured by artist-geologist William Henry Holmes in a chromolithograph, 1882 (Dutton 1882*a*).

Fig. 5. The Grand Canyon and Colorado Plateau from Point Sublime, drawn by artist-geologist William Henry Holmes (Dutton 1882*a*).

formally proposes the term *isostasy* 'for this condition of equilibrium of figure, to which gravitation tends to reduce a planetary body' (Dutton 1889*b*, p. 203) and discusses the great sedimentary bodies of the Appalachian Mountains and the Colorado Plateau which were deposited near sea level and whose great thickness can only be explained by subsidence caused by sedimentation. He then observes that 'wherever broad mountain platforms occur and have been subjected to great erosion the loss of altitude by degradation is made good by a rise of the platform' (Dutton 1889*b*, p. 204). With reference to the western plateaus, he opines that 'the flanks of these platforms ... plainly declare to us that they have been slowly pushed upwards as fast as they were degraded by secular erosion' (Dutton 1889*b*, p. 204). In summary, he states that plateau uplift following erosion and subsidence following deposition 'are, in the main, results of gravitation restoring the isostasy which has been disturbed by denudation on the one hand and by sedimentation on the other' (Dutton 1889*b*, p. 204). He invokes Babbage, Herschel and Pratt in support of his thesis. He explains fold systems (plications) as a result of lateral viscous flow from beneath sediment-loaded littoral basins towards the unloaded continent, invoking as evidence the fold belts found inland from both the Atlantic and Pacific coasts of the United States. But he denies that isostasy is responsible for regional elevation and subsidence, which he attributes, as in 1871, to the expansion and contraction of magmas, respectively (Dutton 1889*b*, p. 211). It is difficult to avoid the conclusion that, with isostasy, Dutton identified and named an interesting concept but applied it to the wrong problem.

Dutton's legacy

Although most of Dutton's work must now be consigned to its historical context, interesting at the time but superseded by later advances, he leaves a legacy that is both literary and scientific.

Literary legacy

Dutton's word pictures of the American West attracted much complimentary interest during his lifetime and have since achieved legendary status. His facility with the English language, influenced early on by Macaulay's essays and Thackeray's novels, was now stimulated by field experiences in the west. His three monographs on the Colorado Plateau are remarkable as much for their colourful prose and aesthetic appreciation of nature as for the striking illustrations of Holmes and Moran. Working in virgin territory, he was not shy at introducing new names for spectacular features. Deeply influenced by eastern culture, he named many features of the Colorado Plateau after Hindu gods, for example Shiva Temple and Vishnu Temple.

Diller, in 1911, paid tribute to Dutton's simple euphonious style and considered him one of the best writers of popular geological science of his day (Diller 1911, p. 137). George Ferdinand Becker's obituary acknowledged Dutton's clarity and simplicity of expression but added, more critically, that 'Ease of expression, however, may be a dangerous gift and Dutton did not always reason soundly' (Becker 1912, p. 387). This may have been due to his methods. Diller reported that Dutton rarely made use of field notes, excepting for figures, read much and discussed ideas with colleagues, but did not prepare a written plan or preliminary draft of his written work. As Diller stated: '... when ready, he penned all his own manuscripts rapidly under the stimulus of an enthusiasm begotten by a consciousness of his comprehensive and complete knowledge of the subject' (Diller 1913, pp. 15–16). Understandably, Dutton's imagination sometimes outdistanced the facts.

Dutton's writings and Holmes' drawings did much to stimulate popular interest in the American West. Wallace Stegner (1909–1993), regarded as the dean of western writers during the mid-twentieth century, later observed with respect to the Grand Canyon that 'though it is Powell's monument to which the tourists walk after dinner to watch the sunset from the South Rim, it is with Dutton's eyes, as often as not, that they see' (Stegner 1953, pp. 173–174).[22]

Scientific legacy

Dutton was among the most prominent American scientists of his time, a reserved person of great integrity and charm, who collaborated easily with others and yet emerged as an intellectual force in his own right. To the modern eye, his scientific legacy is undoubtedly shrouded by colourful prose, rambling digressions, and little regard for economy of style, but his search for explanation still shines through. His geology was largely self-taught and he readily acknowledged the debt he owed to his collaboration with Powell and Gilbert, thus:

> In daily intercourse with Powell and Gilbert, and with the bond of affection and mutual confidence which made the study in a peculiar sense a labor of love, this geological wonderland [the Colorado Plateau] was the never-ending theme of discussion; all the observations and experiences were common stock, and ideas were interchanged, amplified and developed by mutual criticism and suggestion. The extent of my indebtedness to them I do not

know. Neither do they. I only know that it is enormous, and if a full liquidation were demanded it would bring me to bankruptcy (Dutton 1885*a*, p. 113).

Dutton was also generous in praise of earlier scientists. Thus, although later scholars often identify Dutton with isostasy, which he presented in 1889 'in modified form, in a new dress, and in greater detail,' he clearly credited the theory to Babbage and Herschel (Dutton 1889*b*, p. 206). He also gracefully extolled the virtues of his field assistants, such as Diller and Hayden, and the many topographic surveyors on the Colorado Plateau.

Among his individual contributions to volcanic geology, he correctly identified the spatial and temporal relationships between Tertiary volcanic features and the dissection of the rising Colorado Plateau. He saw that Hawaiian volcanoes ranged from active through dormant to extinct as he moved northwestward through the islands. He recognized relationships between the Tertiary and Quaternary volcanics of the Cascade Range, earlier country rocks and later dissection. But time and knowledge have passed on and we now know so much more from subsequent developments in plate tectonics relating to Colorado Plateau uplift, Hawaiian volcanism, and the Cascadia subduction zone. For example, his belief that Hawaiian lavas emanated from shallow reservoirs, perhaps 1–4 km deep, was correct only in defining their immediate source. Subsequent studies of earthquakes beneath Kilauea, for example, suggest that the magma originates some 50–80 km beneath the volcano, in the upper mantle some 30–60 km beneath the oceanic crust, and moves upward through a roughly cylindrical conduit to a shallow magma reservoir only 2–5 km deep (Macdonald *et al.* 1983). Nor did he anticipate the hotspot explanation for sequential Hawaiian volcanism first proposed by Wilson (1963). What Dutton did was to improve the fundamental body of knowledge regarding these features, garnished with sometimes logical, sometimes speculative, observations regarding their origin and significance. In short, like many of his contemporaries, he helped to prepare the way for later scientists by proposing hypotheses that could not be tested at the time.

In similar vein, Dutton's legacy to seismology may be viewed as a way station in a rapidly changing field. Certainly, his 1889 report on the Charleston earthquake of 1886 was in its time a remarkable compilation of facts and theory. This M7.3 event remains enigmatic but a 600-km-long, NNE-trending, east coast fault system beneath the Atlantic Coastal Plain has recently been identified as its probable source (Marple & Talwani 2000). Although this structure affects Cretaceous and Cenozoic sediments, it is not visible at the surface. Thus, with the limited techniques then available, Dutton could neither have inferred such a structure nor its coincidence with the margins of buried Mesozoic rift basins and pre-Mesozoic exotic terranes beneath eastern North America. He did, however, provide a wealth of data as a basis for later work. However, his 1904 book, which some contemporaries viewed as introducing the 'new seismology' to American scientific audiences, soon became dated. In his obituary in 1912, George Becker acknowledged that the book showed Dutton at his best but that it was 'far from exhaustive and cannot be regarded as an adequate presentment of the subject at the time of its publication.' (Becker 1912, p. 388). But to his credit, Dutton continued to probe the unknown, stating in the book's preface that the volcanic and tectonic causes of earthquakes were 'apparently quite distinct, though possibly they may have interrelations not yet recognized.' Shortly before Dutton's death, Diller lauded him as 'the first seismologist of his country' (Diller 1911, p. 137) but in his later obituary toned this down to 'one of the first seismologists of his country' (Diller 1913, p. 10).

With respect to crustal behaviour, Dutton's ideas began to be superseded during his own lifetime and certainly later by developments in plate tectonics. His early contributions during the 1870s were quite bizarre, showing an imperfect understanding of physical processes and an inadequate knowledge of field situations. He erred, for example, in his understanding of magma and sources of volcanism. Yet his arguments against the contraction hypothesis of mountain building were sound enough, reflecting a rising tide of opposition echoed by many of his contemporaries. His recognition that Earth's great fold belts also involved areas of massive sedimentation and subsequent crustal shortening, perhaps reflecting the ideas of his early mentor, James Hall, also heralded later understanding of orogenic systems. Nevertheless, his writings on these topics appear today as archaic legacies of a bygone age.

His contribution to isostasy was different. He introduced the term, even if he did not fully understand its implications. In doing so, he reflected notions regarding Earth's crustal equilibrium that had floated back and forth for some time, certainly since Babbage and Herschel pondered the subject in the 1830s. But they did not give the process a name; Dutton did and in doing so provided a focus for later work in geodesy, geophysics and geology.

The term isostasy was readily accepted during Dutton's lifetime and within a few years a flurry of scientific activity began to emerge under this rubric. Geodesists such as John Hayford (1868–1925), Chief of the Division of Geodesy of the US Coast and Geodetic Survey, recognized the significance of isostatic equilibrium and adopted the Pratt model, which simplified computation, to correct the triangulation surveys of the United States (Hayford 1909). His successor, William Bowie (1872–1940), also invoked the Pratt hypothesis and published an entire book on isostasy in 1927 (Bowie 1927). Indeed, the Pratt hypothesis often came to be known as the Pratt-Hayford model.

Meanwhile in Europe, John Joly (1857–1933), geophysicist of Trinity College, Dublin, followed Dutton in emphasizing the significance of both radioactivity and isostasy to crustal processes in books on *Radioactivity and Geology* (Joly 1909) and *The Surface History of the Earth* (Joly 1925). The latter is essentially a view of Earth's surface from below, 'based directly upon two great recent advances in our knowledge of the earth's crust' (Joly 1925, p. 5). In Finland, geodesist W. A. Heiskanen (1895–1971) of Helsinki's Superior Technical School, favoured the Airy model for his isostatic computations, later referred to as the Airy-Heiskanen model (Heiskanen 1931).

In contrast, many geologists of the early twentieth century were reluctant to enthuse over isostatic models that they considered static and based on mathematical massaging of crustal assumptions that could not be proven at the time. A notable exception was Andrew Lawson (1861–1952), Professor of Geology at the University of California, Berkeley, who published on *The Geological Implications of the Doctrine of Isostasy* in 1924 (Lawson 1924). Lawson described the various ways in which shifts of mass may occur at and beneath the surface, several of which were testable. For these, he invoked both closed-systems, in which isostatic adjustment is achieved by local subcrustal flow from loaded to unloaded regions, and quasi-closed systems in which shifts of load cannot be restored by local compensation and which therefore involve deformation of the geoid.

Lawson's arguments, among others, made little impression on the developing field of geomorphology, dominated as it was by William Morris Davis of Harvard University. It was Dutton's descriptions of the 'great denudation' of the Colorado Plateau, rather than any isostatic implications, that captured Davis' attention. Prolonged denudation was the necessary precursor to peneplanation, the end stage in Davis' cycle of erosion. Furthermore, in an ideal cycle, initiated by rapid uplift and followed by lengthy structural quiescence, isostatic adjustments to denudation were an unwelcome complication. Localized isostatic responses to ice loading, as discussed by Reginald Daly (1871–1957) (Daly 1934), could be tolerated but large-scale responses to denudation were better ignored.

The demise of Davisian geomorphology during the middle decades of the twentieth century, and subsequent advances in plate tectonics, led to a reawakening of interest in isostasy as a concept basic to the explanation of Earth's surface relief. Somewhat belatedly, Dutton's concept of isostasy has once again emerged at the centre of debate regarding denudation and crustal behaviour (Molnar & England 1990). Wherever these debates focus, whether on the American West or the Himalayas or the Andes (Gregory-Wodzicki 2000), they confront a problem fundamental to geology, geodesy and geophysics, namely the extent to which Earth's present surface relief reflects a potential balance between subsurface forces generating uplift, subsidence and mass transfers at depth, and the climate-induced processes responsible for denudation, sedimentation and mass transfers of waste across the surface. In recent years, a combination of geodetic and geophysical methods has done much to restore the significance of isostasy to geomorphology.

Conclusion

Dutton's contribution to geology was always going to be superseded, as befits scientific progress, but his quest for explanation, for seeking to understand the bigger picture from a myriad of local details, is a message for all scientists. As he stated:

> I am fond of viewing the facts observed in the field in their relation to broader and more general facts, and of marshaling them into their proper places (Dutton 1885*a*, p. 198).

He was gifted, perceptive, articulate, highly literate and much travelled, willing to take on challenges for which, as a largely self-taught geologist, he was not always prepared. But he learned quickly. He may not always have been correct but he deserves recognition for addressing these larger issues. His ideas relating denudation to isostasy were remarkably prescient and, although largely ignored by geomorphology for almost a hundred years, have re-emerged at the cutting edge of contemporary scientific enquiry.

I warmly thank K. B. Bork and R. H. Dott, Jr. for their discerning and helpful reviews of this essay.

Notes

[1]Dutton family genealogy, 1177–1923. <freepages. genealogy.rootsweb.com >

[2]Yale College, 1906. *Biographical Record of the Class of Sixty*.

[3]Joseph Silas Diller (1850–1928) was Dutton's principal assistant in the Division of Volcanic Geology, and became its head on Dutton's transfer to the Irrigation Survey in 1888. Diller was responsible for much of the initial geological mapping of the Cascade Range, among other things.

[4]Dutton, C. E. 1871. The causes of regional elevations and subsidences. Address to the American Philosophical Society, 7 April 1871 (published as Dutton 1873).

[5]Dutton, C. E. 1873. A criticism of the contractional hypothesis. Address to the Philosophical Society of Washington, Winter 1873 (published as Dutton 1874).

[6]King 1880, pp. 3–7. In this report, King presents a brief historic background to the various western surveys, including the three surveys that were discontinued following establishment of the USGS.

[7]King, 1880, footnote 8, pp. 6–7. King, describing the ambiguities in the congressional mandate to the USGS, explains that 'I have entirely abandoned that plan, and have divided the region west of the 101st meridian into four geological districts.' These districts were the divisions of the Rocky Mountains, the Colorado, the Great Basin, and the Pacific. Acknowledging that the region has been for many years the field of exploration of Powell, he further states that 'the Division of the Colorado is intended only as a temporary one until this work, already far advanced, can be brought to completion.'

[8]US Statutes at Large, 22 (7 August 1882), 329.

[9]When Diller succeeded Dutton as its head on 1 July 1888, the Division of Volcanic Geology was renamed the Cascade Division.

[10]Letter to Sir Archibald Geikie from Capt. Clarence Edward Dutton, 6 June 1891. Edinburgh University Library Special Collections.

[11]Reprinted in *Journal of Geology*, **14**, 259–268.

[12]In his sympathetic obituary, Diller notes that Dutton smoked vigorously at his work for many years but desisted in later life.

[13]Letter to Archibald Geikie from Capt. Clarence Edward Dutton, 20 January 1881. Edinburgh University Library Special Collections.

[14]Letter to J. D. Dana, Washington, D.C., 8 February 1883. Reprinted as Dutton (1883).

[15]Letter to J. D. Dana, Washington, D.C., 8 February 1883. Reprinted as Dutton (1883, p. 226).

[16]Letter to Archibald Geikie from Capt. Clarence Edward Dutton, 21 July 1884. Edinburgh University Library Special Collections.

[17]The radioactive properties of uranium had been discovered by Henri Becquerel in 1896 and of thorium, polonium and radium by Marie Curie in 1897 and subsequent years. Radioactive decay rates were soon seen by geologists as a means for redefining the age of Earth and its rocks, and for providing a source of heat within the Earth.

[18]This is a brief note by Delaney concerning page 344 of volume 1 of Leonardo's notebooks, translated and edited by Edward MacCurdy.

[19]An excellent summary of early ideas on this theme is provided by Watts (2001, pp. 1–17).

[20]This paper was initially read before the American Philosophical Society in Philadelphia on 7 April 1871.

[21]This paper was presented to the Philosophical Society of Washington during the winter of 1872–1873.

[22]Stegner headed the literary department at Stanford University from 1946 to 1971, and earlier, while a student at the University of Iowa in 1932, had written a thesis on Clarence Dutton's writings.

References

AIRY, G. B. 1855. On the computation of the effect of the attraction of mountain-masses, as disturbing the apparent astronomical latitude of stations of geodetic surveys. *Philosophical Transactions of the Royal Society*, **145**, 101–104.

BECKER, G. F. 1912. Major C.E. Dutton. *American Journal of Science*, 4th Series, **183**, 387–388.

BOWIE, W. 1927. *Isostasy.* E.P. Dutton & Company, New York.

DALY, R. A. 1934. *The Changing World of the Ice Age.* Yale University Press, New Haven.

DANA, J. D. 1849. Geology. *In*: WILKES, C. (ed.) *United States Exploring Expedition, during the years 1838, 1839, 1840, 1842 and 1843, under the command of Charles Wilkes*, **10**. C. Sherman, Philadelphia, with atlas; Putnam, New York.

DELANEY, J. P. 1940. Leonardo da Vinci on isostasy. *Science*, **91**, 546–547.

DILLER, J. S. 1911. Major Clarence Edward Dutton. *Bulletin of the Seismological Society of America*, **1**, 137–142.

DILLER, J. S. 1913. Memoir of Clarence Edward Dutton. *Geological Society of America Bulletin*, **24**, 10–18.

DUTTON, C. E. 1873. The causes of regional elevations and subsidences. *American Philosophical Society Proceedings*, **12**, 70–72.

DUTTON, C. E. 1874. A criticism upon the contractional hypothesis. *American Journal of Science*, 3rd Series, **8** (whole no. 108), No. 44, 113–123.

DUTTON, C. E. 1876. Critical observations on theories of the Earth's physical evolution. *Geological Magazine*, new series, Decade II, **3**, 1876.

DUTTON, C. E. 1877. Report on the lithologic character of the Henry Mountains intrusives. *In*: GILBERT, G. K. (ed.) *Report on the Geology of the Henry Mountains.* US Geographical and Geological Survey of the Rocky Mountain Region, 61–65.

DUTTON, C. E. 1880*a*. Report (from Mount Trumbull, Ariz. September 3, 1880). *In*: KING, C. (ed.) *First Annual Report of the United States Geological Survey*, 29–31.

DUTTON, C. E. 1880*b*. *Report on the Geology of the High Plateaus of Utah.* US Geographical and Geological Survey of the Rocky Mountain Region, **32**.

DUTTON, C. E. 1882*a*. *The Tertiary History of the Grand Cañon District.* US Geological Survey, Monograph 2.

DUTTON, C. E. 1882*b*. Report of Capt. C.E. Dutton, Washington, October 3, 1881. *In*: POWELL, J. W. (ed.) *Second Annual Report of the United States Geological Survey, 1880–81*. Government Printing Office, Washington, 5–10.

DUTTON, C. E. 1882*c*. Review of *Physics of the Earth's Crust*, by the Rev. Osmond Fisher. *American Journal of Science*, 3rd Series, **23**, 283–290.

DUTTON, C. E. 1883. Recent exploration of the volcanic phenomena of the Hawaiian Islands, *Journal of Geology*, Article XXVI, 3rd series, **25**, 219–226.

DUTTON, C. E. 1884. Hawaiian Volcanoes. *In*: POWELL, J. W. (ed.) *Fourth Annual Report of the United States Geological Survey, 1882–83*. Government Printing Office, Washington, 75–219.

DUTTON, C. E. 1885*a*. Mount Taylor and the Zuñi Plateau. *In*: POWELL, J. W. (ed.) *Sixth Annual Report of the United States Geological Survey, 1884–85*. Government Printing Office, Washington, 111–198.

DUTTON, C. E. 1885*b*. Report of Capt. C.E. Dutton, (Mount Shasta, Cal. June 30, 1885). *In*: POWELL, J. W. (ed.) *Sixth Annual Report of the United States Geological Survey, 1884–85*. Government Printing Office, Washington, 59–62.

DUTTON, C. E. 1888. Report of Capt. C.E. Dutton (Ashland, Oreg. July 1, 1886). *In*: POWELL, J. W. (ed.) *Seventh Annual Report of the United States Geological Survey, 1885–86*. Government Printing Office, Washington, 97–103.

DUTTON, C. E. 1889*a*. The Charleston Earthquake of August 31, 1886. *In*: POWELL, J. W. (ed.) *Ninth Annual Report of the United States Geological Survey, 1887–88*. Government Printing Office, Washington, 203–528.

DUTTON, C. E. 1889*b*. On some of the greater problems of physical geology. *Bulletin of the Philosophical Society of Washington*, **11**, 51–64 (printed 1892).

DUTTON, C. E. 1890. Report of Capt. C.E. Dutton, *Tenth Annual Report of the United States Geological Survey*, Part II—Irrigation. Government Printing Office, Washington, 78–108.

DUTTON, C. E. 1891–92. Report on the Nicaragua Canal. United States Senate, 52nd Congress, 1st Session, Miscellaneous Documents 97, Serial 2904, Volume 2.

DUTTON, C. E. 1904. *Earthquakes in the Light of the New Seismology*. G.P. Putnam's Sons, New York.

DUTTON, C. E. 1906. Volcanoes and Radioactivity. *Englewood Times, New Jersey; reprinted in Journal of Geology*, **14**, 259–268.

FISHER, O. 1881. *Physics of the Earth's Crust*. Macmillan and Co., London.

GREGORY-WOSZICKI, K. M. 2000. Uplift history of the Central and Northern Andes. *Geological Society of America Bulletin*, **112**, 1091–1105.

HAYFORD, J. F. 1909. *The Figure of the Earth and Isostasy from Measurements in the United States*. Government Printing Office, Washington, D.C.

HEISKANEN, W. A. 1931. Isostatic tables for the reduction of gravimetric observations calculated on the basis of Airy's hypothesis. *Bulletin Géodésique*, **30**, 110–129.

HERSCHEL, J. 1837. Letter to C. Lyell. *In*: BABBAGE, C. (ed.) *The Ninth Bridgewater Treatise: A Fragment*. John Murray, London, 202–217.

JOHNSTON, A. C. 1996. Seismic moment assessment of earthquakes in stable continental regions—III. New Madrid 1811–1812, Charleston 1886, and Lisbon, 1755. *Geophysical Journal International*, **126**, 314–344.

JOLY, J. 1909. *Radioactivity and Geology: An Account of Radioactive Energy on Terrestrial History*. Constable & Company, London.

JOLY, J. 1925. *The Surface History of the Earth*. Clarendon Press, Oxford.

KING, C. 1880. *First Annual Report of the United States Geological Survey*. Government Printing Office, Washington, 3–7.

LAWSON, A. C. 1924. The geological implications of the doctrine of isostasy. *Bulletin of the National Research Council*, **8**, 46, 1–22.

LONGWELL, C. R. 1958. Clarence Edward Dutton, May 15, 1841–January 4, 1912. *Biographical Memoirs* 32, National Academy of Sciences, Columbia University Press, New York.

MACDONALD, G., ABBOTT, A. T. & PETERSON, F. L. 1983. *Volcanoes in the Sea: The Geology of Hawaii*, University of Hawaii Press, Honolulu.

MARPLE, R. T. & TALWANI, P. 2000. Evidence for a buried fault system in the Coastal Plain of the Carolinas and Virginia—Implications for neotectonics in the southeastern United States. *Geological Society of America Bulletin*, **112**, 200–220.

MOLNAR, P. & ENGLAND, P. 1990. Late Cenozoic uplift of mountain ranges and global climate change: Chicken or egg. *Nature*, **346**, 29–34.

POWELL, J. W. 1882. *Second Annual Report of the United States Geological Survey, 1880–81*. Washington, Government Printing Office.

POWELL, J. W. 1890. *Tenth Annual Report of the United States Geological Survey, 1888–89*; Part I—Geology. Government Printing Office, Washington.

POWELL, J. W. 1891. *Twelfth Annual Report of the United States Geological Survey, 1889–90*. Part I—Geology. Government Printing Office, Washington.

PRATT, J. H. 1855. On the attraction of the Himalaya mountains, and of elevated regions beyond them, on the plumb line in India. *Philosophical Transactions of the Royal Society*, **145**, 53–100.

PRATT, J. H. 1864. Speculation on the constitution on the Earth's crust. *Proceedings of the Royal Society*, **13**, 253–276.

PYNE, S. J. 1999. Dutton, Clarence Edward. *American National Biography*. Volume 7, Oxford University Press, New York.

STEGNER, W. 1953. *Beyond the Hundredth Meridian: John Wesley Powell and the second opening of the West*. Houghton Mifflin Company, Boston, 173–174.

WATTS, A. B. 2001. *Isostasy and Flexure of the Lithosphere*. Cambridge University Press.

WILSON, J. G. & FISKE, J. (eds) 1887–1889. *Appleton's Cyclopedia of American Biography*, 6 Volumes. D. Appleton and Company, New York.

WILSON, J. T. 1963. A possible origin of the Hawaiian Islands. *Canadian Journal of Physics*, **41**, 863–870.

Two Tyrrells cross the Barren Lands of Canada, 1893

D. A. E. SPALDING

1105 Ogden Road, RR#1, Pender Island, British Columbia, Canada, V0N 2M1
(e-mail: david@davidspalding.com)

Abstract: In May 1893 Joseph and James Tyrrell, sons of an Irish immigrant to Canada, left Edmonton in Canada's Northwest Territories with a group of voyageurs, partly aided by an aboriginal map and First Nations guides. By January 1894 the team had reached Winnipeg, Manitoba, completing the first expedition through the last large unexplored area of the mainland of Canada. The travellers suffered from many dangers, and the expedition was more than three months overdue when its members arrived at York Factory.

Some specimens were lost, but records and photographs acquired on the expedition form an important record of northern exploration. Joe Tyrrell identified the Keewatin centre of glaciation, and the post-glacial extension of the Hudson Bay basin is now known as the Tyrrell Sea. There was criticism at the apparent lack of economic benefit, but Tyrrell located iron ore and a possible copper-bearing area, and his work led indirectly to a productive nickel mine and the present working of diamonds. Tyrrell also documented the huge caribou herds, and the lives of the Inuit that depended on them.

J. B. Tyrrell's Barren Lands expedition (1893)

In 1893, a party from the Geological Survey of Canada crossed the Barren Lands west of Hudson Bay. In late May, the two Tyrrell brothers left the railhead[1] at Edmonton and headed north to Lake Athabasca. From there they travelled north and east through dwindling forest, then east to Chesterfield Inlet across the unknown Barren Lands. They journeyed south along the shore of Hudson Bay to Churchill and York Factory, and then to Winnipeg, arriving on 1 January 1894. The party travelled 3200 miles, by wagon, canoe, on foot, and by snowshoe and dog team (Fig. 1).

This journey was the first scientific expedition through this area, the last substantially unknown territory in mainland Canada. It was a turning point in the interesting life of J. B. Tyrrell (Fig. 2) and is exceptionally well documented. Tyrrell's unpublished photographs, field notes, correspondence and other writings have survived in Ottawa[2] and Toronto.[3] His reports were published by the Geological Survey (Tyrrell, J. B. 1896*a*, 1897) and other academic papers by various societies. A classic book on the Barren Lands, written by his brother J. W. Tyrrell, entitled *Across the Sub-Arctics of Canada* (Tyrrell 1898) is largely drawn from the expedition, and J. B. Tyrrell's then fiancée Edith Carey also left an account of personal and public reactions to news of the expedition in her autobiography (Tyrrell 1938). Tyrrell's life has been the subject of three biographical studies (Loudon 1930; Inglis 1978; Martyn 1993), a Ph.D. thesis (Eagan 1972*a*) and is included in a study of Canadian geologists written for children (Shaw 1958). It has also been discussed in histories of surveying in Canada (Thomson 1972), of the Geological Survey (Zaslow 1975), and numerous articles (Eagan 1972*b*). Preparation of a new biography has been announced.

In this examination, I have brought together information from published sources to tell the basic story of the expedition. I have used modern spelling and nomenclature for places, names and groups of people, unless in a direct quotation.[4]

The geographical background

The area of western Canada bounded on the north and east by the Arctic Ocean and Hudson Bay, on the south by the prairies and the west by the Slave and Mackenzie rivers had been but little explored at the time of the expedition. The area was sparsely populated by Dene in the south, and Inuit in the north and east, with a boundary approximately along the tree line (Crowe 1974). Early expeditions generally used native guides or maps drawn by aboriginals.

Hudson Bay (discovered in 1610 by Henry Hudson) provided early access for exploration by sea. The first fur trading post in the region was established in 1688 by des Groseilliers.

Early travellers sent out by the fur traders were Henry Kelsey (1667?–1724) who travelled west to the plains (1690–92), and Anthony Henday (fl 1750–1762) who in 1754–55 was the first European to see the Rocky Mountains. In 1770,

From: WYSE JACKSON, P. N. (ed.) *Four Centuries of Geological Travel: The Search for Knowledge on Foot, Bicycle, Sledge and Camel.* Geological Society, London, Special Publications, **287**, 287–296.
DOI: 10.1144/SP287.22 0305-8719/07/$15.00

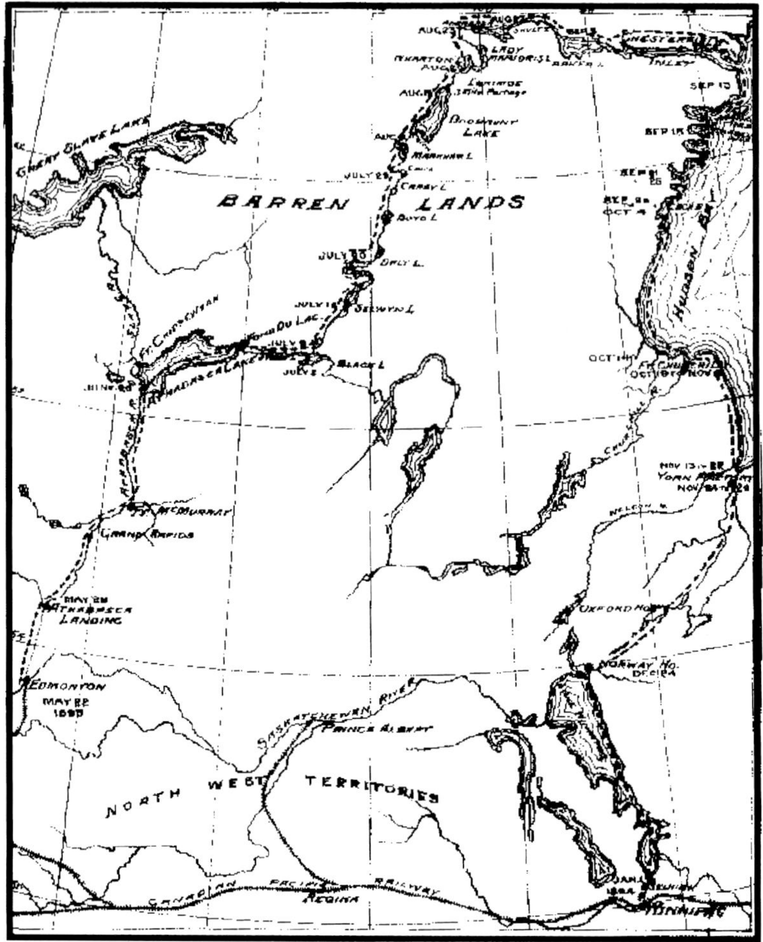

Fig. 1. Route of the Tyrrell expedition of 1893 through northern Canada (from Tyrrell 1898, p. iv).

Samuel Hearne (1745–1792) travelled overland to Great Slave Lake and the Coppermine River (Warkentin 1964; McGoogan 2003). The western rivers had been traversed by Alexander Mackenzie (1764–1820), who in 1789 discovered the river and mountains that now bear his name, and in 1793 he became the first to cross the continent to the Pacific.

The overland naval expeditions led by John Franklin (1786–1847) were supported by the Hudson's Bay Company, and in 1819–22 and 1825–27 they went across the plains and NW to the Arctic Ocean. In 1845 Franklin took two ships into the Arctic islands and disappeared, stimulating a series of expeditions that clarified the geography of the Arctic Islands. Tyrrell himself had undertaken geological exploration of areas to the south of the region in 1892.

The area was first part of Rupert's Land (administered by the Hudson's Bay Company, and transferred to Canada in 1869). Later, it became part of the Northwest Territories, which at the time of Tyrrell's expedition had been divided into the districts of Athabasca and Keewatin. Today, it is largely within the new territory of Nunavut.

The Geological Survey of Canada

The Geological Survey of Canada was founded in 1842, under its first director William Edmond Logan (1798–1875). Canadian born, Logan carried out important work in the coal measures of England and Wales before being appointed to Canada. The Survey's second director Alfred Selwyn (1824–1902) had been trained with

Fig. 2. J. Burr Tyrrell leaving Fort Churchill (from Tyrrell 1898, p. 9).

the Geological Survey in England, then commanded that of Victoria, Australia. He arrived in Canada in 1869 and switched the headquarters of the Survey from Montreal to Ottawa 1881. One reason for this was that Selwyn, initially with only five geologists, was expected to expand the Survey's operations into the new western regions of Canada. The initial priority was to explore the prairies for coal supplies and look for passes through the Rocky Mountains for the projected railway which was to link British Columbia with eastern Canada. Tyrrell was engaged in this work, first under George Mercer Dawson (1849–1901) and then independently. In 1887–1891 he was moved to Saskatchewan and then Manitoba, following which he carried out his two northern explorations.

George Mercer Dawson became the third director of the Survey in 1895 and was responsible for publication of Tyrrell's reports on his work in the north.

The Tyrrell brothers and their canoe men

Both the leader of the 1893 expedition and his chief assistant were sons of William Tyrrell (1816–1904), who was born at Grange Castle, County Kildare, in Ireland and who became a pioneer of Weston, Ontario.

Joseph Burr Tyrrell (1858–1957) was near-sighted, and suffered from scarlet fever in his childhood which left him slightly deaf. These afflictions may have been partly responsible for his reputation which has been described as 'heavy-handed, imperious and ambitious'. He attended Upper Canada College, then took a B.A. at the University of Toronto, becoming more interested in the sciences. In response to his father's wishes, he was articled in a law office, but a lung infection (possibly tuberculosis) allowed him to break free, and with the help of his father's acquaintance with Prime Minister, John A. MacDonald, he joined the Geological Survey in 1881. Here he helped to arrange the geological specimens that had recently been moved from Montreal to Ottawa. This 'practical postgraduate course in Geology and Mineralogy',[5] and later work with Dawson in the field, provided Tyrrell's initial geological training. While at the survey he took an M.A. at the University of Toronto and a B.Sc. at Victoria University, both in 1889.

Tyrrell soon undertook independent surveys in Alberta and east across the prairies, and by 1892 was working east of Lake Athabasca. Here a Chipewyan guide, Ethingo Campbell (Martyn 1993, pp. 20–21), drew him a map of lakes and rivers pointing north to 'Tobon Lake' across the barrens (Martyn 1993, p. 21).

James Williams Tyrrell (1863–1945) was Joe Tyrrell's younger brother. He had been trained as a civil engineer at Toronto's School of Practical Science.[6] He assisted survey geologist Andrew Cowper Lawson (1861–1952) in the field, was a surveyor and weather observer for two expeditions to Hudson Bay, and was then appointed surveyor to the city of Hamilton, Ontario, in 1888. His role on the 1893 expedition was 'assistant, topographer and Eskimo interpreter' (Tyrrell 1896*a*, p. 38A).

Others in the party included three Iroquois brothers from the Caughnawaga Reserve in Quebec: Pierre, Louis and Michel French. Pierre was 'a veteran canoe man, being as much at home in a boiling rapid as in the calmest water' (Tyrrell 1898, p. 9). Louis 'had won some distinction ... through having accompanied Lord Wolseley as a voyageur on his Egyptian campaigns' (Tyrrell 1898, p. 9). Michel, the youngest, was destined to suffer badly from frostbite in the latter part of the expedition.

John Flett, a Chipewyan-speaking part Loucheux 'half breed' from Prince Albert, Saskatchewan, 'was highly recommended ... as an expert portageur of great experience in northern travel, and also an Eskimo linguist' (Tyrrell 1898, p. 9).

James Corrigal (part Cree) and Francois Maurice (part Chipewyan), met the party in Fort McMurray. They 'were also western half-breeds, trained in the use of the pack-straps as well as the paddle, and . . . a pair of fine strong fellows' (Tyrrell 1898, p. 10).

Tyrrell had already learned to work with native guides, and make use of their knowledge of the country. On this expedition local helpers were also employed from time to time, as guides and to provide help with transportation.

Down the Athabasca

Tyrrell had nearly a decade of field experience when he initiated the 1893 expedition. He was no stranger to hardship: he had narrowly escaped drowning, chopped his foot with an axe, and nearly died of typhoid on previous expeditions. Now he drafted 'a carefully worded proposal for an expedition to explore the 'Barren Lands' during the next season. He wrote persuasively of the possibility of valuable minerals being discovered in the area, and argued that the cost of the expedition would be low, since the party would be small and could procure much of its food supply from the caribou herds roaming the region' (Martyn 1993, p. 23). Selwyn gave his approval early in April, and Tyrrell lost no time in making his preparations. He ordered two Peterborough[7] canoes, hired six men, and arranged for letters of credit. Tyrrell had used a camera in the field since 1886 (Martyn 1993, p. 15); now he bought a Hawkeye, a new roll film camera.[8]

J. B. Tyrrell managed to fit in a brief visit to the World Exposition in Chicago on his way west. The main party travelled west to Calgary and north to Edmonton by train. There they obtained supplies, went north on 27 May by wagons to Athabasca Landing, and loaded 22 000 pounds of supplies on the Hudson's Bay Company steamer *Athabasca*. The men tested two canoes by heading downstream, picking up the last two members of the party and the third canoe in Fort McMurray. They reached Fort Chipewyan several days ahead of the steamer, and when it arrived repacked their supplies for canoe travel.

Guided by a 'full-blooded Chippewyan Indian' (Tyrrell 1898, p. 53) named Moberly, the party headed to the east end of Lake Athabasca, where their guide did his best to dishearten the members of the party by warning of the dangers ahead and then refused to go any further. The original party portaged to Black Lake, the farthest point J. B. Tyrrell had reached on previous expeditions, and headed north following Ethingo Campbell's map.

Across the Barrens

Eventually the expedition reached the height of land,[9] naming and crossing Selwyn Lake,[10] and reaching Daly Lake[11] on 18 July. Here 'the forest became thin and intermittent, poplar being seen here for the last time on our journey northward' (Tyrrell 1896*a*, p. 40A). 'After considerable search and some loss of time, we found the outlet of the lake' (Tyrrell 1896*a*, p. 40A), and headed north down the Telzon River 'which either rushed down heavy rapids or widened into lakes thickly studded with islands. In these lakes it was necessary to follow the crooked winding shore in order to find the outlet, while it is essential to land at the head of a rapid in order to decide on the proper channel down which to run the canoes, or to determine where to make the portage' (Tyrrell 1896*a*, p. 40A) (a similar crossing of rapids on the Athabasca River is illustrated here as Fig. 3). Travelling in the forest they had been plagued by mosquitoes, now as they neared the barrens their tormentors were replaced by black fly.

By 29 July the party was running out of food, but they ran into 'an immense herd of caribou' (Tyrrell 1896*a*, p. 40A), shot 60 and spent three days butchering and drying meat. The caribou were so tame that Tyrrell was able to take photographs of the herd from a distance of only a few yards. A nearby lake was named Carey Lake.[12] On 2 August they continued down the river. On the 5th they made camp in 'a small grove of stunted black spruce bushes' (Tyrrell 1896*a*, p. 40A) in a dense fog. 'The next morning, as I looked from the top of a hill behind our camp, I saw a great lake lying before us of which the surface appeared to be almost covered with a solid sheet of ice' (Tyrrell 1896*a*, p. 40A). This was Dubaunt Lake, representing the NW angle of their expedition. They found enough water by the shore to use the canoes, taking two weeks to travel 117 miles along the shore, except for six days when the weather was too 'cold, wet and stormy' (Tyrrell 1896*a*, p. 41A) for them to travel. So far their journey had been over red and grey Laurentian gneisses, with an outlier of fossiliferous Cambro-Silurian limestone, but now 'we found red and grey sandstones and coarse conglomerates, cut and altered by dykes and masses of dark green trap and bright red quartz-porphyry' (Tyrrell 1896*a*, p. 41A), which Tyrrell noted resembled the rocks of the Lake Superior copper district.

Below the lake they bypassed a 'deep narrow gorge' (Tyrrell 1896*a*, p. 41A) over a portage two

Fig. 3. Shooting the Mountain Rapid, Athabasca River (from Tyrrell 1898, p. 41).

and a half miles to another lake. On 19 August they met with a 'solitary Eskimo deer skin tent' (Tyrrell 1896*a*, p. 41A), whose owners were frightened because the survey party was coming from the land of their traditional enemies. But J.W. Tyrrell's command of their language (Inuktitut) and a gift of tobacco soon put the locals at their ease, and the party secured a sketch map of the further course of the river.

The party continued downstream, against head winds and with many portages, reaching the west end of Baker Lake on 2 September. They had traversed 810 miles of unknown country before reaching a point that had been previously reached from Hudson Bay. The next day they shot the last deer of the season. By 12 September they had reached the mouth of Chesterfield Inlet, two months and 22 days from Fort Chipewyan.

Hudson Bay to Winnipeg

The party was now in previously explored territory, but it was still essentially uninhabited; they were 250 miles from Churchill, the first point at which they could obtain supplies, and autumn was well advanced. It took a further three and a half months for all the party to get back to civilization. There is not room to detail their experiences, but they suffered many storms, heavy snowfalls, severe frostbite, and near starvation (Fig. 4). One food shortage was solved by Louis French killing a polar bear. They boiled the meat over a fire of moss, enjoying the unaccustomed food supply until one of the party became ill by eating the bear's poisonous liver.

Reluctantly they cached everything unnecessary for survival, including all the geological specimens that had been collected, and one of the canoes, and headed south. The shore of the bay was very low and flat, and the tide went out several miles, so they could only launch safely or disembark at high tide once every twelve hours. As ice began to build up they were sometimes held away from the shore by uncrossable pack ice.

They reached Churchill on 19 October, recovered their strength and headed south again on 6 November. They had hired James Westascott and his dog team, who carried Michel, whose feet were by now severely frozen. After a ten-day delay before they could cross the running ice of the Nelson River, they reached York Factory on 24 November.

'I need hardly say that the telegraph office was soon found, and messages despatched to anxious friends, who having heard nothing of us for some months, had begun to entertain grave fears for our safety' (Tyrrell 1898, p. 250). 'Expedition very successful. All well.' said one telegram (Quoted in Inglis 1978, p. 51).

Fig. 4. The last of our provisions—a gloomy outlook (from Tyrrell 1898, p. 198).

Publicity and politics

Joe's fiancée, Mary Edith Carey (1876–1945), was perhaps the most anxious to hear news. The expedition was three months overdue, and there was widespread press speculation that the party had perished. Now the family telegrams reached the press, reports of the expedition's success hit the headlines in Canada, the US, and England, and Joe 'returned to Ottawa ... to find himself the toast of North America, with his name spread across many pages of newsprint' (Eagan 1972*b*, p. 18). An admiring reference to the survey's sled parties even appeared in Jack London's novel *The Call of the Wild* (London 1997). One of the Governor General's staff, Dr Gibson, had become engaged to Edith's sister Ellie (Tyrrell 1938, p. 68), and Tyrrell was invited to Government House by the new incumbent Lord Aberdeen (in office 1893–1898). J.B. Tyrrell married Edith Carey in February 1894, and the Tyrrells continued to visit Government House. 'Always a bit of a snob', says one of his biographers, 'Tyrrell bathed in the warmth of the vice-regal smile' (Inglis 1978, p. 109).

The Mining Review on the other hand, was less supportive, and didn't think Tyrrell's expedition was at all worthwhile. Because his travels did not directly provide economic benefit, 'this modern Gulliver', it commented was of 'utter insignificance'.[13] The trip had cost $7000,[14] and had been so criticized in the house that the Prime Minister had had to defend it (Inglis 1978, p. 111). This was only part of the critical lobby against the Survey, for Selwyn was expected to produce miracles from a very small budget. Internally too, many of Tyrrell's colleagues, feeling that their own work was also demanding and dangerous, resented the amount of press coverage he received.

So when Tyrrell planned another Barren Lands trip in 1894, Selwyn was naturally lukewarm, and felt unable to offer any more funds until political influence was brought to bear. Moreover Robert Munro-Ferguson, an aide-de-camp to the Governor-General, was not only anxious to go along, but willing to more than pay his own way, and his chief also contributed. The expedition took a parallel route to the first one, and was carried through even more successfully. By the time Tyrrell returned in January 1895, Selwyn had been retired, and replaced as director by George Dawson.

'Mr Tyrrell did not return to Ottawa in time to furnish a report on his work for publication with others of the same year', says Dawson's formal and rather chilly introduction to Tyrrell's preliminary report of the first expedition (Dawson 1896). Tyrrell had learned much of his practical geology in the field with Dawson, and felt that he 'will be strict, but he will be rational, and if I cannot get along with him it will be largely my own fault'.[15] But before long, he began to feel that Dawson was deliberately holding him back. Married life proved expensive, and his salary had not increased. He was unsuccessful in an attempt to find academic positions,[16] but after a Survey trip to the new Klondike goldfields in the Yukon Territory, he saw an opportunity. He resigned from the Survey in 1898 and began work as a mining engineer in the Klondike.

The Keewatin glaciation and the Tyrrell Sea

Tyrrell's 1893 expedition was in the nature of a geological traverse, covering unknown country to observe and collect evidence of the underlying geology. However, it is remarkable that much geological surveying could be done. Much of the territory traversed was unknown, so a topographical survey of the journey had to be made (based on a boat log, pace surveys, and observations of latitude and longitude). The party had to travel through difficult and dangerous country (covered by snow from early September) and hunt for much of their food. Their scientific duties included the collection of plants, while Tyrrell, the only geologist, carried a camera and took photographs. The abandonment of the substantial geological collections towards the end of the journey did not help in the documentation of their observations.[17]

Tyrrell had studied areas to the immediate south, and some information on the geology of the shores of Hudson's Bay was already available. Tyrrell's summary report (Tyrrell 1896*a*) included numerous comments on geological observations, with preliminary conclusions.

Most of the area visited exposes rocks of the Precambrian shield, which had already been studied by the Survey in eastern Canada; indeed even in his preliminary report Tyrrell made comparisons with the Lake Superior region. Tyrrell reported Huronian rocks at Lake Athabasca (Tyrrell 1896*a*, p. 39A), with Laurentian gneiss underlying the first half of the journey (Tyrrell 1896*a*, p. 41A). 'In Dubaunt Lake the character of the rock suddenly changed', wrote Tyrrell (1896*a*, p. 41A), reporting sandstones and conglomerates cut by dykes, dark green trap and bright red porphyry, similar to the upper copper-bearing Keweenawan Series of Lake Superior.

Few younger rocks were found, though there was a fossiliferous Cambro-Silurian limestone near Dubaunt Lake. Travelling eastwards, they returned to the gneiss, and recorded 'light green Huronian schists' (Tyrrell 1896*a*, p. 43A) near Baker Lake.

Tyrrell's observations on erratics, striae and eskers led to important insights into the glacial history of mainland Canada. He also found raised beaches as much as 500 feet above present sea level, a powerful indication of the weight of ice that had accumulated at the peak of glaciation. He realized that the upland area he was exploring was a height of land on which glaciers had formed and from which they had moved out to south and west. He postulated a centre of glaciation on a low summit north of Dubaunt Lake. He named the ice sheet Keewatin, from the Cree word for the north wind, and mapped directions of ice movement (Tyrrell 1896*b*).

In this work Tyrrell was a leader of the new glacial thinking in Canada. His chief, G. M. Dawson, influenced by his geologist father John William Dawson (1820–1899), had come to terms with glacial explanations in the Rockies by 1890, but still held to a floating ice theory for the prairies (Zaslow 1975, p. 164). Tyrrell was influenced by Warren Upham (1850–1934) of the US Geological Survey, who had worked on this problem with the Canadian Survey, and who was an associate editor with Tyrrell of the *American Geologist*. 'With this [glacial study], and his report on the 1893–1894 expeditions, Tyrrell reached the pinnacle of accomplishment on the Geological Survey ...' says Eagan. 'He had made a first rank contribution to his science and deserves to be recognized for that' (Eagan 1972*b*, p. 21). The inland saltwater sea that developed in the Hudson Bay basin during and after deglaciation is now known as the Tyrrell Sea (Lee 1968, p. 514).

The party also gathered important anthropological and biological data. 'Our knowledge of the Caribou Eskimo begins with the observation of Tyrrell along the Kazan and Dubaunt Rivers', notes anthropologist W. N. Irving (1968, p. 49). The plant collection did reach its destination, and a 21-page list of plants determined by Survey naturalist John Macoun (1831–1920) is appended to J. W. Tyrrell's book, documenting a number of northern and western range extensions. (Some of these specimens survive in the Canadian Museum for Nature). The region west of the Tyrrells' route is now protected by the 56 000 sq km Thelon Game Sanctuary, particularly for protection of musk ox and the Beverly caribou herd (Pelly 1996).

Tyrrell's pioneer nature photographs from the north[18] are commonly reproduced (e.g. Hewitt 1921; Guggisberg 1977), some being early cited as 'the best photographs hitherto taken of this caribou' (Hewitt 1921, p. 63). His photographs of the Inuit were the first to be taken of the Caribou group, and Tyrrell would no doubt be astonished if he could be told that some of these photographs are now on display in Nunavut's Baker Lake Heritage Centre.[19]

Nickel, radium, uranium and diamonds

If, as *The Mining Review* urged, there should be economic benefits arising from survey research, it is appropriate to see what mineral potential was uncovered in the region directly or indirectly as a result of Tyrrell's expedition.

Tyrrell noted 'an extensive deposit of iron ore, consisting of hematite and limonite' on the north shore of Lake Athabasca (Tyrrell 1896*a*, p. 39A). The Dubaunt Lake beds resembled the 'copper-bearing series, of Lake Superior' (Tyrrell 1896*a*, p. 41A). Even if these were of commercial quality, distance from markets precluded any economic benefit in Tyrrell's time. The only successfully workable minerals in the north had to have a high value/weight ratio, and in the subsequent century such resources were indeed discovered in the Barren Lands, by a new generation of geologists exploring with the aid of float planes and helicopters.

A nickel ore body was discovered in 1928 on the shores of Hudson Bay close to Tyrrell's route, and the Rankin Inlet Nickel Mine was operated there from 1957 to 1962 (Whitmore & Liberty 1968, p. 553). Radium had not even been discovered when Tyrrell crossed the Barrens, but from 1933 it was mined at Port Radium in the North West Territories. In 1942, uranium also became of importance with the Manhattan project, and much of the atomic age has been fuelled by uranium discovered north of Lake Athabasca, where Uranium City now stands close to Tyrrell's route.

Perhaps the highest economic value/weight ratio comes from diamonds, and an unexpected and long-delayed result of Tyrrell's work has been the recent discovery of diamonds in the far north (Krajick 2001). In the same year as Tyrrell's expedition, geologist William Herbert Hobbs (1864–1953) of the University of Wisconsin read of a five-year-old boy's discovery of a 3.83-carat diamond in a corn field (Krajick 2001, p. 115). Hobbs documented other finds in the region, going as far back as 1863. He summarized his data in print in 1899, 'demonstrating that all the diamonds lay in a great arc of glacial debris 600 miles long and 200 miles wide—the outer edge of the last great ice sheet that presumably had come down from Canada' (Krajick 2001, p. 116). By this time, many people were looking, and had realized there may be a source in the far north. 'As soon as Tyrrell came home, his maps were torn apart for evidence of where that material came from', says Kevin Krajick.[20]

But the serious search for diamonds required the development of air transportation, and techniques for recognition of indicator minerals for kimberlite pipes. By the early 1990s, the search for diamonds had led 260 corporations to stake 100 000 square miles of the Barren Lands (Krajick 2001, p. 332). The Lac de Gras diamond mine, begun in 1997 (just over a century after Tyrrell's expedition), is now situated in the area north of Athabasca Lake, not far from Tyrrell's route.

Gold and geography

In his later life James Williams Tyrrell surveyed country from Labrador to Great Slave Lake, and in 1900 he led a Dominion Lands Survey through Keewatin. Like his brother, he was later a mine promoter in Red Lake, northern Ontario, and became president of Tyrrell Red Lake Mines. He died in Bartonville, Ontario.

His older brother Joseph Burr Tyrrell became president of the Kirkland Lake Gold Mining Co. in Ontario in 1924, and later a millionaire. But he never forgot the exciting days of travel in unknown lands. In retirement, he edited for publication the journals of important western Canadian explorers, including David Thompson (1770–1857) and his only real predecessor in the Barrens, Samuel Hearne (1745–1792). He was elected a Fellow of the Royal Society of Canada in 1910, served as president of the Champlain Society 1927–1932, and received honorary degrees from Toronto and Queen's universities. He died in his home near Toronto at the age of 98.

Today J. B. Tyrrell is often associated with dinosaurs, for his early discovery of dinosaurs on the Red Deer River has led to his name being given to Canada's major museum of palaeontology, the Royal Tyrrell Museum of Paleontology in Drumheller, Alberta (Spalding 1993, 1999). However, Tyrrell never published research work on dinosaurs, and scientifically his primary claim to fame is still his research on the glacial history of the north.

Tyrrell was among the last to become a field geologist for the Geological Survey of Canada without a degree in the subject. With his two Barren Land expeditions 'one of the last of the great sweep surveys of the Geological Survey of Canada was completed' (Eagan 1972*b*, p. 18). And indeed, with his energy and ambition, his ability to lead parties of diverse talents, and his broad knowledge of science, Tyrrell is remembered scientifically as a great geological explorer. And so he regarded himself, writing in his seventies to his brother James: 'Such days are among my happiest memories, and if I could throw off responsibility and go back to the north where nature had provided a wonderful garden for miles around without any trouble on my part I will be glad to return to it' (Eagan 1972*b*, p. 21).

I am indebted first to J. B. Tyrrell and his relatives for assiduously recording his fieldwork and life and preserving the evidence in print, manuscript, photographs and specimens.

Materials and information from the survey period are preserved by the Geological Survey of Canada, the Canadian Museum for Nature (where information was provided by Mike Shchepanek), and the National Archives. The Thomas Fisher Library of the University of Toronto house Tyrrell's archives and K. Martyn and her colleagues have strenuously preserved and indexed these and published materials in print and on the web featuring Tyrrell.[21]

I am indebted to the Greater Victoria library and the interlibrary loan services of the Dawson and Pender Island libraries for access to obscure published materials. A. Spalding and K. Oke provided assistance by scanning illustrations, and my late friend W. Sarjeant encouraged and assisted my delvings into geological history for nearly half a century.

Notes

[1]'Railhead' is the Canadian term for the current end of the railway line. The line had been built from Calgary in 1891, and later continued to Athabasca Landing.

[2]Geological Survey, National Archives.

[3]Thomas Fisher Rare Book Library, University of Toronto; *The Halcyon*.

[4]The area explored was inhabited by First Nations, formerly known as 'Indians', more specifically Dene, particularly Dogrib and Chipewyan, and Inuit, formerly often known as 'Eskimo', Copper and Caribou. J. W. Tyrrell uses 'Enuit' in his 1898 book. Individuals with parentage of both white and First Nations were then commonly referred to as 'half-breeds'; now usually Métis.

[5]J. B. Tyrrell quoted in Martyn (1993, p. 8).

[6]Different sources indicate James Williams Tyrrell qualified in 1883 or 1889; the first seems most likely.

[7]During the last three decades of the nineteenth century durable canoes from the Peterborough region of Ontario were exported throughout the world (*Canadian Encyclopedia*).

[8]This camera is preserved at the Thomas Fisher Library (Martyn 1993, p. 6).

[9]This term for a watershed has been in use in Canada since 1732.

[10]After the Survey's director.

[11]named for the Minister of the Interior, the Hon. T. M. Daly (1827–1885).

[12]after Tyrrell's prospective father-in-law, Rev. Dr G. W. Carey.

[13]Inglis (1978, p. 52). No more specific source given.

[14]Zaslow (1975, p. 203) says over $6500, about 10% of Survey's annual operations budget.

[15]Letter to Edith, quoted in Inglis (1978, p. 129).

[16]Inquiries and applications to Toronto, McGill and Manitoba in Canada, and Cornell in the USA are referred to by Inglis (1978, p. 141).

[17]Tyrrell was unable to locate the site on his second expedition, but the cache was found by Inuit hunters in the following year, and eventually reached Ottawa in November 1895 (Martyn 1993, p. 39).

[18]Many are housed in the Tyrrell collection at the Thomas Fisher Rare Book Library. University of Toronto.

[19]*The Halcyon*, Issue #26, p. 3 Electronic newsletter of the Friends of the Thomas Fisher Library. http://www.library.utotoronto.ca/development/news/halcyon/nov2000/article1.htm

[20]Krajick, a present day chronicler of the diamond hunt, has presented a history of the diamond discoveries in his Barren Lands.

[21]Martyn has documented her 39 years of work on the Tyrrell Papers to her retirement in 2000 at the web address: www/library/utoronto.ca/fisher. Digital files on Tyrrell may be found at http://digital.library.utoronto.ca/Tyrrell

References

CROWE, K. J. 1974. *A History of the Original Peoples of Northern Canada*. McGill-Queens University Press, Montreal.

DAWSON, G. M. 1896. *Geological Survey of Canada, Annual Report for 1894*.

EAGAN, W. E. 1972*a*. *Dr Joseph Burr Tyrrell, 1858–1957*. Ph.D. thesis, University of Western Ontario.

EAGAN, W. E. 1972*b*. Joseph Burr Tyrrell (1858–1957)—Geologist. *Proceedings of the Geological Association of Canada*, **24**, 17–22. Reprinted *In*: MACQUEEN, R. W. (ed.) *Proud Heritage: People and Progress in Early Canadian Geoscience*. Geological Association of Canada, 119–123.

GUGGISBERG, C. A. W. 1977. *Early Wildlife Photographers*. Taplinger Publishing Company, New York.

HEWITT, G. C. 1921. *The Conservation of the Wildlife of Canada*. Scribners, New York.

INGLIS, A. 1978. *Northern Vagabond. The Life and Career of J.B. Tyrrell—The Man who Conquered the Canadian North*. McClelland & Stewart, Toronto.

IRVING, W. N. 1968. Prehistory—The Barren Grounds. *In*: BEALS, C. S. (ed.) *Science, History and Hudson Bay*. Ottawa, Queen's Printer, **1**, 26–54.

KRAJICK, K. 2001. *Barren Lands. An epic search for diamonds in the North American Arctic*. Henry Holt & Co., New York.

LEE, H. A. 1968. Quaternary Geology. *In*: BEALS, C. S. (ed.) *Science, History and Hudson Bay*. Ottawa, Queen's Printer, **2**, 503–543.

LONDON, J. 1997. *The Call of the Wild*. Annotated and Illustrated. University of Oklahoma Press, Norman. [First published 1903].

LOUDON, W. J. 1930. *A Canadian Geologist*. The Macmillan Company of Canada Ltd., Toronto.

MCGOOGAN, K. 2003. *Ancient Mariner. The Amazing Adventures of Samuel Hearne, The Sailor Who Walked to the Arctic Ocean*. Harper Flamingo Canada, Toronto.

MARTYN, K. 1993. *J.B. Tyrrell: Explorer and Adventurer. The Geological Survey Years 1881–1898*. University of Toronto Library, Toronto.

PELLY, D. F. 1996. Barrens Oasis. *Canadian Geographic*, **116**, 60–63.

SHAW, M. M. 1958. *Canadian Portraits. Geologists and Prospectors. Tyrrell, Camsell, Cross, LaBine*. Clarke Irwin, Toronto.

SPALDING, D. 1993. *Dinosaur Hunters, 150 Years of Extraordinary Discoveries*. Key Porter Books, Toronto and Prima Books.

SPALDING, D. 1999. *Into the Dinosaurs' Graveyard. Canadian Digs and Discoveries*. Doubleday Canada, Toronto.

THOMSON, D. W. 1972. *Men and Meridians. The History of Surveying and Mapping in Canada*. Ottawa: Information Canada, **2**, 1867–1917. [Originally published 1967].

TYRRELL, J. B. 1896*a*. North-west Territories and Keewatin. *Geological Survey of Canada, Annual Report 1894*, **7**, (new series) 38A–46A.

TYRRELL, J. B. 1896*b*. The Genesis of Lake Agassiz. *Journal of Geology*, **6**, 811–815.

TYRRELL, J. B. 1897. Report on the Dubawnt, Kazan and Ferguson Rivers and the NorthWest Coast of Hudson Bay. And on the two overland routes from Hudson Bay to Lake Winnipeg. *Geological Survey of Canada, Annual Report 1896*, 31–34.

TYRRELL, J. W. 1898. *Across the Subarctics of Canada*. T. Fisher Unwin, London. [Republished 1973].

TYRRELL, M. E. 1938. *I Was There; a Book of Reminiscences*. Ryerson Press, Toronto.

WARKENTIN, J. 1964. *The Western Interior of Canada*. McClelland & Stewart, Toronto.

WHITMORE, D. R. E. & LIBERTY, B. A. 1968. Bedrock Geology and Mineral Deposits. *In*: BEALS, C. S. (ed.) *Science, History and Hudson Bay*. Ottawa, Queen's Printer, **2**, 543–557.

ZASLOW, M. 1975. *Reading the Rocks. The Story of the Geological Survey of Canada 1842–1972*. Macmillan of Canada, Toronto.

Investigating the colonies: native geological travellers in the Portuguese Empire in the late eighteenth and early nineteenth centuries

S. F. DE M. FIGUEIRÔA, C. P. DA SILVA & E. M. PATACA

Institute of Geosciences, University of Campinas, P.O. Box 6152, 13083-970 Campinas-SP, Brazil (e-mail: figueiroa@ige.unicamp.br)

Abstract: The historiography of sciences in Brazil frequently analysed the role and the meaning of foreign—predominantly European—travellers in the country, stressing their discoveries and contributions to science. But these texts systematically omitted the 'native' travellers, those who were commissioned by the Portuguese government to survey the territories of the Portuguese empire, especially at the end of the eighteenth century. These expeditions, however, deserve careful attention: they were not sporadic nor dispersed initiatives, but rather co-ordinated, essential parts of a large program of scientific reform and economical reconstruction of the kingdom.

In the second half of the eighteenth century, the Portuguese elite drew upon Enlightenment ideals to reform sectors of the policy considered fundamental to economic growth. The Lisbon Royal Academy of Science (Academia Real de Ciências de Lisboa) was founded in 1779, and from the beginning turned into the focus of a far-reaching policy to promote the more rational exploitation of nature within Portugal and its colonies. The movement reflected a broad awareness of science, conceived as systematic, practical, and useful inquiry.

The academy developed plans to improve agriculture and mining through applied science, and the colonial territories were methodically investigated. The expeditions launched to attend these goals—sometimes entitled '*Viagens Philosophicas*' (Philosophical Travels)—collected thousands of samples (geological, mineralogical, botanical, and zoological), mapped out thousands of linear and square metres of rivers and lands, and produced dozens of *Memoirs*, that filled the Museum of Ajuda and the shelves of the Royal Academy of Science.

Many authors have highlighted the fact that the eighteenth century was a period of major advancement in natural history, and Portugal played its part in this process. At the centre of the actions that marked the Portuguese Empire's adherence to the sciences of the time were the investigations into the natural history of the Kingdom's territories both at home and overseas. The Portuguese reformist policies from the second half of the eighteenth century onwards were born in a context where the Portuguese were trying to overcome problems and recover its international importance and prestige. One of the main goals was to aid economic recovery, that was expected to come about through a rational exploration of the resources provided by nature—this was one of the main reasons for the emphasis laid on natural history.

The Portuguese Government of that time orchestrated a group of actions that mobilized dozens of naturalists—many of whom were actually born in the Portuguese colonies in America, Africa or Asia, who were typical representatives of the Enlightenment—with a view to studying the developments in agricultural and mining sectors. The promotion of investigative and exploratory expeditions within the Kingdom was an essential component of these actions. The overseas expeditions or so-called 'Philosophical Voyages' or 'philosophical research journeys' were organized and carried out by the Portuguese, with material, financial, and political support from the Crown. They became more frequent towards the end of the eighteenth century and the *Memoirs* that were published by the Lisbon Academy of Sciences mark the beginning of this investigation, as they are considered the basis of the process of institutionalization of natural science study in Brazil.

This paper looks at the geological aspects of four of these 'intra-Empire' expeditions—namely, to the Grand-Pará state (Brazil), to Goa and Mozambique, to Angola, and to Cabo Verde Islands, all set off in 1783, in particular those to the Brazilian territory. The wider context in which these studies were framed, and their articulation within the Portuguese project of reforms that were informed by science, as well as their implementation and execution, where the 'Voyage Instructions' played a fundamental role, are important. Our study of the naturalists who travelled in Portuguese America during the transition from the eighteenth to the nineteenth centuries, and an examination of their output, contributes towards a

From: WYSE JACKSON, P. N. (ed.) *Four Centuries of Geological Travel: The Search for Knowledge on Foot, Bicycle, Sledge and Camel.* Geological Society, London, Special Publications, **287**, 297–310.
DOI: 10.1144/SP287.23 0305-8719/07/$15.00

wider understanding of the Luso-Brazilian scientific culture in that period.

Portugal and 'the lights'

In Portugal, as well as in Spain, during the second half of the eighteenth century, various changes that led to a series of reforms, took place as a result of profound changes in the social and economic thinking of the time. The Portuguese elite was inspired by ideals derived from the Enlightenment, and wished to improve those areas considered essential for economic growth, that is, in mining and agriculture (Dias 1968). The reform programme began during the reign of Dom José I (1750–1777) and was conducted by Prime Minister Sebastião José de Carvalho e Melo, the Marquis of Pombal (1699–1782).

For example, one of the trademarks of the reformist action was the restructuring of the University of Coimbra. Models with different origins and inspirations were selected and adapted according to eclectic guidelines that marked the whole process of change. The Marquis of Pombal established various ancient laws that he considered to be good and added other ones, created based on models from universities in England, France and Germany.[1]

The effort undertaken by the State with a view to making an inventory of its wealth using science as a tool is characteristic of this period. It was no accident that the pragmatism that marked the eighteenth century in Europe was so radically expressed in the Portuguese reforms that it reached the point of being considered (alongside eclecticism) its most outstanding characteristic. Also adding to this scenario are the intrinsic characteristics of modern natural sciences, carrying a Baconian profile that took utility and human well-being for granted. Otherwise, why would inventories of the natural world be considered as keys towards a new apprehension of the world? Museums, scientific academies and botanical gardens constituted attempts to administer the explosion of empirical material produced by a wider dissemination of ancient texts, by a greater mobility of people and objects, by exploratory voyages and more systematic forms of communication, exchange and appropriation. By redefining the European view of the world as a measure of 'civilization', albeit a relative one, all these factors also contributed towards the production of new attitudes regarding nature and natural history (Findlen 1996).

Incomprehension or non-incorporation of that theoretical understanding has sometimes led to the whole Portuguese movement of adhesion to modern sciences to be reduced to a simple case of political-strategic 'utilitarianism' or 'pragmatism'. Such cases of restricted understanding are frequently used to suggest the view that Portugal was scientifically backward and was largely unable to participate in wider development of science in Europe at that time.

During the first half of the eighteenth century, during the reign of Dom João V, Portugal had begun to embrace modern sciences and territorial, natural history and human surveys were already being encouraged. In Portugal institutions for education and research, such as the School of Saint Antão in Lisbon where astronomical studies were carried out, existed prior to this period. However, a number of further institutions were established at this time, including the Portuguese Academy of History (1720). These continued to advanced science in Portugal, and so too did the dispute between Spain and Portugal over ownership of a vast territory beyond the Tordesilhas line in Portuguese America, an area that was explored from the seventeenth century onwards—exploration that provided much scientific information. The Portuguese Crown sponsored some expeditions to Brazil with a view to carrying out territorial surveys and producing maps that would support Portugal's international agreements with various nations that bordered Portuguese America i.e. Spain, France, the Netherlands and England.[1]

The Madrid Treaty between Portugal and Spain was signed in 1750 and in the following years two Limit Demarcation Commissions set out to South America. The North Commission was led by the Governor and Captain General of the State of Grand-Pará and Maranhão, Francisco Xavier de Mendonça Furtado (1701–1778) (brother of the Marquis of Pombal). The South Commission was commanded by the governor of Rio de Janeiro, Gomes Freire de Andrada. Various scientific expeditions were carried out in all the border territory, and accounts and maps were produced and various natural products were collected and shipped off to Lisbon. In 1765, during the Pombalin consulate, the Madrid Treaty was annulled. Portugal and Spain again took up negotiations and in 1777 they signed the Saint Ildephonse Treaty. The great geographical and naturalistic surveys carried out during the First Limit Demarcation Commission (1750) served as basis for the agreements between the two powers. Likewise many of the members of the technical commissions that took part in the Limit Demarcation Commission in 1750 also participated in the expeditions that took place during this second phase.

Natural history studies were in the process of being developed in Portugal during the eighteenth century and great progress was made after the reforms were set up in Coimbra; the foundation of the Royal Academy of Sciences of Lisbon was especially important in fostering this development. Portugal's link with its colonial domains in Africa, Asia and America, regions of immense plant, mineral and animal wealth was one of the main factors that drove this advance. Another factor that contributed to the development of natural history was the curiosity that rose in face of the unknown that needed to be inventoried, observed, described and catalogued. This environment involved Portugal 'in the same taste for observing, studying and collecting independently of commercializing.' (Carvalho 1965).

Domingos Vandelli (1735–1816), Félix de Avelar Brotero (d. 1828) and Jose Mayne (stand out amongst those that are important scholars of natural history in Portugal (Calafate 1994). José Correia da Serra (1751–1823), one of the founding members of the Academy, was an important naturalist who was recognized by the international community of that age for his work and publications (Carvalho 1981).

The economic problems that Portugal was facing at that period were worsened by the decline of mining activities in Brazil that concentrated on gold and diamonds. Those close to that crisis had different opinions regarding the issue. Some thought that the mining activity was prejudicial to Portugal. Amongst this group was Bishop José Joaquim de Azeredo Coutinho, founder of the Olinda Seminary who in 1804 published an important tract *Speech on the current state of mines in Brazil*. On the other side was a significant group of men of science dedicated to the fields of mining and mineralogy. They believed that the decline in gold production and also the decline of the mining sector in general was as a result of problems of a technical and scientific nature. With the introduction of reforms guided by reasoning and 'scientific method', they believed that mining output in Brazil could again reach its peak. In general, they argued for the introduction of modern techniques in mineral extraction and for better training for miners and mine administrators. This point of view was shared by the Portuguese Government, whose officials were in permanent contact with men of science connected to various Portuguese institutions such as the University of Coimbra and the Royal Academy of Sciences of Lisbon (Figueirôa 1997).

In the end, this perspective prevailed and resulted in a great governmental effort towards the recovery of this economic sector. This is well documented in numerous official documents found in various archives. There are letters, notifications, regal orders sent by the Crown to different parts of the colony, requiring information about a certain 'mine' recently discovered, appointing naturalists to carry out philosophical observations on the spot, ordering studies to verify the best way to make good use of the resources being discovered; booklists and lists of instruction manuals for mining.[2] Domingos Vandelli[3] wrote of Portugal's great effort towards the investigation and systematization of the natural products from the Portuguese colonies, during the reign of D. José I, in a volume entitled *Natural History of the Colonies*. Indeed Vandelli, a man with a practical spirit and in perfect accordance with the new direction of the University of Coimbra after the Pombalin reforms, took up a project that had been conceived earlier by Diogo de Mendonça Corte Real in 1731 (Carvalho 1987, p. 21).

Vandelli's concern was that the naturalists who had graduated from that institution would be used by the Government, and that they would participate on scientific voyages. For this reason Munteal Filho (1997) highlighted the importance of Vandelli and the group of intellectuals that revolved around him, as they tried to establish a professional group to guide the Crown in its scientific exploration of its Kingdom and possessions. In an attempt to overcome the crisis in the mining sector, the reformist State tried to carry out Vandelli's suggestions by appointing men who were well versed in mining and mineralogy, who were able to produce an inventory of the mineral resources that were found and could also identify how best to exploit them.

The Royal Academy of Sciences of Lisbon

The Lisbon Academy of Sciences was founded on 24 December 1779 with the support of and under the protection of the Portuguese State, which was represented by the Queen, Dona Maria I.

An important study by John Gascoigne (1998) on the subject of science and universities between the sixteenth and the eighteenth centuries puts forward essential thoughts on the relationship between science and universities during that period. Universities are frequently presented as having been obstacles to scientific development because of their strong connections with the Church or are seen as 'medieval relics standing in the way of the rising tide of Enlightenment', in the words of Diderot. According to Gascoigne, philosophers of that period criticized universities, implying a different mode of dealing with knowledge between themselves and university professors. In his article, Gascoigne discards the simplistic explanations that place academies and salons on one side and universities on the other. He observes that in places such as Scotland, Germany and the Iberian countries the determinant role in the dissemination of the Enlightenment lay with the universities. In Portugal, the reform undertaken by the University of Coimbra, that preceded the foundation of the Royal Academy of Sciences of Lisbon, was the element that initiated the dissemination of the Enlightenment within the Kingdom.

The Royal Academy of Sciences of Lisbon was an important centre of production and dissemination of enlightened knowledge. Therefore, in Portugal, the separation between universities and academies in the process of disseminating the Enlightenment did not occur. From its creation in 1779 onwards, the Academy took on the role of institutional coordinator of the reformist actions that were being carried out in various research areas in the Portuguese kingdom. The Academy was the origin of the efforts and guidelines for the production of inventories of the wealth of Portugal and its colonial empire. According to Munteal Filho (1997, p. 89) the articulation between the Portuguese State's policy of promotion and the activities of a speculative and pragmatic approach was one of the objectives of the academic group that conducted the projects to restructure the exploration of overseas domains. Likewise, the delimitation of a useful scientific culture capable of producing a quick return on investments for the Portuguese Crown was necessary.

Thus, the Academy's scientific production was articulated with the State's reformist policies that had as objective a greater knowledge for a better exploration, in the case of the colonial domains, and the establishment of self-sufficiency, in the metropolitan case. Mainly at the end of the eighteenth century, under the guidance of the Academy of Sciences, a truly 'feverish' scientific production was taking place in the Brazilian colony. Munteal Filho (1997) has shown that it was a true process of 'rediscovering the New World', methodically carried out by the Academy.

The 'rediscovering' of Brazil was made possible through the work of naturalists, mainly Brazilians who studied in the reformed University of Coimbra. The scientific output of the naturalists directly associated with the Academy or those that were simply directed by its guidelines are doubtless empirical evidence of how Luso-Brazilian science was made in that age.

The creation of the Academy was a solution that took into consideration not only the interests of scientists who wanted to dedicate themselves to their profession, that was not possible without State financing, but also the interests of the State, who foresaw the profits that science could provide. A process that fostered co-operation of intellectuals and scientists was begun, by means of which the State gained their loyalty. In this 'marriage', scientists maintained their loyalty to the State and supplied it with scientific improvements, while the latter supplied finances for the scientific projects, as long as it could benefit from the outcome. As pointed out by Munteal Filho (1998), within the Academy there was a tacit commitment between bureaucrats and scientific intellectuals so that their speculations had practical objectives and brought to the Crown a return on its investments.

According to the scientific conception that values observation and experience above all, the Academy went about creating establishments that would allow it to bring about this ideal. Apart from having its own printing press, library and cabinet of medals and coins, the Academy possessed a Cabinet of Experimental Physics, a Chemistry Laboratory and a Museum. The members used the laboratories to carry out experiments results of which they published in the *Memoirs* of the Academy. For example, in the Chemistry Laboratory, Domenico Vandelli, Professor of Natural History at Coimbra, Manoel Ferreira da Câmara (1762?–1835) and José Bonifácio de Andrada e Silva (1763–1838) carried out diverse experiments which they described in their texts (Carvalho 1981). These activities point to the nature of the institution as a place of investigation and production of knowledge that afterwards was disseminated in texts such as in the *Memoirs*, that were published, or left in unpublished manuscript form. The Academy's museum was created to collect and conserve animal, plant and mineral species from the Kingdom and the colonies that were considered 'worthy of notice'.

Information was readily shared between these various research areas: the education of natural philosophers and naturalists at university, voyages for surveys, mapping and collecting throughout the Kingdom, the conservation of the collected items in the Museum, experimenting and the dissemination of studies in the form of *Memoirs* of the Academy. The value of these increased according to the degree in which they presented their results based on observation and experiments, had originality and public use. The *Memoirs* were the documents that registered the experiments made by naturalists in Portugal as well as in the colonies and supplied essential information for the Portuguese reformist process.

In the case of mineralogy, the *Memoirs* sent and occasionally published by the Academy, or those simply written according to the institution's guidelines and directions regarding the concept of science and scientific practices contributed to the improvement plans or ideas for overcoming problems associated with the mineral sector that were put forward by the State. Filled with descriptions and details, these documents present a picture of the state of the mining sector at that time and present suggestions for the resolution of the problems that affected it.

The 'Philosophical Voyages' and their instructions

In 1768, Dom José I appointed Domingos Vandelli to be responsible for the establishment of a Botanical Garden at the Royal Palace of Ajuda (Lisbon). The objective of this initiative was to provide the Royal family with a scientific education the would equal that which the prince received in the humanities, which would subsequently transform the king into an enlightened monarch. However, reasons for establishing the Royal Botanical Garden was not limited simply to the education of the Prince. By assisting progress in agriculture, it would also become a place of large-scale experiments on the cultivation of plants that would provide economic benefits for the nation. Thus, as has already been pointed out by Brigola (2000), in the panorama of Portuguese museums in the 1700s, the most relevant initiative was the creation of what that author named as the Museum Complex of Ajuda.

The creation of the Museum Complex of Ajuda enabled a vast project to be undertaken, namely, the *Natural History of the Colonies*, as put into practice by Domingos Vandelli. This project was based on a large survey of the natural products in the plant, animal and mineral kingdoms. The objective was to discover new species and to contribute towards scientific development as well as to evaluate economic potential and carry out geographical observations of the Earth, air and water that might bring new elements to explain the workings of the world. In accordance with Portuguese Imperial policy, natural products were not only to be collected from the colonies and taken back to Portugal, but plant and animal species were also to be acclimatized in all colonies, such as Goa, Macao, Guinea, Mozambique and Angola (Munteal Filho 1998). Thus, the elaboration and execution of scientific voyages in all of the Portuguese Empire constituted obligatory stages in this great project.

Vandelli's first known quotes on expeditions to the colonies date back to 1778. Vandelli's plans were modified with the passing of time and began to grow more and more complex with the unfolding of a series of historical events. The plan that was initially drafted for two naturalists—one who would stay in Rio de Janeiro and the other would travel through the northern parts of Portuguese America while accompanying the Frontier Demarcation Commissions—ended up as a series of scientific expeditions.

In 1778 the naturalists at the Royal Museum of Ajuda were already being prepared for a 'Philosophical Voyage', through the development of scientific instructions and through embarking on various expeditions within Portugal. Essential instruments of control over the production of knowledge, 'Instructions' for scientific expeditions were typical of standardizing the concept of science, according to which the accompaniment of the voyages was made at distance, from the European cabinets (Lopes 1997). These 'Instructions' outlined what the travellers should carry in their luggage. They included detailed directions on how to draw, with objectivity, new examples from mineral, plant, animal and human realms, on how natural and industrial products should be collected and how they should be prepared, the places to be traversed, the ethnographic observations that should be made, the production of diaries and drawings. In short, they presented all the theoretical and practical instructions for the scientific expeditions. In the Instructions detailing exploration on the Portuguese mainland, there are references to people who were intended to participate on the expeditions, the places to be traversed and what should be observed. These documents throw light on how the plans for the execution of voyages were transformed over time.

In 1779, while still in Coimbra, Domingos Vandelli drafted a manuscript meant to be a type of instruction, called *Philosophical voyages* or *Dissertation on the important rules that the naturalist philosopher must mainly observe during his peregrinations*.[4] These instructions are extremely detailed and the author made use of various examples of situations in Brazil, and gave explicit references as to what the naturalist should observe in his expeditions to Portuguese America. Although the instructions are directed to naturalists in general, in various parts of the document there is a special concern regarding Brazil. Undoubtedly the expeditions that set sail after 1783 used this document as their main source of instruction. In the first paragraph of this text Vandelli states that a History of the Earth is very important, that is, the comprehension of how it was formed. Vandelli subscribes to the idea of a reasonably ancient Earth. However, the text is a practical guide and so he prefers to emphasize more general aspects with a view to helping naturalists to extract practical results of interest to Portugal:

> Before dealing with nature's various productions that constitute the globe, it is necessary to say something on the general history of the earth and make it precede the private history of its productions. But this is not the place to present the diverse systems that naturalists have thought up to explain, in their own way, the formation of the earth and the most noteworthy of the phenomena that we observe in the revolutions that occur in the globe, so this is simply the place in which to point out more general things that serve as a lighthouse for the knowledge of natural individuals that can result in favour of Portugal's interests, our voyager is put in charge of everything: while running across most of the world he will encounter at each step irrefutable monuments of ancient times that can suggest ideas on the true Theory of the Earth.[5]

Vandelli then listed in detail what the naturalist must observe. He was to begin by the observation of mountains because they contain 'nature's richest treasures' and they also present infinite uses for humans: they are natural borders between States, pure air can be breathed on top of them, the formation of metal takes place in them and they have the role of water reservoirs. According to Vandelli, all these things were to serve as important reflections for the natural philosopher.

He recommended that the naturalist pay special attention to the primitive mountains[6] as they are supposed to be richer in mineral resources. Regarding mountains, the naturalist was expected to observe amongst other things their height and direction, that 'in America are usually totally different, with mountains ranging more from north to south than from east to west', the display of layers, inclination, arrangement of the veins, types of 'stones' or 'earth' from which they were made of. Vandelli observed that paying attention to the correspondence between the angles of parallel mountains could 'shed great light on the Theory of Earth for Natural History'.[7] A very important task of the naturalist was to verify the type of 'stone' that formed the mountains:

> ... so much because this part of Mineralogy that constitutes lithology is still today so dark and unknown and therefore must be treated with more detail, and observations must be made so as to illustrate the development of the stone and mainly that of the mines and also due to the great utility that we can extract from the crystalline stones and the precious ones. . . .[8]

Vandelli was always concerned with the practical use of the observations made by naturalist-travellers. He suggested the examination of fissures and cavities in mountains 'because besides the crystals they might contain many other precious stones'. The naturalist should proceed with investigations on metal mines by 'starting to examine the cliffs specially where there is a great quantity of spath and quartz, usual sources of these mines'. If traces of a mine were found, the naturalist should obtain a sample and carry out tests on it so as to be able to reach a conclusion regarding its riches or the absence of them. The outcome of tests on a mine should be the calculation of the profit it could bring, 'attending to all the expenses, noting if there is enough wood for its foundry and the distance that separates it from the closest village and between the latter and sea ports'. As regards the veins, their direction (orientation), obliquity, ramification, width, height and depth was to be verified. According to Vandelli, these constituted extremely interesting observations for understanding subterranean physics.

In Brazil, apart from the known mines, the naturalist was to inquire about the existence of silver, platinum, mercury (quicksilver), cinnabar (mercuric sulphide), lead, tin and cobalt, among others. These instructions from Vandelli were in accordance with the Portuguese State's guidelines in relation to the mineral sector, that during this period attempted to expand and diversify mineral exploitation. Apart from the mineralogical tests themselves, the naturalist should also note if exploitation of a specific mine would or would not be of economic benefit. Here Vandelli again refers to Brazilian mines. The naturalist should verify if these mines were being run as they should be, if the machinery in use was adequate or not, and, most importantly, should suggest the means through which the greatest profit could be extracted from the aforementioned mines, at the lowest costs (Fig. 1).

The conclusion that Vandelli gave to this part of his instructions clearly shows what he thought should be the task of the natural philosopher. It also demonstrates the perception the professor has of knowledge. In accordance with the overall view of his time, knowledge for Vandelli had to have a useful nature:

> What is said of the kingdom of the stones comprehends in general all the things that a naturalist should propose to do, so as to produce a complete history of the places he has been to and an assortment of nature's most useful productions that can be of interest to his homeland, the only objective that must guide the fatigues of a philosopher that travels.[9]

In 1779, the arrival at Rio de Janeiro of Luís de Vasconcelos e Sousa, Viceroy of Brazil, led to a modification of Vandelli's plans. Now it was no longer necessary to send a naturalist from the Botanical Garden of Ajuda to Rio de Janeiro as the Viceroy started the process of developing surveys with the help of a resident in the colony. The natural products were prepared by Francisco Xavier Cardoso Caldeira, who was known as 'Xavier of the Birds', and dispatched to Lisbon.

According to documentation relating to the nine year 'Philosophical Voyage' that departed in 1783 for the Brazilian Amazon, the naturalists involved were Alexandre Rodrigues Ferreira (1760–1824),[10] whose duties were to direct ('regulate') the expedition, prepare the diaries, inspect the production of sketches, and to send shipments of natural products; Manuel Galvão da Silva who assisted Ferreira, looked after the 'domestic economy' and inspected the preparation of animals and herbariums; and last, Ângelo Donati (d. 1783) had the duty of making drawings of items and places as determined by Ferreira. In these initial plans each member, naturalist or artist, had a specific duty selected to complement each other (Simon 1983).

The process of preparing for the 'Philosophical Voyages' also comprised the training of the

Fig. 1. Gold exploitation in the River das Velhas and other rivers (unknown artist).

expedition members to carry out excursions within Portugal itself. This activity consisted of the naturalists exploring an already known area, so that they could receive practical training and at the same time be trained in what to expect in an unknown geographical environment. Martin Rudwick (1996) discussed the influence of exposure to well-known, familiar sites to the development of geological theory and practice, and argued that training in familiar geographical regions as well as in unknown ones is essential for theoretical innovation. We establish here that the training of Portuguese naturalists voyagers embodied both of these aspects.

Even before his preparation of Instructions for these expeditions, Vandelli had already discussed the importance of the training of naturalists by taking them on expeditions to known areas. In a letter to Minister Martinho de Melo e Castro dated 22 July 1778, Vandelli commented that the naturalists due to take part in the project the 'Philosophical Voyages' had already been instructed and were about to undertake a trial expedition within the Kingdom:

> The naturalists regarding whom I had the honour of writing to Your Excellency are already prepared for all that regards instructions, lacking only the exercise of a voyage (*apud* Simon 1983, p. 10).

With a view to training the members for the 'Philosophical Voyages', Vandelli chose to visit a coal mine. In 1779, Alexandre Rodrigues Ferreira and João da Silva Feijó (1760–1824)[11] went to the Buarcos coal mine, located close to Cape Mondego, and remained for five days; they produced a journal with some illustrations of the mine (Simon 1983). We have not had access to this journal, but the training took place on an expedition of a predominantly mineralogical nature. Expeditions throughout the Kingdom were a frequently practised exercise in the eighteenth century. Vandelli also made some expeditions to Italy, where he collected various materials that made up his private collection.

The preparatory phase undertaken by the members was essential for the development of cultural, theoretical and practical instruments by the expedition's naturalists and artists. During the naturalists' training, other factors would influence intellectual and perceptive development throughout the voyage, such as the literature read in that period, discussions with professors and colleagues as well as with the scientific community of their network of relationships. But how should a naturalist behave on an expedition in a natural environment that was so different to the one familiar to him? According to Rudwick (1996), without the experience in a familiar environment (initial training in interpretation), the observer of a non-familiar environment would be left only with confusing experiences. The conceptual innovation emerges from the impact of the perception of non-familiar factors that are dealt with using instruments prepared by training the interpretation of familiar factors.

This approach has been used here to interpret the preparations for the overseas expeditions, considering the preparation of the expedition members in familiar environments. Besides the exercise of a field trip to familiar and already explored environments, as proposed by Rudwick, we include in the framework of familiar environments those experiences in other already well known areas as the botanical gardens and natural history museums of Coimbra and Ajuda (in Lisbon), where natural history collections were studied, drawings were observed and executed, and descriptions of natural history species from the Portuguese colonies, especially those from Brazil, were analysed. In the field as well as in these institutions, naturalists and artists also exchanged information, in a process of socialization of experiences that was essential towards this preparatory phase.

The 'Philosophical Voyages' also had the objective of supplying the Royal Museum of Ajuda with natural history collections. Thus, instructions, other than those provided to expedition members were required for those charged with collecting, preparing and dispatching natural products to Lisbon. In 1781 in order to guide the employees of the Royal Museum of Ajuda on issues regarding the preparation and shipping of natural products for the Museum, the naturalists of that institution prepared instructions. This manuscript, which is in the collections of the Bocage Museum, is entitled *Method of collecting, preparing, dispatching and conserving natural products according to the plan that some naturalists have conceived and published for use by curious people that visit the backlands and coastal areas*. In some parts of the manuscript there are annotations in Alexandre Rodrigues Ferreira's handwriting, which possibly indicate that he participated in its preparation. The drawings were made by Joaquim José Codina (d. 1792?) and Ângelo Donati, both of whom were subsequently artists on the 'Philosophical Voyages' (Almaça 1993).

This manuscript corresponds in many points with the Brief instructions for the Lisbon Academy of Sciences' correspondents regarding the dispatching of products and news related to the History of Nature, with a view to forming a National Museum.[12] These instructions were prepared in 1781 by the Lisbon Academy of Sciences with the possible help of naturalists of the Ajuda Museum. There are similarities between the two documents that suggest some collaboration. The Brief Instructions were used on the expeditions and can be used to determine many details related to the dispatch of products and to the various geographical, political, economic and social observations that were made.

The *Brief Instructions* were prepared with the greater objective of supplying a National Museum that was in process of being created. Natural and artificial objects were to be collected in the Kingdom of Portugal and its colonies, to permit the study of natural sciences that would lead to the 'advancement of arts, Commerce, Manufactures and all other branches of Economy'.

The *Brief Instructions* were designed to be used by correspondent members of the Lisbon Academy of Sciences living in Portuguese colonies, such as local administrators or members of the intellectual elite who were not necessarily versed in natural history. Thus the instructions has to be a detailed and thorough text regarding the orientation on methods and techniques of observing, collecting, preparing and dispatching natural history products to Portugal. In addition to the technical guidelines, the *Brief Instructions* contained information on how to observe and record aspects of the 'geographical news of the physical aspect of the country' and the 'morale of the people' (i.e. their practices, costumes and traditions).

The Lisbon Academy instructions were sent to the Governors that sponsored and even carried out great naturalistic surveys. From the 1780s onwards, the colonial administrators started to collect natural products and ship them off to Lisbon. The *Brief Instructions* reached Bahia in 1782. Based on them, some naturalists in that state began to survey, prepare and dispatch natural products to the Royal Botanical Garden of Ajuda. Inácio Ferreira da Câmara became responsible for this task (Damasceno 1969).

The huge task of collecting and shipping colonial products to Portugal was also aided by some naturalists present in the colonies who were hired by the Portuguese Crown. As early as 1781–1782, during the Academy's first years of activity, Domingos Vandelli used all his influence with the State bureaucrats to nominate the philosophy graduates 'that assist in Brazil'. They were:

> Dr Joaquim Veloso, Vila Rica; Francisco Vieira de Couto,[13] Serro Frio; Serafim Francisco de Macedo, Town of São Francisco of the City of Bahia; José da Silva Lisboa, Bahia; Inácio (?) Gularte, Rio de Janeiro [of whom Vandelli wrote:]

> All of these are good and able to observe and collect natural productions. These that follow could be correspondents, such as Antônio da Rocha Barbosa in Rio de Janeiro and Joaquim Barbosa de Almeida in Bahia.[14]

The Viceroy Luís de Vasconcelos e Souza also ordered the promotion of various naturalistic surveys in Rio de Janeiro. In a letter to Minister Martinho de Mello e Castro, dated 17 June 1783, he outlined his pride and appreciation of the natural products sent by him to Portugal through royal naturalists. These products were prepared by 'Xavier of the Birds', caretaker of the Viceroy's private cabinet (Lopes 1997). Besides Luís de Vasconcelos, many other administrators had had an education that included studies in natural philosophy or mathematics, that qualified them to carry out geographical studies and surveys of natural history, as has already been discussed by various authors (Dias 1968; Novais 1989; Gouvêa 2001).

Luís de Vasconcelos also ordered a scientific expedition to carry out natural surveys in the state of Rio de Janeiro, especially a survey of all its flora. The resultant *Botanical Expedition*, that was commanded by Father José Mariano da Conceição Velloso (1741–1811) from 1783 to 1790, a period that coincided with Alexandre Rodrigues Ferreira's expedition to the Amazon, partly contributed to the major project the *Natural History of the Colonies*.

Initially it was intended to carry out only one expedition to the Brazilian Amazon, which was to explore the coast of Pará, the Island of Marajó, the Xingu River, the Amazon River, the Tapajós River and the Madeira River, up to Mato Grosso, and return by the Tocantins River, as explained in the full text of the *Memoir of the Pará Voyage*. However, this scheme was also enlarged. In 1782 this expedition was disbanded and three other 'Philosophical Voyages' were created; these departed Portugal in 1783 and were dispatched to its colonies in Africa, America and Asia respectively.

Alexandre Rodrigues Ferreira, the naturalist from Bahia who from the start was being prepared to command the expedition to the state of Grand-Pará, prepared the initial plans. On 1 September 1783 he set sail from Lisbon and made his way towards Belém on the ships *Águia* (*Eagle*) and *Coração de Jesus* (*Heart of Jesus*). Accompanying the expedition were Agostinho Joaquim do Cabo (d. 1789), a botanical gardener, and the artists José Joaquim Freire (1760–1847) and Joaquim José Codina. Travelling with this expedition was the new Bishop and the Governor of the State of Grand-Pará, Martinho de Sousa e Albuquerque. Once again this was a complementary union between State policy and the enlightened mentality of that period.

Manoel Galvão da Silva (b. 1750), the Bahia-born naturalist who had been prepared to assist Ferreira in the Amazon, left on an expedition to Goa and Mozambique instead. Apart from the task of commanding the expedition, Silva was to also hold the post of Secretary of Government in Mozambique from 1784 to 1793. The botanical gardener José da Costa and the scribe Antônio Gomes accompanied him (Simon 1983).

Ângelo Donati had initially been appointed as the person responsible for the production of drawings on the Amazon expedition, but he set off in 1783 with the same appointment to Angola. As was the case of Manoel Galvão da Silva, the naturalist José Joaquim da Silva, commander of this expedition, also held a similar official position, as the Secretary of Government in Angola from 1783 to 1808. The other member of the party was the artist José Antônio (d. 1784).

And lastly, João da Silva Feijó, who had accompanied Ferreira in 1778 on the preparatory voyage to the Buarcos coal mine, set off in 1783 for the Cabo Verde Islands as naturalist and Secretary of State. He was not accompanied by any artists as he knew how to draw and how to survey and prepare geographical charts. After the end of this expedition, Feijó was sent to Ceará state where he was appointed as naturalist. The same naturalists travelled to various parts of the empire, as highlighted by Gouvêa (2001) in relation to the bureaucrats who performed administrative posts.

The preparation of the expeditions and the collective preparation of the participants reveal various points of continuity between these expeditions, and also reveal the 'successful' results from each of them. The detailed preparations included decisions as to which places had be visited, the tasks each member of the expedition had to carry out, notices about what should be observed and collected, directions as to the type work to be completed *in situ*, and the fieldwork that members of scientific expedition were expected to do. These plans even envisaged the results expected from each expedition.

The four 'Philosophical Voyages', Fr Vellozo's Botanical Expedition and the Frontier Demarcation Commissions are famous for their complementary aspects. However, each expedition was unique in terms of its technical staff, in the particularities of each territory explared, in the vicissitudes of each expedition and in the political, cultural and social interactions that were configured throughout the duration of each one of them.

Once the expedition came to an end, compilation of all the collected data was carried out in the Royal Museum and the Botanical Garden of Ajuda, under the direction of Domingos Vandelli and Alexandre Rodrigues Ferreira, who took up office as the administrator of that institution in 1794. The data collection did not cease after the end of these expeditions, with other scientific voyages being sent off to the colonies, then under the administration of the Minister D. Rodrigo de Sousa Coutinho. We believe that the expeditions carried out after 1796 had a new purpose, that is, they sought to gather complimentary information or more detailing data on the material already collected that was in the process of being described systematically with a view to its being published.

Manuel Arruda da Câmara (1752–1811),[15] for example, carried out expeditions throughout the Brazilian Northeast. From March 1794 to September 1795 the naturalist made a mineralogical voyage through Pernambuco and Piauí, observing various minerals. From December 1797 to July 1799, Câmara was involved in another scientific expedition throughout Paraíba and Ceará. He also carried out expeditions to the São Francisco River. In these expeditions, Manuel Arruda da Câmara carried out mineralogical, botanical and zoological surveys and also sorted the collected data. He wrote some memoirs on agriculture and produced a *Pernambucan Flora* which contained drawings by himself and João Ribeiro de Mello Montenegro (1766–1827) (Mello 1982).

In 1798, with a view to carrying out mineralogical surveys in Minas Gerais, Dom Rodrigo de Sousa Coutinho sent José Vieira Couto (1752–1827)[16] to that State, where he remained until his death in 1827. In a letter from Coutinho to Bernardo José de Lorena, Governor of Minas Gerais,[17] Coutinho directed that José Vieira Couto and José Teixeira da Fonseca be placed in charge of carrying out mineralogical examinations in the State of Minas Gerais, so as to offer 'more substantiated information regarding that country's mines, as well as the profit that can be extracted from them'. In this State, Couto carried out scientific expeditions, where samples were collected and various observations were made that were subsequently incorporated to five memoirs prepared by the naturalist (Silva 2002) (Fig. 2).

On 20 March 1798, Dom Fernando José de Portugal wrote to Dom Rodrigo de Sousa Coutinho in reference to the commission that José de Sá Bittencourt Accioli (1755–1828) had received, that of inspecting Montes Altos' copper and saltpetre mines in the Jacobina district.[18] In this document he describes having informed Accioli of His Majesty's desire that he should travel to Jacobina 'so as to visit the copper and saltpetre mines that exist there' and propose the 'means by which profit can be made from them'.

In the same document, the Minister of State charged João Manso Pereira to make 'a voyage through the States of Rio de Janeiro, São Paulo and that of Minas Gerais, with a view to increasing all knowledge of the riches that exist in the aforementioned States'. For such, João Manso Pereira was to receive a pension of 400 thousand réis per year (roughly US$ 800 today) and also an allowance to pay for expenses and for buying utensils 'that he might find himself in need of for his mineralogical and metallurgical tests'. Joaquim Veloso de Miranda (1742?–1815(1818?)) received the following instructions: 'as His Majesty has been so graceful as to appoint you Secretary of the Government of Minas Gerais, His Majesty expects to see the continuation of your research on the productions of that State'.[19]

The outcome of the four 'Philosophical Voyages' that set sail from Lisbon in 1783 is documented in some mineralogical diaries and memoirs. In 1788, Manoel Galvão da Silva wrote his *Diary or list of Philosophical Voyages that have been carried out according to the orders of His Faithful Majesty on territories of the Villa de Tette Jurisdiction and some of the Massaves'*.[20]

Two years later, in 1790, Alexandre Rodrigues Ferreira wrote his *Philosophical and Political Brochure of Serra de S. Vicente and its establishments* while residing in Vila Bela, capital of Mato Grosso. Up to that point of the expedition, Ferreira had devoted himself mainly to botany, zoology, ethnography, geography, agriculture, shipping and the study of the urbanization process, and had not dedicated any of his texts exclusively to mineralogy and mining. The author's preference for those former subjects had been strengthened by a State policy that conferred special attention to agriculture. Ferreira only dedicated himself to the subjects of mining and mineralogy close to the end of the expedition, after having being urged to do so by Minister Martinho de Mello e Castro, who instructed the naturalist to concentrate on mineralogical tests to supply the museum:

> Taking up again your voyage, always practising the same as is above referred to until you reach Villa Bella, capital of Matto Grosso, immediately waiting on Luís de Albuquerque, Governor and General Captain of that State, who will give you all the help you need to proceed with your tests and exploration of mines, collecting from them the sources that you are able to discover and all other products that belong to Mineralogy that the miners that work there can get hold of for you and when you are certain that the Royal Cabinet is amply provided for regarding everything that belongs to mines from other countries, only those that are part of the domains of His Highness still remain partly drained . . .[21]

The devotion to the subject of mining by Ferreira's expedition of 1788 and 1791, as well as by the other expeditions mentioned above, is clearly connected to the Portuguese Imperial policy that aimed to improve mining economic activities via natural history institutions.

Final considerations

The formulation of an imperial policy aimed at the integration of the diverse Portuguese colonies was clearly present in the project *Natural History of the Colonies*. During the end of the eighteenth century and the beginning of the nineteenth century, that project mobilized various members of the Portuguese scientific community and colonial administrators. Some naturalists kept up a correspondence and developed a kind of network through which they exchanged information on diverse parts of the Portuguese Empire. Frequently the colonial administrators were responsible for dispatching information sent from the colonies to the

Fig. 2. Chart of the 'New Diamond Lorena' (state of Minas Gerais) drawn by José Vieira Couto (1799).

Portuguese mainland, as they maintained a correspondence with the Minister of the Navy and Overseas. Much of this data was used by the Portuguese Crown in the preparation of policies aimed to further the exploration and control of the colonies and were also sent out to the naturalists at the Botanical Garden of Ajuda and at other institutions, such as the Lisbon Academy of Sciences and the University of Coimbra.

The preparation and execution of Vandelli's project must be analysed within the workings of the Portuguese Empire, taking into consideration the complementary aspect between mainland Portugal and her colonies and the close association between the Enlightenment community and State policy.

The 'Instructions' referred to above were documents that guided the 'Philosophical Voyages' sponsored by the Portuguese Crown. These voyages were part of an effort aimed at helping the economic recovery of the Empire by means of a rational exploration of natural resources and they resulted in the recognition and exploitation of natural resources from the colonies, in the delimitation of the frontiers of these colonies, in the collection, description and classification of plants, animals and geological specimens, and their dispatch for Portuguese museums. The 'Instruction' as well as the 'Philosophical Voyages' demonstrate that the Portuguese Kingdom was part of an international movement given to the 'mobilization of worlds' using natural history as means to this end.

Clarete P. da Silva's research is supported by Fapesp, and that of Ermelinda M. Pataca by Capes, support that is gratefully acknowledged.

Notes

[1]Arquiro do instituto de Estudos Brasileiros (IEB), 9th letter, 31 March 1977.

[2]We refer here to some of these explorers (extracted from Sanjad *et al.* 2000, p. 161):

Aloísio Conrado Pfeil (1638–1701): Mathematics teacher at the Colégio do Pará, astronomer, cartographer and explorer. Author of many maps of the North of Brazil, including an important charter of the Amazon River (1684), ordered by the Crown through Padre Antônio Vieira. He explored the State of the North Cape, area of armed conflict between Portugal and France.

Diogo Pinto Gaya: Made a trip to the Tocantins River in 1720.

Francisco de Melo Palheta: In the first half of the eighteenth century, he carried out an expedition to the Madeira and Mamoré rivers, that later became the axis of integration between the central-southern and northern regions of Portuguese domain.

Luís Fagundes Machado: Explored the Madeira River in 1749, in an expedition organized as a model for those to be carried out during the following decade. This voyage produced important accounts regarding the specification of bearings, distances, latitudes, navigation difficulties and the presence of Portuguese and Spanish on the riverbanks.

Besides these explorers, there is also the expedition of the mathematician Jesuits (1729–1748) Diogo Soares and Domingos Capassi throughout the States of Rio de Janeiro, Colônia do Sacramento, Espírito Santo, Minas Gerais, São Paulo and Goiás. Apart from the elaboration of maps, the mathematician priests were also given the task of carrying out a vast inquiry of the communication lines in the Brazilian territory, the economic resources, and of the Indian population (see Sanjad *et al.* 2000).

[3]In the codex 807 of the Rio de Janeiro National Archive (ANRJ) a range of documents that testify this official effort can be found:

Volume 1, sheets 49–52: *Copy of the notice from D. Rodrigo de Sousa Coutinho to Manoel Ferreira da Câmara* dated 14 August 1800, appointing various missions to carry out in Bahia.

Volume 5, sheets 131–8: *Copy of the notice that José de Sá Bittencourt Accioli* sent to D. Rodrigo de Sousa Coutinho, dated 18 May 1799, regarding the economic means of carrying out the extraction of saltpeter from Montes Altos, in the state of Bahia.

Volume 5, sheets 139 and following: *Official letter from Lieutenant Colonel Manuel Martins de Couto Reis* reporting the first steps taken regarding the inquiry into iron mines in the Capitania of Rio de Janeiro, 1795.

Volume 5, sheets 246–7: *Copy of the notice* from D. Rodrigo de Sousa Coutinho to Bernardo José de Lorena, dated 18 March 1797, regarding the Serro Frio mines.

Volume 6, sheet 53: *Notice to Bernardo José de Lorena*, dated 31 October 1798, recommending work in iron, lead and silver mines in Minas Gerais as well as in São Paulo 'making use of Vellozo's lights and the capable metalworker Manso'.

In the Institute of Brazilian Studies (IEB) (Lamego Collection), the codex 82.80, A8, 22 August 1768; 82.87, A8, 27 August 1768; 82.99, A8, 20 October 1768; 82.104, A8, 25 November 1768; 82.148, A8, 6 March 1769 refer to *letters* from D. Francisco Inocêncio de Sousa Coutinho, governor of Angola, to Francisco Xavier de Mendonça Furtado, Secretary of the Navy and Overseas, all regarding 'iron works'. These documents show that other parts of the Portuguese Empire were also part of official concerns.

In the State of São Paulo Archive (in the *Interesting Documents* [Documentos Interessantes—DI] *for History and Customs in São Paulo*, volume 89 1967, pages 47–49; 131–132; 197–199; 209; 218–219; 240–241) there are lists of books and prints sent by the Crown to the state of São Paulo between 1797 and 1802 'with an aim to instruct the people, not only in agricultural issues, but also in other important areas' (DI, volume 89, pages 218–219). Regarding mining, the lists include books such as 'Bergman Mineralogy', 'Memoirs on the saltpeter transferred from Theodoro Dúrtubie [sic]'; 'Extracts from the mode of making saltpeter in the Virginia tobacco factories'; 'Extracts from the method of making potassium nitrate, or saltpeter, by Chaptal'; 'Bergman, T. 2o', 'Potassium, Illuminated'. D. Rodrigo de Sousa Coutinho signs the documents in which these lists can be found.

[4]Physician native of Padova. He was invited by the Marquis of Pombal in 1764 to go to Portugal as lecturer of Natural History and Mineralogy at the University of

Coimbra, where he became master of many Luso-Brazilian naturalists. He created the Botanical Garden of the University of Coimbra and the Royal Botanical Garden of Ajuda (Carvalho 1987).

[5]Manuscript from the Lisbon Academy of Sciences, Manuscript 405 of the red series. Copy of 109 pages, copied in 1796 by Father Franciscano Vicente Salgado.

[6]Manuscript from the Lisbon Academy of Sciences, Manuscript 405 of the red series. Copy of 109 pages, copied in 1796 by Father Franciscano Vicente Salgado, p. 17.

[7]When referring to mountains, scholars of the period discussed here make use of the following adjectives: 'First order', 'Second order' and 'Third order'. These adjectives were in accordance to the nomenclature of that time and respectively correspond to the terms 'primary mountains', 'secondary mountains' and 'tertiary mountains'. They concern the relative age of the rocks, where primary (or primitive) are older and tertiary are younger.

[8]According to Taylor (1988) one of the regularities most subject to discussion by 'geologists' of the eighteenth century was the principle of correspondence of the angles. According to this principle, in cases where two mountains are parallel, the angles of both sides should be correspondent.

[9]Manuscript from the Lisbon Academy of Sciences, Manuscript 405 of the red series. Copy of 109 pages, copied in 1796 by Father Franciscano Vicente Salgado, pp. 21–22.

[10]Manuscript from the Lisbon Academy of Sciences, Manuscript 405 of the red series. Copy of 109 pages, copied in 1796 by Father Franciscano Vicente Salgado, p. 50.

[11]Alexandre Rodrigues Ferreira was born in Bahia in 1756 and died in Lisbon in 1815. It is possible that his father was a slave dealer (Simon, 1983; cf. full reference in next note). He enrolled in the University of Coimbra in 1774 and earned his Bachelor's Degree in Natural Philosophy in 1778. In 1779 he received his Ph.D. title in Philosophy. Subsequently he was hired in the Botanical Gardens of Ajuda, where he made demonstrations in Natural History. In 1794, after the expedition, he was hired as administrator of the Royal Museum and Botanical Gardens of Ajuda, where he later on worked with Domingos Vandelli at the systematization of natural products and the preparation of a *Natural History of the Colonies*.

[12]João da Silva Feijó was born in Rio de Janeiro in 1760 and died in Rio de Janeiro on 9 March 1824. It seems that he graduated in Mathematics from the University of Coimbra, and belonged the Royal Academy of Sciences of Lisbon. He took up a military career, serving in the Engineer Corps and reached the post of colonel.

[13]Lisbon: Regia Officina Typografica. 1781.
This is a mistake made by Vandelli. The correct name is José Vieira Couto.

[14]Arquivo Histórico Ultramarino: Overseas Historical Archive, bundle 26, n. 2722.

[15]Manuel de Arruda Câmara was born in Pernambuco in 1752. In 1783 he took religious vows and entered the Carmelites in the convent of Goyanna (Pernambuco). He later enrolled in the University of Coimbra, studying Natural Philosophy. Afterwards he went to the University of Montpellier, where he received a Ph.D. degree in Medicine. In 1793 he returned to Pernambuco when the Portuguese Crown put him in charge of carrying out a series of naturalistic surveys. He died in Goyanna in 1811.

[16]José Vieira Couto was born in 1752 in Arraial do Tijuco (Minas Gerais) and died in the same town, in 1827. He studied Philosophy and Mathematics at the University of Coimbra, receiving his diploma in the year of 1778.

[17]Arquivo Histórico Ultramarino: Overseas Historical Archive, code 610, p. 202, 18 March 1797.

[18]Arquivo Histórico Ultramarino: Overseas Historical Archive, box 93, doc. 1858.

[19]Arquivo Histórico Ultramarino: Overseas Historical Archive, code. 610, pages 202–3, dated 18 March 1797.

[20]IHGB (Brazilian Historical and Geographical Institute), tin #33, doc. 2.

[21]IHGB (Brazilian Historical and Geographical Institute), tin #282, book 2, document 10, 31 October 1787.

References

ALMAÇA, C. 1993. *Bosquejo histórico da zoologia em Portugal*. Museu Nacional de História Natural, Lisboa.

BRIGOLA, J. C. 2000. Viagem, Ciência, Administração—o complexo museológico da Ajuda (1768–1808). *In: 1o. Congresso Luso-Brasileiro de História da Ciência e da Técnica—Livro de Resumos*. Évora: Universidade de Évora, 49–50.

CALAFATE, P. 1994. *A idéia de natureza no século XVIII em Portugal (1740–1800)*. Imprensa Nacional/Casa da Moeda, (Estudos gerais, série universitária).

CARVALHO, C. T. 1965. Comentários sobre os mamíferos descritos e figurados por Alexandre Rodrigues Ferreira em 1790. *Arquivos de Zoologia*, **12**, 7–70.

CARVALHO, R. DE. 1981. *A atividade pedagógica da Academia das ciências de Lisboa nos séculos XVIII e XIX*. Publicações do segundo centenário da Academia das Ciências de Lisboa, Lisboa.

CARVALHO, R. DE. 1987. *A História Natural em Portugal no século XVIII*. ICALP/Ministério da Educação, Lisboa.

DAMASCENO, D. 1969. Introdução. *In*: SAMPAIO, F. A. DE História dos reinos vegetal, animal e mineral do Brasil, pertencente à medicina. *Anais da Biblioteca Nacional*, **89**, 5–8.

DIAS, M. O. DA S. 1968. Aspectos da ilustração no Brasil. *Revista do Instituto Histórico e Geográfico Brasileiro*, **278** (Jan.–Mar.), 105–170.

FIGUEIRÔA, S. F. DE M. 1997. *As ciências geológicas no Brasil: uma história social e institucional, 1875–1934*. HUCITEC, São Paulo.

FINDLEN, P. 1996. *Possessing Nature. Museums, Collecting, and Scientific Culture in Early Modern Italy.* University of California Press, Berkeley.

GASCOIGNE, J. 1998. The Universities and the Enlightenment. *In*: GASCOIGNE, J. (ed.) *Science, Politics and Universities in Europe, 1600–1800.* Ashgate, Brookfield, 35–53.

GOUVÊA, Mª DE F. 2001. Poder político e administração na formação do complexo atlântico português (1645–1808). *In*: FRAGOSO, J., BICALHO, Mª F. & GOUVÊA, Mª DE F. (org.) *O Antigo Regime nos trópicos: a dinâmica Imperial Portuguesa (séculos XVI–XVIII).* Civilização Brasileira, Rio de Janeiro, 285–315.

LOPES, M. M. 1997. *O Brasil descobre a pesquisa científica: os museus e as ciências naturais no século XIX.* Editora Hucitec, São Paulo.

MELLO, J. A. G. DE. 1982. *Manuel Arruda da Câmara—Obras reunidas. C. 1752–1811.* Secretaria de Educação e Cultura, Recife.

MUNTEAL FILHO, O. 1997. Todo um mundo a reformar: intelectuais, cultura ilustrada e estabelecimentos científicos em Portugal e no Brasil, 1779–1808. *Anais Museu Histórico Nacional*, **29**, 87–108.

MUNTEAL FILHO, O. 1998. Ciência, Natureza e Sociabilidade Intelectual em Portugal no século XVIII: a Academia Real das Ciências de Lisboa e os Caminhos da Ilustração Ibérica (1779–1815). *In*: *Anais do V Seminário Nacional de História da Ciência e da Tecnologia.* Sociedade Brasileira de História da Ciência. São Paulo, 288–295.

NOVAIS, F. 1989. *Portugal e Brasil na crise do Antigo Sistema Colonial (1777–1808).* 5ªed., Hucitec, São Paulo.

RUDWICK, M. 1996. Geological travel and theoretical innovation: the role of 'liminal' experience". *Social Studies of Science*, **26**, 143–159.

SANJAD, N., PATACA, E. M. & LOPES, M. M. 2000. As fronteiras do Império: militares, naturalistas e artistas na Amazônia, século XVIII. *In*: *IV Jornadas Latino-Americanas de Estudos Sociais da Ciência e Tecnologia. Programa e Caderno de Resumos.* IG-UNICAMP, Campinas, 169.

SILVA, C. P. DA. 2002. *O desvendar do grande livro da natureza. Um estudo da obra do mineralogista José Vieira Couto, 1798–1805.* Anna Blume: Fapesp, São Paulo; Campinas: Faep/Unicamp.

SIMON, W. J. 1983. *Scientific expeditions in the Portuguese overseas territories (1783–1808).* Instituto de Investigação Científica Tropical, Lisboa.

TAYLOR, K. 1988. Les Lois naturelles das la géologie du XVIIIème siècle: recherches préliminaires. *Travaux du COFRHIGEO*, 3e série, **2**, 1–28.

Doing and knowing: Charles Darwin and other travellers

S. HERBERT

University of Maryland Baltimore County, Department of History, 1000 Hilltop Circle, Baltimore, MD 21250, USA (e-mail: herbert@umbc.edu)

Abstract: In the 1830s, geology was a young discipline in the process of acquiring uniform standards. This study considers Charles Darwin's work in relation to that of other more practically and less academically oriented travellers. It suggests a continuity exists between the more practically and the more academically-minded groups in such projects as exploring, mining, map and chart making, collecting of specimens, and travel writing. It also highlights the role played by William Fitton as an academically-minded geologist whose instructions on collecting and observing were intended to raise standards for geologists. It suggests that such disciplinary improvements were not unique to geology but reflected a more general willingness at the time to instruct and be instructed.

Permit me to advise the traveller to look into the book of nature, which is always open, and learn what he can (Mawe 1821, p. 55).

A small type and a thin paper have been preferred for the advantage of the traveller; who, it is presumed, will not fail to find the volume an interesting companion of his travels.[1]

Doing and knowing

What counts as travel? Although it is possible to do geological research in one's own garden—consider Darwin and his earthworms—or in one's neighbourhood—consider Darwin and his early exploration of the Llanymynech quarry near Shrewsbury, 'travel' as applied to geology usually means travel at a distance requiring overnight stays away from home (Darwin 1833–1838; Roberts 1996). Because such travel is costly, financially, in the risk to personal health and safety, and in separation from family and friends, a general question comes to mind: who is paying?

As Hugh Torrens has reminded us in regard to William Smith (1769–1839), geologists can be seen as people who do things as well as people who know things (Torrens 2001, p. 61). The question of the doing strongly correlates with the question of who is paying and for what. Thus, for example, mining was the impetus for the efforts of much travel to South America, as indicated in Michael G. Mulhall's classic work *The English in South America* (Mulhall 1878, pp. 445–482).[2] Some ventures failed—such as the South American efforts of Joseph Harrison Freyer (1777–1855) described by Torrens—whereas others prospered not only in business but also published works in which geology played some part (Torrens 2002, XI, pp. 29–46).

John Miers (1789–1879) was the author of *Travels in Chile and La Plata, including accounts respecting the geography, geology, statistics, government, finances, agriculture, manners and customs, and the mining operations in Chile.* Although he was later better known as a botanist, Miers helped to develop English interests in Chilean copper mines. During the voyage of the *Beagle*, Darwin quoted him respectfully on the subject of the paucity of pebbles on the Pampas, and on the subject of the evidence for the rise of land provided by the elevated position of sea shells on the Chilean coast (Miers 1826, quoted in Darwin 1846, p. 35; Barrett *et al.* 1987, p. 37, 53). John Mawe (1766–1829) was another man of business, a dealer in minerals and knowledgeable regarding stratification. As Torrens has shown, Mawe set off to South America in 1804 in the service of the King of Portugal. He travelled as a collector on a 'voyage of commercial experiment'. As Brian Dolan pointed out, Mawe acted as mineral dealer to Edward Daniel Clarke (1769–1822), predecessor to John Stevens Henslow (1796–1861) as Professor of Mineralogy at Cambridge (Dolan 1995, p. 177). Clarke also played a part in the development of the blowpipe for use in mineralogical analysis (Oldroyd 1972).

When Darwin was in Santiago, Chile in 1835, he stayed with Alexander Caldcleugh (d. 1858 in Valparaiso). Caldcleugh had been recommended to Darwin as a 'very accomplished chemist and mineralogist'.[3] At the time when Darwin knew him, Caldcleugh had interests in copper mines at Panuncillo. Darwin visited one of these mines and commented extensively on it in his *Diary* (Keynes 1988, pp. 328–330).[4] Earlier Caldcleugh had served as a private secretary to a British diplomat in Rio de Janeiro and had travelled extensively in

From: WYSE JACKSON, P. N. (ed.) *Four Centuries of Geological Travel: The Search for Knowledge on Foot, Bicycle, Sledge and Camel.* Geological Society, London, Special Publications, **287**, 311–323.
DOI: 10.1144/SP287.24 0305-8719/07/$15.00

Brazil, Chile, and parts of what is now Argentina. Among the 'philosophical instruments' he had with him during his travels was one (a clinometer, a term he did not use) for measuring the 'dip and direction of strata' (Caldcleugh 1825, **1**, pp. 304–305). He published a work of travel narrative that includes some geological observations. He was also the author of a chapter in the innovative 1823 book on meteorology by John Frederic Daniell (1790–1845) (Daniell 1823, pp. 335–348).

In a remarkable essay, written sometime between late December 1833 and mid-April 1834 but unpublished during his lifetime, Darwin developed a theory of the geological formation of the South American continent that included a narrative framework for the history of life on its surface.[5] In this essay, he drew on work by Caldcleugh among others, for example by reminding himself to study Caldcleugh's geological map (Fig. 1). Darwin was interested in correlating recent geological formations with the presence or absence of new species in the area. In his essay he used the traditional term 'Creation' to describe the advent of a new species. He speculated as follows: 'As Patagonia has risen from the waters in so late a period, it may be interesting to consider whence came its organized being[s]. I have conjectured [that] the absence of trees in the fertile Pampas & rich valleys of B. Oriental to be owing to no Creation having taken place subsequently to the formation of the superior Tosca bed'. From the point of view of Darwin's intellectual development, this provides evidence that Darwin was thinking about the subject of species succession over a year before his visit to the Galápagos Islands. Darwin was also drawing on the work of other authors—in the case of Caldcleugh someone attached to the mining aspects of geology and operating at some distance, literally and figuratively, from the centre of academic geology.

Like Darwin, Caldcleugh published an article on the earthquake in Chile of 20 February 1835. Darwin cited this work in his *Beagle* narrative regarding a fall in temperature at a neighbouring spring following the earthquake. Caldcleugh's and Darwin's separate communications to the Geological Society of London on the subject of the evidence for the recent rise of land on the coast of Chile were both read to the Geological Society of London on 4 January 1837 (Caldcleugh 1836).[6]

Geology was also be a sideline to duties of an official nature. The naval captain Basil Hall (1788–1844), son of the geologist James Hall (1761–1832), published articles on his travels around South America. His interpretation of the parallel roads of Coquimbo in Chile became part of the debate over the parallel roads of Glen Roy in Scotland.[7] Of comparable importance was Woodbine Parish (1796–1882), British chargé d'affaires at Buenos Aires (1825–1832), from whose collection Richard Owen (1804–1882) identified and named the *Glyptodon* (Parish 1839).

There was a continuity of interests and methods between these slightly earlier British travellers to South America and Darwin. As a group, they deserve respect for the energy with which they pursued geological and mineralogical subjects, despite working in less than optimum conditions. They are also interesting in that they were not members of a university élite. Even Hall and Parish went directly from school to public service. Of course, the university élite was quite small in those years: annual admissions to the Oxford colleges averaged 400 and the Cambridge colleges 440 from 1820 through 1829. In 1831 the population of England and Wales was 13 994 000, Scotland 2 374 000 and Ireland 7 767 000.[8]

Compared to the mining entrepreneurs and public officials discussed thus far, Darwin had greater freedom to pursue science as a career. In this, he was like other travellers to South America such as his two great continental contemporaries: Alexander von Humboldt (1769–1859), who pursued science at his own expense (including his 1799–1804 research trip through the Spanish colonies) and the French naturalist Alcide Dessalines d'Orbigny (1802–1857), who went to South America on a commission from the Muséum d'Histoire Naturelle. Like Darwin, van Humboldt has long received the attention of scholars and of the public. Recently, through the efforts of those at the Muséum d'Histoire Naturelle, Paris, d'Orbigny's contributions to the history of geology have been acknowledged.[9] Darwin, van Humboldt, and d'Orbigny were similar in that their efforts lay towards the academic end of the knowledge spectrum, as is suggested by each man's substantial publication record. In that sense, with due regard for various practical aspects to each of their labours—von Humboldt, after all, was a prominent authority on mining—their work was overtly academic in style of presentation. But the 'doing and knowing' question is not a matter of either/or, and requires refinement.

The voyage of HMS *Beagle*, 1831–1836

Let us now return to our opening question of doing and knowing and who was paying and apply it to the *Beagle* voyage of 1831–1836. As John Gascoigne has pointed out, the line between public and private funding of science in Britain, at the time of the *Beagle* voyage of 1831–1836, had not been sharply drawn. (Gascoigne 1998, pp. 199–204). Towards the public side of the spectrum was the

Fig. 1. Alexander Caldcleugh's 'Map of the Country between Buenos Ayres and the Pacific Ocean, with a Specification of the different Geological Formations.' The five categories of geological formations listed in the legend include 'Primitive' [pink], 'a very new Stalactiform Limestone' [yellow], 'Red Marl' [green], 'Pebbles & Sand' [salmon], and 'Clay' [blue]. From Caldcleugh, *Travels in South America*. Volume 1.

hydrographical enterprise: the continued surveying of the southern portion of South America, begun on an earlier *Beagle* voyage and circumnavigation using marine chronometers. The results of this operation were greater precision in maps and charts.

As agreed to by Robert FitzRoy (1805–1865), Commander of the *Beagle*, Darwin was officially a supernumerary on the voyage. He wrote to his cousin William Darwin Fox (1805–1880) before departure, 'My appointment is not a very regular affair, as the only thing the Admiralty has done is putting me on the books for Victuals, value 40, per annum.— I have some thoughts of having it taken off again. I should certainly do so, if I thought it would give me a more absolute disposal of my collection, when I return to England'.[10] His father paid his bills, which were substantial (£600 for equipping him for the voyage, £1200 thereafter for five years of travel), rather more than would have been required if his father had simply supported him for the equivalent length of time in England (Browne 1995, p. 229). This left Darwin free to set his own itinerary while on land, and free of any requirements to serve as a mineral prospector or surveyor. The indistinct line between public and private science worked to Darwin's advantage, for example later in his career when he borrowed the British Museum collection of cirripedes to work on at home. In contrast, his American counterpart as world traveller James Dwight Dana (1813–1895) appears not to have had the freedom to loan Darwin specimens from the US Exploring Expedition of 1838–1842.[11]

Many of Darwin's activities as a geological traveller were similar to those of the more practi cally-minded individuals engaged in exploring, map-making, hydrography, mining, and mineral and bone collecting. This is not to suggest that Darwin learned his craft from them; merely a continuum in activity. The more general, and complicated question, of the course of Darwin's geological education is discussed at length in *Charles Darwin, Geologist* (Herbert 2005).

The first example is of a geological overlay Darwin eventually made of the rocks in Tierra del Fuego (Fig. 2). The base map was a chart that had been completed during the first *Beagle* voyage. The Beagle Channel was named. Visually, the map with Darwin's overlay of colour is in the same tradition as the representation produced by Caldcleugh in Fig. 1, but there is an increase in precision at two levels. There is a greater degree of geographical accuracy in the map, a tribute to the hydrographer's work, and a more complex scheme drawn up by Darwin to represent the geology. Thus the hydrographic work by the Admiralty, with its emphasis on precise coastlines, enhanced the work that Darwin was able to perform as a geologist. However, even with the lesser degree of geographical accuracy with which Caldcleugh was working (and, indeed, the interior of the continent was less well known than the coastline during the nineteenth century), Caldcleugh and Darwin were working towards common ends. Darwin was able to achieve more because he was working with greater resources, notably a precise base map.[12]

A second example is in Darwin's approach to the collection of specimens which demonstrates the gradual entrance of rigorous and academically -inspired standards into instructions for travellers. Instructions directed towards travellers on how to collect natural historical specimens were a staple of the literature, a genre sufficiently broad to encompass both Mawe's charming 1821 pocketbook (cited in the epigraph to this article) entitled *Voyager's Companion*, with its frontispiece motto 'Seek and ye shall find' and Darwin's own contribution, a cautionary disquisition in his article on '*Geology*' in the handbook issued for the Admiralty in 1849 (Mawe 1821; Darwin 1849, pp. 156–195; Vaccari 2007).

When Darwin departed on the *Beagle*, he had with him a set of instructions by William Fitton (1780–1861) on how to collect geological specimens. Before the voyage, Darwin had been trained in the field by Henslow and by Adam Sedgwick (1785–1873), and his field technique was formed largely in their company (Secord 1991).[13] For Darwin, Fitton's instructions would have largely repeated what he had learned before his departure on the voyage. Yet, as neither Henslow nor Sedgwick had provided him with written instructions on collecting, Fitton's instructions provided a good documentary record of what a geologist of rank and academic background might advise. Like Henslow and Sedgwick, Fitton was a prominent member of the Geological Society of London, and his instructions reflect a movement towards the improvement and standardization of practice. The act of knowing was taking academic shape.

Fitton had graduated BA from Trinity College, Dublin in 1799 (Wyse Jackson 1994, pp. 16–17, 1998), and thereafter received medical degrees from Edinburgh University and the University of Cambridge. But he was an academic rather than a clinician at heart, and when financially secure, devoted himself full-time to geology, whose study he had begun as an undergraduate in Dublin (Wyse Jackson 1994, pp. 16–17, 1998). Although best-known as a stratigrapher, Fitton also played an important role in shaping the growth of geology as a field. He wrote on its history, served as President of the Geological Society of London (1827–1829) and helped to establish its *Proceedings*, and instructed 'naturalist travellers in the

Fig. 2. Darwin's addition of information on rock types to the 1826–1830 map of southern Tierra del Fuego. From DAR 44 : 13, Darwin Archive, Cambridge University Library. Used with the permission of Syndics of Cambridge University Library.

principles of geology' (Eyles 1972; see also Murchison 1862). Of an open and kindly disposition, and more genuinely welcoming to non-university men than were many of his colleagues in the Geological Society of London, Fitton was able to raise standards in the young field of geology without exciting rancour. His civility is conveyed in his characterization of the verbal exchanges that followed the reading of papers at the Geological Society of London as 'conversations (I will not call them discussions—much less debates)' which were rendered 'both agreeable and instructive' by the 'self-command' of the *participants*.[14]

In 1827, Fitton published his 'Instructions for collecting geological specimens' after examining Australian specimens collected on British naval expeditions of 1801–1805 and 1818–1822 (Fitton 1827*a* (pp. 566–622) *b* (pp. 623–629)). Having looked over these collections made by others than himself, evidently with some dissatisfaction, Fitton advised collectors who had yet to begin their work to 'accompany every specimen' with 'a short note of its geological circumstances' as

> Whether it be found in large shapeless masses, or in strata?
>
> If in strata, what are the thickness, inclination to the horizon, and direction with respect to the compass, of the beds?—. . ..
>
> If the strata be different,—what is the order in which they are placed above each other successively?

Fitton was also firm that labels should be attached to specimens 'immediately, on the spot where they are found', noting that 'this injunction may appear to be superfluous; but so much valuable information has been lost to geology from the neglect of it, that every observer of experience will acknowledge its necessity'. He went on to suggest that

> A *sketch* of a coast or cliff . . . frequently conveys more information respecting the disposition and relations of rocks, than the longest memorandum. If numbers, denoting the situation of the specimens collected, be marked upon such sketches, much time may be saved at the moment of collecting (Fitton 1827*b*, pp. 625–626).

The practice described by Fitton is roughly the one that Darwin followed during the *Beagle* voyage, which would suggest a common standard within the circle that Fitton, Henslow, and Sedgwick operated. Darwin labelled his specimens shortly after he had collected them and interpreted them in their geological context in a set of wonderfully detailed memoranda. On occasion Darwin even wrote the specimen numbers on accompanying sketches, as shown in Figure 3. Particularly note the specimen numbers in the left-hand column of Darwin's page. Everything fits together: specimens, interpretive memoranda, precise identification of location.

Unlike the field notes in pocket-sized notebooks, which Darwin also kept, the complete notes on the Fittonian model lent themselves to relatively easy conversion into published text. Geology was becoming an academic enterprise. The traveller was being schooled. Fitton's standards on note-taking formed part of the educational legacy of the field.

One point on which Fitton's instructions were silent is the dating of notes. Obviously, for historians always, and for participants sometimes, it can be helpful to know when an observation was made. For example, varying opinions on the Sedgwick-Darwin field trip of 1831 have resulted from the fact that neither man systematically dated his notes. The present-day reader of Sedgwick's or Darwin's notes is left to interpolate dates based on those that do appear in the text. In itself this is understandable, for field geologists the emphasis was on the completed picture, often a geological map. Just as a portrait painter would not want to date the application of every stroke, so a geologist might not want to take the trouble to date the collection of every specimen, or the recording of every observation. Fitton's instructions are geared towards the final picture; if the job is done properly, the individual geologist is in effect anonymous and absent. Fortunately for scholars, Darwin kept a Diary, where his entries are often dated. These entries can be used to interpret his formal notes on geology, where the given dates usually refer to the time of his visit to the site rather than to the dates he wrote his notes (Keynes 1988).[15] A few years after the *Beagle* voyage, Darwin became more aware of the practical utility of being able to retrace his steps, and began to date his notes more frequently (Herbert 1977, pp. 207–209).

Methodologically, Fitton was careful to compare the Australian specimens with those from England and other countries. He did this by consulting the private collections held by George Greenough (1778–1855) and Arthur Aikin (1773–1854) and the institutional collection held by the Geological Society of London.[16] The sort of information Fitton was assembling would have been useful to Greenough since he kept maps at the Geological Society for the purposes of recording data from travellers.[17] Overall, the procedure for collection described by Fitton is a good example of what Bruno Latour has characterized as successive missions of exploration acting as 'cycles of accumulation' whose results are formed into scientific knowledge at 'centres of calculation' at home (Latour 1987, pp. 219, 232).

To their credit, early nineteenth-century British travellers of all rank were impressively open to being instructed. To use the vocabulary of the time, this was an 'improving' era. The rapid advancement of the science of geology in the early decades of the nineteenth century was not a

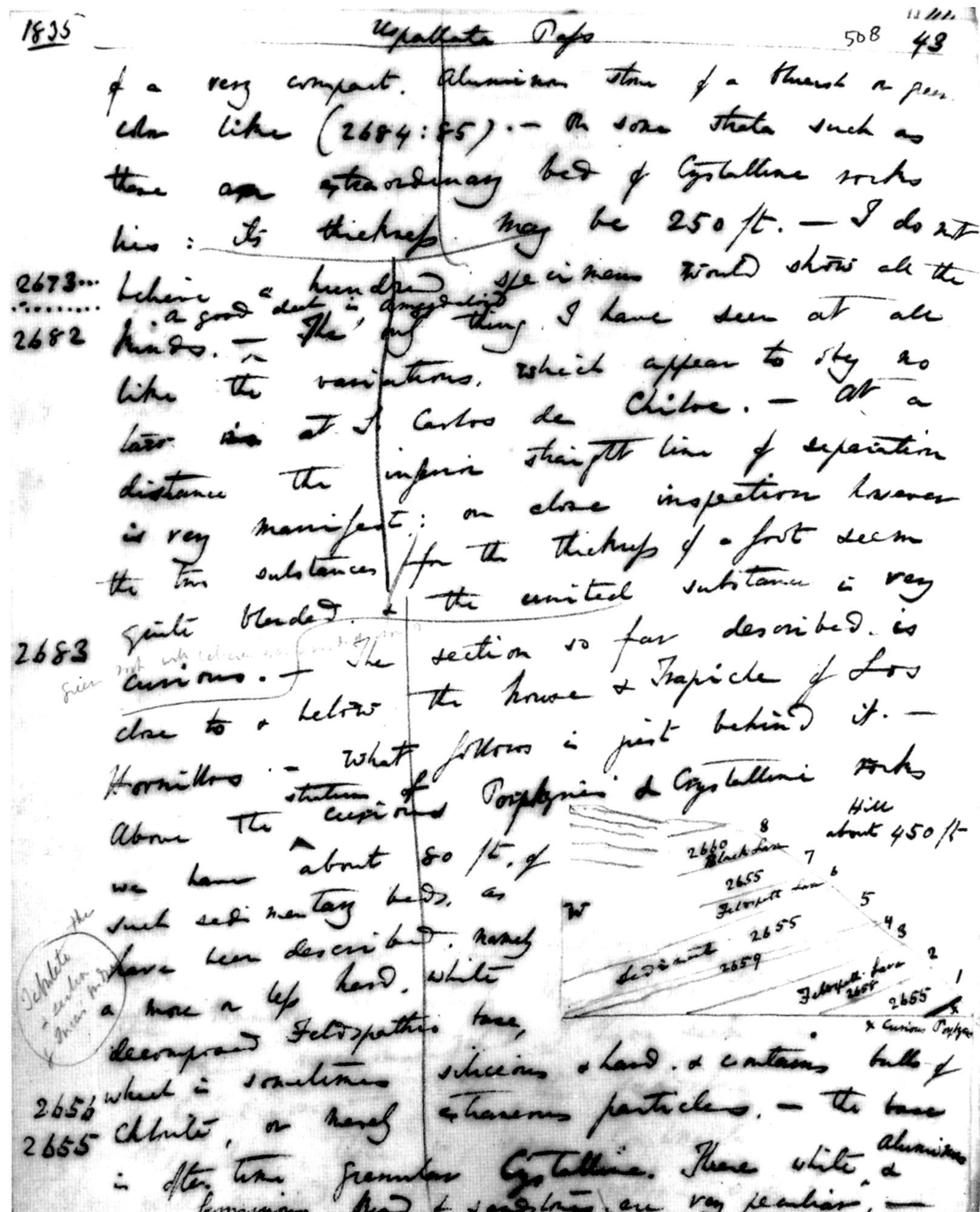
1835 — Uspallata Pass 508 43

of a very compact. Aluminous stone of a bluish or green colour like (2684:85). — On some strata such as these are extraordinary bed of Crystalline rocks lies: Its thickness may be 250 ft. — I do not
2673 .. believe a hundred specimens would show all the
2682 kinds. — The [a good deal in amygdaloid] only thing I have seen at all like the variations, which appear to stay so like [illegible] at S. Carlos de Chiloe. — At a later [illegible] at a distance the [illegible] straight line of separation is very manifest: on close inspection however for the thickness of a foot seem the two substances & the united substance is very quite blended.
2683 curious. — The section so far described is close to & below the house & Trapiche of Los Hornillos. — What follows is just behind it. — Above the stratum of curious Porphyries & Crystalline rocks we have about 80 ft of such sedimentary beds, as have been described, namely a more or less hard, white decomposed Feldspathic base, siliceous & hard & contains balls of which is sometimes
2656
2655 Chlorite, or nearly [illegible] particles. — the base is often time granular Crystalline. These white aluminous & ferruginous kind of sandstones are very peculiar. —

Fig. 3. Page from Darwin's Geological Notes On 'Uspallata Pass' dated 1835. From DAR 36.2:508, Darwin Archive, Cambridge University Library. Used with the permission of the Syndics of Cambridge University Library.

circumstance unique to that field. It would be ahistorical to isolate geology from other disciplines. For example, the publisher Charles Knight (1791–1873) planned to produce a series of volumes intended for the scientific traveller under the general rubric 'How to observe' and including the subjects of 'geology, natural history, agriculture, the fine arts, general statistics, and social manners'. As it happened only two titles appeared: *How to Observe: Geology* (1835) by Henry Thomas

De la Beche (1796–1855), and *How to Observe Morals and Manners* (1838) by Harriet Martineau (1802–1876), a celebrated writer (and traveller) who was a friend of various members of the Darwin family (De la Beche 1835; Martineau 1838). What is interesting in Knight's project is the broad and self-conscious concern with method: on teaching the novice how to observe.

De la Beche's *How to Observe* appeared slightly too late to be useful to Darwin on the *Beagle* voyage. But certainly De la Beche's influence as an instructor must be recognized. During the *Beagle* voyage, Darwin frequently consulted De la Beche's *A Geological Manual* (1831) and his book containing translations of French geological authors (De la Beche 1824, 1831).[18] Perhaps the capstone of De la Beche's career as instructor was his 1845 '*Instructions for the Local Directors of the Geological Surveys of Great Britain & Ireland*' which sought 'briefly to advert to a general mode of observing and recording facts ... by which systematic investigations and uniformity of results may be secured'.[19] In this latter work, De la Beche was addressing the experienced practitioner, rather than the novice, but the didactic impulse was equally present. The 'uniformity of results'—facts—was the goal in each.

The didactic impulse was also present across disciplines, as exemplified by Martineau's work in the '*How to Observe*' series. On its own, it was an early work in sociology, written by Martineau as she embarked on a trip to America (Webb 1960; see also Hoecker-Drysdale 1992).[20] For Darwin, it was a work of importance. Once he became a transmutationist after the voyage, Darwin read widely on the subject of morals (he was particularly interested in whether there was a universal moral sense), and he cited her book in his notebooks. Martineau was part of an intellectual circle that included Thomas Robert Malthus (1766–1834), and, in fact, Darwin read her work immediately before reading Malthus himself.[21] It was reading Malthus that triggered Darwin's development of what he came to call natural selection. As the second author in the *How to Observe* series, Martineau made the case for the connection between the study of geology and the study of society, suggesting that a geological map should cue an observer to the likely occupations of a populace, as one should expect to find miners in Cornwall, and so on.[22] Like Fitton, she also instructed travellers on how to go about taking notes, a somewhat more complicated procedure when one was observing human beings rather than inanimate objects. (She recommended not pulling a notebook out before people's eyes.) So overall this willingness to instruct, be instructed, and to establish 'uniformity of results' (facts) can be seen as a characteristic of that age. Going from reading geology to reading sociology was not trespassing; it was broadening. The motive here was increasingly academic and encompassing of all fields. Geology did not stand alone.

There was also a link between the *Beagle's* mission and the kind of geology Darwin pursued on the voyage. This was his work on the origin and distribution of coral reefs. In his '*Memorandum*' of directions for the voyage, Francis Beaufort (1774–1857), hydrographer, identified the problem as one to be solved. As part of his official duties FitzRoy performed a number of tasks that facilitated Darwin's enquiries. Meteorological measurements fell within the responsibility of the officers of the ship. For example, in explaining the absence of coral reefs at the Galápagos, Darwin credited FitzRoy for the suggestion that the seawater was too cold around the Galápagos Islands to permit the growth of reef-building corals.[23]

Of even greater importance was FitzRoy's charting of the Cocos or Keeling Islands in April 1836, where Darwin was able to explore an actual reef. Citing a letter from FitzRoy to the Admiralty written in February 1836 indicating that the stop at the islands was a possibility rather than a probability, Patrick Armstrong has suggested, plausibly in my view, that Darwin may have influenced FitzRoy's decision (Armstrong 1991, p. 9). In any case, FitzRoy's and Darwin's work at the 'Keelings', as FitzRoy called them, was symbiotic.

The quality of FitzRoy's charting was high; charts of the islands were not remade until the 1940s. In what was for that time a deep measure, FitzRoy showed that there was no sea bottom to be found at 1200 fathoms, measured on the SE side of the larger island at a distance of 2200 yards from the line of breakers. Darwin concluded from FitzRoy's measure that 'the submarine slope of this coral formation is steeper than that of any volcanic cone.'[24] This comment spoke against the presumption of Charles Lyell (1797–1875) that coral atolls grow directly on top of volcanic craters which, in Lyell's view, determined their distinctive shape. Darwin related the doughnut-shape of the atoll, with its interior lagoon, to the inability of the reef-forming organisms to thrive in still water. In other words, the shape of the atoll reflected the parameters of growth of the coral-producing organisms rather than the shape of the base of the platform on which the reef was built.

FitzRoy's plumbing the ocean with a line 7200 feet in length was sufficiently remarkable for its time to require comment. The earliest systematic effort to plumb the deep ocean took place during the 1817–1818 expedition of John Ross (1777–1856) to Baffin's Bay (Rozwadowski 2005, p. 49). Customarily explorers sounded depths that would effect navigation. Once ships were in deep water, and were in no risk of running aground, depth was irrelevant. By performing deep sounding in 'blue water', FitzRoy was in the vanguard of exploration of what would eventually become the science of oceanography. His work at the Keeling Islands permitted Darwin to carry forward his theory within newly expanded limits, though it would not be until the twentieth century that the ocean floor could be studied directly.

For Darwin the question of the origin and distribution of coral reefs was an intellectual puzzle. But it was a puzzle whose solution required thinking about the globe as a whole. The 'doing' of circumnavigations allowed the 'knowing' of coral reef distribution. Particularly important was the very expensive task of bringing of ships and research teams into the Pacific. Here credit must be given to the work of the navigators from many nations. The French naturalist-voyagers Jean-René-Constant Quoy (1790–1869) and Joseph Paul Gaimard (1796–1858), travelling in 1817–1820 under Louis Claude de Freycinet (1779–1842), had observed the key fact that organisms forming coral reefs exist at relatively shallow depths beneath the surface of the water (Quoy & Gaimard 1825). It was on an 1835 French chart, with names altered into English and the longitude reduced to that of Greenwich, that Darwin plotted the elements of his basic theory.

Darwin's theory posited that atolls rested on substantial platforms of reefs. Eventually borings were done that tested and confirmed Darwin's theory, by Australian and British geologists at Funafuti in 1896–1904, and by American geologists at Enewetak Atoll in 1952 (MacLeod 1988; Fautin 2002, p. 447). Understanding the formation and distribution of coral reefs involved the efforts of researchers of several nations. Moreover, academic knowledge regarding reefs depended on the efforts of practical men such as FitzRoy.

From travel to travel narratives

What do travellers with geological interests pursue upon returning home? For the group of travellers under discussion, the answer was often to write a narrative of their travels. In his narrative, Darwin incorporated brief summaries of his individual 'researches', as he termed them, into an account of his travels, which were largely organized chronologically. These summaries, following the pattern of his manuscript notes, were written to the methodologically self-conscious standard of academic geology then evolving (FitzRoy 1839).[25]

In contrast to the academically-oriented Darwin, a number of the authors on which he drew wrote their travel narratives without being grounded in a strong disciplinary tradition for natural history or geology, and in looking over the contributions of his fellow travellers Darwin was, privately, quite harsh in his judgements. He referred to Caldcleugh as 'the author of some bad travels in S. America,' classing him with other hospitable but presumably unacademic 'English merchants'.[26] Of the first volume of the *Beagle's* narrative, by Philip Parker King (1796–1856), he observed that it 'abounds with Natural History of a very trashy nature'.[27] FitzRoy, 'rather old fashioned in habits and ideas' as he described himself, inserted a traditional view of the Noachian flood into his volume of the *Beagle* narrative, thus prompting Darwin's patronizing aside to his sister that, 'You will be amused with FitzRoy's Deluge Chapter—Lyell, who was here to-day, has just read it, & he says it beats all the other nonsense he has ever read on the subject'.[28]

Darwin represented the future of academic geology, but he drew on the work of those whose disciplinary attachments were not so secure: of mining entrepreneurs like Caldcleugh and Miers, of collectors of mineralogical specimens like Mawe, of government officials like Parish, and of naval officers like FitzRoy. To return to Torrens's language, these other travellers were often primarily 'doers', practical men engaged in activities ranging from mining to governing. Their accomplishments were incorporated into academic geology quietly, with significant authors, such as Darwin, carrying forward the disciplinary standard. However, when looking at geology under the rubric of 'travel' their contributions were important.

Conclusion

During the early nineteenth century, contributions to the emerging discipline of geology were made by a wide range of travellers, some of whom were practically rather than academically oriented. During the voyage of HMS *Beagle* (1831–1836), Charles Darwin, a university-trained traveller, used materials provided by a wide range of travellers to develop important ideas in geology. For example, he speculated on the succession of species on the South American continent using material provided by Alexander Caldcleugh, a mining engineer, and on the laws governing the formation and distribution of coral reefs, using material provided by Robert FitzRoy, a naval officer. In considering the subject of 'geological travel' the wide range of travellers needs to be credited in order to understand the various sources of knowledge flowing into the new discipline. As the discipline of geology was just beginning to coalese, identifying the various sources of knowledge was more than a matter of distinguishing basic from applied science, for practical men, as FitzRoy with his plumbing of the ocean's depths, might be carrying out basic science as easily as more academically-trained men. The 'doers' could well be innovators.

At the same time within the circle of geologists active in the Geological Society of London, there were also men such as William Fitton who worked to see common standards applied to the collecting of specimens and the recording of observations. To that end he wrote a set of geological instructions for travellers. Darwin's research practice aboard the *Beagle* was consistent with an on-going emphasis within geology, as within a number of fields of study, on employing proper method. Increasingly the work of all geologically-interested travellers, whatever their background, needed to be pressed into a shape suitable to serve the discipline. Geology was acquiring an academic face.

Notes

[1]Conybeare & Phillips 1822. 'Preliminary Notice'. This volume was part of the library of HMS *Beagle* during its 1831–1836 circumnavigation. See Burkhardt & Smith 1985, **1**, pp. 553–566). 'Books on the *Beagle*'.

[2]Another geologist relevant to this discussion is John Taylor (1779–1863) who owned mines in Mexico though he did not travel there himself. Taylor's financial and managerial skills were important to the early success of the Geological Society of London and the British Association for the Advancement of Science. Taylor's Mexican mining ventures failed even though his British mining ventures flourished. See Burt (1977, pp. 39–47), Thackray (2003, p. 106), Morrell & Thackray (1981, pp. 25–27, 260–265, 461, 478, 502, 539).

[3]Fox, H. S. to Darwin, C., 25 July 1834 (Burkhardt & Smith 1985, **1**, p. 403).

[4]Caldcleugh had purchased the mine for 'one of the English Associations'. (See Darwin 1839, p. 419).

[5]This essay has been published in Herbert (1995). The quotation used in this paragraph is from pp. 32–33. The term 'tosca' had been described by Caldcleugh (1825, **1**, p. 145) 'The upper soil round Buenos Ayres is chiefly of a light nature, approaching to marl, and covering a stiff clay subsoil, called by the inhabitants, tosca'. The short timescale implied by Darwin's remark ('no Creation having taken place subsequently to the formation of the superior Tosca bed') is consistent with Darwin's notes from earlier in the voyage. See Pearson & Nicholas (2007).

[6]Caldcleugh's paper was sent 12 June 1835 from Santiago, and read at the Royal Society on 26 November 1835. Caldcleugh's information on the fall in temperature from 118 °F to 92 °F. in the springs at Cauquenes is quoted in Darwin 1839, p. 321. The papers read 4 January 1837 were printed as follows: Caldcleugh 1833–1838, Darwin 1833–1838. Also see Burkhardt & Smith 1985, **2**, p. 30 for Adam Sedgwick's more positive view of Darwin's paper than of Caldcleugh's, which he found 'diffuse & unnecessarily expanded'. As a university-educated author, Darwin's writing skills were, unsurprisingly, superior to Caldcleugh's.

[7]Hall's work entered the debate over the parallel roads of Glen Roy in Lyell (1833, **3**, pp. 131–132).

[8]Sutherland (1990, **3**, p. 138); see p. 137 for the higher figure for Scottish universities, which had a student population of roughly 4500 in 1825–1826. Mitchell 1988, p. 11. Not quite comparable but of interest is the number of people attending universities was small in relation to general population; about 30 000 men matriculated at Trinity College Dublin from the time of its founding in 1591 to 1846 (Herries Davies 1991, p. 321).

[9]*Colloque International Alcide d'Orbigny*. Paris 1–7 July 2002. The proceedings have been published in *Comptes Rendus Palevol*, 2002, **1**, part 6, Stratigraphie et micropaléontologie, de d'Oribigny à nos jours, and part 7, Voyageur naturaliste et systématicien.

[10]C. Darwin to W. D. Fox, 19 September 1831 (Burkhardt & Smith 1985, **1**, p. 163).

[11]Darwin, C. to Dana, J. D. 8 May 1852 (Burkhardt & Smith 1989, **5**, p. 91) 'I thank you much for your wish for me to have the Cirripedia of the expedition, but I know well how impossible it is'. The editors remark (note 6, p. 93), 'Presumably the specimens were the property of the United States government, sponsor of the expedition on which Dana served as naturalist'.

[12]By way of comparison to the map appearing in Fig. 2, Darwin's earlier freehand geological map of the southern area of South America is instructive as underlining the point that Caldcleugh and Darwin were working within a common tradition. For a reproduction of the map see Herbert (1991, p. 182).

[13]Darwin's notes from his field trip with Sedgwick are published in Barrett (1974). Two important recent reviews of Darwin's 1831 fieldwork in Wales with Sedgwick, both of which revise Secord's and Barrett's views at points, are: Lucas (2002) and Roberts (2001). On the general subject of Darwin's geological education, including his relation to his reading during the *Beagle* voyage see Herbert (2005). I have focused on Fitton's printed instructions in this article because of its relevance to the topic of the conference and because I have not treated it in detail in the book.

[14]Fitton in 1828 quoted in Woodward (1907, p. 76).

[15]Darwin's formal geological notes, as yet unpublished in their entirety, are held at Cambridge University Library. On Darwin's increasing frequency in dating his notes see Herbert (1977, pp. 207–208).

[16]Fitton (1827*a*, **2**, p. 566, 606). Fitton noted that Captain King's collection was presented to the Geological Society, and that duplicates of Brown's collection

were presented to the British Museum. Eventually Greenough's specimens went to University College, London, and some of Brown's material from the *Flinders* voyage went to the Oxford University Museum. Cleevely (1983, pp. 135, p. 71). I thank P.N. Wyse Jackson for this reference.

[17]'George Greenough'. *D.N.B.*

[18]Both are listed in 'Books on the *Beagle*' as having been frequently cited (Burkhardt & Smith 1985).

[19]De la Beche, H.T. 22 May 1845. Instructions for the Local Directors of the Geological Surveys of Great Britain & Ireland. British Geological Survey: GSM 1/4. I thank James Secord for a copy of this document.

[20]I thank Robert Webb for drawing my attention to the self-awareness of method in Martineau, an awareness she shared with geological writers of the period, including De la Beche.

[21]On Darwin's reading of Martineau see Barrett *et al.*, note 7, pp. 537, 555, 558. On the Martineau–Malthus connection to Darwin see Herbert (1977, pp. 211–221).

[22]See Martineau (1838, pp. 144–153) on the value of geological maps to the observer of society.

[23]DAR 37.2, Darwin Archive, Cambridge University Library.

[24]Darwin 1842, p. 8. The location of the measure is taken from Admiralty chart #2510 consulted at the Library of Congress. The material in this section is taken from Herbert (2005).

[25]On the publication history of Darwin's volume see Freeman (1977, pp. 31–54). For an extended treatment of what was entailed by the discipline of geology in Britain during the 1830s Herbert (2005, Chapter 2).

[26]CD to Susan Darwin, 23 April 1835 (Burkhardt & Smith 1985, **1**, p. 446).

[27]CD to Susan Darwin, 1 April 1838 (Burkhardt & Smith 1985, **2**, p. 80).

[28]Robert FitzRoy to CD, 26 February 1838; CD to Caroline Wedgwood, 27 October 1839 (Burkhardt & Smith 1985, **1**, p. 75, p. 236. FitzRoy (1839, **2**, pp. 657–682). On FitzRoy's deluge chapter see Herbert (1992, pp. 77–79).

References

ARMSTRONG, P. 1991. *Under the Blue Vault of Heavens: a Study of Charles Darwin's Sojourn in the Cocos (Keeling) Islands.* Indian Ocean Centre for Peace Studies, Nedlands, Western Australia.

BARRETT, P. H. 1974. The Sedgwick-Darwin geological tour of North Wales. Proceedings of the American Philosophical Society, **118**, 146–164.

BARRETT, P. H., GAUTREY, P. J., HERBERT, S., KOHN, D. & SMITH, S. (eds) 1987. *Charles Darwin's Notebooks, 1836–1844: Geology, Transmutation of Species, Metaphysical Enquiries.* British Museum (Natural History).

BROWNE, J. 1995. *Charles Darwin: Voyaging.* Knopf, New York.

BURKHARDT, F. & SMITH, S. (eds) 1985–1989. *The correspondence of Charles Darwin.* 5 Volumes. Cambridge University Press, Cambridge.

BURT, R. 1977. *John Taylor: Mining Entrepreneur and Engineer, 1779–1863.* Moorland Publishing Company, Buxton.

CALDCLEUGH, A. 1825. *Travels in South America, during the Years 1819–20–21; An Account of Brazil, Buenos Ayres, and Chile.* 2 Volumes. John Murray, London.

CALDCLEUGH, A. 1836. An account of the great earthquake experienced in Chili, 20th February 1835. *Philosophical Transactions of the Royal Society of London*, **126**, 21–26.

CALDCLEUGH, A. 1833–1838. On the elevation of the strata of the coast of Chili; Darwin, C. 1833–1838. Observations of proofs of recent elevation on the coast of Chili, made during the survey of His Majesty's ship Beagle, commanded by Capt. Fitzroy, R. N. *Proceedings of the Geological Society of London*, **2**, 444–449.

CLEEVELY, R. J. 1983. *World Palaeontological Collections.* British Museum (Natural History) and Mansell Publishing Limited, London.

CONYBEARE, W. D. & PHILLIPS, W. 1822. *'Preliminary Notice' to Outlines of the Geology of England and Wales.* William Phillips, London.

DANIELL, J. F. 1823. *Meteorological Essays and Observations.* Underwood, London.

DARWIN, C. 1833–1838. On the Formation of Mould. *Proceedings of the Geological Society of London*, **2**, 574–576.

DARWIN, C. 1839. *Journal of Researches.* Henry Colburn, London.

DARWIN, C. 1842. *The Structure and Distribution of Coral Reefs.* John Murray, London.

DARWIN, C. 1846. *Geological Observations on South America.* Smith Elder and Co. London.

DARWIN, C. 1849. Geology. *In*: HERSCHEL, J. F. D. (ed.) *A Manual of Scientific Enquiry.* John Murray, London, 156–195.

DE LA BECHE, H. T. (translation, notes) 1824. *A selection of the geological memoirs contained in the Annales des Mines, together with a synoptical table of equivalent formations, and M. Brongniart's table of the classification of mixed rocks.* William Phillips, London.

DE LA BECHE, H. T. 1831. *A Geological Manual.* Treuttel and Würtz, London.

DE LA BECHE, H. T. 1835. *How to Observe: Geology.* Charles Knight, London.

DOLAN, B. 1995. Governing matters: the values of English education in the earth sciences, 1790–1830. Unpublished Ph.D. dissertation. University of Cambridge.

EYLES, J. 1972. William Fitton. *Dictionary of Scientific Biography*, **5**, 14.

FAUTIN, D. G. 2002. Beyond Darwin: Coral reef research in the twentieth century. *In*: BENSON, K. R. & REHBOCK, P. F. (eds) *Oceanographic History: The Pacific and Beyond.* University of Washington Press, Seattle, 446–449.

FITTON, W. H. 1827*a*. An account of some geological specimens, collected by Captain P.P. King, in his survey of the coasts of Australia, and by Robert Brown, Esq., on the shores of the Gulf of Carpentaria, during the voyage of Captain Flinders. *In*: KING, P. P. *Narrative of a Survey of the Intertropical and Western Coasts of Australia. 1818–1822*. John Murray, London, Volume 2, 566–622.

FITTON, W. H. 1827*b*. Instructions for collecting geological specimens. *In*: KING, P. P. *Narrative of a Survey of the Intertropical and Western Coasts of Australia. 1818–1822*. John Murray, London, Volume 2, 623–629.

FITZROY, R. (ed.) 1839. *Narrative of the Surveying Voyages of His Majesty's Ships Adventure and Beagle;* 3 volumes + appendix to volume 2. Volume 1. Proceedings of the First Expedition, 1826–1830, under the Command of Captain P. Parker King; Volume 2, Proceedings of the Second Expedition, 1831–1836 under the Command of Captain Robert Fitz-Roy, Appendix; Volume. 3, Journal and Remarks. 1832–1836 by C. Darwin. Henry Colburn, London.

FREEMAN, R. B. 1977. *The Works of Charles Darwin*. Folkestone, Dawson.

GASCOIGNE, J. 1998. *Science in the Service of Empire*. Cambridge University Press, Cambridge.

HERBERT, S. 1977. The place of man in the development of Darwin's theory of transmutation. *Journal of the History of Biology*, **10**, 207–208.

HERBERT, S. 1991. Charles Darwin as a prospective geological author. *British Journal for the History of Science*, **24**, 182.

HERBERT, S. 1992. Between Genesis and geology: Darwin and some contemporaries in the 1820s and 1830s. *In*: DAVIS, R. W. & HELMSTADTER, R. J. (eds) *Religion and Irreligion in Victorian Society*. Routledge, London, 77–79.

HERBERT, S. 1995. From Charles Darwin's portfolio: an early essay on South American geology and species. *Earth Sciences History*, **14**, 23–36.

HERBERT, S. 2005. *Charles Darwin, Geologist*. Cornell University Press, Ithaca, New York.

HERRIES DAVIES, G. L. 1991. Hosce meos filios. *In*: HOLLAND, C. H. (ed.) *Trinity College Dublin and the Idea of a University*. Trinity College Dublin Press, Dublin.

HOECKER-DRYSDALE, S. 1992. *Harriet Martineau, First Woman Sociologist*. Berg, Oxford.

KEYNES, R. D. 1988. *Charles Darwin's Beagle Diary*. Cambridge University Press, Cambridge.

LATOUR, B. 1987. *Science in Action: How to Follow Scientists and Engineers through Society*. Harvard University Press, Cambridge MA.

LUCAS, P. 2002. 'A most glorious country': Charles Darwin and North Wales, especially his 1831 geological tour. *Archives of Natural History*, **29**, 1–26,

LYELL, C. 1833. *The Principles of Geology*. John Murray, London.

MAWE, J. 1821. *The Voyager's Companion*. Longman, London.

MACLEOD, R. 1988. Imperial reflections in the southern seas: the Funafuti expeditions, 1896–1904 *In*: MACLEOD, R. & REHBOCK, P. F. (eds) *Nature in its Greatest Extent: Western Science in the Pacific*. University of Hawaii Press, Honolulu, 159–191.

MARTINEAU, H. 1838. *How to Observe: Morals and Manners*. Charles Knight, London.

MIERS, J. 1826. *Travels in Chile and La Plata, including accounts respecting the geography, geology, statistics, government, finances, agriculture, manners and customs, and the mining operations in Chile*. 2 volumes. Baldwin, Cradock, and Joy, London.

MITCHELL, B. R. 1988. *British Historical Statistics*. Cambridge University Press, Cambridge.

MORRELL, J. & THACKRAY, A. 1981. *Gentlemen of Science: Early Years of the British Association for the Advancement of Science*. Oxford University Press, Oxford.

MULHALL, M. G. 1878. *The English in South America*. "Standard" Office, Buenos Aires.

MURCHISON, R. 1862. William Fitton. *Quarterly Journal of the Geological Society of London*, **18**, xxx–xxiv.

OLDROYD, D. R. 1972. Edward Daniel Clarke, 1769–1822, and his rôle in the history of the blowpipe. *Annals of Science*, **29**, 214–234.

PARISH, W. 1839. *Buenos Ayres, and the Provinces of the Rio de la Plata: their present state, trade, and debt; with some account from original documents of the progress of geographical discovery in those parts of South America during the last sixty years*. John Murray, London.

PEARSON, P. N. & NICHOLAS, C. J. 2007. Charles Darwin's geological observations at Santiago (St Jago) Cape Verde islands. *In*: WYSE JACKSON, P. N. (ed.) *Four Centuries of Geological Travel: The Search for Knowledge on Foot, Bicycle, Sledge and Camel*. Geological Society, London, Special Publications, **287**, 239–253.

QUOY, J. R. C. & GAIMARD, J. P. 1825. Mémoire sur l'accroissement des polypes lithophytes considéré géologiquement, *Annales des Sciences Naturelles*, **6**, 273–290.

ROZWADOWSKI, H. 2005. *Fathoming the Ocean: Discovery and Exploration of the Deep Sea*. Harvard University Press, Cambridge, MA.

ROBERTS, M. 1996. Darwin at Llanymynech: The Evolution of a Geologist. *British Journal for the History of Science*, **29**, 469–478.

ROBERTS, M. 2001. Just before the *Beagle*: Charles Darwin's geological fieldwork in Wales, summer 1831. *Endeavour*, **25**, 33–37.

SECORD, J. 1991. The discovery of a vocation: Darwin's early geology. *British Journal for the History of Science*, **24**, 133–157.

SUTHERLAND, G. 1990. Education. *In*: THOMPSON, F. M. L. (ed.) *The Cambridge Social History of Britain 1750–1950*. Volume 3. Cambridge University Press, Cambridge, 119–169.

THACKRAY, J. C. 2003. *To See the Fellows Fight*. Stanford in the Vale, British Society for the History of Science, p. 106.

TORRENS, H. S. 2001. Timeless order: William Smith (1769–1839) and the search for raw materials 1800–1820. *In*: LEWIS, C. L. E. & KNELL, S. J. (eds) *The Age of the Earth: From 4004 BC to AD 2002*. Geological Society, London, Special Publications, **190**, 61–83.

TORRENS, H. S. 2002. *The Practice of British Geology, 1750–1850*. Ashgate, Aldershot.

VACCARI, E. 2007. The organized traveller: scientific instructions for geological travels in Italy and Europe during the eighteenth and nineteenth centuries. *In*: WYSE JACKSON, P. N. (ed.) *Four Centuries of Geological Travel: The Search for Knowledge on Foot, Bicycle, Sledge and Camel*. Geological Society, London, Special Publications, **287**, 7–17

WEBB, R. K. 1960. *Harriet Martineau, a Radical Victorian*. Heineman, London.

WOODWARD, H. B. 1907. *The History of the Geological Society of London*. Geological Society, London.

WYSE JACKSON, P. N. (ed.) 1994. *In Marble Halls: Geology in Trinity College Dublin*. Department of Geology, Trinity College, Dublin.

WYSE JACKSON, P. N. 1998. William Henry Fitton (1780–1861) and the Wollaston Medal of 1852. *Geoscientist*, **8**, 9.

The quest for limestone in colonial New South Wales, 1788–1825

W. MAYER

Department of Earth and Marine Sciences, Australian National University, Canberra ACT 0200, Australia (e-mail: wolf.mayer@bigpond.com)

Abstract: In the absence of people with scientific qualifications in the newly established colony of New South Wales, Australia, the search for essential commodities, such as limestone, became the task of a few educated laymen, which included military officers and surgeons. The first discovery of calcareous rocks was made in the colony's outpost at Norfolk Island, which was a boon to that settlement but of limited benefit to the larger establishment at Sydney. For many years the major supply of lime for building came from shells collected from beaches and Aboriginal middens. Attempts to cross the Blue Mountains, long unsuccessful, succeeded in 1813 and led to the discovery of significant deposits of limestone. They provided lime for the needs of inland towns but did not supply Sydney until after the introduction of improved quarrying methods, lime manufacture, and building of roads. The first geological observations in New South Wales were made by people with no formal knowledge of geology, who were motivated by the necessity to find basic materials essential to the success of their settlement. Their efforts provided a crude inventory of rock types, which was improved by contributions from occasional visitors with a scientific background.

Limestone was first discovered in mainland New South Wales in 1815, more than 27 years after the founding of the convict colony. The reasons for this long delay lay partly in the geologically unfavourable and topographically difficult terrain that surrounds the site of the first European settlement, and partly in the reluctance of the colonial power to send people with the relevant scientific knowledge to the colony. As a consequence, the first attempts to explore the unknown land for earth resources fell to the initiative of a few military and medical officers, who, at best, had only a rudimentary knowledge of geology. They were motivated, not so much by a desire to make scientific discoveries, but by simple necessity. As the years passed, a small number of travellers with a scientific education, some of them attached to expeditions of discovery, visited New South Wales, and provided preliminary scientific appraisals of its geology.

The main objective of the British Government in 1787 in sending a fleet of 11 ships, carrying more than a thousand convicts, military and government officials to the shores of SE Australia, was to establish a remote settlement for its convicts and thereby ease the pressure on its overflowing gaols and prison hulks. Being despatched with a finite supply of food, seeds, and a range of inadequate tools, it was expected that the land in which they were to settle would supply them with all the remaining necessities of life.

Assorted gentlemen in search of limestone

The new arrivals, in 1788, had fixed on Sydney Cove, in Port Jackson, to establish the first European settlement in the colony of New South Wales, whose boundaries at that time were defined as extending westward as far as 135° of longitude (Fig. 1).[1] The colonists soon realized that most of the trees that grew in abundance around their encampment quickly blunted their saws and axes and did not yield timber suitable for the building of houses. They accepted that they needed to look to the rocks for most of their construction materials. Within weeks of their arrival they were fortunate to find an outcrop of excellent clay for brick making at the head of Long Cove, now Darling Harbour, only about a mile and a half from their settlement (Collins 1798, p. 17). However, their search for limestone to make lime for mortar and to enrich the poor sandy soil of their vegetable gardens remained frustratingly unsuccessful.

Among the pioneers who ventured beyond their tiny establishment at Sydney Cove to examine rocky outcrops and debris for the required substances were Watkin Tench (1759–1833) and William Dawes (1762–1836), both lieutenants of marines, the colony's surgeon-general John White (1757?–1832) and its first Governor, Captain Arthur Phillip (1738–1814) (Cunningham 1996). All were well-educated men by the standards of

From: WYSE JACKSON, P. N. (ed.) *Four Centuries of Geological Travel: The Search for Knowledge on Foot, Bicycle, Sledge and Camel.* Geological Society, London, Special Publications, **287**, 325–342.
DOI: 10.1144/SP287.25 0305-8719/07/$15.00

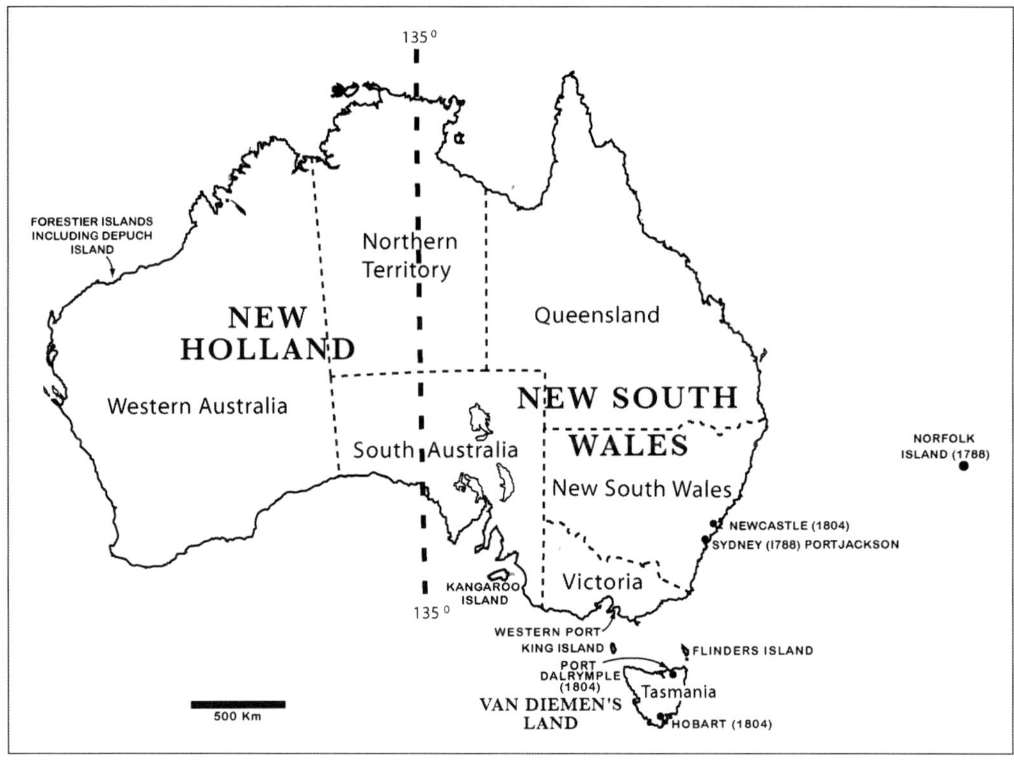

Fig. 1. Map of the Australian continent and adjacent islands. The years in brackets refer to the official establishment of the European settlements.

the time. They possessed the necessary determination and curiosity for the task, though they lacked the skills to interpret the geological nature of their strange new world.

They had few resources to assist them in their surveys. The only previous account of parts of the New South Wales coast appeared in the journals of Captain James Cook (1728–1779).[2] A popularized version of Cook's unpublished journal, intended for a public readership, was published by Hawkesworth in 1773.

As he approached Botany Bay in the *Endeavour* in 1770, the then Lieutenant Cook had recorded 'some white clifts [sic], which rise perpendicular from the sea to a moderate height'.[3] Sailing along the same coast in the convict fleet in 1788, the young naval lieutenant, Phillip Gidley King (1758–1808), whose home was in the south of England, imagined that the white cliffs which Cook had observed were made of chalk.[4] Had King been correct, the colony's requirements for lime could have been amply satisfied.

However, Cook himself had discovered the true nature of the rocks that dominate the coastal area of New South Wales where the first Europeans settled. During a short land excursion from Botany Bay, he noted that 'the stones are sandy and very proper for building'.[5] This journal entry must surely represent the first geological observation recorded in what was to become New South Wales.

It may seem strange that we owe the first description of what are now interpreted as the sandstones of Permo-Triassic Sydney Basin to a sailor, albeit an outstanding one. The surprise is the greater as Cook was accompanied at the time by two men of science, the botanists Joseph Banks (1743–1820) and Daniel Solander (1733–1782), neither of whom seems to have shown much interest in the geology of the newly discovered land. In later years, the influential Banks would give only token support to geological investigations in Australia, to the detriment of the colony's progress (Vallance & Moore 1982, pp. 2–3).

The unexpected appearance at Botany Bay of the French exploring expedition led by the Comte de La Pérouse, only a few days after the British Fleet's arrival, provided the aspiring colonists with some additional geological information. One of the visiting Frenchmen, J. A. Mongez (1751–1788), said to have been an authority on physical and agricultural subjects, suggested that the white clay used by the natives for body paint, would be suitable for the making of china (Vallance 1986, p. 152).

However, Watkin Tench and his fellow explorers soon discovered that the sandstone Cook had observed was the dominant rock type of the land surrounding their little settlement. Governor Phillip notified the Secretary for the Colonies in London that 'The stone of this country is of three sorts: Freestone, which appears equal to Portland stone, a bad firestone, and a stone that appears to contain large proportions of iron. We have good clay for bricks, but no chalk or limestone has yet been found'.[6] Phillip used the term 'freestone' for the thick beds of sandstone that crop out around the harbour, which could be cut with relative ease from the quarry face and worked to provide building stone. The 'bad firestone' he referred to could probably be equated with the white clay that Mongez had observed. According to Lawson, this material represented the leached part of the shale rocks that are interbedded with the sandstone sequence (Lawson 1972). The iron was probably a form of limonite associated with the sandstone.

Tench described the methods the prospectors used in their unsuccessful search for calcareous rocks: 'To find limestone many of our researches were directed. But after repeated assays with fire and chemical preparations on all the different sorts of stone to be picked up, it is still a *desideratum*' (Flannery 1996, p. 230). Surgeon White agreed with Cook and Phillip that 'the stone of this country is excellent for building' but added the proviso 'could any kind of cement be found to keep it together'. His disappointment at the failure of their experiments led him to make the premature prediction that 'there is not any limestone (I believe) in New South Wales' (White 1962, p. 119).

Although the colony's first 'geologists' were unable to recognize limestone on sight, they were clearly familiar with the use of 'chemical preparations', most likely simply the use of acid, to help in the recognition of the stone. It was probably Surgeon White who supplied the chemical substances and who knew how to identify limestone. The medical inventories of surgeons at the end of the eighteenth century included vitriolic (sulphuric) acid, sometimes used in dilute form to treat a malignant fever, as well as nitric acid (Keevil 1957–1963, 2, p. 229).

The effect of acids on limestone was well known. *Chambers Dictionary* of 1787 stated: 'It may easily be determined whether a stone will calcine, by letting fall a few drops of aqua fortis, or any strong acid, on the stone, which will boil and dissolve a part of it, if the stone is calcinable; but will lie upon the stone, like oil, and not ferment, if it will not calcine.'

The lack of rock lime perforce directed the efforts of the newcomers to the use of sea shells. It is not known who among the small community was aware that shells, when burnt, provided excellent lime to make mortar. In England, with its large resources of limestone, there was little need to use shells, but a number of the naval and military personnel at Sydney Cove had been to India or to the West Indies where the burning of shells to make lime was commonly practised. The method must also have been known to a wider public through the pages of the 1771 edition of the *Encyclopaedia Britannica*.

The task of collecting shells from nearby beaches was given to some of the women in the settlement. In the words of Tench, 'the female convicts have hitherto lived in a state of total idleness, except a few who are kept to work in making pegs for tiles and picking up shells for burning into lime' (Flannery 1996, p. 77). This activity marked the beginning of what would become one of the young colony's early industries.

However, such labour was slow and time-consuming and it yielded little raw material for their primitive kiln. As White informed us: 'the governor, notwithstanding that he had collected together all the shells which could be found for the purpose of obtaining from them the lime necessary to the construction of a house for his own residence, did not procure a fourth part of the quantity which was wanted' (White 1962, p. 119). At first, the lack of boats and convict overseers limited excursions to more distant beaches where shells were more plentiful. When such forays became possible, Aboriginal kitchen middens, often found at the mouths of natural caverns, supplemented the shells gathered on the beaches (O'Hara 1817, pp. 50–51). Even with this additional supply lime remained a scarce resource.

Frustrated by the lack of lime for the building of adequate houses and public stores, White blamed Captain Cook 'for having led them to suppose that oyster and cockle shells might be procured in such quantities as to make a sufficiency of lime ... but this is by no means the case. That great navigator, notwithstanding his usual accuracy and candour, was certainly too lavish of his praises on Botany Bay' (White 1962, p. 145). The criticism was undeserved. Cook's journal entries on this matter were accurate but had been embroidered by Hawkesworth[7] in his account of the navigator's voyages. It was this embellished version that White had read.

The shortage of lime forced the builders to use puddled mud instead of mortar to lay their bricks. The walls had to be made of great thickness, and no higher than one story, to stop them collapsing. Even then, rain would weaken the bond between the bricks so that within a year or two chimneys toppled over and walls crumbled.

The great value that the colonists attached to the possession of lime is well illustrated by Lieutenant Ball's report that 'twenty pounds of chalk have been saved from the wreck' following the loss of the *Sirius* in 1790 (Ball, quoted in Hunter 1795, p. 382). Ball was presumably referring to a consignment of lime from the small stores in Sydney intended for use on Norfolk Island, where the ship had struck a reef.

A quantity of coral and other testaceous substances

It was from distant Norfolk Island, late in 1791, that news arrived at Sydney Cove of the first discovery of calcareous stone. An excited Lieutenant King, who had established an outpost of the colony on the island in 1788, conveyed the good tidings to Governor Phillip. This time, after his earlier blunder off the New South Wales coast, King was not mistaken. He described the find as 'a quantity of coral and other testaceous substances ... [which] produced a very fine white lime ... and excels any lime made of chalk ... and proves to be a very tough cement'.[8]

It is somewhat surprising that the discovery was not made earlier, as the outcrop clearly differs in appearance from the Tertiary basalts and tuffs which make up most of the island. The wave-cut platform and some low cliffs where the sedimentary rock crops out are within sight of the first settlement established on the island. The stone has since been described as a cross-bedded and massive calcarenite, containing fossils of Quaternary age (Abell & Falkland 1991). Chemical analyses showed it to have a calcium carbonate content of 85.65% (Carne & Jones 1919, p. 4). The untrained eyes of the settlers may not have suspected that the light-coloured rock resembling sandstone contained lime. The usefulness of the stone was probably revealed by accident when fragments were placed in a fire and turned into a white powder.

'That valuable acquisition' as King referred to it, provided him with both a building stone and an abundance of lime for mortar. The quarry also supplied headstones for graves. It did not, however, solve the building problems in Sydney. With only one very small government vessel, the *Supply*, available at that time, the shipment of food from the fertile island to a hungry population in the larger settlement had to be given priority. Only small quantities of lime reached Sydney from Norfolk Island.

Limestone ballast

As early as September 1788 Governor Phillip had suggested to the authorities in London that ships bound for the colony carry limestone as ballast.[9] In 1791, appalled by 'the ruinous state of the dwellings and store houses', he made the same request in letters to the Secretary for the Colonies and to the Commissioners of the Navy. When several supply ships arrived at Sydney in 1791, carrying ballast of valueless stone, he made his point more forcefully in a letter to Lord Grenville (1759–1834): 'Your Lordship will readily conceive of how much consequence it would have been to the settlement had two or three hundred tons of limestone been sent out, and which might have been done, if those ships found it necessary to bring so much shingle ballast; for the limestone might with very little trouble have been changed for the stone of this country.'[10]

Governor Phillip's urgent pleas did not go unheard in London business circles. A shipping contractor, William Richards, lobbied Joseph Banks, who had considerable influence at the Colonial Office, with a proposal to send Dorking limestone, which he described as superior to any other in the country and of which one ton would go nearly as far as two tons of any other. He offered to supply it at a cost of 16s. to 18s. per ton and added that this was 'so trifling an expense compared to the magnitude of the object for which it is wanted'.[11]

The Commissioners of the Navy issued instructions in 1792 that limestone be sent as ballast in transport ships, but it took more than another year for the first of a very few shipments to arrive at Sydney. The *Sugar Cane*, transporting Irish convicts to the penal colony, brought a miserly 44 tons of the stone, shipped from Cork (Collins 1798, p. 262). Surgeon-Superintendent Bell, travelling on that vessel, opportunistically augmented the treasured cargo with another 10 tons at Rio de Janeiro, for which he paid a total of 10 shillings.[12]

Before leaving his post in late 1792, Governor Phillip once more reported the crumbling of the buildings for want of lime, a complaint that would be echoed by his successors over a period of more than twenty years.

First inland surveys

The task of exploring the country was made more difficult for the small band of settlers who attempted it by the ruggedness of their physical environment, which they encountered some way beyond their base at Sydney Cove, and by their limited resources and skills. The land behind the first settlement at

Sydney Cove occupies a roughly semicircular area of undulating lowland, the Cumberland Plain, which extends for about 60 km or 37 miles westwards before rising to form the formidable barrier of the Blue Mountains (Fig. 2). In order to expand their tiny territory and to search for resources such as limestone, it seemed necessary to explore and cross these mountains. Several tentative incursions into this rugged range during the 1790s came to a halt in the mountains' thickly wooded valleys and against the steeply rising sandstone cliffs.

In the absence of offers from members of the established community to engage in the tiresome and seemingly unrewarding task of inland exploration, John Hunter (1737–1821), who in 1795 had succeeded Phillip as Governor, allowed John Wilson (?1770–1800), an illiterate former convict who had served his time, to lead a small party on two journeys to the interior in January and March of 1798. The small band of adventurers included John Price, one of the Governor's servants, who kept a journal of their travels.[13] Instead of heading west, as most of the early explorers had done, Wilson chose to follow a southwesterly direction along a more easily traversed land corridor between the coastal ranges and the Blue Mountains (Fig. 2). During the second of these journeys he and his party reached a point about 125 miles (200 km) to the SW of Sydney. Theirs was the first inland expedition to travel beyond the near horizontal strata of the Sydney Basin terrain onto the deformed basement rocks of what is today referred to as the Palaeozoic Lachlan Fold Belt. Their path took them across the Blue Mountains at their easily negotiated southern end. They believed that they 'saw plenty of coal and limestone'[14] during their first journey, and that they crossed over ground that was 'covered with limestone and a kind of marble'[15] on their second excursion.

Although Wilson was an excellent bushman, his scientific credentials, and those of the other members of his party, were negligible. Nevertheless, it seems reasonable to assume that a person, said to have been born in Lancashire, England, would recognize coal. It is highly likely that this band of untutored adventurers discovered the first inland outcrop of coal in New South Wales, a short distance to the west of Colo Vale (Fig. 2). The coal seams that crop out there are part of the Permian Illawarra Coal Measures which had first been described in the preceding year in coastal cliff exposures, at what is now Coal Cliff, by the explorer and surgeon George Bass (1771–1803) (Branagan 1972, p. 12). The party's record of this find has gone largely unremarked in the literature. Cambage (1920, p. 6) was sceptical of their claim and suggested that their route may have taken them a few miles to the east of the locality where the coal appears at the surface. Given the sketchy record in their journal with regard to direction and distances travelled, this issue is debatable.

With respect to the claim by Wilson's party to have found limestone, it is clear that they were mistaken. Cambage suggested that the specimens they brought back to Sydney may have been samples of trachyte (Cambage 1920, p. 25). This rock type, of Tertiary age, crops out along their probable routes a short distance to the south of Mittagong and near Colo Vale (Fig. 2). It is possible that the greyish colour of the weathered trachyte pebbles and outcrops reminded the collectors of a similar coloration they had seen in limestone in England. Their misidentification of the rock may have persuaded Governor Hunter that their claim to have found coal, of which they do not seem to have collected specimens, might be equally false. He took no action to check out their discovery and, in doing so, may have deprived some of the least educated among the colony's subjects of some minor fame in the annals of geological exploration in Australia.

Mounds of shells and musings about their origin

Meanwhile, the citizens of Sydney continued to depend on sea shells as a source of lime. With the continued arrival of convicts, as well as some free settlers, their resources and collecting skills increased. They became better at locating shell mounds at beaches around Sydney and had more small boats for transporting them to their kilns. Governor Hunter was able to report that shells had been found 'all around the harbour in considerable quantities' and to declare that workmen were covering the decaying and crumbling houses with lime to 'ensure their lasting at least twenty years to come'.[16] His statement proved to be wildly optimistic. A coating of lime did not for long prevent the poorly built houses from further deterioration, nor did the supply of shells prove equal to the demand.

In 1800, the former Lieutenant King, now a post-captain, succeeded Hunter as Governor of New South Wales. The colony's growing population, that had now reached 5000, required increasing amounts of raw materials for building. King recognized the urgent need to explore beyond the settled surroundings of Sydney for these essential commodities.

First, in 1801, King sent his Lieutenant-Governor, Colonel Paterson (1755–1810), in a small naval vessel, the *Lady Nelson*, to the mouth of the Coal River, today's Hunter River and the site of Newcastle. Coal had accidentally been discovered in the vicinity by escaping convicts in

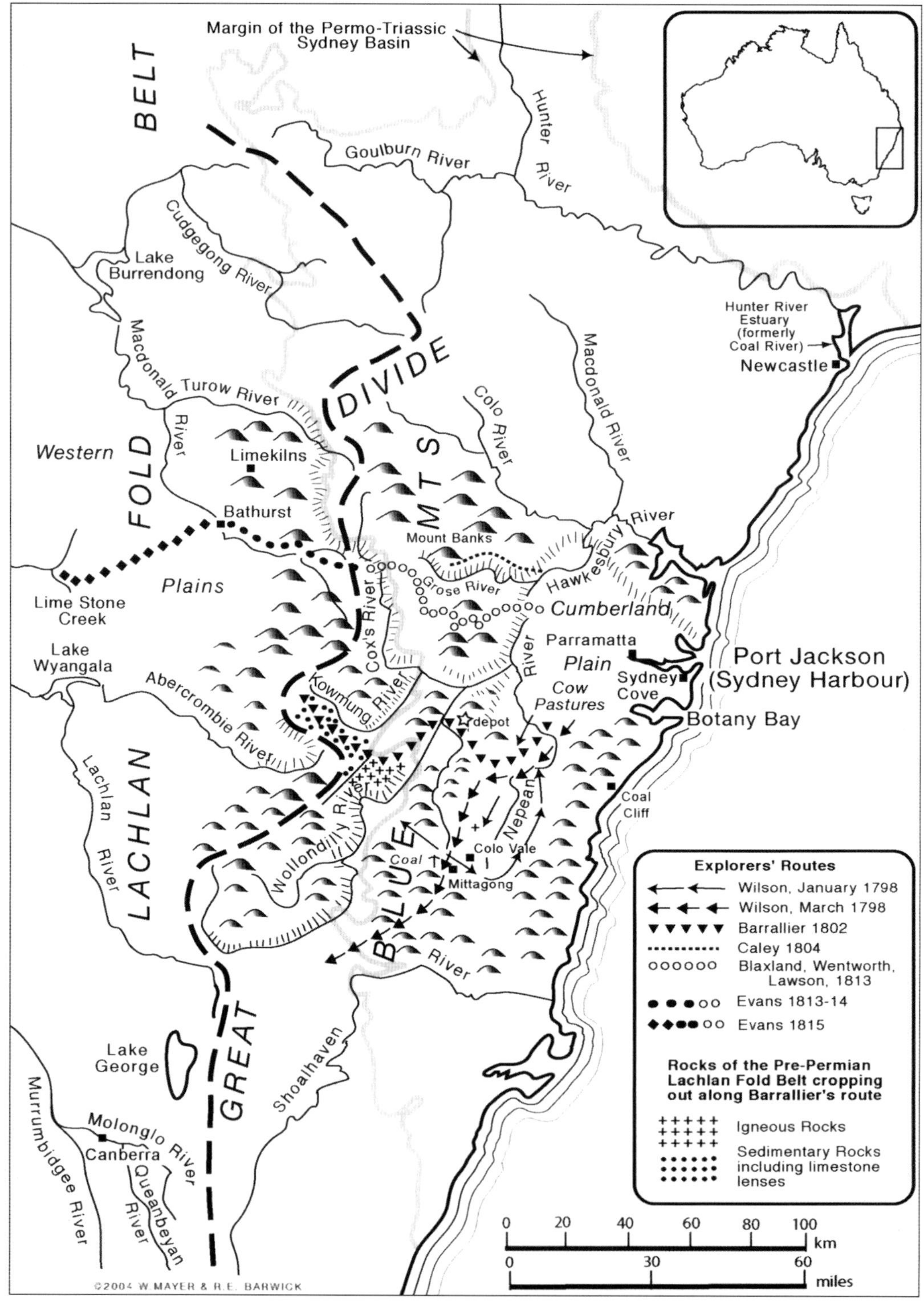

Fig. 2. Map showing an outline of the Sydney Basin and the country near its western margin. The routes followed by some of the colony's first European explorers are indicated. Note that topographic relief is not indicated in the northern and southern parts of the map.

1791 and at the mouth of the river itself in 1797 (Branagan 1972, pp. 11–14). King instructed his deputy to pay particular attention, amongst other things, to the occurrence of limestone or shells. Paterson, a Scot, who had arrived in the colony in 1791, was well suited for this assignment. He was a keen amateur botanist with a basic knowledge of geology, and had explored in parts of South Africa before he joined the army. He had also made collections of plant and rock specimens near Sydney and on Norfolk Island, many of which he had sent to Joseph Banks.

Following his detailed examination of the Coal River estuary and the adjacent country, Paterson was able to report: 'I have not as yet discovered anything like limestone, but the quantity of oyster shells on the beaches inland is beyond conception; they are in some places for miles. These are four foot deep without either sand or earth. Vessels might lie within a few yards of where they are found'.[17] These extensive accumulations of shells on the banks of this river would, for a time, become the major source of lime for Sydney.

The origin of these vast shell mounds became the subject of discussion among some settlers, an increasing number of which had been educated and took an intellectual interest in their surroundings. W. C. Wentworth (1790–1872), in his account of New South Wales, published in 1819, referred to their early speculations over the origin of the shell mounds:

> How they came there has long been a matter of surprise and speculation to the colonists. Some are of opinion that they have been gradually deposited by the natives in those periodical feasts of shell fish, for the celebration of which they still assemble at stated seasons in large bodies: others have contended, and I think with more probability, that they originally were large natural beds of oysters, and that the river has on some occasion or other, either changed its course or contracted its limits, and thus deserted them (Wentworth 1819, pp. 55–57).

W. C. Wentworth, who was born in the colony and became one of its most prominent public figures, provided us here with an early view of the formation of river terraces, though he was unaware of what produced the fluctuations in water level.

Later scientific investigations suggest that there was truth in both of the popular opinions referred to by Wentworth. Recent archaeological surveys (Dyall 1971) have revealed the presence of large clusters of Aboriginal campsites, marked by heaps of discarded shells, along ocean beaches and in the Hunter River estuary. The extensive shell accumulations, 'without either sand or earth', which Paterson encountered near the water's edge, represented most likely a series of overlapping middens. They would have been the first and most easily obtainable deposits of shells which, within a few years of their discovery, were collected by convicts and carted to nearby kilns for burning into lime.

In 1890, the geologists David and Etheridge published an account of raised beaches in the Hunter estuary, which include layers rich in marine shells. Thom and Murray-Wallace (1988) mapped some of the terraces in the upper part of the estuary and ascribed a Late Pleistocene (Last Interglacial) age to them. A section at Largs, about 20 miles upstream from Newcastle, shows shell beds associated with layers of clay and muddy sand.

Most of the shells the convict labourers burnt undoubtedly came from the loose accumulations of the midden mounds. It is likely that, as the supply from middens diminished, some naturally formed shell deposits were also used. This is supported by a report into the state of the colony which states that the shells 'are found in a loose and rich deposit of sandy loam, that is easily removed' (Bigge 1822, pp. 114–118).

Blue Mountains surveys and a first scientific assessment of the geology

Governor King's next venture was much more ambitious. Reminding himself again that the future growth of the settlement depended on access to both land and resources beyond the limits prescribed by the encircling mountain barrier, he sanctioned another expedition in 1802 to try to conquer this troublesome range.[18] For its leader, he chose the French-born Ensign Francis Barrallier (1773–1853), the son of an *émigré* from Napoleon's France, who had fled his native country in 1793 and subsequently spent several years in England. He came to New South Wales in the Governor's company in 1800. Little is known of his professional training in Europe; his biographer merely states that he was instructed by his father, who had been an engineer in the French Navy, in mathematics, fortifications and drawing (McQueen 1993). Whether or not he had acquired formal professional qualifications, Barrallier was clearly a very talented and versatile person. In 1801 he had surveyed Western Port, now part of the state of Victoria (Fig. 1), and had accompanied Paterson to the Hunter River where he produced a survey of the estuary. He was in charge of the improvement and maintenance of Sydney's defences and, at the time of his mission to cross the mountains, was aide-de-camp to the Governor. It is likely that he gained his first rudimentary understanding of geology during his earlier surveys and from his association with Paterson.

Barrallier's party of 1802 was better prepared than any that had been sent out previously. He was given the opportunity in October that year to make a preliminary journey to determine the best route into the mountains and to select a site for a supply depot (Fig. 2). The exact location of this staging post is not known; however Cunningham (1996, p. 96) believes that it was near the head of Sheahy's Creek just short of the more deeply incised valleys where Permian rocks are exposed below the Triassic sandstones and shales over which Barrallier had travelled from Sydney. After his return from this reconnaissance, the Sydney community was soon abuzz with rumours, repeated later in English newspapers, that Barrallier had discovered an abundance of limestone.

The botanist George Caley (1770–1829), who had arrived in the colony on board the same ship as Barrallier in 1800, and had also been one of the exploring party to Western Port, received these speculations with scepticism. In a letter to Sir Joseph Banks, dated 1 November 1802, he expressed his doubts about such discoveries based on his belief that Barrallier had not penetrated any further into the mountains than he himself had done on an earlier occasion.[19] Caley seems to have seen Barrallier as a competitor in the exploration of the mountains and believed himself to be better suited for the task.[20] While he was correct in his claim that the explorer had not passed beyond the western boundary of the Sydney Basin, Caley could not have based this statement on any firm evidence. He himself had, at that time, only entered the fringes of the chain (Cunningham 1996, p. 113; Webb, J. 1995). Caley was to be given the opportunity to make his own attempt at crossing the range some two years later in 1804, when he managed to advance as far as Mount Banks (Fig. 2), named by him after his patron, before being forced to return.

Governor King also corrected the popular opinion of Barrallier's discoveries by informing Banks that the specimens brought back included 'imperfect limestone' only.[21] Later, in a letter to Lord Hobart, King further reduced their value to 'very imperfect limestone'.[22]

It is somewhat puzzling how the notion of a major discovery of limestone took hold among the people of Sydney, given the insignificance of Barrallier's find during his short scouting trip. What can be of little doubt is the source of King's confident statement regarding the low calcium carbonate content of the specimens. From the middle of 1802, a French expedition led by Nicolas Baudin (1754–1803) had been anchored in Port Jackson. Its large scientific staff included Louis Depuch (1774–1803) and Joseph Charles Bailly (1777–1844), two scientists with professional training in geology.

Before arriving at Sydney the expedition had already explored along the coastlines of West Australian and Tasmania. They were to complete their charting of the southern margin of the Australian continent on their return voyage. The Baudin expedition carried out the first geological surveys by professionally trained scientists in Australia (Mayer 2005).

Despite being at war with France, the Government in London, in a spirit of scientific cooperation, had issued Baudin with a passport allowing him to conduct scientific investigations in Australia. Soon after his arrival at Sydney he received news that a peace treaty between the two countries, which was to last only a few months, had been concluded at Amiens.

Governor King generously entertained and assisted the French explorers in the repair and re-provisioning of their ships. He was less generous in granting them freedom to explore by land. A request by the zoologist François Péron (1775–1810) to accompany Barrallier on his first journey into the mountains met with a refusal (Péron & de Freycinet 1807–1816, **1**, p. 394). King seems to have retained some suspicion of French intentions in the region and preferred that they should confine their research to the ground that was already known. Nevertheless, not having a qualified mineralogist of his own in the colony, he asked Depuch to examine the specimens Barrallier had sent back from his first journey. The French geologist prepared a short report on the nature of the rock samples, now in the Mitchell Library, in Sydney.[23] The report's existence became known to us indirectly from a letter written three years later by King to Earl Camden, the then Colonial Secretary, in which he gave a brief account of Barrallier's first journey and an appraisal of the geological discoveries made. In his letter King merely referred to 'the Mineralogist' as the source of his geological information.[24] This vague attribution led Vallance (1975, p. 31) to assume, incorrectly, that King's 'mineralogist' was A. W. H. Humphrey (1782–1829) (Fig. 3), who had been appointed as the colony's first government geologist in 1803.

Depuch, stated in his report that among the samples Barrallier had sent back, only some specimens of shale were *légèrment calcaire* (slightly calcareous). It was this information on which King had based his statement that 'very impure limestone' had been found. The very mention of the word limestone may possibly have been the source of the wildly exaggerated rumours that had spread through the Sydney community.

The rock samples that Depuch was given to examine, and based on the locations given in Barrallaier's journal, were probably specimens of Triassic sandstone and shale. They also enabled the French geologist to draw some conclusions regarding the extent to which the explorer had penetrated

Fig. 3. Adolarius William Henry Humphrey (1782–1829), His Majesty's Mineralogist in New South Wales, from 1803 to 1812 (Photograph courtesy National Library of Australia, Canberra).

into the mountains. During one of his own brief excursions to the eastern margin of the Blue Mountains, Depuch in the company of fellow geologist Bailly had reached the point where the Grose and Nepean Rivers join to form the Hawkesbury River (Fig. 2). There they observed pebbles and boulders of a 'large variety of granite' which the rivers had brought down from the highlands.[25] Vallance and Branagan have also commented on the observations at this locality by the two Frenchmen (Vallance & Branagan 1968, Branagan 2002). Although they applied the term granite to a wide range of crystalline rocks (Plomley *et al.* 1990), Depuch's assumption, that the higher and more central parts of the range included igneous rocks of differing composition, proved to be correct. We are aware today that even pebbles from granite exposed on the western side of the Blue Mountains are transported to the eastern side by some of the rivers, including the Kowmung and Cox's Rivers (Fig. 2). In his report to the Governor, Depuch made use of his own field observations to declare that Barrallier had not reached the centre of the chain of mountains as, had he done so, he could not have missed the 'numerous varieties of granite' which must crop out there.[26]

The views held by the French geologists concerning the formation and classification of rocks reflected, in large part, those of Abraham Gottlob Werner (1749–1817). In his best known and most influential work, published in 1786 (Werner 1786, 1971), Werner grouped crustal rocks into four categories. The Primitive, or first-formed rocks, including the majority of igneous and metamorphic rocks, were deposited by precipitation in a primeval ocean and contained no fossils. Overlying these, and also deposited in an aqueous environment, were sequences of stratified rocks, often fossiliferous, the Flötzgebirge of Werner and referred to as the Secondary by Depuch. The younger alluvial deposits consisted largely of unconsolidated sediments and were derived mainly from the erosion of older rocks. In what Werner considered a minor category, he included rocks of volcanic origin, whose extent he thought to be limited. When exploring the West Australian coastal regions in 1801, Depuch had already recognized these four categories of rocks and believed them to constitute all of the 'exterior part of the globe'.[27] He accepted therefore that the 'granites' brought down by the rivers from the Blue Mountains, belonged to the Primitive or oldest rocks in the region and formed the core of the range. The layered sedimentary rocks of the Sydney Basin, that flanked this central core, were of a younger age and part of the Secondary, a term already in use for such rock types before Werner proposed his system of classification (Oldroyd 1996, pp. 80–82). Depuch had attended the lectures of Déodat de Dolomieu (1750–1801) at the Ecole des Mines in Paris. The celebrated geologist followed the neptunian views of Werner in explaining the origin and formation of the major crustal rock types (Mayer 2005). With respect to the formation of basalt, however, Dolomieu disagreed with Werner, ascribing their origin instead to the outpouring of subterranean lavas. Depuch closely followed Dolomieu's teaching in his interpretation and classification of the rocks he examined in Australia.

On his second and much longer journey, during November and December 1802, Barrallier did indeed encounter and collect specimens of 'granite'. Unfortunately, by the time of his return, the French ships had left Port Jackson. For a description of the rocks and minerals the expedition brought back we depend on Barrallier's journal entries, as none of the specimens is known to have been preserved.[28] The young explorer had been able to greatly increase his knowledge of geology during his conversations with the French geologists and with the zoologist Péron, who had a wide understanding of natural science. A comment in Barrallier's journal, stating that he believed he had passed the centre of the mountains as he did not find any more granite but only sandstone,[29] is evidence of how well he attended to the teaching of his mentors, though it is hard to know today what might have been his 'centre'.

Barrallier penetrated further westward into the mountains than anyone before him (Fig. 2).

He passed beyond the western rim of the Sydney Basin, and traversed some of the pre-Permian rocks of the Lachlan Fold Belt. As was the case with previous attempts at crossing the mountains, the extreme ruggedness of the terrain exhausted and finally defeated the party. With provisions running low and more difficult country lying ahead, he decided to return. The irony is that had he been able to continue for one or two more days he would have crossed the Great Divide and reached the gently undulating ground on the margins of the western plains. During his arduous passage through this difficult terrain he also passed tantalizingly close to deposits of limestone that he had so fervently sought to discover (Fig. 2).

Barrallier collected many rock specimens on his journeys. Pebbles of sandstone containing sea shells, which he picked up along the banks of the Kowmung River, represent the first discovery in New South Wales of what are now regarded as Devonian fossils. In his journal entry Barrallier noted that 'these stones seemed as if they had been carried from the summit of the mountain by this stream ...' and, further, that 'these discoveries which seemed to me interesting for a naturalist, strengthened my perseverance, as I hoped from day to day that I would be able to make more useful ones'.[30]

Among the discoveries Barrallier hoped to make, limestone rated highly. However, in this endeavour he was unfortunately unsuccessful. On the return journey, back among the rocks of the Sydney Basin, he found 'in a kind of slate ... various solid bodies, assuming the form of lozenges ...' which 'vitriolic [sulphuric] acid causes to effervesce'.[31] It is likely that the 'solid bodies' he refers to were calcareous concretions. It might be assumed here that the calcareous rocks Barrallier collected belonged to the Berry Siltstone, a formation within the dominantly marine Shoalhaven Group, that crops out below the Permian coal beds in the deeper river valleys of the Blue Mountains. This formation is in part calcareous and even contains minor limestone (Bembrick 1980, p. 137). However, Barrallier collected the above samples after he had returned to the eastern side of the mountains where the terrain is made up of Triassic rocks.

The positive reaction of the rocks to the acid may have given the explorer a moment of encouragement before he realized that their presence did not point to outcrops of limestone. It is interesting to note here that many of the exploring parties sent out, since the start of settlement, carried supplies of acid to test for limestone. As late as 1820, instructions issued to a party of explorers exhorted them 'to observe what sort of stone there is in the country you pass through, any that looks like limestone, you are to try with the acid you are provided with'.[32]

Péron, when writing the official account of the Baudin expedition, summarized the views of the French scientists with regard to the search for limestone in the Sydney area as follows: 'Our mineralogists did not discover in any of the lands occupied by the English [he was referring to the settlements in and around Sydney] the slightest vestige of carbonated lime, and the inhabitants are every where reduced to make use of the small quantity only which they procure by the calcination of shells, particularly by those of oysters, which latter are found at Botany Bay in vast abundance. The governor has in vain offered large rewards to those inhabitants who may discover any bank of limestone; and there is no reason to believe that such a discovery will be made.' (Péron 1809, p. 310). Péron and his fellow scientists had seen enough of Sydney Basin geology to convince them that the likelihood of limestone being found among its rocks was very low. However, they showed that such discoveries could be made elsewhere in Australia when they discovered limestone on Kangaroo Island (Fig. 1).

Barrallier's failure to cross the Blue Mountains had a more serious consequence. It created a general belief in the community that the range was insurmountable and must for ever remain a barrier to the extension of the settlement. This may have discouraged any would-be adventurers from making further attempts to explore the western country for several years. Governor King himself seemed to have decided that further exploration was not warranted. In a letter to Banks he stated that 'I cannot help thinking that persevering in crossing the mountains, which are a confused and barren assemblage of mountains with impossible chasms in between, would be as chimerical as useless'.[33] As Cunningham has pointed out, King modified his view in advice given to William Bligh (1754–1817), his successor as Governor, when he opined that crossing the mountains may be justified, but only if the settlement expanded.[34]

The reader may be puzzled by the persistent attempts by the explorers to cross the mountains in a westerly direction, given that Wilson had found a much easier route in 1798 which avoided the more rugged parts of the mountains and led the way to open country to the southwest. The reasons for this apparently illogical behaviour are twofold. Governor Phillip, in 1788, had landed two bulls and five cows on arrival in the colony which a few months later wandered off into the bush. They were found again in 1795 at what became known as the Cow Pastures, west of Sydney. The herd had increased in number and, at the time Wilson commenced his journeys, it contained well over 100 beasts. As the cattle were considered the property of government, neither Governor Hunter nor his successor, Governor King, who had been given Phillips' original one-third share in the herd, wanted the animals to be interfered with. The presence of the cattle therefore effectively limited the colony's expansion to the southwest. After King's departure in 1806, prominent settlers such as the influential John Macarthur (1766–1834) who had been granted large leases in the area, discouraged movement through his land (Cunningham 1996, pp. 87–88). Although a few individuals moved into the new country and established small farms, it was not until 1820 when Governor Macquarie journeyed to Lake George, first sighted by Europeans earlier in the same year, that official settlement commenced in that area. Less than two months after Macquarie's visit Charles Throsby Smith (1798–1876) found small deposits of limestone at the site of present-day Canberra. In the following year his uncle, the surgeon, explorer and major landholder, Charles Throsby (1777–1828), discovered larger deposits on the Queanbeyan River, to the south of Lake George (Fig. 2) (Mayer 2000, pp. 20–23).

Appointment of a mineralogist

The prospects for a more coordinated appraisal of the colony's resources received a boost with the surprise appointment, in 1803, of A. H. W. Humphrey as government mineralogist. Lord Hobart in a letter to Governor King[35] saw the appointment as 'the most efficacious aid in ascertaining the mineral productions of your Government'. He described Humphrey, who was given the title of 'His Majesty's Mineralogist', 'as a person in every respect well qualified for the duty' and 'that the results of his labour will be productive of considerable benefit to the settlement'. King was, however, to be disappointed both by the ability and the commitment of Humphrey. Vallance (1981) gives a thorough account of the mineralogist's background and his work in the colony. It appears that, rather than being trained in science, Humphrey derived his knowledge of mineralogy from his involvement in his family's business activities as collectors and dealers in minerals.

The Colonial Office in London, aware of the discoveries of coal and of reports that deposits of iron had been found in New South Wales, started to take a greater interest in the resource potential of its distant colony. The understandable fears expressed by Governor Phillip in 1788, when the tiny community was fighting for its survival, that the finding of minerals and precious metals 'would be the greatest evil that could befall the settlement',[36] had given way to a belief that such discoveries might benefit both the colonial power and its distant colony.

Officialdom in London seemed, however, to have formed the view that the search for Earth resources would be just as well accomplished by a knowledgeable mineral collector as by a trained scientist. This is born out by the yearly salary of only £95-5-0 paid to Humphrey, one that might have been appropriate for a prospector but was well below that due to a scientifically trained mineralogist (Vallance 1981, p. 109).

The establishment of a new settlement in Van Diemen's Land by Lieutenant-Governor Collins, with whom Humphrey had arrived in the colony, meant that the mineralogist commenced his official duties at the newly founded town of Hobart, in 1804. Governor King impatiently awaited his arrival in Sydney where he hoped Humphrey would give particular attention to making 'observations' in the Blue Mountains, in which endeavour he [King] would 'assist his researches to the utmost in my power'.[37] It was not until 1805, close to the end of King's tenure as governor, that Humphrey arrived in Sydney. His main activity there seems to have been selecting specimens of iron ore for shipping to England. Nothing came of the plans King had for him. By 1807 Humphrey was back in Hobart.

It is clear that Humphrey found life as a landholder and settler in Tasmania more congenial and much more financially rewarding than the duties expected of 'His Majesty's Mineralogist'. During the nine years he spent in the employment of government he displayed little enthusiasm for geological work, apart from a brief period after his arrival in the company of the botanist Robert Brown (1773–1858) (Vallance 1981, p. 114; Vallance & Moore 1982, pp. 32–33; Vallance *et al.* 2001, p. 462). In a letter to England written in 1803, he revealed his awareness of the colony's lack of limestone. He noted that it 'has not been discovered at Port Jackson, though no pains, as I am informed, have been spared by the Governor of that Colony to find so valuable and useful a Substance' (Vallance 1981, p. 129). But Humphrey seemed little inclined to contribute towards this effort. One historian comments: 'It is impossible even to figure the role of A. W. H. Humphrey, the mineralogist, at this time. We have seen that he was capable of active bush exploration, but nothing remains to mark his own specialised operations. In later years he did useful work as a police magistrate' (Giblin 1939, **2**, pp. 33–34).

The mineralogist had got himself into trouble with the government as early as 1806 when, in Sydney, he had entered into a liaison with Harriet Sutton, a convict girl whom he persuaded to come with him to Tasmania. Despite a Government

Order for her return and, some five years later, a direct verbal order from visiting Governor Macquarie (1762–1824) to the same effect, Humphrey refused to give her up (Vallance 1981, p. 119). When he offered his resignation as mineralogist in 1812, Governor Macquarie accepted it without regret and referred to him as 'being naturally an indolent man, and of a weakly and sickly constitution [who] has never made any discoveries in this country that are worthy of notice'.[38] To his credit, Humphrey married Harriet Sutton in the same year he resigned from his government position.

It is ironic that during Humphrey's initial stay in Van Diemen's Land it was a naval officer, Captain Kent (1751–1812), who found limestone and freestone for building in the northern part of the island in 1804, an area the mineralogist had briefly visited in the previous year. Kent, the nephew of the former Governor Hunter, joined the navy at the age of twelve. He arrived in Sydney in command of the *Supply* in 1795 in convoy with the *Reliance* which brought his uncle to the colony. Kent was sent in command of a number of different ships to Cape Town, India and the East Indies to obtain supplies for the colony. On one of these journeys he discovered and charted Port St Vincent in New Caledonia. In 1804 he took Lieutenant-Governor Paterson's party from Sydney to the northern coast of Tasmania where the latter was to establish the island's second settlement at Port Dalrymple, today's Launceston (Pike 1967, pp. 46–47). Paterson reported Kent's timely discovery to Governor King with great satisfaction.[39] Soon after, however, with his experience at the Hunter River to guide him, Paterson found 'a very large bank of shells on a nearby beach extending for near a quarter of a mile, and from 3 to 4 feet deep'. When trials established that the shells produced a superior kind of lime, the beach deposits became, in the early years, the preferred source of supply. As the settlers' primitive technology improved, the limestone which Captain Kent had discovered provided for the needs of the growing town (Hosking & Hueber 1954).

Sydney continues to depend on shell lime

The lack of limestone to provide a reliable and sufficient supply of lime continued to be felt in Sydney. Surrounding beaches were still being scoured for shells but yielded an inadequate supply for the kilns at Sydney and nearby Parramatta. The readily mined strata of coal and the large shell accumulation found by Paterson three years earlier at the Hunter River, persuaded Governor King to form a new settlement there which grew into the town of Newcastle, the second major settlement in New South Wales. From 1804, sporadic shipments of lime reached Sydney from the Hunter River to supplement local production. Ships returning to Sydney from Norfolk Island also brought some limestone as ballast (Turnbull 1806, p. 83). While supplies from these sources increased over the years they still did not meet all the needs of the Sydney region.

In 1806, King handed over the governorship of the colony to William Bligh. It was not long before the new governor despatched letters to London with the now familiar comments regarding the ruinous state of the buildings. The unfortunate Bligh, twice deposed from command, first as captain by the mutineers of the *Bounty* and, in 1808, by his own military officers as Governor, joined the chorus of his predecessors when he wrote that 'no marl, chalk or limestone has been seen' in the colony.[40] Following Bligh's departure, Lieutenant-Governor Foveaux (1765–1846) made an even more grim assessment of the town's buildings which 'are in a state of deplorable decay and dilapidation—so much so that I am decidedly of opinion that most of them must be rebuilt'.[41] He was, however, over optimistic about being able to obtain sufficient supplies of lime when he wrote: 'for so great is the plenty of shells at this settlement and its neighbouring coasts that no scarcity of lime can be apprehended for several years'.[42]

During Bligh's short but turbulent reign and during the period that followed it until 1810, when a new Governor restored stability in the colony, virtually no new efforts were made to assess the mineral potential of New South Wales. A gentleman of some renown, Dr Robert Townson (1763–1827), who had written books on mineralogy and natural history, and had carried out considerable work on limestone in Hungary, arrived in Sydney in 1807. Unfortunately, however, although he had brought with him chemicals and other equipment for rock and mineral testing, he received no encouragement from Governor Bligh (Vallance 1986, p. 160) and did little scientific work in the colony. In time, he became a wealthy land holder.

The arrival of Governor Macquarie in 1810 ushered in an era of expansion and greater prosperity for the colony. The population, now exceeding 10 000, and including a large number of free settlers and former convicts who had served their sentences, demanded leases of cultivable land that was now scarce. It was imperative for the colony to expand beyond the mountains that ringed Sydney on the inland side.

The production of shell lime at Newcastle and closer to Sydney was greatly increased to cater for the new Governor's major building programme. However, much more was needed, prompting

Macquarie to request, as the first governor had done, that all transport ships coming to the colony be ballasted with limestone.[43] Like that of his predecessor, Governor Phillip, Macquarie's call met with little success. Following the discovery of limestone near Hobart, he also urged the Colonial Secretary to send a skilful mineralogist to the colony to find limestone in New South Wales.[44] This request was also ignored. It was not until 1823 that John Busby (1765–1857) was appointed as civil engineer and mineral surveyor.

There is no doubt that most of the colony's inhabitants were as aware as the Government itself of the urgent need for limestone. Encouraged by the prospect of a reward, many must have kept their eyes open for likely looking rocks as they travelled through the countryside. Sir H. Hayes Browne, one of the colony's gentlemen convicts, who had been transported for the interesting crime of abducting an heiress, was one who believed he had succeeded. Writing to Governor Macquarie he informed him that he would take him to the spot where he had found limestone and modestly added that he would think himself 'amply rewarded by discovering so valuable an acquisition to a rising colony should it prove successful'.[45] As the spot he referred to appears to have been close to Sydney, his geological knowledge seems to have been wholly deficient.

Fig. 4. George William Evans (1780–1852), surveyor and explorer (Photograph courtesy Mitchell Library, Sydney).

Limestone find brings praise and censure to discoverer

The major breakthrough came in 1813, when three enterprising colonists succeeded in crossing the Blue Mountains in a westerly direction and descended to open country of undulating topography. The explorers Gregory Blaxland (1778–1853), William Lawson (1774–1850) and William Charles Wentworth (1790–1872) had chosen a route, long used by Aborigines, which followed the ridges of the range rather than the valleys, as previous expeditions had done. At the end point of their journey they found themselves on granite rocks of the Lachlan Fold Belt and close to the Great Divide.

In the same year, the Governor dispatched his assistant-surveyor, George William Evans (1780–1852) (Fig. 4) with a small party to explore the newly discovered land. He was the first European to make a complete crossing of the mountains and to reach the western plains. Following the success of this expedition, Macquarie had a road built over the mountains. In 1815, the hands-on Governor, in the company of Evans and the Surveyor-General John Oxley (1783–1828) travelled to the western plains himself to assess the new country. He was pleased by the sight of the vast stretches of much needed grazing land and selected a site for what was to become the city of Bathurst. His one complaint was 'that neither coals nor limestone have been discovered in the western country; articles in themselves of so much importance, that the want of them must be severely felt, whenever that country shall be settled'.[46] His concerns with regard to limestone soon proved to be unfounded.

Surprisingly, the Governor chose George Evans rather than the latter's superior, John Oxley, to continue the exploration of the country west of Bathurst. It may be that the success of Evans' first journey persuaded the Governor to entrust him with the performance of this new task. Some 45 miles SW of the new town site the surveyor, following along the banks of a creek, found 'pure lime stone of a misty grey colour'. He named the small stream Lime Stone Creek.[47] The deposit he discovered there (Fig. 2) is now recognized as part of the Cliefden Caves Limestone of Ordovician age (Lishmud *et al.* 1986, p. 101).

The long search, which had begun with the tentative efforts of the first colonists and had been pursued, almost obsessively at times, by both private individuals and by those who explored on behalf of the Government, had at last brought results. A grateful Governor complimented Evans on his discovery and rewarded him with £100 and

1000 acres of land in Tasmania.[48] The Secretary for the Colonies in London was less gracious. While conceding that Evans deserved praise for his perseverance in the field and for his achievements, the style of writing in his journal suggested to Lord Bathurst that he did not appear to be 'qualified by his education for the task of giving the information respecting this new country'. He proposed that in 'the future prosecution of these discoveries' Evans should be accompanied by 'some person of more scientific observation and of more general knowledge' whom he thought Macquarie would find 'among the officers of the regiment ... or among the medical officers'.[49]

Bathurst's comments do Evans a grave injustice and are undeserved. Although his short apprenticeship in engineering and architecture and some elementary training in surveying[50] did not prepare him for scientific exploration, and his modest social background (his father was secretary to the Earl of Warwick) did not recommend him to those in charge at the Colonial Office, he was nevertheless a careful and accurate observer in the field who recorded his findings in a simple, workmanlike style. It seems that the conservative nobleman strongly believed that military and medical officers, with a more traditional education and higher social standing, were more capable and better suited to the task of leading expeditions into unknown territory. When he approved, grudgingly it seems, the reward Macquarie had paid to his assistant-surveyor, he did so 'in consideration of his [Evans'] fatigue and extra expenses on his journey into the interior' rather than in recognition of his discovery of limestone.[51] Later, the appreciative citizens of the rapidly growing town of Bathurst, named by Macquarie after the very same critical aristocrat, erected a handsome statue, which today still graces the town's central park, to commemorate the explorer and his discoveries. His name is also attached to several topographic features in the Blue Mountains.

When, two years later, another expedition prepared to explore the interior, the Governor, who by nature tended to be less conservative in his choice of personnel, nevertheless paid lip service to his superiors in London and appointed the surveyor-general, John Oxley, to lead the party. Although not from the officer class, Oxley had served in the forces as a sailor. The capable Evans went as his assistant (Oxley 1820).

The discovery of a route to the interior of New South Wales had another consequence. Lord Bathurst sensed that the vast expanse of land to the west, in addition to its potential for agriculture, might also yield valuable mineral products. He sent out a detailed set of instructions for future explorers, requesting that in their travels they also look for and record the presence of minerals, precious metals or stone and bring back carefully labelled specimens. Their observations should also include the shape and direction of mountains.[52]

Within a few years after the first discovery of limestone in New South Wales other deposits were found, mainly on the western margin of the Sydney Basin. In 1817, Oxley was able to write to Macquarie that the country 'abounded in limestone' (Oxley 1820, pp. 373–374). The colony, for so long hampered and frustrated in its progress by the lack of lime, now seemed to possess an abundance of riches.

Continued shortage in the midst of plenty

If there was jubilation and satisfaction in the ranks of the Government in Sydney at this turn in fortune, the practicalities of exploiting the new-found resource proved more problematical. Skilled workers to quarry the rock and to build and operate lime kilns were in short supply. The distance from Sydney to Bathurst of about 125 miles (200 km), with a further 45 miles (72 km) to Lime Stone Creek, coupled with the steep and difficult ascent on the western side of the range, proved too great an impediment to the transport of bulk material to Sydney. For the time being, at least, the main settlement had to continue its reliance on supplies of shell lime from Newcastle and from beaches closer to the town, and on the occasional arrival of limestone from Norfolk Island.

Only a year after he received Oxley's favourable letter, Macquarie issued a General Order making it an offence to take sea shells from some of Sydney's beaches subject to a fine of not less than £5.[53] The great limestone deposits beyond the Blue Mountains, although they would soon provide ample supplies of lime for the building of new towns in the west, proved, as yet, of little benefit to Sydney.

Ironically, the deposit that Evans had discovered was considered to be too far away to exploit even for the Bathurst market, and in its early years the new town still imported lime from Sydney (Pearson 1981). However, in 1821, a local farmer and explorer, James Blackman Jnr (?1792–1868), found limestone some 16 miles to the north of Bathurst, at an area now known as Limekilns (Fig. 2). Blackman and his family arrived in Sydney as free settlers in 1801. He was one of the first farmers settled by Governor Macquarie at Bathurst in 1818, where he later also took on the duties of superintendent of convicts. His journeys of exploration in the western country led to the opening up of large tracts of land for settlement (Pike 1966, pp. 110–111). The limestone he discovered on one of these journeys occurs in scattered deposits of what is now known as the Early to Middle Devonian Limekilns Group (Lishmud *et al.* 1986, p. 247) and this produced the first rock lime in New South Wales. Production may have started soon after the discovery was made, but had certainly commenced by 1823.[54]

Another Frenchman, René Lesson (1799–1849), the surgeon of the visiting French vessel *Coquille*, visited the area in 1824 and commented on the significance of these new finds: 'They have discovered ... the mineral substance that seems to be lacking in New Holland, limestone, the greatest need of which is experienced by the English for the construction of their buildings, since the coast does not produce sufficient shells to meet requirements. The lime derived from it is very adhesive and consequently rated highly; only it is very expensive' (Mackness 1965, part 2, p. 71).

It had taken some 35 years since European settlement for the first rock lime to be produced in New South Wales. With the spread of agricultural settlements and the growth of new towns west of the ranges from the 1820s onwards, other deposits of limestone were discovered in the rocks around Sydney outside the strata of the Sydney Basin. Although the cost of transport prevented their full exploitation for some time to come, they eventually met all the needs of local communities.

By the middle of the nineteenth century, advances in technology and transport and the availability of persons skilled in the industry, made it possible to utilise the vast limestone deposits beyond the Blue Mountains for the requirements of Sydney. This led to the phasing out of burning shells that had sustained the colony's building trade for so many years.

Conclusion

For nearly half a century after the foundation of the colony of New South Wales the casual search for essential Earth materials was largely undertaken by a small number of its citizens, who had little or no geological learning. The military and naval officers, surveyors, surgeons, settlers, and even convicts, who engaged in this work sometimes succeeded in their endeavour by diligent application and a keen interest in the natural world, and sometimes by luck. It was not until the 1830s and 1840s that professionally-trained geologists, some of them permanent residents, carried out more extensive mineral surveys of the then much more widely settled continent of Australia.

The literature rarely mentions the names of the mostly untutored pioneers who made the first observations of a geological nature in New South Wales. This article gives some recognition to the geological work carried out by the dedicated early observers who explored the unknown land of New South Wales in search of limestone. Although their efforts were often motivated by need rather than by the nobler objective of scientific enquiry, their finds stimulated further interest in Earth science and made important economic contributions to the progress of the colony.

I am very grateful to D. Branagan and to D. Oldroyd for their comments and helpful suggestions from which I have greatly benefited. The latter has generously supplied me with a copy of the typescript of his paper published in this book. I also thank P. Wyse Jackson for help with editorial matters. I am greatly indebted to R. Barwick, in the Department of Earth and Marine Sciences at the Australian National University, for giving freely of his time in preparing the graphics for the two maps, Figures 1 and 2. I also thank G. Macauley for his generous help with the illustrations. G. Baglione, the curator of the Lesueur collection at the *Le Havre Muséum d'histoire naturelle*, has kindly sent me a number of manuscripts relating to the Baudin expedition. It is a pleasure to acknowledge here the professional help received from the staff of both the Australian National Library in Canberra and the Mitchell Library in Sydney. The former institution granted permission to publish Figure 3, and the latter Figure 4.

Notes

[1]Governor Philip's first commission. *Historical Records of Australia*, **1**, pt.1, 1–2.

[2]Cook, J. Journal on H.M.S. *Endeavour*. MS in National Library of Australia, Canberra.

[3]Cook, Journal on H.M.S. *Endeavour*, 26 April 1770, p. 241; Hawkesworth (1773) has quoted this observation unchanged.

[4]King, P.G. 1788. Journal, 14 January. *Historical Records of New South Wales*, **2**, Appendix C, pp. 538–589. King's own observations recorded in his journal leave no doubt that he had read Hawkesworth's book.

[5]Cook, Journal on H.M.S. *Endeavour*, 3 May 1770, p. 246. Hawkesworth (1773) has changed this to 'the stone is sandy and might be used with advantage for building'.

[6]Phillip, A. to Lord Sydney, 9 July 1788. *Historical Records of New South Wales*, **1**, 146.

[7]Hawkesworth 1773, p. 248; Hawkesworth had changed Cook's journal entry, Journal on H.M.S. *Endeavour*, p. 248: 'On the sand and mud banks are oysters, mussels and cockles...' to 'On the banks of sand and mud there are great quantities of oysters, mussels and cockles...'.

[8]King, P.G. to Phillip, A. 29 December 1791. *Historical Records of New South Wales*, **1**, 574–575.

[9]Phillip, A. to Lord Sydney, 28 September 1788. *Historical Records of New South Wales*, **1**, 189–190.

[10]Phillip, A. to Lord Grenville, 8 November 1791. *Historical Records of New South Wales*, **1**, 547; Phillip, A. to Commissioners of the Navy, 9 November 1791. *Historical Records of New South Wales*, **1**, 550.

[11]Richards, W. to Banks, J., 17 July 1792. *Historical Records of New South Wales*, **1**, 633. Richards is clearly referring to a particular deposit in the Cretaceous Chalk and probably compares it with the products from other quarries in the south of England. The various units making up the Chalk in the south of England have a calcium carbonate content that varies from as low as 63% in the Lower Chalk to as high as 98% in the upper Chalk. The more distant Jurassic and Carboniferous limestone deposits have a calcium carbonate content of 80% and 98% respectively. See in: Minerals Planning Guide 10: Provision of Raw Materials for the Cement Industry. *Office of the Deputy Prime Minister.* http:/www.odpm.gov.uk/stellant/groups/odpm_planning/documents/page/odpm_plan_606846-08.hcsp

[12]Bell, D. W. to Nepean E., 12 July 1793. *Historical Records of New South Wales*, **2**, 57.

[13]An account of two journeys into the interior of New South Wales, written by John Price. *Historical Records of New South Wales*, **3**, Appendix C, 819–828; On the first journey Wilson and Price were accompanied by an Irish convict named only as Roe; their companion on the second journey was a man named Collins.

[14]Price, *Historical Records of New South Wales*, **3**, Appendix C, p. 821.

[15]Price, Historical Records of New South Wales, **3**, Appendix C, p. 825.

[16]Hunter, J. to Duke of Portland, 12 November 1796. *Historical Records of New South Wales*, **3**, 175–176.

[17]Paterson, W. to King, P. G., 25 June 1801. *Historical Records of New South Wales*, **4**, 414–415.

[18]King, P. G. to Banks, J., 2 October 1802. *Historical Records of New South Wales*, **4**, 845.

[19]Caley, G. to Banks, J., 1 November 1802. *Historical Records of New South Wales*, **4**, 881.

[20]Cunningham 1996, p. 112. Caley, 1 November 1802. *Historical Records of New South Wales*, **4**, pp. 881–882.

[21]King, P. G. to Banks, J., 2 October 1802. *Historical Records of New South Wales*, **4**, 845.

[22]King, P. G. to Lord Hobart, 30 October 1802. *Historical Records of New South Wales*, **4**, 875.

[23]Depuch, L. 1802. Monsieur Depuch's opinion of the items lately sent from the mountains by M. Barrallier, Nov 13 [1802]. In *King Papers*, **8**: Further papers, 71 (in French). MS in Mitchell Library, Sydney, CY 906.

[24]King, P. G. to Earl Camden, 1 November 1805. *Historical Records of Australia*, **5**, 589.

[25]Péron & de Freycinet 1807–1816, **1**, pp. 439–440. The account of this journey was written by Bailly and incorporated by Péron in his official history of the French expedition.

[26]Depuch 1802. Monsieur Depuch's opinion of the items lately sent from the mountains by M. Barrallier, Nov 13 [1802]. In *King Papers*, **8**: Further papers, 71 (in French). MS in Mitchell Library, Sydney, CY 906.

[27]Depuch, L. 1801. *Quelque observations sur l'île des Amiraux.* MS n° 08015 (transcription by Jacqueline Bonnemains). *Le Havre, Muséum d'histoire naturelle.* The island named in the title is now known as Depuch Island and forms part of the Forestier Group lying off the coast of Western Australia (see Fig. 1 for location). Depuch does not seem to have been aware that Werner had introduced the Transitional as a fifth category into his system of rock classification, separating the Primitive from his Floetzgebirge or stratified rocks (the Secondary in Depuch's writings). Based on various manuscripts seen by the author, both Depuch and Bailly appear to have accepted Werner's neptunist views of an aqueous origin for both the Primitive and the rocks of the Secondary. With respect to volcanic rocks, however, they did not follow Werner in that they considered basalt to be of volcanic origin (Mayer 2005).

[28]Barrallier, F. 1802. Journal of journeys of exploration into the Blue Mountains, in *Historical Records of New South Wales*, **5**, Appendix A, pp. 749–825.

[29]Barrallier, F. 1802. Journal of journeys. *Historical Records of New South Wales*, **5**, Appendix A, p. 799.

[30]Barrallier, F. 1802. Journal of journeys. *Historical Records of New South Wales*, **5**, Appendix A, p. 801. David Branagan remarked to this author that the Devonian fossils Barrallier found were probably silicified.

[31]Barrallier, F. 1802. Journal of journeys. *Historical Records of New South Wales*, **5**, Appendix A, p. 813.

[32]Mayer 2000, Throsby, C. 1820. Instructions to exploring party. MS in New South Wales State Records, Reel 6034, pp. 167–177.

[33]King. P. G. to Banks J., 2 November 1805. *Historical Records of New South Wales*, **5**, 225–226.

[34]Cunningham 1996, p. 121; King, P. G. to Bligh, W., 22 August 1806. *Historical Records of New South Wales*, **6**, 171.

[35]Lord Hobart to King, P. G., 24 February 1803. *Historical Records of New South Wales*, **5**, 47–48.

[36]Phillip, A. to Lord Sydney, 28 September 1788. *Historical Records of New South Wales*, **1**, 190.

[37]King, P. G. to Lord Hobart, 1 March 1804. *Historical Records of New South Wales*, **5**, 331.

[38]Macquarrie, L. to Lord Liverpool, 17 November 1812. *Historical Records of Australia*, **7**, 587.

[39]Paterson, W. to King, P. G., 26 November 1804. *Historical Records of New South Wales*, **5**, 485.

[40]Bligh, W. to Windham, W., 31 October 1807. *Historical Records of Australia*, **6**, 145.

[41]Foveaux, J. to ____, 10 September 1808. *Historical Records of Australia*, **6**, p. 662. Probably a private letter to Under Secretary Chapman (see note 188, p. 745 in above volume).

[42]Foveaux, J. to Macquarrie, L., 17 January 1810. *Historical Records of New South Wales* **7**, 274.

[43]Macquarie, L. to a Gentleman, 30 June 1813. *Colonial Secretary's Index*, in Mitchell Library, Sydney, Reel 6043; 4/1728, pp. 86–87.

[44]Macquarie, L. to Lord Bathurst, 28 April 1814. *Historical Records of Australia*, **8**, 159.

[45]Sir Hayes Browne, H. to Macquarrie, L., 11 July 1810. *Colonial Secretary's Index*, in Mitchell Library, Sydney, Reel 6042; 4/1725, p. 250.

[46]Macquarie, L. to Lord Bathurst, 24 June 1815. *Historical Records of Australia*, **8**, 575.

[47]Evans, G. W. 1815. Journal of his expedition. *Historical Records of Australia*, **8**, 611–619.

[48]Macquarie, L. to Lord Bathurst, 18 March 1816. *Historical Records of Australia*, **9**, 60–61.

[49]Lord Bathurst to Macquarie, L., 18 April 1816. *Historical Records of Australia*, **9**, 114–117.

[50]Weatherburn (1966). Evans left England at the age of eighteen for the Cape of Good Hope where he worked in the Naval Storekeeper's Department. He arrived at Sydney in 1802 and commenced work as a surveyor in the following year. Given his short period of formal education it is likely that he built on his limited skills by practical experience.

[51]Lord Bathurst to Macquarie, L., 30 January 1817. *Historical Records of Australia*, **9**, 203.

[52]Lord Bathurst to Macquarie, L., 18 April 1816. *Historical Records of Australia*, **9**, 115–116.

[53]Macquarie, L., 24 October 1818. General Order. *Colonial Secretary's Index*, in Mitchell Library, Sydney, Reel 6047, 4/1741, p. 67.

[54]Cunningham, A. 1823. Journal and account of my daily travels from Bathurst to Liverpool Plains in New South Wales. MS in Mitchell Library, Sydney.

References

ABELL, R. S. & FALKLAND, A. C. 1991. The hydrology of Norfolk Island, South Pacific Ocean. *Bureau of Mineral Resources, Geology and Geophysics, Bulletin* **234**.

BEMBRICK, C. 1980. Geology of the Blue Mountains, Western Sydney Basin. *In*: HERBERT, C. & HELBY, R. A guide to the Sydney Basin. *Geological Survey of New South Wales, Department of Mineral Resources*, Bulletin **26**, 137.

BIGGE, J. T. 1822. *Report of the Commissioner of Inquiry into the State of the Colony of New South Wales*. London.

BRANAGAN, D. F. 1972. *Geology and Coal Mining in the Hunter Valley 1791–1861*. Newcastle History Monograph **6**, Newcastle Public Library, Newcastle.

BRANAGAN, D. F. 2002. Australian stratigraphy and palaeontology: the nineteenth century French contribution. *Comptes rendus, Palevol*, **1**, 657–662.

CAMBAGE, R. H. 1920. Exploration beyond the Upper Nepean in 1798. *The Royal Australian Historical Society*, **6**, 6.

CARNE, J. E. & JONES, L. J. 1919. The limestone deposits of New South Wales. *Mineral Resources New South Wales*, **25**, 4.

COLLINS, D. 1798. *An Account of the English Colony in New South Wales*. T. Cadell and W. Davies, London. [Australian edition 1975. A.H. and A.W. Reed, Sydney].

CUNNINGHAM, C. 1996. *The Blue Mountains Rediscovered—Beyond the Myth of Early Australian Exploration*. Kangaroo Press, Kenthurst.

DAVID, T. W. E. & ETHERIDGE, R. 1890. The raised beaches of the Hunter River delta. *Geological Survey of New South Wales, Records*, **2**, 37–52.

DYALL, L. K. 1971. Aboriginal occupation of the Newcastle coastline. *Hunter Natural History*, **3**, 154–157.

FLANNERY, T. (ed.) 1996. *Watkin Tench 1788*. The Text Publishing Company, Melbourne.

GIBLIN, R. W. 1939. *The Early History of Tasmania*. Methuen, London.

HAWKESWORTH, J. 1773. *An Account of the Voyages Undertaken . . . for Making Discoveries in the Southern Hemisphere*, 3 Volumes. Printed for W. Strahan & T. Cadell, London.

HOSKING, J. S. & HUEBER, H. V. 1954. The limestones of Tasmania and their industrial development. *Commonwealth Scientific Research Organisation*. Melbourne.

HUNTER, J. 1795. *An Historical Journal of the Transactions at Port Jackson and Norfolk Island*. Stockdale, London.

KEEVIL, J. J. 1957–1963. *Medicine and the Navy, 1200–1900*, 4 Volumes. E. & S. Livingstone, Edinburgh.

LAWSON, W. 1972. A history of industrial pottery production in New South Wales. *Journal of the Royal Australian Historical Society*, **57**, 17–39.

LISHMUD, S. R., DAWOOD, A. D. & LANGLEY, W. V. 1986. *The Limestone Deposits of New South Wales*. 2nd edn. Geological Survey of New South Wales, Mineral Resources, **25**.

MCQUEEN, A. 1993. *Blue Mountains to Bridgetown—The Life and Journeys of Barrallier, 1773–1853*. Published privately.

MACKNESS, G. (ed.) 1965. *Fourteen Journeys over the Blue Mountains 1813–1841*. D.S. Ford, Sydney.

MAYER, W. 2000. In search of a 'considerable river': Retracing the paths of the early explorers to Lake George, the Limestone Plains and the Murrumbidgee. *Canberra Historical Journal*, **45**, 15–24.

MAYER, W. 2005. Deux géologues français en Nouvelle-Hollande (Australie): Louis Depuch et Charles Bailly, membres de l'expédition Baudin (1801–1803). *Travaux du Comité français d'histoire de la géologie*, 3e série, **19**, 95–112.

O'HARA, J. 1817. *The History of New South Wales*. Printed for the author and sold by J. F. Hamilton. London.

OLDROYD, D. 1996. *Thinking about the Earth: A History of Ideas in Geology*. Athlone, London.

OXLEY, J. 1820. *Journal of Two Expeditions into the Interior of New South Wales, 1817–8*. John Murray, London.

PEARSON, M. 1981. Simple lime-burning methods in nineteenth century New South Wales. *Archaeology Oceania*, **19**, 112–115.

PÉRON, F. 1809. *Voyage to the Southern Hemisphere*. Printed for Richard Phillips by B. McMillan, London.

PÉRON, F. & DE FREYCINET, L. 1807–1816. *Voyage de decouvertes aux terres australes, exécuté par ordre de sa majesté l'Empereur et Roi, sur les corvettes* Le Géographe, *et* Le Naturaliste, *et la goëlette* Le Casuarina *pendant les années 1800, 1801, 1802, 1803, et 1804* ... 2 Volumes. Imprimerie Impériale, Paris.

PIKE, D. (ed.) 1966. *Australian Dictionary of Biography*, **1**, 1788–1850. Melbourne University Press, Melbourne.

PIKE, D. (ed.) 1967. *Australian Dictionary of Biography*, **2**, 1788–1850. Melbourne University Press, Melbourne.

PLOMLEY, B., CORNELL, C. & BANKS, M. 1990. François Péron's Natural History of Maria Island. *Records of the Queen Victoria Museum*, **99**.

THOM, B. G. & MURRAY-WALLACE, C. V. 1988. Last interglacial (Stage 5e) estuarine sediments at Largs, New South Wales. *Australian Journal of Earth Sciences*, **35**, 571–574.

TURNBULL, J. 1806. *Voyage Around the World.* August Campe, Hamburg. [German translation].

VALLANCE, T. G. 1975. Origins of Australian geology (Presidential Address). *Proceedings of the Linnean Society of New South Wales*, **100**, 31.

VALLANCE, T. G. 1981. The start of government science in Australia: A.W.H. Humphrey, His Majesty's Mineralogist in New South Wales, 1803–12. *Proceedings of the Linnean Society*, **105**, 107–146.

VALLANCE, T. G. 1986. Sydney Earth and after: Mineralogy of colonial Australia 1788–1900. *Proceedings of the Linnean Society of New South Wales*, **108**, 152.

VALLANCE, T. G. & BRANAGAN, D. F. 1968. New South Wales geology—its origin and growth. *In: A Century of Scientific Progress.* Royal Society of New South Wales, Sydney, 265–279.

VALLANCE, T. G. & MOORE, D. T. 1982. Geological aspects of the voyage of H.M.S. *Investigator* in Australian Waters, 1801–5. *Bulletin of the British Museum of Natural History, (historical series)* **10**, 2–3.

VALLANCE, T. G., MOORE, D. T. & GROVES, E. W. 2001. *Nature's Investigator: The Diary of Robert Brown in Australia, 1801–1805.* Australian Biological Resource Study, Canberra.

WEATHERBURN, A. K. 1966. *George William Evans—Explorer.* Angus and Robertson, Sydney.

WEBB, J. 1995. *George Caley: Nineteenth Century Naturalist: A Biography.* Surrey Beatty and Sons, Chipping Norton, New South Wales.

WENTWORTH, W. C. 1819. *Statistical, Political and Historical Description of the Colony of New South Wales.* Whittaker, London.

WERNER, A. G. 1786. Kurze Klassifikation und Beschreibung der verschiedenen Gebirgsarten. *Abhandlungen der Böhmischen Gesellschaft der Wissenschaften, auf das Jahr 1786*, 272–297, Prag und Dresden: In der Waltherischen Hofbuchhandlung.

WERNER, A. G. 1971. *Short classification and description of the various rocks.* Translated by A. M. OSVOPAT. Hafner, New York.

WHITE, J. 1962. *Journal of a Voyage to New South Wales.* Angus and Robertson, Sydney.

In the footsteps of Thomas Livingstone Mitchell (1792–1855): soldier, surveyor, explorer, geologist, and probably the first person to compile geological maps in Australia

D. OLDROYD

School of History and Philosophy, The University of New South Wales, Sydney, New South Wales 2075, Australia (e-mail: doldroyd@optushome.com.au)

Abstract: The Scotsman, Thomas Mitchell, was one of Australia's leading pioneers and explorers. Trained as a military surveyor, he travelled to Sydney in 1827, and was soon in charge of the mapping of New South Wales, an important position as it was vital to get the new colony surveyed, so that land could be apportioned correctly and fairly. In addition, Mitchell had responsibility for laying out the plans for the early road system of New South Wales, in which work he displayed excellent judgement, selecting a viable route for a road across the Blue Mountains, and directing the construction work. But Mitchell also had ambitions to be a proper explorer and to make original scientific contributions. He undertook four major expeditions, and later wrote them up in what are now classics of early Australian geographical writing. His Department's masterly 'Nineteen Counties' topographic map of the Sydney region and its hinterland, produced in 1834, was a milestone for the development of NSW. Mitchell had no formal training in geology, but on learning that he was to take up an appointment in NSW he rapidly made the acquaintance of London geologists and acquired the rudiments of the subject. While in the field in 1834, he was fortunate to be an early visitor to Wellington Caves, west of the Blue Mountains, where he excavated, collected, and dispatched many new mammalian specimens to experts in Britain. These attracted great interest in Edinburgh, London, and Paris, where the study of cave remains had been attracting attention for several years. Also in 1834, Mitchell coloured the outcrops of the main rock types of the Sydney region onto his 'Nineteen Counties' map, thus producing what was probably the first geological map of any part of Australia. This is reproduced here in colour for the first time. His later, published, geological maps of the Wellington area and a locality near Bathurst are also reproduced. The paper discusses the possible sources of information for the 1834 map. Mitchell wrote up his Wellington observations and submitted them to the Geological Society of London, but only an abstract was published and the full paper was refused publication (for unknown reasons). He eventually wrote up his findings, with illustrations, in a book describing his explorations (1838); but following his 'rejection' he did little further work in geology other than a hasty examination of a goldfield near Bathurst at the time of the NSW gold rush (1852). The present paper seeks to give a comprehensive account of Mitchell's geological work, which was an important adjunct to his career as a 'traveller/explorer' and cartographer.

Thomas Livingstone Mitchell (1792–1855), was a Scotsman who made a permanent mark on the history of Australia through his explorations, his survey work and mapping, and his laying out of lines of roads and towns. He oversaw the actual construction of roads and was responsible for three important pieces of road engineering. Besides this, he made many geological observations during his travels and some significant contributions to the early development of Australian geology. But Mitchell was essentially a military man, so it is interesting to know why and how he equipped himself to carry out geological work. In the 1830s, he compiled what were probably the first geological maps of Australia, one published, the other one only existing in manuscript form. He also published a geological map of an inland area in the 1850s, when he was engaged to make a report on the New South Wales goldfields; and this map had an interesting geological section of the main divide of eastern Australia. Most notably, Mitchell made contributions by collecting and sketching mammalian fossils from caves west of Sydney, and bringing them to the attention of European experts. He was a fine artist, and also had literary interests and abilities. As an explorer, he was unquestionably a traveller, and thus he fits well into the framework of this book. By the standards of his day, he was also a geologist. In addition, Mitchell made contributions to the study of Australian botany and zoology.

Mitchell was not exceptional among Australian explorers in having eclectic scientific interests. Others, like John Oxley (1783–1828) and Ludwig Leichhardt (1813–1848), also made significant geological observations, even in the early phases

From: WYSE JACKSON, P. N. (ed.) *Four Centuries of Geological Travel: The Search for Knowledge on Foot, Bicycle, Sledge and Camel.* Geological Society, London, Special Publications, **287**, 343–373.
DOI: 10.1144/SP287.26 0305-8719/07/$15.00

of Australian exploration. But Mitchell lived on in the colony to continue his work over a long period, even though his 'audience' was to a considerable extent in Europe, as might be expected according to George Basalla's (1967) views on the several stages of colonial science.

According to this model, science in colonial countries such as Australia passed through three stages. First, in a period that was simply an extension of geographical exploration, there were visitors, such as the botanist Robert Brown (1773–1858), Charles Darwin (1809–1882) or François Péron (1775–1810), who made observations in the 'peripheral' colonies and then took them back to their home countries, for presentation to colleagues at the 'centre'. In the second stage, which Basalla dubbed 'colonial science', there were those such as Frederick M'Coy (1817–1899) who trained overseas but stayed in the colonies, and did their work there. However, they were likely, where possible, to send their work back to the 'centre' for publication and were to a considerable extent dependent on the 'central' scientific culture. The 'colonial scientists' read the literature at the 'centre' and had institutional contacts there, looking as much to the 'centre' for reputation and distinctions as in their adopted homelands. They had access to many novel sources of information, but could not easily belong to the 'invisible college' or 'core set' at the 'centre'. Third, there came the innumerable people who trained in the colonies (or members of the Commonwealth), made their careers there, published there, and established an independent scientific tradition. Among geologists, one might mention, for example, Reginald Sprigg (1919–1994) of Adelaide University, famous for his work on the Ediacara fauna. The three stages overlap and there have been several modifications of the model and attempts to vary it according to the different colonial situations, such as India, Australia, or wherever. But as an approximate description the model works quite well, and Mitchell fits into it neatly. He may appropriately be categorized as a 'colonial scientist'. Incidentally, although science has long since passed into the third stage, the professional study of the history of science is, for the most part, even now mostly in the second stage in Australia.

To assist comprehension, a sketch-map showing the locations of the more important places referred to in the present paper is provided in Figure 1.

Mitchell's early career[1]

Born in Scotland in 1792, Mitchell's legal father John Mitchell was harbour-master at Grangemouth near Edinburgh and his mother was Janet (née Wilson); but he may have been the offspring of a local laird, Thomas Livingstone (died 1809). After his legal father's death, Thomas Mitchell was supported by his 'uncle', or likely father, Thomas Livingstone, who paid for him to attend Edinburgh University for a time. The young Mitchell also managed a coal-mine belonging to his 'uncle' for a brief period while but a teenager, and this experience may have stimulated some geological interests.[2]

However it was decided that Mitchell should have a military career. He obtained a commission as a junior officer in 1811 and went off to serve in the Peninsular Wars. By 1825 he was a captain. He served on the staff of Wellington's Quartermaster, the Scotsman Sir George Murray (1772–1846), who became Mitchell's patron.[3] Mitchell's role in the campaigns was to act as a kind of forward scout, getting information about the lie of the land and reporting on enemy troop movements and dispositions to headquarters. This involved the preparation of quickly drafted sketches and maps, and he trained as a surveyor in the army under the guidance of Major Charles Pierrepoint (died 1812). By the end of the campaigns (in which Mitchell saw considerable action), he was regarded as one of the army's best draftsmen and surveyors, so that when the battles were over he had (with Murray's support) the honour of being chosen to prepare the official maps for the wars on the Iberian Peninsula and spent five years on this task. While in Portugal, in 1818, he married Mary Blunt (1800–1883), daughter of a general in the British army. They eventually had twelve children.

Mitchell's post-war map-work was partly undertaken at Sandhurst, where he took courses in chemistry and geology under Richard Phillips (1778–1851),[4] as well as in painting and drawing; and he also was reading classical literature. He was promoted to major in 1826, and thereafter was generally known as Major Mitchell, though in 1839, after suitable lobbying, he obtained a knighthood, chiefly for his survey work and explorations in New South Wales. He was promoted to colonel in 1841. Not long before he left England, Mitchell wrote a book on surveying techniques for use by the military (Mitchell 1827).

But from 1826, as a half-pay army officer, Mitchell sought better opportunities, and again with the assistance of Murray he obtained the position of Assistant Surveyor-General in New South Wales, in the knowledge that he would soon succeed the Surveyor-General John Oxley, who was in poor health. There was an understanding that Mitchell would have the opportunity to undertake work as an explorer in addition to his duties as a surveyor. He arrived in Australia on 23 September 1827, and the following year,

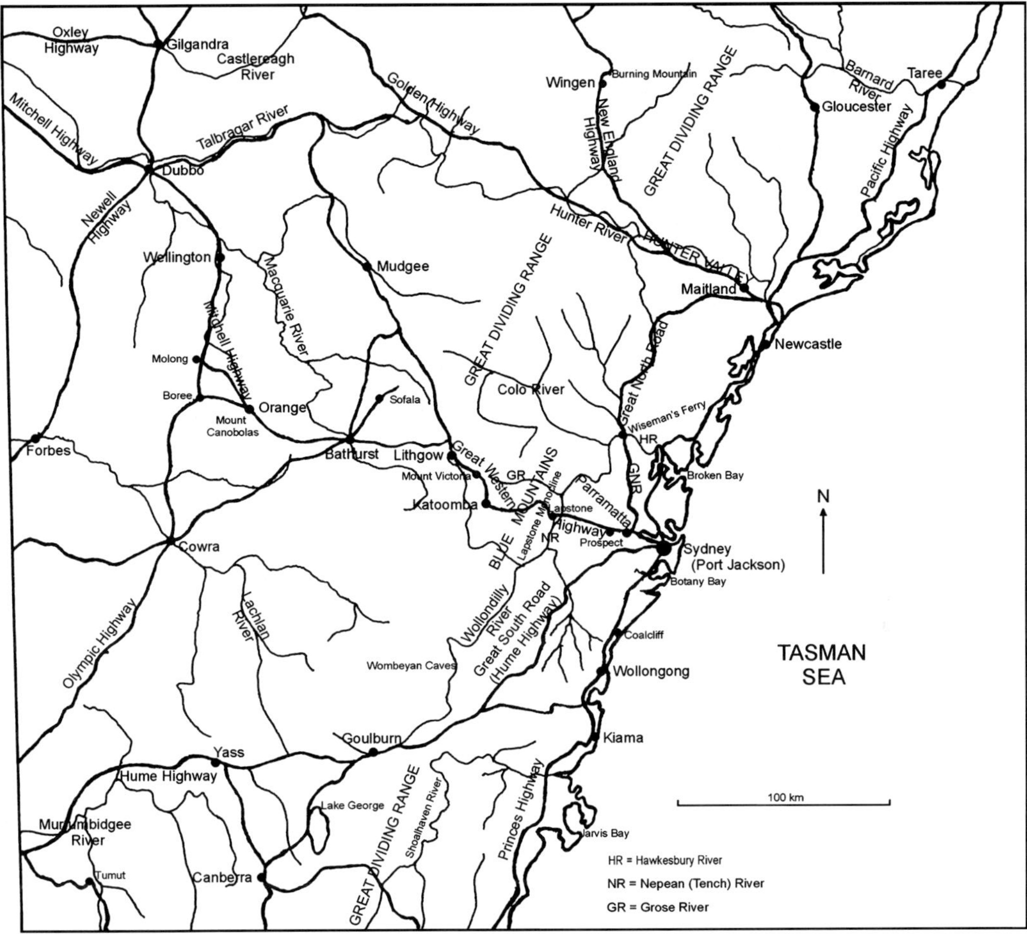

Fig. 1. Locality map for places referred to in the present paper.

on John Oxley's death, he succeeded him as Surveyor-General.

The post was one of importance for the well-being of the colony. The country needed to be mapped, and surveyed for the purpose of the distribution of land to the clamouring settlers. Lines of roads had to be selected, following the most convenient routes; and then the roads and bridges needed to be constructed. The task was huge, but the convicts provided free, or essentially slave, labour. Mitchell entered into this work with zeal and competence.

Early geological interests

There are records of Mitchell's seemingly sudden early interest in geology, not long before he left England, as can be seen from his diary of the period.[5] In February 1827, he was busy getting acquainted with members of the Geological Society, apparently starting with Roderick Murchison (1792–1871), which was natural in that both were retired but young Scottish soldiers, who had fought in the Peninsular Wars. (They may even have met in the army.) Mitchell was rapidly elected a Fellow of the Society, on 20 April 1827, being proposed by William Fitton (1780–1861), William Buckland (1784–1856), and the conchologist and zoologist (and subsequently a magistrate in London) William John Broderip (1789–1859) of Oriel College, a younger fellow-worker with Buckland in Oxford, who was with him when the first marsupial fossils were examined from the Stonesfield Slate near Oxford (now regarded as Jurassic). It was probably through Buckland's work, widely known amongst geologists in the 1820s, that Mitchell became interested in cave deposits in Australia, and perhaps with fossil marsupials.

Mitchell recorded for 20 February 1827, that he called on the Scotsman John Macculloch (1773–1835), who was then pioneering the geological mapping of Scotland, and 'beseeched' him to give him 'some instruction in Geology as to the manner of going Geologically over a country'. He was shown a table of 'the strata in the different formations on the world's surface' and was advised to examine stream sections. He should 'measure the thickness of the strata[,] the order—and the quality—and particularly the angle of inclination with the horizon'. '[S]uch materials were enough when sufficiently numerous to form a plan [map]'. Macculloch recommended Arran as a good place to start observing, it being, as he rightly put it, 'an epitome of the world'. But Mitchell had no time to follow up this advice.

On 22 February, Mitchell was introduced by Murchison to Dr Fitton, who showed him his geological library. The previous year, Fitton had published an account of all the geological knowledge of Australia then available, an offprint of which he gave to Mitchell (Fitton 1826). He recommended the Isle of Wight as a good place for a novice, but again Mitchell had no time to go there, so far as is known. Fitton drew Mitchell's attention to the common direction of the coastline and the ranges of eastern Australia, which parallelism was also found in other parts of the world. Fitton also advised him about the purchase of portable field equipment, suitable for the use of a travelling geologist (blow pipe, etc.) and available for two guineas.

The following day, Fitton introduced Mitchell at breakfast to John Herschel (1792–1871), who gave advice on the use of astronomical observations for surveyors. Joined by Murchison, Fitton and Mitchell then went to call on the botanist Robert Brown, who talked to him about Australian botany. Brown had made geological as well as botanical collections in Australia while accompanying the voyage of exploration of Matthew Flinders (1774–1814), and had journeyed westwards from Sydney to the foot of the Blue Mountains, and north to the Hunter Valley, where he had found fossil leaves, indicative of what is now termed the *Glossopteris* flora (Vallance & Moore 1982; Vallance *et al.* 2001, p. 16).

On 2 March, Mitchell attended a lecture on volcanoes by George Scrope (1797–1876) at the Geological Society and was introduced by Murchison to Colonel Thomas Colby (1753–1852), who was in charge of the trigonometric survey of Britain. He advised about the establishment of a base-line in New South Wales, as a basis for the overall mapping of the colony.[6] On the 4th, Fitton showed him how to use sections to help compile maps, and recommended sections of coastal cliffs as being especially useful. He explained a technique he used whereby a slip of paper was divided, each division representing fifty paces. Then the strata observed at fifty-pace intervals were entered on the section. As will be seen later, Mitchell used something like this technique when surveying in the New South Wales goldfields as late as 1851. Fitton recommended that places of special interest should be sketched in more detail, and collected specimens were to be noted on the section. Specimens should be labelled with gummed paper, indicating the place and date of collection. Fitton also showed Mitchell characteristic specimens of the older types of rock, according to the Wernerian sequence of granite, schist, clay-slate, etc.

Also on the 4th, Mitchell dined at the Murchison household where, among others, he met George Scrope, the Oxford chemist and geologist Charles Daubeny (1795–1867), the President of the Royal Society, Davies Gilbert (1767–1839), Fitton and the military man Captain Henry Kater (1777–1835), a multi-skilled instrument maker, metrologist, surveyor, and geodesist, who had worked on the great Indian Survey. On 26 March, Mitchell attended a lecture at the Geological Society by Macculloch. Given these last-minute activities, and the elementary information imparted to Mitchell by Fitton, it is unlikely that Mitchell had learnt much geology at Sandhurst from Richard Phillips, whose major interests were chemical. But he was now evidently making every effort to 'gen up' on the subject in a few months.

On 28 April, Mitchell called on a 'Mr Martin' (perhaps the well-known 'romantic' artist John Martin), who showed him the techniques of engraving mezzotints: cover a steel sheet with a warm thin layer of wax and rosin; then the drawing should be made on the wax, and the steel etched with *aqua fortis* (nitric acid). The well-known English landscape artist Anthony Vandyke Copley Fielding (1787–1855) advised on artistic representations of scenes and useful pigments. On 2 May, Mitchell was taking advice from the well-known taxidermist Benjamin Leadbeater[7] on the skinning and preservation of bird specimens (which skill doubtless became useful to Mitchell during his Australian explorations—he was much interested in the Continent's birds). On the 16th he was in discussions with the publisher John Murray about his book on military surveying, though it was eventually issued by Samuel Leigh. Mitchell sailed for Australia with his family on 10 June, in the *Prince Regent*, which arrived in Sydney on 27 September. He took up his duties on 6 October, having met John Oxley shortly before.

Mitchell was an enigmatic character. On the one hand, he could be exceedingly kind and efficient, and he was certainly a manic worker. On the other hand, he had a filthy temper and quarrelled with a succession of Governors and subordinates. He was brave, competent, and ambitious, and desirous of obtaining fame (which he achieved). The department that he was to take over had got into a mess through Oxley's illness and Mitchell pulled it together. But his real goal was to make a name for himself as an explorer and discoverer of new scientific facts, areas suited for agriculture and valuable mineral deposits. So he was forever off on exploring trips rather than attending to the routines of survey work. He was on leave in Britain from 1837 to 1841, from 1847 to 1848, and again from 1853 to 1854. As a colonial scientist, he was always eager for leave, seeking to maintain his contacts with Britain, publish his work there, and get it known at the 'centre'. Mitchell came close to dismissal on more than one occasion for insubordination and exceeding his leave, but with good contacts he was able to appeal successfully to the higher authorities in Britain, over the heads of the local Governors. So he was perhaps a mixed blessing to Australia. He certainly contributed much by his mapping, exploration, and road construction, even if he was frequently out of the country and his department eventually fell into disarray.

Geological knowledge in Australia at the time of Mitchell's arrival; and his geological map of 1834

In 1838, Mitchell presented a hand-coloured geological map of the area of Sydney and its hinterland to the Geological Society of London (see Fig. 2), and a similar one was probably donated to the Royal Geographical Society.[8] The Geological Society's copy (catalogued at 994/4), a fragile, folded document, is dated on the outside of the cloth cover in Mitchell's hand: 'New South Wales T. L. Mitchell *1834*'. This is presumably the date when the map was geologically coloured.

This was very likely the first general geological map compiled in Australia,[9] and it may, for the most part, be reconciled satisfactorily with the modern map of the region.[10] However, how much of the information was actually collected by Mitchell? And was it a great achievement to produce such a map? To answer these questions, one must consider the knowledge already available from previous observers.

First the massive sandstones, which are obvious everywhere in the Sydney area: one does not have to be a geologist to recognize them. They mostly form a gently folded basin, but are folded up in the steep Lapstone Monocline, as it is commonly called today, west of Sydney, to form the Blue Mountains, in which range they form the famous orange-yellow cliffs that tower over the valleys cut into them by fluvial erosion. Modern geologists divide the sandstones into an upper Hawkesbury Sandstone and a lower Narrabeen Group, but this distinction was not made by Mitchell.

The coal seams depicted on Mitchell's map had been known from the early days of settlement. The Newcastle coal was first discovered by a group of escaped convicts in March 1791, only three years after the Europeans established a settlement at Sydney. In June 1797, some shipwrecked sailors, making their way northwards along the coast to get to Sydney, discovered a coal seam cropping out on the coast almost at sea level between Sydney and Wollongong, among the softer sediments that crop out at the edge of the basin, below the ubiquitous Sydney sandstones.[11] The Newcastle coal was soon being exploited and mining began in a small way in 1801, with convicts as labourers.[12]

By the time that Mitchell was in the Hunter Valley/Newcastle area in 1829, arrangements for substantial commercial mining had been established, with the lease allocated to the Australian Agricultural Company. Though the company did not actually start shipping out coal until 1831, a report (dated 24 March 1827) was prepared by the Company's mining superintendent, John Henderson (*c.* 1784–1835),[13] who examined the Newcastle area's coal outcrops and reported favourably on the prospects. In addition, he prepared a fine coloured east–west geological section, showing the coal seams passing under (or through) the hill to the north of which Newcastle's city centre is now situated. Following Henderson's recommendation, the coal mines were later opened to the west of the city a mile or two inland—and later further west, up the Hunter Valley. His 1827 section showed that vertical pits had already been dug on either side of the hill, down to and through, the coal seams. Henderson also looked for coal in other areas in the neighbourhood of Newcastle (and also at Paramatta), so he could have provided Mitchell with a good deal of information.

I am informed by Dr Pennie Pemberton archivist at the Noel Butlin Archives Centre of the Australian National University, Canberra, who has made a close study of the Agricultural Company's archives, that Mitchell would have been familiar with the nature and extent of the area of the Newcastle coalfield, as one of his surveyors marked out the Company's area of lease and Mitchell and Henderson probably met one another (pers. comm., 13 July 2004). Also, in his duties as Surveyor-General, Mitchell was involved with issues concerned with

Fig. 2. Thomas Mitchell's hand-coloured geological map of the Sydney area and surrounding region, dated 1834: probably the first geological map of any part of Australia. Reproduced by courtesy of the Geological Society, London. Key: (1) a central area of 'Upper Shales' (represented in brown by Mitchell); (2) a large area of 'Sydney Sandstone' (orange-yellow) and alternations of sandstone and shale; (3) Coal beds (black/grey) to the south of Sydney, near the present coal mines of the Wollongong area; and to the north near the seaward end of the Hunter Valley, in the area of the modern industrial town of Newcastle; (4) 'Calcareous Sandstone' (blue) along the south coast near Wollongong and to the north of the Newcastle coal measures; [5] Trap (dark green) cropping out on the coast south of Wollongong at a place called Kiama; and west of Sydney, in two masses at places now called Prospect and Dundas. Top of map as printed here is north.

the Company's pastoral leases further north. Dr Pemberton states that Henderson prepared the section in May 1827, and it was dispatched to London, where it was later archived, but subsequently returned to Australia in 1966. Being fragile, it was separated from other documents and was only located in Canberra in 2004.[14]

Fragments of coal and kerosene shale were also collected from the beds of rivers where they exit the Blue Mountains on their eastern side, suggesting some western exposures, but, as can be seen, Mitchell's map gives no indication of western coals (which in fact crop out in the valleys of the mountains and at the bottom of their west-facing scarp, and were known in the Lithgow area as early as 1824). The Wollongong and Newcastle exposures were inconveniently distant from Sydney, so in 1799 rock-boring equipment was sent out from England through the good offices of Joseph Banks (1743–1820), and at least one trial, probably more, was made near the centre of the basin. Some indications of coal were found within the upper shales, but it was not until many years later that mining began in the coal measures proper, under a spot not far from the centre of Sydney, at a much greater depth (nearly 3000 feet) than could be reached with augers in the early nineteenth century.[15] Still, the fact that they were looking for coal close to Sydney suggests that there was awareness, early on, of a basin structure, such as is suggested by Mitchell's map of 1834.

The area coloured brown by Mitchell corresponds to what is known today as the Wianamatta Group, which consists chiefly of shales, some containing thin and poor-quality coal seams. The Wianamatta unit was recognized by the important colonial scientist, the Reverend William Branwhite Clarke (1798–1878), being mentioned in his correspondence in the 1840s, but was not formally introduced at that time.[16] It produces a much better soil than the sandstones, despite being saltier. When the first fleet arrived in Sydney in 1788 the settlers had severe food problems, as vegetables could not be grown satisfactorily on the sandstone soils. So they hunted around for better soils, and soon found them at Parramatta on the plains to the west of Sydney. A second settlement was established there and the region was rapidly examined for the occurrence of the shale areas for agricultural purposes. Shale beds, within the Wianamatta Group and close to Sydney, were also later used shortly after for brick making.

The shales were described by François Péron (1775–1810) in 1807, in his account of the Nicolas Baudin (1754–1803) expedition of 1800–1804, reporting a fairly short journey, inland from Sydney, by the two mineralogists that accompanied that expedition: Louis Depuch (died 1802) and Joseph Charles Bailly (1772–1844) (Vallance 1981–1982). Bailly referred to the shales near Parramatta as *schistes bitumineux* and mentioned that they contained fossilized fern leaves. He also mentioned that there could be coal at depth in the Parramatta area, which suggests that he too may have had the idea of a basin structure for the strata around Sydney. Near the foot of the Blue Mountains the *schistes* were found to burn with a 'lively flame', but these were only specimens of kerosene shale carried eastwards by the rivers (Péron 1807).[17]

Depuch also wrote a brief report[18] on some specimens collected by Francis Luis Barrallier (1773–1853), a French military engineer and surveyor who had left France because of the French Revolution and had settled in Australia, joining the military establishment. Barrallier was directed to try to cross the Blue Mountains in 1802. To this day there is some controversy as to how far he penetrated to the west. It seems that he did not achieve a complete crossing to the western plains, though he probably reached the Palaeozoic rocks in the valleys to the west of the main outcrop of sandstones. He brought back some rock specimens that were examined by Depuch (see Mayer 2007). But it is unlikely that Mitchell knew much, if anything, about Barrallier's useful geological observations, and they do not appear to have influenced his map-making.

Mitchell's mapped outcrop of the shales corresponds fairly well with that shown on modern maps, except that, for some reason, he did not record some of the rocks now known to crop out near Sydney as belonging to that unit. This is a little odd, considering that the area was not then mostly covered with buildings as it is today. And the shale areas have characteristic 'turpentine' trees, which one might have expected Mitchell to recognize as an indication of the shale outcrops. Outcrops of this rock-type to the SW of Sydney are not particularly well mapped either.

The road west from Sydney beyond Parramatta passes a substantial laccolithic intrusion of dolerite at a place called Prospect. Even in Mitchell's day, this rock was being quarried for road metal, as it is to this day. This intrusion is one of the outcrops depicted as 'trap' on Mitchell's map. One also finds 'trap' at a nearby locality, Dundas, where he obtained road metal. There has been so much subsequent quarrying there that almost no igneous rock now remains, and the area is a football pitch. Significantly, however, it is called Mitchell Park, and thus his work there is remembered.

Trap is also marked to the south of Wollongong at Kiama where it crops out most obviously on the coast at the famous Kiama blowhole, first recorded by George Bass in 1797 (see note 11). This fertile area was already settled by the time Mitchell arrived and the volcanic rocks would have been well known. What Mitchell may have thought about the age relations of the volcanics cannot, however, be deduced from his map.

Finally, on Mitchell's map there is the 'calcareous sandstone' north of the Newcastle coal measures. The geology is much more complex in this area than in most other parts of the Sydney Basin, and it is difficult to know precisely which rocks Mitchell had in mind. But certainly there are some calcareous sandstones in that area, though not distributed as he represented them, as a single widespread unit. Mitchell's unit appears to correspond with a variety of Carboniferous and Permian sandstones, some calcareous and tuffaceous, and some fossiliferous. The overlying coals are now regarded as Permian, though in the nineteenth century there was heated controversy as to whether they were Mesozoic ('oolitic') or Carboniferous (Vallance 1981).

So it is likely that a lot of the geological information in Mitchell's map would have been available to him from what one might call 'general knowledge'. But the base map itself was another thing altogether. This map, the so-called 'Nineteen Counties Map', was published in 1834, as a result of Mitchell's assiduous efforts to get the area of Sydney and its hinterland properly mapped (see Fig. 3).[19] The job was not done single-handed of course. Mitchell was chiefly the 'centre of calculation' for the map's compilation (Latour 1987). Two base-lines were marked out accurately on the shores of Botany Bay in 1828 and the mapping proceeded by triangulation and chaining from there. The data were supplied by his team of surveyors, mostly young men of good family, supported by gangs of convicts, who would, for example, do the chaining and provide the labour for cutting trees, getting water, or driving the bullock carts that served as the main means of transport. But much of the terrain in the mountains behind Sydney is exceedingly rugged, so a good deal of the travel had to be done on horseback or on foot. The assistant surveyors were encouraged to supply information about rocks and soils in their reports, and they did this to a limited extent; so Mitchell had some information from his staff that would have found its way into his geological map.

The general artistic style of the map was one that Mitchell had developed from his work in Portugal and Spain, and which can also be seen in his notebooks. These commonly show the lines of roads that he surveyed, and for which he was responsible for finding the best line of route and laying out the

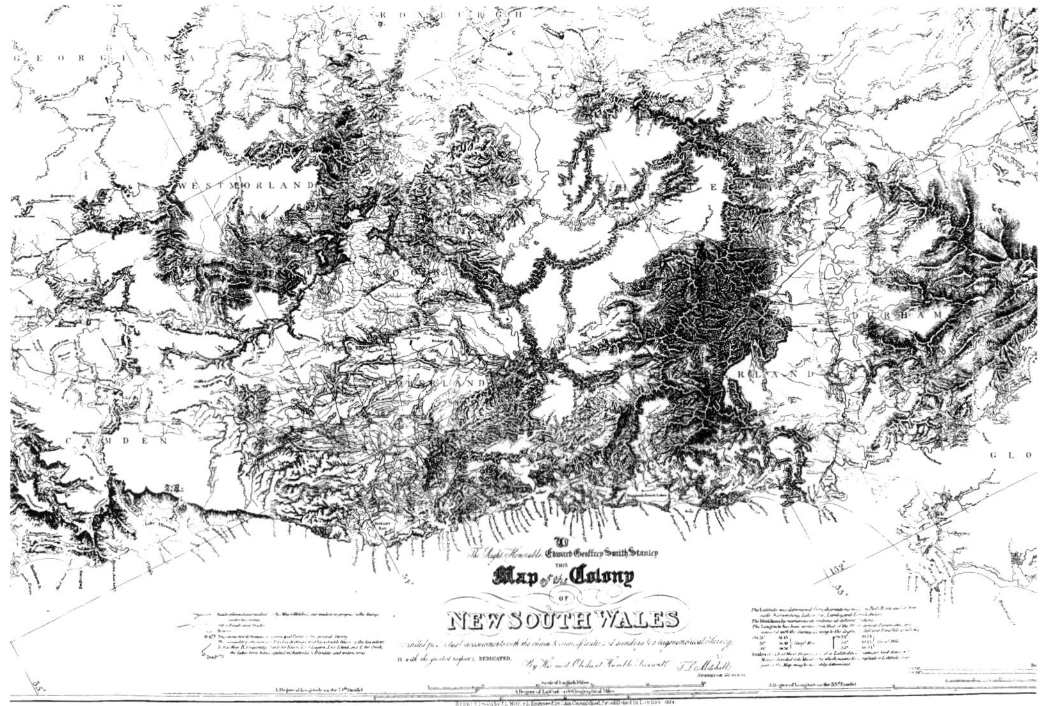

Fig. 3. A portion of Mitchell's *Map of the Colony of New South Wales Compiled from actual measurements with the chain & circumferenter, and according to a trigonometrical Survey* [Sydney, 1834].

lines for construction. He also undertook some of the actual engineering design work and overseeing the work of the convict stonemasons. Early on, Mitchell drove the Great North Road through the rugged hills north of Sydney to Newcastle.[20] In fact it proved to be something of a white elephant, for almost as soon as it was finished it was realized that it was still cheaper to transport goods such as coal by sea rather than by the new land route. Even so, it was a masterpiece of engineering, though now disused except by hikers. Apart from this now disused road and prior to the building of motorways, the main lines of road construction spreading out from Sydney, were laid out by Mitchell. One of them, running out towards the NW from Bathurst, in the direction of Mitchell's first exploration, is now called the Mitchell Highway. There was also the Great South Road (now called the Hume Highway), to Goulburn and Yass, the route of which is that now followed for the northern portion of the main Sydney–Melbourne highway.

As Superintendent of Roads as well as Surveyor-General, Mitchell was also responsible for getting a road built over the mountains to the agricultural land to the west, around settlements such as Bathurst and Orange. The Blue Mountains form a substantial barrier to the west. The early explorers had tried to cross them by following the valleys, but this took them into a terrible tangle of vegetation and huge boulders in the river beds. Eventually, in 1813, the settlers learnt to do what the aborigines had long done, namely follow the ridge tops. But this had pitfalls too, as the first road, which was opened in 1815 followed a ridge as far as it could to the most westerly outpost of the Blue Mountains at Mount York, and then made a hair-raising 1-in-4 descent. So in 1827 a reward of land was offered to anyone who could prove a better route. One suggested by the explorer Hamilton Hume (1783–1872) brought him a reward, and another one was made to a settler named James Collets or Collits (1806–1880), with a new route off Mount York. In 1830, Mitchell was dispatched with a gang of convicts to construct a road along that proposed route. But while there he decided there was a better route off the mountains at a place somewhat to the south, now called Mount Victoria. He requested permission from Sydney to change the plan. It was refused, but Mitchell was on the spot and simply went ahead and began constructing the road where he thought best. It was completed in 1832 (having received retrospective approval) and was another masterpiece of engineering (see Fig. 4). The modern road still uses his route. It was such high-handedness that so often led to Mitchell's quarrels with his superiors, but he always emerged the winner.

Getting up onto the Blue Mountain plateau from the plains to the west of Sydney was also a substantial problem, involving the ascent of that important geological structure: the Lapstone Monocline. The old route made a zig-zag ascent, but following the success of the work at Victoria Pass Mitchell recommended a passage up a small pass (now called Mitchell's Pass), which, with the help of a fine stone bridge (the oldest that survives in mainland Australia) constructed by the master Scottish mason David Lennox (1788–1873), allowed a more direct and simpler ascent. This route was used from 1833 until 1926, when a new road was constructed along the approximate route of the first line of ascent.

The explorations

Mitchell is best known in Australia for his work as an explorer, and he undertook four major expeditions, the routes of the first three of which are shown in Figure 5. The first was in 1831–1832 when he went NW from the Hunter Valley, hoping to corroborate a report by an escaped convict that there was a river leading towards the Gulf of Carpentaria, or even an inland sea. This proved illusory but much new ground was crossed and Mitchell's footsteps were soon followed by graziers and settlers. The second trip (1835) took him down the course of the Darling River in western New South Wales into arid country, though again pastoralists soon followed him. The third expedition (1836) was a great success, in which he penetrated as far south as the rich agricultural lands of Western Victoria and the Grampian Mountains, reaching the southern ocean, to the west of where Melbourne now stands, and meeting some new settlers on the coast, to their mutual surprise. Mitchell called this rich and fertile land 'Australia Felix', and it soon became one of the wealth centres for Australian agriculture. These explorations were written up in two volumes in 1838, while Mitchell was on leave in England, as *Three Expeditions into the Interior of Eastern Australia*. It had two editions and there is a modern facsimile reprint of the second one (Mitchell 1838). The book is one of the main texts recounting early Australian exploration. A fourth journey from Sydney into tropical Queensland was undertaken in 1845–1846 (Mitchell 1848).

Mitchell received much applause for these exertions. The Department of Lands building, completed in 1881, is one of the fine old Sydney sandstone edifices, and its external walls are decorated with statues of Australia's major explorers. Mitchell is there in his niche along with the rest, and he thus has his niche in history (see Fig. 6).[21]

MOUNT VICTORIA
in its Original state

The PASS as completed

Fig. 4. Mitchell's road construction at Victoria Pass, for the descent of the western scarp of the Blue Mountains (from: Anon. 1833, pp. 323–326).

The explorations used convicts as manpower, but Mitchell also had some educated technical assistants. On his second journey, he was accompanied by the botanist Allan Cunningham (1791–1839), who got lost and was killed and eaten by aborigines. The explorers used several carts or wagons, drawn by numerous bullocks or horses respectively. Portable boats were also carried, but they were of limited use. Additionally, they brought a flock of sheep and a pack of hunting dogs, and the men were armed. It was all run with military skill and discipline. Each night, the wagons were drawn up and the tents erected with the same arrangement, appropriate for defence.

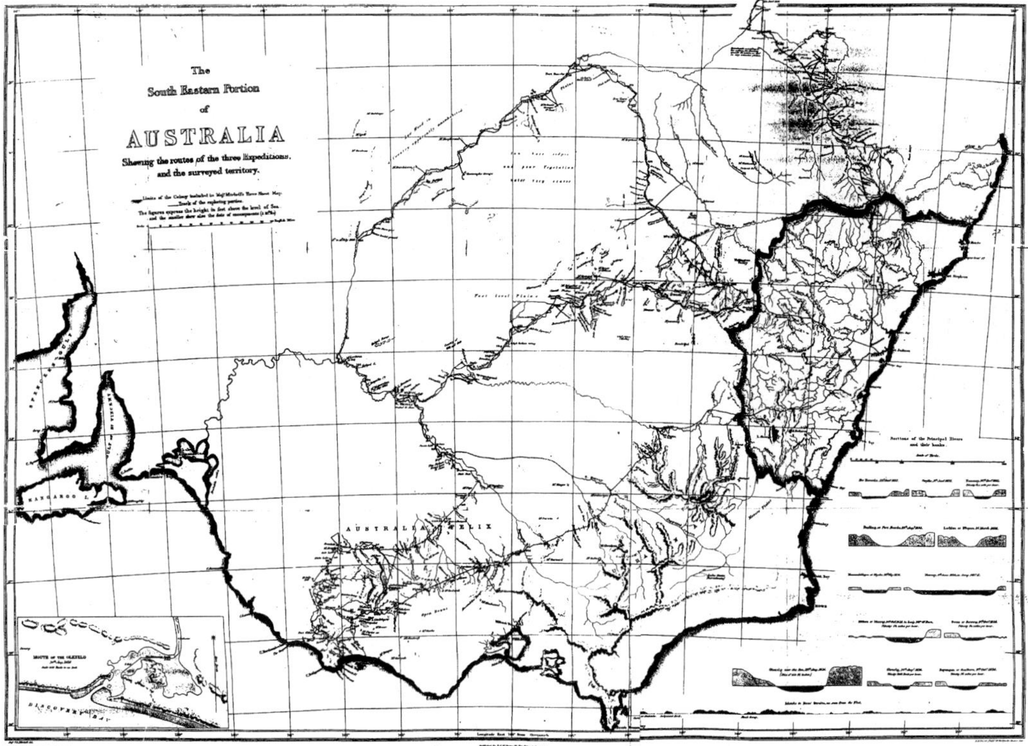

Fig. 5. Route maps for Mitchell's first three inland expeditions (*Three Expeditions*, 2nd edition, 1839, **1**, plate following p. 337).

The whole journey was chained, as was necessary for the accurate survey of the flat inland areas; and subsequent analyses of Mitchell's results show that he was extremely accurate in determining positions, almost miraculously so. For his southern exploration, after a journey of nearly 2500 miles, the closing error was a mere 1.75 miles. Mitchell would frequently go off from the main party on various forays, spying out the land, making topographic survey observations, and looking for water, which was nearly always in short supply. To keep track of his position, he relied on the supposition that his horse's forefeet averaged 950 steps per mile (a technique also used by travellers on camel in the Sahara Desert). He moved a counter from a pocket on one side to one on the other at every hundred paces, and kept a note in his pocketbook of every change of direction, as determined with his compass.

At times, the men suffered considerably from thirst but there was not much sickness, apart from some scurvy. Mitchell's narrative of the expeditions reveals that he was constantly examining soils, vegetation, and fauna, and he was making geological observations and collecting specimens. 586 of the geological specimens were subsequently taken back to England, listed by William Lonsdale (1794–1871), and published by Mitchell in 1851, but the locations and identifications of these were rather imprecise, which was probably Mitchell's fault, not Lonsdale's. Thus there are entries such as 'Dividing Range E: Quartz Rock'; 'From Towrang: Ironstone'. Others were more precise, such as 'Mount Lawson: Grey trap, consisting principally of granular feldspar. Feldspar the base of all trap'. Presumably these notes represent information pasted onto the specimens by Mitchell himself. If so, they suggest that he collected in a somewhat amateurish manner, and not in accord with what Fitton had attempted to teach him.

Mitchell also collected fossils; some from the Hunter Valley area were figured and named in his 1838 book, according to the proposals of Lonsdale or the invertebrate palaeontologist James de Carl(e) Sowerby (1787–1871), whom Mitchell consulted when he was back in London (see Fig. 7). Later, they were cited by W. B. Clarke (1878, p. 113) as being 'Carboniferous[,] and Devonian?' The forms shown in Figure 7 are now regarded as Permian.[22] But Clarke was convinced

Fig. 6. Statue of Thomas Mitchell, Department of Lands Building, Sydney. Photographed by the author, 2003.

Fig. 1. Pl. 2.

Fig. 2.

MEGADESMUS GLOBOSUS.

Fig. 7. Samples of fossils collected by Mitchell, figured by J. de C. Sowerby, *Three Expeditions*, 2nd edition, 1839, **1**, Plate 2, facing p. 14.

that the Australian coal measures were Carboniferous. Mitchell himself expressed no firm opinion about their ages. However, corals that he found near the confluence of the Tumut and Coodradigbee Rivers with the Murrumbidgee, SW of Sydney, on the return leg of his third expedition, were recorded by him as *Favosites*, *Stromatopora concentrica*, and *Helipora pyriformis*; and crinoid remains were also found. These were suggested as belonging to 'Mr. Murchisson's [sic] Silurian system'. Here Mitchell first used the term 'Silurian' in reference to Australian strata. The rocks were subsequently interpreted as Devonian (which System had not yet been proposed in 1838).[23] Mitchell also commented on what we now regard as Silurian rocks in the Bungonia area further north towards Sydney, though he did not then call them Silurian. But given that Murchison's *Silurian System* was only published in 1839, and the concept did not exist at all when Mitchell left England in 1827, it is clear that Mitchell was endeavouring to catch up with the progress of geology when he was back in Britain on leave and finalizing the text of his *Three Expeditions* for publication.

One of the interesting results of Mitchell's first expedition was a product of his visit to a place called Mount Wingen at the upper end of the Hunter Valley, NW of Sydney, also called the 'Burning Mountain'. This geological curiosity was discovered by an unknown settler in 1828 and reported in the Sydney newspaper *The Australian* on 28 March that year. It was described by a Mr Mackie, of Cockle Bay in Sydney,[24] as an active volcano (*The Australian*, 30 July). It was again reported as such in Edinburgh in 1830 (Anon. 1830), although the idea was soon corrected by a local geologist, the Reverend Charles Pleydell Wilton (1795–1859, MA Cantab. and Member of the Cambridge Philosophical Society), who ministered in Parramatta and then in Newcastle. He gave a careful description of the site, interpreting the phenomenon as being due to a burning coal seam (Wilton 1830). This, of course, would be a rare case that would have been agreeable to the Wernerian theory of volcanoes, but that issue was largely defunct by 1830. The coal seam was thought by Wilton to have been ignited thousands of years ago, presumably by a bush fire.

Mitchell visited the site in 1829, rushing there on a two-day trip away from his work on the Great North Road, but his written description published in his *Three Expeditions* was chiefly the result of a one-day visit on 2 December 1831, during his first journey of exploration. In his book, he published an accurate plan of the area of the fire (see Fig. 8)[25] and described the rocks within which the fire appeared to be burning, noting the occurrence of *Spirifer* impressions in the sandstone there. Plant impressions could also be seen. He noted the occurrence of basalt and 'trap rock' on adjacent hills, but evidently agreed with Wilton's interpretation of the unusual phenomenon. Mitchell's accurate plan has allowed the rate of combustion of the coal seam to be gauged: the site of the still on-going combustion has moved up the hill by about 200 metres since Mitchell's day. Subsequent estimates of the duration of the combustion have varied from 5000 to 500 000 years but the smaller figure is today thought more probable.

Mount Wingen is an interesting locality. Few examples of such a phenomenon are known in the world, though Mitchell referred to an example, described by Buckland, at Holworth near Weymouth on the south coast of England, thought to have been ignited by the action of rain-water on the iron pyrites in a bituminous shale. In the Australian case, the combustion may have been started by a bush fire, as Wilton suggested, or by some chemical action of the pyritized coal.

Fossil megafauna[26]

Mitchell's major contribution to geology was undoubtedly his investigation and collection of fossil bones from Wellington Valley to the west of the Blue Mountains, NW of Bathurst and towards the modern town of Dubbo. Mitchell was evidently very interested in the remains and took time off from his road building work in 1830, and again from his first expedition, to examine the famous caves. He would have gathered from his contacts in England immediately before his departure for Australia that cave deposits were a 'hot topic' in geology at that time, and Mitchell wanted to make a name for himself by obtaining specimens in the southern hemisphere, should this be possible.

The Wellington Valley was first explored by Oxley and some naturalists in 1817, and in 1820 Oxley recorded limestone in the region, an important matter for the new settlement in Sydney, which urgently needed cement and plaster (Mayer 2007). Oxley called the fertile valley the 'Vale of Tempe' (meaning beautiful valley or delightful spot). Settlement began there in 1823, with a few convicts and their guards. The artist Augustus Earle (1793–1838) was in the area in 1826 or 1827 and evidently entered the caves, for pictures by him of both the interior and the exterior are known. His painting of the farms in the valley at that early date shows a surprisingly ordered and pleasant rural scene (Hackforth-Jones 1980, pp. 100, 102, and 103). In 1830, Mitchell met a fellow Scot, George Ranken (1793–1860), who had migrated to Australia in 1821 and taken up land as the second settler near Bathurst in 1822. He was a prominent agriculturalist, and became magistrate of Bathurst. Evidently he had an interest in natural history, and also some engineering knowledge, for he constructed a bridge over the Macquarie River near Bathurst in 1856. Mitchell and Ranken examined the Wellington Caves together, as we shall see.

The Wellington Caves were first discovered by white men in the 1820s, perhaps by an escaped convict. As said, the caves were visited by Earle, who made paintings of both the exterior and interior of one of the caves, and in 1828 by the explorers, Hamilton Hume and Charles Sturt (1795–1869). In May 1830, a newspaper article in the *Sydney Gazette*, signed 'L', reported that Ranken had been in the caves, and had discovered fossil animal bones inside. He had made a small collection and taken them to Sydney. From there they were transferred to Robert Jameson (1774–1854) in Edinburgh by the Reverend John Dunmore Lang (1799–1878), a prominent Scottish settler, intellectual, and subsequent New South Wales politician, who had almost certainly written the *Sydney Gazette* report.[27] Jameson republished the article in the *Edinburgh New Philosophical Journal*. The fact that Jameson's journal was chosen for many of the publications relevant to the Wellington Caves, and that the people with whom Mitchell was involved at Wellington were Scots, reflects the rather close intellectual and social connection between New South Wales and Scotland at that time.

Ranken penetrated deeper into the caves than others had done. He attached a rope to a protruding object in order to lower himself into a cave. The

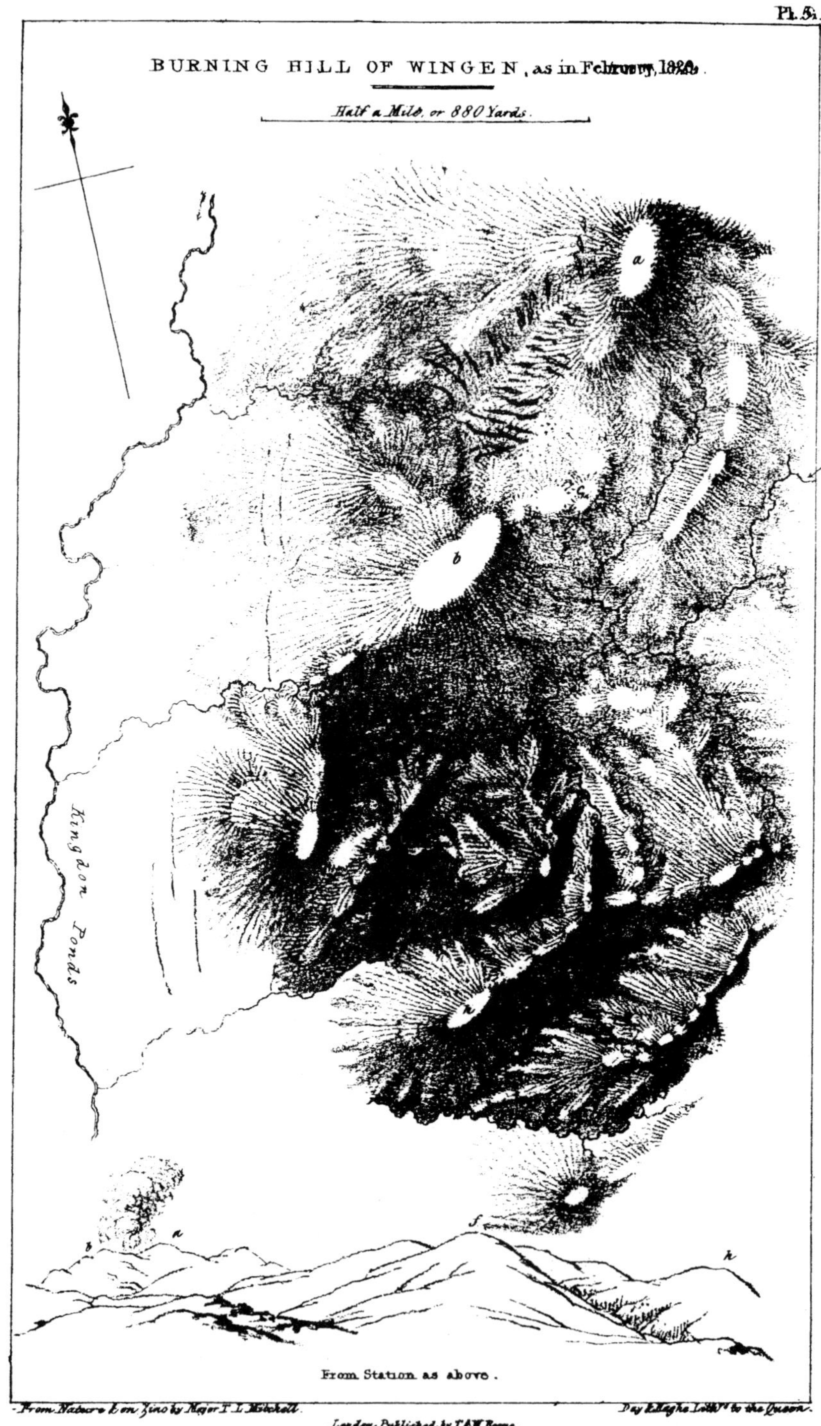

Fig. 8. Plan of the area of Mount Wingen, or Burning Mountain, published by Mitchell: *Three Expeditions*, 2nd edition, 1839, **1**, Plate 5, facing p. 23.

object broke and Ranken had a tumble. It was only then he realized that the protruding thing was a bone. In the *Gazette* report it was suggested that it might be some large animal like an ox, an Irish elk, or even an elephant. Later it was recognized as one of the Australian megafauna: probably a *Diprotodon*.

In early June 1830, Mitchell was in the area between the Blue Mountains and Bathurst, collecting various geological specimens, but seemingly not in a systematic way. On the 5th, he went with Ranken to the Wellington Caves, and subsequently also to some neighbouring caves at Molong. They met up with Alexander Kinghorne (1770–1846), who was responsible for discovering and exploring some of the caves at Wellington.[28] Various soldiers from Wellington were also involved, and some convicts were used to assist with the excavation of the cave deposits.

On 3 July 1830, Mitchell recorded his surveying of the Wellington Valley, following the Bell River northwards toward its junction with the Macquarie River, where the township of Wellington is now located. He made extensive notes on his fieldwork (or cave-work) and mentioned the common occurrence of red earth in the caves, as well as a white 'ash', into which Ranken sank up to his waist. Subsequently, various other caves were visited on horseback, some at the suggestion of local aborigines. Mitchell and Kinghorne had with them a man carrying a theodolite, so it is clear that he was engaged in survey work in the course of his cave investigations.

In fact, Mitchell's notes show that he was combining business (surveying) with pleasure (fossil hunting in the caves). He worked till late at night in one of the caves, finding bones which he thought were human. Kinghorne supplied boxes for the transport of specimens to Sydney, which were wrapped in wool, appropriately for Australian palaeontology. Some excellent drawings were made (see Figs 9[29] and 10). 'Rankin [sic] read a poem he had written in which he noticed most flatteringly my survey, drawings &c at Wellington'. Further caves were visited near Boree. During one night, Mitchell slept in a cave 'tolerable well' under a kangaroo cloak. Near Molong they met a Dr John Henderson (died 1836) from Tasmania, a 'very odd personage' who was 'making a section of the strata & c'. But they do not appear to have collaborated.[30] Further caves were visited by Mitchell and his associates and they found more bone breccias in caves, the deposits always being associated with the red earth found in the caves. By July, they were back at Ranken's property near Bathurst. On 10 July, Mitchell recorded that he had finished his plan of the Wellington valley, and a sketch of a plan of the bone cave was started on the 11. He set off back to Sydney on 24 July.

As mentioned, Mitchell thought that some of the cave remains were human, which is not impossible, though there have been no subsequent authenticated finds. He mused in his diary:

> Like the lost remains of a shipwreck . . . [the remains] lay a melancholy vestige of a tremendous storm; and I could not behold these vestiges of a being once animated like myself, but which had existed long prior to the earliest Egyptian mummy without the most elevated and interesting reflections. Could this but be reanimated, what light could it not throw on the most puzzling question[:] how and whence comes the red earth always containing bones? This might have been a body from Asia, and at least as ancient as Noah![31]

These words suggest that Mitchell was thinking, even if subconsciously, about a former catastrophic flood in the southern hemisphere, back in the time of Noah, and perhaps washing in bones from as far away as Asia. After all, it was a geological commonplace in the early 1830s that the Noachian flood had been a geological agent and a universal event. Therefore, it was pleasing to Mitchell that he was finding evidence that diluvial agencies had been at work in Australia, as Buckland and Cuvier might have supposed. Later (1838) Mitchell published a geological map of the Wellington Valley (the first geological map ever published for any part of

Fig. 9. Interior of one of the Wellington Caves, drawn by Mitchell, *Three Expeditions*, 2nd edition, 1839, **2**, Plate 43, facing p. 360. An original of this picture is preserved at the *Muséum d'Histoire Naturelle*, Paris, Cuvier MSS 640, Folio 153.

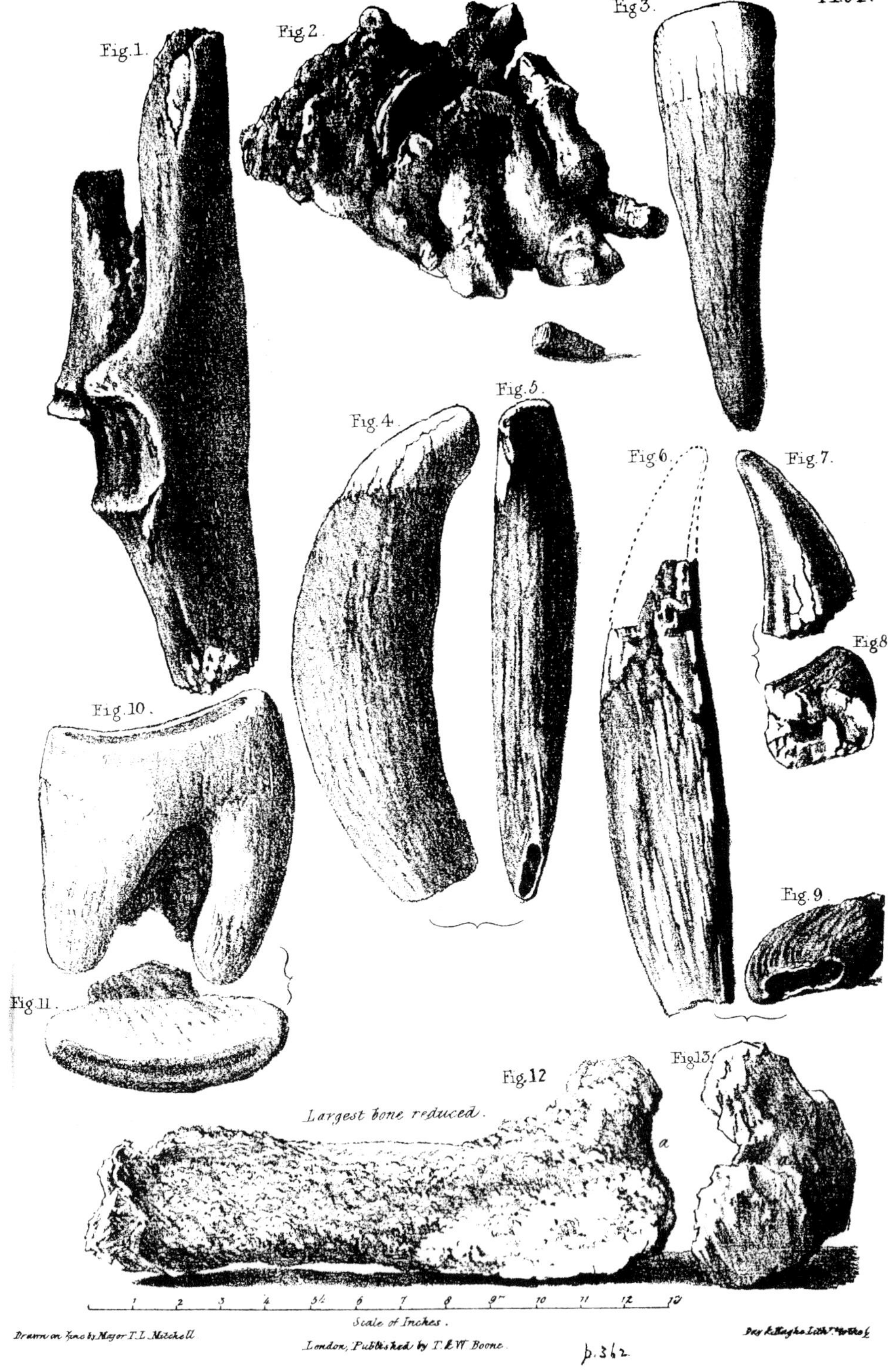

Fig. 10. Mitchell's drawings of some bones and teeth excavated from the Wellington Caves, *Three Expeditions*, 2nd edition, 1839, **2**, Plate 51, facing p. 372.

Australia). It is significant that the map[32] made a clear distinction between 'alluvium' and 'diluvium' (see Fig. 11).

On his return to Sydney, Mitchell prepared a paper of twenty-nine hand-written pages entitled '*An Account of the Limestone Caves at Wellington Valley, and of the Situation, near one of them, where Fossil Bones have been Found*', dated 14 October 1830.[33] It travelled to England via the ship *Gilmore*, along with three boxes of specimens and ten pages of drawings, and was delivered to the Geological Society. It was read before the Society on 13 April 1831, and an abstract, prepared by Fitton, subsequently appeared in its *Proceedings* for that year (Mitchell 1834). Fitton informed Mitchell[34] that his specimens and drawings were also tabled on 13 April, and that a paper by Mitchell had recently appeared in Jameson's *Edinburgh New Philosophical Journal* ([Mitchell] 1831)[35] The Geological Society abstract, as published by Fitton, contained identification of bones by the fossil mammal expert at the Royal College of Surgeons, William Clift (1759–1849),[36] as belonging to kangaroo, wombat, dasyurus (Tasmanian devil), koala, and possum. A large bone might, he thought, belong to an elephant, and one that was 'obscure and imperfect' might have belonged to a dugong. The dust in the caves had been analysed by the chemist Edward Turner (1796–1837), Professor of Chemistry at University College, London, who found carbonate of lime and phosphate of lime.[37]

In fact, the first announcement in a British scientific journal of the Wellington discoveries had been made in a brief note by Jameson, relaying information, dated 22 April 1831, by 'Colonel [Patrick] Lind[e]say of the 39th Regiment, a very active and intelligent inquirer', who commented on the large size of the animals compared with modern Australian forms (Anon. [Lind(e)say, P.] 1831).[38] Lind[e]say (1776–1839), yet another Scot, was at the time supporting Mitchell in his controversies with Governor Ralph Darling (1775–1858), and sanctioned his first inland exploration.[39] He had a collection of Mitchell's material dispatched to Europe, which was examined by Georges Cuvier (1769–1832) and his co-worker, the Irish zoologist and palaeontologist Joseph Pentland (1797–1873) (Delair & Sarjeant 1976; Sarjeant & Delair 1979) in Paris. But Lind[e]say played no further part in the story. The specimens were passed around between the experts in Edinburgh, London, and Paris, and it is not always easy to know who was looking at what, and when. But clearly the experts at the 'centre' were interested in the information received from Mitchell at the periphery and regarded it as important.

Some revisions to Mitchell's Geological Society paper were made in Sydney on 2 September 1831, and a copy was sent to Cuvier in Paris, where it survives in the Cuvier archives (MSS 640). At the time of the initial presentation, Mitchell had been puzzled as to how caves were formed, but by 1831 he had evidently read Charles Lyell's (1797–1875) *Principles of Geology*, and now added a prefatory page in a version of the paper that was intended for Cuvier, along with additional specimens:

> It is no longer difficult to understand how the large cave at Wellington may have been gradually hollowed out or at least enlarged in this manner [by action of acidic water] until its' [sic] roof nearly reached the surface as at present since the last inundation, or to suppose that the other class of caves containing bones which now fill them to the surface are but the ruins of antediluvian caves which had been hollowed out in a similar manner and at length broken in and finally filled up with the wreck of the animal world under the operation of that great catastrophe which moulded the Hill now partly composed of these remarkable remains into its' [sic] present smooth external form: and that the same process continuing in the lower parts of the caves, thus choked up with osseous breccia, has occasioned that subsidence, by which this substance has been discovered and which distinguishes so remarkably all the caves where it occurs by an appearance of disruption and disorder never seen in the other Limestone Caves where no breccia is to be found.
>
> I now avail myself of the opportunity kindly afforded me by Captain Laplace [a French sea captain then returning from Sydney to France] to submit a copy of that description and drawings with some specimens to M. Le Baron Cuvier from a desire to afford some information respecting the fossil remains we have found in New South Wales and in hopes to profit in common with the rest of the world by the speculations of that distinguished individual.[40]

This text suggests that though Mitchell had now read Lyell he was still a Cuvierian catastrophist. Alternatively, his draft new introductory page may have been worded the way it was because he was sending materials to Cuvier himself.

Be this as it may, several sets of specimens were sent to Europe in 1830–1831: Lind[e]say's to Jameson and Cuvier; Ranken's or Lang's to Jameson; Mitchell's to the Geological Society; and Mitchell's (Lind[e]say's?) to Cuvier. The bones in Edinburgh were examined by Jameson and his 'friend and former pupil' Dr Adam of the Royal College of Physicians, as Jameson (1831) reported. They recognized kangaroo and wombat bones, but a large one was thought to be a kind of *hippopotamus*. The specimens then went to London, conveyed by Jameson's friend Dr Turnbull Christie, where they were re-examined by Clift in July 1831, after the Geological Society presentation. A letter from Clift to Fitton, dated 29 July 1831,[41] indicates that Clift only had time to make a cursory examination of the specimens, but that he marked Mitchell's drawings with some names,

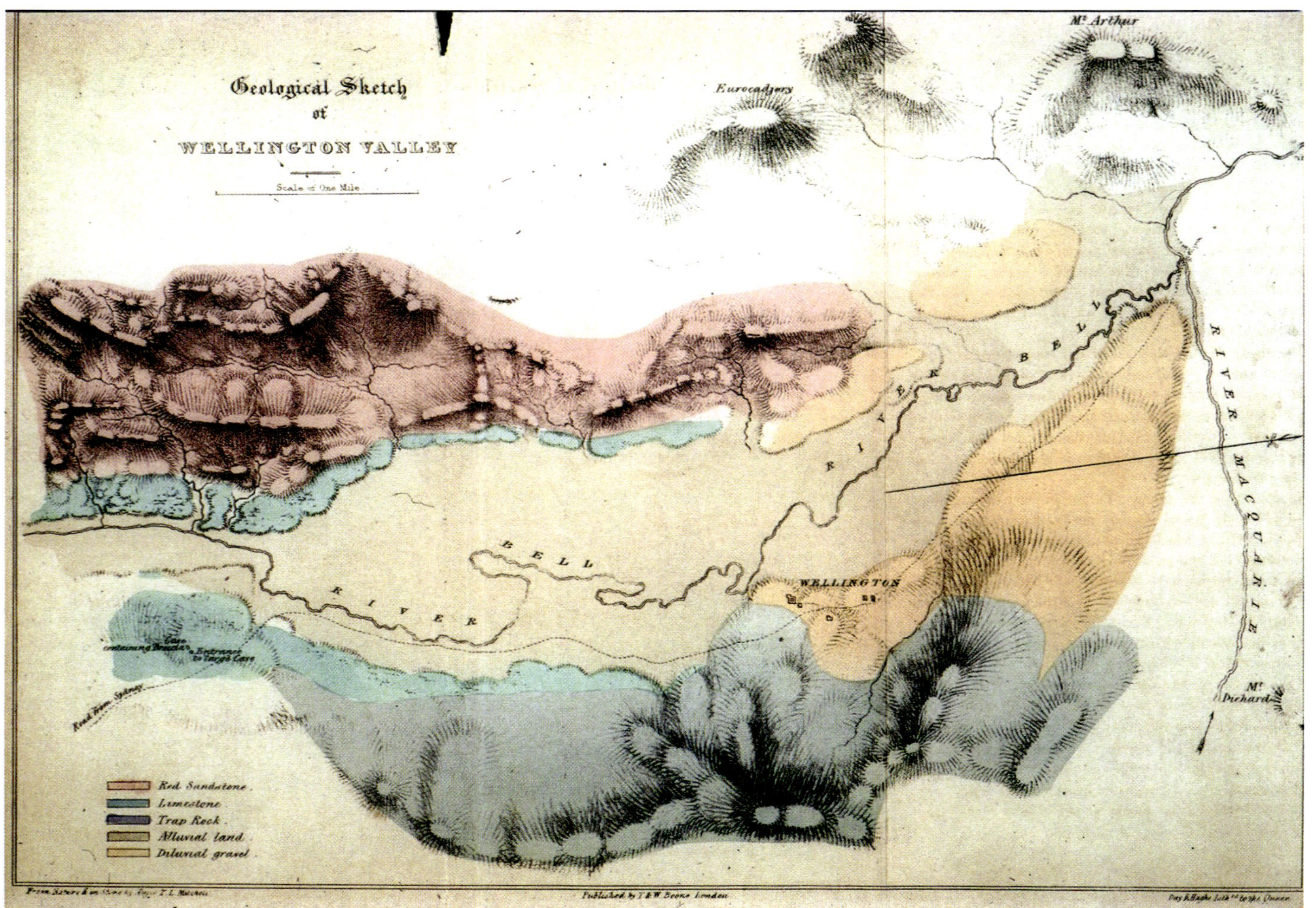

Fig. 11. Geological map of Wellington Valley, New South Wales, from Mitchell, *Three Expeditions*, 2nd edition, 1839, **2**, Plate 42, facing p. 359.

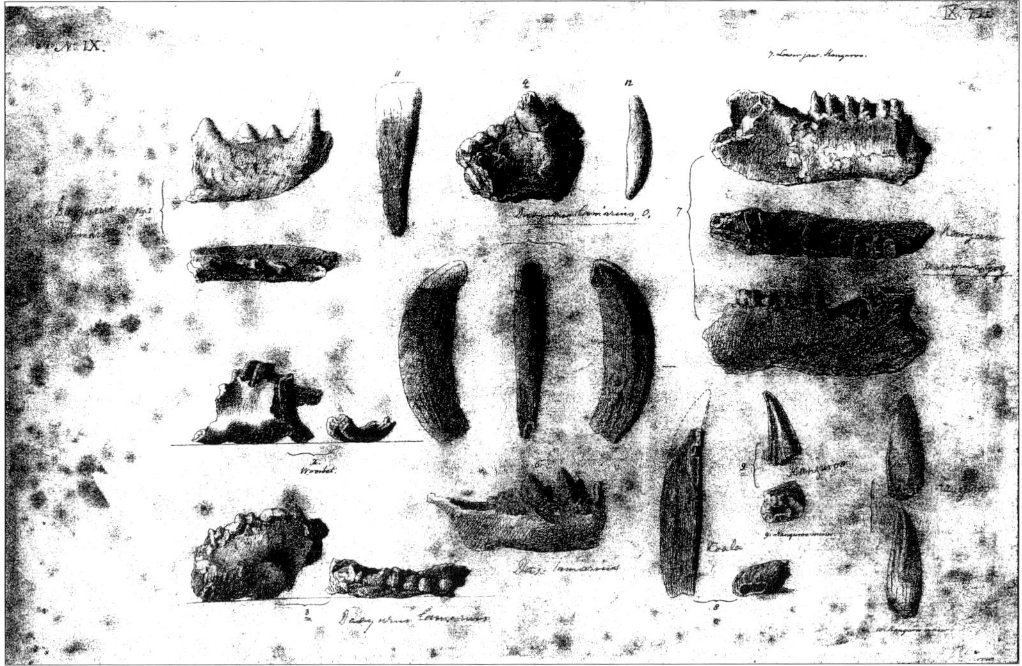

Fig. 12. Figures of mammalian remains from Wellington Caves, figured by Mitchell and annotated by Clift (Geological Society archives, LDGSL 99/1; reproduced by courtesy of The Geological Society).

as can be seen from the drawings now held in the Geological Society's archives (see Fig. 12). Later that year, Clift himself published a note on the fossils in Jameson's journal, now suggesting that the mysterious large bone might have belonged to an ox or hippopotamus, but not an elephant (Clift 1831).

As mentioned, Mitchell himself added further information about the appearance of the caves, both inside and out, and their contents. The bone breccia could be found on the walls of the cave, and bones could also be found in the layered limestone material on the cave floors, which had some bones standing upright and encased in more than one layer. The bones were of large and small animals, mixed together, but did not appear to be water-worn. The bone breccias occurred outside the caves as well as within. He described one cave as being 'stopped up with diluvial earth' but did not speculate further.

Great interest was shown in the megafaunal remains, both in Britain and Australia, as they were evidently different from anything still living in Australia. In a note, presumably written by Jameson, added to Clift's Edinburgh paper (1831), it was stated that the caves and their bone breccias were analogous to ones found in Europe; that the ancient fauna of Australia had the same distinctive character as Australian animals alive today; but they were much larger. The agencies that had brought together the bones in Australia were supposedly similar to those that had operated in Europe.

The specimens that went to Paris for Cuvier's consideration were never formally described by him, as he died in May 1832, though he is reported to have suggested that the largest bones belonged to some kind of *elephant*.[42] This was also the view published by Pentland. He wrote two accounts of the Australian specimens for Jameson's *Edinburgh New Philosophical Journal*: one in 1832, describing specimens deposited in the Edinburgh College Museum and others seen at the Geological Society (Pentland 1832); and the other, dated 15 November 1833, referring to the further collections sent to Cuvier by Mitchell (Pentland 1833). Pentland regarded many of the types as new to science, but belonging to the same genera as those still extant in Australia. However, perhaps under the influence of the deceased Cuvier's opinion of the matter, Pentland also opined that the large bone sent by Ranken to Jameson belonged to an elephant, *not* a hippopotamus as Clift had suggested. The remains, Pentland thought, were similar to those of a fossil elephant type found in the Arno Valley in Italy, about one-third smaller than the Asiatic elephant.

While Mitchell's Geological Society paper was, according to Fitton's letter to Mitchell, received with 'all the attention it deserved' by Buckland and others, the paper did not get printed in full in the Society's *Transactions*: there was only the 'abstract' in the *Proceedings*. This may partly be explained by the publication problems that the Society was experiencing in the 1830s, prior to the establishment of the *Quarterly Journal* in 1845. However, Mitchell himself was preoccupied with many other matters in the mid-1830s, notably with his own explorations and topographic surveys, previously discussed. An undated note in Mitchell's papers in Sydney records that the paper was 'referred' on 27 April 1831, a fortnight after its presentation by Fitton. But it was only on 16 March 1834, that it was 'reported' on. A ballot on its publication was held on 6 January 1836, when it was resolved 'not to be printed'.[43]

One can only speculate about the paper's refusal, and it may have been to do with the publication problems that were unsettling the Society at that period. But a letter from Mitchell to Ranken dated 24 July 1833, suggests that Buckland could have had a hand in the matter:

> I understand Buckland's nose is put completely out of joint by the bones from Australia, their not being those of lions and hyenas is, I find, a fact that is considered in England to entirely upset his theory. And I have now heard from the best authority that the fact of their fossil bones belonging to animals similar to those now existing has worked a great change in all their learned speculating on such subjects at home. The big bone is neither part of an elephant or dugong, but a nondescript animal, perhaps extinct, though peculiar to New South Wales.[44]

This, perhaps, may hint at the reasons why the work of the Australian traveller was subsequently 'suppressed' Mitchell himself doubted whether the European 'experts' were correct in thinking that the large bones from Wellington were some such animal as an elephant.[45]

Moreover, the fact that the Wellington fossils were mostly similar in type (at the genus level) to modern Australian forms might not have been pleasing to those who had in mind universal catastrophes followed by subsequent creations of new forms. If that was the way things had happened, why should the new forms have any relation to those that they superseded? By contrast, Darwin subsequently used the similarity between the South American megafauna and modern types as one of his strands of evidence to argue for transformism. He also referred in *The Origin of Species* to Clift's paper (based on Ranken and Mitchell's material), among other examples, in favour of what Darwin called 'the law of the succession of types' (For discussion of this issue see Duggan 1980). Possibly such considerations militated against acceptance of Mitchell's paper. Or maybe Richard Owen (1804–1892) wanted to publish the Australian information? In keeping with our representation of Mitchell as a 'colonial scientist', he made no attempt to issue his work in Australia. His goal had been to publish his full account in the *Transactions of the Geological Society*. Failing that, a London-published book was evidently the best option, there being no very suitable outlet for his work in a New South Wales journal at that time.[46]

Leave in England

Following his arduous explorations, Mitchell was granted leave to write up his results, and he sailed for England in March 1837. While in England he was chiefly occupied with completing his maps for the Peninsular War and writing up his three expeditions. But in June that year he had requested the return of his 1831 memoir and the associated drawings and map of the Wellington area;[47] and these were then published as an addendum to the second volume of his travel book (1838), along with reproductions of some of his drawings of the caves and their bones and a geological map of the valley (see Figs 11 and 12). The map appears to have been added to the book late in its production, as it was not referred to in the text. Mitchell presented his manuscript geological map of Sydney and its environs to the Geological Society in November 1838.[48]

In England, Mitchell made direct contact with Owen for the first time. Owen examined and described some of Mitchell's fossil collection, naming a new and large wombat *Phascolomys mitchelli*, in honour of the explorer and collector. The *Diprotodon* appeared on the scene in a letter from Owen to Mitchell, dated 8 May 1838, published in his *Three Expeditions* (Owen in Mitchell 1838), which formally named the controversial large bone as belonging to the new genus. As the name implied, *Diprotodon* was a large creature with two distinctive protruding teeth. The teeth had affinities to those of modern wombats, though a section of a tooth had, Owen thought, some similarity to that of a hippopotamus. It was definitely not a dugong, but the skeleton was very incomplete, so no complete description of the animal could be given. It is now indeed regarded as a kind of giant wombat-like creature, and a marsupial, as Owen himself subsequently suggested.

Although Mitchell wrote that he did not pretend 'to account for the phenomena presented by the caverns' he did speculate a bit in his book. It seemed to him that the caves had been temporarily inundated at least twice and he suggested that this might have been due to overall changes in sea level. He correctly asserted that evidence was provided by coastal topographies such as lakes and the form of Sydney harbour, which suggested that it was a flooded valley. The occurrence of inland salt lakes suggested to Mitchell former marine inundations. So, he proposed, the plains of the interior could formerly have been arms of the sea. At the time of marine incursions, the bone beds might have been washed in, but it was suggested that the land surface had been altered since it was first inhabited, such operations having taken place subsequent to the extinctions (Mitchell 1838, pp. 371–375). However, this did not explain why the bones did not display water abrasion; nor did it explain why the bone breccias were found only in or near the caves, but not at other locations. (Later theorists believe that the red matrix to the cave bones and bone breccias is wind-born dust, a loess-type deposit.)

Mitchell continued to send Owen materials, more particularly ones received by him from the Condamine River, Darling Downs, in southern Queensland, inland from Brisbane.[49] Mitchell was regarded as the local expert on mammalian fossils, so specimens were channelled through him. In 1841, Mitchell also asked his son Roderick (1823–1851) and one of his surveyors named James Charles Burnett (1815–1854) to collect materials in northern New South Wales, in the areas of the old localities of Wellington and Boree, and SW of Sydney at the Wombeyan Caves, near Goulburn, and towards what is now Canberra. Initially, Owen thought that the fossils he had been sent from the Darling Downs were those of a mastodon or a *Dinotherium*, the large elephant-like creature with downward-pointing tusks, and indicated this in a paper in *Annals and Magazine of Natural History* (Owen 1843*a*). Following the receipt of a letter from Mitchell containing some new figures, Owen wrote in the same issue that the remains belonged to *Dinotherium* (Owen 1843*b*); but on receipt of yet more material from Mitchell he opted for *Diprotodon* (Owen 1845 for 1844, p. 225). With teeth found actually in the jaw sockets they could be seen to be procumbent, not curving downwards, as in *Dinotherium*.

Mitchell himself, writing to Owen, emphasized the gigantic size of the kangaroo specimens that he was forwarding. He saw a 'gradation of several intermediate sizes from *Macropus* [=kangaroo] *Atlas* ... up to the *Macropus Elephantus* (or whatever you may [like to] call it)'.[50] Rather implausibly, Mitchell suggested that the 'energies of animal nature' appeared 'to be on the wane' in Australia; but this could be 'a wise provision of providence for the introduction of those other large animals by man's agency which have been found better suited to his wants'.

Basing his claims on some molar teeth in mutilated mandibles, supplied to him by Mitchell in 1842 from the Condamine River area, Owen again honoured his correspondent, at the British Association (BA) in 1844, by naming what came to be regarded as a large rhinoceros-resembling marsupial as *Nototherium mitchelli* (Owen 1845 for 1844, p. 232).[51] Initially, Owen thought he was looking at the remains of a young *Diprotodon*, but closer examination of the teeth revealed that they belonged to a different and smaller species. In the BA paper, Owen discussed the evidence for the fossil forms in given regions ('particular provinces') as being similar to those found there at present. In general, he found the evidence confirmed the idea that the ancient and modern forms of a 'province' were of similar types, but he declined to go so far as to call the generalization a 'law'. Whereas (as mentioned) the generalization became an important consideration for Darwin's transformist views, Owen spoke of the 'last-created class of animals'. Although appreciating the significance of evidence that might be construed in favour of Darwin's 'law of succession', Owen was not a transmutationist.

By 1844–5, Owen no longer suggested that *Diprotodon* or *Nototherium* were anything but marsupials, but belief in an Australian pachyderm lingered on in the form of some kind of *Mastodon*, supported by a claim by the Polish explorer Paul de Strzelecki (1797–1873) that he had acquired a bone of the animal from an Aboriginal at Boree (or Buree) (Strzelecki 1845, p. 312). Owen named this as *Mastodon australis*, and continued to support the claim of the former existence of elephant-like creatures in Australia, even after it was challenged by the vertebrate palaeontologist Hugh Falconer (1808–1865) in 1863, who maintained that the Strzelecki *Mastodon* bone resembled a South American type, and that the matter could best be explained by supposing that specimens had somehow become switched or confused (Falconer 1863).

When Owen published his final synthesis of the mammalian palaeontology of Australia, he produced a good figure of the skull of a well preserved *Diprotodon*, and also a representation of nearly the whole of the skeleton, as shown in Figure 13.[52] This specimen was collected at King's Creek, southern Queensland, and was described and figured for Owen by his correspondent and collector in Sydney, the physician Dr George Bennett (1804–1893), in a letter dated 18 September 1863.[53]

Mitchell's fourth expedition, to tropical Queensland (1845–1846)

This journey did not yield important new geological results, and the intention of getting through to the

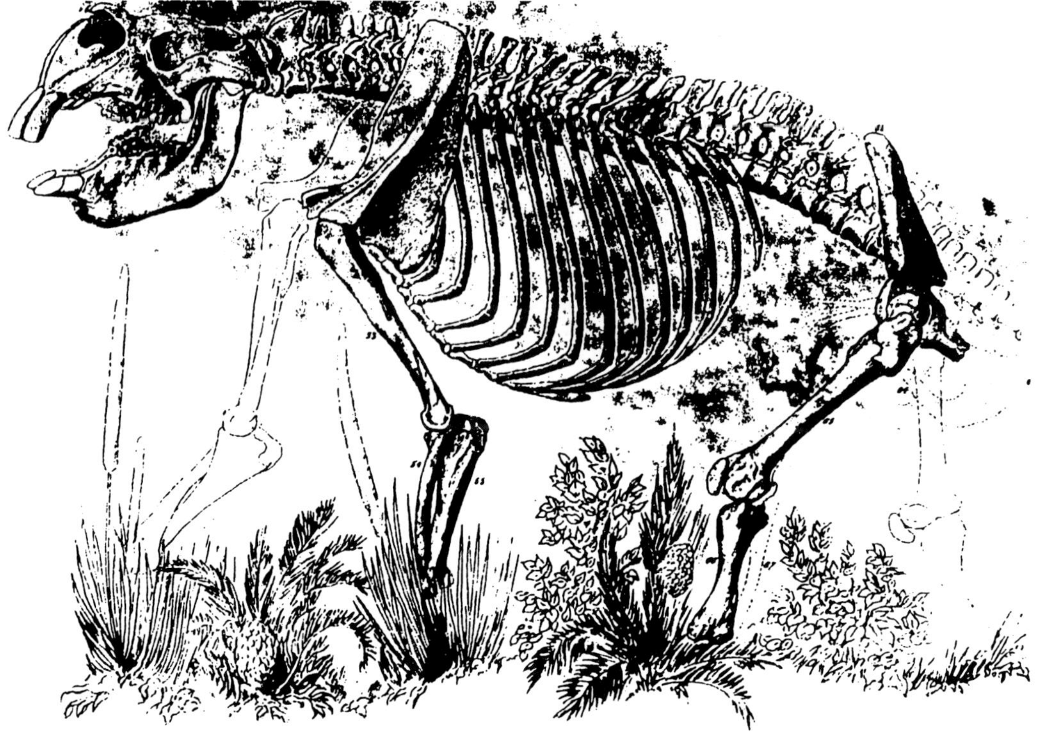

Fig. 13. Representation of skeleton of *Diprotodon*, according to Richard Owen (1877, **2**, plate XXXV).

Gulf of Carpentaria was not realized; but Mitchell made extensive botanical observations and collections. Pieces of fossil wood were recorded as being found loose on the ground on the plains of central Queensland (then called North Australia) (Mitchell 1848, p. 293).[54]

In 1847, Mitchell was back in England again and he arranged for the publication of his narrative of his fourth expedition. His illustrations were now made with the help of a *camera lucida*, but there was no substantial improvement in quality as a result, compared with his illustrations of 1838. Mitchell also went to Spain during the course of this leave, which, on his return to Sydney, led to the publication of a book on viticulture; but he had difficulty in explaining to the authorities why he was jaunting on the Continent.

Mitchell's work on the goldfields of New South Wales

On his return to Australia in July, 1848, Mitchell was again combating the authorities, at a time when his survey department was in disarray, losing funding as the economic circumstances deteriorated in New South Wales in the 1840s. But things improved in the 1850s with the announcement of the discovery of payable gold in New South Wales in April, 1851. There was an immediate gold rush, and the Government at last began to take an interest in geological matters. Samuel Stutchbury (1798–1859) was appointed Government Geologist and began survey work in the goldfields area west of the Blue Mountains, and then worked his way up into Queensland. W. B. Clarke (who was probably the first to discover gold in New South Wales in the 1840s, but did not announce the fact) was also in the field, and having examined the new goldfield area near Bathurst he turned south towards the highest mountains of Australia (the Snowy Mountains) and studied the geology of that large region, while continuing to look for gold. Additionally, the Government turned to Mitchell, now getting on in years, to examine the new western field in the area north and west of Bathurst.

Mitchell set off with his son Livingston on 31 May 1851, along with throngs of people travelling to the gold fields. When they reached the settlement of Molong (whose limestones and caves he had previously studied), not far from Wellington, they were joined by an assistant, Walker Davidson, one of the surveyors in Mitchell's department. The work was hard, as it was an exceptionally wet and cold winter, but they were a quite well-equipped official party, and the topography of the region was by then fairly well known. Stutchbury, working with much less assistance, crossed Mitchell's path briefly and wrote rather deprecatingly of the scale and performance of Mitchell's expedition.[55] Be this as it may, Mitchell's survey was completed and the party back in Bathurst by 13 August. His report was submitted on 16 October 1851, and published as a separate pamphlet with accompanying coloured map in 1852 (Mitchell 1852*a*, 1853). This work was quite an accomplishment for a man of fifty-nine, approaching old age after a tumultuous life.

Mitchell's goldfield report was, in a way, the only substantial piece of geological work he did without the assistance of experts. His earlier work consisted largely in making and reporting observations, collecting samples, and sending them to Europe for expert evaluation. Mitchell had his own ideas but he did not publish them, except in his *Three Expeditions*. As we have seen, his main paper was, for whatever reason, rejected after peer assessment.

The goldfields report was something different. Mitchell was essentially on his own, and for the first time dealing in a serious way with something other than mammalian remains.[56] He prepared two versions of a map. The first, dated 1851, was evidently a draft for the second, which was the officially published map, issued in 1852 (Mitchell 1852*b*).[57] The second Mitchell map (but not the first) was accompanied by two interesting east–west sections (see Fig. 14), which used the junction of the rivers Bell and Macquarie (at Wellington) as the datum line, this horizon being taken to be the 'lowest point of diluviatile [river] action in the country thus attempted to be represented'.

Mitchell's map shows most obviously the outline of the trachytic rock of Mount Canobolas (1395 m), south of the town of Orange and west of Bathurst. The main area of the map, marked in pink, he designated as 'gneiss and schist', but in the report it was referred to as 'chlorite or quartziferous schist', which was a reasonable description of the rocks of the region where the gold was being found. As can be seen from the map, Mitchell extended his mapping westwards to the Catombal Range, where he noted and mapped sandstone and conglomerate (orange), and to the Currambenya Range (red: granite). Mitchell did not know the age of the sandstone and conglomerate, but thought they might correspond to the 'oldest rock of the same kind, found in the formations of Europe'. This was drawing a long bow. No fossil evidence was cited. The sandstones appeared to strike approximately north–south, as did outcrops of limestone (blue), which included the limestone caves south of Wellington, which Mitchell knew so well.

Mitchell noted that the new settlement (or encampment) of Ophir, where alluvial gold was being found, was at the junction of two streams, which provided a favourable locality for the accumulation of gold. Both streams were sourced at sites close to the contact of the Canobolas trap and the quartz rock and schist. Mitchell thought, from his reading of Andrew Ure (1778–1857), that such a contact might be favourable for the production of gold, the idea being that the heat associated with igneous activity had driven up gold from the earth's interior.[58] Mitchell noted the high-angle dip of the schists, such that their line of strike was little affected by changes in topography. However, it is doubtful that he distinguished between bedding and cleavage, for the high angled planes represent cleavage rather than bedding. The Mullions Range at the centre of Mitchell's map (designated 'primary') was said to provide the 'auriferous ridges', whereas the rarely exposed granite was 'the basis of the whole'. At the time of his visit to Ophir, all the gold obtained was alluvial. Later, mines were cut into the schists in a hunt for the source of the gold in quartz veins. Mining continued in the region until the 1950s and one still sees fossickers there armed with metal detectors.

Mitchell's main effort was devoted to examining the streams and their courses and declivities, and their relations to the underlying rocks, for it was in the streams that the prospectors were finding gold. He also considered whether gold might be found *in situ*, but found that such occurrences were rare, though there were quartz veins (which Mitchell called dykes) aplenty.[59] Writing to Owen in 1852,[60] Mitchell bemoaned the fact that he had described quartzose rocks in his *Three Expeditions* that looked very like those that now turned out to be auriferous. But he had not suspected that possibility at the time. He suggested that the famous quartzites at Holyhead (Anglesey) looked similar and might prove auriferous.[61] Thus did he strain to find analogies between geological features in different continents. Of course, Sir Roderick Murchison had 'successfully' predicted that the mountains of the main divide in Australia would be auriferous, based on analogy with the Ural Mountains gold deposits. Such far-flung speculation as Mitchell offered was not regarded as absurd at that time. On the contrary, geologists were keen to establish global correlations, as they are today, though more in terms of time. Moreover, it was sometimes thought in the nineteenth century that the orientation of mountain ranges could be correlated with their ages.

Mitchell's northern section suggested a general anticlinal structure for his schistose rocks, but

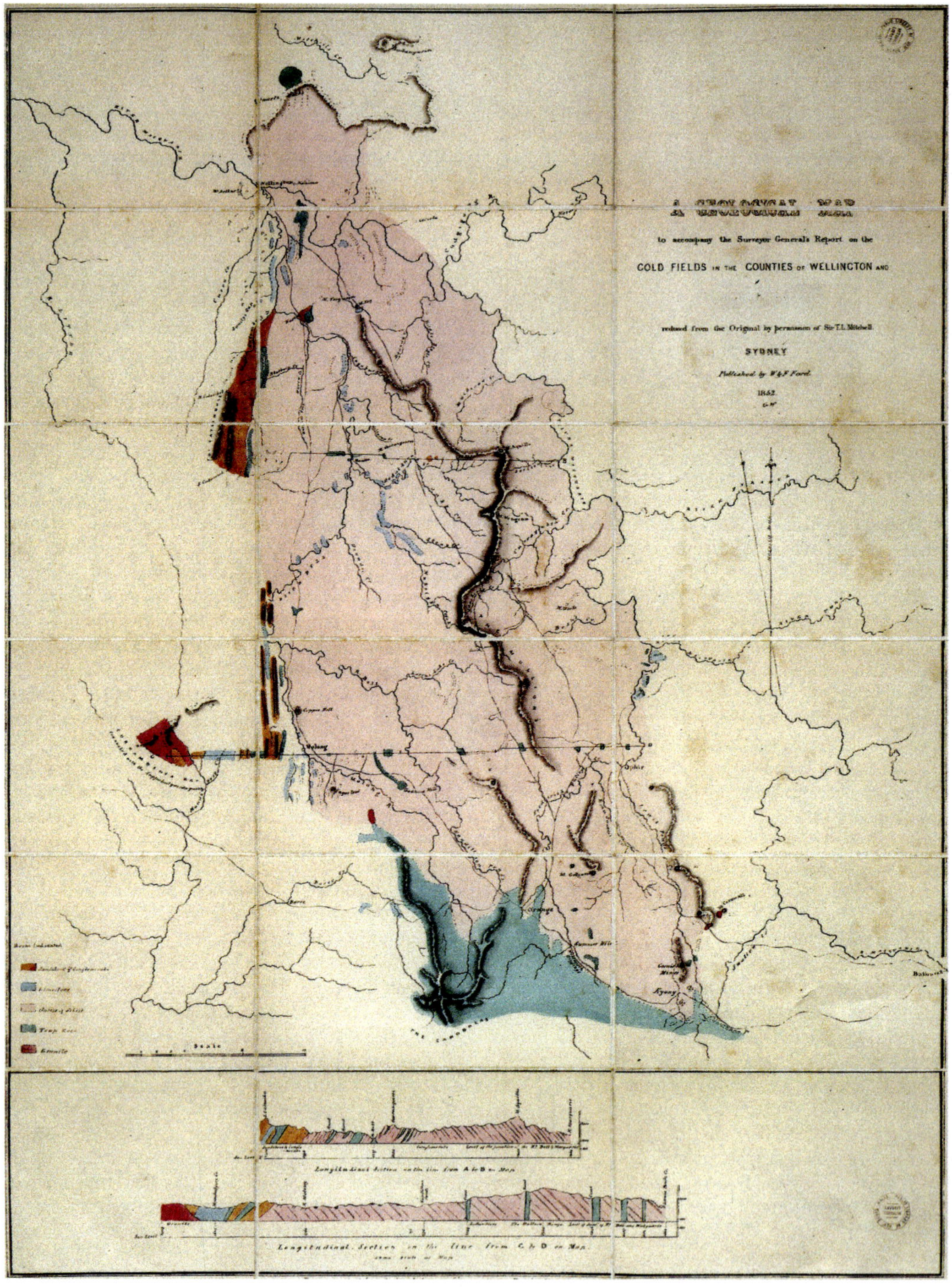

Fig. 14. Geological map and east–west sections of the district of Orange, Molong, Wellington, Ophir, Sofala, etc. (NW of Bathurst), published by Mitchell to accompany his Goldfields Report, 1852. The lines of section are marked on the map. (Reproduced by courtesy of the Mitchell Library, State Library of New South Wales, Sydney, Ref. 553.41/M.)

whereas the strike in the area is indeed generally north–south as his map implied, the actual structure of the region is vastly more complex than that suggested by his section. We are in the region known today as the Lachlan fold-belt. Mitchell remarked the apparent association of quartz conglomerate in a ferruginous matrix as being associated with the occurrence of gold. And on that basis, remembering the rocks he had observed in inland regions during his expeditions of the 1830s he suggested that the goldfields might be of much wider extent than that of the area of diggings at Ophir and Sofala. In the event, such expectations proved to be over-sanguine. In any case, he cautioned that access to the gold (if any) in more remote areas would be difficult for want of water. However, his expectation that gold might be found in 'Australia Felix' (Victoria) proved correct, and the rushes to Ballarat and Bendigo down south soon followed. Likewise, his prediction of the occurrence of gold in the Snowy Mountain region near Tumut proved correct.

A manuscript topographic map and profile drawn, approximately north–south, from Bathurst northwards to the goldfield town of Sofala also survives, with some geological information entered by Mitchell on the profile of the line of route.[62] Mitchell was apparently adopting the procedure suggested to him many years before by Fitton when Mitchell had been looking for geological advice in London, prior to his first voyage to Australia. Mitchell was hardly able to take much advantage of this procedure for his goldfields map. The Sofala–Bathurst profile was taken along a road approximately parallel to the strike, whereas sections required for the published map needed to cut across the strike, and for the most part, his map was based on what could be observed in occasional stream beds where gold might be looked for. The area is geologically complex, involving a thrust fault, and the greywackes and volcanics show many variations. In any case, Mitchell did not have sufficient time to make a good map of the area as a whole. Nor did he have the requisite geological knowledge. Also, much of the ground was then wooded and exposures were less common then than now. There are some reasonable outcrops that would probably have been visible in Mitchell's day, but his Sofala–Bathurst profile does not show high-angle dips, such as are evident in some places in the modern road cuttings. Mitchell also missed an easily observed exposure of granite in one of the stream beds along the route. (But perhaps the water level was high at that time.) As the published geological map does not cover the Sofala area, one may suspect that Mitchell simply made a detour from Bathurst to Sofala during his return journey to Sydney but did not collect enough information to add significantly to his map.

It is instructive to compare Mitchell's map (Fig. 14) with one by Stutchbury (taken from his diary,[63] not a finished map), working in the same area in the same season, in the same bad weather, with fewer resources. There isn't a great deal to choose between them in terms of quality, though Mitchell had the advantage of an accurate topographic map, whereas Stutchbury's appears to be hand-drawn, albeit probably based on the same base map as that which Mitchell used.[64] They do not agree well, even in the matter of the outcrop of the Mount Canobolas volcanics. So this early work did not really lay the basis for the subsequent geological mapping of New South Wales, and the early maps of Mitchell and Stutchbury were largely ignored by subsequent workers. But Stutchbury's published reports to the New South Wales Government gave much more detail than did Mitchell's.

The closing years and closing remarks

Things were bad for Mitchell on his return from the goldfields. He got into more bureaucratic and political wrangles and his son Roderick was drowned at sea in August 1851, having been washed overboard from a vessel sailing south from Moreton Bay to Sydney. Mitchell even fought a duel on 27 September 1851, with a politician who accused him of being responsible for excessive expenditure in his department, but fortunately both parties were still unscathed before they were separated by their seconds.

Mitchell was in England again in 1853–1854, but he died back in Australia on 5 October 1855, from a chill that turned to pneumonia, acquired while surveying a line of road to the SW of Sydney. The new Governor Sir William Denison (1804–1871) had wished to be rid of him and his department was being made the subject of an official inquiry. But death called Mitchell to account before his political masters could do so.

Though he had quarrelled with many people, Mitchell's activities, abilities, and accomplishments were substantial and varied. He had numerous admirers for his support of the small settlers. His long funeral cortege went all the way through Sydney to his burial place at St Stephen's Church in the suburb of Camperdown, there were so many mourners. His eldest son, Thomas Livingstone, inherited his substantial estate, but the four other sons died relatively young. His daughters married well and Lady Mitchell lived into her old age. Several of Mitchell's descendants were involved in survey work or served in the army.

With his experienced 'eye for country', and manifest abilities as a traveller, Mitchell might well have been a major figure in the history of Australian geology. As it was, he was chiefly a provider of information from the 'periphery' or colony to the 'centre' or London, and his attempts to sustain his geological contacts in London, during extended periods of leave, along with his failure to get his major work published by the Geological Society, mark him as a 'Basalla Stage 2' geologist, with the difficulties associated with such a condition. Mitchell's geological map of 1838 went unpublished and, stuck in London, it was of little use to those people in Australia interested in geology.[65] His real work in Australia lay in topographic mapping, exploration, and road building. He would have had greater renown as a geologist had he remained in Britain, and might well have become a second Murchison. That was not to be. But Mitchell made a permanent mark on Australian history by his explorations and mapwork, and his laying out of roads for the country's first transport network. Moreover, W. B. Clark, the country's leading geologist in the mid-nineteenth century, knew Mitchell's abilities well enough to write that his 'intelligence and observation would have made him a first-rate geologist could he but have dedicated his time to the necessary studies' (Clarke 1860, p. 244). In fact, most people in Australia today know Mitchell not for his geological work, his roads, or even his explorations, but by the name given to one the country's most beautiful birds: the 'Major Mitchell Cockatoo', which was figured in *Three Expeditions*.

I am much indebted to D. Branagan for extensive assistance and advice. P. Wyse Jackson provided numerous perceptive and appreciated editorial recommendations. W. Mayer kindly made available to me a copy of the typescript of his paper in the present volume. A. Mussell provided a copy of Mitchell's map held at the Geological Society, and figures of mammalian remains drawn by Mitchell; and W. Cawthorne from the Society's library also assisted. Thanks are further due to P. Taquet, of the *Muséum d'Histoire Naturelle*, Paris, who has provided information about the Mitchell material in the Cuvier archive. It has been possible to consult relevant correspondence held at the Natural History Museum, London, by the good offices of P. Tucker. P. Pemberton provided valuable information about the early coal investigations in the Newcastle area (NSW). In Sydney, the help of the staff of the Mitchell Library[66] has been greatly appreciated, and permission to publish Figure 14 is gratefully acknowledged.

Notes

[1]An excellent biography of Mitchell is: Foster (1985). General information about Mitchell's career, as described in this paper, may be found in this volume, though other sources have been consulted.

[2]An invoice, dated 11 March 1811, drawn up in London by Mitchell, requested payment of £36 9s 10d for management of the Rumford Colliery from 11 to 30 June 1809 (Mitchell Papers, **1**, 1811–1819, 36, Mitchell Library, Sydney).

[3]Following his retirement from the army, Murray became an influential politician and secretary for the colonies, from which position he was able to exert favorable influence on Mitchell's behalf.

[4]Phillips was a founder Member of the Geological Society of London, and subsequently became Curator of the Museum of Practical Geology in London.

[5]Mitchell, T. L. 1827. Science note-book and diary. Mitchell Papers, Mitchell Library, Sydney, C 39.

[6]I have been informed by Geoffrey Larminie (pers. comm., July 2003), a former Director of the British Geological Survey who worked for a time in Australia, that Mitchell's base-line and meridian were still the basis of mapping in New South Wales in the second half of the twentieth century. However, other base-lines were also laid out at Lake George near Canberra and at Richmond, on the plains between Sydney and the Blue Mountains.

[7]His dates are not known to me, but photographs of items for sale of specimens of his fine work may be found on the internet.

[8]This cannot be checked at the time of writing as its archives are currently being relocated and the map they hold is not available for inspection.

[9]David Branagan has pointed out to me that the Polish physician and naturalist Johann Lhotsky (1795–1866), who was in Australia between 1832 and 1838 and travelled to the alpine regions of New South Wales in the early months of 1834, may have prepared a geological map of the areas he visited. But it is not known to have survived. On Lhotsky's geological work in Australia, see Vallance (1977).

[10]See, for example, New South Wales Department of Mines, 1969. *Sydney Basin 1: 250 000*, Government Printer, New South Wales.

[11]The locality is now called Coalcliff. The shipwrecked sailors' discovery was confirmed the same year by the naval surgeon and traveller, George Bass (1771–1803), who also ascertained that Tasmania was an island, so that his name is given to Bass Strait, which separates Tasmania from the Australian mainland. The coals were examined geologically by W. B. Clarke and J. D. Dana in the late 1830s, but mining did not begin in the area until the 1850s. The basin-like structure for the Sydney region was vaguely hinted at very early on by Bass in 1797: letter to William Paterson, 20 August, 1797, *Historical Records of New South Wales*, Charles Potter, Sydney, 1895, **3**, 289–290: 'it is probable that they [the coal seams] extend a considerable way, for I am much mistaken if it will not be found that the Blue Mountains wind round to

this place [Coalcliff near Wollongong], and of course end there. If so, this structure of coal may possibly run through the whole range. Coals, you know, have been found washed down the Grose and the Tench [rivers, running eastward from the Blue Mountains]'.

[12]For a good account of the early history of coal mining in the Newcastle area, see: Branagan (1972). See also: Branagan (1990). Preceding the first page of this paper there is a black-and-white photocopied reproduction of the map reproduced in Figure 2 of the present publication.

[13]Henderson had been manager of a mine in Dunfermline, Scotland, and had sound experience in surveying coal districts.

[14]For Henderson's section, see Noel Butlin Archives Centre, Australian National University, Australian Agricultural Company, 78/1/2, p. 196; now imaged as Ref D128). See also: Pemberton (2004).

[15]For early boreholes for coal in the Sydney basin, see Branagan (1990), p. 5–6.

[16]Moyal (2003, **1**, 150) (Clarke to William Sharpe Macleay, 15 April, 1844).

[17]The report by Bailly and Depuch was prepared after their journey west of Sydney to the foot of the Blue Mountains in October, 1802.

[18]Depuch, L. 1802. Monsieur Depuch's opinion of the items lately sent from the mountains by M. Barrallier Nov. 13 [n.d. 1802 or 1803?]. In *King Papers*, **8**: Further Papers, 71, Mitchell Library, CY 906.

[19]There is a detailed account of the surveying and its culmination in the production of the map in: Andrews (1992). The map was reprinted in 1884 and subsequently republished by the Central Mapping Authority of New South Wales (1977). The scale is 50 English miles to 15 cm. A circumferenter (or circumferentor) was 'an instrument consisting of a flat brass bar with sights at the ends and a circular brass box in the middle, containing a magnetic needle, which play[ed] over a graduated circle; the whole being supported on a tripod' (*Shorter Oxford Dictionary*, 1969).

[20]The road was under construction from 1826 to 1836. It involved a steep descent to and ascent from the Hawkesbury River north of Sydney, where a ferry was located, and then northwards across rugged sandstone hills. The main work was completed by 1832, with Mitchell being a major driving force behind the operation. He had to reroute the road for its ascent from the northern side of the River and move the location of the ferry. For an account of the road history and construction, see: Upton (n.d.).

[21]The statues decorating the building were the work of J. White, W. P. McIntosh, and an Italian sculptor, but I have no information as to who carved the Mitchell statue. See: Anon. (n.d.).

[22]*Megadesmus globosus*, a bivalve mollusc, is now regarded as a Permian fossil. Clarke (1878, p. 113) stated that such fossils were known to be found at Harpur's Hill (near Maitland in the Hunter Valley). For a modern photograph of a specimen of the species, held in the Australian Museum, see: ⟨www.amonline.net.au/collections/palaeontology/record.cfm?id=7⟩

[23]The several older systems of the stratigraphic column were named as follows: Cambrian (by Sedgwick 1835); Ordovician (by Lapworth 1879); Silurian (by Murchison 1835); Devonian (by Sedgwick & Murchison 1839); Carboniferous (by Conybeare & Phillips 1822), Permian (by Murchison 1841); Triassic (by Alberti 1834), Jurassic (by Boué 1829); Cretaceous (by Omalius d'Halloy 1822).

[24]Now called Darling Harbour, the heart of the tourist area of Sydney.

[25]The indistinct date on this figure is 1829, but there appears to have been a correction.

[26]For previous discussion of Mitchell's work on megafauna, see: Foster (1936); Lane & Richards (1963); Dawson (1985); Osborne (1991); Branagan (1992); and Holland (1992).

[27]L[ang, John Dunmore], Interesting discovery. To the Editor of the *Sydney Gazette*. *Sydney Gazette*, 21 May, 1830. Lang thought that the bones must have been taken into the caves by 'beasts of prey'. It was suggested that the cave at Wellington might be analogous to the one made famous by William Buckland in Yorkshire: the Kirkdale cavern, which Lang called the Kirkham cave. He opined that the remains were antediluvian, were different in kind from animals presently found in Australia, and that there had been a 'physical convulsion that destroyed the various races of animals' upwards of 4000 years ago. Like Buckland, Lang thought there was evidence of the Deluge. Divine design could be recognized in the elimination of dangerous large animals which would have made Wellington Valley not 'so eligible a place for the residence of man as it presently is'. Lang very likely gave the specimens to Jameson in person as he visited Britain in 1830.

[28]Kinghorne had reached Australia in 1822 and was appointed superintendent of the Government agricultural establishment at Emu Plains, near the Hawkesbury River below the eastern scarp of the Blue Mountains, before being allocated a farm as a 'squatter'. He eventually acquired extensive pastoral interests.

[29]I thank Professor Philippe Taquet for informing me of the location of the original of this picture.

[30]John Henderson was a surgeon in the Bengal Army with a considerable interest in botany. He served in various places in India between 1815 and 1829, but was on leave in Australia in 1829–1831, trying to recover his health after his extended sojourn in Asia. Initially in Tasmania, he sought to establish a scientific society there, and largely through his efforts the Van Diemen's Land Scientific Society was formed in 1830, building on the first efforts towards scientific education in Hobart through a mechanics institute there. But the Society was short-lived, with Henderson soon taking off for New South Wales and residing in Sydney for several months, planning to compare the geology of New South Wales and Van Diemen's Land

(Tasmania). He first went to Wellington and spent some time exploring the caves there, and tried to get backing from Governor Darling for travel further west to search for an inland sea, but having failed to obtain Government support for this enterprise he decided to return by striking across wild country to the NE towards the Hunter Valley, accompanied only by an Indian servant and without even a compass to assist navigation. His journey involved considerable hardships, but the two travellers got back to civilization safely. Henderson wrote a report for Darling about his observations at Wellington, but they added little to what was known from the work of Mitchell and others. It appears, however, that some specimens he collected were to be transported to London for examination by Fitton and Buckland. It is not known whether they ever arrived. The report (eventually published in Calcutta) contained a rudimentary diagram of the cave systems at Wellington and Boree, and very simple drawings of three teeth and a fossil 'coral' (seemingly a crinoid). His report listed the fossils collected, which he thought included a thylacine (Tasmanian 'tiger'), opossums, kangaroo rats, ducks, a carnivorous animal the size of a leopard, possibly a human upper jawbone, a large animal thought to be 'pisciverous', seeming pigs' teeth, and some other bits and pieces. Henderson clearly did not have the knowledge to make a good collection. He returned to India in 1831, where he undertook immense journeys, and did much to try to promote public education and health; but he died before his travels were published. For Henderson's Australian work, see: Henderson (1832; republished 1965), and also Hoare (1968). I thank David Branagan for these references.

[31]Mitchell, T. L. 1830. Diary entry, 3 July, 1830. Mitchell Library, Sydney.

[32]Several of the originals for *Three Expeditions*' illustrations of the Wellington Caves and their fossils survive at the *Muséum d'Histoire Naturelle*, Paris (Cuvier MSS 640), but the original map is not in the Paris collection. A map could have been hand-coloured in the early 1830s, but it is likely that the version, as it appears in the book, was prepared in the year of publication, 1838.

[33]Mitchell Papers, A 292, Mitchell Library, Sydney, 14 October, 1830.

[34]Fitton to Mitchell, 11 August, 1831. Mitchell Papers, A 292, Mitchell Library, Sydney, 137–140.

[35]When this paper was first published, Jameson erroneously supposed that the author was Lang, not Mitchell. Jameson subsequently published a correction. There is no evidence that Lang visited the caves in person at the time in question, though he was interested in geology.

[36]Keeper of the Hunterian Museum at the Royal College of Surgeons, Clift was a leading expert at the time on fossil bones and frequently described and figured specimens for other persons (including Darwin's South American specimens).

[37]Turner had studied medicine at Edinburgh, providing another Scottish connection. Presumably the analysis was organized by Fitton.

[38]Lind[e]say was commander of the 39th Regiment and Acting Governor of New South Wales for six weeks in 1831, probably during which time he wrote a note about the finds to Jameson.

[39]Mitchell named a peak for Lind(e)say in the Nandewar Range, near Narrabri, northern New South Wales. Lind(e)say was another soldier-Scot. He was for a time commander of the garrison at Sydney, and was Acting-Governor of the Colony for a short period in 1831. He also transmitted Australian bird-skins to Edinburgh and was generally interested in natural history. On his return to Britain in 1836, he was promoted to major-general and knighted.

[40]Mitchell Papers, Mitchell Library, Sydney, **8**, 189.

[41]Mitchell Papers, 1830–1839 (CYA 292), Mitchell Library, Sydney.

[42]Fitton, writing to Mitchell, 11 August 1831, stated that Cuvier was of the opinion that one of the large bones belonged to an elephant (Mitchell Papers, A 292, 137–140, Mitchell Library, Sydney).

[43]Mitchell Papers, Mitchell Library, Sydney, **8**, 224.

[44]Mitchell, letter to Ranken, 24 July 1833, in: Ranken (1916, p. 28).

[45]Mitchell, letter to Ranken, 30 October 1831, in: Ranken (1916, p. 25).

[46]Much of W. B. Clarke's early work appeared in the *Sydney Morning Herald* newspaper at that period. In this respect, he was not a 'pure' colonial scientist.

[47]Mitchell to Lonsdale, 2 June 1838, Geological Society archives, LR3/310.

[48]Mitchell to William Lonsdale, 6 November 1838, Geological Society Archives, LR4/49.

[49]Mitchell to Owen, 6 April 1842. Owen Correspondence, British Museum of Natural History, **19**. See also reference to letter from Mitchell to Owen, 3 January 1842, quoted in: Owen (1877, p. 240). Mitchell described the river as consisting of basins in some parts, swamp areas in others, and cutting channels through deep beds of alluvium in yet others. The bones were discovered in the alluvium.

[50]Mitchell to Owen, 28 January 1843. Owen Correspondence, British Museum of Natural History, **19**.

[51]*Nototherium* = 'southern beast'.

[52]Owen lacked the bones of part of the fore-limbs, and the feet; and that these deficiencies were covered up by the imagined vegetation. (The plants shown are recognizable as belonging to the Australian flora.)

[53]Bennett to Owen (Owen 1877, **1**, p. 190).

[54]There are records of fossil wood specimens, collected by Mitchell, in the collections of the Queensland Museum, Brisbane, and the British Museum of Natural History. See: Cleevely 1983, p. 205. I am informed by Susan Turner that the Brisbane specimens cannot now be located in the Museum.

[55]Stutchbury wrote in his diary: 'Sir Thomas Mitchell's mode of taking sections is by observing an Aneroid Barometer upon heights and in the bottoms fixing his distance by horse paces!! The geology of a country he ascertains by knocking off a piece of rock here and

there' (27 June 1851); and 'This morning Sir Thomas Mitchell moved his camp, and in so doing experienced great trouble with his horses. His party consist[ed] of himself, his son (Mr Livingstone Mitchell), and three men, together with Mr Davidson, Assistant Surveyor, and six men—twelve persons in all. He had with him three horses and yet was obliged to hire a team of eight bullocks to assist the removal of his camp and baggage' (28 June 1851) (Mitchell Library, A 2839). Stutchbury, by contrast, recorded in his diary (1 May 1852) that he only had two assistants and three horses. Astonishingly, he also asserted (3 August 1852) that he differed substantially from Mitchell about the general direction of strike of the rocks in the goldfield. Stutchbury had them striking 11°–20° west of north, whereas Mitchell had them running east of north; and the surveyors did not allow for magnetic variation.

[56]In a sense, he was now just edging towards Basalla's (1967) Phase 3.

[57]The topographic mapwork is attributed in the State Records to a surveyor named J. S. Adams. (There is a word missing in the printed map's title.)

[58]Mitchell did not give a reference for Ure's idea, but cited him as follows on page 5 of the goldfield report: 'That silica and its associated bases, which are oxydised at the surface of the earth, and thus deprived of their elementary activity exist at a moderate depth beneath the surface, in a state of simple combustibles, there is little reason to doubt'. On this basis, Mitchell thought that 'the nearer the quartz is found to the igneous rock and places of irruption, the greater will be the probability that so heavy a metal as gold, or indeed any other metal, may be found with it'.

[59]In fact, mining from quartz veins was later carried out extensively in the goldfield at Ophir, and subsequently further north at Hill End.

[60]Mitchell to Owen, 5 January, 1852. Owen Correspondence, The British Museum of Natural History, **19**.

[61]I have no reason to think that Mitchell visited Holyhead personally.

[62]The map, prepared by Mitchell in 1851, shows the outcrops of a small number of different rock-types, while the profile depicts the supposed underground extensions of some of the strata. The document is held in the State Archives of New South Wales, Kingswood, as: Surveyor General. CGS 13859, Maps and Plans, 1792–1880, Map No. 5132.

[63]Stutchbury, S. Undated coloured geological map. Manuscript diary. Mitchell Library, A 2639.

[64]David Branagan, who has done extensive work on Stutchbury, informs me that Stutchbury, the official Government geologist, had difficulty in getting access to the official Government maps prepared by Mitchell's department.

[65]No copy of the map is known to me in Australia, though presumably Mitchell kept one for himself.

[66]The Mitchell Library in Sydney is named after the nineteenth-century bibliophile, David S. Mitchell, not Thomas L. Mitchell.

References

ANDREWS, A. E. J. 1992. *Major Mitchell's Map 1834: The Saga of the Country of the Nineteen Counties.* Blubber Head Press, Hobart.

ANON. [Lind(e)say P.] 1831. Bone caves discovered in New Holland. *Edinburgh New Philosophical Journal*, **10**, 179.

ANON. 1830. Volcano in New South Wales. *Cheek's Edinburgh Journal of Natural and Geographical Science*, **2**, 62.

ANON. 1833. Description of Mount Victoria. *The New South Wales Calendar and General Post Office Directory*, Sydney, 323–326.

ANON. n.d. *Lands Department Building: Bridge Street Sydney*. Department of Main Roads, Sydney.

BASALLA, G. 1967. The spread of Western science. *Science*, **156**, 611–622.

BASS, G. 1797. Letter to William Paterson, 20 August, 1797. *Historical Records of New South Wales*, Charles Potter, Sydney, 1895, **3**, 289–290.

BRANAGAN, D. F. 1972. *Geology and Coal Mining in the Hunter Valley 1791–1861*. Newcastle Public Library, Newcastle History Monographs No. 6, Newcastle.

BRANAGAN, D. F. 1990. A history of New South Wales coal mining. *In*: BRANAGAN, D. F. & WILLIAMS, K. L. (eds) *Coal in Australia*. The Third Edgeworth David Symposium, The University of Sydney, Sydney, 1–15.

BRANAGAN, D. F. 1992. Richard Owen in the antipodean context [a review]. *Journal and Proceedings of the Royal Society of New South Wales*, **125**, 95–102.

CENTRAL MAPPING AUTHORITY OF NEW SOUTH WALES. 1977. *Map of the Colony of New South Wales Drawn by Sir Thomas Mitchell 1834* (facsimile reprint), Bathurst (NSW).

CLARKE, W. B. 1860. *Researches in the Southern Gold Fields of New South Wales*. Reading and Wellbank, Sydney.

CLARKE, W. B. 1878. *Remarks on the Sedimentary Formations of New South Wales: Illustrated by References to other Provinces of Australia*. 4th edn. Government Printer, Sydney.

CLEEVELY, R. 1983. *World Palaeontological Collections*. British Museum (Natural History) and Mansell Publishing Limited, London.

CLIFT, W. 1831. Report by Mr Clift, of the College of Surgeons, London, in regard to the fossil bones found in the caves and bone-breccia of New Holland. *Edinburgh New Philosophical Journal*, **10**, 394–396.

DAWSON, L. 1985. Marsupial fossils from Wellington caves, New South Wales; the historic and scientific significance of the collections in the Australian Museum, Sydney. *Records of the Australian Museum*, **37**, 55–69.

DELAIR, J. B. & SARJEANT, W. A. S. 1976. Joseph Pentland: a forgotten pioneer in the osteology of fossil marine reptiles. *Proceedings of the Dorset Natural History and Archaeological Society*, **97**, 12–16.

DUGGAN, K. G. 1980. Darwin and *Diprotodon*: the Wellington Caves fossils and the Law of Succession. *Proceedings of the Linnean Society of New South Wales*, **104**, 265–272.

FALCONER, H. 1863. On the American fossil elephant of the regions bordering the Gulf of Mexico (*E. Columbi*,

Falc.); with general observations on the living and extinct species. *The Natural History Review: A Quarterly Journal of Biological Sciences*, **3** (new series), 43–114 and plates.

FITTON, W. H. 1826. An account of some geological specimens, collected by Captain P. P. King, in his survey of the coasts of Australia, and by Robert Brown, Esq., on the shores of the Gulf of Carpentaria, during the voyage of Captain Flinders. *Tilloch's Philosophical Magazine*, **68**, 14–34, 132–147.

FOSTER, W. 1936. Colonel Sir Thomas Mitchell, D.C.L., and fossil mammalian research. *Journal and Proceedings of the Royal Australian Historical Society*, **22**, 433–443.

FOSTER, W. C. 1985. *Sir Thomas Livingston Mitchell and his World 1792–1855: Surveyor General of New South Wales 1828–1855.* The Institution of Surveyors, Australia, Sydney.

HACKFORTH-JONES, J. 1980. *Augustus Earle Travel Artist: Paintings and Drawings in the Rex Nan Kivell Collection, National Library of Australia.* National Library of Canberra, Camberra.

HENDERSON, J. 1832. *Observations on the Colonies of New South Wales and Van Diemen's Land.* Baptist Mission Press, Calcutta (republished in facsimile by Libraries Board of South Australia, Adelaide, 1965).

HOARE, M. E. 1968. Doctor John Henderson and the Van Diemen's Land Scientific Society. *Records of the Australian Academy of Science*, **1**, 7–24.

HOLLAND, J. 1992. Thomas Mitchell and the origins of Australian vertebrate paleontology. *Journal and Proceedings of the Royal Society of New South Wales*, **125**, 103–106.

JAMESON, R. 1831. On the fossil bones found in bone-caves and bone-breccia of New Holland. *Edinburgh New Philosophical Journal*, **10**, 393.

LANE, E. A. & RICHARDS, A. M. 1963. The discovery, exploration and scientific investigation of the Wellington Caves, New South Wales. *Helectite: Journal of Australasian Cave Research*, **2**, 3–53 and plates.

LATOUR, B. 1987. *Science in Action: How to Follow Scientists and Engineers through Society.* Open University Press, Milton Keynes.

MAYER, W. 2007. The quest for limestone in colonial New South Wales, 1788–1825. *In*: WYSE JACKSON, P. N. (ed.) *Four Centuries of Geological Travel: The Search for Knowledge on Foot, Bicycle, Sledge and Camel.* Geological Society, London, Special Publications, **287**, 325–342.

MITCHELL, T. L. 1827. *Outlines of a System of Surveying, for Geographical and Military Purposes: Comprising the Principles on Which the Surface of the Earth May be Represented on Plans.* Samuel Leigh, London.

MITCHELL, T. L. 1831. Additional information illustrative of the natural history of the Australian bone-caves and osseous breccia. Communicated by Dr. Lang. *Edinburgh New Philosophical Journal*, **10**, 368–371.

MITCHELL, T. L. 1834 (for 1831). An account of the limestone caves at Wellington Valley, and of the situation, near one of them, where fossil bones have been found. *Proceedings of the Geological Society of London*, **1**, 321–322.

MITCHELL, T. L. 1838. *Three Expeditions into the Interior of Eastern Australia; With Descriptions of the Recently Explored Region of Australia Felix, and of the Present Colony of New South Wales.* T. and W. Boone, London (2nd edn, T. and W. Boone, London, 1839; reprinted in facsimile, Libraries Board of South Australia, Adelaide, 1965).

MITCHELL, T. L. 1848. *Journal of an Expedition into the Interior of Tropical Australia, in Search of a Route from Sydney to the Gulf of Carpentaria*, Longman, Brown, Green and Longmans, London.

MITCHELL, T. L. 1851. Catalogue of 586 specimens of Australian rocks, now in the collection of the Geological Society of London, with their descriptions, by William Lonsdale, Esquire, F.G.S. Tabled at the Legislative Council [of New South Wales] 2 December 1851. Appendix to 'Report of the Surveyor General on the Gold Fields of Bathurst, Wellington, &c.', 40–47.

MITCHELL, T. L. 1852*a*. *Geological Report on the Goldfield, in the Counties of Wellington and Bathurst.* W. and F. Ford, Sydney.

MITCHELL, T. L. 1852*b*. *A Geological Map to Accompany the Surveyor General's Report on the Gold Fields in the Counties of Wellington and [—] Realised from the Original by Permission of Sir T. L. Mitchell.* W. and F. Ford, Sydney.

MITCHELL, T. L. 1853. Geological Report on the gold fields, in the Counties of Wellington and Bathurst. *Further Papers Relating to the Recent Discovery of Gold in Australia (In Continuation of Papers Presented to Parliament June 14, 1852.) Presented to both Houses of Parliament by Command of Her Majesty, February 28, 1853.* HMSO, London.

MOYAL, A. 2003. *The Web of Science: The Scientific Correspondence of the Rev. W. B. Clarke, Australia's Pioneer Geologist.* 2 Volumes. Australian Scholarly Publishing, Melbourne.

OSBORNE, R. A. L. 1991. Red earth and bones: the history of cave sediment studies in New South Wales, Australia. *Earth Sciences History*, **10**, 13–28.

OWEN, R. 1838. Letter, Owen to Mitchell, 8 May, 1838. Description of bones collected from Wellington Caves. *In*: MITCHELL, T. L. *Three Expeditions into the Interior of Eastern Australia; With Descriptions of the Recently Explored Region of Australia Felix, and of the Present Colony of New South Wales.* T. and W. Boone, London, **2**, 365–369.

OWEN, R. 1843*a*. On the discovery of the remains of a mastodontoid pachyderm in Australia. *Annals and Magazine of Natural History*, **11** (new series), 7–12.

OWEN, R. 1843*b*. Additional evidence proving the Australian Pachyderm to be a Dinotherium, with remarks on the nature and affinities of that genus. *Annals and Magazine of Natural History*, **11** (new series), 329–332.

OWEN, R. 1845 *for* 1844. Report on the extinct mammals of Australia, with descriptions of certain fossils indicative of the former existence in that continent of large marsupial representatives of the Order *Pachydermata*. *Report of the Fourteenth Meeting of the British Association for the Advancement of Science; Held at York in September 1844.* John Murray, London, 223–240.

Owen, R. 1877. *Researches on the Fossil Remains of the Extinct Mammals of Australia; with a Notice of the Extinct Marsupials of England.* 2 Volumes. J. Erxleben, London, **1**.

Pemberton, P. 2004. 'Coal from Newcastle: The Australian Agricultural Company's collieries in the nineteenth century' 12th Biennial National Conference of the Australian Historical Association, 5–9 July, 2004. *Abstracts*, 104.

Pentland, J. 1832. On the fossil bones of Wellington Valley, New Holland, or New South Wales. *Edinburgh New Philosophical Journal*, **13**, 301–308.

Pentland, J. 1833. Observations on a collection of fossil bones sent to Baron Cuvier from New Holland. *Edinburgh New Philosophical Journal*, **14**, 120–121.

Péron, F. 1807. *Voyages Découvertes aux Terres Australes, Exécuté par Ordre de Sa Majesté l'Empereur et Roi, sur les Corvettes le Géographe, le Naturaliste et la Goëlette le Casuarina, pendant les Années 1800, 1802, 1803 et 1804; publié par Décret Impériale, sous le Ministère de M. de Champagny, et rédigé par M. F. Péron.* 2 Volumes. L'Imprimerie Impériale, Paris.

Ranken, W. B. 1916. *The Rankens of Bathurst.* S. D. Townsend and Co., Sydney.

Sarjeant, W. A. S. & Delair, J. B. 1979. An Irishman in Cuvier's laboratory. The letters of Joseph Pentland, 1820–1832. *Bulletin of the British Museum (Natural History), Historical Series*, **6**, 245–319.

Strzelecki, P. E. de, 1845. *Physical Description of New South Wales and Van Diemen's Land. Accompanied by a Geological Map, Sections, and Diagrams, and Figures of the Organic Remains.* Longman, Brown, Green, and Longmans, London.

Upton, T. H. n.d. *The Establishment of Direct Communication between Sydney and Newcastle as Published in the Journal of the Institution of Engineers, Australia, Vol. 4 Nos 5, 6 and 7, May, June and July 1932.* The Department of Main Roads [of] New South Wales, Sydney.

Vallance, T. G. 1977. John Lhotsky and geology. *In*: Kruta, V. *et al.* (ed.) *Dr. John Lhotsky: The Turbulent Australian Writer Naturalist and Explorer.* Australia Felix Literary Club, Documentary and Historical Series, No. 2, Melbourne, 41–56.

Vallance, T. G. 1981. The fuss about coal: troubled relations between paleobotany and geology. *In*: Carr, D. J. & Carr, S. J. M. (eds) *Plants and Man in Australia.* Academic Press, Sydney, 133–176.

Vallance, T. G. 1981–1982. Lamarck, Cuvier and Australian geology. *Histoire et Nature*, No. **19–20**, 133–136.

Vallance, T. G. & Moore, D. T. 1982. Geological aspects of the voyage of HMS *Investigator* in Australian waters, 1801–5. *Bulletin of the British Museum (Natural History), historical Series*, **10**, 1–43.

Vallance, T. G., Moore, D. T. & Groves, E. W. 2001. *Nature's Investigator: The Diary of Robert Brown in Australia, 1801–1805.* Australian Biological Resources Library, Canberra.

Wilton, C. P. N. 1830. Account of the Burning Mountain in Australia, called Mount Wingen, near Hunter's River. *Edinburgh Journal of Science*, **2** (new series), 270–273.

Nineteenth-century observations of the Dun Mountain Ophiolite Belt, Nelson, New Zealand and trans-Tasman correlations

M. R. JOHNSTON

395 Trafalgar Street, Nelson 7010, New Zealand (e-mail: mike.johnston@xtra.co.nz)

Abstract: The Nelson Mineral Belt, part of the Dun Mountain Ophiolite Belt of Permian age, in the north of the South Island of New Zealand, was the subject of intense interest soon after the European settlement of New Zealand owing to the discovery of copper and chromite mineralization. The first geologist to survey the Mineral Belt in east Nelson was Thomas Ridge Hacket (*c.* 1830–1884), who arrived from Britain in 1857, although he was primarily interested in the mineralization. The first scientific description of the belt followed the visit in 1859 of the geologist Ferdinand von Hochstetter (1829–1884) of the Austrian Geological Survey who regarded the belt as a sill. He recognized that Dun Mountain was not serpentinite, like the rest of the belt, but an olivine-rich rock that he named dunite. von Hochstetter tentatively considered that thick mafic rocks of the Brook Street Volcanics, that were then considered to be intrusives rather than sedimentary, may have been a correlative of the Mineral Belt. Between the belt and the Brook Street rocks he mapped the Wooded Peak Limestone and Maitai Slates. von Hochstetter tentatively assigned a Mesozoic age to all of the above rocks and for nearly 100 years Mesozoic sedimentary rocks in many parts of New Zealand were regarded as a correlative of the Maitai Slates.

Both Hacket and von Hochstetter confirmed an abundance of chromite in the Mineral Belt, a finding that resulted in the building of New Zealand's first railway. Mining was short-lived but hopes were raised that gold might be found following Hacket's move to Queensland in the late 1860s. He recognized that the Gympie gold-fields had many similarities with Dun Mountain and reasoned that east Nelson might also be auriferous. This was taken up by men in Nelson such as Sir David Monro (1813–1877) and William Wells (1810–1893) who were keen geological observers, and by Monro's son in law Dr, later Sir James, Hector (1834–1907) who was director of the New Zealand Geological Survey. In 1870, Hector arranged for Edward Heydelbach Davis (1845–1871) to undertake a detailed examination of the Permo-Triassic rocks of east Nelson. Davis, like Hacket, regarded the Mineral Belt as metamorphosed Maitai rocks. As well as demonstrating that the east Nelson rocks did not contain economic concentrations of gold, copper or chromite, Davis showed that they were distinct from what had been thought to be similar rocks in the Coromandel gold-fields in North Island of New Zealand. This was the first, if not fully appreciated, step to limit the Maitai rocks. Hacket's correlation of the Gympie and Dun Mountain rocks was forgotten until the advent of plate tectonics. The various major rock units in Nelson are now recognized as terranes that accumulated as a result of convergent plate tectonics along the Mesozoic margin of Gondwanaland.

To the east of Nelson city (Fig. 1), in the north of New Zealand's South Island, is the classic section through the Permo-Triassic rocks of New Zealand. The section is not simple because the rocks are now known to comprise a number of fault-bounded strips or terranes (Rattenbury *et al.* 1998). Exposed in the Maitai River, which rises on Dun Mountain to join The Brook before discharging into the sea at Nelson, are the Dun Mountain–Maitai and Brook Street terranes in the east and west respectively. The eastern terrane consists of the Early Permian Dun Mountain Ophiolite Belt and the unconformably overlying sedimentary Maitai Group of Late Permian–Early Triassic age. Except for an absence of pelagic sediments and pillow lava, the ophiolite belt is substantially complete. However, in any one section, parts of it may be thin or absent due to subsequent faulting so that, for example, adjacent to Dun Mountain it is dominated by cumulate ultramafics (Johnston 1981). The Brook Street terrane is composed of Permian volcanogenic breccia and tuff, with minor volcanics, that were regarded as igneous in the nineteenth century. To the SW of the Maitai valley the Murihiku terrane, composed of the Triassic sedimentary Richmond Group, intervenes between the Brook Street and Dun Mountain–Maitai terranes (Fig. 2). The terranes, along with others forming the basement rocks of eastern New Zealand, accumulated as a result of convergent plate tectonics on the Mesozoic margin of Gondwanaland. Beginning in Late Cretaceous times, the accreted terranes separated from the former continent with the opening of the Tasman Sea between Australia and New Zealand. The propagation of the boundary between the Pacific and Indo-Australian plates

From: WYSE JACKSON, P. N. (ed.) *Four Centuries of Geological Travel: The Search for Knowledge on Foot, Bicycle, Sledge and Camel.* Geological Society, London, Special Publications, **287**, 375–387.
DOI: 10.1144/SP287.27 0305-8719/07/$15.00

Fig. 1. Nelson City between Nelson Haven and the eastern ranges drained by the Maitai River (top left) and the Brook (top right). The low conical hills just beyond the city are of Brook Street Volcanics whereas Maitai Group underlies the smoother intermediate hills. The Dun Mountain Ophiolite Belt forms the skyline and Dun Mountain (1129 m) is the broad mountain top right. Photo: Lloyd Homer, courtesy of GNS Science, New Zealand.

into the South Pacific in the Early Miocene resulted in the Alpine Fault and the buckling of the Gondwanaland crust to form the islands of New Zealand.

The ophiolite belt is up to 10 km wide and extends for 150 km through the mountains of east Nelson. It also crops out in the south of the South Island, on the other side of the Alpine Fault, and has been traced magnetically to the NW and SE of New Zealand. Because of the high magnesium content of the soils, which prevents the uptake of calcium, plant growth on the ultramafic part of the belt is stunted so that it contrasts strongly with the forested mountains developed on the enclosing mafic and sedimentary rocks (Fig. 3). With its numerous brown weathering outcrops, higher peaks within it were bestowed names such as Red Hills and Dun Mountain. Because it was of no use for farming, the European settlers ignored the serpentinitic rocks of the ophiolite belt until copper and chromite were found near Dun Mountain. The serpentinitic rocks soon acquired the name of the 'Mineral Belt' and for nearly a century Nelsonians adhered to the belief that it would yield great mineral wealth (Johnston 1987). From 1857, a number of geologists began to assess this perceived wealth. Although economic interest subsequently waned, the advent of plate tectonics and the recognition of ophiolites as a slice of mantle tectonically emplaced into continental crust ensured that scientific investigation of the belt continued.

Maori and European mineral discoveries

The Mineral Belt was of interest to the first people to occupy New Zealand. The Maori discovered that mudstone incorporated as blocks into a serpentinitic melange along the eastern margin of the Mineral Belt (Fig. 3) could be worked into stone implements. The mudstone, metasomatized to an albite and tremolite-rich rock known to the Maori as pakohe, is very tough because of the high tensile strength of the tremolite fibres. Nevertheless it had a conchoidal fracture that allowed it to be fashioned into a variety of tools, particularly adzes. Extensive quarries, including several in the Maitai valley, were developed and the resulting tools were traded widely throughout New Zealand. With increasing European contact, and the resulting introduction of iron and steel implements, the quarries were quickly abandoned and by the time of organized European settlement they were forgotten.

To the Europeans who arrived in New Zealand, understanding the basement rocks of New Zealand was to prove a formidable challenge. From 1840, early European settlement, largely from the British Isles, was organized by companies whose principal goal was to place farmers on the land. Consequently arable lands adjacent to natural ports were the main areas of interest. Nelson was the first planned settlement in the South Island but in 1841 an advance party chose the extensive plains of the east coast of the island. The colonial administrators vetoed this and instead insisted that the new settlement had to be closer to existing settlements in the North Island. Consequently Nelson was established in the north of the South Island, which was largely mountainous with only scattered lowlands and few conveniently situated natural harbours. From the start, the Nelson settlement was constrained by limited areas suitable for farming; however, indications of mineral wealth looked as though they might compensate for this deficiency. Traces of alluvial gold were found as early as 1842, although it was not until after the Australian rushes of the early 1850s that New Zealand's first payable field was discovered in 1856 in west Nelson, in what the Dutch explorer Abel Tasman had, in 1642, named Murderers Bay. The area was soon to be known by its present appellation of Golden Bay. Prior to the Golden Bay rush, traces of copper were found in 1852 in the Mineral Belt close to the town of Nelson, the port for the scattered Nelson settlement. The following year it became the administrative centre for the newly promulgated Nelson Province.

The discovery of copper created widespread interest among the settlers who instantly recognized the benefits that a copper mine would bring to

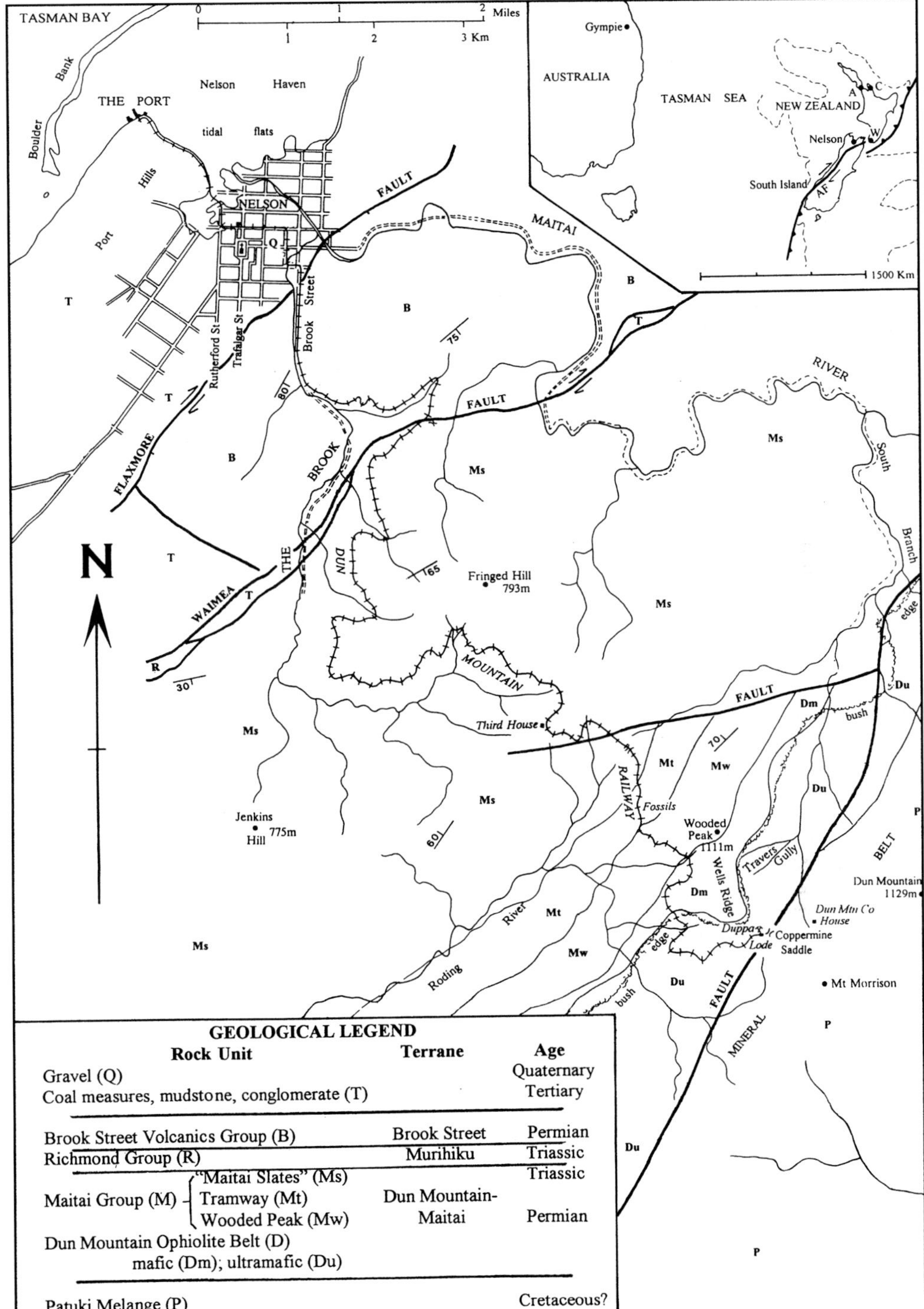

Fig. 2. Simplified geological map of the district between Nelson and Dun Mountain with inset map showing New Zealand, Tasman Sea and eastern Australia. AF, Alpine Fault and boundary between Pacific and Indo-Australian plates; W, Wellington; A, Auckland; C, Coromandel peninsula. The dashed line is 3000 m isobath, which approximates to the edge of the New Zealand continental crust. The coast at Nelson is as it was in 1870.

Fig. 3. View NE along the Nelson Mineral Belt (Dun Mountain Ophiolite Belt) which is characterized by a lack of forest cover on the ultramafic rocks. Dun Mountain, the type locality of dunite, is the prominent relatively flat-topped peak top centre. Chromite and copper mineralization is in the vicinity of Coppermine Saddle between Dun Mountain and the broad bush capped Wooded Peak and Wells Ridge (centre left). The Dun Mountain Railway crosses the Mineral Belt on the SE flank of Wells Ridge. Extending north and SW from the saddle respectively are the valleys of the Maitai and Roding rivers. Except for a narrow band of mafic rocks, comprising the upper part of the ophiolite belt, adjacent to the NW (left) margin of the ultramafic rocks, the bush-covered slopes are underlain by sedimentary rocks of the Maitai Group. The hummocky topography to the SE (right) of the ultramafics is on Patuki Melange. Photo: Lloyd Homer, courtesy of GNS Science, New Zealand.

Nelson with its limited pastoral lands. Seams of bituminous coal were already known in Golden Bay and the stage was set, so it was confidently felt, for the establishment of a copper smelting industry. One immediate drawback was that New Zealand, with a population of only 110 000 people, would not be able to contribute the necessary capital. Also, although it was not recognized at the time, there was no one in the colony with the expertise to assess the copper mineralization. The need for a thorough assessment of the mineralization was largely neglected because several Nelson settlers had worked in the copper and tin mines of SW England. There was, however, one significant difference between the Cornish mines and Nelson. The Cornish mines were in granite whereas in Nelson the copper had been found in serpentinite. This was not considered to be of any great significance to one former Cornish miner, William Long Wrey (*c.* 1792–1886), who soon determined that the copper, in the form of malachite, cuprite and native copper, was present at 24 locations within a 3 km length of the Mineral Belt. He had no hesitation in describing these occurrences as copper lodes.

Thomas R. Hacket—mining engineer and geologist

With assurances from the more influential members of the Nelson community as to Wrey's integrity, the capital to mine the copper was put forward when in 1857 the economically powerful Dun Mountain Copper Mining Company Ltd was registered in London. The company wasted no time in purchasing the land containing the copper and shipping to Nelson the materials necessary to build a railway from the port to the mines some 870 m above sea level. Despite this, no steps were taken to corroborate Wrey's statements on the mineralization. Neither the company's directors, nor the members of the Nelson community who had vouched for Wrey's integrity, had the expertise to judge whether in fact sufficient work had been carried at to verify that lodes existed. Nevertheless some of these more influential men were interested in science and had a passing knowledge of geology. For example, Dr David Monro (1813–1877) (Wright-St Clair 1971) had trained as a doctor in Edinburgh University, where his father Alexander Monro *tertius* was Professor of Anatomy. Amongst his pupils was palaeontologist Richard Owen who described New Zealand's giant fossil bird the moa (*Dinornis*). In New Zealand, David Monro rarely practised medicine but became an influential pastoral farmer and politician and, for his services as Speaker of the New Zealand parliament he was knighted in 1866. Another of Nelson's leading settlers with an interest in geology was wealthy landowner, politician and educationalist William Wells (1810–1893) (Lash 1992) who, along with Monro, was subsequently appointed to a local management committee of the Dun Mountain Company. As well as the Dun Mountain Company, both men were investing in, and taking a leading role in, the management of several other companies that had been formed to exploit a number of mineral occurrences discovered in Nelson Province. Although undoubtedly their prime objective was to benefit financially, they ardently observed the geology that was exposed in the mine workings. Paradoxically this interest became more acute when the mines failed to live up to expectations and reasons for why this should be were actively sought. Despite their interest in geology, neither Monro, Wells, nor others were qualified to report on the Dun Mountain copper and were reliant on what Wrey enthusiastically told them. In any case it was up to the company to confirm that its property did indeed contain rich copper lodes.

It was not until the Dun Mountain Company belatedly sent mining engineer Thomas Ridge Hacket (*c.* 1830–1884) to Nelson that a professional evaluation of the copper, and the first geological appraisal of the Mineral Belt, was undertaken. Hacket had received his professional training in London but had worked in Germany prior to his Dun Mountain appointment. Following his arrival in Nelson on 31 October 1857, Hacket wasted no time venturing into the Mineral Belt and the results of his evaluation were soon known to the 3000 residents of Nelson. In his reports to London, he described the lodes as 'moonshine' a finding that Wrey vigorously disputed. Nelsonians quickly divided into two factions: the 'Wreyites' willing to believe the old and certainly experienced, if not in mineral assessment, Cornishman and the 'Hackettites' who accepted the professional advice from the relatively young mining engineer. The matter would undoubtedly have concluded with the Dun Mountain Company abandoning the Mineral Belt if Hacket had not also examined a nearby deposit of chromite called the Duppa Lode that had been found in 1853. Chromite was a relatively new industrial mineral that was little known in Nelson and with its black colour it did not have the same allure to fire the imagination as the bright green and red copper ores. As a consequence the Duppa Lode had been ignored as work progressed on opening up the supposed copper lodes. In the late 1850s, it was recognized that chromite could produce a range of brilliant colours, particularly chrome green that replaced dangerously unstable arsenic-based dye. The demand for chromite increased dramatically and the price rose to £15 a ton. This was considered sufficiently high to justify building a railway from the Duppa Lode to Nelson where the ore could be shipped to England for processing.

Ferdinand von Hochstetter and the naming of dunite

Having learned from the fiasco over the copper, the Dun Mountain Company directors sought confirmation of Hacket's opinion of the Duppa Lode before they committed themselves to the expense of building the railway. They were not the only ones interested in what Dun Mountain might contain. With the controversy over the copper still raging in Nelson, attempts to mine Golden Bay coal not progressing smoothly, and declining returns from the Golden Bay gold-field, the provincial government of Nelson became aware of the visit to New Zealand of the Austrian research frigate *Novara*. On board *Novara* was the geologist Ferdinand von Hochstetter (1829–1884) who had been born in Germany but had joined the Austrian Geological Survey in 1853 (Fleming 1990). While the ship was in Auckland, von Hochstetter had been engaged by the Auckland Provincial Council to undertake a geological survey of that province, including in the east the nascent gold-fields on the Coromandel peninsula (Fig. 2). Accompanying von Hochstetter was German geologist Julius von Haast (1822–1887) who had just arrived independently in Auckland as an immigration agent (Maling 1990). Under pressure from local merchants and other men of influence, the Nelson Provincial Council engaged von Hochstetter who, with von Haast, arrived in Nelson on 4 August 1859 and remained for two months. During that time he visited much of the north of the province which encompassed the northern third of the South Island. He delegated examination of the eastern part of the province, which was soon to become Marlborough, to von Haast. Despite wet weather, which resulted in flooded rivers making work in the mountains difficult, von Hochstetter was able to visit most areas of interest.

High among von Hochstetter's, and many others, priorities was the Dun Mountain mineralization. On 29 August 1859, von Hochstetter, accompanied by von Haast, Wrey, Monro, and others, proceeded up the Maitai valley recognizing and naming the Brook Street and Maitai rocks, as well as the Wooded Peak Limestone at the base of the Maitai Slates. Two days later, after completing an examination of most of the copper and chromite deposits, von Hochstetter climbed the barren rusty slopes of Dun Mountain and, much to his surprise, found that its summit was not serpentinite like the rest of the Mineral Belt. Instead beneath its dun coloured weathering, the rock was a beautiful green colour being composed of olivine with grains of chromite. von Hochstetter named the rock dunite. As for the Mineral Belt as a whole he interpreted it as a giant sill that might have been associated with the Brook Street Volcanics which he described as diabasic intrusive rocks. Both the Mineral Belt and Brook Street rocks were regarded as intruding the Maitai rocks; all three units being of Mesozoic age (von Hochstetter 1864). Not until about 1900 was it realized that the Brook Street rocks were largely sedimentary. On outskirts of Nelson, von Hochstetter collected Late Triassic *Monotis* and other fossils from the Richmond Sandstone, now part of the Richmond Group.

Before leaving Nelson on 1 October 1859, von Hochstetter gave a public lecture on the geology of Nelson. He diplomatically downplayed the controversy over the Dun Mountain copper, and stated that if the chromite could be mined economically 'the abundance of it must prove a source of much wealth'. Privately, he had conferred with the Dun Mountain management committee in Nelson, which comprised Hacket, Monro and Wells, and it was arranged that he would also report to the company's directors in London. The directors were assured that their property contained chromite in abundance.

von Hochstetter, with the assistance of von Haast and others, was able to compile geological maps of Auckland and Nelson provinces, the first regional maps of any part of New Zealand. The Nelson map, at a scale of 1:1 300 000, looks remarkably similar to present day maps in that the distribution of the various rock units is accurately portrayed. Although this was a notable achievement, and without diminishing von Hochstetter and von Haast's accomplishments, the geology of Nelson was in fact reasonably well known prior to the arrival of the geologists. Hacket had provided advice in a professional capacity to a number of individuals and small local companies that were, mostly unsuccessfully, attempting to open up mineral prospects. A colliery engineer, James Burnett (1826–1872), had undertaken mineral surveys, particularly of the extensive coal measures in the west of the province. Even old Wrey was advertising himself as a mineral surveyor. Although no one was employing him as such, Wrey was vigorously promoting Nelson as a rich mineral province and had a large following which collectively had considerable political clout.

The land surveyors, largely employed under the auspices of the provincial government, were searching the mountainous interior for land suitable for farming. Although not trained geologists, the surveyors could identify basic rocks whose presence they recorded. In New Zealand's youthful landscape, rock type had a direct influence on soil fertility which, along with possible mineral wealth had a bearing on whether the provincial council should sell the land, and if so for how much, or retain it for mining purposes. This, coupled with information gleaned from prospectors with experience of the Californian and Australian gold-fields, meant that Nelson Province was not a complete geological unknown at the time of von Hochstetter's visit. What many were looking for from von Hochstetter was an independent assessment of the known and potential mineral wealth. After von Hochstetter left New Zealand, von Haast undertook geological explorations of the Nelson west coast and this data was added to the maps before they were published in 1864.

Chromite and New Zealand's first railway

On receiving von Hochstetter's confirmation of Hacket's assessment of the quantity of chromite, the directors of the Dun Mountain Company authorized construction of the railway to Dun Mountain. Even before the route of the railway was decided, Monro expressed doubts. On 1 May 1860 he wrote to von Hochstetter that although there was a large quantity of chromite 'it does not appear to preserve its continuity in the same manner as metaliferous lodes in Europe'. Nevertheless the railway (with a gradient of 1:18) was laid out from Brook Street through steep heavily forested mountains to the mines. After inspecting the route Monro and others joined Hacket on 12 January 1861 and again Monro had misgivings, for he wrote in his diary that the chromite did not occur as a lode but was irregularly mixed with the serpentinite. This does not seem to have perturbed Hacket and construction of the railway soon commenced. As the workers were excavating on the side of Wooded Peak a bed of prismatic shelled bivalves was revealed in part of the Maitai Group that was much later to be named the Tramway Sandstone (Waterhouse 1959).

The 21 km long railway, New Zealand's first, commenced carrying chromite on 3 February 1862 (Fig. 2). In its first year of operation the railway transported over 3000 tons of chromite but this was not maintained. As the total amount of chromite used in Britain was only a few thousand tons a year, the market was soon saturated leading to a drop in price of the ore. The outbreak of the American Civil War in 1861 interrupted cotton supplies to Lancashire, which in turn lessened the demand for dyes. Further compounding the lack of demand was the discovery of aniline dyes that could be produced cheaply as a by-product of the coal-gas industry that was expanding rapidly as industrialization swept Britain. When the market stabilized at the end of the American Civil War and the Dun Mountain Company was able to increase production to the level of its first year of operation, it found that, despite the earlier assurances of Hacket and von Hochstetter, it had run out of chromite.

It was only after the chromite at and close to the surface had been extracted by open cast mining that tunnelling into the hillside below the mines failed to encounter any economic deposits. Hacket and von Hochstetter had assumed that the chromite would extend to considerable depth. For this they could not be blamed too much in that the nature of the mineralization in serpentinite was not understood at that time although Hacket, particularly in the light of Monro's concerns, should have ensured that the underground exploration was sufficient to confirm the inferred reserves. Hacket was saved potential embarrassment as he had left the company in mid-1861 after he opened his own small chromite mines to the SW of Dun Mountain. Hacket and von Hochstetter were also led astray by the quantities of loose chromite that littered the hillside. Rather than coming from a single large lode it was derived from a number of lenses and, being more resistant than the host serpentinite, it formed a lag deposit on the hillside. With no one in its employment to assess accurately whether its property could again be worked profitably in 1864 the company appointed an experienced mining manager, Joseph Cock (*c.* 1817–1877), from Devon. In common with other similarly qualified mining professionals he was accorded the title of Captain.

Before Cock could arrive in Nelson, the chromite mines were examined by Dr James Hector (1834–1907), the Otago provincial geologist (Dell 1990). Otago, in the south of the South Island, had extensive alluvial gold-fields and was wealthy enough to employ its own geologist. Hector had obtained a medical degree from Edinburgh University, as this was the only way he could train as a geologist. His dual training enabled him to join the Palliser Expedition to western Canada in 1857 as surgeon and geologist and this in turn set him on an illustrious scientific career in New Zealand beginning with his appointment in Otago in 1861. The visit to Nelson of Hector, assisted by Hacket, was to collect geological samples for the New Zealand Exhibition that was held in Otago in 1865. The geologists, accompanied by Wells, Monro and others, were taken up the railway in one of the wagons. Hector, being an Edinburgh man, had much in common with Monro. He was to have an even closer link as four years later he married Monro's elder daughter Maria Georgiana. Although Hector collected dunite and visited the mines there is no record of what he or Hacket thought about the company's prospects. However, this was soon determined when Cock arrived in Nelson on 5 July 1864. By late 1864, Cock had completed his examination of the company's property and found that although chromite was widespread, it was both patchy and highly variable in quality, and that it could not be worked profitably. On receiving this not entirely unexpected news, the company suspended most of its operations. However, the company was reluctant to abandon its Dun Mountain property as it looked as though gold might restore its fortunes.

The search for gold

In April 1864 alluvial gold had been discovered in the Wakamarina Valley only 20 km east of Dun Mountain. The company's initial annoyance at losing almost its entire workforce to the gold-rush was momentarily offset by the hope that its property would also prove auriferous. To the disappointment of many, including the Dun Mountain Company, the Wakamarina gold-field had all but petered out by 1865, but gold was being found in considerable quantities elsewhere in New Zealand. This in turn reinforced the opinion of the directors that it would be premature to abandon Dun Mountain. A further assessment was forthcoming when in 1866 Hector again visited east Nelson. In the previous year, Hector had been appointed the inaugural director of the New Zealand Geological Survey, a diminutive department of central government based in Wellington. As well as being the only geologist, Hector had a number of other responsibilities including the Colonial Museum and Colonial Laboratory (Burton 1965). Hector provided no information on what he referred to as 'a short exploration of the mineral ranges lying to the east of Nelson'. Either on this visit or the one with Hacket in 1864, he collected fossils from the shell bed exposed on the Dun Mountain Railway. The fossils he identified as *Inoceramus* thereby apparently confirming a Mesozoic age for the Maitai Group and, with time, for similar looking sediments throughout New Zealand. Unfortunately this misidentification of an atomodesmatinid fossil marked the start of what was to become the 'Maitai Problem'. The Maitai Problem was not finally resolved until the mid-twentieth century (Waterhouse 1965, 1967; Nicholson 2003).

Although Hector was the only geologist in the Geological Survey, he had Thomas Hacket assisting him on a demanding eight-month exploration of the West Coast of the South Island coal-fields that immediately followed his visit to east Nelson in 1866. Hacket was keen to join the Geological Survey, and certainly had the ability. Furthermore, both men had the greatest respect for each other and treated each other as equals. Despite Hacket's undoubted attributes, the offer of employment was never forthcoming. Initially there was not enough money to employ two geologists full time but later, when Hector was able to take on a permanent assistant, Hacket was never given the opportunity of applying. Perhaps Hector considered that Hacket regarded himself as too much of an equal. Certainly Hacket was a dogmatic individual and was somewhat intolerant of people, such as von Haast, with differing views from himself. At the end of the West Coast survey, with no offer of employment Hacket arrived in Melbourne in January 1868 to seek work in Victoria. As for Hector, he returned to Nelson in late December 1868, primarily to marry 'Georgie' Monro. Nevertheless, prior to the wedding he examined traces of coal in the Maitai Valley, and gave a public lecture on Nelson's geology and potential mineral wealth. He also discussed these matters with Cock and other representatives of the Dun Mountain Company. One particular focus was whether the 'greenstones' signified the presence of economic deposits of gold. At the time the term 'greenstone' was a field name that included any green altered rock, such as diorite, dolerite or gabbro, that commonly contained chlorite or other green minerals. Of greenstones Nelson had an abundance.

Greenstones and Queensland correlations

Although in the late 1860s gold was being found in ever increasing amounts in many parts of New Zealand, most of it was from alluvial deposits. A notable exception was in the Coromandel peninsula in Auckland Province where fabulous returns were being obtained from veins in volcanic rocks that were underlain by a sequence of sedimentary rocks. Hector and others considered that small intrusions of 'greenstones', containing iron and copper sulphides, were responsible for the gold in the younger parts of the sedimentary sequence. The overlying volcanic rocks were young, being Cenozoic in age, and as the sedimentary sequence was believed to have affinities with the Maitai Group in Nelson it was thought to be Mesozoic in age. Furthermore the Brook Street Volcanics looked as though they might represent a southern equivalent of the Coromandel 'greenstones'. If these correlations were correct, then the possibility of gold being present in economic amounts in East Nelson was considerably strengthened.

Hacket further reinforced the association of 'greenstones' with gold. In the winter of 1868 he had obtained employment with the Queensland Geological Survey and recognized that the rocks at Gympie and Kilkivan, in the SE of the state, had much in common with those in east Nelson. Even the hill above the town of Kilkivan he likened to Dun Mountain itself. As well as mapping gold-bearing quartz reefs, which Hacket discovered were not continuous but tended to crop out in an en echelon manner, he critically examined the host rocks and collected a very large number of rocks and fossils. Hacket incorporated all of his data in a comprehensive report. He was also at odds with Queensland geologists, such as Richard Daintree (1831–1878), because he considered the serpentinite, as at Dun Mountain, to be sedimentary in origin. This was not an uncommon view at the time, but Daintree and others favoured emplacement as igneous dykes. Hacket corresponded regularly with Hector whilst he was in Gympie. His letters were comprehensive and, as well as geology, included comments on such matters as Queensland politics along with numerous hints about working in New Zealand. In addition to samples, Hacket sent maps and reports, including a number that were unpublished and had been forwarded without the knowledge of the Queensland

Survey which was about to be abolished. Hacket also wrote to William Wells in Nelson and informed him that it was more than likely that the chromite and copper of Dun Mountain had diverted attention from the real wealth of east Nelson that was its potentially auriferous nature. Hacket also provided an important link between Hector and the Australian geologists, who were all working in areas that were geologically unexplored and which commonly bore little resemblance to the areas of Europe where they had been trained. It was a service for which Hacket has received little recognition (Fleming 1987).

Despite the superficial similarities, there were differences between the geology around Nelson and the gold-fields of Coromandel and Queensland. The most obvious from an economic viewpoint was that Coromandel, although it contained minor copper shows, completely lacked chromite. As for Queensland the fossils included several of Late Palaeozoic age whereas the rocks around Nelson on the basis of the Dun Mountain '*Inoceramus*' and the Richmond Group fossils, were regarded as Triassic. The evidence for a Triassic age was further strengthened because, to the SW of Dun Mountain, the Richmond group appeared to dip under the Maitai rocks.

Dun Mountain revisited

With Hacket's observations, along with the results of recent work in Coromandel by Captain Frederick Wollaston Hutton (1845–1871) who had joined the Geological Survey as assistant-geologist, Hector in October 1869 confirmed to the Dun Mountain Company that he still considered it possible that east Nelson might prove auriferous. With Hector's endorsement, in October 1869 the Dun Mountain Company sent Cock to Coromandel to compare its rocks with those in Nelson. In addition, samples from Thames (Coromandel Peninsula) were donated by the Geological Survey to the Nelson Museum so that prospectors would have a better guide to potentially auriferous rocks. Although only prospecting would determine if economic gold deposits were present in east Nelson, a geological appraisal of the Dun Mountain area was now warranted. This task was entrusted to Edward Hydelbach Davis (1845–1871) who had recently arrived in New Zealand.

Davis had been exposed to mining from an early age as his father owned iron mines and foundries near Dalton-in-Furness in the north of England. After working for his father, Davis attended lectures at the London School of Mines and in 1866 was elected to fellowships in the Chemical Society and the Geological Society of London. After graduation he was employed as chemist and geologist with the Portugal Iron and Coal Company in Portugal. Following his marriage, his next job was with the Mariquita Mining Company in Colombia but the tropical climate affected his health so badly that he returned to England. Davis arrived in New Zealand in 1870, on behalf of London investors, to investigate the smelting of extensive deposits of iron sands on the west coast of the North Island. After examining the iron sands he met Hector in Auckland in April 1870 and was temporarily engaged to assist in fieldwork in the Coromandel. In June, he was appointed as assistant-geologist in the Geological Survey in place of Hutton who had resigned at the beginning of the year. His first fieldwork in this post was to undertake an evaluation of Dun Mountain area. From his temporary work with the Geological Survey he was familiar with the Thames rocks and, courtesy of Hacket, there was a reference collection from Queensland. As part of the evaluation, Davis was instructed to determine the relationship between the Maitai rocks and the Mineral Belt and Brook Street rocks flanking them to the east and west respectively. Hector, unlike Hacket, regarded both the Mineral Belt and Brook Street rocks as igneous. Hector wanted an opinion as to whether the age of the Maitai Group was contemporaneous with the igneous rocks. As for economic geology, Davis was to look for auriferous quartz veining in the Maitai Group and the presence of gold generally in east Nelson. In addition, he was to reassess the old copper and chromite workings in the somewhat forlorn hope that economic deposits might exist. To achieve this Hector instructed Davis to map east Nelson in detail and to construct sections based on the Maitai and Roding rivers and the intervening Dun Mountain Railway (Fig. 2).

On 26 September 1870 Davis arrived in Nelson from Wellington and from the boat would have seen the rusty brown slopes of Dun Mountain rising amongst the bush covered hills beyond the city which he described as 'a quiet little place'. Two days later Davis and Cock went up the Dun Mountain Railway to the Mineral Belt observing on the way the Maitai Group, including the *Inoceramus* shell bed and the Wooded Peak Limestone. He devoted the next few days to a thorough examination of the old copper and chromite mines (Davis 1871). Although hampered by the collapse of most of the underground workings, Davis confirmed that there were plenty of traces of copper, in the form of the native metal, silicates and carbonates, and which he correctly attributed to precipitation from groundwater moving through fractures and joints in the serpentinite. This process was still continuing, for in an old drive he observed that copper had accumulated on the walls and floor since the miners had abandoned it only a few years earlier. Further proof was soon forthcoming for Davis spent an uncomfortable night from copper poisoning after drinking water from one of the drives. As for true lodes, he doubted that any would be found in the serpentinite, which he interpreted as metamorphosed Maitai Group sedimentary rocks stratigraphically below the Wooded Peak Limestone. He attributed the metamorphism to a granite intrusion whose existence at depth was apparently confirmed by what he termed elvan dykes in the Roding River. The dykes were later shown to be a variety of a calcium garnet that was named rodingite in 1911 (Bell *et al.* 1911). If any copper lodes existed, Davis reasoned they would be in the granite but appreciated that there was little evidence to support this. He wisely concluded that further mapping was required, particularly in the granite country to the west of Nelson, 'before spending a penny more on Dun Mountain'.

As for the chromite, it was obvious to Davis that no large deposits existed. Instead, he confirmed that the mineralization was patchy, a view that Monro had reached in May 1860 and reiterated seven months later. Davis also carefully mapped Dun Mountain and concluded that the olivine-rich rock, in his opinion unnecessarily named dunite by von Hochstetter, had a very restricted distribution being confined to the SW corner of the peak. Its apparently wider distribution was due to extensive scree deposits mantling the western slopes of the mountain. Around the margins of the mountain, Davis recorded a mixture of rocks including siltstone and what he called felstone but was in fact rodingite. Davis found this part of the Mineral Belt difficult to interpret, writing that 'It is not possible to make out any distinct boundaries for these rocks, as they are mixed up inextricable confusion'. The rocks are now mapped as the Patuki Melange that forms the southeastern boundary of the Dun Mountain Ophiolite Belt.

The demise of Davis and the Dun Mountain Company

From Dun Mountain, Davis extended his mapping into the Maitai River, noting that the sequence was similar to that exposed on the railway. As well as locating the *Inoceramus* shell bed in the upper part of the river, he also collected the first ammonite from the Maitai Group but unfortunately it was mislaid at some later date. It was not until the 1970s that further ammonites were collected from the group enabling a Triassic age to be assigned to the rocks above the shell bed. While in the Maitai Valley, Davis was puzzled that a branch of the river was full of smooth boulders the size of small houses. Although mapping them as being derived from a drift deposit, he correctly concluded that the boulders could not possibly have been water worn. They are complete tectonically rounded inclusions transported from the Patuki Melange in an earth flow. Further down the river, Davis mapped the well-bedded Maitai Group rocks, comprising the Maitai Slates of von Hochstetter and which are now referred to the Greville and Waiua formations, before passing into the Brook Street Volcanics.

The contact between the Maitai and Brook Street rocks was not exposed and, without comprehending the possibility of major faulting, he was unable to reach any conclusion as to its nature (Fig. 4). Although he was puzzled about the contact, the presence nearby of Cenozoic coal measures does not seem to have troubled him. It appears that Davis, like his contemporaries, regarded the Cenozoic rocks as being deposited on the existing topography rather than the mountains post-dating the coal measures. To complete his east Nelson survey, Davis again went up the railway to Dun Mountain and then followed the Roding River to the SW. The rocks were similar to those he had already seen and did not add materially to the information he had already acquired. From his Dun Mountain work, and a brief look at some of the quartz reefs being opened out in Golden Bay and elsewhere in west Nelson, Davis dismissed the possibility that the Nelson rocks could be correlated with those of the Coromandel Peninsula. Writing from Nelson to Hector, Davis advised that the latter rocks were, in contrast to those further south, magnesium deficient and 'to my mind this separates the two as far as light and darkness'.

Davis returned to Wellington and in November 1870 two preliminary reports, lacking much geological data, were released. For those hoping that Nelson might have a prosperous mining future, they did not make agreeable reading. Before he finished his main report, Davis returned to Nelson in January 1871 and extended his mapping further SW from Dun Mountain to the fossiliferous Richmond Group rocks, before proceeding to map the Grey River district on the West Coast of the South Island. It was there, on 9 February 1871 that, while attempting to cross the mouth of the flooded Ten Mile Creek, he was washed off his horse into the surf and drowned. He has the melancholy distinction of being the only employee of the New Zealand Geological Survey to be killed on duty. His geological report on Nelson was published posthumously and, being based on his November reports, it was understandably not as complete as if he had finished it himself. Davis' reports, and posthumous analyses of ultramafic and other rocks

Fig. 4. View SW from above the Maitai valley to The Brook (centre). The conical hills (right) are of Brook Street Volcanics and are separated by the Waimea Fault (arrowed) from hills of Maitai Group (far left). Photo: Lloyd Homer, courtesy of GNS Science, New Zealand.

he had collected, showed that economic deposits of gold did not exist and that economically Dun Mountain would not be another Gympie. As for the Dun Mountain Company, it was wound up in 1872 having spent about £100 000 for a return of a third of this figure.

A faulted sequence

Davis had been content to accept a Triassic age for the east Nelson rocks, based on Hector's *Inoceramus* identification and the Triassic fossils in the Richmond Sandstone. Alexander McKay (1841–1917), who examined the Dun Mountain area in May 1878, was suspicious of the age of the 'Dun Mountain *Inoceramus*' and was careful to distinguish it from other Mesozoic inoceramitid fossils in New Zealand (McKay 1878). The following year McKay discovered Late Palaeozoic fossils in the Maitai rocks to the SW of Dun Mountain but this still did not resolve their age or the relationship with fossiliferous Richmond rocks. However, McKay did recognize that the contact between the two was a significant one and considered the possibility that it could be faulted. It was not until

Charles Taylor Trechmann (1885–1964) examined the rocks closely, taking a particular interest in the fossils, that it was demonstrated that the Dun Mountain *Inoceramus* belonged to the atomadesmatiniid group, had close affinities with Australian forms, and was Permian in age. The fossil is now placed in the genus *Maitaia*. He also recognized that the fossiliferous Triassic rocks were separated from the Maitai rocks by a fault that was later named the Eighty-eight Fault.

In 1911, geologists of the Geological Survey had recognized the Waimea Fault and what is now the Flaxmore Fault on either side of the Brook Street Volcanics even though they did not fully appreciate their significance. The Richmond Group rocks, now included in the Murihiku terrane, are pinched out along the fault as it merges with the Waimea Fault adjacent to Nelson City (Fig. 2). Ironically the Permian fossils found by McKay are now known to be in allochthonous limestone blocks in Triassic rocks and the Maitai Group thus straddles the Permian–Triassic boundary. The Brook Street rocks are of Permian age. Once the question of the age of the Nelson rocks was resolved, one of the main supposed differences between them and some of the Gympie rocks was removed.

The Mineral Belt reborn

Despite the reports of Hacket, Cock and Davis, further unsuccessful attempts were made to mine chromite and copper in the Mineral Belt. By early in the twentieth century the Mineral Belt was finally abandoned by the miners, except for the quarrying of serpentinite. However, it came to the attention of geologists, particularly with the advent of plate tectonics and the recognition of the origin of ophiolites. The outcrop of the Dun Mountain Ophiolite Belt and its overlying Maitai Group were also to lead to the identification of major horizontal movement on the Alpine Fault within the South Island. Recognizing that Dun Mountain/Maitai rocks were present in both the north and south of the South Island, Harold Wellman of the Geological Survey in 1949 put forward the hypothesis that they had been formerly continuous but had been displaced 480 km by dextral strike-slip movement on the Alpine Fault. The fault constitutes the boundary between the Indo-Australian and Pacific plates within the fragment of Gondwanaland continental crust that is New Zealand.

Conclusions

Although Hacket was the first geologist to examine the Dun Mountain Ophiolite Belt or Mineral Belt, it was von Hochstetter who provided the first geological map and a description, including the recognition and naming of dunite. Hacket noted the similarities between east Nelson and Gympie, and by communicating between Australia and New Zealand played a valuable role in the exchange of information and ideas across the Tasman Sea. In conveying his observations concerning Queensland to Hector, Monro and Wells, Hacket helped to ensure that the next geological appraisal of east Nelson, by Davis, was undertaken. One of the prime objectives for Davis was to determine the relationships between the 'greenstones' and the sedimentary Maitai Group as well as to ascertain whether these rocks could be correlated with rocks containing 'greenstones' at Coromandel and Gympie. Davis remained perplexed about the Brook Street–Maitai contact, now known to be a major fault that is considered to separate two terranes.

Davis was more positive concerning the contact of the Maitai Group with the serpentinite of the Mineral Belt. Like Hacket, he interpreted the serpentinites of the Mineral Belt as metamorphosed sedimentary rocks rather than, as favoured by von Hochstetter and Hector, intrusive rocks. Although neither of these interpretations was correct, Hacket and Davis rightly considered that the unmetamorphosed rocks, comprising the Maitai Slates and the Wooded Peak Limestone of von Hochstetter, were stratigraphically above the serpentinite. It is now known the Maitai Group rests unconformably on the Dun Mountain Ophiolite Belt. Furthermore, Davis conclusively demonstrated that the Nelson rocks could not be correlated with those of Coromandel. Paradoxically, if his ammonite had not been lost, then it would ultimately have been identified as Triassic in age and therefore apparently confirmed a similar age for the Dun Mountain *Inoceramus*.

Davis offered no opinion on the Gympie connection advocated by Hacket and the latter's correlation of Gympie and Nelson was forgotten. It was not until the late twentieth century following fieldwork in Gympie by Larry Harrington, who had worked in Nelson for the New Zealand Geological Survey in the 1940s and 1950s, that the similarity was again noted (Harrington 1983). Thus east Nelson and Gympie rocks provide an important link between the New Zealand and Australian fragments of Gondwanaland.

The writer is indebted to P. Ballance of Nelson, L. Harrington of Canberra and H. Nicholson of Auckland for constructive criticism of the manuscript. In addition to the references cited, considerable use has been made of the New Zealand Geological Survey records particularly relating to Davis and held by National Archives (Series 7214, Box 1 GS 1870/15A). Inwards and Outward

Correspondence of the Geological Survey is held by the Museum of New Zealand Te Papa Tongarewa Archives and the author thanks J. Twist, archivist at the museum, for making them available. The writer acknowledges the use of the excellent aerial oblique photographs taken by L. Homer and held by GNS Science, Lower Hutt.

References

BELL, J. M., CLARK, E. DE C. & MARSHALL, P. 1911. The geology of the Dun Mountain Subdivision, Nelson. *New Zealand Geological Survey Bulletin*, **12**.

BURTON, P. 1965. *The New Zealand Geological Survey 1865–1965*. Information Series 52. Department of Scientific and Industrial Research, Wellington.

DAVIS, E. H. 1871. On the geology of certain districts of the Nelson Province. *New Zealand Geological Survey Reports of Geological Explorations 1870–1*, 103–135.

DELL, R. K. 1990. Hector, James. *In*: *The Dictionary of New Zealand Biography*. Volume 1, 1769–1869. Allen & Unwin, Department of Internal Affairs, Wellington, 183–184.

FLEMING, C. A. 1987. Some nineteenth century trans-Tasman influences in geology. *Australian Journal of Earth Sciences*, **34**, 261–267.

FLEMING, C. A. 1990. Hochstetter, Christian Gottlieb Ferdinand von. *In*: *The Dictionary of New Zealand Biography*. Volume 1. Allen & Unwin, Department of Internal Affairs, Wellington, 199–200.

HARRINGTON, H. J. 1983. Correlation of the Permian and Triassic Gympie terrane of Queensland with the Brook Street and Maitai terranes of New Zealand. *In*: MURRAY, C. G. (pref.) Permian Geology of Queensland. *Geological Society of Australia, Queensland Division, Brisbane*, 431–436.

HOCHSTETTER, F., VON, 1864. *Geology of New Zealand: Contributions to the Geology of the Provinces of Auckland and Nelson*. Translated and edited by FLEMING, C. A. 1959. Government Printer, Wellington.

JOHNSTON, M. R. 1981. *Sheet O27AC—Dun Mountain. Geological Map of New Zealand 1:50 000*. Department of Scientific and Industrial Research, Wellington.

JOHNSTON, M. R. 1987. *High Hopes–The History of the Nelson Mineral Belt and New Zealand's first Railway*. Nikau Press, Nelson.

LASH, M. D. 1992. *Nelson Notables 1840–1940*. Nelson Historical Society, Nelson, 149–150.

MCKAY, A. 1878. Report on the Wairoa and Dun Mountain districts. *New Zealand Geological Survey Reports of Geological Explorations 1877–78*, 119–159.

MALING, P. B. 1990. Haast, Johann Franz Julius von. *In*: *The Dictionary of New Zealand Biography*. Volume 1, 1769–1869. Allen & Unwin, Department of Internal Affairs, Wellington, 167–169.

NICHOLSON, H. H. 2003. *The New Zealand Greywackes*. Unpublished Ph.D. thesis, University of Auckland.

RATTENBURY, M. S., COOPER, R. A. & JOHNSTON, M. R. 1998. *Geology of the Nelson Area. Institute of Geological & Nuclear Sciences 1:250 000 geological map 9*. 1 sheet + 67 p. Institute of Geological & Nuclear Sciences Limited, Lower Hutt, New Zealand.

WATERHOUSE, J. B. 1959. Stratigraphy of the lower part of the Maitai Group, Nelson. *New Zealand Journal of Geology and Geophysics*, **2**, 944–953.

WATERHOUSE, J. B. 1965. A historical survey of the pre-Cretaceous geology of New Zealand Part 1. *New Zealand Journal of Geology and Geophysics*, **8**, 931–998.

WATERHOUSE, J. B. 1967. A historical survey of the pre-Cretaceous geology of New Zealand Part 2—Metamorphic and igneous rocks. *New Zealand Journal of Geology and Geophysics*, **10**, 923–981.

WRIGHT-ST CLAIR, R. E. 1971. *Thoroughly a Man of the World—a Biography of Sir David Monro*. Whitcombe & Tombs, Christchurch.

Franz Hilgendorf (1839–1904): introducer of evolutionary theory to Japan around 1873

M. YAJIMA

College of Liberal Arts and Sciences, Tokyo Medical and Dental University, 2-8-30 Kohnodai, Ichikawa-shi, Chiba 272-0827, Japan (e-mail: yajima-michiko@gupi.jp)

Abstract: Franz Martin Hilgendorf (1839–1904) arrived in Japan in March, 1873 and stayed in Tokyo until October 1876 as a 'foreign employee' (a Westerner employed in the modernization of Japan). In the beginning of the Meiji Era, the Japanese Government invited many Westerners to introduce facets of Western civilization. Hilgendorf's doctoral dissertation at Tübingen University (1863) dealt with molluscan evolutionary lineages of Miocene age from Steinheim, Germany. Hilgendorf's ideas already included the concept that fossils had evolved. Charles Darwin (1809–1881) himself had mentioned Hilgendorf's findings in the sixth edition of his *Origin of Species*, published in 1872. Hilgendorf lectured on evolution at the Tokyo Medical School (the former University of Tokyo) well before the American, Edward Sylvester Morse (1838–1925), whom it was thought had introduced Darwinian theory to Japan in 1877. The priority of Hilgendorf is proved by contemporary notes taken by the famous Japanese novelist, Mori Ougai (1862–1922), who had attended Hilgendorf's lectures whilst a student.

During my studies of the systematics of fossil and Recent Ostracoda (Crustacea), I discovered that the aquatic firefly, *Vargula hilgendorfii*, a Recent ostracod, was first described and named by Gustav Wilhelm Müller (1857–1940) in 1890, based on specimens collected by Franz Martin Hilgendorf (1839–1904) (Fig. 1) in Japan. (Aquatic fireflies are usually used for enzyme experiments in Japanese high schools.) So, in 1994, I visited the Museum für Naturkunde, Berlin, in order to check the type specimens of *Vargula hilgendorfii*, and I found there more specimens comprising many other taxa, mostly fishes, which Hilgendorf had collected in Japan. I decided to hold a Hilgendorf Exhibition in order to demonstrate his contribution to the development of Japanese ichthyology and fishery sciences. During preparations for the Hilgendorf Exhibition, I also researched Hilgendorf's work which demonstrated that he had a very modern view of evolution, and predated Edmund Morse in bringing Charles Darwin's evolutionary theory to Japan.

Hilgendorf has until recent been largely forgotten in both Japan and Germany. In Japan, he is remembered as the person who introduced Western biology, this is based on information given in an obituary notice written by Weltner in 1906 (Weltner 1906; Ueno 1968; Isono 1986). My own research uncovered Hilgendorf's own *curriculum vitae*, written by himself, his marriage registration, and the notice of his death at the Museum für Naturkunde, Berlin. These documents were also displayed during the Hilgendorf Exhibition, in Japan in 1997 and 1998.

Franz Martin Hilgendorf was born on 5 December 1839 in Neudamm in the region (Mark) of Brandenburg, the third son of Johan Hilgendorf, a merchant. He was not a strong child, but nevertheless was very intelligent. He liked reading books, skating and fencing. In October 1859, at the age of 19, he entered the University of Berlin, initially to study philology. He soon became interested in biology when he took a part-time job in the Museum für Naturkunde in Berlin. He probably first read Darwin's *Origin of Species* in 1859 or 1860. After four semesters in Berlin, Hilgendorf continued his studies with two more semesters at the University of Tübingen, where he joined Professor Friedrich August Quenstedt (1809–1889), a specialist in Jurassic ammonites and their use in stratigraphy.

During the summer of 1862, Hilgendorf undertook his first scientific journey to Steinheim under the leadership of Quenstedt. In May 1863, Hilgendorf took his doctoral degree, submitting his work as *Beiträge zur Kenntnis des Süßwasserkalkes von Steinheim.*[1] The thesis included a discussion of the problem of the phylogeny of *Planorbis multiformis*, a gastropod, and Hilgendorf included the idea that these organisms had evolved (Fig. 2). Unfortunately, his thesis was not published, probably because the idea of such evolutionary transformations seemed too daring for the time. Charles Darwin (1809–1881) himself mentioned Hilgendorf's findings in the sixth edition of his *Origin of Species*, published in 1872. Darwin was mistaken in stating that Hilgendorf's work had been carried out in Switzerland, probably because Steinheim is in Schwaebischen Alps.

From: Wyse Jackson, P. N. (ed.) *Four Centuries of Geological Travel: The Search for Knowledge on Foot, Bicycle, Sledge and Camel.* Geological Society, London, Special Publications, **287**, 389–393.
DOI: 10.1144/SP287.28 0305-8719/07/$15.00

Fig. 1. Franz Martin Hilgendorf (1839–1904).

Hilgendorf continued his studies in Berlin, mainly focusing on organic chemistry, and he worked as an analyst in the laboratory there before switching his attention to zoology. Under the influence of Wilhelm Karl Hartwig Peters (1815–1883), Professor of Zoology and Director of the Museum für Naturkunde, Berlin, he was appointed to a job at this institute. After successive positions as the head of Hamburg Zoo, the Head of the library at Dresden, and Lecturer in Biology at the Dresden Gymnasium, Hilgendorf went to Japan in 1873. He arrived in Japan on 2 March 1873, and stayed in Tokyo until 24 October 1876, as a 'foreign employee', or Westerner employed in the modernization of Japan. In the early years of the Meiji Era, the Japanese Government invited many Westerners to introduce various facets of Western civilization to the country.[2]

After his return from Japan to Berlin, Hilgendorf worked as the Curator of the Museum für Naturkunde from 1887 until his death on 5 July 1904. He married Juria Anting on 19 October 1880 and they had two sons and one daughter. He was promoted to professorship in 1893. He continued his systematic study of vermes, crustaceans, and fish. But he never forgot his Steinheim snail studies. His planorbid studies may be considered to be his most important scientific work, conducted from

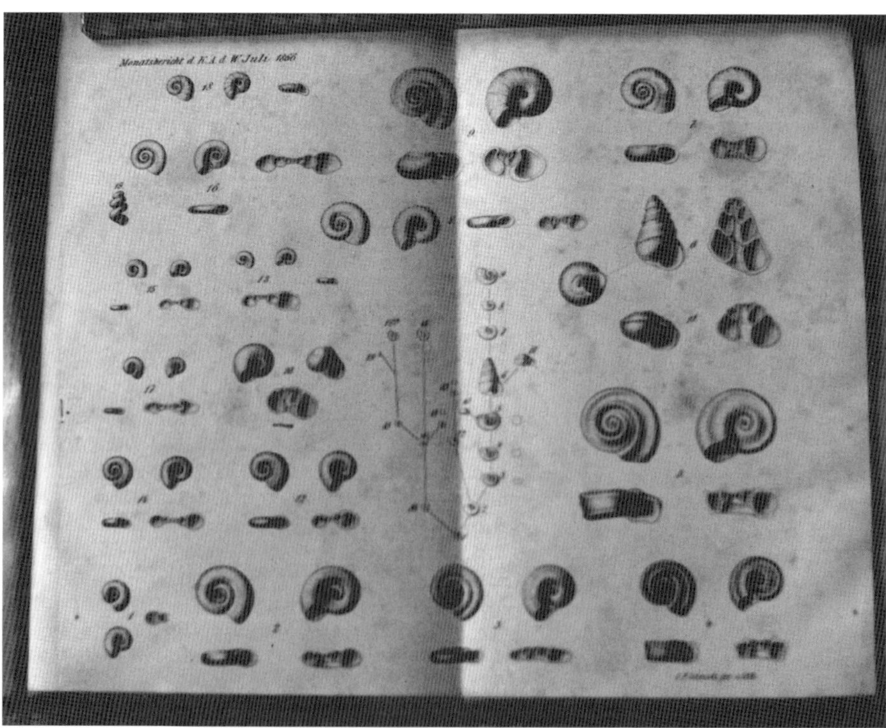

Fig. 2. Hilgendorf's planorbid phylogenetic tree of 1866.

his first publication in 1866 (Hilgendorf 1866) until his last contribution on this subject in 1901. Hilgendorf published his ideas on the evolution of *Planorbis* in a paper in 1879 (Hilgendorf 1879) in the journal *Kosmos*, which had been founded at Leipzig two years previously, for the express purpose of promoting the concept of natural evolution. On the editorial board of the journal appear the names of Charles Darwin and Ernst Haeckel (1834–1919), who was the most prominent exponent of Darwinism in Germany.

Hilgendorf's main work in Japan

Hilgendorf was appointed to the Tokyo Medical School (the former University of Tokyo) where he delivered 24 lectures a week on natural history and the natural sciences. He also collected many samples of fishes, mammals, reptiles, gastropods and other animals in Japan, and studied their systematics. He proposed 36 new species of Japanese fish, of which 25 are still valid (Yajima 1997). Among the fish species, he was especially interested in the genus *Gobius* about which the current Emperor of Japan is a leading authority.[3]

Hilgendorf found the mollusc *Pleurotomaria berichii* (Fig. 3) along the coast near Tokyo and described it as a 'living fossil' (Hilgendorf 1877). The term 'living fossils' was first used by Darwin in Chapter 4 'Natural Selection' of his *Origin of Species* (Darwin 1859). After Hilgendorf's paper, the British Museum (Natural History) in London asked colleagues in Japan to collect and donate specimens of *Pleurotomaria berichii* for its collections. Skilled Japanese collectors were able to find it and shells of this 'living fossil' were expensive to purchase. In Japan *Pleurotomaria berichii* has been given the Japanese name 'the millionaire shell'.

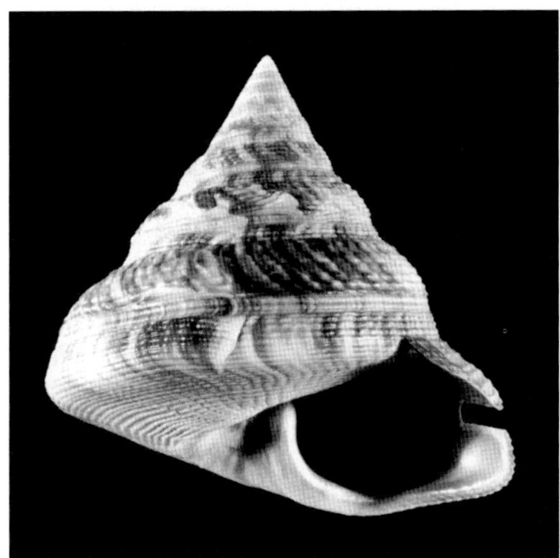

Fig. 3. *Pleurotomaria berichii* Hilgendorf, 1877.

In Japan, Hilgendorf travelled throughout Hokkaido (northern Japan) to collect fish samples and to study the Ainu (a Japanese indigenous and minority race), and around Tokyo he travelled to Nikko, Hakone, Chiba to collect biological samples. In 1874, Hilgendorf visited Ludwig Haber in Hokkaido, the uncle of Nobel Prize-winning chemist Fritz Haber who was later murdered by Japanese Samurai. Hilgendorf narrowly escaped assassination. He later described Japanese salmon and dedicated the species to Ludwig Haber.

After returning to Berlin, Hilgendorf continued his systematic description of Japanese animals, and he helped the Japanese in the International Fishery Exhibition held at Berlin in 1880.

Hilgendorf's lectures in Japan

Hilgendorf lectured on evolution at the Tokyo Medical School well before the American Edmund Morse (1838–1925), who has long been thought to have been the person who introduced the Darwinian theory to Japan in 1877. Hilgendorf's priority in this regard is proved by the notes taken at his lectures by the famous Japanese novelist, Mori Ougai (1862–1922), who was one of Hilgendorf's students. These notes are now at the Ougai Memorial Hongo Library, Tokyo.

Ougai's notes show that Hilgendorf's course comprised zoology (nearly 50%), botany (25%), and then crystallography (25%). The part on zoology consists of a title page (p. 1), an introduction to the units of classification (p. 2), sections on the systematics of vertebrates (pp. 3–22) and the systematics of invertebrates (pp. 22–24). He did not discuss the concept of phylum; in the 1860s biologists still thought of varieties as units of classification. In the section on the systematics of vertebrates, there are notes on mammals, birds, reptiles and fish. In some places there are details concerning fish of which Hilgendorf was particularly interested. There are also many anatomical details scattered throughout. In the section on man, Hilgendorf introduced Haeckel's classification of humans based on hair type. When considering men and apes, Hilgendorf duly introduced Darwin's evolutionary theory. In the section on the systematics of invertebrates, Hilgendorf dealt only with insects. In the section on botany, there is a list of 15 classes, species, their scientific names and Japanese equivalents. There are also 19 fine sketches of plants. In the section on crystallography there are also many figures of crystal types.

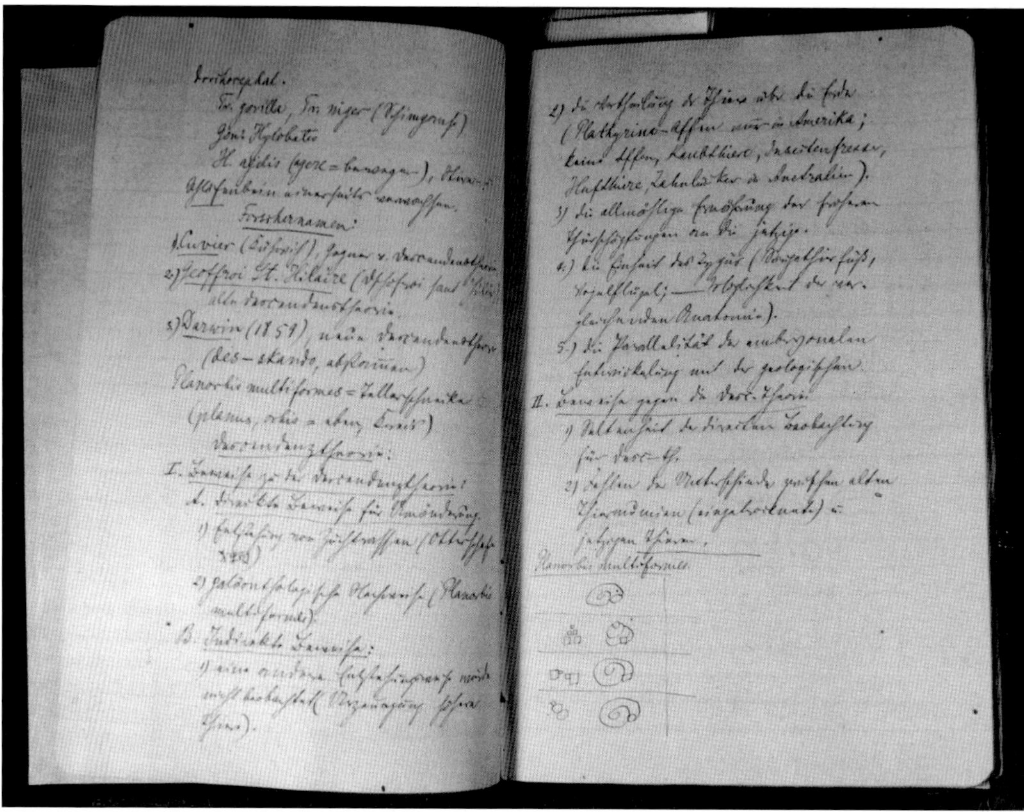

Fig. 4. Mori Ougai's note on Hilgendorf's lecture on evolution in Tokyo. On the left page are written the names of Cuvier, Geoffroy St Hilaire, and Darwin. On the right page, there are figures of planorbids (The Ougai Memorial Hongo Library, Tokyo).

Hilgendorf introduced Darwin's evolutionary theory in the following manner (Fig. 4). He first mentioned the names of scholars like Georges Cuvier (1769–1832), who had been hostile to evolutionary ideas, and Geoffroy St Hilaire who was noted to be an early evolutionist (although Darwin's (1859) theory was substantially different from those of his predecessors). Hilgendorf then mentioned the evidence for and against Darwinian evolutionary theory. As direct evidence for evolution, he mentioned the development of cultivated species, and his own palaeontological evidence for his study of *Planorbis multiformis*. As indirect evidence, he mentioned homology, zoogeographical evidence, and the parallelisms between embryonic development and systematics. As evidence against evolutionary theory, he mentioned the rarity of any direct observations of evolution and the lack of differences between mummified and recent animals. He gave a figure of the morphological changes in planorbids.

Concluding remarks

Hilgendorf was the first person to give a lecture on Darwinian evolution in Japan; his lecture of *c.* 1873–76 being earlier than that of the American biologist E. Morse who lectured on evolution in 1877 (Isono 1988). Matsubara Shinnosuke (1879), the Japanese collaborator during Hilgendorf's stay in Japan, had stated that Hilgendorf had given lectures on evolution before Morse and that he had first introduced the notion of evolution to Japan. But this was mentioned in passing in one sentence in the introduction to his paper of 1879 (Matsubara 1879). No-one paid any attention to Matsubara's paper, first, because it was not widely distributed and second, because Matsubara was not an academic, 'merely' one of the administrators of the Japanese Fishery Association.

To discuss the direct influence of Hilgendorf on the novelist Mori Ougai is now difficult, because his novels contain nothing about evolution and Darwin, unlike those of the other great Japanese novelist of the Meiji Era, Natsume Souseki (1867–1916). Souseki wrote a novel about *Shumino iden [Inheritanceof favoured persons]*, and mentioned his own view of Herbert Spencer's evolutionary ideas.

Hilgendorf's planorbid tree and study was highly praised by W. E. Reif (1983), J. Levinton (1988), and S. J. Gould (2002). The planorbid strata at Steinheim are now recognized as being of Miocene age that were deposited in the meteorite crater at the locality (Mensink 1984).

After considerable research, I found type specimens of *Vargula hilgendorfii* in the Museum für Naturkunde, Berlin in 1996. I organized the Hilgendorf Exhibition in 1997 to exhibit these specimens at five museums in Japan between April 1997 and May 1998, in cooperation with many German and Japanese researchers (Yajima 1998). The Hilgendorf Exhibition was the first attempt to develop a travelling exhibition about natural history in Japan. No less than 33 000 people, including the Emperor and Empress, visited the Exhibition. This exhibition and the Hilgendorf Symposium held on 19 October 1997 at the University Museum of the University of Tokyo have helped to make the name of Franz Martin Hilgendorf and his pioneering work and his introduction of evolutionary ideas into Japan more widely known.

I am most grateful to D. Oldroyd for help during the preparation of this paper. I am much indebted to H. Landsberg and H.-J. Paepke of the archives department of the Museum für Naturkunde, Berlin and H. J. of Tübingen University for the research of Hilgendorf's work (Janz 1999).

Notes

[1]Hilgendorf, F. M. 1863. *Beiträge zur Kenntnis des Süßwasserkalkes von Steinheim.* 42 pp., Unpublished Ph.D. thesis, Philosophical Faculty, University of Tübingen.

[2]Edmund Naumann (1854–1927), one of 'Foreign Employees', was famous as the 'Father of Geology' in Japan. More than 500 Westerners came to Japan as 'Foreign Employees' during 1867–1885. Japanese government stopped the programme of 'Foreign Employees' because of the high cost of their salaries.

[3]The Emperor has written some scientific papers and contributed parts of Japanese Fish Catalogues. Most Japanese people know that the Emperor studies *Gobius*.

References

DARWIN, C. 1859. *Origin of Species.* J. Murray, London.

GOULD, S. J. 2002. *The Structure of Evolutionary Theory.* Belknap, Harvard.

HILGENDORF, F. M. 1866. Planorbis multiformis *im Steinheimer Süßswasserkalk: Ein Beispiel von Gestaltveränderung im Laufe der Zeit.* W. Weber, Berlin.

HILGENDORF, F. M. 1877. Vorlegung aus seinen in Japan gemachten Sammlungen ein Exemplar einer *Pleurotomaria. Sitzungs- Bericht der Gesellschaft Naturforschender Freunde zu Berlin*, **72**, 73.

HILGENDORF, F. M. 1879. Zur Streitfrage des *Planorbis multiformis. Kosmos*, **5**, 10–22, 90–99.

ISONO, N. 1986. German 'Foreign Employees'. *Proceedings of Keio University, Natural Sciences*, **2**, 24–27.

ISONO, N. 1988. Introducing evolutionism to Japan. *In*: MORIYA, T. (ed.) *E.S. Morse and Japan*. Shogakkan, Tokyo, 295–327.

JANZ, H. 1999. Hilgendorf's planorbid tree—the first introduction of Darwin's Theory of Transmutation into palaeontology. *Palaeontological Research*, **3**, 287–293.

LEVINTON, J. 1988. *Genetics, Paleontology, and Macroevolution.* Cambridge University Press, Cambridge.

MATSUBARA, S. 1879. *Seibutu Shinronn [New theory of Biology].* Bansuido, Tokyo.

MENSINK, H. 1984. Die Entwicklung der gastropoden im miozaenen See der Steinhimer Beckens (Sueddeutschland). *Palaeontographica, Abteilung A*, **183**, 1–63.

REIF, W. E. 1983. Hilgendorf's (1863) dissertation on the Steinheim planorbids (Gastropoda; Miocene): the development of a phylogenetic research program for paleontology. *Palaeontologische Zeitsrift*, **57**, 7–20.

UENO, M. 1968. Franz Martin Hilgendorf. *In*: UENO, M. *ET AL.* (eds) *Foreign Employees*. Kajima Publishing Co., Tokyo, 100–112.

WELTNER, W. 1906. Franz Hilgendorf. *Archiv für Naturgeschichte*, **72**, 1–12.

YAJIMA, M. 1997. *Franz Hilgendorf: the Father of Japanese Ichthyology and Fishery Sciences.* The Committee of the Hilgendorf Exhibition, Tokyo.

YAJIMA, M. 1998. Reviews of the Hilgendorf Exhibition: A Trial of the Traveling Exhibition. *Journal of Geography*, **107**, 872–878.

Geophysical travellers: the magneticians of the Carnegie Institution of Washington

G. A. GOOD

History Department, West Virginia University, Morgantown, WV, 26506-6303, USA (e-mail: greg.good@mail.wvu.edu)

Abstract: Between 1904 and World War II, a group of researchers ranged the world over in an effort to understand the Earth's magnetism. They called themselves 'magneticians' and they worked for the Department of Terrestrial Magnetism at the Carnegie Institution of Washington. Directed by Louis Agricola Bauer (1865–1932) and John Adam Fleming (1877–1956), these investigators followed carefully selected routes through Africa, Asia, South America, and other remote regions. They carried with them a heavy complement of instruments, camp gear, and evening wear, for those times when they reached outposts of European civilization.

L. A. Bauer and motivations for the World Magnetic Survey

This article focuses on the geophysical travellers of the Carnegie Institution of Washington's Department of Terrestrial Magnetism (the DTM): their scientific goals and methods, their experiences in the field, and a consideration of how their experience compared with that of geological travellers. Their extensive expeditions in the early twentieth century were not the first aimed primarily at understanding Earth's magnetism. Edmond Halley (1656–1742) had undertaken two expeditions into the Atlantic Ocean to chart declination, winds, and ocean currents, around 1700 (Chapman 1941). Alexander von Humboldt (1769–1859) and Carl Friedrich Gauss (1777–1855) inspired and coordinated a network of a few dozen magnetic observatories, but did not try to send out expeditions to map geomagnetism. The British, led by Edward Sabine (1788–1883), extended von Humboldt and Gauss's efforts and sent out several expeditions in the mid-nineteenth century, most notably that of James Clark Ross (1800–1862), primarily to make magnetic measurements (Morrell & Thackray 1981, pp. 353–370 and 523–531; Good 1988). And some polar expeditions in the late nineteenth century had included magnetic work. These expeditions, however, were sporadic and uncoordinated.

These efforts paled before the ambitious plans of the DTM. The DTM sent out so many expeditions between 1904 and the Second World War that it is no exaggeration to speak of the DTM as an 'expedition factory.' (Good 1994). Certainly, later activities spurred on by the International Geophysical Year, major military interests, and the development of satellites and other new technologies, increased geomagnetic survey investigations by several more orders of magnitude (Zmuda 1971*a*, *b*).[1] Nevertheless, the mass production of geophysical expeditions by the DTM marked a critical development in the history of scientific travel.

Louis Agricola Bauer (1865–1932) (Fig. 1), an American of German parents with a Ph.D. in physics from Berlin (1895*a*, *b*), was the individual who built the factory and orchestrated the mass production of expeditions (Good 1999*c*). He was one of the most ambitious investigators of the Earth's magnetism. His doctoral dissertation provided a mathematical analysis of the slow, long-term changes (secular changes) taking place in Earth's main magnetic field.[2] Bauer saw these secular changes as a gateway toward understanding 'a connection between heavenly bodies and our Earth' beyond the gravitational link. Only through geophysical research, he argued, could still more mutual interactions be revealed.[3] As an analyst, Bauer was critically aware of the need for more magnetic measurements around the globe. The gaps meant that mathematical analysis could not be based on evenly distributed measurements of the magnetic elements (declination, inclination, and intensity); measurements had been made in an opportunistic, apparently haphazard way. Some countries such as Britain, France, and the United States, had systematic magnetic surveys underway in the nineteenth century, but little was being done in South America, Africa, Australia, or Asia. Bauer's solution in 1895 was to work from maps of isomagnetic lines,[4] from which he 'read' values at 126 points, equally distributed around the globe. Although this made his calculations easier, these values were derivative and the 'chart error' built into these interpolations left Bauer ultimately unsatisfied (Bauer 1895*a*, *b*). The task of investigating geomagnetic theory, he realized, required systematic magnetic

From: Wyse Jackson, P. N. (ed.) *Four Centuries of Geological Travel: The Search for Knowledge on Foot, Bicycle, Sledge and Camel.* Geological Society, London, Special Publications, **287**, 395–408.
DOI: 10.1144/SP287.29 0305-8719/07/$15.00

Fig. 1. Louis Agricola Bauer, observing the horizontal intensity of Earth's magnetic field onboard the DTM vessel *Galilee* in 1906. (Courtesy of Carnegie Institution of Washington, Department of Terrestrial Magnetism Archives. Photo 0.4.)

measurements on a global scale never before attempted. Theory required thousands of well distributed measurements. Theory required expeditions.

The founding of the Carnegie Institution of Washington (CIW) in 1902 presented Bauer with a means to tackle this monumental problem. The Institution aimed to support new areas of research beyond the usual problems. Moreover, the Institution intended to supplement work being undertaken elsewhere. To Bauer, this was an opportunity to move into this subject that was not quite physics, certainly not geology, and really had no home. It also offered a chance for him to be active on an international stage as a representative of a non-governmental organization. He argued that' ... to become an expert in terrestrial magnetism requires a lifetime of exclusive devotion and singleness of purpose ... ' (Bauer 1904, pp. 1–2). He thought it would soon be necessary to specialize in particular problem areas in geomagnetism, as in perhaps secular variation '[an] area sufficiently broad and extensive to occupy one's *sole* attention.' (Bauer 1904, p. 2).

The establishment of the DTM independent of other departments within the CIW was seen by Bauer as an advance towards establishing the independent status of the field. With this security and the ability to develop an unencumbered research programme, Bauer saw the DTM's scope of operations 'embracing the whole globe.' Bauer saw two hurdles facing geomagnetic research: the insufficiency of data and the undeveloped state of mathematical analyses of these data. The first, necessitating extensive fieldwork, is the more romantic, the job of the theoretician being deskbound. In every problem area in geomagnetism, he said, 'the observational data in general have not the requisite extent and proper distribution either in time or space or both' (Bauer 1904, p. 2). The 'lack of requisite data,' was a problem, for example, in the central problem of the overall intensity of the magnetization of the Earth, which in Bauer's time was discussed in terms of the determination of the magnetic potential of the Earth. He wrote:

> Considering the earth as a whole, very little advance in our knowledge of the distribution of the earth's permanent magnetism was made during the second half of the past century. Chiefly on this account, it is found that the accuracy of the determination of the magnetic potential of the earth has in no wise been increased by the most refined and elaborate of the modern calculations .... Such an important question as whether the earth, like any other magnet, is gradually losing its magnetism or not cannot be definitely answered, because of the lack of sufficient and accurate data (Bauer 1904, p. 3).

Bauer argued that it was the task of his generation to gather that data and begin the process of its reduction so that a later generation could develop the theory. Bauer was himself a theoretical physicist by training, intimately familiar with mathematical analysis of magnetic phenomena, so this was no call for Baconian fact gathering. The grand scheme of observations he envisioned was driven by the needs of theory. Every observer he ultimately sent out was, unwittingly or not, contributing to that grand scheme.

Bauer the theoretician thus became Bauer the director of an expedition factory. Perhaps the complexity of this enormous task ultimately contributed to the loss of his sanity. He spent his last years under psychiatric care and committed suicide in 1932. In the intervening years, however, he presided over an ambitious programme of observation which involved two ships circumnavigating the globe, hundreds of land expeditions ranging between the poles and the equatorial zone, two geophysical observatories in the most remote places, and a central office in Washington, D.C., which not only standardized instruments and reduced data, but which also conducted both laboratory and theoretical research. This paper focuses just on the expedition enterprise.

The expedition factory

Although the DTM continues to exist in the twenty-first century, its commitment to Bauer's expeditionary enterprise ended with the Second World War. A fitting symbol of this closure is the publication of two volumes in 1947 by Ernest Harry Vestine (1906–1968), the last of the managers of the enterprise (Good 1999*b*). One volume encapsulated the data that described Earth's main magnetic field and its secular variation and the other pushed forward the analysis of that data.[5] The enterprise began around 1900 with the vision that enough money, organization, and personnel could be devoted to this single activity to complete a magnetic survey of the entire planet in one generation. It had never been done before, even once, over any period of time. It took a little more than one generation.

How to present an overview of so extensive an enterprise? One way is to start with numbers. Altogether the DTM employed at least 199 magnetic observers on Land Magnetic Survey teams between 1905 and 1944. Some may have been missed in slip-ups in record keeping. Of these, 66 are not listed in the records as having worked on a team with another observer. They might have been an incidental member of a team at some point and not received credit for observations made. At least some of them travelled alone. On the other hand, 110 observers were listed as having worked on a team with one or two other observers. Sometimes both (or all three) members were experienced and perhaps had executed a number of solo expeditions. There are also 23 observers who are in the lists only as members of teams. Interpretation of these numbers is difficult. Perhaps some of these 23 were local assistants, or youth on a single job assignment. Others might have been much more significant individuals who merely moved on to greener pastures than this. After all, by the 1930s, the magnetic survey work run by the DTM was winding down. Ambitious people would have been looking elsewhere.[6]

Bauer directed all the surveys in the beginning and even participated in a few (Good 1991). John Adam Fleming (1877–1956) succeeded him in this role around 1910 (Good 1999*a*). Harlan Fisk (1869–1932) became the Chief of Land Magnetic Survey for the work that produced the 1921 and 1927 volumes. How many expeditions were there? I haven't yet figured out how to count them. Fleming and Fisk reported 178 land expeditions between 1905 and 1926 (Fig. 2) (Fleming & Fisk 1928, pp. 27–36, table 1). But there are ambiguities. Individual observers joined each other and split off as their paths crossed. In 1908, James Percy Ault (1881–1929) and C. C. Stewart travelled extensively together north of Winnipeg, Canada, but when Ault went on to other duties in the US, Stewart proceeded to 14 more stations in Manitoba, Saskatchewan, and Alberta (Bauer 1912). Sometimes other duties intervened and an observer had to divide an expedition into two. For example, Harlan Fisk made measurements at over 80 stations in the Bermuda Islands in July and August 1907, but did not continue on to Jamaica, Barbados, and the north coast of South America until July through December 1908 at 38 stations. Moreover, members of the cruises of the DTM's research vessels *Galilee* and the *Carnegie* made measurements ashore on both islands and continents, sometimes leaving the ship for extended periods for a mini-expedition. An even more ambiguous case is presented by the *Maud* expedition conducted by Roald Amundsen (1872–1928) for the DTM, in which he purposefully froze his ship into the arctic ice pack over Asia from 1922 to 1925. Fleming and Fisk explained: 'Although [Amundsen's observations were] made over the ocean, the instruments used and the methods of observing were the same as those employed at land stations, so that they are appropriately published in the list of land magnetic results.' (Fleming & Fisk 1928, p. 28). Land expeditions continued after 1928, but on a diminishing schedule. There were more than 200 DTM expeditions. A better measure of the activity of the DTM expeditionaries might be the number of stations where magnetic measurements were made. Fleming and Fisk reported a total of 5685 stations occupied by 1928, or an average of 271 occupations per year (Fleming & Fisk 1928, table 2). The difficulty of this work will become clear shortly with a discussion of the nature of the magnetic measurements being made, as well as of the conditions of travel.

In any case, there were a great many expeditionaries out and about at any one time. In 1907, John Fleming was in Central America, Fisk in Bermuda, J. C. Pearson in Western Canada, and the *Galilee* carried William John Peters (1863–1942), J. P. Ault, H. E. Martyn, and Pearson around the north Pacific Ocean (Bauer 1912, **1**, 54, pp. 107–108). Magneticians who were regular employees, such as Ault, also performed multiple duties. He established magnetic stations in the Caribbean in 1905, in Mexico in 1906–1907, served on the first two cruises of the *Galilee* either side of this expedition, and travelled 1800 miles through northern Canada in 1908 making magnetic measurements at nineteen stations. He shipped aboard the first cruise of the *Carnegie* in 1909 and then was assigned to data reduction work at the main office in Washington (Bauer 1912, **1**, 54, pp. 109, 112–113).

Although the magneticians frequently were called for tours of solo duty, they were part of a

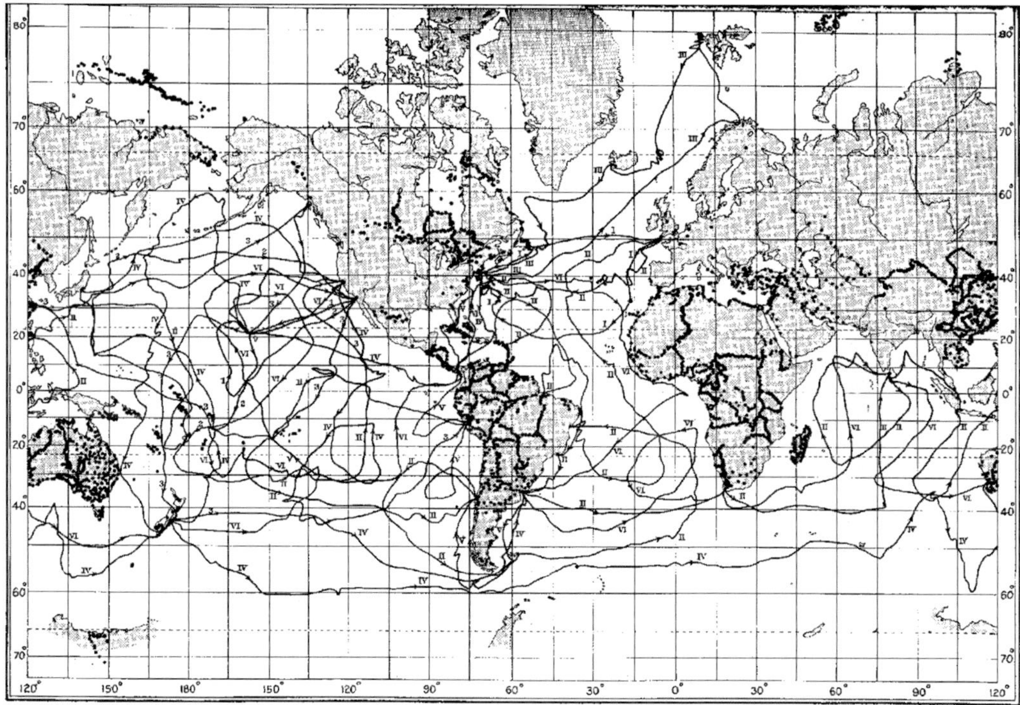

Fig. 2. Map of geomagnetic survey expeditions and observations of the Department of Terrestrial Magnetism, 1904 to 1926. The cruises of the *Galilee* are indicated by Arabic numbers, those of the *Carnegie* by Roman numerals; black dots show the land stations (*Source*: Fleming & Fisk 1928, Fig. 1.)

tightly organized group of scientific workers. They were trained in using the most arcane of instruments: magnetometers, dip circles, induction coils (Fig. 3). They learned the basics of electricity and magnetism. But more central to their future role, they learned the mathematical techniques of data reduction, the first stages of which they would some day perform in a tent in the Rio Negro in Brazil or perhaps at sea on the *Carnegie*. A first field season was a testing time with one of the best workers. Some assistants never had a second season, others went on to a full career. Through this winnowing the most dedicated individuals attached themselves to this select group. Some proceeded to assignments of months or years at one of the DTM's geophysical observatories, or to a cruise of the *Carnegie*, or to more advanced data reduction, instrumentation, or laboratory research at the central office in Washington.

Another aspect of central control of these expeditionaries emerges from the plan of research for the DTM. Every track was meticulously planned, based on the need for data from particular areas of the globe. In those pre-electronic computer days, simplified mathematical models were especially important. Some periods of intensive, prolonged investigations, certain days, were designated to coordinate observations around the globe. The political, material, and social issues of travel in foreign lands was coordinated and balanced against these scientific needs. No wonder magneticians were repeatedly suspected of being spies!

Running such an expedition factory was as daunting a task as that faced by each individual observer. A central office needed to be organized to manage the many tasks. In 1919 more than fifty employees were divided into seven main divisions: administrative, research in terrestrial magnetism, experimental work, magnetic survey, the *Carnegie*, observatories, and instruments and properties. The Magnetic Survey Division included an overall chief (Fleming), a chief of ocean work (Ault), a chief of land work (Fisk), three observers (Harry E. Sawyer, Allen Sterling, and Herbert R. Grummann), and a junior observer (Robert R. Mills). Also within this division were people normally assigned to the DTMs two geomagnetic observatories in Peru and Australia, but who were periodically assigned to survey work as well: two 'magneticians-in-charge' (Harry M. W. Edmonds, 1862–1945, and William F. Wallis), one observer (Wilfred C. Parkinson, 1884–1964), one assistant

Fig. 3. A 'universal' magnetometer, developed and constructed at the Department of Terrestrial Magnetism in 1910. This instrument was widely used on the most remote land expeditions. An observer could measure magnetic declination, inclination, and intensity with this one instrument, thus greatly saving in weight of instrumentation (*Source*: Multhauf & Good 1987).

observer (Federico G. Rosemberg), and one observatory aid (Theodore C. Kiesel). The Washington staff included Bauer as chief, two expert computers (Charles R. Duvall and Henry B. Hedrick), a librarian-translator (Harry D. Harradon, 1883–1949), along with instrument makers, stenographers, and other support staff.[7]

A budget needed to be raised to keep so large an operation going. Institutional politicians needed to be cultivated. From \$20 000 in 1904, the budget grew to \$76 920 in 1910, and \$231 625 in 1921. Moreover, in 1909 the *Carnegie* was built for \$115 000, in 1913–14 a research building was built in Washington, and from 1914 to 1918 magnetic observatories were under construction in Washington, Peru, and Australia, all projects beyond the operational budget. Although all of this funding was raised within the CIW, it nevertheless required active advocacy. The DTM had a larger budget in 1921 than any other CIW department.[8]

Overseas connections needed to be maintained through the US State Department and many other contacts. The files of the DTM are replete with special requests to ambassadors, factors of economic concerns, and representatives of foreign governments. Requests asked for duty-free entrance of instruments, considerations for bills of credit, and passport assistance. The Carnegie Institution even issued its own passport, which sometimes worked when the American passport did not!

Public relations were also crucial. Scrapbooks full of newspaper accounts of expeditions kept at the Carnegie Institution evidence both the level of interest of the press and the public and the value that the Institution placed in these accounts. Articles in popular magazines, such as *National Geographic*, indicate both active grooming of image and also the high level of acceptance of the activity of the DTM. Glossy magazines and newspapers were supplemented by radio talks as time went on.[9] Bauer and Fleming realized, too, the special importance of cultivating public opinion in Washington, D.C., where their political and financial support centred. The DTM sponsored annual 'exhibits' and lectures. In 1914, the exhibit included a model of the ship *Carnegie*, photographs, magnetic instruments, and a globe in which pins were stuck for each magnetic station. As the author of a report on the exhibit noted, the globe was getting crowded and 'As the globe is more and more covered, a smaller sized head might be better'.[10] The 640 attendees included professionals, technical government employees, naval officers, teachers, and 'ladies.'

The distribution of the expeditions around the globe was planned around the principle that the DTM would supplement the existing efforts of nations and individuals. Because the US and European countries generally undertook their own magnetic survey work, there was no need for the DTM to be active there. In some places, qualified individuals only needed material support to undertake the work, such as in South Africa, where Professors J. C. Beattie and J. T. Morrison were able to use leave from their universities if the DTM supplied instruments and travel expenses (Bauer 1912, **1**, table 5, pp. 101–106). In Canada, the government was unable to meet the full needs of magnetic surveying and the DTM found itself negotiating a difficult middle way, undertaking the survey work while still encouraging the national government (Good 1991). When any competent observer was to be part of a polar expedition, the DTM cooperated by loaning or standardizing instruments and by training observers. In one case the DTM even loaned the observer, when DTM magnetician Richard H. Goddard was signed over to the MacMillan Arctic Expedition of 1923.[11]

In the first five years of land survey work (1905–1910), the only magnetic measurements made by DTM personnel within the US (52 stations) or Europe (36 stations) were for standardizing instruments or establishing critical baseline data. Africa had the highest concentration of stations with 386, followed by North America with 328 (mostly in Canada, Central America, Newfoundland), Asia with 308 (mostly in China), and South America with 111 (Bauer 1912, **1**, table 5, p. 55). The islands of the Pacific Ocean (51 stations) and the Atlantic Ocean (68 stations) outpaced Australia and New Zealand, with only 10 stations. Effort re-focused more strongly on South America (247 stations), Australia and New Zealand (284 stations), and Africa (207 stations) in 1911–1913 (Bauer 1912, **2**, table 5, p. 25). From 1914 to 1920, 1747 stations were occupied mostly in Africa, Asia, Australia and New Zealand, and South America (Bauer *et al.* 1921, table 1, p. 5). Lastly, for comparison, 1443 stations were occupied between 1921 and 1926, mainly in Asia and South America (Fisk 1927, table 8, p. 33). It's clear where the Carnegie Institution concentrated its expeditions, at least in the biggest features. It is also notable that the average number of expeditions in these four periods did not vary much, respectively being 217 per year, 327, 249, and 206 per year. Over this 21-year period, 1143 stations were established by DTM magneticians in Asia, 1053 in Africa, 939 in South America, 826 in Australia and New Zealand, ranging down to 120 in Europe and 30 in Antarctica. In 1928, Fleming and Fisk proclaimed that:

> The general magnetic survey of the globe, to the accomplishment of which the Carnegie Institution of Washington, through its Department of Terrestrial Magnetism under the energetic guidance of Director Louis A. Bauer, devoted its energies for many years, has been completed for the major part of the Earth (Fleming & Fisk 1928, 27).

A lot of stories hide behind these statistics, but before we turn to some of the individual magneticians' experiences, a few more questions about how this complex enterprise was regulated must be addressed. For the work of these magneticians to be of any use extraordinary consistency was required. After all, this was the main reason for this enterprise to be conducted by a single institution rather than by many.

This consistency was maintained through standards and a rigorous local culture. Bauer ran a tight organization, with what was sometimes seen by others as an obdurate, obsessive subjugation of the individual. In 1906 he sent a memo to all members of the DTM, stating:

> Gentlemen: The work entrusted to us is of such magnitude and importance and marks such an epoch in magnetic science that we must all esteem it a privilege to be associated with it.
>
> The successful execution of duties assigned in every case demands entire devotion and a whole-hearted interest.
>
> The difficulties to be surmounted and the work demanded are frequently such that no adequate financial compensation can be made for services rendered so that the chief satisfaction must be derived from the knowledge of duties well and faithfully performed.[12]

The memorandum dictated disciplined attention to clarity, careful reporting of expenses, the necessity of following instructions, stewardship of instruments, habits of mind, record keeping, and the priority of the enterprise over the individual. Bauer himself was a driven and disciplined worker and he expected the same of other members of the department.

Bauer's expectation of commitment made itself known in ways ranging from detailed, typed instructions, to short notes correcting what he saw as inappropriate behaviour. This original 1906 memorandum noted that observers must keep all instructions and model calculations they received so that methods would be rigorously consistent.

> You may be called upon at any time either to observe independently or to train others and to take charge of work. Failure to do this has at times caused confusion in methods and entailed unnecessary additional work upon the Office.

He continued that careful habits were essential in handling and packing instruments, the alternative being damage, expense, and 'expenditure of time which could have been put to better advantage.' Bauer commented on the importance of entering data and descriptions immediately on the prescribed form developed by the DTM. Records, he said, should allow an 'office computer to make intelligently an entirely independent computation .... A neat arrangement of observations or of computations should be an object of pride.' He also cautioned observers that when they travelled in foreign lands they represented the Carnegie Institution and that 'a judgment of the entire work may rest upon [your] work and [your] actions.'

Every magnetician received three types of instructions before setting out on an expedition. First were the instructions regarding itinerary, foreign governmental contacts, etc. Observers were given freedom to decide in the field if these directions needed to be changed, but only if political or other contingencies absolutely demanded it. Changes were to be communicated and approved at the first opportunity. These instructions also sometimes included specific recommendations for coping with local customs or diseases.[13]

The second type of instructions regarded methods of observation. Every observer was well trained in the use of the instruments before going into the field and most also trained as assistant observers for a tour before taking off on their own. Nevertheless, each observer carried a booklet describing in detail the use and care of each instrument. The managers found that observers also needed special instructions 'applicable to conditions that pertain to magnetic surveys generally outside the United States and in places without considerable cultural development (sic).'[14] The use of many scientific instruments entails much knowledge that is tacitly learned or which comes through demonstration and occasional discussion. A 6-page memo attempted to spell out some such matters for observers before they went to places where close communication was impossible. Because the re-location of occupation sites was important but difficult in the wilds, photographs from several vantage points were recommended. The memo recommended that magneticians carry three chronometers in rough country. It gave directions for determining latitude using the *Nautical Almanac* and star observations; means of correcting for local magnetic or electrical disturbances; and ways of dealing with finicky earth inductors and galvanometers. Some observers received special instructions directing unusual regimens of observation not undertaken by all observers, as for example when Harlan Fisk was detailed to make magnetic observations during a total solar eclipse in May 1918.[15]

The third type of instructions related to record keeping, computations, and quality control. Bauer was repeatedly frustrated when he detected errors in calculations several stages after they should have been detected and corrected. Observers were provided with forms that reminded them of critical data. Instructions explicitly directed proper methods of recording. They were required to perform a first level of computation in the field. In the office, two computers each re-calculated the base data and then checked each other. And yet, errors crept in. In 1915 Bauer directed that both the computer and reviser must initial and date each sheet of calculations and briefly discuss anything unusual.[16] As Bauer had said in a similar memo in 1909, 'Bear in mind that *somebody* must sooner or later attend to the things *you* neglect.'[17] Neatness, system, and order were essential.

Bauer effected oversight of the magneticians through his 'lieutenants'. This list remained consistent throughout the period of 1905 to 1926, when the main thrust of the world magnetic survey was underway: J. P. Ault, John Fleming, Harlan Fisk, and William Peters. Several of these men had been with him at the US Coast and Geodetic Survey or were part of the Cosmos Club membership in Washington, D.C. They shared an interest in exploration, geoscience of the mathematical and physical variety, and disciplined work. Fisk and Peters provide a glimpse of this second echelon at the DTM. Their backgrounds and their careers within the DTM are among the most completely detailed for those who spent most of their adult lives within the department.

Rather than rehearse the stories of the travels of these two magneticians, who did indeed travel extensively on both land and ocean expeditions during their full careers at the DTM, I will instead take a closer look at their roles in the main office. Harlan Wilbur Fisk was born in Geneva, Kansas, of parents with a long New England lineage. Fisk obtained a B.S. at Carleton College, Minnesota, in 1896 and took only a few graduate classes at the University of Wisconsin in 1901 and Carleton College in 1902–1903. He married in 1897 and fathered three sons and a daughter. His first job after college was as high school principal in Grand Meadow, Minnesota, from 1896–1899. He then became a professor of mathematics at Fargo College, North Dakota, a post he held from 1899 to 1906. During summer breaks in 1904 to 1906 Fisk worked as a magnetic observer for Bauer with the Coast and Geodetic Survey. Apparently this work suited him and in October 1906 he gave up his professorship to become first a Magnetic Observer (1906), then a Magnetician (1907), then Chief of Section of Land Magnetic Survey (1919) at the DTM.[18] He conducted a few field studies for the DTM after his permanent hire, visiting the Bermuda Islands in 1907 and Central and northern South America in 1908. Afterwards he worked mainly in Washington.

W. J. Peters was a field scientist by inclination and by training. Born in Oakland, California, he spent only a short time at the University of California, then shifted to topographic surveys and boundary surveys of California in the 1880s. From 1884 until 1902 he climbed through the ranks at the US Geological Survey. He developed the skills of a topographic surveyor, a practical astronomer, and a computer. He was chief of party from 1898 to 1902 in extensive surveys in Alaska and Utah. At this point he shifted to Arctic, ocean-based exploration with the Ziegler Expedition of 1903–1905. He commanded the scientific party of the expedition (he had four assistants), which returned home with extensive records of geomagnetism, auroras, meteorology, and tides, despite the loss of its ship. Having proven himself as a leader and as a teacher, Peters was hired by the DTM to command first its leased ship *Galilee* and then in 1909 the *Carnegie*.[19]

In the home office, Fisk designed and tested instruments and he helped train new observers and assistant observers for several decades. He guaranteed the regimentation and the common background in instruments, departmental expectations, mathematical skills, etc., on which the enterprise depended. He developed a special interest in secular variation of geomagnetism, which led him to play a prominent role in selecting the most important locations for 'repeat-stations'. Hence, he helped to decide where observers would go. Fisk helped the office in an efficiency study of observers, producing a system in the 1920s that rated how many stations one could expect an observer to occupy in a given time, based on actual performance of observers from 1914 to 1924. He noted in one memo that while the average in earlier expeditions had been one locality per 5.17 days, the average had decreased to one per 7.44 days because the emphasis was shifting to repeat stations for secular variation work. Repeat stations were further apart. This analysis was completed for each observer, and for each class of observation he made, for each month of fieldwork. In this way, each observer could be managed by appeal to an objective measure.[20]

Peters was another source of continuity for the DTM, since many of its land observers also served on the ships and often were transported to their new expeditions by them. After 1914, Peters was mainly assigned to Washington, where he worked primarily in data reduction and wrote reports on the results of various expeditions. He was Chief Computer and Assistant to the Director through the 1920s and until 1931. In his role as Chief Computer he trained new computers and guaranteed the quality of all calculations. He was well known at the DTM for directing others in these essential activities. He retired in 1934.

The DTMs world magnetic survey also assumed a careful standardization of instruments, both in their design and in their regulation. Suffice it to say here that the DTM was extremely diligent in this regard, building its own instruments or carefully supervising contractors such as Bausch and Lomb. They also compared each instrument to a standard of its type both before and after each expedition.[21]

Overall, the Department of Terrestrial Magnetism was a model of corporate scientific research in the first generation of its existence. It was built around one main goal and all activities of its dozens of employees were coordinated as a part of reaching that goal. The DTM was, because of that goal, an expedition factory that turned out seemingly endless volumes of magnetic data, all of which added up to a magnetic snapshot of the Earth for the early twentieth century.

The magneticians, their scientific work and their travels

To accurately and fully portray the magneticians of the Department of Terrestrial Magnetism would require detailed analysis that is beyond this article. Ideally, each of the several hundred employees of the DTM should be followed from their childhood through their education to the culmination of their careers. For now it must suffice to provide a quick overview and then to pull out a few individuals and let them stand as indicators of the group.

A few initial points about the group: some made careers at the DTM; only Ault, Bauer, Fleming, Fisk, Peters, and six other observers spent most of their careers there. The other 188 magneticians were more transient: 100 of them were with the DTM for only one 'tour of duty'; most magneticians were employed for only a short while.

Although many of the magneticians were not much more than technicians, with very thorough but limited training, some did indeed go on to great things. Some received advanced degrees in physics, electrical engineering, or other areas. Some went on to become famous: Edward Crisp Bullard (1907–1980) for his work on geodynamo theory and plate tectonics; Harald Sverdrup (1888–1957) in oceanography at the Scripps Institution; Scott Forbush (1904–1984) for his work in cosmic rays; Roald Amundsen, Thomas C. Poulter (1897–1978), and Paul Siple (1908–1968) for their participation in Antarctic exploration; Edward Kidson (1882–1939) in Australia because of the Kidson Track that he blazed; and Andrew Thomson (1894–1974) and Frank T. Davies (1904–1981) for upper atmospheric research in Canada. Others, though, laboured in relative obscurity. The following individuals were at the heart of the expedition factory.

Charles Keynes Edmunds and Frederick Crawford Brown in China

Consider C. K. Edmunds. Edmunds was born in 1876 and, at age 21, received his Ph.D. in geology from Johns Hopkins University in 1897, and may well have met Bauer when he was a visiting instructor there. Edmunds entered the DTM with the first class of magneticians in 1905. He was already at that time the president of Canton Christian College in China, which had been founded by the American Presbyterian Board of Foreign Missions in 1888. Edmunds proved to be both a competent observer and later a contact point for other DTM magneticians entering China. From 1906 to 1910, he dedicated three or more months each year to magnetic survey work. At first he borrowed a Kew magnetometer and dip circle from the British Hong Kong Observatory, where he standardized his instruments before and after his first expedition. That expedition was to Hainan Island, 'a region magnetically unexplored, and at some points which had never before been visited by a white man'. He also tested French magnetic instruments borrowed from Zi-Ka-Wei Observatory and used them at a number of stations on the coast in 1907. In 1908, the DTM sent him a set of its own instruments so that he could compare them in the field against the British and French instruments. Magnetician D. C. Sowers, on his way to an expedition in northern China and Turkestan, visited Edmunds and compared their instrument sets, too. Edmunds then conducted a series of forays throughout SE China and into Indo-China as well. By 1910 he had observed at over 100 stations.[22]

Edmunds described these expeditions in detail in reports printed in 1915 by the DTM (Bauer & Fleming 1915).[23] He fitted in the magnetic work during holidays and furloughs from his college, and his expenses were paid by the DTM. This allowed him to use Canton as a base of operations, where he finished his data reduction. Between January 1906 and March 1912 he conducted seven separate expeditions, three of them lasting four to five months. When he travelled to the US on college business in 1910, he spent September at the Washington office of the DTM comparing his instruments with the DTM standards. Altogether, Edmunds was very familiar with efforts to standardize magnetic measurements.

Like all DTM observers, Edmunds was as scrupulous in reporting logistical and budget details as he was with his scientific measurements. He reckoned that he had travelled 26 000 miles in SE Asia during these seven expeditions, 15 000 of those in the field. With no salary from the DTM he was a definite bargain to Bauer, costing only $5335 in field expenses. Altogether in this time he occupied 116 stations.

Edmunds tended to travel in well-settled regions. He made extensive use of the network of missionary compounds, especially of his Presbyterian colleagues. He also visited western scientists at observatories, for example the Jesuit priest Joseph de Moidry (1858–1937) at Zi-Ka-Wei, and hospitals and actively engaged western military, customs, and consular officials. He was also helped by western commercial factors. Although he travelled extensively by steamer and rail, he also did not shy away from hiring local junks or walking, although he hired 'coolies' to carry his heavy load of instruments and baggage (Bauer & Fleming 1915, 96–97, for example).

On his 1911 travels Edmunds ran up against political reality. The intended loop was to go via Haiphong and Hanoi up into Burma and on to the Yangtse. As he said, 'the reports of trouble on the border between Yunnan and Burma were quite alarming'. So he cabled Washington and obtained permission instead to organize an expedition through French Indo-China and Siam. This expedition across the peninsula was accomplished mainly by foot and canoe, portaging around numerous rapids, and staying in 'native huts'. Once in Bangkok, the US consul helped arrange free use of the railroads to extend the magnetic measurements (Bauer & Fleming 1915, 104–107).

In 1914 plans were afoot for Edmunds to participate in an ambitious and more complicated survey of northern and NW China and Mongolia (Bauer *et al.* 1921, 154–160).[24] He provided the expertise in outfitting a campaign in China and planned the routes for himself and another DTM magnetician, Frederick Brown (Fig. 4). Brown, hired 30 May 1913, had trained as an assistant observer with

Fig. 4. Frederick Brown travelled around China and Mongolia by various means, including camel and sedan chair. Here Brown stands alongside his chair and porters in Kweichow in April 1915. (Courtesy of Carnegie Institution of Washington, Department of Terrestrial Magnetism Archives. Photo DTM 5769.)

Edward Kidson in Australia (Bauer & Fleming 1915, 81–84). He advanced quickly and led two expeditions across northern Queensland and north-west Australia in 1913 and another in Northern Territory in 1914 (Bauer *et al.* 1921, 100–105). Brown arrived in Canton in January 1915, where Edmunds was his Chief of Party. After two months of organizing, Brown left Canton for his first tour, taking Mr N. K. Ip as his assistant, along with a cook. For the next year and a half, Brown and Edmunds met up periodically, travelled together a little, but mostly apart. When separate, each took along an assistant able to speak the local dialect, a cook, and porters. The story of this cluster of expeditions is one of the more complicated, since it lasted over a year and ranged over 12 000 miles for Edmunds and likewise, by different routes, for Brown.

The manuscript correspondence extant in the DTM archives tells a slightly different story of Brown's activities in Mongolia than that published in the *Land Magnetic Observations*. On 9 June 1915, Brown wrote that an official had not trusted him because he was British. An official prohibited Brown from making any maps and said his passport didn't allow him to make observations. In one town he had to be surrounded by soldiers to keep curious crowds at bay. As he proceeded he also repeatedly faced bandits. The published account underplayed these problems, a stratagem that was consciously employed to minimize offence to the countries in which DTM observers travelled.[25]

Although Edmunds had turned to other duties in 1916, Brown continued in Manchuria throughout 1917, then he moved on to Africa from 1919 to 1921 (Bauer *et al.* 1921, 105–152). He was assigned around the world in Africa, South America, and elsewhere. He returned to China in 1922 and stayed there again from 1931 to 1936, during which time he worked extensively with Dr Paul C. T. Kwei (1895–1961), a graduate of Canton Christian College. Kwei worked at the DTM in Washington in 1936 and became a prominent ionospheric physicist in China (Wallis & Green 1947; Wang Shen 1994).[26]

Despite all of their expeditioning for the DTM, Edmunds and Brown were both primarily committed to education and missionary work. Edmunds presided over Canton Christian College for over twenty years and he published *Modern Education in China* for the US Government Printing Office in 1919. He also served as president of Pomona College from 1928 to 1941. Brown ran a mission school in Changsha in the 1920s. Their training was in science, and they took advantage of the needs of the DTM to devote themselves to field work that took them throughout Asia, from Indo-China to Mongolia. It was a good arrangement for both sides, but perhaps Edmunds and Brown were not such typical magneticians as, say, Harlan Fisk and William J. Peters.

Earl Parker Hanson

The other magnetician to be highlighted was somewhat different to others in his line. Earl Parker Hanson (1899–1978) was born outside the US (in Berlin), of an expatriate father and a Danish mother. He graduated in engineering from the University of Wisconsin in 1922 and went to work as an engineer in Chile until 1925. At that point he became an international traveller, a writer, and an editor. He was a bit of a bum, priding himself on being able to get anywhere with next to nothing in his pocket. He was an 'irregular', which marked him out from most of the earlier magneticians. But by the early 1930s, the main world magnetic survey was complete and the goals of the DTM were shifting, Bauer was out of the picture, and a person like Hanson could find a niche.

In fact, Hanson fitted well the new needs of the DTM. Although the DTM overall was shifting its research to the laboratory and to nuclear physics, the programme still called for making 'repeat measurements' for the study of secular variation. Fleming, Fisk, and several other top officials saw the need to send out a diminishing number of observers to make these observations, but they no longer needed to maintain magneticians on permanent staff. Hanson, along with other adventuring types, was just right. In 1931 he was the Executive Secretary of the Explorers' Club of New York. He had experience as a solo traveller, he was cultured, flexible, and he understood the use of measuring instruments.

Hanson also had another advantage. As a writer, he left one of the best records available by any magnetician of his daily activities. He published *Journey to Manaos* (1938) as a partial record of his travels for the DTM in South America between 1931 and 1933. More valuable, though, are the many manuscripts of letters and reports to and from him still at the DTM. For this account, though, I will draw mainly on his book (Hanson 1938, 47–48). The book provides an account of the difficult challenges faced in simply getting from one point to another and provides a few simplified glimpses of the measurements that made the travel important. True, the book is a somewhat sanitized version of events, since Hanson and the Department both had reasons to submerge embarrassing events. For current purposes, however, that is unimportant.[27]

Hanson travelled up the Orinoco River in Venezuela, through the canal connecting the headwaters of this river with those of the Rio Negro in the Amazon basin, and on to Bolivia and Peru after visiting numerous repeat stations along the Amazon (Wallis & Green 1947).[28] He noted his magnetic observations:

> Specifically my task was simple. I was given a set of delicate instruments, a tent to shelter them while I was making observations, proper forms, credentials, letters of credit, and the like, and was instructed to transport those things to some eight stations in the West Indies and South America. These stations were places, carefully described in the Institution's records, where my predecessors had worked and where my successors will work some day.
>
> I was to set up my instruments at the exact spot where some other man had set his up anywhere from five to twenty years before, and I was to do the same things he had done. I was to measure the 'declination' of the earth's magnetic field, ... the strength of the earth's field ..., and its 'inclination' ...
>
> I was to measure those things at a number of places listed, along a route of some twenty thousand miles, so that the scientists in Washington could compare my observations with those of my predecessors in order to determine how the earth's magnetic field had changed in the interim in the areas covered. At certain points I was to make 'Class A' observations of that throbbing, ever-changing field ...
>
> After doing that work for a year and a half I could go home again (Hanson 1938, 3–4).

This glib account made it all sound too simple. Later in the book, Hanson discussed complicating factors encountered in making measurements. First, he had to find the monuments left behind by previous magneticians. This was made difficult by the vigorous tropical vegetation, by the abandonment of villages after the end of the rubber boom, and by other vagaries. J. T. Howard, for example, had observed in Santa Isabel, Brazil, in 1925 and left behind two monuments with the letters CIW on them. The one in the centre of town had disappeared. The other was reported to be on Itacoatiari island in the river. No one knew of any such island. The local store owner said 'Heaven knows what name some Indian gave the island when your Mr Howard was here. People use different names for the same places here'. (Hanson 1938, 263–266). Ultimately, Hanson found the monument on the island with the help of a young girl who recognized an odd tree in Howard's photographs.

Hanson reported that his own monuments were often dug up soon after he left them, locals thinking that he was marking treasures that he had discovered using his mysterious instruments. Hanson learned to place several prominent but false monuments and to make the real monuments small and inconsequential.

The observations were demanding, requiring close attention for hours on end, and sometimes for several days. Hanson had troubles with curious on-lookers who wanted to buy pencils, corks, or needles, and who wanted to have their pictures taken (Hanson 1938, 143–144). He also was greatly hindered by malaria, which revisited him several times and sapped his strength and concentration. He said that he had to centre his daily life on the need to make magnetic observations, and not allow himself to use his limited strength for anything else. Once, in a particularly severe flare up, he wrote

> I thought that a magnetic expedition would have been a lot of fun but for the need for observing the earth's magnetism ... (Hanson 1938, 301).

He fretted at several points that he might have to give up the expedition and break his contract. As he said to himself, 'If you don't do your magnetic work, why are you here?' (Hanson 1938, 257).

Hanson repeatedly touched on the need for concentration in carrying out the measurements. He had to measure declination, inclination, latitude, longitude, and time the swinging magnet as it indicated the intensity of the magnetic field. A certain amount of illness, he said, sharpened the senses. He was methodical. After making his measurements, he would pack the instruments in their cases, listen to a short-wave radio for the time signal for the longitude, and do his preliminary data reduction. The people at the DTM counted on this (Hanson 1938, 271–272).

Hanson proved his value even more when he returned to the States. Not only did he write the book about his journey, but he published several articles, gave numerous interviews to the press, and he organized a lecture tour.[29] The Institution definitely got their money's worth from Hanson. Hanson never worked directly for the DTM again, which was characteristic of the last generation of magneticians. He went on to a career in regional development in Latin America, advised the US military on jungle warfare in World War II, served the State Department in Liberia, and chaired the Geography Department at the University of Delaware. His final position was as Counselor for Economic Development for the State Department in Puerto Rico.

Conclusion and contrasts: the end and meanings of Bauer's enterprise

Every factory is driven by both a need and by a dynamo. These indirectly shape the experiences of every worker in the enterprise. The need behind the DTM's magnetic survey of the world was at the heart of the magnetic phenomena themselves: their time-scale. Unlike most geological processes, which require eons to occur, the magnetic elements being measured change constantly and in a number of different ways. In particular, there are daily changes due to the Earth rotating beneath the gaze of the Sun; irregular changes called magnetic storms due to events on the Sun itself; and there are secular changes, due to movements deep within the Earth. The latter presents the most interesting difficulty for terrestrial magnetic science, at least in the days before air surveys and satellites. Because the things being measured are constantly drifting in value, it is critical that the measurements be made quickly relative to those changes. Geologists have no problem accomplishing this. Geophysicists do. So, the goal of producing a 'snap shot' within one generation originated in the phenomena and the science. The shutter on the magnetic camera was open from 1905 to 1926. That snap shot continues to be of use to researchers today, as they incorporate such historical data into ongoing investigations (Jonkers *et al.* 2003).

As for the dynamo, that was Bauer himself. Bauer felt an intense inner urge, the source of which may never be penetrated, to understand Earth's magnetism. Nothing could stand in his way. He was undeterred by the disinterest of both geologists and physicists, although their occasional disdain wounded him. The efforts of petty bureaucrats in government science departments to make geomagnetic needs subservient to, for example, geodetic needs, drove him to seek a new patron in a private institution. The bickering and meddling of antagonistic nationalists in other international endeavours like meteorology led him to define 'international science' in a new and different way, too. Bauer spoke of zeal, devotion, and duty in memos and publications. He meant it. It drove him and the DTM on. Those who didn't understand this, moved on.

Bauer several times claimed that the DTM magneticians were the first white men to be seen in certain remote areas of the world. While this was certainly true in a few cases, that is, that a magnetician was the first contact of an indigenous people with representatives of modern society, in most cases it was not. Although travel in Amazonas challenged the traveller, in fact the magneticians there followed in the tracks of the Venezuelan military, wholesale merchandisers from Manaos, the rubber economy, the missionaries, and the professional explorers. On the Asian steppes they followed the ancient Silk Road. Just as the lines of DTM ocean cruises tended to follow and interlink commercial travel lanes, the dots jumping up the Congo River or across China followed old trading routes. The magneticians took advantage of this travel infrastructure, these lanes of cultural interaction, and so exposed the furthest tendrils of European and American interaction with indigenous societies. These magneticians generally were not pioneers, but that is not to say they were timid. Their travels nonetheless challenged their ingenuity and self-reliance.

Indeed, most of the challenges faced by the magneticians were social or cultural ones. Whether magneticians travelled alone or in a team, they were frequently the only outsiders in a village or river valley. They were always hiring local porters and assistants, arranging canoes or camels, beds or meals or medicines. This side of geophysical travel engaged much of the magneticians' time. The actual scientific measurements were precise but repetitive. Calming the vibrating needle of a magnetometer was second nature to them. That is, the scientific part of these expeditions became routine, almost a ritual.

How did geophysical travel differ from geological travel? The logistical challenges were similar, both in the heavy loads and in the need for the traveller to deal effectively with social and cultural problems. The geophysicist and the geologist frequently went into equally challenging environments, from the poles to the equator, from mountain tops to jungles. The magnetician, however, could concentrate much of the work in settled places, whereas geologists more often have to be away from towns.

There are more basic contrasts, though. The exact surroundings are of less importance to the geophysicist. What counts are longitude, latitude, and the magnetic measurements. For much (though not all) of geomagnetic research, the local rocks were not that important. For the geologist, the rocks were exactly the point. The magnetician travelled through the real world of rocks, towns, and people, but did his work in an abstract world of measurements, data reduction, and mathematical theory. Imagine the *Carnegie* in the middle of the Pacific Ocean, with not a feature to mark the horizon and the ship cruising along undetectably while the observers measure the magnetic variables unconcerned about their surroundings. This trackless, featureless landscape symbolizes the whole project.

Secondly, although a geologist might make a discovery with a single rock sample, or with a small collection made on a single trip, this could never happen to a magnetic traveller. Each reading taken was but a small act that only made sense within the whole collective action of the DTM and of all the magnetic researchers that ever have and ever will take readings. Bauer's DTM was producing a snapshot of Earth's magnetic field and each reading was akin to a single pixel in a larger picture. Alone, it was useless and uninteresting. A fitting conclusion to this article is a quote from Earl Hanson,

> No one field observer can make 'discoveries' in magnetism ... All we can do is go back again and again, to measure, purely as a long-time program of routine-research, *how* the earth's magnetic field has changed since the work of our predecessors in the region covered, collecting the facts that our colleagues in Washington need for their studies. It is a program that should never be ended ... (Hanson 1938, 332)

Notes

[1]Zmuda (1971*a*, *b*). The latter volume includes short historical essays in addition to scientific reports.

[2]Bauer (1895*a*). Bauer dedicated his dissertation to Max Planck, Wilhelm von Bezold, and Wilhelm Foerster, all of whom he studied with in Berlin.

[3]Bauer (1895*a*), unpaginated sheet, Thesis IV: 'Es wird mehr und mehr anerkannt, dass die Gravitation nicht das einzige Band zwischen den Himmelskörpern und unserer Erde ist, sondern dass noch andere Wechselwirkungen vorhanden sind, über welche man nur durch geophysikalische Erforschung ins Klare kommen wird'.

[4]Neumayer (1891). *Atlas des Erdmagnetismus* with 5 maps. No. 1 is a map of equal declinations. The maps Bauer read were contained in this marvellous volume, a colourful visual summary of observations made by numerous observers over the better part of the nineteenth century.

[5]Vestine *et al.* 1947*a*, *b*. As Vestine (1947*b*) notes in his Preface (p. iii), these two works also represent the culmination of work undertaken during World War II for the US Navy. This work related the detonation of harbour mines to variation in magnetic intensity.

[6]These preliminary counts are based on charts of observers in five volumes published by the Carnegie Institution of Washington under the title: *Land Magnetic Observations*. Vol. 1 covered the period 1905–1910, vol. 2 1911–1913, vol. 4 1914–1920, vol. 6 1921–1926, and vol. 8 1927–1944. These were published by the Carnegie Institution of Washington, in Washington, D.C., in 1912, 1914, 1921, 1927, and 1946.

[7]Department of Terrestrial Magnetism, Archives. File: Personnel. Fisk. Organization. Dated 10 January 1919. Asterisks beside half a dozen names designate these individuals as on flexible assignments.

[8]The various dollar figures are taken from Carnegie Institution of Washington *Yearbooks*, published by the institution.

[9]See, for example, Bauer (September 1907); and Fleming (1931). Fleming's talk was broadcast over the Columbia Broadcasting System.

[10]Carnegie Institution of Washington. Archives. File: DTM. Miscellaneous, 1904–1917 (II). Edmonds, Harry M. W. Report on Exhibit of Department of Terrestrial Magnetism at Administrative Building, 14–19 December 1914. 5 p., on 4.

[11]Carnegie Institution of Washington. Archives. File: DTM. Arctic Expedition. This file contains letters and contracts over a three-year period.

[12]Department of Terrestrial Magnetism. Archives. File: Director's Memoranda. This one is dated 10 December 1906.

[13]Department of Terrestrial Magnetism. Archives. File: Special Instructions – Observations. This file includes two memos: 'To Reduce Seasickness to a Minimum' and 'Suggestions for Care of Health in the Tropics'.

[14]Department of Terrestrial Magnetism. Archives. File: Unnamed file. No title or date on memorandum, but after 1928 from internal evidence.

[15]Department of Terrestrial Magnetism. Archives. File: Instructions. Fisk.

[16]Department of Terrestrial Magnetism. Archives. File: Director's Memoranda. Dated 31 March 1915. This memo detailed eight 'precepts' to minimize errors.

[17]Department of Terrestrial Magnetism. Archives. File: Director's Memoranda. Dated 5 April 1909.

[18]Department of Terrestrial Magnetism. Archives. File: Personnel. Fisk. Organization. These details come from a biographical data form Fisk completed for the Carnegie Institution in 1922. A slightly more complete account is Harradon (1933).

[19]Department of Terrestrial Magnetism. Archives. File: General Files to 1934. Peters, W. J.

[20]Department of Terrestrial Magnetism. Archives. File: Observers–Score.

[21]Multhauf & Good (1987). A series of DTM instruments is shown and discussed in Figures 43–49, pp. 57–63. It is possible to trace the 'life story' of each instrument just as it is of each magnetic observer, through the detailed information in published land magnetic results.

[22]Edmunds is mentioned in several web sites on the history of missionary work. See Haas (1996) on p. 231; and Bauer (1912), p. 119. The 1909 expedition of Sowers in northern China is detailed in Bauer (1912), pp. 117–118.

[23]Bauer & Fleming (1915). No correspondence or other archival material from this early period of Edmund's career is known to survive.

[24]Bauer *et al.* (1921), 154–160. Copies of Brown's letters to Edmunds during these travels are in the archives of the DTM. File: General Files to 1934. C. K. Edmunds. Mongolian Expedition; and Russian Expedition.

[25]Archives. DTM. File: General Files to 1934. C. K. Edmunds. Mongolian Expedition.

[26]Wallis & Green (1947), and Wang Shen (1994). See also Archives. DTM. File: China. This file contains correspondence from Fleming to Brown and Kwei regarding survey work in China, from 1929 to 1935.

[27]Numerous files at the DTM provide a more candid view and will be used at a later time.

[28]Wallis & Green (1947). Hanson observed at 89 repeat stations.

[29]Department of Terrestrial Magnetism. Archives. File: Hanson, Earl. Correspondence; and Carnegie Institution of Washington. Archives. File: President. Hanson.

References

BAUER, L. A. 1895*a*. *Beiträge zur Kenntniss des Wesens der Säcular-Variation des Erdmagnetismus.* Inaugural -Dissertation zur Erlangung der Doctorwürde von der philosophischen Fakultät der Friedrich-Wilhelms-Universität zu Berlin. Mayer and Müller, Berlin.

BAUER, L. A. 1895*b*. On the Distribution and the Secular Variation of Terrestrial Magnetism, No. I. *The American Journal of Science*, (3rd ser.) **50**, 109–115; No. 2. *The American Journal of Science*, (3rd ser.) **50**, 189–205; and No. 3. *The American Journal of Science*, (3rd ser.) **50**, 314–325.

BAUER, L. A. 1904. The Present Problems of Terrestrial Magnetism. *In*: ROGERS, H. J. (ed.) *International Congress of Arts and Sciences, Universal Exposition, St. Louis*. 8 vols. Houghton, Mifflin, and Co., Boston and New York. **4**, 1–7.

BAUER, L. A. 1907. The Work in the Pacific Ocean of the Magnetic Survey Yacht "Galilee." *National Geographic*, September 1907, 601–611.

BAUER, L. A. 1912. *Land Magnetic Observations 1905–1910*. Researches of the Department of Terrestrial Magnetism. Carnegie Institution of Washington Publication No. 175. Washington, D. C., **1**, 113.

BAUER, L. A. & FLEMING, J. A. 1915. *Land Magnetic Observations 1911–1913*, 95–107.

BAUER, L. A., FLEMING, J. A., FISK, H. W. & PETERS, W. J. 1921. *Land Magnetic Observations 1914–1920*. Researches of the Department of Terrestrial Magnetism. Carnegie Institution of Washington Publication No. 175. 4, Table 1, p. 5.

CHAPMAN, S. 1941. *Edmond Halley as Physical Geographer and the Story of his Charts*. Royal Astronomical Society, London.

FISK, H. W. 1927. *Land Magnetic Observations 1918–1926*. Researches of the Department of Terrestrial Magnetism. Carnegie Institution of Washington Publication No. 175. **6**.

FLEMING, J. A. 1931. The Magnetism of the Earth. *Science Service Radio Talks*, 74–77.

FLEMING, J. A. & FISK, H. W. 1928. Summary of Magnetic-Survey Work by the Carnegie Institution of Washington. *Terrestrial Magnetism and Atmospheric Electricity*, **33**, 27–36.

GOOD, G. A. 1988. The study of geomagnetism in the late nineteenth century. *EOS*, **69**, 218–228.

GOOD, G. A. 1991. Scientific Sovereignty: Canada, The Carnegie Institution, and the Earth's Magnetism in the North. *Scientia Canadensis*, **38**, 3–37.

GOOD, G. A. 1994. Vision of a global physics: The Carnegie Institution and the First World Magnetic Survey. *History of Geophysics*, **5**, 29–36.

GOOD, G. A. 1999*a*. John Adam Fleming. *In*: GARRATY, J. A. & CARNES, M. C. (eds) *American National Biography*. 24 vols. Oxford University Press, New York, **8**, 102–104.

GOOD, G. A. 1999*b*. Ernest Harry Vestine. *In*: GARRATY, J. A. & CARNES, M. C. (eds) *American National Biography*. 24 vols. Oxford University Press, New York, **22**, 343–344.

GOOD, G. A. 1999*c*. Louis Agricola Bauer. *In*: GARRATY, J. A. & CARNES, M. C. (eds) *American National Biography*. 24 vols. Oxford University Press, New York, **2**, 349–351.

HAAS, W. J. 1996. *China Voyager. Gist Gee's Life in Science*. M.E. Sharpe, Armonk, New York and London, England.

HANSON, E. P. 1938. *Journey to Manaos*. Reynal and Hitchcock, New York.

HARRADON, H. D. 1933. Harlan Wilbur Fisk, 1869–1932. *Terrestrial Magnetism and Atmospheric Electricity*, **38**, 55–58.

JONKERS, A. R. T., JACKSON, A. & MURRAY, A. 2003. Four centuries of geomagnetic data from historical records. *Reviews of Geophysics*, **41**, 2.1–2.36.

MORRELL, J. & THACKRAY, A. 1981. *Gentlemen of Science: Early Years of the British Association for the Advancement of Science*. Clarendon Press, Oxford.

MULTHAUF, R. P. & GOOD, G. 1987. *A Brief History of Geomagnetism and a Catalog of the Collections of the National Museum of American History*. Smithsonian Studies in History and Technology No. 48. Smithsonian Institution Press, Washington, D.C.

NEUMAYER, G. 1891. *Atlas des Erdmagnetismus. Berghaus' Physikalisher Atlas, Abteilung IV*. Justus Perthes, Gotha.

VESTINE, E. H., LAPORTE, L., COOPER, C. *ET AL*. 1947*a*. *Description of the earth's main magnetic field and its secular change, 1905–1945*. Carnegie Institution of Washington, Washington, D.C., Publication No. 578.

VESTINE, E. H., LAPORTE, L., LANGE, I. & SCOTT, W. E. 1947*b*. *The Geomagnetic Field, Its Description and Analysis*. Carnegie Institution of Washington, Washington, D.C., Publication No. 580.

WALLIS, W. F. & GREEN, J. W. 1947. *Land and Ocean Magnetic Observations, 1927–1944*. Researches of the Department of Terrestrial Magnetism, **8**, 38–44.

WANG SHEN. 1994. Dr. C. T. Kwei and the Carnegie in China in 1930s and 1940*s*. *History of Geophysics*, **5**, 171–172.

ZMUDA, A. J. 1971*a*. The World Magnetic Survey 1957–1969. *EOS*, **52**, 60–67.

ZMUDA, A. J. (ed.) 1971*b*. *The World Magnetic Survey 1957–1969*. IAGA Bulletin No. 28, International Union of Geodesy and Geophysics, Association of Geomagnetism and Aeronomy. p. 206.

Index

Note: Page numbers for figures are denoted in *italics*. Page numbers followed by a lower case 'n' denote references to the notes.

Abich, Hermann Willhelm 177–81, *178*, 180n
Africa 184, 186, 191–2
Agassiz, Louis 255–69, *256*
  and Desor 258–9
  funding 256, 257
  and Girard 261
  and Guyot 265
  and Lesquereux 263
  Neuchatel 255–6
  and Pourtales 265–6
Airy, George Biddell 279
Airy hypothesis 279, 280
Allan, Thomas 149, 157
Alpine Fault 386
altitude, calculating 58
America 196, 210, 255, 275
  American West 273–4
  Cascade Range 275, 283
  Civil War 261, 271, 272
  immigrants 258, 267
  Mount Taylor 275, 280
  slavery 261–2
  surveys 267, 274
  Zuni Plateau 275, 280
  see also Colorado Plateau; United States Geological Survey; Washington
Amundsen, Roald 397
Angola 297, 305
Antônio, José 305
Aouelloul crater 199–200, 201
Armenia 177–81, *179*
Arran 31–47, *35*, 210
  Glen Cloy 37–9, *38*
  Glen Rosa 39–40
  Goatfell 36–7, 39–40
  Kilmichael 36–7
  Lochranza unconformity 41–2, 46n
  Sannox 42
  Tormore dykes 42–5, *44*, *45*, 46n
  Witch's Step 40–1
Atlantic Ocean, dives 188, 192
Ault, J.P. 401
Australia
  European settlements *326*
  geological maps 343–73
  geology [1930s] 347–51
  limestone 325–42
  megafauna 357, 361
  surveys 332, 343
  see also Blue Mountains; Nelson; New South Wales; Queensland; Sydney
Auvergne, France
  [1751–1800] 73–96
  geological pilgrimage 73–4, 79, 81n
  inventory of Auvergnats 86–8
  inventory of travellers 82–5
  map of *75*, *80*
  mineral resources 78, 81n
  motivation for travel 77–8
  province definition 76, 81n
  regional details 74–6
  visitors and locals 78–9, 81n
Azerbaijan 177, 178
Azevedo, Antonio 211–12, 213, 214
Azores 229–38, 237n

Bailly, Joseph Charles 332, 333, 349, 369n
Baird, Spencer F. 261
Bakewell, Robert 73
Banks, Joseph 326, 332, 334, 349
baobab tree, age 244–5
Barrallier, Francis 331–4, 349
Barrande, Joachim 122, 132n
Barringer, Daniel Moreau 196
Barrios, Charles 124–5, 126
Baudin, Nicolas 332, 349
Bauer, Louis Agricola 395–6, *396*, 400, 401–2, 405–6
Beaumont, Léonce Élie de 207, 211, 225, 226–7
Benot, Eduardo 129–31
Berthoud, Fritz 260
Blackman, James 338–9
Bland, Phil 202–3
Blaxland, Gregory 337
Bligh, William 334, 336
Blue Mountains 325, 331–5
  crossing 329, 334–5, 337, 338, 343, 349, 351
Boissieu, Jean-Jacques de *78*, 82
Bonpland, Alexandre Aimé Goujaud 164
Born, Ignaz von 49, 50–4, *50*, *52*, 151, 155
Boscovich, R. G. 279
Bosler, Jean 195, 196, 200
botany 7, 56, 172, 207
Boué, Ami *12*, 13–14, *13*
Bouguer, Pierre 279
Brard, Cyprien Prosper 12
Brazil 297, 301, 305, 311–12, 314
  mining 299, 308n
  Rio de Janeiro 304–5
  surveys 298, 302, 305
Bresnard, Wladimir 192
Brewster, David 97, 99, 106n
Bright, Richard 140
Britain
  Royal Society of London 19
  science funding 312–14
Broderip, William John 345
Brook Street Volcanics 375, 380, 382, 383, 384, 386
Brown, Frederick Crawford 402–4, *403*
Browne, H. Hayes 337
Buch, Leopold von 63, 64–5, 177, 180, 207, 210–11, 218, 221, 226
  upheaval theory 229, 232, 235
Buckland, William 345–6, 362
Buckman, Sydney Savory 136
Buitrago, Jesus 124
Burkhardt, Jacques 258

Burnett, James 380
Burning Mountain 354–5, *356*
Busby, John 337
Butler, Gerard W. 139, *140*

Cabo, Agostinho Joaquim do 305
Caldcleugh, Alexander 311–12, *313*, 314, 319, 320n
Caley, George 332
Câmara, Manuel Arruda da 305–6, 309n
Canada 287–96, *288*
  expedition dangers 287, 290, 291, *292*, 293
  expedition records 287–8, 293
  geographical background 287–8
  Geological Survey of 288–9, 292, 294
  Lake Athabasca 290, *291*
Canary Islands 207–28, 229, 231
Cape Verdi Islands 297, 305
  Flag Staff Hill 245, 248, 249–50, *249*, *250*, 251
  Quail Island 242, 243, *243*, *244*, 245
  Red Hill 245, 249, 251
  Santiago 239–53, *242*
  valley form 245–6, *246*, 247–8, *247*
carbonatite 123, 132n
Carnegie Institute, DTM 395–408
  employees 397, *398–9*, 402
  expedition factory 395, 396, 397–402
  expedition locations 397, *398*, 399–400, 402, 404
  expedition training 398, 399, 400–1
  funding 399, 403
  results 397, 401, 403
Carpathian Mountains 49–61
Caspian Sea 167–9
Cassidy, William A. 200–1, 203
catastrophist v uniformitarian 241, 243, 245, 248
Caucasian Mineral waters 179
Caucasian ridge 178–9
Caucasus 177–81
  Ararat volcano 178
Charleston Earthquake 276, *277–8*, 283
Charpentier, Jean de 209
China 402–4
Chinguetti 191–205, *193*
  Aouelloul crater 199–200, 201
  guelb Aouinet 201, 202, 203, *203*
  locals 195, 198–9, 200
chromite 375, 379, 380–1, 383–4, 386
Chudeau, René 183, 184–5, *185*
Clarke, William Branwhite 349, 364
Clerk, John 31, 32
Clift, William 359–61, 362
coal 40, 177, 178, 329, 336, 347–9, 369n, 378
Cock, Joseph 381, 383
Codina, Joaquim José 304, 305
Cole, Blanche (née Vernon) 137–8, *137*, 142–3
Cole, Grenville Arthur James 135–47, *136*, *137*, *140*, *142*
  As We Ride 142–3
  bicycles 138, *144*, 145
  biography 136–8
  European travels 139–43, *141*
  Ireland 143–5, *144*
  publications 138, 143, 145
  The Gypsy Road 138, 139–42, *139*
Colorado Plateau 274, 279, 280, *281*, 283, 284
Cook, James 326, 327, 339n
Copenhagen 154–5
copper 375, 376–9, 380, 383–4, 386
coral reefs 318–19, 328
Coromandel gold fields 375, 380, 382–3, 386
Costa, Pereira da 120
Coutinho, Dom Rodrigo de Souza 306
Couto, José Vieira 306, *307*, 309n
craters of elevation 210–11, 225, 226, 233
cryolite 155, 159
crystal form 26, *27*
Cuvier, Georges 209, 359, 361
cyclic processes 20, 21–5, 26

Daly, Reginald 284
Dana, James Dwight 225
Darwin, Charles 32, 239–53, 311–23, 362, 389
  catastrophism 239, 245–8
  evolutionary theory 239, 312, 391–2
  fieldwork 241–2, *242*, 314–16
  geological education 239, 314
  geological notes 252, 316, *317*, 320n
  Lyellian geology 240, 247, 250–2
  Quail Island 242–3, *244*, 245
Daubeny, Charles 240, 248, 249, 251
Davis, Edward Hydelbach 375, 383–5, 386
Davis, William Morris 145, 284
Dawes, William 325
Dawson, George Mercer 289, 292–3
De la Beche, Henry Thomas 317–18, 321n
Delgado, Nery 119–34, *120*
  correlations 121–2, 124, 127, 128, 130
  diplomacy and negotiation 122–4, 130, 132n
  Huelva 123, 130
  hydrothermal activity 123, 126, 130
  primordial fauna 125, 126
  Spain 120–1, 126, 129–30
Depuch, Louis 332–3, 340n, 349, 369n
Desmarest, Nicolas 78, *78*, *80*, 82
Desor, Edouard 255, 256, 258–60, *258*, 267
diamonds 294, 295n
Diller, Joseph Silas 273, 275, 282, 284n
diluvium 239–40, 241, 244, 252
dinosaurs 184, 187, 294
Dolomieu, Déodat de 333
Donati, Angelo 302, 304, 305
Drake, Raleigh 201
Dun Mountain 375–87, *377*, *378*
dunite 379–80, 381, 384, 386
Dutton, Clarence Edward 271–86, *272*
  biography 272–4
  crustal deformation hypothesis 279–80
  scientific legacy 282–4
  volcanics 274–6
dykes 39, 43, *44*, *45*, 213, 218, 220

Earle, Augustus 355
earthquakes 276, *277–8*, 283, 312
Edmunds, Charles Keynes 402–4
Ehrenberg, Christian Gottfried 161, 165, 169, 170, 172
Enlightenment geologists 74, 297, 298, 299–300
erosion 24, *25*, 41, 245–6, 279–82, 284
Etna, Mount 225–6, 227, 233

Europe 7–17, 139–43
Evans, George William 337–8, *337*, *338*, 341n
Everest, George 279
evolution 21, 29n, 122, 389–91

Faeroe Islands 154
Feijó, Joao da Silva 305, 309n
Fer de Dieu meteorite 191–205
  misprint theory 203–4
  searches for 194–6, 199–201, 202–4
Ferreira, Alexandre Rodrigues 302, 303, *303*, 304–6, 309n
Fichtel, Johann Ehrenbert von 49, *53*, 54–5
field instruments 346, 399–400, 401, 402–3
fieldtrips 109, 116, 304
  cycling 143–4
  funding 145, 146n, 297
  students 143–5
fieldwork 4, 7, 49, 58, 79, 120, 121, 131n
  instructions 8–10, 11, 15n, 58, 138, 302, 316, 334
Filho, Munteal 300
Fisher, Osmond 280
Fisk, Harlan 397, 401–2
Fitton, William 311, 320, 345–6
  field techniques 314–16
  identification of bones 359–61
  instructions 345–6
Fitzroy, Robert 314, 318, 319, 320
Fleming, John Adam 397, 401
fold belts 282, 283
Forbes, James David 97–107, *98*
  itinerary 97–9
  Naples 105–6
  Physical Notices 99–105
fossils 21, 70, 243–4, 343, 353–4, 380, 389, 391
  ammonites *23*
  Diprotodon 363, *364*
  identification of bones 362–3
  Inoceramus 383, 384, 385–6
  megafauna 355–62
  organic origin 21–2, 25, 26
  Pleurotomaria berichii 391
  primordial fauna 122, 124–6, 127, 130, 132n
  Vargula hilgendorfii 389, 393
Foveaux, Lieutenant-Governor 336
France 81n, 109–17, 207–8
  see also Auvergne
Freiberg mining 154, 162
Freihaus Theatre company 149, 151, 158
Freire, José Joaquim 305
Fric, Antonín 140

Gagel, Curt 235
Galápagos Islands 318
Gascoigne, John 299–300
Gauss, Carl Friedrich 395
Geikie, Archibald 73
geological time 25–6, 28, 41, 70, 208, 209, 242–5
geological travel 1–6, 8, 120, 129, 130, 319, 403
  definition 10, 311
  funding 145, 146n, 165, 170, 229, 297, 311
  instructions 7–17, 314
  literature 49, 319
  local guides 76, 79, 111
  motivation for 1, 2–3, 77–8, 325
geologists 7, 68–9, 76–7, 320, 365
  academically minded 311, 319
  philosophy and economics of 63–72
  practically minded 311, 319–20
  women 109–10, 116
geology
  advancement of 316–18, 319
  international cooperation 119, 120, 312
  mapping 120, 183
  physical 271, 276
  practical 120, 138, 303, 319–20
  as a science 7, 10, 49, 68–9
  textbooks 12–14, 138
  training 116, 312, 320n
geomorphology 142, 145, 271, 284
geophysics 4–5, 171, 396, 406
Georgia 177, 178
German Association of Graduates and Friends of Moscow Lomonossow University (DAMU) 173–5, *174*
Germany 153–4, 231–2
  and Russia 161–2, 172, 173, 174
  Scheibenberg locality 149, 154
Giesecke, Charles Lewis 149–60, *152*, *157*
  Berlin 153–4
  biography 149–50, 153, 158
  Dublin 157–8
  Greenland 155–7
  mineralogy 149, 153, 155, 157–8, 159
  pseudonym 149–50, 153
  setbacks 155–7
  theatre 149, 150–3
  Vienna 151–2
Gilbert, Grove Karl 271, 273, 282–3
Girao, Cabo 213
Girard, Charles 255, 256, 258, 260–2, 260, 267
glaciation 177, 293–4
Goa 297, 305
Goethe, Johann Wolfgang von 63–72, 157, 248
  correspondence 63, 69
  scientific approach 66–70, 71
  volcanism 65–6
gold 165, 294, 303, 365, 367, 375, 381–2, 386
  alluvial 365, 376, 381–2
Gran Canaria *216*, 217–18, *217*, 226, 234–5
Grand Canyon 274, 280, *281*
granite 31, 39, 42, 167
Greenland 149, 155–7, 159
  Ahnighito iron 197
  Disko Island 155
  Godhavn basalts *156*
  Holstenborg District *156*
  Ivigtut location 155
  North West Passage 159
Gresley, Amanz 256
Guettard, Jean-Etienne 78, 82
Guyot, Arnold 255, 256, 258, 264–5, *264*, 266, 267

Haast, Julius 380
Hacket, Thomas Ridge 375, 379, 380, 381, 382–3, 386
Hacquet, Belsazar de la Motte 49, 55–8, *56*, *58*
Halley, Edmond 395

Hanson, Earl Parker 404–5, 406
Hartung, Georg 212–13, 214, 229–38
  Azores 232, 233–5
  biography 230–1
  craters of explosion 234, 237n
  Germany 231–2
  glacial geology 235–6
  Gran Canaria *216*, 217–18, 234–5
  La Palma 218–22, *219*, *220*
  letters to Lyell 235, 236, 236n
  Madeira *212*, *215*, 231, 233–5
  publications 229, 230–1, 232, 233–5, 237n
  Tenerife 222–3, *232*
  travel problems 229, 232
Hawaii 275, 276, 283
Hayford, John 283–4
Hector, James 375, 381, 382–3, 386
Heer, Oswald 230, 231, 235, 236n
Heiskanen, W. A. 284
Henderson, John 347–9, 369-70n
Henslow, John Stevens 239, 251
Herschel, John 279
Hilgendorf, Franz 389–93, *390*
  biography 389–91
  evolution 391, *392*
  Exhibition 389, 393
  Japan 391–2, 392
  Ougai, Mori 392–3
  phylogeny 389–91, *390*, 393
HMS *Beagle* voyage 312–19
Hobbs, William Herbert 294
Hochstetter, Ferdinand von 375, 379–80, 381, 384, *385*, 386
Hoff, Adolf von 66
Hoffmeister, Franz Anton 153
Hooke, Robert 19–30, *20*
  artist 19, 22, *23*
  fossils 21, 22
  polar wander 21
Horner, Mary 209
Hubert, Henry 192–4
Humboldt, Alexander von 63, 64–5, 101, 161–75, *163*, 177, 179, 180, 256, 312, 395
  biography 161, 163–5
  correspondents 173–4
  expedition route 166–70
  funding 165, 170
  publications 164–5, 170–1, *173*
  reviews 170–1, 172–3
  Russia 165–70, 173–5
Humphrey, Adolarius William Henry 332, *333*, 335–6
Hungary, volcanic deposits 140–1
Hunter, John 329
Hutton, Fredreick Wollaston 383
Hutton, James 21, 43, 45, 63, 64, 249
  field observations 33, 41–3, 45
  influence of Hooke 26–8
  Theory of the Earth 31–47

Ibero, Carlos Ibañez de 129–31
International Geological Congress 119
Ireland 143–5, 157–8
  Royal Dublin Society 157–8
Isle of Wight 19–30, *25*
  cyclic processes 21–5
  Mermaid Arch and Stag 22–4, *24*
  Middle Chalk 22
  Needles 21, *22*
isostasy 271, 274, 279–82, 283–4
Italy 98–105, 109–17, 177, 208, 211
  eighteenth and nineteenth century 7–17
  Pozzuoli Temple of Serapis 102–3, *107*, 248–9
  see also Etna; Naples; Pompeii; Sicily; Stromboli; Vesuvius

Jameson, Robert 31–47, 97, 99, 106n, 159, 355, 359–61
  Arran journals 34–45, *35*
  fieldwork 31, 34–6, 45
  lithological relationships 38–9
  Tormore dykes 43, *44*, *45*
  Wernerian approach 33, 42
Japan, evolutionary theory 389–93
Joly, John 284
Judd, John Wesley 45

Kammerbühl, Bohemia 64–7
Keeling Islands 318
Keewatin glaciation 293–4
Kilian, Conrad 183, 185–7
  itinerary *186*
  Tassilis des Ajers 186
  unknown mountains 186–7
King, Clarence 273
King, Governor 328, 329–31, 332, 334, 335, 336
Kinghorne, Alexander 357
Kulik, Leonid A. 196
Kwei, Paul C. T. 404

La Palma 218–22, *219*, 226
  Caldera de Taburiente 218–21, *220*, *221*, *222*
  volcanic activity 222, 225
Lacroix, Alfred 192–4
Lang, John Dunmore 355
Lapaz, Jean 200
Lapaz, Lincoln 200
Lapparent, Albert Félix de 183, 187, *188*
lava, slope angle 211, 212, 213, 214, 218–20, 224, 225–6, 227
Lawson, Andrew 284
Lawson, William 337
Leohnard, Karl Caesar 13
Lesley, J. Peter 255
Lesquereux, Léo 255, 258, 260, 262–4, *262*, 267
Lesson, René 338–9
limestone 40, 57, 248, *249*, 252, 325–42, 355
  ballast 328, 336, 337
  exploiting resource 338–9
  identification of 329, 334
  search for 327, 337–8
  shells 243–4, *244*, 327, 329–31, 334
  see also Praya limestone; Wooded Peak Limestone
Lindesay, Patrick 359, 370n
Lisbon Royal Academy of Science 297, 299–300
  brief instructions 304
  Memoirs 297, 300
Logan, William Edmond 288
Lomonosov, Michail Vasil'evic 162

Lyell, Charles 63, 73, 207–28, 229, 240, 318, 359
biography 207
craters of elevation 210–11, 225, 226
and Darwin 250–2
France 109–17
geology 109–10, 209, 239
Gran Canaria *216*, 217–18
and Hartung 230
Italy 109–17, 211, 225–6
La Palma 218–22, *219*, *220*
Madeira *212*, *213*, *215*, 224, 231
and Murchison *110*, 111–15, *112*
Porto Santo 214
publications 223–4
Tenerife 222–3, *232*
travel route *110*, 207–10
upbuilding theory 232, 233–4, 235, 248–9
Lyell, Mary (née Horner) 211, 214

Macculloch, John 346
McKay, Alexander 385
Macpherson, José 124, 130
Macquarie, Governor 335, 336–7
Madeira 207–28, 229–38, 236n
cross section *215*
structure of 218, 233–4
volcanics 216, 226
Madrid Treaty 298
Maffiotte, Pedro 217
Magic Flute, The 149, 151, 158, 159
magnetic surveys 200, 202, 395–6, 405–6
magneticians 395–408
magnetometer *399*
Maitai Group 375, 380, 382, 383, 384, 385–6, *385*, 386
Malavoy, Jean 194–5, 197
Mallada, Lucas 124
Malthus, Thomas Robert 318
Martineau, Harriet 318, 321n
Mauritania *193*, 201
Mawe, John 311
Mensenin, Nicolai Stepanovic 172
meteorite 191–205, *193*
craters 196–7, 199–200, 201
impact glasses 199, 201
iron 196–7
see also Fer de Dieu meteorite
Miers, John 311
mineralogy 9–10, *15*, 49–61, 149, 151, 300, 332, 335–6
mining 1, 9–10, 51, 53, 162, 165, 300, 311–12, 347
Mitchell, Thomas Livingstone 343–73, *354*
assistants 350, 352
Australia 347–51, *348*, 363–7
biography 343, 344–5, 347, 362, 367
Burning Mountain 354–5, *356*
England 362–3
expeditions 343, 351–5, *353*
fossils 343, 353–4, *354*, 362
geology 343, 345–7, 349–50, 368
goldfields 364–7, *366*, 371n
map work 343, 344, *348*, 349, 350–1, 350
Mount Victoria Pass 351, *352*
New South Wales *350*, 364–7, *366*
publications 343, 351, 362–3, 368
Queensland 363–4
road construction 343, 351, 369n
specimens 346, 353–4, 370n
Wellington Caves 355–63, 370n
Mitscherlich, Eilhard 207, 225, 227
Monod, Théodore Andre 183, 187–9, 201, *203*
biography *189*, 191–2
Fer de Dieu meteorite 191–205
meteorite search 189, 199–201, 202–4
publications 189, 191
Monro, David 375, 379, 381, 386
Morse, Edward Sylvester 389, 391
Mossier, Jean Baptiste 77
mountain formation 53–5, 57, 279–80, 283, 302, 309n
Mozambique 297, 305
Münteer, Friederich 155
Murchison, Charlotte 109–17, *110*, 207–8
Murchison, Roderick Impey 109–17, 207–8, 345–6, 365
and Lyell *110*, 111–15, *112*

Namibia, Hoba iron 197
Naples 97–107
Astroni 102
Cappomazza 103
geology 105, 106
Monte Barbaro 103
Monte Epomeo 104
Pozzuoli molluscan damage theory 103
navigation 202, 319, 353
Necker, Louis Albert 209
Nelson *376*
geological map 375, *377*
Mineral Belt 375, 376, *378*, 383, 384, 386
Neptunist theory 32–4, 45, 54, 57, 63, 64, 67
New South Wales
colonial 325–42
goldfields 343, 364–7
limestone 325, 337, 338–9
mapping *330*, 343, *345*, *350*
surveys 326–7, 328–9, 335, 336, 338, 339
New Zealand 375–87
chromite 380–1, 383–4
coal 378, 379, 382
copper 383–4
first railway 375, 380–1
geological map 380
Geological Survey 382, 383
gold 379, 381–2
mineral discoveries 376–9
Newton, Isaac 19–20, 21–2
Norfolk Island 325, 328
Nuovo, Monte 103–4

oceanography 266, 318–19
Olivier, Charles P. 196
ophiolite belts 375–6, 386
Orbigny, Alcide Dessalines d' 312
Ougai, Mori 389, 391, 392
Oxley, John 337, 338, 344, 346–7, 355

palaeontology 120, 294
Pallas, Peter Simon 162
Paterson, Lieutenant-Governor 31, 329, 336
Pentland, Joseph 359, 361
Péron, François 334, 349
Peters, William 401, 402
Phillip, Arthur 325, 327, 328, 334, 335
Piccard, Auguste 192
Pickova, Olga 192
plate tectonics 27, 283, 375–6
Playfair, John 46n, 63, 106–7n
Plutonist theory 32–4, 45, 63, 64
polar wander 21, 27–8, 29n
Pompeii 99, 100–1
Portugal
  Ajuda Museum 301, 304, 305
  correlations with Spain 124, 127, 128, 130, 131
  geological map 120, 123–4, 130, 133n
  Geological Survey of 119, 120, 130, 132–3n
  Sao Domingos 129, 130, 132n
  science 131n, 298–9
Portugese Empire
  expeditions 297, 301–6, 307
  native travellers 297–310
  Philosophical Voyages 297, 301, 303, 304, 307
  reformist policies 297, 298
Pourquié, Andre 199
Pourtales, Louis François de 255, 256, 258, 261, 265–6, *265*, 267
Powell, John Wesley 271, 273–4, 282–3
Prado, Casiano 124, 128
Pratt hypothesis 279, 280, 284
Pratt, John Henry 279
Praya limestone 245, 248, 249, *250*, 252
Price, John 329
Prokop, Karel 202
Prussian Mining and Metallurgy Administration 164

Queensland
  greenstones 382–3, 386
  Gympie goldfields 375, 382–3, 386
Quenstedt, Friedrich August 389

radioactivity 201, 276, 285n
Ranken, George 355–7, 359, 362
Renovantz, Hans Michael 167, *168*
Reub, Franz Ambros 66
Ribeiro, Carlos 119–20, 121, 131n
Richât crater 201
Richmond Group 386
Ripert, Gaston 193–5, 197, 199, 200, 202, 203–4
Robilant, Benedetto Spirito Di 9–10
Roemer, Carl-Ferdinand von 124
Rose, Gustav 161, 165, 169, 170, 171–2, 171, 172
Ross, James Clark 395
Rossi–Forel scale 276
Russell, Sara 202–3
Russia 161–75, 177, 179
  Altai mountains 167
  and Germany 161–2, 172, 173, 174
  Kolyvan 167, 169
  mining 165, 170, 172
  science in 161–3, 173

Sabine, Edward 395
Sahara 183–90, 191
  inland sea 183–4
  safety 183, 184, 190
  trans-Saharan railway 184
Saint Ildephonse Treaty 298
Saussure, Horace Bénédict de 10–12, *10*
Scandinavia 209, 210, 233
Schulz, Guillermo 125
science
  and agriculture 297, 298
  funding 312–14
  and mining 297, 298, 299
  and universities 299–300
scientific instruments 8, 9, *11*, 12–13, *12*, *13*, 15n, 165, 265
  see also field instruments
Scrope, George Poulett 225
Sedgwick, Adam 239
seismology 271, 276–8, 283
  isoseismal map 276, *277–8*
Selwyn, Alfred 288–9, 292
Semisayat crater 201
Shetland Islands 31, 32, 46n
Siberia 165, 167, 196
Sicily, Valle del Bove 207, 208, 211, 225–6
Silva, José Jaoquim da 305
Silva, Manuel Galvao da 302, 305, 306
Släpfer, Ernesto 200–1
Smith, Charles 223, 335
Smith, James 211
Smith, William 122
Solander, Daniel 326
Somma, Mount 99–100
Sonrel, Auguste 258
Souseki, Natsume 393
Spain 119–34
  Almadén 131
  Asturias 121, 124–6, 127, 130
  Cambrian 123–6, 130
  Cantabria 121, 124–7
  Carboniferous 123
  Ciudad–Real 121, 128–9
  geological culture 129–30
  Huelva 121, 122–4
  Léon 121, 124–6, 127
  Silurian 123–4, 128–9, 131
Spanish Geological Survey 120, 129–30
Spencer, Leonard J. 203
Spinoza, Baruch de 63, 67–8, 71
Sternberg, Caspar Maria Count 63–72
  correspondence 63, 70
  inductive approach 69–70, 71
  volcanism 65–6
stratigraphy 120, 183
Stromboli, lava flows 225, 227
Studer, Bernhard 209
Stutchbury, Samuel 364, 365, 367, 370–1n
subsidence 248–9, 271, 282
Suess, Eduard 180
Switzerland 209, 257, 262, 266
  Neuchatel 255–69
Sydney
  basin structure 349, 368n
  shell lime 336–7

Tarin, Gonzalo y 121, 123
Temimichat Ghallaman crater 201
Tench, Watkin 325, 327
Tenerife 222–3, *223*, *224*, 226
Tenoumer crater 201
Throsby, Charles 335
Tierra del Fuego, map 314, *315*
Townson, Robert 336
Tozzetti, Giovanni Targioni 8–9
transport 287
  bicycles 135–47, *140*, *142*, *144*
  camel 183–90, 191, 231
  canoe 287, 290, 295n
  cycling 135–47
  walking 56–7, 184, 188, 191, 287
Trechmann, Charles Taylor 386
Tyrell, Joseph & James 287–96, *289*
  expedition 287, *288*, 290–3
  expedition dangers 290, 291, *292*, 293
  gold mining 294
  native guides 289–90, 291
  photographs 293–4
Tyrell Sea 287, 293–4

unconformity 20, 41–2, 46n
United States Geological Survey 271, 273–4
uplift 248–50, 252, 271, 280, 282
Urals 166, 167, 172

Vallisneri, Antonio 8
Vandelli, Domingos 299, 301–2, 304, 307
Vasconcelos, Luís de 305
Verneuil, Edouard 124
Vestine, Ernest Harry 397
Vesuvius, Mount 97–8, *98*, 99–100, *100*, *101*, 225
Vienna 151–2, 180
Vogt, Carl 256–7, *257*, 258–9
volcanoes 64–5, 74, 97, 101, 177, 271, 273, 274–6
  origin 54–5, 100, 236, 237n, 276
  uplift 208, 249–50

Waimea fault 384–6, *385*
Wakamarina Valley 381–2
Walker, John 32–4, 39–41
Waller, Richard 19
Washington, Carnegie Institute 395–408
Wellington Caves 343, 355, *357*, 359, 361, *361*
  bones *358*, 369n, 370n
Wellington Valley 355, 357–9, *360*
Wellman, Harold 386
Wells, William 375, 379, 386
Wentworth, William Charles 331, 337
Werner, Abraham Gottlob 33, 54, 63, 64, 149, 154, 163, 333, 340n
Wernerian principles 31, 39, 41, 45, 159
White, John 325, 327
white phonolite 140–1
Wianamatta Group 349
Wilson, John 329
Wilton, Charles Pleydell 355
Wooded Peak Limestone 375, 380
Woodward, John 7, *8*, 25
World Magnetic Survey 395–6
Wrey, William Long 378–9, 380
writing styles 49, 50, 52, 55, 57, 109, 143
  diaries 69, 109, 171
  letters 52, 69, 70, 109

Zanda, Brigitte 203, 204
zoology 172, 191, 239